AF712947

Dirichlet Branes and Mirror Symmetry

Clay Mathematics Monographs
Volume 4

Dirichlet Branes and Mirror Symmetry

Paul S. Aspinwall
Tom Bridgeland
Alastair Craw
Michael R. Douglas
Mark Gross
Anton Kapustin
Gregory W. Moore
Graeme Segal
Balázs Szendrői
P.M.H. Wilson

American Mathematical Society
Clay Mathematics Institute

Clay Mathematics Institute Monograph Series

Editors in chief: S. Donaldson, A. Wiles
Managing editor: J. Carlson
Associate editors:

B. Conrad	I. Daubechies	C. Fefferman
J. Kollár	A. Okounkov	D. Morrison
C. Taubes	P. Ozsváth	K. Smith

2000 *Mathematics Subject Classification.* Primary 14J32, 18E30, 53D12, 53D40, 53D45, 53C38, 53C29, 53C25, 81T30, 81T40, 81T45, 81T75.

For additional information and updates on this book, visit
www.ams.org/bookpages/cmim-4

Library of Congress Cataloging-in-Publication Data

Dirichlet branes and mirror symmetry / Paul Aspinwall... [et al.].
p. cm. — (Clay mathematics monographs ; v. 4)
Includes bibliographical references and index.
ISBN 978-0-8218-3848-8 (alk. paper)
1. D-branes. 2. Mirror symmetry. 3. Superstring theories. I. Aspinwall, Paul, 1964–

QC794.6.S85D57 2009
539.7′258—dc22 2009014258

Copying and reprinting. Individual readers of this publication, and nonprofit libraries acting for them, are permitted to make fair use of the material, such as to copy a chapter for use in teaching or research. Permission is granted to quote brief passages from this publication in reviews, provided the customary acknowledgment of the source is given.

Republication, systematic copying, or multiple reproduction of any material in this publication is permitted only under license from the American Mathematical Society. Requests for such permission should be addressed to the Acquisitions Department, American Mathematical Society, 201 Charles Street, Providence, Rhode Island 02904-2294, USA. Requests can also be made by e-mail to reprint-permission@ams.org.

© 2009 by the authors. All rights reserved.
Published by the American Mathematical Society, Providence, RI,
for the Clay Mathematics Institute, Cambridge, MA.
Printed in the United States of America.

∞ The paper used in this book is acid-free and falls within the guidelines
established to ensure permanence and durability.
Visit the AMS home page at http://www.ams.org/
Visit the Clay Mathematics Institute home page at http://www.claymath.org/

10 9 8 7 6 5 4 3 2 1 14 13 12 11 10 09

Contents

Preface

The 2002 Clay School on Geometry and String Theory was held at the Isaac Newton Institute for Mathematical Sciences, Cambridge, U.K., from 25 March through 19 April 2002. It was run jointly by the organizers of two concurrent workshops at the Newton Institute: one on Higher Dimensional Complex Geometry organized by Alessio Corti, Mark Gross and Miles Reid, and the other on M-theory organized by Robbert Dijkgraaf, Michael R. Douglas, Jerome Gauntlett and Chris Hull, in collaboration with Arthur Jaffe, then president of the Clay Mathematics Institute.

This is the second of two books that provide the scientific record of the school. The first book, *Strings and Geometry* [**131**], edited by Michael R. Douglas, Jerome Gauntlett and Mark Gross, was a proceedings volume and largely focused on the topics of manifolds of special holonomy and supergravity.

The present volume, intended to be a monograph, covers mirror symmetry from the homological and torus fibration points of view. We hope that this volume is a natural sequel to *Mirror Symmetry*, [**242**], written by Hori, Katz, Klemm, Pandharipande, Thomas, Vafa, Vakil and Zaslow, which was a product of the first Clay School in the spring of 2000. We shall refer to it as MS1. A familiarity with the foundational material of MS1 can be viewed as a prerequisite for reading this volume, and we shall often refer to MS1 for background.

The overall goal of this volume is to explore the physical and mathematical aspects of Dirichlet branes. The narrative is organized around two principal ideas: Kontsevich's Homological Mirror Symmetry conjecture and the Strominger-Yau-Zaslow conjecture. While Kontsevich's conjecture predates the introduction of D-branes into physics, we will explain how the conjecture really is equivalent to the identification of two different categories of D-branes. In particular, we examine how the physics leads us naturally to mathematical concepts such as derived categories and Fukaya categories. We explore the ramifications and the current state of the Strominger-Yau-Zaslow conjecture. We relate these ideas also to a number of active areas of research, such as the McKay correspondence, topological quantum field theory, and stability structures.

As with mirror symmetry in general, these areas have benefited from a remarkably fruitful interaction between mathematicians and physicists. And, over the six year gestation period of the book, a great deal of progress has been made in clarifying and in understanding, and in some cases proving, the original conjectures.

It seems fair to say that to fully appreciate the resulting picture requires having some understanding of both mathematical and physical points of view. Conveying both in the same book has been a challenge and an opportunity. We were not satisfied to simply tell the story twice, once from each point of view. Rather, we attempted a unified presentation, in which both mathematics and physics have their essential insights to provide, explained in a way that physicists and mathematicians can follow without necessarily having all of the foundations of both subjects at their fingertips.

Part of the difficulty in doing this stems from the numerous differences in background and language between physicists and mathematicians; while we feel we have done a great deal to bridge these gaps, it is all the more obvious to us how many gaps remain.

Of course there is a more essential difficulty, which is that the breadth of topics needed to tell the entire story is such that none of the authors are experts in all of them. We have thus divided the main part of the writing while nevertheless striving to unify the book by extensive editing and cross-referencing. The task was carried out by Michael R. Douglas and Mark Gross on the basis of cross-reading and comments made by all of the authors. Michael R. Douglas and Mark Gross take responsibility for the book's success or failure on this level.

Chapter 1 is intended to give a largely physical overview of the topics of the book. Chapter 2, on topological open string theory, is due to Greg Moore and Graeme Segal. An earlier draft of this material appeared as arXiv:hep-th/0609042v1.

Chapters 3 and 5, on the physics of Dirichlet branes, are largely due to Paul Aspinwall, Michael R. Douglas and Anton Kapustin. Parts of this material appeared in arXiv:hep-th/0403166, while §§5.7 and 5.8 are heavily based on Tom Bridgeland's published work on stability structures.

Chapter 4, on representation theory, is largely due to Tom Bridgeland, Alastair Craw, and Balázs Szendrői.

Chapters 6 and 8 are due to Mark Gross, while Chapter 7 is due to Mark Gross and Pelham Wilson.

The entire manuscript was read by Robert Karp and Arthur Greenspoon, both of whom caught numerous imprecisions and unclear points. We also benefited from discussions with and comments by Mohammed Abouzaid, Gary Gibbons, Akira Ishii, Dmitri Orlov, and Bernd Siebert. Several of the authors would also like to thank the hospitality of the IHES, where portions of the book were completed.

Let us again repeat our thanks to those who made the 2002 school possible: H. Keith Moffatt and John Kingman, the directors of the Newton Institute; and its staff, Wendy Abbott, Tracey Andrew, Caroline Fallon, Jackie Gleeson, Louise Grainger, Rebecca Speechley and Christine West.

Finally, let us express our thanks to Jim Carlson and to the staff of the Clay Mathematics Institute, especially Vida Salahi, in helping with the preparation and production of this book. We especially thank Lori Lejeune for providing the artwork in the book.

Michael R. Douglas and Mark Gross
October 2008

CHAPTER 1

Overview and physical background

This book is an introduction to a collection of topics at the interface between theoretical physics and mathematics, referred to collectively as "mirror symmetry." The concept of mirror symmetry evolved in the late 1980's out of the study of superstring compactification, and received its first precise formulation in the 1991 work of Candelas, de la Ossa, Green, and Parkes [**85**] conjecturing (on the basis of solid physical arguments) a formula for the number of rational curves of given degree on a quintic Calabi-Yau manifold, in terms of the periods of the holomorphic three-form on another "mirror" Calabi-Yau manifold. Further developments along these lines included Batyrev's general mirror symmetry construction for hypersurfaces in toric varieties and Givental's and Lian, Liu and Yau's proof of the validity of the instanton number predictions of Candelas et al. In physics terms, these developments all concern the relation between the A- and B- topologically twisted $N = 2$ sigma models with Calabi-Yau target space, and relate to the theory of closed strings in these spaces. These topics are covered in depth in the prequel to this volume [**242**], as well as in [**101, 458**].

In the mid-nineties, two bolder developments emerged, inspired by the physics of open string theory: Kontsevich's 1994 proposal of homological mirror symmetry [**309**], and the geometric picture put forth by Strominger, Yau and Zaslow [**433**] in 1996. These ideas lifted mirror symmetry beyond the somewhat specialized domains of enumeration problems in algebraic geometry and two-dimensional sigma models in physics to a broader picture with more wide-ranging importance in both fields. These two developments and the work they inspired are the subject of our book.

We begin by recalling some of the general physical background from string theory, and give an intuitive description of string compactification, Dirichlet branes, T-duality and the other physical concepts we will discuss in more depth, primarily in Chapters 2, 3 and 5. We then summarize the mathematics behind homological mirror symmetry and SYZ, which we will discuss in depth in Chapters 4, 6 and 7. In Chapter 8 we give a precise formulation of Kontsevich's original homological mirror conjecture, and the worked example of the elliptic curve.

1.1. String theory and sigma models

The central physical object which motivates both proposals, explicitly in Strominger-Yau-Zaslow, and which (as emerged later) lies behind Kontsevich's proposal as well, is the Dirichlet brane, introduced in 1995 by Polchinski.[1] A Dirichlet brane is defined physically as an allowed end point for an open string, or equivalently a boundary condition in two-dimensional conformal field theory.

What does this mean? While various useful mathematical definitions and explanations of conformal field theory and Dirichlet branes have been made, at present none of them provides a completely satisfactory starting point for our purposes. Thus, our general approach in this book will be to explain this physics on an intuitive level, extract the parts we need, and then provide mathematical definitions which can serve as the basis of a more precise discussion.

In general terms, a string theory describes the motion of one-dimensional strings, topologically loops (closed strings) or segments (open strings), in some *target space*, a Riemannian manifold M. To represent the motion of a string through time, one uses a map from a two-dimensional Riemannian manifold Σ, the *world-sheet*, into the target space-time (a product $M \times \mathbb{R}$, where the $\mathbb{R}$ factor represents time).

To specify a quantum theory of strings, we must define a Hilbert space $\mathcal{H}$ of "string wavefunctions," and various linear and multilinear operations on this space. In very rough terms, one can think of $\mathcal{H}$ as a space of functionals on the loop space of M; although in detail this picture is not really right (wave functions have support on discontinuous loops) it gives a reasonable intuitive starting point.

The linear operations correspond to particular world-sheets, or operations on world-sheets. Given a world-sheet Σ without boundary, the quantum theory produces a number, the *partition function*. On each boundary of Σ, one specifies a "boundary condition," an element of $\mathcal{H}$, and in return gets a number. For example, the sphere with three boundaries (or "pair of pants") corresponds to a linear functional on $\mathcal{H}^{\otimes 3}$. Other operators acting on $\mathcal{H}$ correspond to varying the metric on Σ or to other physical observables.

The resulting structure, quantum field theory and conformal field theory, is comparable to and in a sense a generalization of an algebra of functions on M. While we will give a flavor of this subject in Chapter 3, as with almost all work on mirror symmetry, our primary discussion will be based on a simplified but still very rich subset of the problem, called *topological string theory* and *topological quantum field theory*.

[1] Actually, *re-introduced*; see [**393, 461**] for the history.

We will introduce topological string theory in Chapter 2 with the following approach. The correspondences between world-sheets and linear operations satisfy "sewing relations," coming from the fact that a world-sheet Σ can be decomposed into a connected sum of smaller world-sheets in a variety of ways, and the corresponding compositions of linear operations must all lead to the same results. Some simple examples appear in Figure 3 in Chapter 2.

These sewing relations can be summarized as follows:

DEFINITION 1.1. A string theory is a functor from a geometric category to a linear category.

We discuss the simplest example in Chapter 2, that of topological string theory. Here we choose the geometric category to be the category whose objects are oriented $(d-1)$-manifolds, and whose morphisms are oriented cobordisms. The corresponding linear category can then be understood in terms of an associated finite dimensional algebra and its modules.

One can discuss physical quantum field theories using the same language, by now constructing the geometric category out of manifolds with metric. Now the Hilbert space $\mathcal{H}$ is infinite-dimensional, and the morphisms depend on the metric on Σ. The resulting structure has only been made explicit in a few cases, the "exactly solvable" or "integrable" theories. Since we will need more general results, we must discuss the physics definitions of these theories. The standard approach is in terms of a functional integral over maps $\Phi : \Sigma \to M \times \mathbb{R}$, the corresponding "perturbative" graphical expansion, or in some cases using representation theory of infinite-dimensional algebras. We will describe these approaches in Chapter 3.

Another important ingredient in this physics is supersymmetry. Physically, supersymmetry produces much better behaved quantum theories, in which many of the problematic divergences which require renormalization in fact cancel between fermions and bosons. Supersymmetry is also at the heart of many of the connections with mathematics, starting with Witten's famous works of the early 1980's connecting supersymmetry, Morse theory and index theory [**466, 465, 464, 473**].

If one assumes extended supersymmetry, meaning a symmetry algebra with several supercharges with a compact Lie group action (called R symmetry), one gets even stronger constraints on the theory. This structure is at the root of most of the connections with algebraic geometry. The case of primary interest for our book is conformal theory with "$(2,2)$" supersymmetry (§3.1.4 and §3.3.2). In this case, M must be a complex Kähler manifold. There are several other cases, surveyed (for example) in [**153**].

The central new ingredient in quantizing these theories is the renormalization group, as outlined in §3.2.5 and §3.2.6. This leads to conditions on the metric of the target space M (and the other couplings if present) which are necessary for conformal invariance. For the closed string (and at leading order in a sense we describe shortly), this is the condition of Ricci flatness of the metric, and more generally the equations of supergravity.

While not rigorous, the physics analyses give strong evidence that a wide variety of two-dimensional conformal field theories exist. One general class takes M to be a complex Kähler manifold with a Ricci-flat metric. By Yau's proof of the Calabi conjecture, such a metric will exist if $c_1(M) = 0$, and a large number of such "Calabi-Yau manifolds" have been constructed, for example as hypersurfaces in toric varieties.

We also know from Yau's theorem that the Ricci-flat metric is uniquely determined by a choice of complex structure on M, and a choice of Kähler class. Physics arguments show that these CFT's admit deformations which are in one-to-one correspondence with infinitesimal variations of complex structure, and variations of a *complexified Kähler class.* The additional deformations correspond to those of an additional two-form B, satisfying the condition that it is harmonic (this agrees with the equations of supergravity).

Other general classes of CFT's include the "Landau-Ginzburg models" and "gauged linear sigma models." These can be thought of as physics versions of the operations of restriction to the zeroes of a section, and of quotient by a holomorphic isometry.

For any of these models, physics defines a "spectrum of operators" and "correlation functions," and techniques for computing these in an expansion around an exactly solvable limit. The basic such limit for the sigma model is the "large volume" limit[2], in which the operator spectrum and correlation functions reduce to geometric invariants. A basic example is the algebra of harmonic forms, while supersymmetric theories based on complex target spaces can make contact with more subtle concepts, such as variation of Hodge structure.

1.1.1. Stringy and quantum corrections. While the sigma model approach emphasizes the relations between quantum field theory and geometry, there is an opposing strain in the physics discussion, which focuses on the differences between string theory and conventional ideas of geometry. These can be seen by computing the corrections to the large volume limit, by using other more algebraic approaches to conformal field theory, and by "semiclassical" arguments that include additional contributions to the functional integral from instantons and solitons. Many have suggested that these differences will ultimately find their proper understanding in some new, "stringy" form of geometry.

Let us begin with an example of the first phenomenon, that of corrections to the large volume limit. One can show that, in the supersymmetric sigma

[2]Also called the $\alpha' \to 0$ limit, or for euphony as well as historical reasons, the zero-slope limit.

model, the conformal invariance condition on the target space metric coming from the renormalization group analysis is actually not Ricci flatness, but rather a deformation of this,

$$0 = R_{ij} + l_s^6[R^4]_{ij} + \mathcal{O}(l_s^8 R^5). \tag{1.1}$$

Here $[R^4]_{ij}$ is a symmetric tensor constructed from four powers of the Riemann curvature tensor, given explicitly in [**198**], and l_s is a real (dimensionful) deformation parameter called the "string length." In the limit that $l_s \sim 0$ compared to the curvature length, this condition reduces to Ricci flatness. The corrections are defined by quantum field theoretic perturbation theory, and are believed to continue to all orders in l_s^2.

Almost all of the correlation functions obtain similar corrections and these might be regarded as defining a deformation of each of the geometric structures seen in the large volume limit, for example, the algebra of harmonic forms on M. However, little is known in this generality; almost all results in this direction at present come from topological string theory, as we discuss below.

Besides the string length, there is a second "parameter" in string theory,[3] the string coupling, denoted g_s. The defining property of the string coupling is that it controls an expansion whose terms arise at different world-sheet genera: for example, the Einstein equations, which arise from computations involving a genus zero (sphere) world-sheet, could get a correction at genus one of order g_s^2, at genus two of order g_s^4, and so on.

While a fair amount is known about mirror symmetry at higher genus, regrettably the topic will not appear in this book. Perhaps it will receive its due in a Mirror Symmetry III.

Our second source of information about "stringy geometry" comes from world-sheet or "non-geometric" approaches to conformal field theory. These are largely based on the representation theory of Kac-Moody and related infinite-dimensional algebras, such as the Virasoro and super-Virasoro algebra. A famous example is the "Gepner model" §3.3.6, which provides an independent (and in principle rigorous) definition of certain Calabi-Yau sigma models.

One of these topics will play a central role in our discussion, namely the theory of the $N = 2$ superconformal algebra (§3.3.3). This is the basis for the primary physical argument for mirror symmetry (§3.4.3) and will lead to most of the specific physical conclusions we draw in Chapters 3 and 5.

We finally turn to information from semiclassical methods. These incorporate extended field configurations, which in general fall into two broad

[3]We put the word parameter in quotes because one can show that its value can be changed by varying a space-time field, called the dilaton, and thus all of the theories obtained by starting with different values of g_s are physically equivalent. This is somewhat analogous to the fact that the string length l_s is not a parameter, because a different choice of l_s could always be compensated by an overall scale transformation.

classes, instantons and solitons. Both of these are nontrivial critical points of the action functional used in the functional integral definition of a quantum field theory, where nontrivial means that the field configuration (in a sigma model, the map $\Phi : \Sigma \to M$) is nonconstant on Σ. Typically (though not always) such critical points exist for topological reasons.

An instanton is a field configuration which is "concentrated" or associated with a point in the underlying space-time Σ (in particular, at an "instant" in time). As one goes to infinity in any direction, it asymptotes to a constant. It is used in approximate evaluations of the functional integral as a "saddle point;" thus the integral is regarded as a sum of contributions from each critical point. As we will see in Chapter 3, a nontrivial critical point will lead to a correction which, unlike the power-like corrections in (1.1), is exponentially small in the deformation parameter (here l_s).

In the case at hand, the basic example is to consider $\Sigma \cong S^2$ and a target space M with nontrivial π_2. These are called "world-sheet instantons" and lead to corrections in many correlation functions. We will review these corrections and their by-now familiar role in mirror symmetry in §3.4. In the case of open string theory, an analogous role will be played by maps from Σ a disk.

A soliton is a nontrivial solution associated to a line which extends through time, but is concentrated in space. In other words, as one goes to infinity in any spatial direction, the field configuration approaches a constant. The basic example of a soliton for us will be the "winding string" which underlies T-duality, as explained shortly in §1.3.

For Σ of dimension greater than two, one can go on to consider a solution which asymptotes to a constant in some but not all of the spatial directions. These are referred to as "branes" (short for membranes). The Dirichlet brane we are about to discuss is an example, if we consider it in "space-time" (i.e., ten-dimensional) terms.

The upshot of this very brief overview is that there are a variety of physical effects which can make stringy geometry differ significantly from conventional geometry, but all are controlled by two parameters, the string length and the string coupling. The important parameter in our subsequent discussion will be the string length l_s; when a geometric scale (curvature length, injectivity radius, volume of cycle) is small compared to l_s, stringy geometry (whatever it is) is relevant.

Note that in places (and commonly in the string literature), an alternate convention $\alpha' = l_s^2$ is used for this parameter.

1.1.2. Topological string theory, twisting and mirror symmetry. A fully general treatment of "stringy geometry" probably awaits a more complete and satisfactory mathematicization of quantum field theory. However there is a significant portion of the problem which can be satisfactorily understood within our current frameworks, namely the part which can be framed within topological string theory.

A good primary definition of topological string theory, or topological quantum field theory more generally, is as a geometric functor from a category of topological manifolds and cobordisms to a linear category. On this level, the subject is essentially mathematics, by which we mean that physics techniques do not have much to say.

Physics techniques become more valuable when we can relate a topological field theory to a physical quantum field theory, defined by a functional integral. There are two ways in which this can work. One is for the quantum field theory to be independent of the metric on Σ, as in Chern-Simons theory.[4] The other, which is relevant here, is "cohomological topological field theory," in which the theory contains a nilpotent operator Q such that the stress tensor (the operator generating infinitesimal variations of the metric) is Q-exact.

We discuss this construction for $(2,2)$ superconformal theory in §3.3, going into many details which we will need for the open string case. There are two possibilities, the A- and B-twists, which isolate different, essentially independent subsectors of the physical theory. Correlation functions in the A twisted theory (§3.4.1) depend only on complexified Kähler moduli, while those in the B-twisted theory (§3.4.2) depend only on complex structure moduli. The physics discussion is very asymmetric between the two theories – whereas the B-model can be completely understood in terms of standard geometry (variation of Hodge structure), instanton corrections in the A-model modify the algebra of operators from the classical de Rham cohomology ring to a new "quantum cohomology ring."

In terms of our discussion of stringy geometry, what makes the topological theory tractable is that almost all of the power-like (perturbative) corrections are absent, leaving (in the A-model) an interesting series of instanton corrections. These can be computed, for example by using localization in the functional integral, and summed to provide an explicit "invariant of stringy geometry."

As discussed in detail in MS1, closed string mirror symmetry equates the A-model on a Calabi-Yau manifold X to the B-model on a mirror Calabi-Yau manifold Y, usually with a fairly simple relation to X. We outline that part of the story which is essential for us in §3.4.3; to a good extent one can take the techniques of closed string mirror symmetry (localization, Picard-Fuchs equations, mirror maps and so forth) as a "black box" which will be called on at specific points in the open string story.

[4]It might have some minimal sort of dependence, such as the framing dependence of Chern-Simons theory.

1.1.3. Dirichlet branes. We can now explain our definition of a Dirichlet brane as an allowed end point for an open string. In an open string theory, the Hilbert space $\mathcal{H}$ should roughly look like a space of functionals on maps from the interval to M. Of course, an interval has two distinguished points, its start and end. The image of either of these points traces out a one-dimensional trajectory (or "world-line") in M. To complete the definition of open string, we must state boundary conditions for these endpoints.

The most general definition of these boundary conditions is phrased in terms of conformal field theory, and need not have any obvious interpretation in terms of a target space geometry. However, if our conformal field theory is a sigma model with target M, it is natural to look for such a picture. As we explain in §3.5, this leads to

DEFINITION 1.2. A *geometric* Dirichlet brane is a triple (L, E, ∇_E) – a submanifold $L \subset M$, carrying a vector bundle E, with connection ∇_E.

The real dimension of L is also often brought into the nomenclature, so that one speaks of a Dirichlet p-brane if $p = \dim_{\mathbb{R}} L$.

An open string which stretches from a Dirichlet brane (L, E, ∇_E) to a Dirichlet brane (K, F, ∇_F), is a map X from an interval $I \cong [0, 1]$ to M, such that $X(0) \in L$ and $X(1) \in K$. An "open string history" is a map from $\mathbb{R}$ into open strings, or equivalently a map from a two-dimensional surface with boundary, say $\Sigma \equiv I \times \mathbb{R}$, to M, such that the two boundaries embed into L and K.

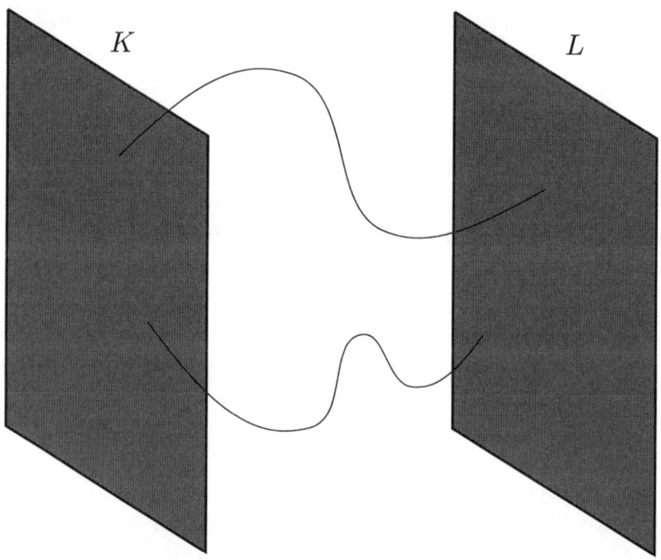

FIGURE 1. Open strings ending on D-branes.

The quantum theory of these open strings is defined by a functional integral over these histories, with a weight which depends on the connections

the submanifold M to be volume minimizing in the metric on X, and the connection ∇ to satisfy the Yang-Mills equations.

Like the Einstein equations, the equations governing a submanifold of minimal volume are highly nonlinear, and their general theory is difficult. This is one motivation to look for special classes of solutions; the physical arguments favoring supersymmetry are another.

Just as supersymmetric compactification manifolds correspond to a special class of Ricci-flat manifolds, those admitting a covariantly constant spinor, supersymmetry for a Dirichlet brane will correspond to embedding it into a special class of minimal volume submanifolds. Since the physical analysis is based on a covariantly constant spinor, this special class should be defined using the spinor, or else the covariantly constant forms which are bilinear in the spinor.

The standard physical arguments leading to this class are based on the kappa symmetry of the Green-Schwarz world-volume action, for which a good introduction is [**172**]. We will not explain this, but begin at its penultimate step, in which one finds that the subset of supersymmetry parameters ϵ which preserve supersymmetry, both of the metric and of the brane, must satisfy

$$\phi \equiv \operatorname{Re} \epsilon^t \Gamma \epsilon|_M = \operatorname{Vol}|_M. \tag{1.2}$$

In words, the real part of one of the covariantly constant forms on M must equal the volume form when restricted to the brane.

Clearly $d\phi = 0$, since it is covariantly constant. Thus,

$$Z(M) \equiv \int_M \phi$$

depends only on the homology class of M. Thus, it is what physicists would call a "topological charge," a "central charge" or a "BPS central charge," depending on context.

If in addition the p-form ϕ is dominated by the volume form Vol upon restriction to any p-dimensional subspace $V \subset T_x X$, i.e.,

$$\phi|_V \le \operatorname{Vol}|_V, \tag{1.3}$$

then ϕ will be a calibration in the sense of Harvey and Lawson [**226**]. This condition can be checked locally, but implies the global statement

$$\int_M \phi \le \int_M \operatorname{Vol} \tag{1.4}$$

for any submanifold M. Thus, the central charge $|Z(M)|$ is an absolute lower bound for $\operatorname{Vol}(M)$.

A calibrated submanifold M is now one satisfying (1.2), thereby attaining the lower bound and thus of minimal volume. Physically these are usually called "BPS branes," after a prototypical argument of this type due to Bogomol'nyi and Prasad-Sommerfield, for magnetic monopole solutions in nonabelian gauge theory.

For a Calabi-Yau X, all of the forms ω^p can be shown to be calibrations, and it is not hard to show that the corresponding calibrated submanifolds are p-dimensional holomorphic submanifolds. Furthermore, the n-form $\mathrm{Re}\, e^{i\theta}\Omega$ for any choice of real parameter θ is a calibration, and the corresponding calibrated submanifolds are called *special Lagrangian.*

The previous discussion generalizes to the presence of a general connection on M, and leads to the following two types of BPS branes for a Calabi-Yau X. Let $n = \dim_{\mathbb{R}} M$, and let F be the ($\mathrm{End}(E)$-valued) curvature two-form of ∇.

The first kind of BPS D-brane, based on the ω^p calibrations, is (for historical reasons) called a "B-type brane." Here the BPS constraint is equivalent to the following three requirements:

(1) M is a p-dimensional complex submanifold of X.
(2) The 2-form F is of type $(1,1)$, i.e., (E, ∇) is a holomorphic vector bundle on M.
(3) In the supergravity limit, F satisfies the Hermitian Yang-Mills equation:
$$\omega|_M^{p-1} \wedge F = c \cdot \omega|_M^p$$
for some real constant c.

Taking into account the l_s corrections of §1.1.1, the Hermitian Yang-Mills equation is deformed to the "MMMSL" equation [**347, 330**],

(3′) F satisfies $\mathrm{Im}\, e^{i\phi}(\omega|_M + il_s^2 F)^p = 0$ for some real constant ϕ.

Actually, this statement is not precise either, but the further corrections require a lengthier discussion, which we give in Chapter 3.

The second kind of BPS D-brane, based on the $\mathrm{Re}\, e^{i\theta}\Omega$ calibration, is called an "A-type" brane. The simplest examples of A-branes are the so-called special Lagrangian submanifolds (SLAGs), satisfying

(1) M is a Lagrangian submanifold of X with respect to ω.
(2) $F = 0$, i.e., the vector bundle E is flat.
(3) $\mathrm{Im}\; e^{i\alpha}\Omega|_M = 0$ for some real constant α.

More generally, one also has the "coisotropic branes." In the case when E is a line bundle, such A-branes satisfy the following four requirements:

(1) M is a coisotropic submanifold of X with respect to ω, i.e., for any $x \in M$ the skew-orthogonal complement of $T_xM \subset T_xX$ is contained in T_xM. Equivalently, one requires $\ker \omega_M$ to be an integrable distribution on M.
(2) The 2-form F annihilates $\ker \omega_M$.

(3) Let $\mathscr{F}M$ be the vector bundle $TM/\ker \omega_M$. It follows from the first two conditions that ω_M and F descend to a pair of skew-symmetric forms on $\mathscr{F}M$, which we denote by σ and f. Clearly, σ is nondegenerate. One requires the endomorphism $\sigma^{-1}f : \mathscr{F}M \to \mathscr{F}M$ to be a complex structure on $\mathscr{F}M$.
(4) Let r be the complex dimension of $\mathscr{F}M$. One can show that r is even and that $r + n = \dim_{\mathbb{R}} M$. Let Ω be the holomorphic trivialization of K_X. One requires that $\mathrm{Im}\, e^{i\alpha}\Omega|_M \wedge F^{r/2} = 0$ for some real constant α.

Coisotropic A-branes carrying vector bundles of higher rank are not fully understood.

Physically, one must also specify the embedding of the Dirichlet brane in the remaining (Minkowski) dimensions of space-time. The simplest possibility is to take this to be a time-like geodesic, so that the brane appears as a particle in the visible four dimensions. This is possible only for a subset of the branes, which depends on which string theory one is considering. Somewhat confusingly, in the type IIA theory, the B-branes are BPS particles, while in IIB theory, the A-branes are BPS particles (the notations were introduced before this relationship was known).

1.1.5. String theory and mirror symmetry. Of the various ways one can formulate mirror symmetry, perhaps the most useful for string theory is

CONJECTURE 1.3. *Type IIA string theory compactified on a Calabi-Yau threefold X is dual to type IIB string theory compactified on a mirror Calabi-Yau threefold Y.*

The word "dual" more or less means that, for any theory of the first type, there exists some isomorphic theory of the second type; in particular all physical predictions of the two theories are the same.

All of the known mathematical consequences of mirror symmetry can be derived from this conjecture, by matching the various physical observables. We will not go into all of its ramifications, as this would require going far deeper into the physics than we want for this book. Rather, we will focus on the consequence of central importance for us, namely,

CONJECTURE 1.4. *The set of BPS D-branes in type IIA theory compactified on X is isomorphic to the set of BPS D-branes in type IIB theory compactified on Y.*

Since BPS D-branes are particles which could be produced and detected by a hypothetical observer living in one of these space-times, this certainly follows from the main conjecture. Having made this conjecture, one might go on to test it by comparing lists of BPS Dirichlet branes in pairs of dual theories. Of course, this is difficult. A better approach might be to look for some simple construction which, given a BPS D-brane on the first list, produces its counterpart on the second.

On the other hand, a skeptic might ask the following question. All of the concepts we just introduced are easily explained in the standard language of differential geometry, and had been the focus of mathematical attention for some time. If mirror symmetry had a simple explanation in these terms, why should it have come as any surprise to mathematicians? This suggests that we need something new from string theory to motivate or explain mirror symmetry. While indeed much of our book will be devoted to explaining just what this new input is, in actual fact there was a mathematical proposal predating the physics we just outlined, so let us begin with that.

1.2. The homological approach

In his 1994 ICM talk, Kontsevich made the prophetic proposal that mirror symmetry could be explained through an equivalence between the bounded derived category of coherent sheaves $D^b(X)$ on a Calabi-Yau manifold X and the (derived) Fukaya category of its mirror Y. Objects in the Fukaya category are Lagrangian submanifolds of Y, while morphisms are elements of Floer cohomology. A derived equivalence between these two categories is a deeper version of an isomorphism between (in the odd-dimensional case) the even and odd cohomology of X and Y, respectively. Kontsevich predicted that such an equivalence lay behind the enumerative predictions of mirror symmetry.

This rather abstract proposal took some time to be appreciated by either mathematicians or physicists – when it was made the Dirichlet brane was almost unknown,[5] and the categories being equated are not part of the general working knowledge of mathematicians. Furthermore, while the proposal again has the great advantage of going beyond conventional differential and algebraic geometry, this means that motivating it requires somewhat more knowledge of string theory. For both of these reasons, our discussion must start with more background material.

Thus, in Chapter 3, we provide a review of the ingredients we will need from superconformal field theory. Since a good introduction can be found in MS1, we will not aim for completeness here, but rather focus on the following points. First, there is a close analogy between CFT and quantum mechanics, and many of the relations between quantum mechanics and mathematics (in particular, spectral geometry and Hodge theory) have simple generalizations to CFT. We then discuss the general theory of the $N = 2$ superconformal algebra, and the operation of topological twisting, which makes contact with our discussion in Chapter 2.

---

[5]What was known at this point was the definition of A- and B-type boundary conditions in topological sigma models [**463**], and this was Kontsevich's starting point.

Chapter 3 ends with an overview of supersymmetric boundary conditions. We explain the origin of the A- and B-type BPS conditions from this point of view, and the physics of T-duality.

We then switch to mathematics. In Chapter 4, we review algebraic preliminaries: homological algebra, coherent sheaves and their derived categories and derived equivalences.

The theory of quiver representations provides an ideal motivating example, in Chapter 4. Next we turn to the specific context in which homological mirror symmetry operates, beginning with a review of the notion of coherent sheaf on a variety X. Coherent sheaves form an abelian category: every morphism can be extended to an exact sequence using kernels and cokernels. However, the category of sheaves is not particularly well-behaved under certain natural operations such as pullback and push-forward. For example, pulling back or tensoring non-locally-free sheaves is not a pleasant operation, often leading to loss of information. A solution is offered by the derived category $\mathrm{D}(X)$ of the variety X. Objects in the derived category are complexes of coherent sheaves, but the notion of morphism is subtler than the notion of a morphism between complexes, making the derived category non-intuitive at first sight. Functors such as pullbacks, push-forwards, tensor products, and global sections have an extension to the derived category, with very natural properties. The notion of an exact sequence is replaced by that of an exact triangle, providing the analogue of the long exact sequence for the various functors. This theory is explained in a down-to-earth way in Chapter 4, with ample references to the literature where the technical details can be consulted.

A fundamental operation involving the derived category is the Fourier-Mukai transform. Given varieties X and Y and an object $P \in \mathrm{D}^b(X \times Y)$, with p_1, p_2 the projections, we obtain a functor $\mathrm{D}^b(X) \to \mathrm{D}^b(Y)$ via $E \mapsto p_{2*}(P \otimes p_1^* E)$. This can be viewed as a sheaf-theoretic analogue of the Fourier transform. Such a transform is most interesting when it is an equivalence of categories, when it is called a Fourier–Mukai functor. It was initially used by Mukai to prove that the derived categories of dual abelian varieties were equivalent. In particular this shows that the derived category $\mathrm{D}^b(X)$ does not necessarily determine the variety X. In Chapter 4 we recall the original functor of Mukai, and subsequent extensions to other (relative) Calabi–Yau contexts: flops, elliptic fibrations etc. As another illustration, we show how a Fourier–Mukai functor can be used to study the derived category of projective space $\mathbb{P}^n$ in terms of linear algebra using the simple set of generators $\mathcal{O}, \ldots, \mathcal{O}(-n)$ (Beilinson's trick). This theory is explained in Chapter 4.

Another area where the Fourier-Mukai transform has proved useful is in the McKay correspondence. A celebrated observation of John McKay states that the graph of ADE type associated to the quotient of $\mathbb{C}^2$ by a finite subgroup $G \subset \mathrm{SL}(2, \mathbb{C})$ can be constructed using only the representation theory of G. This establishes a one-to-one correspondence between exceptional prime divisors of the well known minimal resolution $Y \to \mathbb{C}^2/G$ and the nontrivial irreducible representations of G. If G is a finite subgroup of $SL(n, \mathbb{C})$ acting on $\mathbb{C}^n$, the natural generalisation of these ideas involves Nakamura's moduli space G-Hilb$(\mathbb{C}^n)$ of G-clusters on $\mathbb{C}^n$. He proved that G-Hilb$(\mathbb{C}^3)$ is a crepant resolution of $\mathbb{C}^3/G$ when G is Abelian and conjectured that the same holds for all finite subgroups $G \subset \mathrm{SL}(3, \mathbb{C})$. More generally, whenever $Y = G\text{-Hilb}(\mathbb{C}^n) \to \mathbb{C}^n/G$ is a crepant resolution, the universal bundle on Y determines locally free sheaves $\mathcal{R}_\rho$ on Y in one-to-one correspondence with the irreducible representations ρ of G, which, according to a proposal of Reid, should generate the K-theory or the derived category of Y analogously to the Beilinson generators of the derived category of $\mathbb{P}^n$. The McKay conjectures of both Nakamura and Reid were proved simultaneously for a finite subgroup $G \subset \mathrm{SL}(3, \mathbb{C})$ by Bridgeland, King and Reid, who established an equivalence of derived categories $\Phi\colon \mathrm{D}(Y) \to \mathrm{D}^G(\mathbb{C}^n)$ between the bounded derived category of coherent sheaves on Y and that of G-equivariant coherent sheaves on $\mathbb{C}^n$. Recent work of Craw and Ishii establishes a derived equivalence similar to Φ for any crepant resolution Y of $\mathbb{C}^3/G$, at least for finite Abelian subgroups of $\mathrm{SL}(3, \mathbb{C})$. In Chapter 4, we explain the ideas behind this circle of results.

At this point, we are now prepared to plunge into the physical origins and explanation of homological mirror symmetry which takes up Chapter 5.

1.2.1. Stability structures. To explain the basic point, we compare the objects entering homological mirror symmetry with the original geometric descriptions of BPS branes. Recall that the A-type branes are special Lagrangian manifolds; in homological mirror symmetry these are represented as isotopy classes of Lagrangians, a related but more general class of objects. And, while at first the B-type branes look more similar, being holomorphic objects (submanifolds or sheaves) in both cases, on looking at the connections we realize that a geometric B-type brane carries not just a holomorphic bundle or sheaf, but the more specific Hermitian Yang-Mills connection, satisfying an additional differential equation.

The precise relation between the two classes of B-type branes follows from the Donaldson and Uhlenbeck-Yau theorems. These state the necessary and sufficient conditions for a holomorphic bundle to admit an irreducible Hermitian Yang-Mills connection – it is that the bundle be μ-stable, i.e., stable in the sense of [**372**]. This is often true but not always, so the geometric B-type branes form a subset of the holomorphic objects. Furthermore, μ-stability (in complex dimensions two and higher) depends on the Kähler class of the underlying manifold, and thus the set of geometric B-type branes is **not** invariant under deformations of the Kähler data. Similar geometric considerations on the A side, first due to Joyce, show that the special Lagrangian submanifolds form a subset of the Lagrangian submanifolds. Each isotopy class of Lagrangian submanifolds contains either zero or one special Lagrangian, depending on a stability condition which varies with the complex structure of the underlying manifold.

Thus, we have a correspondence between stability conditions on the A and B sides, as required by open string mirror symmetry. However, from the point of view of topological string theory this is rather surprising, as the A- and B-twisted topological string theories only depend on the Kähler (for A) and complex structure (for B) moduli; they are not supposed to depend on the other moduli at all. This implies that the concept of BPS brane **cannot** be defined strictly within the topological string theory; one must bring in more information to make this definition.

But what makes this problem particularly accessible is that the stability conditions only depend on a small subset of the geometric information in the problem, the Kähler class for μ-stability and the period map for Joyce's stability condition. Thus, we can hope to formulate a single stability condition which includes both of the known geometric stability conditions as special cases, and use this to conjecture that a BPS brane is a stable object in either of the equivalent categories of topological boundary conditions. Such a stability condition, often called Pi-stability, was developed in work of Douglas, Aspinwall and Bridgeland.

To develop this picture, one must identify the elements of Kontsevich's proposal in the conformal field theory underlying the A- and B-twisted topological string theories. It turns out that only a few additional ingredients are necessary, mostly originating in the structure of boundary conditions for the $U(1)$ current algebra sector of the $N = 2$ superconformal algebra. This sector is the physical construct which underlies the grading of Dolbeault cohomology, and by taking these physical choices into account one can construct graded complexes from the original topological boundary conditions, leading directly to the derived category.

The main physical consequence of this construction is that the grading is not static but dynamic, in that while varying the "missing" moduli (say the Kähler moduli in the B-twisted model) preserves all the previously existing structure of topological string theory, the grading structure introduced at this point can vary. Understanding this "flow of gradings" leads fairly directly to the proposal of Pi-stability.

From a practical point of view, the most effective way to use the resulting framework is to use the B-model definition of topological branes, and take the information entering the stability condition from the **mirror** B-model, as in both cases the definitions and computations can be phrased in standard algebraic geometric terms (there are no world-sheet instanton corrections). While the derived category of coherent sheaves on a general compact Calabi-Yau manifold is not yet well understood, for hypersurfaces in projective spaces many sheaves can be obtained by restriction, providing material for simple examples. Furthermore, the techniques of Chapter 4 provide complete and explicit quiver descriptions of the derived category in a large class of "local" Calabi-Yau manifolds, such as those obtained by the resolution of quotient singularities. On the closed string side, computing periods is a highly developed art, because of its applications in closed string mirror symmetry, allowing us to exhibit many simple examples of Pi-stability and its variation, which can be checked against other physical constructions of BPS branes.

Following Bridgeland, the further analysis of Pi-stability requires introducing several additional concepts, such as a generalized Harder-Narasimhan filtration. It is more convenient at this point not to strictly follow the physics but instead to axiomatize the concept of "stability structure" as the most general realization of these concepts. One can then show that the space of stability structures forms an open manifold, which includes the physical examples as a submanifold.

1.2.2. Comparison of A_∞ structures. The discussion we just made combines A- and B-models and in this sense goes beyond topological string theory. If we ask about the consequences of mirror symmetry for topological string theory, it is natural to look for a quantity analogous to the prepotential which encodes the variation of Hodge structure and correlation functions in the closed string case.

Physically, this is provided by the superpotential, which can be regarded as generating open string correlation functions in a very analogous manner. In this language, the basic enumerative prediction of mirror symmetry is to count disks with specified homology class and bounding specified special Lagrangian manifolds in terms of the series expansions of suitable open string B-model correlation functions.

While this approach has been pursued successfully, it ignores some crucial differences between the closed and open string cases, which in some ways are more interesting than the actual enumerative predictions. The first of these, in some ways elementary but still significant, is that – as indicated in §1.1.3 – the superpotential is best thought of as a function of noncommuting variables. This is because an open string correlation function corresponds to a set of operators on the boundary of a disk, which comes with an ordering.

A deeper difference is that whereas the moduli spaces which appear in the closed string theory are unobstructed, deformations of vector bundles on a Calabi-Yau can be obstructed. This obstruction theory turns out to be precisely governed by the superpotential – an unobstructed deformation is one for which all gradients of the superpotential vanish. As a consequence of this, the spectrum of operators of the topological open string theory can vary under deformation.

Such a structure is more naturally described, not by an associative category, but by an A_∞ category. Such a structure also emerges naturally from the construction of correlation functions on the boundary of a disk, and indeed this is how it was first seen, in Fukaya's construction of a category based on Lagrangian manifolds and Floer homology. This structure was developed before the physics concept of Dirichlet brane and was the direct motivation for Kontsevich's proposal.

In retrospect, some modifications to the open string A- and B-models are required to fully realize Kontsevich's proposal. On the A-model side, the Fukaya category did not realize the triangulated structure of the derived category of coherent sheaves; this can be remedied by the so-called "twist construction" of Bondal and Kapranov [**50**]. On the B-model side, one has to work a bit to see the A_∞ structure; we explain this (following Kontsevich and Soibelman [**311**]) in §8.2.1.

Finally having a precise formulation of open string mirror symmetry, we illustrate it by working out the basic relations for the case of M an elliptic curve in §8.4.

1.3. SYZ mirror symmetry and T-duality

Although the formalism of homological mirror symmetry is very powerful, one may reasonably ask for other explanations of mirror symmetry which lie closer to classical differential and algebraic geometry. This brings us to the proposal of Strominger, Yau and Zaslow.

The central physical ingredient in this proposal is T-duality. To explain this, let us consider a superconformal sigma model with target space (M, g), and denote it (defined as a geometric functor, or as a set of correlation functions), as

$$\mathrm{CFT}(M, g).$$

In physics terms, a "duality" is an equivalence

$$\mathrm{CFT}(M, g) \cong \mathrm{CFT}(M', g')$$

which holds despite the fact that the underlying geometries (M, g) and (M', g') are not classically diffeomorphic. Rather, one must use the "stringy" features outlined in §1.1.1 to see the equivalence.

T-duality is a duality which relates two CFT's with toroidal target space, $M \cong M' \cong T^d$, but different metrics. In rough terms, the duality relates a "small" target space, with noncontractible cycles of length $L < l_s$, with a "large" target space in which all such cycles have length $L > l_s$.

This sort of relation is generic to dualities and follows from the following logic. If all length scales (lengths of cycles, curvature lengths, etc.) are greater than l_s, string theory reduces to conventional geometry. Now, in conventional geometry, we know what it means for (M, g) and (M', g') to be non-isomorphic. Any modification to this notion must be associated with a breakdown of conventional geometry, which requires some length scale to be "sub-stringy," with $L < l_s$.

To state T-duality precisely, let us first consider $M = M' = S^1$. We parameterise this with a coordinate $X \in \mathbb{R}$ making the identification $X \sim X + 2\pi$. Consider a Euclidean metric g_R given by $ds^2 = R^2 dX^2$. The real parameter R is usually called the "radius" from the obvious embedding in $\mathbb{R}^2$. This manifold is Ricci-flat and thus the sigma model with this target space is a conformal field theory, the "$c = 1$ boson." Let us furthermore set the string scale $l_s = 1$.

As discussed in elementary textbooks on string theory [**395**], and as we will prove in §3.2.3.6, there is a complete physical equivalence

$$\mathrm{CFT}(S^1, g_R) \cong \mathrm{CFT}(S^1, g_{1/R}).$$

Thus these two target spaces are *indistinguishable* from the point of view of string theory.

Just to give a physical picture for what this means, suppose for sake of discussion that superstring theory describes our universe, and thus that in some sense there must be six extra spatial dimensions. Suppose further that we had evidence that the extra dimensions factorized topologically and metrically as $K_5 \times S^1$; then it would make sense to ask: What is the radius R of this S^1 in our universe? In principle this could be measured by producing sufficiently energetic particles (so-called "Kaluza-Klein modes"), or perhaps measuring deviations from Newton's inverse square law of gravity at distances $L \sim R$. In string theory, T-duality implies that $R \geq l_s$, because any theory with $R < l_s$ is equivalent to another theory with $R > l_s$. Thus we have a nontrivial relation between two (in principle) observable quantities, R and l_s, which one might imagine testing experimentally.

Returning to the general discussion, let us now consider the theory $\mathrm{CFT}(T^d, g)$, where T^d is the d-dimensional torus, with coordinates X^i parameterising $\mathbb{R}^d/2\pi\mathbb{Z}^d$, and a constant metric tensor g_{ij}. Then there is a complete physical equivalence

$$\mathrm{CFT}(T^d, g) \cong \mathrm{CFT}(T^d, g^{-1}). \tag{1.5}$$

In fact this is just one element of a discrete group of T-duality symmetries, generated by T-dualities along one-cycles, and large diffeomorphisms (those not continuously connected to the identity). The complete group is isomorphic to $\mathrm{SO}(d, d; \mathbb{Z})$.[6]

While very different from conventional geometry, T-duality has a simple intuitive explanation. This starts with the observation that the possible embeddings of a string into X can be classified by the fundamental group $\pi_1(X)$. Strings representing non-trivial homotopy classes are usually referred

[6]For comparison, the group $\mathrm{Diff} / \mathrm{Diff}_0 \cong \mathrm{SL}(d, \mathbb{Z})$.

to as "winding states." Furthermore, since strings interact by interconnecting at points, the group structure on π_1 provided by concatenation of based loops is meaningful and is respected by interactions in the string theory. Now $\pi_1(T^d) \cong \mathbb{Z}^d$, as an abelian group, referred to as the group of "winding numbers" for evident reasons.

Of course, there is another $\mathbb{Z}^d$ we could bring into the discussion, the Pontryagin dual of the $U(1)^d$ of which T^d is an affinization. An element of this group is referred to physically as a "momentum," as it is the eigenvalue of a translation operator on T^d. Again, this group structure is respected by the interactions. These two group structures, momentum and winding, can be summarized in the statement that the full closed string algebra contains the group algebra $\mathbb{C}[\mathbb{Z}^d] \oplus \mathbb{C}[\mathbb{Z}^d]$.

In essence, the point of T-duality is that if we quantize the string on a sufficiently small target space, the roles of momentum and winding will be interchanged. This can be seen by a short functional integral argument which will appear in §3.2.3.6. But the main point can be seen by bringing in some elementary spectral geometry. Besides the algebra structure we just discussed, another invariant of a conformal field theory is the spectrum of its Hamiltonian H (technically, the Virasoro operator $L_0 + \bar{L}_0$). This Hamiltonian can be thought of as an analog of the standard Laplacian Δ_g on functions on X, and it is easy to see that its spectrum on T^d with metric g as above is

$$\text{Spec}\, \Delta_g = \{\sum_{i,j=1}^{d} g^{ij} p_i p_j;\ p_i \in \mathbb{Z}^d\}. \tag{1.6}$$

On the other hand, the energy of a winding string is (as one might expect intuitively) a function of its length. On our torus, a geodesic with winding number $w \in \mathbb{Z}^d$ has length squared

$$L^2 = \sum_{i,j=1}^{d} g_{ij} w^i w^j. \tag{1.7}$$

Now, the only string theory input we need to bring in is that the total Hamiltonian contains both terms,

$$H = \Delta_g + L^2 + \cdots$$

where the extra terms $\cdots$ express the energy of excited (or "oscillator") modes of the string. Then, the inversion $g \rightarrow g^{-1}$, combined with the interchange $p \leftrightarrow w$, leaves the spectrum of H invariant. This is T-duality.

There is a simple generalization of the above to the case with a non-zero B-field on the torus satisfying $dB = 0$. In this case, since B is a constant antisymmetric tensor, we can label CFT's by the matrix $g + B$. Now, the basic T-duality relation becomes

$$\mathrm{CFT}(T^d, g+B) \cong \mathrm{CFT}(T^d, (g+B)^{-1}).$$

Another generalization, which is considerably more subtle, is to do T-duality in families, or fiberwise T-duality. The same arguments can be made, and would become precise in the limit that the metric on the fibers varies on length scales far greater than l_s, and has curvature lengths far greater than l_s. This is sometimes called the "adiabatic limit" in physics.

While this is a very restrictive assumption, there are more heuristic physical arguments that T-duality should hold more generally, with corrections to the relations we discussed proportional to curvatures $l_s^2 R$ and derivatives $l_s \partial$ of the fiber metric, both in perturbation theory and from world-sheet instantons. These corrections have not been much studied, which is unfortunate as they would probably shed much light on the subtleties involved in making the SYZ conjecture precise, as we will discuss below.

1.3.1. T-duality and Dirichlet branes. The discussion we just made was for closed strings. Clearly maps from an interval to a manifold are not classified by π_1 and indeed there is no analogous choice. How then can the T-duality relation hold for open strings?

This is the question which led to the original discovery of Dirichlet branes [**106**]. Suppose we start with open strings which are free to propagate anywhere in T^d, in modern terms with a Dirichlet d-brane wrapping T^d. While there is no winding number in this case, since such a string state is a function on T^d, we can still apply Pontryagin duality and conclude that this open string algebra contains $\mathbb{C}[\mathbb{Z}^d]$. In physics terms, there is a d-dimensional conserved momentum. Furthermore, the open string Hamiltonian will still contain a piece which looks like the Laplacian on T^d, whose spectrum will still be (1.6).

If we apply the inversion $g \to g^{-1}$, clearly the simplest way to recover the original spectrum is to identify a *new* open string sector in which the spectrum is again possible values of (1.7). While this is not true for open strings on a d-brane wrapping T^d, it could be true if we forced the two endpoints of the open strings to coincide, as the minimal length of a geodesic satisfying this condition is again (1.7). Furthermore, since π_1 is commonly defined using based loops (of course here it will not depend on whether or not there is a base point), a sector of open strings which are forced to begin and end at a specific point p will again contain $\mathbb{C}[\mathbb{Z}^d]$ as an algebra.

Thus, the simple proposal, which will be justified by functional integral arguments in §3.5.4, is that the T-dual to theory $\mathrm{CFT}(T^d, g)$ containing a Dirichlet d-brane is the theory $\mathrm{CFT}(T^d, g^{-1})$ containing a Dirichlet 0-brane. Such a brane is defined by a choice of zero-dimensional submanifold, i.e., a point $p \in X$, and its open strings must begin or end at p.

This reproduces the spectrum, but now one must ask what the choice of p corresponds to in the original d-brane theory. The beautiful answer to this question is that to complete the specification of the d-brane, we must also specify a bundle E and connection ∇. It is plausible (and correct) that the T-dual of the 0-brane is a d-brane with trivial bundle E, and a flat connection. But the moduli space of flat connections on T^d is itself a torus; denote this by $\tilde{T}^d$. Then a slightly non-trivial statement, which one can check, is that the natural metric on $\tilde{T}^d$, obtained by restricting the natural metric on the space of connections on T^d (with metric g) to the flat connections, is the flat metric g^{-1}. While the overall scale of the metric on the space of connections is undermined, string theory determines this relation to be precisely (1.5). Thus the moduli spaces of the proposed pair of T-dual Dirichlet brane theories coincide as metric spaces.

1.3.2. Mirror Symmetry and Special Lagrangian Fibrations. We are now ready to explain the Strominger-Yau-Zaslow proposal [**433**]. Consider a pair of compact Calabi-Yau 3-folds X and Y related by mirror symmetry. By the above, the set of BPS A-branes on X is isomorphic to the set of BPS B-branes on Y, while the set of BPS B-branes on X is isomorphic to the set of BPS A-branes on Y.

The simplest BPS B-branes on X are points. These exist for all complex structures on X, even nonalgebraic ones. Their moduli space is X itself. Let us try to determine which BPS A-branes on Y they correspond to. The conditions on BPS A-branes described above imply that an A-brane can have real dimension 5 or 3. However, the conditions for the existence of 5-dimensional A-branes depend sensitively on the the symplectic structure; for example, rescaling the symplectic form by a constant factor, in general, will eliminate such A-branes, because the curvature 2-form of a line bundle has quantized periods. In contrast, special Lagrangian submanifolds remain special Lagrangian if we rescale the symplectic form by an arbitrary constant factor. This matches the properties of points on X. Thus the mirror of a point on X must be a three-dimensional A-brane (N, E, ∇). BPS conditions imply in this case that $F = 0$ (we assume for simplicity that $B = 0$) and that N is a special Lagrangian submanifold of Y. Thus we conclude that there exists a family of SLAGs on Y parametrized by points of X. Moreover, this family is the moduli space of a single SLAG N regarded as a BPS A-brane on Y.

What else can we say about this special family of SLAGs on Y? According to McLean's theorem (§6.1.1), the moduli space of a special Lagrangian submanifold N is locally smooth and has dimension $b_1(N)$. A BPS A-brane is a SLAG equipped with a Hermitian line bundle E and a flat connection ∇ on E. The moduli space of flat connections on a fixed N is a torus of real dimension $b_1(N)$. Thus the total dimension of the moduli space of (N, E, ∇) is $2b_1(N)$. Since this moduli space is X, we must have $b_1(N) = 3$. It follows that X is fibered by tori of dimension 3. Exchanging the roles of Y and X we conclude that Y is also fibered by three-dimensional tori.

The T^3-fibrations of X and Y obtained in this way cannot be smooth everywhere (otherwise the Euler characteristic of both X and Y would be zero). Singularities occur when the deformed SLAG ceases to be smooth.

Next we would like to argue that the smooth fibers of both T^3 fibrations are themselves SLAGs. Let us imagine that the induced metric on the fibers of the T^3 fibration of X is flat. This assumption is unrealistic, but we may hope that in the large volume limit this is a good approximation away from the singular fibers. Then we may perform T-duality on the (nonsingular) fibers and obtain a (noncompact) Calabi-Yau manifold X' which is mirror to (an open piece of) X. T-duality maps a point $p \in X$ sitting in a fiber M into a D-brane of the form (M', E, ∇), where the 3-torus M' is dual to M and ∇ is a flat connection determined by the location of p on M. Varying p on X will deform M' as well as the flat connection ∇. This strongly suggests that X' is an open piece of Y, and that (M', E, ∇) is the A-brane N mirror to p. Then the T^3-fibrations of X and Y are (approximately) T-dual to each other. Furthermore, if we consider a point on $N = M'$, then its T-dual is M. But since any point on Y is a BPS B-brane, its T-dual M must be be a BPS A-brane on X. Therefore each nonsingular fiber of the T^3 fibration of X is a SLAG. Reversing the roles of X and Y, we conclude that nonsingular fibers of the T^3 fibration of Y are also SLAGs.

We summarize with the following

CONJECTURE 1.5. *(Strominger-Yau-Zaslow): For any mirror pair of compact simply-connected Calabi-Yau 3-folds X and Y, there should exist T^3 fibrations of X and Y which have the following two properties:*

- *Their nonsingular fibers are special Lagrangian submanifolds.*
- *If one takes the large volume limit for X (and the corresponding "large complex structure" limit for Y), the two fibrations are T-dual to each other.*

Since the physical arguments are based on genus zero CFT, there is no evident restriction to six real dimensions and $\hat{c} = 3$. Thus, we may further conjecture that the SYZ proposal is valid not only for 3-folds, but also for Calabi-Yau mirror pairs of arbitrary dimension d, now predicting T^d fibration structures.

1.3.3. Mathematics of the SYZ conjecture. As we saw, physics suggests a fairly simple picture of mirror symmetry; for non-singular fibres, it follows by applying the operation of T-duality to the tori. On the other hand, the work of SYZ did not explain how to deal with singular fibres, nor have subsequent physical developments really filled this gap.

But since the initial paper of 1996, there has been much mathematical progress in understanding the conjecture. Important work of Hitchin described natural structures which appear on the base of special Lagrangian fibrations. Suppose that $f : X \to B$ is the special Lagrangian fibration anticipated by Conjecture 1.5. If $B_0 = \{b \in B | f^{-1}(b) \text{ is a non-singular torus}\}$, then B_0 carries a so-called Hessian or affine Kähler manifold structure. These structures for dual fibrations should then be related by a natural duality procedure which is essentially just a Legendre transform.

We will describe these structures in detail, and use this to give a complete picture of the structures which arise in *semi-flat* mirror symmetry in §6.2. This is the situation where $B = B_0$, and the Ricci-flat metric on X restricts to a flat metric on each fibre of f. In this case, all of the information about X can be recovered from structures on B, and one begins to realize that the crucial objects in mirror symmetry are not the Calabi-Yau manifolds themselves, but rather the affine Kähler bases. We will explore this in detail. We will be able to see the mirror isomorphism between complex and Kähler moduli promised to us by mirror symmetry, and in addition describe mirror symmetry at a deeper level as a phenomenon which interchanges certain Lagrangian submanifolds on X with vector bundles or coherent sheaves on Y. This shows how the SYZ conjecture is related to the homological mirror symmetry conjecture. This material will be covered in §6.3.

To more fully explore approaches to the SYZ conjecture, we will take several approaches. First, to conclude the initial discussion of the SYZ conjecture in Chapter 6, we will show how to construct torus fibrations which compactify to non-trivial Calabi-Yau manifolds, such as the quintic in $\mathbb{P}^4$. This requires an understanding of singular fibres, which we cover in Chapter 6. In this case, we are able to construct topological torus fibrations where the tori degenerate over a codimension two locus $\Delta \subseteq B$, called the discriminant locus. This discriminant locus turns out to be a trivalent graph in the case of the quintic. However, this is a purely topological approach, and does not address metric aspects of the conjecture.

The issue of metrics is addressed in Chapter 7, in which we will abandon the comfort of the semi-flat case. We will begin with a number of points of view for describing and producing examples of Ricci-flat and other special metrics, drawing on recent work of Hitchin.

For compact Calabi-Yau manifolds, it is a deep theorem of Yau that there exists a Kähler Ricci-flat metric for each Kähler class on the manifold. However, there is not a single, explicit, non-trivial example known. On the other hand, in the non-compact case, in the presence of symmetries, it is often possible to reduce the complicated partial differential equation governing Ricci-flatness to an ordinary differential equation. There are some general ansätze applicable in such a situation, such as the Gibbons-Hawking ansatz. We discuss a selection of examples in Chapter 7.

In Chapter 7 we also address some of the key recent discoveries of Joyce on special Lagrangian fibrations. While the semi-flat and topological SYZ pictures discussed in Chapter 6 fit well with the original SYZ picture, Joyce developed a picture of general special Lagrangian fibrations which does not fit with this picture. In particular, he dispelled hopes that special Lagrangian fibrations were necessarily C^∞, and as a result, one does not expect that the discriminant locus of a special Lagrangian fibration need be a nice codimension two object. Thus the topological picture for the quintic developed in Chapter 6, with a graph as discriminant locus, seems to be only an approximation to the hypothetical special Lagrangian fibration, where we would now expect to have some fattening of this graph as discriminant locus. In addition, evidence developed by Joyce suggests that the dualizing procedure should actually change this fattened discriminant locus. Thus, the original strong version of SYZ, in which there is a precise duality between the fibres, cannot hold, for one might have a situation where $f^{-1}(b)$ is non-singular but $\check{f}^{-1}(b)$ is singular, for dual fibrations. Thus we are forced to revise the metric version of the SYZ conjecture.

This revision has been developed in work by Kontsevich and Soibelman, and Gross and Wilson. The basic idea is that we expect SYZ to hold only near a so-called *large complex structure limit point.* Essentially, this means that we are given a degenerating family $\pi : \mathcal{X} \to S$ of Calabi-Yau manifolds, with some point $0 \in S$ having $\pi^{-1}(0)$ being an extremely singular Calabi-Yau variety, where "extremely singular" can be made precise. Then we expect that for $t \in S$ near 0, the fibre $\mathcal{X}_t$ will carry a special Lagrangian fibration, though perhaps with some quite bad properties, and perhaps only on some open subset of $\mathcal{X}_t$. However, as $t \to 0$, we hope that this special Lagrangian fibration will improve its behavior, and "tend" towards some well-behaved limit. We will make this precise in Chapter 7. Current evidence suggests that this is closely related to the global behaviour of the Ricci-flat metric on $\mathcal{X}_t$ as $t \to 0$. In particular, we can define a precise notion of limit of metric spaces, and one can conjecture that $\mathcal{X}_t$ actually exhibits collapsing as $t \to 0$, namely the special Lagrangian tori shrink to points, resulting in a limit which is a manifold of half the dimension of that of $\mathcal{X}_t$. This limit manifold should in fact coincide with the base of the special Lagrangian fibration. This allows one to formulate a limiting form of the SYZ conjecture, and currently this looks like the most viable differential geometric form of the conjecture.

CHAPTER 2

D-branes and K-theory in 2D topological field theory

Let us begin our study of Dirichlet branes in the simplest possible context, that of two dimensional *topological* field theory (TFT). As explained in MS1 and as we will review, a 2d closed TFT is a finite dimensional commutative Frobenius algebra $\mathcal{C}$. Our goal in this chapter is to generalize this result to 2d topological open and closed TFT.

Starting with a geometric category of cobordisms between 1-manifolds with boundary, we shall show in §2.1 that describing the sewing relations and their solutions is a non-trivial but tractable problem. In rough terms, the result is that D-branes (boundary conditions) correspond to modules over the closed string algebra $\mathcal{C}$.

One corollary of this result is a general relation between D-branes and K-theory. This relation will play a central role throughout our book, and later we will give a variety of mathematical and physical arguments for it, particularly in §5.1. However we see here that its origin is far more primitive; ultimately it follows just from the sewing constraints.

If $\mathcal{C}$ is semisimple, we can go on and completely classify the D-branes, and do so in §§2.2, 2.3. This captures the physics of a zero-dimensional target space.

In §§2.4, 2.5 we begin working up to higher-dimensional examples, for which $\mathcal{C}$ is not semisimple. Perhaps the simplest examples are the "Landau-Ginzburg models," in which $\mathcal{C}$ is a Jacobian algebra of functions modulo an ideal generated by gradients. To go further, one can look at natural algebras based on the cohomology of the target space.

In §2.6 we will extend these results to the equivariant case, where we are given a finite group G, and the worldsheets are surfaces equipped with G-bundles. This is relevant for the classification of D-branes in orbifolds.

2.0.1. Summary of results. Our main results are the following two theorems, which we state now but whose meanings will become clearer in what follows. To state the first we must point out that a semisimple commutative Frobenius algebra[1] $\mathcal{C}$ is automatically the algebra of complex-valued

[1] A Frobenius algebra $\mathcal{C}$ — commutative or otherwise — means an algebra over the complex numbers equipped with a linear map $\theta : \mathcal{C} \to \mathbb{C}$ such that the pairing $(a, a') \mapsto \theta(aa')$ is a non-degenerate bilinear form on $\mathcal{C}$. The classical example is the complex

functions on the finite set $X = \mathrm{Spec}(\mathcal{C})$ of algebra homomorphisms from $\mathcal{C}$ to $\mathbb{C}$. (For a finite dimensional algebra this agrees with the definition of Spec which is usual in algebraic geometry.) We think of X as a "space-time" which is equipped with a "volume-form" or "dilaton field" θ which assigns the measure θ_x to each point $x \in X$.

THEOREM A. *For a semisimple 2-dimensional TFT, corresponding to a finite space-time (X, θ), the choice of a maximal category of D-branes fixes a choice of a square root of θ_x for each point x of X. The category of boundary conditions is equivalent to the category Vect(X) of finite-dimensional complex vector bundles on X. The correspondence is, however, not canonical, but is arbitrary up to composition with an equivalence Vect(X) $\to$* Vect(X) *given by tensoring each vector bundle with a fixed line bundle (i.e., one which does not depend on the particular D-brane).*

Conversely, every semisimple Frobenius category $\mathcal{B}$ is the category of boundary conditions for a canonical 2-dimensional TFT, whose corresponding commutative Frobenius algebra is the ring of endomorphisms of the identity functor of $\mathcal{B}$.

We shall explain in the next section the sense in which the boundary conditions form a category. The theorem will be proved in §2.2. In §2.2.4 we shall describe an analogue of the theorem for spin theories.

The second theorem relates to "G-equivariant" or "G-gauged" TFTs, where G is a finite group. Turaev has shown that in dimension 2 a semisimple G-equivariant TFT corresponds to a finite space-time X on which the group G acts in a given way, and which is equipped with a G-invariant dilaton field θ and as well as a "B-field" B representing an element of the equivariant cohomology group $H^3_G(X, \mathbb{Z})$.

THEOREM B. *For a semisimple G-equivariant TFT corresponding to a finite space-time (X, θ, B) the choice of a maximal category of D-branes fixes a G-invariant choice of square roots $\sqrt{\theta_x}$ as before, and then the category is equivalent to the category of finite-dimensional B-twisted G-vector bundles on X, up to an overall tensoring with a G-line bundle.*

In this case the category of D-branes is equivalent to that of the "orbifold" theory obtained from the gauged theory by integrating over the gauge fields, and it does not remember the equivariant theory from which the

group-algebra $\mathbb{C}[G]$ of a finite group G, where for a linear combination $a = \sum \lambda_g g$ of group elements we define $\theta(a) = \lambda_1$. Another example is the cohomology algebra of a compact oriented manifold with complex coefficients. For basic material on Frobenius algebras see, for example, ch.9 of [**104**], or [**142**].

orbifold theory arose. There is, however, a natural enrichment which does remember the equivariant theory.

This will be explained and proved in §2.6.

When these theorems apply, they provide a complete answer to our main questions. But, as we will see, the restriction to the semisimple case makes them of limited applicability, essentially only to the case of a zero-dimensional target space. Nevertheless, it is worth discussing them in detail, because this is by far the simplest way to understand the essential structure of the theory. In subsequent chapters, we will develop the formalism required to go beyond the semisimple case. Here, we foreshadow this in §2.1 and §2.5, explaining how the category of boundary conditions is naturally an A_∞ category in the sense of Fukaya, Kontsevich, and others.

2.1. The sewing theorem

2.1.1. Definition of open and closed 2D TFT. Roughly speaking, a d-dimensional quantum field theory is a particular kind of rule which assigns a number — called the *partition function* of the theory — to each closed d-dimensional manifold with appropriate structure. What makes the rule a quantum field theory is the way the partition function behaves when the closed manifold is subdivided. Mathematically, the structure can be conveniently formalized as a *functor* from a geometric category to a linear category. The simplest example is a *topological field theory*, where we choose the geometric category to be the category whose objects are closed, oriented $(d-1)$-manifolds, and whose morphisms are oriented cobordisms (two such cobordisms being identified if they are diffeomorphic by a diffeomorphism which is the identity on the incoming and outgoing boundaries). The linear category in this case is simply the category of complex vector spaces and linear maps, and the only property we require of the functor is that (on objects and morphisms) it takes disjoint unions to tensor products. A closed d-manifold can be regarded as a cobordism from the empty $(d-1)$-dimensional manifold to itself, and the tensoring axiom implies that the vector space assigned to the empty $(d-1)$-manifold is just the complex numbers, so the theory assigns a 1×1 matrix — i.e., a complex number — to each closed d-manifold. This is the partition function of the theory. The case $d = 2$ is of course especially well known and understood.

There are several natural ways to generalize the geometric category. One may, for example, consider manifolds equipped with some additional structure such as a Riemannian metric or a spin structure. Bringing in the metric will quickly involve us in quantum field theory in all of its complexity, so we postpone this to Chapter 3. One can however go partway by incorporating spin structure, which we do in §2.1.6.

The main focus in this chapter is on a different kind of generalization, where the partition function is defined not just for a closed manifold but for an oriented d-manifold with a boundary whose connected components have been labelled with elements of a fixed set $\mathcal{B}_0$, called the set of *boundary conditions.* This is formalized by taking the objects of the geometric category to be oriented $(d-1)$-manifolds with boundary, with each boundary component labelled with an element of the set $\mathcal{B}_0$. In this case a cobordism from Y_0 to Y_1 means a d-manifold X whose boundary consists of three parts, $\partial X = Y_0 \cup Y_1 \cup \partial_{\mathrm{cstr}} X$, where the "constrained boundary" $\partial_{\mathrm{cstr}} X$ is a cobordism from ∂Y_0 to ∂Y_1. Furthermore, we require the connected components of $\partial_{\mathrm{cstr}} X$ to be labelled with elements of $\mathcal{B}_0$ in agreement with the labelling of ∂Y_0 and ∂Y_1.

Thus when $d = 2$ the objects of the geometric category are disjoint unions of circles and oriented intervals with labelled ends. A functor from this category to complex vector spaces which takes disjoint unions to tensor products will be called an *open and closed topological field theory*: such theories will give us a "baby" model of the theory of D-branes. We shall always write $\mathcal{C}$ for the vector space associated to the standard circle S^1, and $\mathcal{O}_{ab}$ for the vector space associated to the interval with ends labelled by $a, b \in \mathcal{B}_0$ oriented from b to a (i.e., so that it is a cobordism from the point b to the point a, and NOT the other way round).

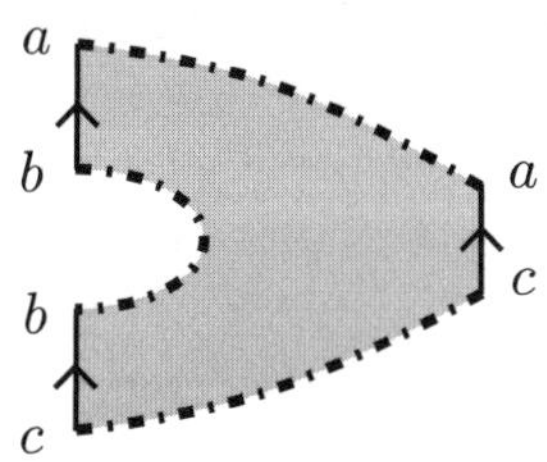

FIGURE 1. Basic cobordism on open strings. In this and all following open string diagrams we indicate the constrained boundary by a dot-dash line.

The cobordism of Figure 1 gives us a linear map $\mathcal{O}_{ab} \otimes \mathcal{O}_{bc} \to \mathcal{O}_{ac}$, or equivalently a bilinear map

$$\mathcal{O}_{ab} \times \mathcal{O}_{bc} \to \mathcal{O}_{ac}, \tag{2.1}$$

which we think of as a composition law. In fact we have a $\mathbb{C}$-linear *category* $\mathcal{B}$ whose objects are the elements of $\mathcal{B}_0$, and whose set of morphisms from b to a is the vector space $\mathcal{O}_{ab}$, with composition of morphisms given by (2.1). (To say that $\mathcal{B}$ is a $\mathbb{C}$-linear category means no more than that the bilinear composition (2.1) is associative in the obvious sense, and that there is an identity element $1_a \in \mathcal{O}_{aa}$ for each $a \in \mathcal{B}_0$; we shall explain presently why these properties hold.)

For any open and closed TFT we have a map $e : \mathcal{C} \to \mathcal{C}$ defined by the cylindrical cobordism $S^1 \times [0,1]$, and a map $e_{ab} : \mathcal{O}_{ab} \to \mathcal{O}_{ab}$ defined by the square $[0,1] \times [0,1]$. Clearly $e^2 = e$ and $e_{ab}^2 = e_{ab}$. If all these maps are identity maps we say the theory is *reduced.* There is no loss in restricting ourselves to reduced theories, and we shall do so from now on.

2.1.2. Algebraic characterization. The most general 2D open and closed TFT, formulated as in the previous section, is given by the following algebraic data:

1. $(\mathcal{C}, \theta_{\mathcal{C}}, 1_{\mathcal{C}})$ is a commutative Frobenius algebra.

2a. $\mathcal{O}_{ab}$ is a collection of vector spaces for $a, b \in \mathcal{B}_0$ with an associative bilinear product

$$\mathcal{O}_{ab} \otimes \mathcal{O}_{bc} \to \mathcal{O}_{ac}. \tag{2.2}$$

2b. The $\mathcal{O}_{aa}$ have non-degenerate traces

$$\theta_a : \mathcal{O}_{aa} \to \mathbb{C}. \tag{2.3}$$

In particular, each $\mathcal{O}_{aa}$ is a not necessarily commutative Frobenius algebra.

2c. Moreover,

$$\mathcal{O}_{ab} \otimes \mathcal{O}_{ba} \to \mathcal{O}_{aa} \stackrel{\theta_a}{\to} \mathbb{C} \tag{2.4}$$

is a perfect pairing, and

$$\theta_a(\psi_1\psi_2) = \theta_b(\psi_2\psi_1) \tag{2.5}$$

for $\psi_1 \in \mathcal{O}_{ab}, \psi_2 \in \mathcal{O}_{ba}$.

3. There are linear maps:

$$\begin{aligned} \iota_a : \mathcal{C} &\to \mathcal{O}_{aa} \\ \iota^a : \mathcal{O}_{aa} &\to \mathcal{C} \end{aligned} \tag{2.6}$$

such that

3a. ι_a is an algebra homomorphism

$$\iota_a(\phi_1\phi_2) = \iota_a(\phi_1)\iota_a(\phi_2). \tag{2.7}$$

3b. The identity is preserved

$$\iota_a(1_{\mathcal{C}}) = 1_a. \tag{2.8}$$

3c. Moreover, ι_a is central in the sense that

$$\iota_a(\phi)\psi = \psi\iota_b(\phi) \tag{2.9}$$

for all $\phi \in \mathcal{C}$ and $\psi \in \mathcal{O}_{ab}$.

3d. ι_a and ι^a are adjoints:

$$\theta_{\mathcal{C}}(\iota^a(\psi)\phi) = \theta_a(\psi\iota_a(\phi)) \tag{2.10}$$

for all $\psi \in \mathcal{O}_{aa}$.

3e. The "Cardy conditions."[2] Define ${\pi_b}^a : \mathcal{O}_{aa} \to \mathcal{O}_{bb}$ as follows. Since $\mathcal{O}_{ab}$ and $\mathcal{O}_{ba}$ are in duality (using θ_a or θ_b), if we let ψ_μ be a basis for $\mathcal{O}_{ba}$ then there is a dual basis ψ^μ for $\mathcal{O}_{ab}$. Then we define

$$ {\pi_b}^a(\psi) = \sum_\mu \psi_\mu \psi \psi^\mu, \tag{2.11} $$

and we have the "Cardy condition":

$$ {\pi_b}^a = \iota_b \circ \iota^a. \tag{2.12} $$

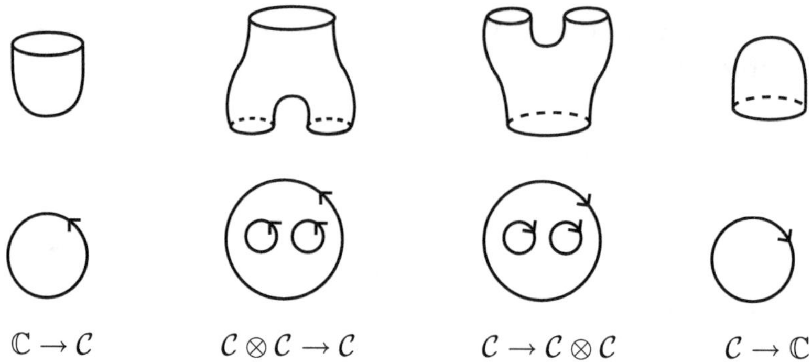

FIGURE 2. Four diagrams defining the Frobenius structure in a closed 2d TFT. It is often more convenient to represent the morphisms by the planar diagrams in the second row. In this case our convention is that if a circle is oriented so that the surface lies on its right then it is an ingoing circle.

2.1.3. Pictorial representation.

2.1.3. Pictorial representation. Let us explain the pictorial basis for these algebraic conditions. The case of a closed 2d TFT is very well-known. The data of the Frobenius structure is provided by the diagrams in Figure 2. The consistency conditions follow from Figure 3.

In the open case, entirely analogous considerations lead to the construction of a not necessarily commutative Frobenius algebra in the open sector. The basic data are summarized in Figure 4. The fact that (2.4) are dual pairings follows from Figure 5. The essential new ingredients in the open and closed theory are the open-to-closed and closed-to-open transitions. In 2d TFT these are the maps ι_a, ι^a. They are represented by Figure 6. There are five new consistency conditions associated with the open/closed transitions. These are illustrated in Figures 7 to 12. (In checking the topological assertions about these diagrams, it is usually best to imagine the surfaces as "flattened out": thus the two surfaces of Figure 9 are both annuli, with one

[2]These are actually generalizations of the conditions stated by Cardy. One recovers his conditions by taking the trace. Of course, the factorization of the double twist diagram in the closed string channel is an observation going back to the earliest days of string theory.

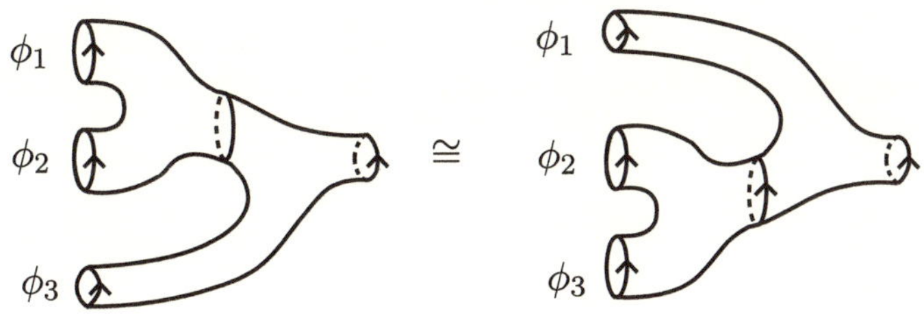

$$(\phi_1 \cdot \phi_2) \cdot \phi_3 = \phi_1 \cdot (\phi_2 \cdot \phi_3)$$

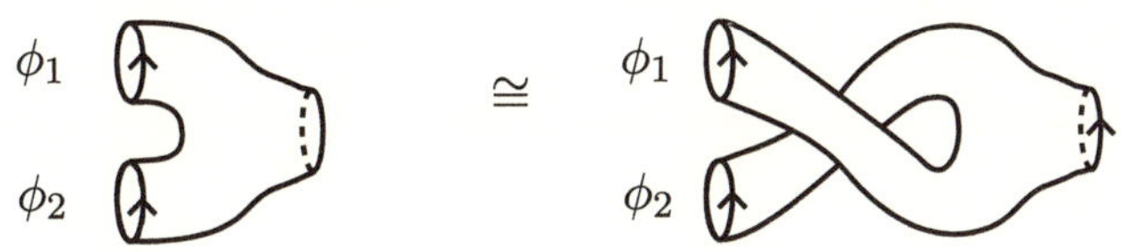

$$\phi_1 \cdot \phi_2 = \phi_2 \cdot \phi_1$$

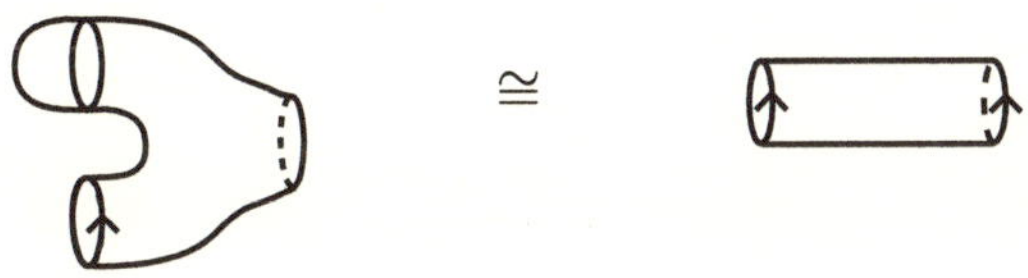

$$1_{\mathcal{C}} \cdot \phi = \phi$$

FIGURE 3. Associativity, commutativity, and unit constraints in the closed case. The unit constraint requires the natural assumption that the cylinder correspond to the identity map $\mathcal{C} \to \mathcal{C}$.

boundary circle being the incoming closed circle, while the other boundary circle is subdivided into an outgoing interval and an interval of constrained boundary. Similarly, the two surfaces in the Cardy diagram 12 are annuli, while the surfaces of 7 are each discs with two holes — i.e., discs from which two open subdiscs have been removed.)

2.1.4. Sewing theorem. Geometrically, any oriented surface can be decomposed into a composition of morphisms corresponding to the basic data defining the Frobenius structure. However, a given surface can be decomposed in many different ways. The above sewing axioms follow from the consistency of these decompositions. The sewing theorem guarantees that

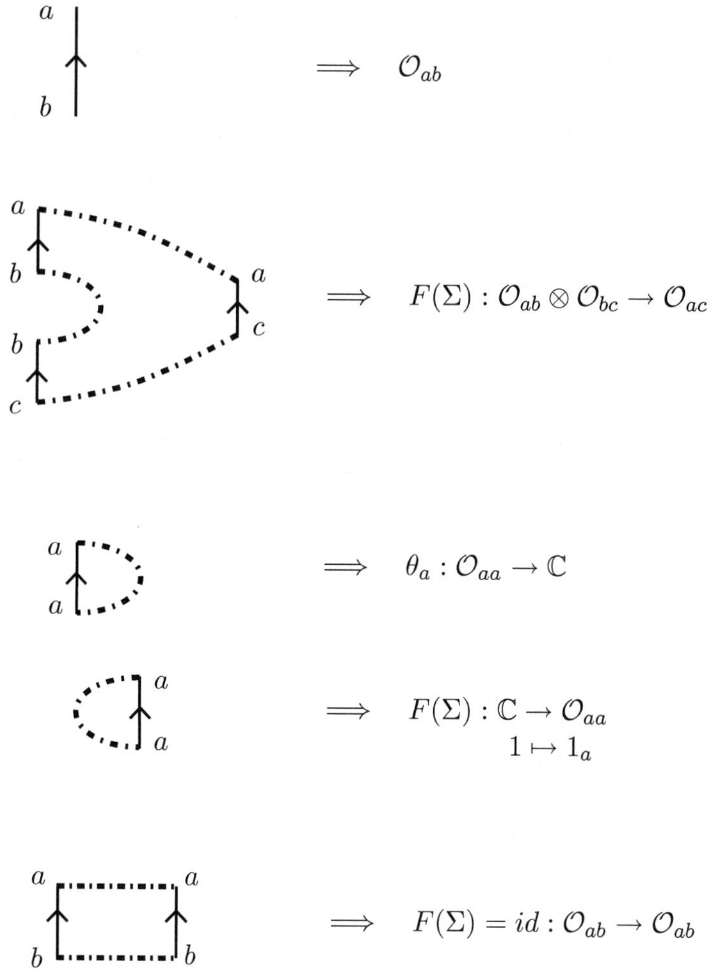

FIGURE 4. Basic data for the open theory. Constrained boundaries are denoted with dot-dash lines, and carry a boundary condition $a, b, c, \cdots \in \mathcal{B}_0$.

there are no further relations on the algebraic data imposed by consistency of sewing.

THEOREM 2.1. *Conditions 1,2,3 above are sufficient to ensure that the algebraic data give rise to a well-defined open and closed topological field theory.*

The proof is in §2.7.

2.1.5. The category of boundary conditions. The category $\mathcal{B}$ of boundary conditions of an open and closed TFT is a $\mathbb{C}$-linear category. We can adjoin new objects to it in various ways. For example, if the category

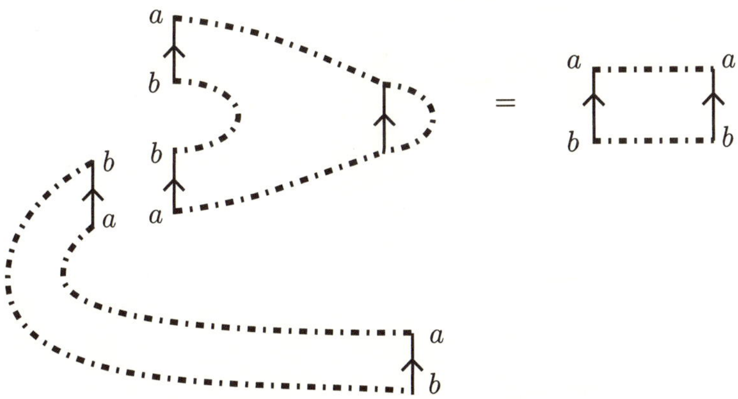

FIGURE 5. Assuming that the strip corresponds to the identity morphism we must have perfect pairings in (2.4).

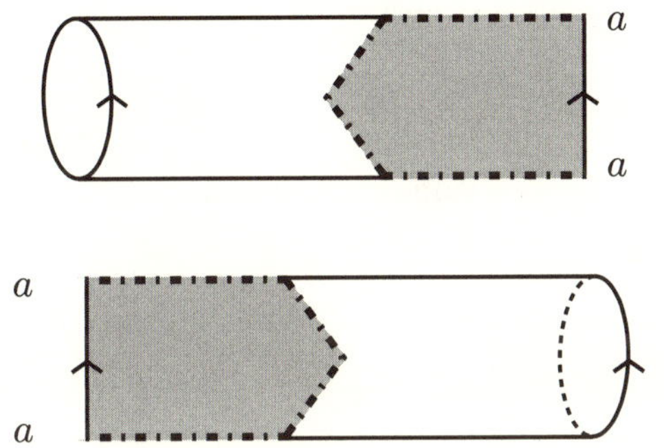

$\iota_a : \mathcal{C} \to \mathcal{O}_{aa}$

$\iota^a : \mathcal{O}_{aa} \to \mathcal{C}$

FIGURE 6. Two ways of representing open to closed and closed to open transitions.

does not possess direct sums, we can define for any two objects a and b a new object $a \oplus b$ by

$$\mathcal{O}_{a\oplus b,c} := \mathcal{O}_{ac} \oplus \mathcal{O}_{bc} \tag{2.13}$$

$$\mathcal{O}_{c,a\oplus b} := \mathcal{O}_{ca} \oplus \mathcal{O}_{cb}, \tag{2.14}$$

and hence[3]

$$\mathcal{O}_{a\oplus b,a\oplus b} := \begin{pmatrix} \mathcal{O}_{aa} & \mathcal{O}_{ab} \\ \mathcal{O}_{ba} & \mathcal{O}_{bb} \end{pmatrix}, \tag{2.15}$$

with the obvious composition laws, and

$$\theta_{a\oplus b} : \mathcal{O}_{a\oplus b,a\oplus b} \to \mathbb{C} \tag{2.16}$$

[3]The matrix notation here is intended to help with understanding the composition of the morphisms: as a vector space, $\mathcal{O}_{a\oplus b,a\oplus b}$ is simply the sum of the four spaces $\mathcal{O}_{aa}, \mathcal{O}_{ab}, \mathcal{O}_{ba}, \mathcal{O}_{bb}$.

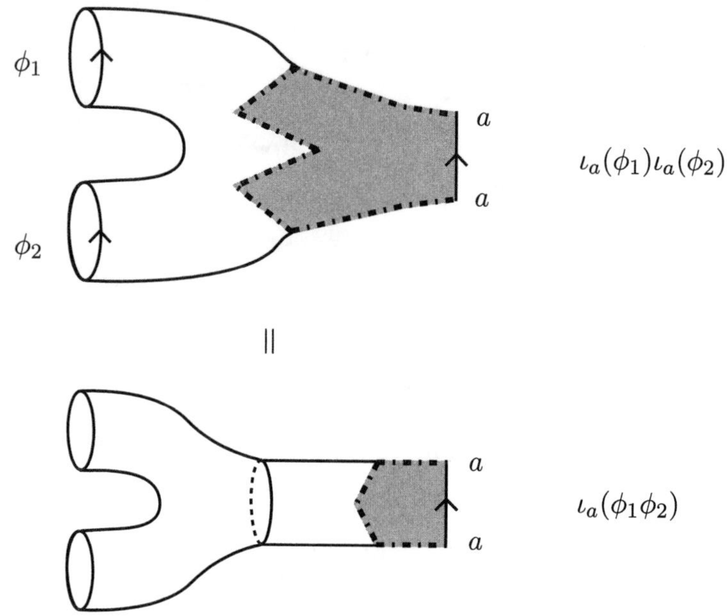

FIGURE 7. ι_a is a homomorphism.

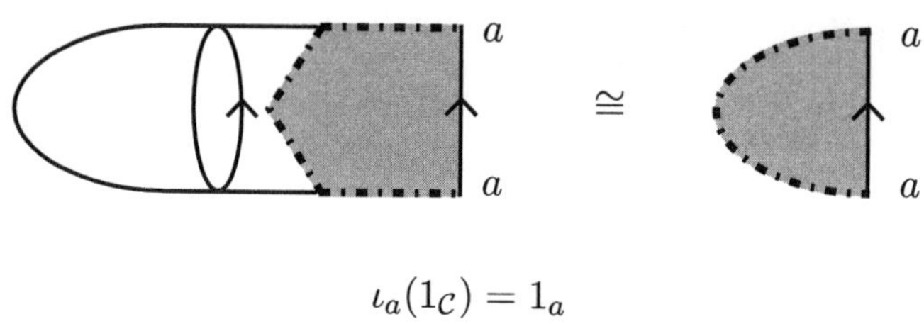

FIGURE 8. ι_a preserves the identity.

given by

$$\theta_{a\oplus b}\begin{pmatrix}\psi_{aa} & \psi_{ab}\\ \psi_{ba} & \psi_{bb}\end{pmatrix} = \theta_a(\psi_{aa}) + \theta_b(\psi_{bb}). \tag{2.17}$$

The new object is the direct sum of a and b in the enlarged category of boundary conditions. If there was already a direct sum of a and b in the category $\mathcal{B}$ then the new object will be canonically isomorphic to it. In the opposite direction, if we have a boundary condition a and a projection $p \in \mathcal{O}_{aa}$ (i.e., an element such that $p^2 = p$) then we may as well assume there is a boundary condition $b = \text{im}(p)$ such that for any c we have $\mathcal{O}_{cb} =$

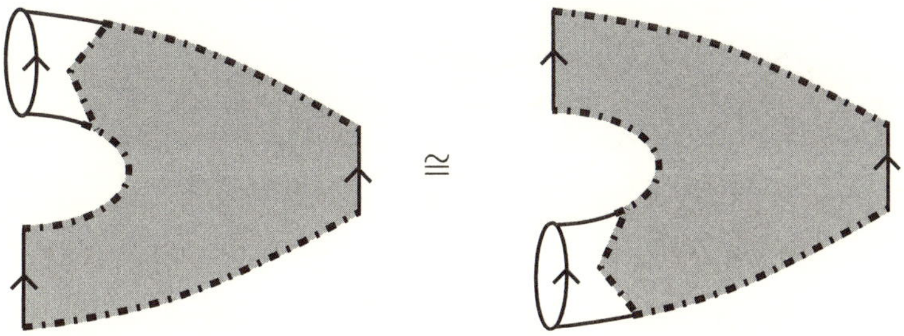

$$\iota_*(\phi) \cdot \psi = \psi \cdot \iota_*(\phi)$$

FIGURE 9. ι_a maps into the center of $\mathcal{O}_{aa}$.

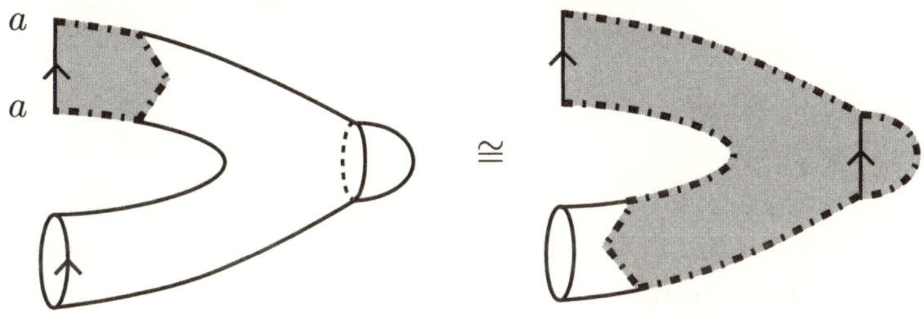

$$\theta_{\mathcal{C}}(\iota^a(\psi) \cdot \phi) = \theta_a(\psi \cdot \iota_a(\phi))$$

FIGURE 10. ι^a is the adjoint of ι_a.

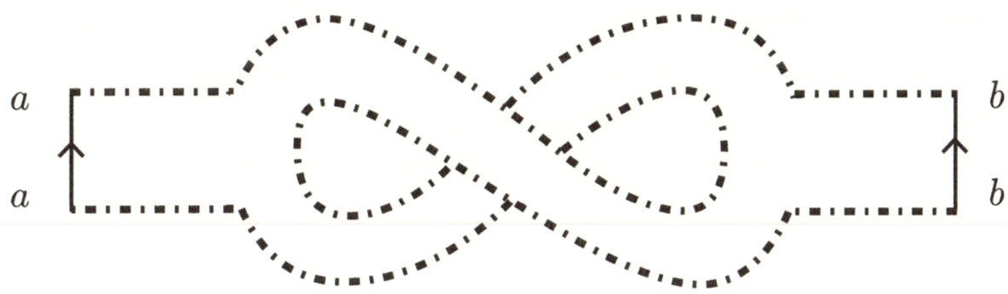

$${\pi_b}^a : \mathcal{O}_{aa} \to \mathcal{O}_{bb}$$

FIGURE 11. The double-twist diagram defines the map ${\pi_b}^a : \mathcal{O}_{aa} \to \mathcal{O}_{bb}$.

$\{f \in \mathcal{O}_{ab} : pf = f\}$ and $\mathcal{O}_{bc} = \{f \in \mathcal{O}_{ba} : fp = f\}$. Then we shall have $a \cong \mathrm{im}(p) \oplus \mathrm{im}(1-p)$.[4]

[4]A linear category in which idempotents split in this way is often called *Karoubian*. See the brief related discussion at the end of §8.3.4.

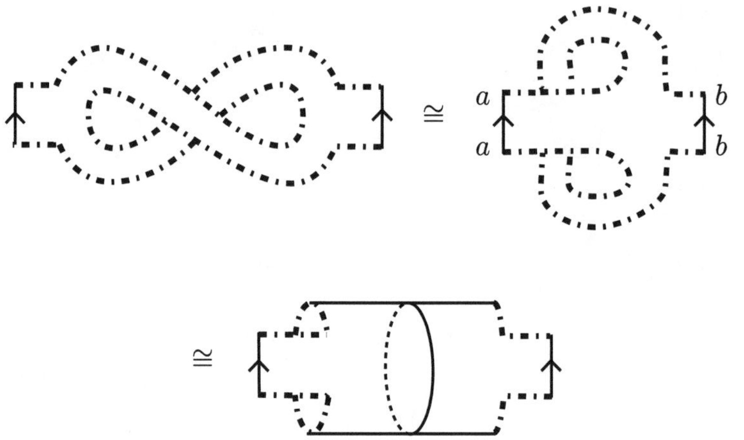

FIGURE 12. The (generalized) Cardy condition expressing factorization of the double-twist diagram in the closed string channel.

One very special property that the category $\mathcal{B}$ possesses is that for any two objects a and b the space $\mathcal{O}_{ab}$ of morphisms is canonically dual to $\mathcal{O}_{ba}$, by a pairing which factorizes through the composition in either order. It is natural to call a category with this property a *Frobenius* category, or a *Calabi-Yau* category.[5] It is a strong restriction on the category: for example the category of finitely generated modules over a finite dimensional algebra does not have the property unless the algebra is semisimple.

EXAMPLE 2.2. Probably the simplest example of an open and closed theory of the type we are studying is one associated to a finite group G. The category $\mathcal{B}$ is the category of finite dimensional complex representations M of G, and the trace $\theta_M : \mathcal{O}_{MM} = \mathrm{End}(M) \to \mathbb{C}$ takes $\psi : M \to M$ to $\mathrm{trace}(\psi)/|G|$. The closed algebra $\mathcal{C}$ is the center of the group-algebra $\mathbb{C}[G]$, which maps to each $\mathrm{End}(M)$ in the obvious way. The trace $\theta_{\mathcal{C}} : \mathcal{C} \to \mathbb{C}$ takes a central element $\sum \lambda_g g$ of the group-algebra to $\lambda_1/|G|$.

In this example the partition function of the theory on a surface Σ with constrained boundary circles $C_1, \ldots, C_k$ labelled $M_1, \ldots, M_k$ is the weighted sum over the isomorphism classes of principal G-bundles P on Σ of

$$\chi_{M_1}(h_P(C_1)) \cdots \chi_{M_k}(h_P(C_k)),$$

[5]The latter terminology comes from the case of coherent sheaves on a compact Kähler manifold, where for two sheaves E and F the dual of the morphism space $\mathrm{Ext}(E,F)$ is in general $\mathrm{Ext}(F, E \otimes \omega)$, where ω is the canonical bundle. This coincides with $\mathrm{Ext}(F,E)$ only when ω is trivial, i.e., in the Calabi-Yau case. (For details see §4.3.) We shall discuss this example further in §2.5.

where $\chi_M : G \to \mathbb{C}$ is the character of a representation M, and $h_P(C)$ denotes the holonomy of P around a boundary circle C. Each bundle P is weighted by the reciprocal of the order of its group of automorphisms.

Returning to the general theory, we can now ask three basic questions.

(i) If we are given a "closed" TFT, can we enlarge it to an open and closed theory, and, if so, is the enlargement unique?

(ii) If we are given the category $\mathcal{B}$ of boundary conditions of an open and closed theory, together with the linear maps $\theta_a : \mathcal{O}_{aa} \to \mathbb{C}$ which define the Frobenius structure, can we reconstruct the whole theory, i.e., can we find the closed Frobenius algebra $\mathcal{C}$?

(iii) Is an arbitrary Frobenius category the category of boundary conditions for some closed theory?

For the first question to be well-posed, we should assume that the category of boundary conditions is *maximal*, in the sense that if $\mathcal{B}'$ is an enlargement of it then any object of $\mathcal{B}'$ is isomorphic to an object of $\mathcal{B}$. Even so, we shall see that there are subtleties which prevent any of these questions from having a simple affirmative answer.

2.1.6. Generalizations: Spin theories. We can obtain many interesting generalizations of the above structure by modifying either the geometrical or the linear category. We shall now survey a few kinds of examples by way of illustration, sometimes giving only a sketch of the details.

The most general target category we can consider is a symmetric tensor category: clearly we need a tensor product, and the axiom $\mathcal{H}_{Y_1 \sqcup Y_2} \cong \mathcal{H}_{Y_1} \otimes \mathcal{H}_{Y_2}$ only makes sense if there is an involutory canonical isomorphism $\mathcal{H}_{Y_1} \otimes \mathcal{H}_{Y_2} \cong \mathcal{H}_{Y_2} \otimes \mathcal{H}_{Y_1}$.

A very common choice in physics is the category of *super vector spaces*, i.e., vector spaces V with a mod 2 grading $V = V^0 \oplus V^1$, where the canonical isomorphism $V \otimes W \cong W \otimes V$ is $v \otimes w \mapsto (-1)^{\deg v \deg w} w \otimes v$. One can also consider the category of $\mathbb{Z}$-graded vector spaces, with the same sign convention for the tensor product.

In either case the closed string algebra is a graded-commutative algebra $\mathcal{C}$ with a trace $\theta : \mathcal{C} \to \mathbb{C}$. In principle the trace should have degree zero, but in fact the commonly encountered theories have a *grading anomaly* which makes the trace have degree $-n$ for some integer n.[6] The formulae (2.5), (2.9), and (2.11) must be replaced by their graded-commutative analogues. In particular if we choose a basis ψ_μ and its dual ψ^μ so that

$$\theta_{\mathcal{C}}(\psi^\mu \psi_\nu) = \delta^\mu_\nu \tag{2.18}$$

[6] It is easy to see that, up to an overall translation of the grading, the most general anomaly assigns an operator of degree $\frac{1}{2}n(i - o - \chi)$ to a cobordism with Euler number χ and i incoming and o outgoing boundary circles.

then

$$\pi_b{}^a(\psi) = \sum_\mu (-1)^{\deg\psi_\mu \deg\psi} \psi_\mu \psi \psi^\mu. \tag{2.19}$$

We can also obtain interesting structures by changing the geometrical category of manifolds and cobordisms by equipping them with extra structure.

EXAMPLE 2.3. We define *topological-spin theories* by replacing "manifolds" with "manifolds with spin structure."

A *spin structure* on a surface means a double covering of its space of non-zero tangent vectors which is non-trivial on each individual tangent space. On an oriented 1-dimensional manifold S it means a double covering of the space of positively-oriented tangent vectors. For purposes of gluing it is useful to note that this is the same thing as a spin structure on a ribbon neighbourhood of S in an orientable surface. Each spin structure has an automorphism which interchanges its sheets, and this will induce an involution T on any vector space which is naturally associated to a 1-manifold with spin structure, giving the vector space a mod 2 grading by its ± 1-eigenspaces. We define a topological-spin theory as a functor from the cobordism category of manifolds with spin structures to the category of super vector spaces with its graded tensor structure. The functor is required to take disjoint unions to super tensor products, and we also require the automorphism of the spin structure of a 1-manifold to induce the grading automorphism $T = (-1)^{\text{degree}}$ of the super vector space. We shall see presently that this choice of the supersymmetry of the tensor product rather than the naive symmetry which ignores the grading is forced on us by the geometry of spin structures if we want to allow the possibility of a semisimple category of boundary conditions. There are two non-isomorphic circles with spin structure: S^1_{ns}, with the Möbius or "Neveu-Schwarz" structure, and S^1_r, with the trivial or "Ramond" structure. A topological-spin theory gives us state spaces $\mathcal{C}_{ns}$ and $\mathcal{C}_r$, corresponding respectively to S^1_{ns} and S^1_r.

There are four cobordisms with spin structures which cover the standard annulus. The double covering can be identified with its incoming end times the interval $[0,1]$, but then one has a binary choice when one identifies the outgoing end of the double covering over the annulus with the chosen structure on the outgoing boundary circle. In other words, alongside the cylinders $A^+_{ns,r} = S^1_{ns,r} \times [0,1]$ which induce the identity maps of $\mathcal{C}_{ns,r}$ there are also cylinders $A^-_{ns,r}$ which connect $S^1_{ns,r}$ to itself while interchanging the sheets. These cylinders $A^-_{ns,r}$ induce the grading automorphism on the state spaces. But because $A^-_{ns} \cong A^+_{ns}$ by an isomorphism which is the identity on the boundary circles — the Dehn twist which "rotates one end of the cylinder by 2π" — the grading on $\mathcal{C}_{ns}$ must be purely even. The space $\mathcal{C}_r$

can have both even and odd components. The situation is a little more complicated for "U-shaped" cobordisms, i.e., cylinders with two incoming or two outgoing boundary circles. If the boundaries are S^1_{ns} there is only one possibility, but if the boundaries are S^1_r there are two, corresponding to $A^{\pm}_r$. The complication is that there seems no special reason to prefer either of the spin structures as "positive". We shall simply *choose* one — let us call it P — with incoming boundary $S^1_r \sqcup S^1_r$, and use P to define a pairing $\mathcal{C}_r \otimes \mathcal{C}_r \to \mathbb{C}$. We then choose a preferred cobordism Q in the other direction so that when we sew its right-hand outgoing S^1_r to the left-hand incoming one of P the resulting S-bend is the "trivial" cylinder A^+_r. We shall need to know, however, that the closed torus formed by the composition $P \circ Q$ has an *even* spin structure. Note that the Frobenius structure θ on $\mathcal{C}$ restricts to 0 on $\mathcal{C}_r$.

There is a unique spin structure on the pair-of-pants cobordism of Figure 2 which restricts to S^1_{ns} on each boundary circle, and it makes $\mathcal{C}_{ns}$ into a commutative Frobenius algebra in the usual way. If one incoming circle is S^1_{ns} and the other is S^1_r then the outgoing circle is S^1_r, and there are two possible spin structures, but the one obtained by removing a disc from the cylinder A^+_r is preferred: it makes $\mathcal{C}_r$ into a graded module over $\mathcal{C}_{ns}$. The chosen U-shaped cobordism P, with two incoming circles S^1_r, can be punctured to give us a pair of pants with an outgoing S^1_{ns}, and it induces a graded bilinear map $\mathcal{C}_r \times \mathcal{C}_r \to \mathcal{C}_{ns}$ which, composing with the trace on $\mathcal{C}_{ns}$, gives a non-degenerate inner product on $\mathcal{C}_r$. At this point the choice of symmetry of the tensor product becomes important. Let us consider the diffeomorphism of the pair of pants which shows us in the usual case that the Frobenius algebra is commutative. When we lift it to the spin structure, this diffeomorphism induces the identity on one incoming circle but reverses the sheets over the other incoming circle, and this proves that the cobordism must have the same output when we change the input from $S(\phi_1 \otimes \phi_2)$ to $T(\phi_1) \otimes \phi_2$, where T is the grading involution and $S : \mathcal{C}_r \otimes \mathcal{C}_r \to \mathcal{C}_r \otimes \mathcal{C}_r$ is the symmetry of the tensor category. If we take S to be the symmetry of the tensor category of vector spaces which ignores the grading, this shows that the product on the graded vector space $\mathcal{C}_r$ is graded-symmetric with the usual sign; but if S is the graded symmetry then we see that *the product on $\mathcal{C}_r$ is symmetric in the naive sense.* (We must bear in mind here that if ψ_1 and ψ_2 do not have the same parity then their product is in any case zero, as we have seen that $\mathcal{C}_{ns}$ is purely even.)

There is an analogue for spin theories of the theorem which tells us that a two-dimensional topological field theory "is" a commutative Frobenius algebra. It asserts that a spin-topological theory "is" a Frobenius algebra $\mathcal{C} = (\mathcal{C}_{ns} \oplus \mathcal{C}_r, \theta_{\mathcal{C}})$ with the properties just mentioned, and with the following additional property. Let $\{\phi_k\}$ be a basis for $\mathcal{C}_{ns}$, with dual basis $\{\phi^k\}$ such that $\theta_{\mathcal{C}}(\phi_k\phi^m) = \delta_k^m$, and let β_k and β^k be similar dual bases for $\mathcal{C}_r$. Then the Euler elements $\chi_{ns} := \sum \phi_k\phi^k$ and $\chi_r = \sum \beta_k\beta^k$ are independent of the choices of bases, and the condition we need on the algebra $\mathcal{C}$ is that $\chi_{ns} = \chi_r$. In particular, this condition implies that the vector spaces $\mathcal{C}_{ns}$ and $\mathcal{C}_r$ have the same dimension.[7] In fact, the Euler elements can be obtained from cutting a hole out of the torus. There are actually four spin structures on the torus. The output state is necessarily in $\mathcal{C}_{ns}$. The Euler elements for the three even spin structures are equal to $\chi_e = \chi_{ns} = \chi_r$. The Euler element χ_o corresponding to the odd spin structure, on the other hand, is given by $\chi_o = \sum(-1)^{\deg \beta_k}\beta_k\beta^k$.

We shall omit the proof that the general spin theory is what we have just described, but it is almost identical with the proof we shall give in §2.7 of the theorem of Turaev about G-equivariant theories in the simple case when the group G is $\mathbb{Z}/2$. Indeed a spin theory is very similar to — but not the same as — a $\mathbb{Z}/2$-equivariant theory, which is the structure obtained when the surfaces are equipped with principal $\mathbb{Z}/2$-bundles (i.e., double coverings) rather than spin structures. We shall discuss equivariant theories in §2.6. (One difference is that in the equivariant case the $\mathbb{Z}/2$ action is nontrivial in the sector $\mathcal{C}_1$ and trivial in $\mathcal{C}_g$, precisely the opposite of what we have found in the spin case.) Comparing with the equivariant theory, the surprising result that the product on $\mathcal{C}_r$ is naive-symmetric can be understood as twisted anticommutativity.

It seems reasonable to call a spin theory *semisimple* if the algebra $\mathcal{C}_{ns}$ is semisimple, i.e., is the algebra of functions on a finite set X. Then $\mathcal{C}_r$ is the space of sections of a vector bundle E on X, and it follows from the condition $\chi_{ns} = \chi_r$ that the fibre at each point must have dimension 1. Thus the whole structure is determined by the Frobenius algebra $\mathcal{C}_{ns}$ together with a binary choice at each point $x \in X$ of the grading of the fibre E_x of the line bundle E at x.

We can now see that if we had not used the graded symmetry in defining the tensor category we should have forced the grading of $\mathcal{C}_r$ to be purely even. For on the odd part the inner product would have had to be skew, and that is impossible on a 1-dimensional space. And if both $\mathcal{C}_{ns}$ and $\mathcal{C}_r$ are purely even then the theory is in fact completely independent of the spin structures on the surfaces.

A concrete example of a two-dimensional topological-spin theory is given by $\mathcal{C} = \mathbb{C} \oplus \mathbb{C}\eta$ where $\eta^2 = 1$ and η is odd. The Euler elements are $\chi_e = 1$ and $\chi_o = -1$. It follows that the partition function of a closed surface with spin structure is ± 1 according as the spin structure is even or odd. (To prove this it is useful to compute the Arf invariant of the quadratic refinement of

[7]Thus, in a sense, the theory has "space-time supersymmetry."

the intersection product associated to the spin structure and to note that it is multiplicative for adding handles.)

The most common theories defined on surfaces with spin structure are not topological: they are 2-dimensional conformal field theories with $\mathcal{N} = 1$ supersymmetry. The general features of the structure are still as we have described, but it should be noticed that if the theory is not topological then one does not expect the grading on $\mathcal{C}_{ns}$ to be purely even: states can change sign on rotation by 2π. If a surface Σ has a conformal structure then a double covering of the non-zero tangent vectors is the complement of the zero-section in a two-dimensional real vector bundle L on Σ which is called the *spin bundle.* The covering map then extends to a symmetric pairing of vector bundles $L \otimes L \to T\Sigma$ which, if we regard L and $T\Sigma$ as complex line bundles in the natural way, induces an isomorphism $L \otimes_{\mathbb{C}} L \cong T\Sigma$. An $\mathcal{N} = 1$ superconformal field theory is a conformal-spin theory which assigns a vector space $\mathcal{H}_{S,L}$ to the 1-manifold S with the spin bundle L, and is equipped with an additional map

$$\Gamma(S, L) \otimes \mathcal{H}_{S,L} \to \mathcal{H}_{S,L} \tag{2.20}$$

$$(\sigma, \psi) \mapsto G_\sigma \psi, \tag{2.21}$$

where $\Gamma(S, L)$ is the space of smooth sections of L, such that G_σ is real-linear in the section σ, and satisfies $G_\sigma^2 = D_{\sigma^2}$, where D_{σ^2} is the Virasoro action of the vector field σ^2 related to $\sigma \otimes \sigma$ by the isomorphism $L \otimes_{\mathbb{C}} L \cong T\Sigma$. Furthermore, when we have a cobordism (Σ, L) from (S_0, L_0) to (S_1, L_1) and a holomorphic section σ of L which restricts to σ_i on S_i we have the intertwining property

$$G_{\sigma_1} \circ U_{\Sigma,L} = U_{\Sigma,L} \circ G_{\sigma_0}. \tag{2.22}$$

EXAMPLE 2.4. We define *topological-spin^c theories*, which model 2d theories with $\mathcal{N} = 2$ supersymmetry, by replacing "manifolds" with "manifolds with spin^c structure".

A *spin^c structure* on a surface with a conformal structure is a pair of holomorphic line bundles L_1, L_2 with an isomorphism $L_1 \otimes L_2 \cong T\Sigma$ of holomorphic line bundles. A spin structure is the particular case when $L_1 = L_2$. On a 1-manifold S a spin^c structure means a spin^c structure on a ribbon neighbourhood of S in a surface with conformal structure. An $\mathcal{N} = 2$ superconformal theory assigns a vector space $\mathcal{H}_{S;L_1,L_2}$ to each 1-manifold S with spin^c structure, and an operator

$$U_{S_0;L_1,L_2} : \mathcal{H}_{S_0;L_1,L_2} \to \mathcal{H}_{S_1;L_1,L_2} \tag{2.23}$$

to each spin^c-cobordism from S_0 to S_1. To explain the rest of the structure we need to define the $\mathcal{N} = 2$ Lie superalgebra associated to a spin^c

1-manifold $(S; L_1, L_2)$. Let $\mathcal{G} = \mathrm{Aut}(L_1)$ denote the group of bundle isomorphisms $L_1 \to L_1$ which cover diffeomorphisms of S. (We can identify this group with $\mathrm{Aut}(L_2)$.) It has a homomorphism onto the group $\mathrm{Diff}^+(S)$ of orientation-preserving diffeomorphisms of S, and the kernel is the group of fibrewise automorphisms of L_1, which can be identified with the group of smooth maps from S to $\mathbb{C}^\times$. The Lie algebra $\mathrm{Lie}(\mathcal{G})$ is therefore an extension of the Lie algebra $\mathrm{Vect}(S)$ of $\mathrm{Diff}^+(S)$ by the commutative Lie algebra $\Omega^0(S)$ of smooth real-valued functions on S. Let $\Lambda^0_{S;L_1,L_2}$ denote the complex Lie algebra obtained from $\mathrm{Lie}(\mathcal{G})$ by complexifying $\mathrm{Vect}(S)$. This is the even part of a Lie superalgebra whose odd part is $\Lambda^1_{S;L_1,L_2} = \Gamma(L_1) \oplus \Gamma(L_2)$. The bracket $\Lambda^1 \otimes \Lambda^1 \to \Lambda^0$ is completely determined by the property that elements of $\Gamma(L_1)$ and of $\Gamma(L_2)$ anticommute among themselves, while the composite

$$\Gamma(L_1) \otimes \Gamma(L_2) \to \Lambda^0 \to \mathrm{Vect}_{\mathbb{C}}(S) \tag{2.24}$$

takes (λ_1, λ_2) to $\lambda_1\lambda_2 \in \Gamma(TS)$.

In an $\mathcal{N} = 2$ theory we require the superalgebra $\Lambda(S; L_1, L_2)$ to act on the vector space $\mathcal{H}_{S;L_1,L_2}$, compatibly with the action of the group $\mathcal{G}$, and with a similar intertwining property with the cobordism operators to that of the $\mathcal{N} = 1$ case. For an $\mathcal{N} = 2$ theory the state space always has an action of the circle group coming from its embedding in $\mathcal{G}$ as the group of fibrewise multiplications on L_1 and L_2. Equivalently, the state space is always $\mathbb{Z}$-graded.

An $\mathcal{N} = 2$ theory always gives rise to two ordinary conformal field theories by equipping a surface Σ with the spin^c structures $(\mathbb{C}, T\Sigma)$ and $(T\Sigma, \mathbb{C})$. These are called the "A-model" and the "B-model" associated to the $\mathcal{N} = 2$ theory. In each case the state spaces are cochain complexes in which the differential is the action of the constant section 1 of the trivial component of the spin^c-structure.

2.1.7. Generalizations: Cochain level theories. The most important "generalization", however, of the open and closed topological field theory we have described is of a more fundamental kind. Our topological theories are intended to be a toy model of the conformal field theories that arise in string theory. In closed string theory the central object is the vector space $\mathcal{C} = \mathcal{C}_{S^1}$ of states of a single parametrized string. This has an integer grading by the "ghost number", and an operator $Q : \mathcal{C} \to \mathcal{C}$ called the "BRST operator" which raises the ghost number by 1 and satisfies $Q^2 = 0$. In other words, $\mathcal{C}$ is a *cochain complex*. If we think of the string as moving in a spacetime M then $\mathcal{C}$ is roughly the space of differential forms defined along the orbits of the action of the reparametrization group $\mathrm{Diff}^+(S^1)$ on the free loop space $\mathcal{L}M$ (more precisely, square-integrable forms of semi-infinite degree). Similarly, the space $\mathcal{C}$ of a topologically-twisted $\mathcal{N} = 2$ supersymmetric theory, as just described, is a cochain complex which models the space of semi-infinite differential forms on the loop space of a Kähler manifold — in this case, *all* square-integrable differential forms, not just those along the orbits of $\mathrm{Diff}^+(S^1)$. In both kinds of example, a cobordism Σ from p circles to q circles gives an operator $U_{\Sigma,\mu} : \mathcal{C}^{\otimes p} \to \mathcal{C}^{\otimes q}$ which depends on a conformal structure μ on Σ. This operator is a cochain map, but its crucial feature is that changing the conformal structure μ on Σ changes the operator $U_{\Sigma,\mu}$ only by a cochain homotopy. The cohomology $H(\mathcal{C}) = \ker(Q)/\mathrm{im}(Q)$ — the "space of physical states" in conventional string theory — is therefore the state space of a topological field theory. (In the usual string theory situation the topological field theory we obtain is not very interesting, for the BRST cohomology is concentrated in one or two degrees, and there is a "grading anomaly" which means that the operator associated to a cobordism Σ changes the degree by a multiple of the Euler number $\chi(\Sigma)$. In the case of the $N = 2$ supersymmetric models, however, there is no grading anomaly, and the full structure is visible.)

A good way to describe how the operator $U_{\Sigma,\mu}$ varies with μ is as follows. If $\mathcal{M}_\Sigma$ is the moduli space of conformal structures on the cobordism Σ, modulo diffeomorphisms of Σ which are the identity on the boundary circles, then we have a cochain map

$$U_\Sigma : \mathcal{C}^{\otimes p} \to \Omega^*(\mathcal{M}_\Sigma, \mathcal{C}^{\otimes q}) \tag{2.25}$$

where the right-hand side is the de Rham complex of forms on $\mathcal{M}_\Sigma$ with values in $\mathcal{C}^{\otimes q}$. The operator $U_{\Sigma,\mu}$ is obtained from U_Σ by restricting from $\mathcal{M}_\Sigma$ to $\{\mu\}$. The composition property when two cobordisms Σ_1 and Σ_2 are concatenated is that the diagram

(2.26)

$$\begin{array}{ccc}
\mathcal{C}^{\otimes p} & \longrightarrow & \Omega(\mathcal{M}_{\Sigma_1}, \mathcal{C}^{\otimes q}) \\
\downarrow & & \downarrow \\
\Omega(\mathcal{M}_{\Sigma_2 \circ \Sigma_1}, \mathcal{C}^{\otimes r}) & \longrightarrow & \Omega(\mathcal{M}_{\Sigma_1} \times \mathcal{M}_{\Sigma_2}, \mathcal{C}^{\otimes r}) = \Omega(\mathcal{M}_{\Sigma_1}, \Omega(\mathcal{M}_{\Sigma_2}, \mathcal{C}^{\otimes r}))
\end{array}$$

commutes, where the lower horizontal arrow is induced by the map $\mathcal{M}_{\Sigma_1} \times \mathcal{M}_{\Sigma_2} \to \mathcal{M}_{\Sigma_2 \circ \Sigma_1}$ which expresses concatenation of the conformal structures.

Many variants of this formulation are possible. For example, we might prefer to give a cochain map

$$U_\Sigma : C_\bullet(\mathcal{M}_\Sigma) \to (\mathcal{C}^{\otimes p})^* \otimes \mathcal{C}^{\otimes q}, \tag{2.27}$$

where $C_\bullet(\mathcal{M}_\Sigma)$ is, say, the complex of smooth singular chains of $\mathcal{M}_\Sigma$. We may also prefer to use the moduli spaces of Riemannian structures instead of conformal structures.

There is no difficulty in passing from the closed string picture just presented to an open and closed theory. We shall not discuss these cochain level theories in any depth in this work, but it is important to realize that they are the real objective. We shall now point out a few basic things about them. A much fuller discussion can be found in Costello's work [**100**].

For each pair a, b of boundary conditions we shall still have a vector space — indeed a cochain complex — $\mathcal{O}_{ab}$, but it is no longer the space of morphisms from b to a in a category. Rather, what we have is, in the terminology of Fukaya, Kontsevich, and others, an A_∞-*category.* The notion of A_∞-category will appear a number of times throughout the book, but is discussed in greatest detail in Chapter 8. Summarizing briefly here, this means that instead of a composition law $\mathcal{O}_{ab} \times \mathcal{O}_{bc} \to \mathcal{O}_{ac}$ we have a *family* of ways of composing, parametrized by the contractible space of conformal structures on the surface of Figure 1. In particular, any two choices of a composition law from the family are cochain homotopic. Composition is associative in the sense that we have a contractible family of triple compositions $\mathcal{O}_{ab} \times \mathcal{O}_{bc} \times \mathcal{O}_{cd} \to \mathcal{O}_{ad}$, which contains all the maps obtained by choosing a binary composition law from the given family and bracketing the triple in either of the two possible ways.

REMARK 2.5. This is not the usual way of defining an A_∞-structure. According to Stasheff's original definition, an A_∞-structure on a space X consists of a sequence of choices: first, a composition law $m_2 : X \times X \to X$; then, a choice of a map

$$m_3 : [0,1] \times X \times X \times X \to X$$

which is a homotopy between

$$(x,y,z) \mapsto m_2(m_2(x,y),z) \text{ and } (x,y,z) \mapsto m_2(x, m_2(y,z));$$

then, a choice of a map

$$m_4 : \overline{S_4} \times X^4 \to X,$$

where $\overline{S_4}$ is a convex plane polygon whose vertices are indexed by the five ways of bracketing a 4-fold product, and $m_4|((\partial\overline{S_4}) \times X^4)$ is determined by m_3; and so on. There is an analogous definition — in fact slightly simpler — applying to cochain complexes rather than spaces. (See §8.1 for the precise definitions.) These definitions, however, are essentially equivalent to the one above coming from 2-dimensional field theory: the only important point is to have a *contractible* family of k-fold compositions for each k. (A discussion of the relation between the definitions can be found in [**412**].)

Apart from the composition law, the essential algebraic properties we have found in our theories are the non-degenerate inner product, and the commutativity of the closed algebra $\mathcal{C}$. Concerning the latter, when we pass to cochain theories the multiplication in $\mathcal{C}$ will of course be commutative up to cochain homotopy, but, unlike what happened with the open string composition, the moduli space $\mathcal{M}_\Sigma$ of closed string multiplications, i.e., the moduli space of conformal structures on a pair of pants Σ, modulo diffeomorphisms of Σ which are the identity on the boundary circles, is not contractible: it has the homotopy type of the space of ways of embedding two copies of the standard disc D^2 disjointly in the interior of D^2 — this space of embeddings is of course a subspace of $\mathcal{M}_\Sigma$. In particular, it contains a natural *circle* of multiplications in which one of the embedded discs moves like a planet around the other, and there are two *different* natural homotopies between the multiplication and the reversed multiplication. This might be a clue to an important difference between stringy and classical space-times. The closed string cochain complex $\mathcal{C}$ is the string theory substitute for the de Rham complex of space-time, an algebra whose multiplication is associative and (graded-)commutative on the nose. Over the rationals or the real or complex numbers, such cochain algebras are known by the work of Sullivan [**434**] and Quillen [**400**] to model[8] the category of topological spaces up to homotopy, in the sense that to each such algebra $\mathcal{C}$ we can associate a space $X_\mathcal{C}$ and a homomorphism of cochain algebras from $\mathcal{C}$ to the de Rham complex of $X_\mathcal{C}$ which is a cochain homotopy equivalence. If we do not want to ignore torsion in the homology of spaces we can no longer encode the homotopy type in a strictly commutative cochain algebra. Instead, we must replace commutative algebras with so-called E_∞-algebras, i.e., roughly, cochain complexes $\mathcal{C}$ over the integers equipped with a multiplication which is associative and commutative up to given arbitrarily high-order homotopies. An arbitrary space X has an E_∞-algebra $\mathcal{C}_X$ of cochains, and conversely one can associate a space $X_\mathcal{C}$ to each E_∞-algebra $\mathcal{C}$. Thus we have a pair of adjoint functors, just as in rational homotopy theory. A long evolution in algebraic topology has culminated in recent theorems of Mandell [**346**] which show that the actual homotopy category of topological spaces is more or less equivalent to the category of E_∞-algebras. The cochain algebras of closed string theory have less higher commutativity than do E_∞-algebras, and this may be an indication that we are dealing with non-commutative spaces in Connes's sense: that fits in well with the interpretation of the B-field of a string background as corresponding to a bundle of matrix algebras on space-time. At the same time, the non-degenerate inner product on $\mathcal{C}$ — corresponding to Poincaré duality — seems to show we are concerned with *manifolds*, rather than more singular spaces.

For readers not accustomed to working with cochain complexes it may be worth saying a few words about what one gains by doing so. To take the simplest example, let us consider the category $\mathcal{K}$ of cochain complexes of finitely generated *free* abelian groups and cochain homotopy classes of cochain maps. This is called the *derived category* of the category of finitely generated abelian groups. (Derived categories will be discussed in detail in

[8]In this and the following sentence we are overlooking subtleties related to the fundamental group.

Chapter 4.) Passing to cohomology gives us a functor from $\mathcal{K}$ to the category of $\mathbb{Z}$-graded finitely generated abelian groups. In fact the subcategory $\mathcal{K}_0$ of $\mathcal{K}$ consisting of complexes whose cohomology vanishes except in degree 0 is actually equivalent to the category of finitely generated abelian groups.[9] But the category $\mathcal{K}$ inherits from the category of finitely generated free abelian groups a duality functor with properties as ideal as one could wish: each object is isomorphic to its double dual, and dualizing preserves exact sequences. (The dual C^* of a complex C is defined by $(C^*)^i = \mathrm{Hom}(C^{-i}, \mathbb{Z})$.) There is no such nice duality in the category of finitely generated abelian groups. Indeed, the subcategory $\mathcal{K}_0$ is not closed under duality, for the dual of the complex C_A corresponding to a group A has in general *two* non-vanishing cohomology groups: $\mathrm{Hom}(A, \mathbb{Z})$ in degree 0, and in degree $+1$ the finite group $\mathrm{Ext}^1(A, \mathbb{Z})$ Pontrjagin-dual to the torsion subgroup of A. This follows from the exact sequence (not to be confused with the cochain complex):

$$0 \to \mathrm{Hom}(A, \mathbb{Z}) \to \mathrm{Hom}(F_A, \mathbb{Z}) \to \mathrm{Hom}(R_A, \mathbb{Z}) \to \mathrm{Ext}^1(A, \mathbb{Z}) \to 0 \tag{2.28}$$

derived from an exact sequence

$$0 \to R_A \to F_A \to A \to 0$$

The category $\mathcal{K}$ also has a tensor product with better properties than the tensor product of abelian groups (which does not preserve exact sequences), and, better still, there is a canonical cochain functor from (locally well-behaved) compact spaces to $\mathcal{K}$ which takes Cartesian products to tensor products. (The simplicial, Čech, and other candidates for the cochain complex of a space are *canonically* isomorphic in $\mathcal{K}$.)

We shall return to this discussion in §2.5.

2.2. Solutions of the algebraic conditions: The semisimple case

2.2.1. Classification theorem. We now turn to the question : given a closed string theory $\mathcal{C}$, what is the corresponding category of boundary conditions? In our formulation this becomes the question: given a commutative Frobenius algebra $\mathcal{C}$, what are the possible $\mathcal{O}_{ab}$'s?

We can answer this question in the case when $\mathcal{C}$ is semisimple. We will take $\mathcal{C}$ to be an algebra over the complex numbers, and in this case the most useful characterization of semisimplicity is that the "fusion rules"

$$\phi_\mu \phi_\nu = N_{\mu\nu}^{\ \ \lambda} \phi_\lambda \tag{2.29}$$

[9] To an abelian group A one can associate the cochain complex

$$C_A = (\ \cdots\ \to 0 \to R_A \to F_A \to 0 \to\ \cdots\ ),$$

where F_A is a free abelian group (in degree 0) with a surjective map $F_A \to A$, and R_A is the kernel of $F_A \to A$. The choice of F_A is far from unique, but nevertheless the different choices of C_A are *canonically* isomorphic objects of $\mathcal{K}$.

are diagonalizable.[10] That is, the matrices $L(\phi_\mu)$ of the left-regular representation, with matrix elements $N_{\mu\nu}^\lambda$, are simultaneously diagonalizable.

Equivalently, there is a set of basic idempotents ε_x such that

$$\begin{aligned} \mathcal{C} &= \oplus_x \mathbb{C}\varepsilon_x \\ \varepsilon_x\varepsilon_y &= \delta_{xy}\varepsilon_y. \end{aligned} \tag{2.30}$$

Equivalently, yet again, $\mathcal{C}$ is the algebra of complex-valued functions on the finite set $X = \mathrm{Spec}(\mathcal{C})$ of characters of $\mathcal{C}$.

The trace $\theta_\mathcal{C} : \mathcal{C} \to \mathbb{C}$, which should be thought of as a "dilaton field" on the finite space-time $\mathrm{Spec}(\mathcal{C})$, is completely described by the unordered set of non-zero complex numbers

$$\theta_x := \theta_\mathcal{C}(\varepsilon_x) \tag{2.31}$$

which is the only invariant of a finite dimensional commutative semisimple Frobenius algebra.

It should be mentioned that the most general finite dimensional commutative algebra over the complex numbers is of the form $\mathcal{C} = \oplus \mathcal{C}_x$, where x runs through the set $\mathrm{Spec}(\mathcal{C})$, and $\mathcal{C}_x$ is a local ring, i.e., $\mathcal{C}_x = \mathbb{C}\varepsilon_x \oplus m_x$, with ε_x as in (2.30) , and m_x a nilpotent ideal. If $\mathcal{C}$ is a Frobenius algebra, then so is each $\mathcal{C}_x$, and there is some ν_x for which $\theta_\mathcal{C} : m_x^{\nu_x} \to \mathbb{C}$ is an isomorphism, while $m_x^{\nu_x+1} = 0$. Let us write $\omega_x \in m_x^{\nu_x}$ for the element such that $\theta_\mathcal{C}(\omega_x) = 1$. The element ω of $\mathcal{C}$ with components ω_x can be regarded as a "volume form" on space-time. (A typical example of such a local Frobenius algebra $\mathcal{C}_x$ is the cohomology ring — with complex coefficients — of complex projective space $\mathbb{P}^n$ of dimension n. The cohomology ring is generated by a single 2-dimensional class t which satisfies $t^{n+1} = 0$. The trace is given by integration over $\mathbb{P}^n$, and takes t^k to 1 if $k = n$, and to 0 otherwise. Thus $\omega_x = t^n$ here.)

A useful technical fact about Frobenius algebras — not necessarily commutative — is that, in the notation of (2.11) , the "Euler" element $\chi = \sum \psi_\mu \psi^\mu$ is invertible if and only if the algebra is semisimple[11], which in the general case means that the algebra is isomorphic to a sum of full matrix

[10]The structure constants $N_{\mu\nu}^\lambda$ need not be integral, though in many interesting examples there is a basis for the algebra in which they are integral.

[11]To see this, one observes that for any element ψ of the algebra we have $\theta(\psi\chi) = \mathrm{tr}(\psi)$, where $\mathrm{tr}(\psi)$ denotes the trace of ψ in the regular representation. (This holds because $\theta(\psi^\mu\psi\psi_\nu) = \theta(\psi\psi_\nu\psi^\mu)$ is the (μ,ν) matrix element of the matrix representing ψ in the regular representation.) As the pairing $(\psi_1,\psi_2) \mapsto \theta(\psi_1\psi_2)$ is non-degenerate, it follows that the trace-form $(\psi_1,\psi_2) \mapsto \mathrm{tr}(\psi_1\psi_2)$ is non-degenerate if and only if χ is invertible, and non-degeneracy of the trace-form is well-known to be a criterion for a finite dimensional algebra to be semisimple. There are several definitions of semisimplicity, and their equivalence amounts to the classical theorem of Wedderburn. For our purposes, a semisimple algebra is just a sum of full matrix algebras.

algebras. The element χ always belongs to the center of the algebra; in the commutative case it has components $\dim(\mathcal{C}_x)\omega_x$.

In the semisimple case we have the following complete characterization of the possible open algebras $\mathcal{O}_{aa}$ compatible with a fixed closed algebra $\mathcal{C}$.

THEOREM 2.6. *If $\mathcal{C}$ is semisimple then $\mathcal{O} = \mathcal{O}_{aa}$ is semisimple for each a and necessarily of the form $\mathcal{O} = \mathrm{End}_{\mathcal{C}}(W)$ for some finite dimensional representation W of $\mathcal{C}$.*

PROOF. The images $\iota_a(\varepsilon_x) = P_x$ are central simple idempotents. Therefore $\mathcal{O}_x = P_x\mathcal{O} = P_x\mathcal{O}P_x$ is an algebra over the Frobenius algebra $\mathcal{C}_x = \varepsilon_x\mathcal{C} \cong \mathbb{C}$, and so it suffices to work over a single space-time point. Then $\iota^a(1_{\mathcal{O}_x}) = \alpha 1_{\mathcal{C}_x}$ for some element $\alpha \in \mathbb{C}$. By the Cardy condition

$$\alpha 1_{\mathcal{O}_x} = \chi_{\mathcal{O}_x} = \sum \psi_\mu \psi^\mu. \tag{2.32}$$

Applying θ we find $\alpha = \dim \mathcal{O}_x$, and hence $\chi_{\mathcal{O}_x}$ is invertible if $\mathcal{O}_x \neq 0$. It follows that $\mathcal{O}_x$ is semisimple at each point x, i.e. a sum of matrix algebras $\oplus_i \mathrm{End}(W_i)$. In fact, the Cardy condition shows that there can be at most one summand W_i at each point, i.e. the algebra is simple. For the map $\pi : \mathcal{O}_x \to \mathcal{O}_x$ must take each summand $\mathrm{End}(W_i)$ into itself, and cannot factor through the 1-dimensional $\mathcal{C}_x$ if more than one W_i is non-zero. □

According to Theorem A, the most general $\mathcal{O}_{aa}$ is obtained by choosing a vector space $W_{x,a}$ for each basic idempotent ε_x, i.e., a *vector bundle* on the finite space-time $X = \mathrm{Spec}(\mathcal{C})$, and forming:

$$\mathcal{O}_{aa} = \oplus_x \mathrm{End}(W_{x,a}). \tag{2.33}$$

But let us notice that when we have an algebra of the form $\mathrm{End}(W)$ the vector space W is determined by the algebra only up to tensoring with an arbitrary complex line: any irreducible representation of the algebra will do for W.

Elements $\psi \in \mathcal{O}_{aa}$ will be denoted $\psi = \oplus\psi_x$. We have seen above that the projection operator P_x onto the x-th summand is given by

$$\iota_a(\varepsilon_x) = P_x. \tag{2.34}$$

From the adjoint relation and the Cardy condition we readily deduce the relations:

$$\theta_a(\psi) = \sum_x \sqrt{\theta_x}\,\mathrm{Tr}_{W_{x,a}}(\psi_x) \tag{2.35}$$

$$\iota^a(\psi) = \oplus_x \mathrm{Tr}_{W_{x,a}}(\psi_x)\frac{\varepsilon_x}{\sqrt{\theta_x}} \tag{2.36}$$

(one must use the same square root in the formula for $\theta_{\mathcal{O}}$ and ι^a.) Note that $\theta_{\mathcal{C}}(\frac{\varepsilon_x}{\sqrt{\theta_x}}\frac{\varepsilon_y}{\sqrt{\theta_y}}) = \delta_{x,y}$, i.e., the elements $\frac{\varepsilon_x}{\sqrt{\theta_x}}$ form a natural orthonormal basis for $\mathcal{C}$. Thus, a boundary condition a gives us a tuple of *positive integers* $w_x = \dim W_x$, one for each basic idempotent, as well as a choice of the square root $\sqrt{\theta_x}$. The relation (2.5), however, shows that these square roots are an intrinsic property of the Frobenius category $\mathcal{B}$, and do not depend on which particular object in it we are considering.

Let us now determine the $\mathcal{O}_{aa} \times \mathcal{O}_{bb}$ bimodules $\mathcal{O}_{ab}$ associated to a pair of boundary conditions a, b. These are again fixed by the Cardy condition.

LEMMA 2.7. *When $\mathcal{C}$ is semisimple we have*

$$\mathcal{O}_{ab} \cong \oplus_x \mathrm{Hom}(W_{x,b}, W_{x,a}) \tag{2.37}$$

PROOF. Restricting to each $\mathcal{O}_{aa}$ we can invoke Theorem 2.6. Then the

$$\iota_a(\varepsilon_x)\mathcal{O}_{ab} = \mathcal{O}_{ab}\iota_b(\varepsilon_x)$$

are bimodules for the simple algebras $\mathcal{O}_{x,aa}$ and $\mathcal{O}_{x,bb}$. We restrict to a single idempotent and drop the x, that is, we take $\mathcal{C} = \mathbb{C}$. The only irreducible representation of $\mathcal{O}_{aa} = \mathrm{End}(W_a)$ is W_a itself, and the only irreducible $\mathcal{O}_{aa} \times \mathcal{O}_{bb}$-bimodule is $W_a \otimes W_b^*$. Therefore, $\mathcal{O}_{ab} \cong n_{ab} W_a \otimes W_b^*$, where n_{ab} is a nonnegative integer. Let us work out the Cardy condition. If v_m is a basis for W_a and w_n is a basis for W_b then a basis for $\mathcal{O}_{ab}$ is $v_{m,\alpha} \otimes w_{n,\alpha}^*$ where $\alpha = 1, \ldots, n_{ab}$. Then $\pi_b{}^a(\psi) = n_{ab}\mathrm{Tr}_{W_a}(\psi)P_b/\sqrt{\theta_x}$. Comparing to $\iota_b\iota^a(\psi)$ we get $n_{ab} = 1$. □

A consequence of this Lemma is that

$$\pi_b{}^a(\psi) = \oplus_x \frac{1}{\sqrt{\theta_x}} \mathrm{Tr}_{W_{x,a}}(\psi_x)P_{x,b} \tag{2.38}$$

We can now describe the maximal category $\mathcal{B}$ of boundary conditions. We first observe that if $p \in \mathcal{O}_{aa}$ is a projection —i.e., $p^2 = p$ —we can assume that $a = b \oplus c$ in $\mathcal{B}$, where b is the image of p. For we can adjoin images of projections to any additive category in much the same way as we adjoined direct sums. If the closed algebra $\mathcal{C}$ is semisimple we can therefore choose an object a_x of $\mathcal{B}$ for each space-time point x so that a_x is supported at x — i.e., $\iota_{a_x}(\varepsilon_x)\mathcal{O}_{a_x a_x} = \mathcal{O}_{a_x a_x}$ — and is simple, i.e., $\mathcal{O}_{a_x a_x} = \mathbb{C}$. For any object b of $\mathcal{B}$ we then have a canonical morphism

$$\oplus_x \mathcal{O}_{ba_x} \otimes a_x \to b, \tag{2.39}$$

where on the left we have used the possibility of tensoring any object of a linear category by a finite dimensional vector space. Furthermore, it follows from the lemma that the morphism (2.39) is an isomorphism, for both sides have the same space of morphisms into any other object c. Finally, notice that a_x is unique up to tensoring with a line L_x, for if a'_x is another choice then $a'_x \cong a_x \otimes L_x$, where $L_x = \mathcal{O}_{a'_x a_x}$.

THEOREM 2.8. *Suppose $\mathcal{C}$ is semisimple, corresponding to a space-time X. Then*

(i) *the category $\mathcal{B}$ of boundary conditions is equivalent to the category* $\mathrm{Vect}(X)$ *of vector bundles on X, by the inverse functors*

$$W \mapsto \oplus_x W_x \otimes a_x, \tag{2.40}$$

where W is a vector bundle on X, and W_x is its fibre at $x \in X$, and

$$a \mapsto \{\mathcal{O}_{a_x a}\}, \tag{2.41}$$

where the right-hand side denotes the vector bundle on X whose fibre at x is $\mathcal{O}_{a_x a}$.

(ii) *The equivalence of $\mathcal{B}$ with* $\mathrm{Vect}(X)$ *is unique up to transformations*

$$\mathrm{Vect}(X) \to \mathrm{Vect}(X)$$

given by tensoring with a line bundle $L = \{L_x\}$ on X.

(iii) *The Frobenius structure on $\mathcal{B}$ is determined by choosing a square root $\{\sqrt{\theta_x}\}$ of the dilaton field. It is therefore unique up to multiplication by an element $\sigma \in \mathcal{C}$ such that $\sigma^2 = 1$.*

REMARK 2.9. (1) A boundary condition a has a *support*

$$\mathrm{supp}(a) = \{x \in X : W_x \neq 0\} \tag{2.42}$$

contained in $X = \mathrm{spec}(\mathcal{C})$. If two boundary conditions a and b have the same support then $\mathcal{O}_{ab}$ is a Morita equivalence bimodule between $\mathcal{O}_{aa}$ and $\mathcal{O}_{bb}$. The reader might wish to compare this discussion to §6.4 of [**418**]. Note that it is necessary to invoke the Cardy condition to draw this conclusion.

(2) Examples of semisimple Frobenius algebras in physics include:

(a) The fusion rule algebra (Verlinde algebra) of a RCFT.

(b) The chiral ring of an $\mathcal{N} = 2$ Landau-Ginzburg theory for generic superpotential W (that is, as long as all the critical points of W are Morse critical points). This is the case when the IR theory is massive.

(c) Generic quantum cohomology of manifolds.

2.2.2. Comment on B-fields. We can see from this discussion just where the idea of a B-field would appear,[12] though in fact on a 0-dimensional space-time any B-field must be trivial. We showed that there is a category of boundary conditions associated to each point of space-time, and that it is isomorphic to the category of finite dimensional vector spaces, though not canonically. More precisely, it contains minimal — i.e., irreducible —

[12] We are going to discuss the standard physics definitions of the B-field in §§3.2.6, 3.3.2 and 3.5.2.7.

objects from which any other object can be obtained by tensoring with a finite dimensional vector space.

Now a B-field is in essence a bundle of *categories* on space-time in which the fibre-categories are all isomorphic but not canonically. We can suppose that each fibre is isomorphic to the category of finite dimensional vector spaces. The crucial feature is that the ambiguity in identifying each fibre with the standard fibre is a "group" — in this case actually a category — of equivalences whose elements are complex lines and in which composition is given by the tensor product. Our category of boundary conditions is precisely the category of "sections" of a bundle of categories with this structural group.

It may be helpful to think of this in the following way. An electromagnetic field is a line bundle with connection on space-time. It is something we can think of as part of the structure of space-time, and makes sense in the absence of fermions. But in a theory with fermions there is a spinor space at each point of space-time, and the electromagnetic field is "really" the information about how the spinor spaces are connected together from point to point of space-time. In this sense the electromagnetic field "is" the spinor bundle with its connection. A B-field similarly "is" the bundle of boundary conditions.

On a general topological space X the classes of B-fields are classified by the elements of the cohomology group $H^3(X, \mathbb{Z})$, which can be understood as $H^1(X, \mathcal{G})$, where $\mathcal{G}$ is the "group" of line bundles under tensor product, which in algebraic topology is an Eilenberg-MacLane object of type $K(\mathbb{Z}, 2)$. We shall return to this topic in §2.6.

2.2.3. Reconstructing the closed algebra. When we have an open and closed TFT each element ξ of the closed algebra $\mathcal{C}$ defines an endomorphism $\xi_a = i_a(\xi) \in \mathcal{O}_{aa}$ of each object a of $\mathcal{B}$, and $\eta \circ \xi_a = \xi_b \circ \eta$ for each morphism $\eta \in \mathcal{O}_{ba}$ from a to b. The family $\{\xi_a\}$ thus constitutes a natural transformation from the identity functor $1_{\mathcal{B}} : \mathcal{B} \to \mathcal{B}$ to itself. (See also Definition 4.10).

For any $\mathbb{C}$-linear category $\mathcal{B}$ we can consider the ring $\mathcal{E}$ of natural transformations of $1_{\mathcal{B}}$. It is automatically commutative, for if $\{\xi_a\}, \{\eta_a\} \in \mathcal{E}$ then $\xi_a \circ \eta_a = \eta_a \circ \xi_a$ by the definition of naturality. (A natural transformation from $1_{\mathcal{B}}$ to $1_{\mathcal{B}}$ is a collection of elements $\{\xi_a \in \mathcal{O}_{aa}\}$ such that $\xi_a \circ f = f \circ \xi_b$ for each morphism $f \in \mathcal{O}_{ab}$ from b to a. But we can take $a = b$ and $f = \eta_a$.) If $\mathcal{B}$ is a Frobenius category then there is a map $\pi_a{}^b : \mathcal{O}_{bb} \to \mathcal{O}_{aa}$ for each pair of objects a, b, and we can define $j^b : \mathcal{O}_{bb} \to \mathcal{E}$ by $j^b(\eta)_a = \pi_a{}^b(\eta)$ for $\eta \in \mathcal{O}_{bb}$. In other words, j^b is defined so that the Cardy condition $\iota_a \circ j^b = \pi_a{}^b$ holds. But the question arises whether we can define a trace $\theta : \mathcal{E} \to \mathbb{C}$ to make $\mathcal{E}$

into a Frobenius algebra, and with the property that

$$\theta_a(\iota_a(\xi)\eta) = \theta(\xi j^a(\eta)) \tag{2.43}$$

for all $\xi \in \mathcal{E}$ and $\eta \in \mathcal{O}_{aa}$. This is certainly true if $\mathcal{B}$ is a semisimple Frobenius category with finitely many simple objects, for then $\mathcal{E}$ is just the ring of complex-valued functions on the set of classes of these simple elements, and we can readily define $\theta : \mathcal{E} \to \mathbb{C}$ by $\theta(\varepsilon_a) = \theta_a(1_a)^2$, where a is an irreducible object, and $\varepsilon_a \in \mathcal{E}$ is the characteristic function of the point a in the spectrum of $\mathcal{E}$. Nevertheless, a Frobenius category need not be semisimple, and we cannot, unfortunately, take $\mathcal{E}$ as the closed string algebra in the general case. If, for example, $\mathcal{B}$ has just one object a, and $\mathcal{O}_{aa}$ is a commutative local ring of dimension greater than 1, then $\mathcal{E} = \mathcal{O}_{aa}$, and so $\iota_a : \mathcal{E} \to \mathcal{O}_{aa}$ is an isomorphism, and its adjoint map j^a ought to be an isomorphism too. But that contradicts the Cardy condition, as ${\pi_a}^a$ is multiplication by $\sum \psi_i \psi^i$, which must be nilpotent. In §2.6 we shall give an example of two *distinct* closed string Frobenius algebras which admit the same open string algebra $\mathcal{O}_{aa}$.

The commutative algebra $\mathcal{E}$ of natural endomorphisms of the identity functor of a linear category $\mathcal{B}$ is called the *Hochschild cohomology* $HH^0(\mathcal{B})$ of $\mathcal{B}$ in degree 0. The groups $HH^p(\mathcal{B})$ for $p > 0$, whose definition will be given in a moment, vanish if $\mathcal{B}$ is semisimple, but in the general case they appear to be relevant to the construction of a closed string algebra from $\mathcal{B}$. Let us notice meanwhile that for any Frobenius category $\mathcal{B}$ there is a natural homomorphism $K(\mathcal{B}) \to HH^0(\mathcal{B})$ from the Grothendieck group[13] of $\mathcal{B}$, which assigns to an object a the transformation whose value on b is ${\pi_b}^a(1_a) \in \mathcal{O}_{bb}$. In the semisimple case this homomorphism induces an isomorphism $K(\mathcal{B}) \otimes \mathbb{C} \to HH^0(\mathcal{B})$.

For any additive category $\mathcal{B}$ the Hochschild cohomology is defined as the cohomology of the cochain complex in which a k-cochain F is a rule that to each composable k-tuple of morphisms

$$Y_0 \xrightarrow{\phi_1} Y_1 \xrightarrow{\phi_2} \cdots \xrightarrow{\phi_k} Y_k \tag{2.44}$$

assigns $F(\phi_1, \ldots, \phi_k) \in \operatorname{Hom}(Y_0, Y_k)$. The differential in the complex is defined by

$$\begin{aligned} (dF)(\phi_1, \ldots, \phi_{k+1}) &= F(\phi_2, \ldots, \phi_{k+1}) \circ \phi_1 + \\ &\quad + \sum_{i=1}^{k} (-1)^i F(\phi_1, \ldots, \phi_{i+1} \circ \phi_i, \ldots, \phi_{k+1}) + \\ &\quad + (-1)^{k+1} \phi_{k+1} \circ F(\phi_1, \ldots, \phi_k). \end{aligned} \tag{2.45}$$

[13]I.e., the group formed from the semigroup of isomorphism classes of objects of $\mathcal{B}$ under $\oplus$.

(Notice, in particular, that a 0-cochain assigns an endomorphism F_Y to each object Y, and is a cocycle if the endomorphisms form a natural transformation. Similarly, a 2-cochain F gives a possible infinitesimal deformation $F(\phi_1, \phi_2)$ of the composition law $(\phi_1, \phi_2) \mapsto \phi_2 \circ \phi_1$ of the category, and the deformation preserves the associativity of composition if and only if F is a cocycle.)

In the case of a category $\mathcal{B}$ with a single object whose algebra of endomorphisms is $\mathcal{O}$ the cohomology just described is usually called the Hochschild cohomology of the algebra $\mathcal{O}$ with coefficients in $\mathcal{O}$ regarded as a $\mathcal{O}$-bimodule. This must be carefully distinguished from the Hochschild cohomology with coefficients in the dual $\mathcal{O}$-bimodule $\mathcal{O}^*$. But if $\mathcal{O}$ is a Frobenius algebra it is isomorphic as a bimodule to $\mathcal{O}^*$, and the two notions of Hochschild cohomology need not be distinguished. The same applies to a Frobenius category $\mathcal{B}$: because $\mathrm{Hom}(Y_k, Y_0)$ is the dual space of $\mathrm{Hom}(Y_0, Y_k)$ we can think of a k-cochain as a rule which associates to each composable k-tuple (2.44) of morphisms a linear function of an element $\phi_0 \in \mathrm{Hom}(Y_k, Y_0)$. In other words, a k-cochain is a rule which to each "circle" of $k+1$ morphisms

$$\cdots \xrightarrow{\phi_0} Y_0 \xrightarrow{\phi_1} Y_1 \xrightarrow{\phi_2} \cdots \xrightarrow{\phi_k} Y_k \xrightarrow{\phi_0} \cdots \tag{2.46}$$

assigns a complex number $F(\phi_0, \phi_1, \ldots, \phi_k)$.

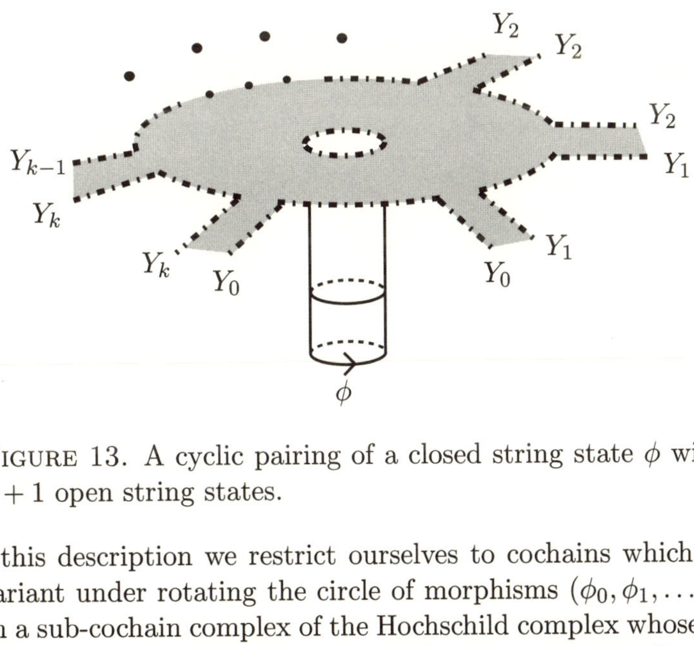

FIGURE 13. A cyclic pairing of a closed string state ϕ with $k+1$ open string states.

If in this description we restrict ourselves to cochains which are cyclically invariant under rotating the circle of morphisms $(\phi_0, \phi_1, \ldots, \phi_k)$ then we obtain a sub-cochain complex of the Hochschild complex whose cohomology is called the *cyclic cohomology* $HC^*(\mathcal{B})$ of the category $\mathcal{B}$. The cyclic cohomology — which evidently maps to the Hochschild cohomology — is a more natural candidate for the closed string algebra associated to $\mathcal{B}$ than is the Hochschild cohomology (because, for example, a state represented by the vector (2.46) pairs in a cyclically invariant way with a closed string state to give a number, in virtue of Figure 13. In our baby examples the cyclic and Hochschild cohomology are indistinguishable, but it is worth pointing out[14] that while $HH^2(\mathcal{B})$ is, as indicated above, the space of infinitesimal deformations of $\mathcal{B}$ as a category, the group $HC^2(\mathcal{B})$ is its space of infinitesimal deformations as a Frobenius category.

A very natural Frobenius category on which to test these ideas is the category of holomorphic vector bundles on a compact Calabi-Yau manifold: that example will be discussed in §2.5.

2.2.4. Spin theories and mod 2 graded categories. Let us give a brief outline, without proofs, of the modifications of the preceding discussion which are needed to describe the category of boundary conditions for a topological-spin theory as defined in §2.1.6.

There is just one spin structure on an interval, and its automorphism group is (± 1), so for each pair of boundary conditions a, b the vector space $\mathcal{O}_{ab}$ will have an involution, i.e. a mod 2 grading. The bilinear composition $\mathcal{O}_{ab} \otimes \mathcal{O}_{bc} \to \mathcal{O}_{ac}$ will preserve the grading. There is a non-degenerate trace $\theta_a : \mathcal{O}_{aa} \to \mathbb{C}$ which satisfies the commutativity condition (2.5) (without signs).

If the closed theory is described by a Frobenius algebra $\mathcal{C} = \mathcal{C}_{ns} \oplus \mathcal{C}_r$, as in §2.1.6, there will be adjoint maps

$$\begin{aligned} \iota_a^{ns} &: \mathcal{C}_{ns} \to \mathcal{O}_{aa} \\ \iota_{ns}^{a} &: \mathcal{O}_{aa} \to \mathcal{C}_{ns} \\ \iota_a^{r} &: \mathcal{C}_{r} \to \mathcal{O}_{aa} \\ \iota_{r}^{a} &: \mathcal{C}_{aa} \to \mathcal{C}_{r} \end{aligned} \tag{2.47}$$

which preserve the grading. Moreover, the maps ι_a^{ns} and ι_a^r fit together to define a homomorphism of algebras $\mathcal{C} \to \mathcal{O}_{aa}$. The centrality condition becomes

$$\begin{aligned} \iota_a^{ns}(\phi)\psi &= \psi\iota_a^{ns}(\phi) \\ \iota_a^{r}(\phi)\psi &= (-1)^{\deg\phi\deg\psi+\deg\psi}\psi\iota_a^{r}(\phi). \end{aligned} \tag{2.48}$$

Thus, ι^{ns} maps into the naive center of the algebra $\mathcal{O}_{aa}$. The reason we get the naive centre here, rather than the graded-algebra centre, and also the reason that the trace is naively commutative, is the same as that given in §2.1.6 for the naive commutativity of the algebra $\mathcal{C}$. The sign for ι^r is obtained by carefully following the choices of sections of the spin bundle one chooses under the diffeomorphism in Figure 9.

There are two Cardy conditions

$$\begin{aligned} \iota_a^{ns}\iota_{ns}^{b}(\psi) &= \pi_b^a(\psi) := \sum(-1)^{\deg\psi_\mu\deg\psi}\psi_\mu\psi\psi^\mu \\ \iota_a^{r}\iota_{r}^{b}(\psi) &= \tilde{\pi}_b^a(\psi) := \sum(-1)^{\deg\psi_\mu(\deg\psi+1)}\psi_\mu\psi\psi^\mu. \end{aligned} \tag{2.49}$$

[14]As we learned from Kontsevich.

If we assume the closed algebra is semisimple then, just as before, we can assume that $\mathcal{C}_{ns}$ is the algebra of functions on a finite set X, and we can determine the category of boundary conditions point-by-point. In other words, we can assume that $\mathcal{C} = \mathbb{C}[\eta]$, where the generator η of $\mathcal{C}_r$ satisfies $\eta^2 = 1$, but may have either even or odd degree. In either case, the argument we have already used shows (by means of the first Cardy formula) that the algebra $\mathcal{O}_{aa}$ is the full matrix algebra of a vector space W. If the degree of η is even then $\iota^r(\eta) = P$ with P even, $P^2 = 1$, and $P\psi P = (-1)^{\deg \psi}\psi$. In this case the category of boundary conditions at the point is equivalent to the category of mod 2 graded vector spaces. If, on the other hand, the degree of η is odd, then P is odd, $P^2 = 1$ and P is (naive) central. The involution of the algebra $\mathcal{O}_{aa}$ corresponds to an involution of the module W, and the action of P is an isomorphism between the two halves of the grading. The even subalgebra of $\mathcal{O}_{aa}$ is a full matrix algebra. Thus the category of boundary conditions is equivalent to the category of graded representations of the superalgebra $\mathbb{C}[\eta]$, which in turn is equivalent simply to the category of ungraded vector spaces. The Frobenius structure of the open algebra determines that of the closed algebra by taking the square, as in the ungraded case. The two cases $\deg \eta = 0$ and $\deg \eta = 1$ are roughly analogous to the distinction between the even and odd degree Clifford algebras over the complex numbers.

Suppose, conversely, that we have an arbitrary semisimple mod 2 graded category $\mathcal{B}$, i.e., a linear category equipped with an involutory functor S which one thinks of as the flip of the grading. Such a category has two kinds of simple object P: those such that $S(P) \cong P$, and those for which this is not true. The first kind of object generates a subcategory of $\mathcal{B}$ isomorphic to the category of vector spaces, and the second kind generates a subcategory isomorphic to the category of graded vector spaces. Thus any semisimple graded category $\mathcal{B}$ is the category of boundary conditions for a unique topological-spin theory.

2.3. Vector bundles, K-theory, and "boundary states"

In the semisimple case there is a nice geometrical interpretation of the category $\mathcal{B}$ of boundary conditions: the possible objects correspond to the vector bundles over the "space-time" $X = \mathrm{Spec}(\mathcal{C})$ associated to $\mathcal{C}$, which is just a finite set of points. The fibre above a point x is just the vector space W_x.

Let us now make some comments on "boundary states". As we will discuss in §3.5, in conformal field theory one associates to a boundary condition a a corresponding "state" B_a in the closed string state space. (Strictly, B_a is an element of the algebraic dual.)

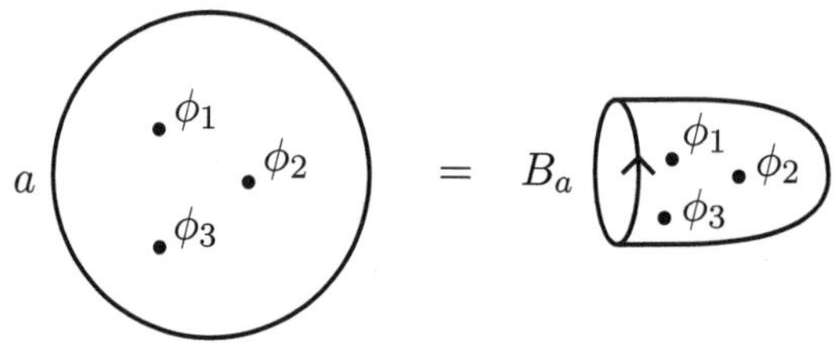

FIGURE 14. Correlations on the upper half plane with boundary condition a are the same as the closed string amplitude for an insertion of a boundary state B_a.

Translated to the present context, $B_a \in \mathcal{C}$. The defining property of the boundary state is that the correlation functions of operators on a disk with the boundary condition a are equal to the correlation functions of the closed theory on the sphere obtained by capping off the disk with another disk and inserting the state B_a at the center of the cap. This is illustrated in Figure 14.

In equations,

$$\theta_a\big(\iota_a(\phi_1)\cdots\iota_a(\phi_n)\big) = \theta_{\mathcal{C}}\big(B_a\phi_1\cdots\phi_n\big) \tag{2.50}$$

for all $\phi_1,\ldots,\phi_n$. Using the adjoint relation and non-degeneracy of the trace we find that

$$B_a = \iota^a(1_{\mathcal{O}_{aa}}) \tag{2.51}$$

The map $a \mapsto B_a$ is a natural homomorphism

$$K(\mathcal{B}) \to \mathcal{C}. \tag{2.52}$$

More generally, if χ is the Euler element of $\mathcal{C}$, we can think of the element $\chi^g\prod(B_a)^{h_a}$ of $\mathcal{C}$ as the operator which "adds g handles and $h = \sum h_a$ holes, where h_a of the holes have the boundary condition a", in the sense that the correlation functions of a collection of operators on a surface of genus g with h holes with the given boundary conditions are the same as the correlation functions of the same operators on a sphere with the additional insertion of $\chi^g\prod(B_a)^{h_a}$.

Let us record one simple property of these boundary states. First, using the Cardy condition we have

$$\begin{aligned}\theta_{\mathcal{C}}(B_aB_b) &= \theta_a(\iota_a(B_b))\\ &= \theta_a(\iota_a\iota^b(1_b))\\ &= \theta_a(\pi_a{}^b(1_b))\\ &= \dim\mathcal{O}_{ab}.\end{aligned} \tag{2.53}$$

In the semisimple case the formulae (2.35) give an explicit formula for the "boundary state" in terms of the basic idempotents:

$$B_a = \iota^a(1_{\mathcal{O}_{aa}}) = \sum_x (\dim W_x) \frac{\varepsilon_x}{\sqrt{\theta_x}}. \tag{2.54}$$

The formula shows that the boundary states form a positive cone in the unimodular lattice $\mathcal{L}_B$ spanned by the orthonormal basis $\frac{\varepsilon_x}{\sqrt{\theta_x}}$ in the closed algebra $\mathcal{C}$. In particular it follows from (2.54) that boundary states can only be added with positive integral coefficients. They are therefore *not* like quantum mechanical states of branes. The fundamental integral structure is a result of the Cardy condition.

It is natural to speculate whether there should be an operation of "multiplication" of boundary conditions. There are arguments both for and against. The original perspective on D-branes, according to which they are viewed as "cycles" in space-time on which open strings can begin and end, suggests that there should be a multiplication, corresponding to the intersection of cycles. As no multiplication seems to emerge from the toy structure we have developed in this chapter one may wonder whether an important ingredient has been omitted. Against this there are the following considerations. Our boundary conditions seem to correspond more closely to vector bundles — i.e., to K-theory classes — on space-time than to homology cycles: that will be plainer when we consider the equivariant situation in §2.7. Now the K-theory classes of a ring have a product, coming from the tensor product of modules, only when the ring is commutative; and we have already remarked that the B-fields which are part of the closed string model of space-time seem to encode a degree of noncommutativity. More precisely, D-branes seem to define classes in the *twisted* K-theory of space-time, twisted by the B-field, and the twisted K-theory of a space does not form a ring: the product of two twisted classes is a twisted class corresponding to the sum of the twistings of the factors. But in string theory there is no concept of "turning off" the B-field to find an underlying untwisted space-time. For example, the conformal field theory corresponding to a torus with a non-zero B-field can be isomorphic by "T-duality" to a theory coming from another torus with no B-field.

Another reason for not expecting a multiplication operation on D-branes also comes from T-duality in conformal field theory. There the closed string theories defined by a Riemannian torus T and its dual T^* are isomorphic, and we do indeed have a K-theory isomorphism $K(T) \cong K(T^*)$, but it is not compatible with the multiplication in K-theory. On the other hand, in some examples of TFTs coming from $\mathcal{N}=2$ supersymmetric sigma models the category of boundary conditions does seem to be a tensor category.

The formula (2.54) for the boundary state shows that the lattice $\mathcal{L}_B$, which is picked out inside $\mathcal{C}$ by the dilaton field θ, is not closed under multiplication in $\mathcal{C}$ unless $\theta_x = 1$ for all points x; but the lattices corresponding to different dilaton fields multiply into each other just as happens with twisted K-classes. Nevertheless, in the semisimple case, if we define an element $S := \sum_x \sqrt{\theta_x}\varepsilon_x$, then the operation

$$(B_1, B_2) \mapsto SB_1B_2 \tag{2.55}$$

does define a multiplication on boundary states, though its significance is unclear.

2.3.1. "Cardy states" versus "Ishibashi states". The formula for the boundary state (2.54) is reminiscent of the relation between the "Cardy states" and the "Ishibashi states" in boundary conformal field theory [**86, 87**]. For readers familiar with this relation, let us comment briefly on this resemblance, as the two sides of (2.54) do in fact correspond to these two bases.

As we will review in §3.5.2, the Cardy and the Ishibashi states are two natural bases for the boundary states. The Cardy states are physical boundary states, for which all correlation functions are single-valued. On the other hand, the "Ishibashi states" are defined as the simplest solutions of the consistency condition (3.136), and are in direct one-to-one correspondence with the closed string primary fields. The distinction between the two is in how left-moving and right-moving components of the boundary state are glued together; diagonally for the Ishibashi states, and satisfying the Cardy relations for the Cardy states.

Our analogy is that ε_i, $i = 1, \ldots, N$, correspond to Ishibashi states while the basis ϕ_μ is analogous to a basis of primary fields of definite conformal weight and is characterized by

$$\phi_\mu\phi_\nu = N_{\mu\nu}^{\lambda}\phi_\lambda \tag{2.56}$$

with positive integral $N_{\mu\nu}^{\lambda}$.

The analogy should not be pushed too far since in the topological theory there is no chiral algebra, and we should think of *every* element of $\mathcal{C}$ as a solution to $(T - \bar{T}) = 0$, and its generalizations. There are no left-movers or right-movers. Nevertheless, using these formulae we recover, essentially, Cardy's formula for Cardy boundary states in terms of character boundary states. Note that there is no need to use any relation to the modular group.

We close with one further brief remark. It is nice to see the standard relation that the closed string coupling is the square of the open string coupling in the present context. If we scale $\theta_{\mathcal{C}} \to \lambda^{-2}\theta_{\mathcal{C}}$ then $\chi_{\mathcal{C}} = \sum_\mu \phi_\mu\phi^\mu \to \lambda^{+2}\chi_{\mathcal{C}}$. We may therefore interpret λ^2 as the closed string coupling. On the other hand, the square root of θ_i in $B_{\mathcal{O}}$ shows that $B_{\mathcal{O}} \to \lambda B_{\mathcal{O}}$, and therefore λ is the open string coupling. Indeed, the partition function for a surface with g handles and h holes is $Z(\Sigma) = \theta_{\mathcal{C}}((\chi_{\mathcal{C}})^g(B_{\mathcal{O}})^h)$, and therefore scales as $Z(\Sigma) \to \lambda^{-\chi(\Sigma)}Z(\Sigma)$, as expected, where $\chi(\Sigma) = 2 - 2g - h$ is the Euler number of Σ.

2.4. Landau-Ginzburg theories

D-branes can be defined in general two-dimensional $\mathcal{N} = 2$ Landau-Ginzburg theories [**241**]. Such theories can be topologically twisted, producing topological Landau-Ginzburg theories. It is interesting to compare with the D-branes obtained from our results applied to the resulting closed topological theory. Here we confine ourselves to a few very elementary remarks. In the past few years, following an initial suggestion by Kontsevich, an elaborate theory of categories of topological Landau-Ginzburg branes has been developed. See §3.6.8 and references therein for details. These categories are thought to capture more physical information about the D-branes. In the case when all the critical points of the superpotential are Morse there is a functor to the category of branes we construct.

Let us recall the definition of a topological LG theory. One begins with a superpotential $W(X_i)$ which is a holomorphic function of chiral superfields $X_1, \ldots, X_n$. When W is a polynomial the Frobenius algebra is simply the Jacobian algebra

$$\mathcal{C} = \mathbb{C}[X_1, \ldots, X_n]/(\partial_1 W, \ldots, \partial_n W). \tag{2.57}$$

The Frobenius structure is defined by a residue formula. For example, in the one-variable case we define

$$\theta(\phi) := \mathrm{Res}_{X=\infty} \frac{\phi(X)}{W'(X)}. \tag{2.58}$$

If the critical points of W are all Morse critical points then the algebra (2.57) is semisimple. Physically, Morse critical points correspond to massive theories, while non-Morse critical points renormalize to nontrivial 2d CFT's in the infrared.

If all the critical points are Morse then the trace is easily written in terms of the critical points p_a as

$$\theta(\phi) = \sum_{dW(p_a)=0} \frac{\phi(p_a)}{\det(\partial_i\partial_j W|_{p_a})}. \tag{2.59}$$

In the semisimple one-variable case we can construct the basic idempotents as follows. Let

$$W' = \prod_{\alpha=1}^{n}(X - r_\alpha) \tag{2.60}$$

where we assume all the roots are distinct. Then it is easy to check that

$$\varepsilon_\beta := \prod_{\alpha:\alpha\neq\beta} \frac{(X-r_\alpha)}{(r_\beta - r_\alpha)} \tag{2.61}$$

are basic idempotents. (To prove this, write $(X-r_\alpha) = (X-r_\beta)+(r_\beta - r_\alpha)$).

EXAMPLE 2.10. $W = \frac{1}{3}t^3 - qt$. For $n = 2$ we can explicitly write

$$\begin{aligned} \varepsilon_1 &= \frac{\sqrt{q}+t}{2\sqrt{q}}, \\ \varepsilon_2 &= \frac{\sqrt{q}-t}{2\sqrt{q}} \end{aligned} \tag{2.62}$$

Note that $\theta_1 = 1/(2\sqrt{q})$ and $\theta_2 = -1/(2\sqrt{q})$.

Then from the general result above one finds $\mathcal{O}_{aa} = \mathrm{End}(W_1)\oplus\mathrm{End}(W_2)$ and

$$\theta_{\mathcal{O}}(\Psi) = \sqrt{\theta_1}\mathrm{Tr}(\Psi_1) + \sqrt{\theta_2}\mathrm{Tr}(\Psi_2) \tag{2.63}$$

$$\iota^a(\Psi) = \frac{1}{\sqrt{\theta_1}}\mathrm{Tr}(\Psi_1)\varepsilon_1 + \frac{1}{\sqrt{\theta_2}}\mathrm{Tr}(\Psi_2)\varepsilon_2. \tag{2.64}$$

Thus, the general boundary state is

$$B = w_1 \frac{\varepsilon_1}{\sqrt{\theta_1}} + w_2 \frac{\varepsilon_2}{\sqrt{\theta_2}} \tag{2.65}$$

where w_1, w_2 are integers. It is interesting to work out the monodromy in the boundary states as q circles counterclockwise around the origin along the curve $q(t) = |q|e^{2\pi i t}$, $0 \le t \le 1$. Under this operation ε_1 and ε_2 are exchanged. Moreover, $\sqrt{\theta_1} = e^{-i\pi t/2}/\sqrt{2}$ and $\sqrt{\theta_2} = ie^{-i\pi t/2}/\sqrt{2}$. Thus branes of type (w_1, w_2) evolve into branes of type $(w_2, -w_1)$.

Clearly there will be similar phenomena for general Landau-Ginzburg theories. The space of superpotentials W has a codimension one "discriminant locus" where it has non-Morse critical points. Analytic continuation around this locus will permute the ε_i, but will only permute the $\sqrt{\theta_i}$ up to sign. One may understand in this elementary way some of the brane permutation/creation phenomena discussed in numerous places in the literature.

The "vector bundles on space-time" that we have found can be taken quite literally in the context of the theory of strings moving in less than one dimension which was worked out in 1988-1991. (For reviews [**180, 122, 118**].) Strings moving in a space-time of n disjoint points can be modelled by matrix chains or by topological field theory. The latter point of view is described in, for example, [**122, 118**]. In the latter point of view, one considers topological gravity coupled to topological matter. For n space-time points the topological matter can be taken to be the $\mathcal{N} = 2$ Landau-Ginzburg theories associated to W given by the unfolding of A_n singularities:

$$W = \frac{x^{n+1}}{n+1} + a_n x^n + \cdots + a_0. \tag{2.66}$$

For generic W we find vector bundles on n space-time points. This is of course what we expect for the branes in such space-times!

It is worth mentioning that in these simplest of string theories (the "minimal string theories") considerable progress has been made in recent years in understanding the full spectrum of D-branes, going beyond the topological field theory truncation. See [**417**] for a review.

2.5. Going beyond semisimple Frobenius algebras

The examples of topological field theories coming from $\mathcal{N} = 2$ conformal field theories — Landau-Ginzburg models and the quantum cohomology rings of Calabi-Yau manifolds — suggest that it is of interest to understand the possible solutions of the algebraic conditions in the case when $\mathcal{C}$ is not semisimple. In this section we shall make some partial progress with this problem, and we shall also explain how it should perhaps be viewed in a wider context.

2.5.1. Examples related to the cohomology of manifolds. A natural example of a graded commutative Frobenius algebra is the cohomology with complex coefficients of an even-dimensional compact oriented manifold X. Thus, for $\mathcal{C}$ we can take the algebra $\mathcal{C} = H^*(X, \mathbb{C})$ with trace $\theta(\phi) = \int_X \phi$. What are the corresponding $\mathcal{O}$'s?

A natural guess, which turns out to be wrong, but for interesting reasons, is that we should take $\mathcal{O} = \mathcal{C} \otimes \mathrm{Mat}_N(\mathbb{C}) = \mathrm{Mat}_N(\mathcal{C})$ for some $N > 0$, together with

$$\theta_{\mathcal{O}}(\psi) = \int_X \mathrm{Tr}(\psi). \tag{2.67}$$

While $\mathcal{O}$ is indeed a Frobenius algebra, the only natural candidate for the map ι_* is $\iota_*(\phi) = \phi \otimes 1_N$. However, this fails to satisfy the Cardy condition: one computes $\iota^*(\psi) = \mathrm{Tr}(\psi)$ from the adjoint relation, and hence $\iota_*\iota^*(\psi) = \mathrm{Tr}(\psi) \otimes 1_N$. On the other hand, one also finds

$$\begin{aligned} \pi(\psi) &= \sum (-1)^{\deg \omega_i (\deg \psi + \deg \omega^i)} (\omega_i \otimes e_{lm}) \psi (\omega^i \otimes e_{ml}) \\ &= (\chi(TX)\mathrm{Tr}(\psi)) \otimes 1_N. \end{aligned} \tag{2.68}$$

Here ω_i and ω^i are dual bases for $H^*(X, \mathbb{C})$ with respect to the Poincaré inner product, e_{ml} are matrix units, and $\chi(T_X) \in H^{\mathrm{top}}(X, \mathbb{C})$ is the Euler class of the tangent bundle T_X, which is given by $\chi(T_X) = \sum \omega_i \omega^i$, and, finally, we have used the matrix identity $\sum e_{lm}\psi e^{ml} = \mathrm{Tr}(\psi) \otimes 1_N$. The map π annihilates forms of positive degree, and cannot agree with $\iota_*\iota^*$.

This example can be modified to give an open and closed theory by taking $\mathcal{O}$ to be associated with a *submanifold* of X. This is, after all, the standard picture of D-branes! Let us work in the algebraic category of $\mathbb{Z}$-graded vector spaces, and continue to take $\mathcal{C} = H^*(X, \mathbb{C})$, with X a compact connected oriented n-dimensional manifold, and the trace $\theta_{\mathcal{C}}(\phi) = \int_X \phi$, of degree $-n$ as above. Let us look for an open algebra of the form $\mathcal{O} = \mathrm{Mat}_N(\mathcal{O}_0)$, with $\mathcal{O}_0$ commutative. Then $\mathcal{O}_0$ is a Frobenius algebra, and we may as well assume that it is $H^*(Y, \mathbb{C})$ for some compact oriented manifold[15] Y of dimension m, and that $\iota_* : \mathcal{C} \to \mathcal{O}_0$ is f^* for some map $f : Y \to X$.

Thus $\mathcal{O} = H^*(Y, \mathbb{C}) \otimes \mathrm{Mat}_N(\mathbb{C})$ with open string trace

$$\theta_{\mathcal{O}}(\Psi) = \theta_o \int_Y \mathrm{Tr}(\Psi) \tag{2.69}$$

of degree $-m$, where θ_o is a constant. This is a non-commutative Frobenius algebra.

The adjoint relation determines ι^*:

$$\iota^*(\Psi) = \theta_o f_*(\mathrm{Tr}(\Psi)), \tag{2.70}$$

where f_* is the adjoint of the ring homomorphism $f^* : H^*(X) \to H^*(Y)$ with respect to Poincaré duality. Thus ι^* has degree $n - m$. On the other hand, one sees at once that $\pi : \mathcal{O} \to \mathcal{O}$ has degree m, so if the Cardy condition is to hold we must have $n = 2m$. If that is true, then we can assume, by making a small generic perturbation of f, that f is an immersion of Y in X. We can now make the adjoint map f_* more explicit:

$$f_*(\psi) = \mathrm{pr}^*(\psi) \wedge \Phi_{N_Y}, \tag{2.71}$$

where $\mathrm{pr} : N_Y \to Y$ is the projection of the normal bundle (identified with a tubular neighbourhood of Y in X) and Φ_{N_Y} is the Thom class of the bundle, compactly supported in the tubular neighbourhood, which represents the cohomology class of Y in X. One easily finds that

$$\iota_*\iota^*(\Psi) = \theta_o \chi(N_Y) \wedge \mathrm{Tr}(\Psi) \otimes 1. \tag{2.72}$$

where $\chi(N_Y)$ is the Euler class of the normal bundle of $Y \hookrightarrow X$, i.e. the homological self-intersection of Y in X.

On the other hand, from (2.68) and (2.70), we have

$$\pi(\Psi) = \frac{1}{\theta_o} \mathrm{Tr}(\Psi) \chi(T_Y) \otimes 1_N, \tag{2.73}$$

where $\chi(T_Y) \in H^{\mathrm{top}}(Y, \mathbb{C})$ is the Euler class of the tangent bundle T_Y, whose integral is the Euler number of Y.

---

[15] In fact we need to allow Y to have orbifold singularities to ensure this.

Evidently the Cardy conditions are satisfied if we choose θ_o so that $\chi(T_Y) = \theta_o^2\chi(N_Y)$. This is always possible if $\chi(N_Y)$, which is the self-intersection number of Y in X, is non-zero, and also possible if Y is a Lagrangian submanifold of a symplectic manifold X, for then $N_Y \cong T_Y$. The boundary state is $B = \theta_o N\Phi_{N_Y}$.

One immediate consequence of this discussion is that if we start, say, with $\mathcal{O} = H^*(\mathbb{C}P^2, \mathbb{C})$ as our open algebra then we can easily find two different closed algebras compatible with it, by regarding Y as a submanifold either of $X = \mathbb{C}P^4$ or of $X' = \mathbb{H}P^2$.

Unfortunately we do not know how to describe the *category* of boundary conditions for $\mathcal{C} = H^*(X, \mathbb{C})$. But it seems likely, in any case, that to get a significant result one would have to consider the theory on the cochain level. We next turn our attention to that case.

2.5.2. The Chas-Sullivan theory. There is an interesting example — due to Chas and Sullivan [**91**] — on the cochain level of a structure a little weaker than that of our open and closed theories which may illuminate the use of cochain theories. Let us start with a compact oriented manifold X, which we shall take to be connected and simply connected. We can define a category $\mathcal{B}$ whose objects are the oriented submanifolds of X, and whose vector space of morphisms from Y to Z is $\mathcal{O}_{YZ} = \mathrm{Ext}^*_{H^*(X)}(H^*(Y), H^*(Z))$ — the cohomology, as usual, has complex coefficients, and $H^*(Y)$ and $H^*(Z)$ are regarded as $H^*(X)$-modules by restriction. The composition of morphisms is given by the Yoneda composition of Ext groups. With this definition, however, it will not be true that $\mathcal{O}_{YZ}$ is dual to $\mathcal{O}_{ZY}$. (To see this it is enough to consider the case when $Y = Z$ is a point of X, and X is a product of odd-dimensional spheres; then $\mathcal{O}_{YZ}$ is a symmetric algebra, and is not self-dual as a vector space.)

We can do better by defining a cochain complex $\hat{\mathcal{O}}_{YZ}$ of "morphisms" by

$$\hat{\mathcal{O}}_{YZ} = \mathcal{B}_{\Omega(X)}(\Omega(Y), \Omega(Z)), \tag{2.74}$$

where $\Omega(X)$ denotes the usual de Rham complex of a manifold X, and $\mathcal{B}_A(B, C)$, for a differential graded algebra A and differential graded A-modules B and C, is the usual cobar resolution

$$\mathrm{Hom}(B, C) \to \mathrm{Hom}(A \otimes B, C) \to \mathrm{Hom}(A \otimes A \otimes B, C) \to \cdots, \tag{2.75}$$

in which the differential is given by

$$\begin{aligned} df(a_1 \otimes \cdots \otimes a_k \otimes b) &= a_1 f(a_2 \otimes \cdots \otimes a_k \otimes b) \\ &+ \sum (-1)^i f(a_1 \otimes \cdots \otimes a_i a_{i+1} \otimes \cdots \otimes a_k \otimes b) \\ &+ (-1)^k f(a_1 \otimes \cdots \otimes a_{k-1} \otimes a_k b) \end{aligned} \tag{2.76}$$

whose cohomology is $\mathrm{Ext}_A(B, C)$. This is different from

$$\mathcal{O}_{YZ} = \mathrm{Ext}^*_{H^*(X)}(H^*(Y), H^*(Z)),$$

but related to it by a spectral sequence whose E_2-term is $\mathcal{O}_{YZ}$ and which converges to $H^*(\hat{\mathcal{O}}_{YZ}) = \mathrm{Ext}_{\Omega(X)}(\Omega(Y), \Omega(Z))$. But more important is that $H^*(\hat{\mathcal{O}}_{YZ})$ is the homology of the space $\mathcal{P}_{YZ}$ of paths in X which begin in Y and end in Z. To be precise, $H^p(\hat{\mathcal{O}}_{YZ}) \cong H_{p+d_Z}(\mathcal{P}_{YZ})$, where d_Z is the dimension of Z. On the cochain complexes the Yoneda composition is associative up to cochain homotopy, and defines a structure of an A_∞-category $\hat{\mathcal{B}}$. The corresponding composition of homology groups

$$H_i(\mathcal{P}_{YZ}) \times H_j(\mathcal{P}_{ZW}) \to H_{i+j-d_Z}(\mathcal{P}_{YW}) \tag{2.77}$$

is the composition of the Gysin map associated to the inclusion of the codimension d_Z submanifold $\mathcal{M}$ of pairs of composable paths in the product $\mathcal{P}_{YZ} \times \mathcal{P}_{ZW}$ with the concatenation map $\mathcal{M} \to \mathcal{P}_{YW}$.

Let us try to fit a "closed string" cochain algebra $\mathcal{C}$ to this A_∞ category. The algebra of endomorphisms of the identity functor of $\mathcal{B}$, denoted $\mathcal{E}$ in §2.2, is easily seen to be just the cohomology algebra $H^*(X)$. We have mentioned in §2.1 that this is the Hochschild cohomology $HH^0(\mathcal{B})$.

The definition of Hochschild cohomology for a linear category $\mathcal{B}$ was given at the end of §2.2. In fact the definition of the Hochschild complex makes sense for an A_∞ category such as $\hat{\mathcal{B}}$, and it is one candidate for the closed algebra $\mathcal{C}$.

In the present situation $\mathcal{C}$ is equivalent to the usual Hochschild complex of the differential graded algebra $\Omega(X)$, whose cohomology is the homology of the free loop space $\mathcal{L}X$ with its degrees shifted downwards[16] by the dimension d_X of X, so that the cohomology $H^i(\mathcal{C})$ is potentially non-zero for $-d_X \leq i < \infty$. This algebra was introduced by Chas and Sullivan in precisely the present context — they were trying to reproduce the structures of string theory in the setting of classical algebraic topology. There is a map $H^i(X) \to H^{-i}(\mathcal{C})$ which embeds the ordinary cohomology ring of X in the Chas-Sullivan ring, and there is also a ring homomorphism $H^i(\mathcal{C}) \to H_i(\mathcal{L}_0 X)$ to the Pontrjagin ring of the based loop space $\mathcal{L}_0 X$, based at any chosen point in X.

The other candidate for $\mathcal{C}$ mentioned in §2.1 was the cyclic cohomology of the algebra $\Omega(X)$, which is well-known [**176**] to be the *equivariant* homology of the free loop space $\mathcal{L}X$ with respect to its natural circle action. This may be an improvement on the non-equivariant homology.

The structure we have arrived at is, however, *not* a cochain-level open and closed theory, as we have no trace maps inducing inner products on

---

[16] Thus the identity element of the algebra, in $H^0(\mathcal{C})$, is the fundamental class of X, regarded as an element of $H_n(\mathcal{L}X)$ by thinking of the points of X as constant loops in $\mathcal{L}X$.

$H^*(\hat{\mathcal{O}}_{YZ})$. When one tries to define operators corresponding to cobordisms it turns out to be possible only when each connected component of the cobordism has non-empty outgoing boundary. (A theory defined on this smaller category is often called a *non-compact* theory.) The nearest theory in our sense to the Chas-Sullivan one is the so-called "A-model" defined for a symplectic manifold X. There the A_∞ category is the *Fukaya category* (see Chapter 8 for many more details), whose objects are the Lagrangian submanifolds of X equipped with bundles with connection, and the cochain complex of morphisms from Y to Z is the Floer complex which calculates the "semi-infinite" cohomology of the path space $\mathcal{P}_{YZ}$. In good cases the cohomology of this Floer complex has a vector space basis indexed by the points of intersection of Y and Z, and the cohomology of the corresponding closed complex is just the ordinary cohomology of X. From our perspective the essential feature of the Floer theory is that it satisfies Poincaré duality for the infinite dimensional manifold $\mathcal{L}X$.

2.5.3. Remarks on the B-model. Let X be a complex variety of complex dimension d with a trivialization of its canonical bundle. That is, we assume there is a nowhere-vanishing holomorphic d-form Ω. The *B-model* [**468**] is a $\mathbb{Z}$-graded topological field theory arising from the $\mathcal{N}=2$ supersymmetric σ-model of X. The natural boundary conditions for the theory are provided by holomorphic vector bundles on X.

The category of holomorphic vector bundles is not a Frobenius category. There is, however, a very natural $\mathbb{Z}$-graded Frobenius category associated to X: the category $\mathcal{V}_X$ whose objects are the vector bundles on X, but whose space of morphisms from E to F is

$$\mathcal{O}_{EF} = \mathrm{Ext}^*_X(E,F) = H^{0,*}(X, E^* \otimes F). \tag{2.78}$$

The trace $\theta_E : \mathcal{O}_{EE} \to \mathbb{C}$, of degree $-d$, is defined by

$$\theta_E(\Psi) = \int_X \mathrm{Tr}(\Psi) \wedge \Omega. \tag{2.79}$$

This is non-degenerate by Serre duality, but the category is still not semisimple — in fact the non-vanishing of the groups Ext^i for $i > 0$ precisely expresses the non-semisimplicity of the category. (A non-zero element of $\mathrm{Ext}^1_X(E,F)$ corresponds to an exact sequence $0 \to F \to G \to E \to 0$ which does not split, i.e., to a vector bundle G with a subbundle F with no complementary bundle.)

What are the endomorphisms of the identity functor of $\mathcal{V}_X$? Multiplication by any element of $H^{0,*}(X)$ clearly defines such an endomorphism. A holomorphic vector field ξ on X also defines an endomorphism of degree 1, for any bundle E has an "Atiyah class"[17] $a_E \in \mathrm{Ext}^1_X(E, E \otimes T^*_X)$ — its

[17]Corresponding to the extension of bundles $0 \to E \otimes T^*_X \to J^1E \to E \to 0$, where J^1E is the bundle of 1-jets of holomorphic sections of E.

curvature — which we can contract with ξ to give $e_\xi = \iota_\xi a_E \in \mathrm{Ext}^1_X(E;E)$. More generally, a class

$$\eta \in H^{0,q}(X, \bigwedge^p T_X) = \mathrm{Ext}^q_X(\bigwedge^p T^*_X, \mathbb{C})$$

can be contracted with $(a_E)^p \in \mathrm{Ext}^p_X(E, E \otimes (T^*_X)^{\otimes p})$ to give

$$e_\eta = \iota_\eta (a_E)^p \in \mathrm{Ext}^{p+q}_X(E,E).$$

Now Witten has shown in [**468**] that $H^{0,*}(X, \bigwedge^* T_X)$ is indeed the closed string algebra of the B-model. To understand this in our context we must once again pass to the cochain-level theory of which the Ext groups are the cohomology. A good way to do this is to replace a holomorphic vector bundle E by its $\bar{\partial}$-complex $\hat{E} = \Omega^{0,*}(X,E)$, which is a differential graded module for the differential graded algebra $A = \Omega^{0,*}(X)$. Then we define $\hat{O}_{EF}$ as the cochain complex $\mathrm{Hom}_A(\hat{E}, \hat{F})$, whose cohomology groups are $\mathrm{Ext}^*_X(E,F)$. If we are going to do this, it is natural to allow a larger class of objects, namely all finitely generated projective differential graded A-modules. Any coherent sheaf E on X defines such a module: one first resolves E by a complex $\mathcal{E}$ of vector bundles, and then takes the total complex of the double complex $\hat{\mathcal{E}}$. The resulting enlarged category is essentially the *bounded derived category* (one of the main topics of Chapter 4) of the category of coherent sheaves on X. In this setting, we find without difficulty that the endomorphisms of the identity morphism are given, just as in the topological example above, by the Hochschild complex

$$\hat{\mathcal{C}} = \{A \to A \otimes A \to A \otimes A \otimes A \to \cdots\},$$

whose cohomology is $H^*(X, \bigwedge^* T_X)$. There is still, however, work to do to understand the trace maps on $\hat{\mathcal{C}}$, and the adjoint maps ι_E and ι^E. We feel that this has not yet been properly elucidated in the literature. For some progress on this question see [**348, 81, 82**], as well as more recent progress in [**80**].

2.6. Equivariant 2-dimensional topological open and closed theory

An important construction in string theory is the "orbifold" construction. Abstractly, this can be carried out whenever the closed string background has a group G of automorphisms. There are two steps in defining an orbifold theory. First, one must extend the theory by introducing "external" gauge fields, which are G-bundles (with connection) on the world sheets. Next, one must construct a new theory by summing over all possible G-bundles (and connections).

We begin by describing carefully the first step in forming the orbifold theory. The second step — summing over the G-bundles — is then very easy in the case of a finite group G.

2.6.1. Equivariant closed theories. Let us begin with some general remarks. In d-dimensional topological field theory one begins with a category $\mathcal{S}$ whose objects are oriented $(d-1)$-manifolds and whose morphisms are oriented cobordisms. Physicists say that a theory admits a group G as a *global symmetry group* if G acts on the vector space associated to each $(d-1)$-manifold, and the linear operator associated to each cobordism is a G-equivariant map. When we have such a "global" symmetry group G we can ask whether the symmetry can be "gauged", i.e., whether elements of G can be applied "independently" — in some sense — at each point of space-time. Mathematically the process of "gauging" has a very elegant description: it amounts to extending the field theory functor from the category $\mathcal{S}$ to the category $\mathcal{S}_G$ whose objects are $(d-1)$-manifolds equipped with a principal G-bundle, and whose morphisms are cobordisms with a G-bundle.[18] We regard $\mathcal{S}$ as a subcategory of $\mathcal{S}_G$ by equipping each $(d-1)$-manifold S with the trivial G-bundle $S \times G$. In $\mathcal{S}_G$ the group of automorphisms of the trivial bundle $S \times G$ contains G, and so in a gauged theory G acts on the state space $\mathcal{H}(S)$: this should be the original "global" action of G. But the gauged theory has a state space $\mathcal{H}(S, P)$ for each G-bundle P on S: if P is non-trivial one calls $\mathcal{H}(S, P)$ a "twisted sector" of the theory. In the case $d = 2$, when $S = S^1$ we have the bundle $P_g \to S^1$ obtained by attaching the ends of $[0, 2\pi] \times G$ via multiplication by g. (The fibre of P_g at the basepoint of S^1 is by definition the group G.) Any bundle is isomorphic to one of these, and P_g is isomorphic to $P_{g'}$ if and only if g' is conjugate to g. But note that the state space depends on the bundle and *not* just its isomorphism class, so we have a twisted sector state space $\mathcal{C}_g = \mathcal{H}(S, P_g)$ labelled by a group element g rather than by a conjugacy class.

We shall call a theory defined on the category $\mathcal{S}_G$ a *G-equivariant* TFT. It is important to distinguish the equivariant theory from the corresponding "gauged theory," described below. In physics, the equivariant theory is obtained by coupling to nondynamical background gauge fields, while the gauged theory is obtained by "summing" over those gauge fields in the path integral.

An alternative and equivalent viewpoint which is especially useful in the two-dimensional case is that $\mathcal{S}_G$ is the category whose objects are oriented $(d-1)$-manifolds S equipped with a map $p : S \to BG$, where BG is the classifying space of G. In this viewpoint we have a bundle over the space $\mathrm{Map}(S, BG)$ whose fibre at p is $\mathcal{H}_p$. To say that $\mathcal{H}_p$ depends only on the G-bundle p^*EG on S pulled back from the universal G-bundle EG on BG by p is the same as to say that the bundle on $\mathrm{Map}(S, BG)$ is equipped with a flat connection allowing us to identify the fibres at points in the same connected component by parallel transport; for the set of bundle isomorphisms

[18]We are assuming here that the group G is *discrete*: if G is a Lie group we should define $\mathcal{S}_G$ as the category of manifolds equipped with a principal G-bundle with a connection.

$p_0^*EG \to p_1^*EG$ is the same as the set of homotopy classes of paths from p_0 to p_1. When $S = S^1$ the connected components of the space of maps correspond to the conjugacy classes in G: each bundle P_g corresponds to a specific point p_g in the mapping space, and a group element h defines a specific path from p_g to $p_{hgh^{-1}}$.

The second viewpoint makes clear that G-equivariant topological field theories are examples of "homotopy topological field theories" in the sense of Turaev [**452**]. We shall use his two main results: first, an attractive generalization of the theorem that a two-dimensional TFT "is" a commutative Frobenius algebra, and, secondly, a classification of the ways of gauging a given global G-symmetry of a semisimple TFT. We shall now briefly review his work.

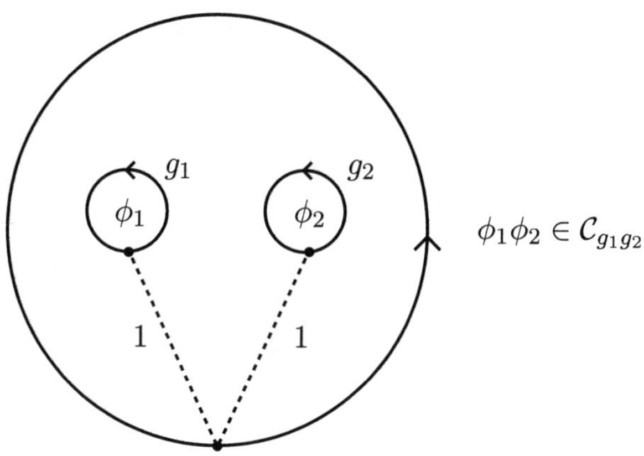

FIGURE 15. Definition of the product in the G-equivariant closed theory. The heavy dot is the basepoint on S^1. To specify the morphism unambiguously we must indicate consistent holonomies along a set of curves whose complement consists of simply connected pieces. These holonomies are always along paths between points where by definition the fibre is G. This means that the product is not commutative. We need to fix a convention for holonomies of a composition of curves, i.e., whether we are using left or right path-ordering. We will take $h(\gamma_1 \circ \gamma_2) = h(\gamma_1) \cdot h(\gamma_2)$.

A G-equivariant TFT gives us for each element $g \in G$ a vector space $\mathcal{C}_g$, associated to the circle equipped with the bundle P_g whose holonomy is g. The usual pair-of-pants cobordism, equipped with the evident G-bundle which restricts to P_{g_1} and P_{g_2} on the two incoming circles, and to $P_{g_1g_2}$ on the outgoing circle, induces a product

$$\mathcal{C}_{g_1} \otimes \mathcal{C}_{g_2} \to \mathcal{C}_{g_1g_2} \tag{2.80}$$

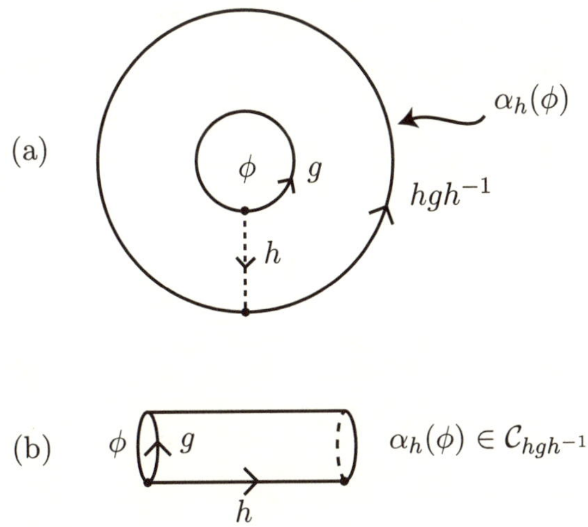

FIGURE 16. (a) The action of α_h on a state $\phi \in \mathcal{C}_g$. This can also be represented by the cylinder as in (b).

making $\mathcal{C} := \oplus_{g\in G}\mathcal{C}_g$ into a G-graded algebra, as shown in Figure 15.

As in the usual case there is a trace $\theta : \mathcal{C}_1 \to \mathbb{C}$ defined by the disk diagram with one ingoing circle. Note that the holonomy around the boundary of the disk must be 1. Making the standard assumption that the cylinder corresponds to the unit operator we obtain a non-degenerate pairing $\mathcal{C}_g \otimes \mathcal{C}_{g^{-1}} \to \mathbb{C}$.

A new element in the equivariant theory is that G acts as an automorphism group on $\mathcal{C}$. That is, there is a homomorphism $\alpha : G \to \mathrm{Aut}(\mathcal{C})$ such that

$$\alpha_h : \mathcal{C}_g \to \mathcal{C}_{hgh^{-1}}. \tag{2.81}$$

Diagramatically, α_h is defined by the surface in Figure 16.

Now let us note some properties of α. First, if $\phi \in \mathcal{C}_h$ then $\alpha_h(\phi) = \phi$. The reason for this is explained in Figure 17.

Next, while $\mathcal{C}$ is not commutative, it is "twisted-commutative" in the following sense. If $\phi_1 \in \mathcal{C}_{g_1}$ and $\phi_2 \in \mathcal{C}_{g_2}$ then

$$\alpha_{g_2}(\phi_1)\phi_2 = \phi_2\phi_1. \tag{2.82}$$

The necessity of this condition is illustrated in Figure 18.

The last property we need is a little more complicated. The trace of the identity map of $\mathcal{C}_g$ is the partition function of the theory on a torus with the bundle with holonomy $(g, 1)$. Cutting the torus the other way, we see that this is the trace of α_g on $\mathcal{C}_1$. Similarly, by considering the torus with a bundle with holonomy (g, h), where g and h are two commuting elements of G, we see that the trace of α_g on $\mathcal{C}_h$ is the trace of α_h on $\mathcal{C}_{g^{-1}}$. But we need a strengthening of this property. Even when g and h do not commute we

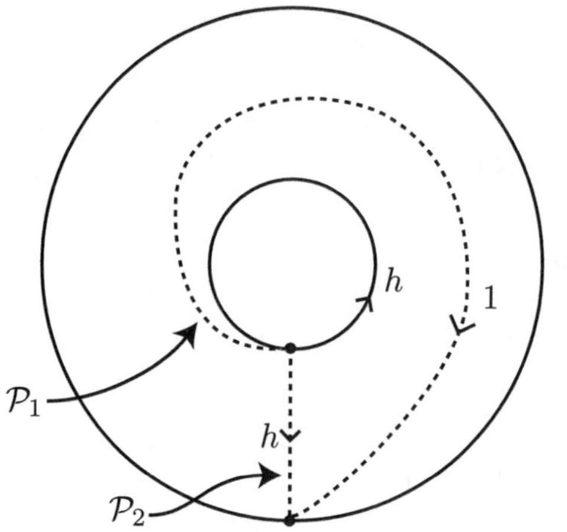

FIGURE 17. If the holonomy along path $\mathcal{P}_2$ is h then the holonomy along path $\mathcal{P}_1$ is 1. However, a Dehn twist around the inner circle maps $\mathcal{P}_1$ into $\mathcal{P}_2$. Therefore, $\alpha_h(\phi) = \alpha_1(\phi) = \phi$, if $\phi \in \mathcal{C}_h$.

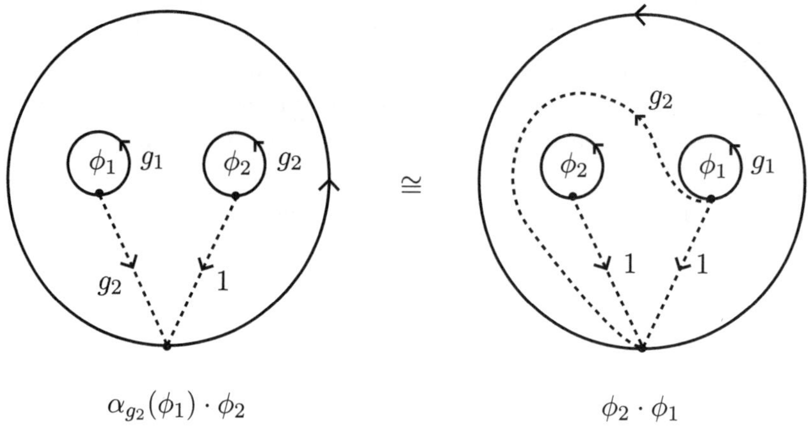

FIGURE 18. Demonstrating twisted centrality.

can form a bundle with holonomy (g, h) on a torus with one hole, around which the holonomy will be $c = hgh^{-1}g^{-1}$. We can cut this torus along either of its generating circles to get a cobordism operator from $\mathcal{C}_c \otimes \mathcal{C}_h$ to $\mathcal{C}_h$ or from $\mathcal{C}_{g^{-1}} \otimes \mathcal{C}_c$ to $\mathcal{C}_{g^{-1}}$. If $\psi \in \mathcal{C}_{hgh^{-1}g^{-1}}$ let us introduce two linear transformations L_ψ, R_ψ associated to left- and right-multiplication by ψ. On the one hand, $L_\psi \alpha_g : \phi \mapsto \psi \alpha_g(\phi)$ is a map $\mathcal{C}_h \to \mathcal{C}_h$. On the other hand $R_\psi \alpha_h : \phi \mapsto \alpha_h(\phi)\psi$ is a map $\mathcal{C}_{g^{-1}} \to \mathcal{C}_{g^{-1}}$. The last sewing condition states that these two endomorphisms must have equal traces:

$$\mathrm{Tr}_{\mathcal{C}_h}\Big(L_\psi \alpha_g\Big) = \mathrm{Tr}_{\mathcal{C}_{g^{-1}}}\Big(R_\psi \alpha_h\Big). \tag{2.83}$$

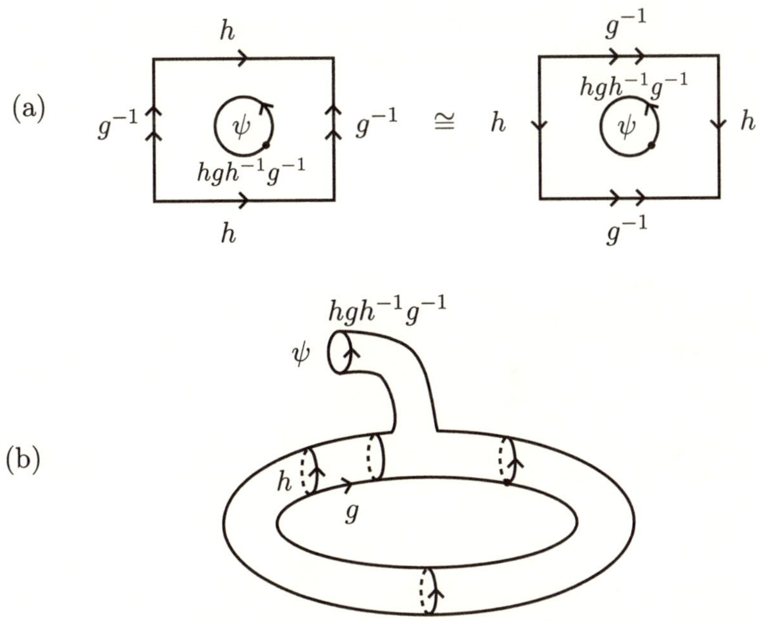

FIGURE 19. Deforming the LHS of (a) into a space-time evolution diagram yields (b), whose value is $\mathrm{Tr}_{\mathcal{C}_h}(L_\psi \alpha_g)$. Similarly deforming the RHS of (a) gives a diagram whose value is $\mathrm{Tr}_{\mathcal{C}_{g^{-1}}}(R_\psi \alpha_h)$.

The reason for this can be deduced by pondering the diagram in Figure 19.

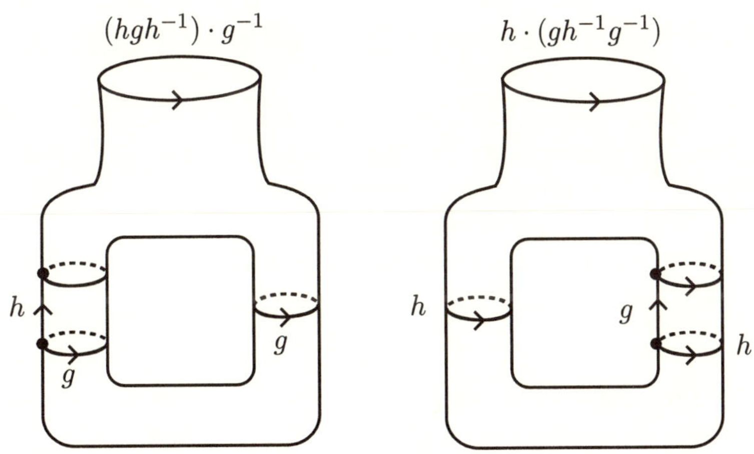

FIGURE 20. A simpler axiom than Turaev's torus axiom.

The equation (2.83) was taken by Turaev as one of his axioms. It can, however, be reexpressed in a way that we shall find more convenient. Let $\Delta_g \in \mathcal{C}_g \otimes \mathcal{C}_{g^{-1}}$ be the "duality" element corresponding to the identity cobordism of (S^1, P_g) with both ends regarded as outgoing. We have $\Delta_g = \sum \xi_i \otimes \xi^i$, where ξ_i and ξ^i run through dual bases of $\mathcal{C}_g$ and $\mathcal{C}_{g^{-1}}$. Let us also write

$$\Delta_h = \sum \eta_i \otimes \eta^i \in \mathcal{C}_h \otimes \mathcal{C}_{h^{-1}}.$$

Then (2.83) is easily seen to be equivalent to

$$\sum \alpha_h(\xi_i)\xi^i = \sum \eta_i \alpha_g(\eta^i), \tag{2.84}$$

in which both sides are elements of $\mathcal{C}_{hgh^{-1}g^{-1}}$. This equation is illustrated by the isomorphic cobordisms of Figure 20.

In summary, the sewing theorem for G-equivariant 2d topological field theories is given by the following theorem:

THEOREM 2.11. ([**452**]) *To give a 2d G-equivariant topological field theory is to give a G-graded algebra $\mathcal{C} = \oplus_g \mathcal{C}_g$ together with a group homomorphism $\alpha : G \to \mathrm{Aut}(\mathcal{C})$ such that*

(1) There is a G-invariant trace $\theta : \mathcal{C}_1 \to \mathbb{C}$ which induces a non-degenerate pairing $\mathcal{C}_g \otimes \mathcal{C}_{g^{-1}} \to \mathbb{C}$.
(2) The restriction of α_h to $\mathcal{C}_h$ is the identity.
(3) For all $\phi \in \mathcal{C}_g, \phi' \in \mathcal{C}_h$, $\alpha_h(\phi)\phi' = \phi'\phi$.
(4) For all $g, h \in G$ we have

$$\sum \alpha_h(\xi_i)\xi^i = \sum \eta_i \alpha_g(\eta^i) \in \mathcal{C}_{hgh^{-1}g^{-1}}, \tag{2.85}$$

where $\Delta_g = \sum \xi_i \otimes \xi^i \in \mathcal{C}_g \otimes \mathcal{C}_{g^{-1}}$ and $\Delta_h = \sum \eta_i \otimes \eta^i \in \mathcal{C}_h \otimes \mathcal{C}_{h^{-1}}$ as above.

REMARK 2.12. (1) We will give a proof of the sewing theorem in §2.7.3.
(2) Warning: Turaev calls the above a *crossed G Frobenius algebra*, but it is not a crossed-product algebra in the sense of C^* algebras (see below). We will refer to an algebra satisfying the conditions of the theorem as a *Turaev algebra*.
(3) Axioms 1 and 3 have counterparts in the non-equivariant theory, but axioms 2 and 4 are new elements.

2.6.2. The orbifold theory. Before going any further, let us describe how we obtain the orbifold theory from the Turaev algebra.

Let us return to the general discussion at the beginning of §2.6.1, where we outlined the definition of an equivariant theory. Roughly speaking, the gauged theory is obtained from the equivariant theory by summing over the gauge fields. More precisely, the state space which a gauged theory associates to a $(d-1)$-manifold S consists of "wave-functions" ψ which associate to each G-bundle P on S an element ψ_P of the state space $\mathcal{H}_{S,P}$ of the equivariant theory. The map ψ must be "natural" in the sense that when $\theta : P \to P'$ is a bundle isomorphism the induced isomorphism $\mathcal{H}_{S,P} \to \mathcal{H}_{S,P'}$ takes ψ_P to $\psi_{P'}$. This is often referred to as the "Gauss law." In the two-dimensional case, the Gauss law amounts to saying that the state space $\mathcal{C}_{\text{orb}}$ for the circle is the G-invariant part of the Turaev algebra $\mathcal{C} = \oplus \mathcal{C}_g$. In other words,

$$\mathcal{C}_{\text{orb}} = \oplus \{\mathcal{C}_g\}^{Z_g}, \tag{2.86}$$

where now g runs through a set of representatives for the conjugacy classes in G, and we take the invariant part of $\mathcal{C}_g$ under the centralizer Z_g of g in G. The algebra $\mathcal{C}_{\text{orb}}$ is not a graded algebra if G is non-abelian. One must check that the product in $\mathcal{C}_{\text{orb}}$ is simply the restriction of the product in $\mathcal{C}$. The trace $\mathcal{C}_{\text{orb}} \to \mathbb{C}$ is the restriction of the trace $\mathcal{C} \to \mathbb{C}$ which is the given trace on $\mathcal{C}_1$ and is zero on $\mathcal{C}_g$ when $g \neq 1$. Then $\mathcal{C}_{\text{orb}}$ is a commutative Frobenius algebra which encodes the orbifold theory.

2.6.3. Solutions of the closed string G-equivariant sewing conditions. Having found the sewing conditions in the G-equivariant case we can try to classify examples of the structure. The Frobenius algebra $\mathcal{C}_1$ with its G-action corresponds to a topological field theory with a global G-symmetry. In the case when $\mathcal{C}_1$ is a semisimple Frobenius algebra — and therefore the algebra of functions on a finite G-set X — Turaev finds a nice answer: ways of gauging the symmetry, i.e., of extending $\mathcal{C}_1$ to a Turaev algebra, correspond to *equivariant B-fields* on X, i.e., to equivariant 2-cocycles of X with values in $\mathbb{C}^\times$. Furthermore, two such B-fields define isomorphic Turaev algebras if and only if they represent the same class in $H^2_G(X, \mathbb{C}^\times) \cong H^3_G(X, \mathbb{Z})$. We now review this result and take the opportunity to introduce a more geometric picture of Turaev's algebra $\mathcal{C}$ (in the semisimple case). We shall first recall some very general constructions.

2.6.3.1. *General constructions.* Whenever a group G acts on a set X we can form a category $X//G$, whose objects are the points x of X, and whose morphisms $x_0 \to x_1$ are

$$\text{Hom}(x_0, x_1) := \{g \in G : gx_0 = x_1\}. \tag{2.87}$$

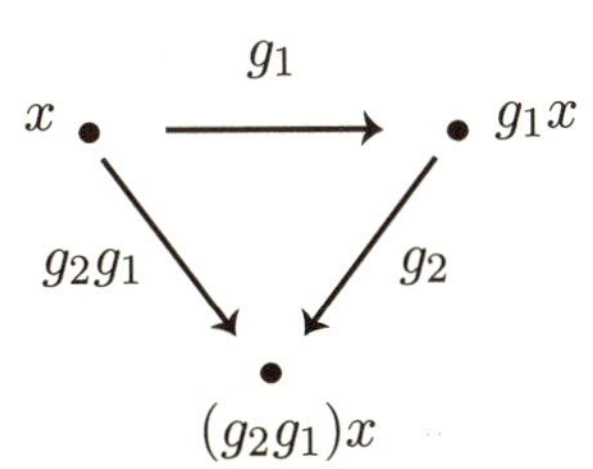

FIGURE 21. An oriented two-simplex Δ_{x,g_1,g_2} in the space $|X//G|$.

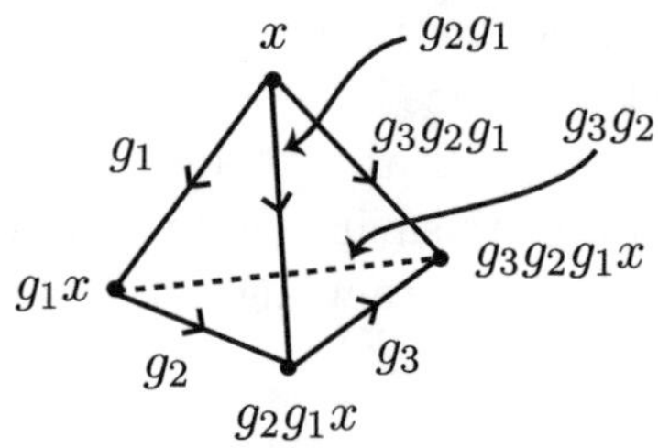

FIGURE 22. An oriented 3-simplex in $|X//G|$.

Next, for any category C, one can form the *space of the category*, denoted $|C|$. This is an oriented simplicial complex whose p-simplices are in 1-1 correspondence with the composable p-tuples of morphisms in the category. To be specific, the vertices are the objects of the category. The edges are the morphisms. Triples of morphisms (f_1, f_2, f_3) with $f_3 = f_2 \circ f_1$ correspond to 2-simplices, and so forth. In the present case, when we form the simplicial complex $|X//G|$ the 2-simplices are the triples (g_1, g_2, x) illustrated in Figure 21. Three-simplices are shown in Figure 22, etc.

The space $|X//G|$ is a model for $(X \times EG)/G$. Hence the (cellular) cohomology of this space $H^*(|X//G|, \mathbb{C}^\times)$ is the equivariant cohomology $H^*_G(X, \mathbb{C}^\times)$.

Another object which we can associate to any category C is its *algebra* $A(C)$ over the field $\mathbb{C}$. This has a vector space basis $\{\varepsilon_f\}$ indexed by the morphisms of C, and the product is given by $\varepsilon_{f_1}\varepsilon_{f_2} = \varepsilon_{f_1 \circ f_2}$ when f_1 and f_2 are composable, and $\varepsilon_{f_1}\varepsilon_{f_2} = 0$ otherwise. For the category $X//G$ the algebra $A(X//G)$ is the usual *crossed-product* algebra $A(X) \ltimes G$ in the sense of operator algebra theory, where $A(X)$ is the algebra of complex-valued functions on the set X.[19]

The construction of the category-algebra $A(C)$ can be generalized. A *B-field* on a category C is a rule which associates a complex line L_f to each morphism f of C, and associative isomorphisms

$$L_{f_1} \otimes L_{f_2} \to L_{f_1 \circ f_2}$$

to each pair (f_1, f_2) of composable morphisms . In concrete terms, to give such a product is to give a 2-cocycle on the space $|C|$. Indeed, choosing basis elements $\ell_f \in L_f$, we must have

$$\ell_{f_1} \cdot \ell_{f_2} = b(f_1, f_2, f_3)\ell_{f_3} \tag{2.89}$$

[19]For any commutative algebra A with G-action, $A \ltimes G$ is spanned by elements $a \otimes g$ with $a \in A$ and $g \in G$, and the product is given by

$$(a_1 \otimes g_1)(a_2 \otimes g_2) = a_1 g_1(a_2) \otimes g_1 g_2. \tag{2.88}$$

The isomorphism $A(X//G) \to A(X) \ltimes G$ takes $\varepsilon_{g,x}$ to $\chi_{gx} \otimes g$, where χ_x is the characteristic function supported at x.

where $b(f_1, f_2, f_3) \in \mathbb{C}^\times$ defines a 2-cochain on $|C|$. (We choose values in $\mathbb{C}^\times$ rather than $\mathbb{C}$ so the product is non-degenerate.) Associativity of (2.89) holds iff b is a 2-cocycle. A change of basis of the L_f modifies b by a coboundary. Hence the isomorphism classes of B-fields on C are in 1-1 correspondence with cohomology classes $[b] \in H^2(|C|, \mathbb{C}^\times)$. When we have a B-field b on C we can form a *twisted* category-algebra $A_b(C)$, which as a vector space is $\oplus L_f$, and where the multiplication is defined by means of the associative maps $L_{f_1} \otimes L_{f_2} \to L_{f_1 \circ f_2}$.

Applying the above construction to the category $X//G$, an associative product on the lines $L_{g,x}$ is the same thing as a 2-cocycle in $H^2_G(X, \mathbb{C}^\times)$. In terms of the basis elements $\ell_{g,x}$ for the lines $L_{g,x}$ we shall write the multiplication

$$\ell_{g_2,x_2}\ell_{g_1,x_1} = \begin{cases} b_{x_1}(g_2, g_1)\ell_{g_2g_1,x_1} & \text{if } x_2 = g_1x_1 \\ 0 & \text{otherwise} \end{cases} \tag{2.90}$$

Here $b_{x_1}(g_2, g_1) = b(\Delta_{x,g_1,g_2})$ is the value of the cocycle on the oriented 2-simplex of Figure 21.

Notice that if G_x is the isotropy group of some point $x \in X$ then restricting (2.90) to the elements $\ell_{g,x}$ with $g \in G_x$ shows that b_x defines an element of the group cohomology $H^2(G_x, \mathbb{C}^\times)$, corresponding to the central extension of G_x by $\mathbb{C}^\times$ whose elements are pairs (g, λ) with $g \in G_x$ and $\lambda \in L_x - \{0\}$. This central extension of the isotropy group G_x does not in general extend to any central extension of the whole group G. It does so, however, in the particular case when the B-field b is pulled back from a 2-cocycle of G by the map $X \to$ (point), i.e. when $b_x(g_2, g_1)$ is independent of x. In general the cocycle $b : G \times G \times X \to \mathbb{C}^\times$ can be regarded as a cocycle of the group G with values in the abelian group $A(X)^\times = \mathrm{Map}(X, \mathbb{C}^\times)$ with its natural G-action. Thus it defines a (non-central) extension

$$1 \to A(X)^\times \to \tilde{G} \to G \to 1.$$

One technical point to notice is that for any B-field we have $L_f = \mathbb{C}$ canonically when f is an identity morphism. Thus $L_{g,x} = \mathbb{C}$ when $g = 1$. We shall always choose $\ell_{g,x} = 1$ when $g = 1$, thereby normalizing the cocycle so that $b_x(g_1, g_2) = 1$ if either g_1 or g_2 is 1.

The algebra $A_b(X//G) = \oplus_{g\in G, x\in X} L_{g,x}$ with the multiplication rule defined by (2.90) can be identified with the twisted crossed-product algebra $A(X) \rtimes_b G$ via

$$\ell_{g,x} \mapsto \chi_{gx} \otimes g,$$

where χ_x is the characteristic function supported at x. The twisted crossed-product is defined by

$$(f_1 \otimes g_1)(f_2 \otimes g_2) = \alpha_{g_1g_2}(b(g_1, g_2))f_1\, \alpha_{g_1}(f_2) \otimes g_1g_2, \tag{2.91}$$

where $b(g_1, g_2)$ denotes the function $x \mapsto b_x(g_1, g_2)$ in $A(X)^\times$, and the group G acts on $A(X)$ in the natural way

$$\alpha_g(f)(x) = f(g^{-1}x),$$

so that $g \cdot \chi_x = \chi_{gx}$.

If we wish to apply these considerations to the spin case described in §2.1.6 and §2.2.4 then we must consider the lines L_f to be $\mathbb{Z}/2$ graded. In this case the theory will admit a further twisting by $H^1(|C|, \mathbb{Z}/2)$. However, we will not discuss this generalization further.

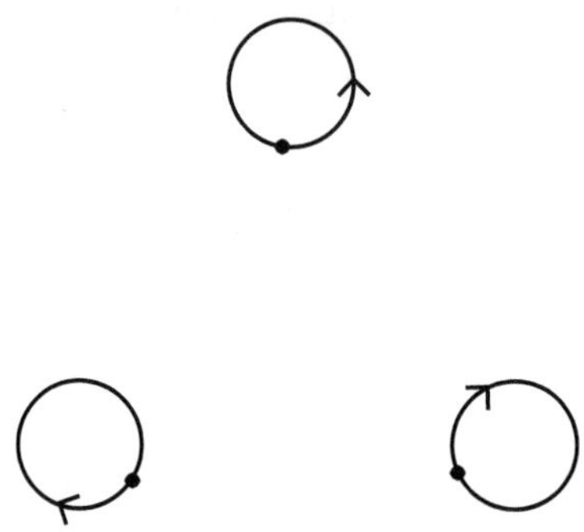

FIGURE 23. The algebra of little loops for $X = S_3/S_2$, where S_n is the permutation group on n letters.

2.6.3.2. *The Turaev algebra associated to a G-space.* The algebra

$$A_b(X//G)$$

does *not* satisfy the sewing conditions and is not a Turaev algebra. In particular (2.82) is usually not satisfied for a crossed-product algebra. However, the subcategory defined by the morphisms with the same initial and terminal object does lead to a Turaev algebra for any B-field b on $X//G$. We call this the "algebra of little loops". Thus we define $\mathcal{C} = \oplus_g \mathcal{C}_g \subset A_b(X//G)$ by

$$\mathcal{C}_g := \oplus_{x:gx=x} L_{g,x} \tag{2.92}$$

and define the trace by

$$\theta(\ell_{g,x}) = \delta_{g,1}\theta(\varepsilon_x) \tag{2.93}$$

where on the right θ is the given G-invariant trace on $\mathcal{C}_1$, and the ε_x are the usual idempotents in the semisimple Frobenius algebra $\mathcal{C}_1 = A(X)$, i.e., $\varepsilon_x = 1 \in L_{1,x} = \mathbb{C}$. The algebra of little loops can be visualized as in Figure 23.

An equivalent way to describe $\mathcal{C}$ is as the commutant of $\mathcal{C}_1 = A(X)$ in

$$A_b(X//G) = A(X) \rtimes_b G.$$

As $A(X)$ is in the centre of $\mathcal{C}$, it is natural to think of $\mathcal{C}$ as the sections of a bundle of algebras on X; the fibre of this bundle at $x \in X$ is the twisted group algebra $\mathbb{C}_{b_x}[G_x]$, where G_x is the isotropy group of x. Furthermore, the bundle of algebras has a natural G-action, covering the G-action on X. To see this, notice that the extension $\tilde{G} = \{(f,g) : f \in A(X)^\times, g \in G\}$ of the group G by the multiplicative group $A(X)^\times$ defined by the B-field sits inside the multiplicative group of $A(X) \ltimes_b G$, normalizing the subalgebra $A(X)$. As $A(X)$ is in the center of $\mathcal{C}$, this means that G acts by conjugation on the algebra $\mathcal{C}$. Notice, however, that only $\tilde{G}$, and not G, acts on the larger algebra $A_b(X//G)$.

In terms of explicit formulae, the action of G on the algebra $\mathcal{C}$ is given by

$$\alpha_{g_1}(\ell_{g_2,x}) = \ell_{g_1,x}\ell_{g_2,x}\ell_{g_1,x}^{-1} = z_x(g_2,g_1)\ell_{g_1g_2g_1^{-1},g_1x}, \tag{2.94}$$

where

$$z_x(g_2,g_1) = \frac{b_x(g_1,g_2)b_x(g_1g_2,g_1^{-1})}{b_x(g_1,g_1^{-1})}. \tag{2.95}$$

In this way we obtain a Turaev algebra, which we shall denote by $\mathcal{C} = T(X,b,\theta)$. The only non-trivial point is to verify the "torus" axiom (2.83). But in fact it is easy to see that both sides of the equation are equal to

$$\sum_x \ell_{h,x}\ell_{g,x}\ell_{h,x}^{-1}\ell_{g,x}^{-1},$$

where x runs through the set $\{x \in X : hx = gx = x\}$.

Turaev has shown that the above construction is the most general one possible in the semisimple case.

THEOREM 2.13. ([**452**], *Theorem 3.6)* *Let $\mathcal{C}$ be a Turaev algebra. If $\mathcal{C}_1$ is semisimple then $\mathcal{C}$ is the twisted algebra $T(X,b,\theta)$ of little loops on $X = \operatorname{Spec}(\mathcal{C}_1)$ for some cocycle $b \in Z^2_G(X,\mathbb{C}^\times)$.*

PROOF. If $\mathcal{C}_1$ is semisimple we may decompose it in terms of the basic idempotents ε_x. Then $\mathcal{C}_g$ is a module over $\mathcal{C}_1$, and hence it should be identified with the cross sections of the vector bundle over the finite set X whose fibre at x is $\mathcal{C}_{g,x} = \varepsilon_x\mathcal{C}_g$. (This is a trivial case of what is called the Serre-Swan theorem.) Now we consider the torus axiom (2.83) in the case $h = 1$. We have $\Delta_1 = \sum \theta(\varepsilon_x)^{-1}\varepsilon_x \otimes \varepsilon_x$, and hence

$$\sum \theta(\varepsilon_x)^{-1}\alpha_g(\varepsilon_x)\varepsilon_x = \sum \theta(\varepsilon_x)^{-1}\varepsilon_x,$$

where the second sum is over x such that $gx = x$. On the other hand we readily calculate that if $\{a_{x,i}\}$ is a basis of $\mathcal{C}_{g,x}$ and $\{a^*_{x,i}\}$ is the dual basis of $\mathcal{C}_{g^{-1},x}$ then $a_{g,i}a^*_{g,i} = \theta(\varepsilon_x)^{-1}\varepsilon_x$, so that the other side of the torus axiom is

$$\sum \theta(\epsilon_x)^{-1}\dim(\mathcal{C}_{g,x})\varepsilon_x.$$

Thus the axiom tells us that $\mathcal{C}_{g,x}$ is a one-dimensional space $L_{g,x}$ when $gx = x$, and is zero otherwise. The multiplication in $\mathcal{C}$ makes these lines into a G-equivariant B-field on the category of small loops in $X//G$. Finally, it is not hard to show that the category of B-fields on $X//G$ is equivalent to the category of G-equivariant B-fields on the category of small loops; but we shall omit the details. □

Let us now consider the orbifold theory coming from the gauged theory defined by the Turaev algebra $\mathcal{C} = T(X, b, \theta)$. We saw in §2.6.2 that it is defined by the commutative Frobenius algebra $\mathcal{C}_{\text{orb}}$ which is the G-invariant subalgebra of $\mathcal{C}$. In the case of the Turaev algebra of a G-space X we have

THEOREM 2.14. *The orbifold algebra $\mathcal{C}_{\text{orb}}$ is the center of the crossed-product algebra $A(X) \ltimes_b G$. It is the algebra of functions on a finite set $(X/G)_{\text{string}}$ which is a "thickening" of the orbit space X/G with one point for each pair ξ, ρ consisting of an orbit ξ and an irreducible projective representation ρ of the isotropy group G_x of a point $x \in \xi$, with the projective cocycle b_x defined by the B-field.*

PROOF. The Turaev algebra $\mathcal{C}$ consists of the elements of $A(X) \ltimes_b G$ which commute with $A(X)$. But an element of $A(X) \ltimes_b G$ belongs to its centre if and only if it commutes with $A(X)$ and also commutes with the elements of G, i.e., is G-invariant.

Now we saw that $\mathcal{C}$ is the product over the points $x \in X$ of the twisted group-algebras $\mathbb{C}_{b_x}[G_x]$. The invariant part is therefore the product over the orbits ξ of the G_x-invariant part of $\mathbb{C}_{b_x}[G_x]$, i.e. of the centre of $\mathbb{C}_{b_x}[G_x]$, which consists of one copy of $\mathbb{C}$ for each irreducible representation ρ with the cocycle b_x. □

The Turaev algebra $\mathcal{C} = T(X, b, \theta)$ sits between $\mathcal{C}_{\text{orb}}$ and $A(X) \ltimes_b G$. We shall see in Theorem 2.16 that $A(X) \ltimes_b G$ is semisimple, and hence Morita equivalent[20] to its centre $\mathcal{C}_{\text{orb}}$. But the Turaev algebra retains more information than the orbifold theory: it encodes X and its G-action. The difference is plainest when G — of order n — acts freely on X; then $A(X) \ltimes G$ is the product of a copy of the algebra of $n \times n$ matrices for each G-orbit in X, and provides us with no way of distinguishing the individual points of X. We shall see in §2.6.5 that the category of boundary conditions for the gauged theory $\mathcal{C}$ is a natural enrichment of the category for the orbifold theory, at least in the semisimple case.

[20]This means that the category of representations of $A(X) \ltimes_b G$ is equivalent to the category of representations of $\mathcal{C}_{\text{orb}}$, uniquely up to tensoring with a "line bundle" — a representation L of $\mathcal{C}_{\text{orb}}$ such that $L \otimes_{\mathcal{C}_{\text{orb}}} L' \cong \mathcal{C}_{\text{orb}}$ for some L'.

It might come as a surprise that the crossed-product algebra of space-time $A(X) \rtimes G$ is not the appropriate Frobenius algebra for G-equivariant topological field theory, in view of the occurrence of the crossed-product algebra as a central concept in the theory of D-branes on orbifolds developed in [**141, 308, 352**]. In fact, this fits in very nicely with the philosophy of this chapter. The Turaev algebra remembers the points of X, and so allows only the "little loops" above. In this way the sewing conditions - which are meant to formalize worldsheet locality - also encode a crude form of *space-time* locality.

We shall conclude this section by making contact with the usual path integral expression for the orbifold partition function on a torus. To do this we compute $\dim \mathcal{C}_{\mathrm{orb}}$ by computing the projection onto G-invariant states in $\mathcal{C}$. Note that $\alpha_g(\ell_{h,x})$ is only proportional to $\ell_{h,x}$ when $[g,h] = 1$ and $gx = x$, and then

$$\alpha_g(\ell_{h,x}) = \frac{b_x(g,h)}{b_x(h,g)}\ell_{h,x} \tag{2.96}$$

where we have combined (2.95) with the cocycle identity. Thus we find

$$\dim \mathcal{C}_{\mathrm{orb}} = \frac{1}{|G|}\sum_{gh=hg}\sum_{x=gx=hx}\frac{b_x(g,h)}{b_x(h,g)}. \tag{2.97}$$

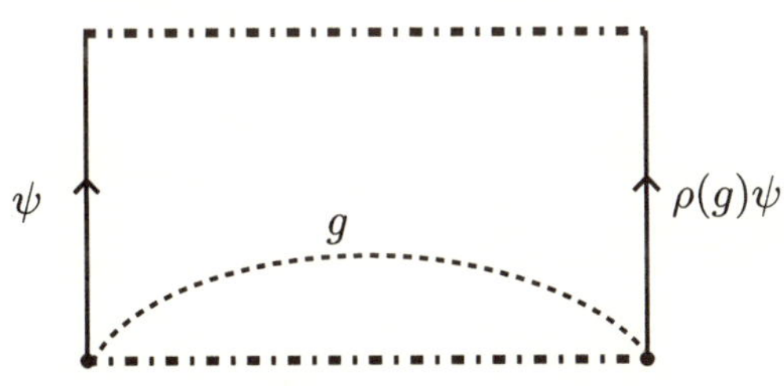

FIGURE 24. The wavy line is a constrained boundary. If there is holonomy g along the dotted path $\mathcal{P}$ then this morphism gives the G-action on $\mathcal{O}$.

2.6.4. Sewing conditions for equivariant open and closed theory. Let us now pass on to consider G-equivariant open and closed theories. We enlarge the category $\mathcal{S}_G$ so that the objects are oriented 1-manifolds with boundary, with labelled ends, equipped with principal G-bundles. The morphisms are the same cobordisms as in the non-equivariant case, but equipped with G-bundles. Up to isomorphism there is only one G-bundle on the interval: it is trivial, and admits G as an automorphism group. So an equivariant theory gives us for each pair a, b of labels a vector space $\mathcal{O}_{ab}$ with a G-action. The action of $g \in G$ on $\mathcal{O}_{ab}$ can be regarded as coming from

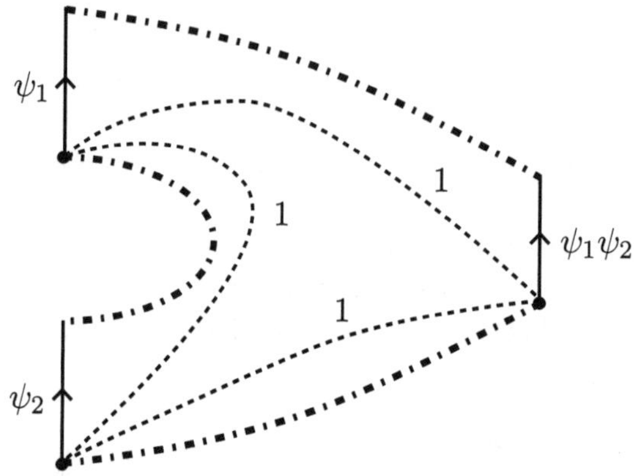

FIGURE 25. The definition of the multiplication in $\mathcal{O}$. The holonomy on all dotted paths is 1. Note the order of multiplication.

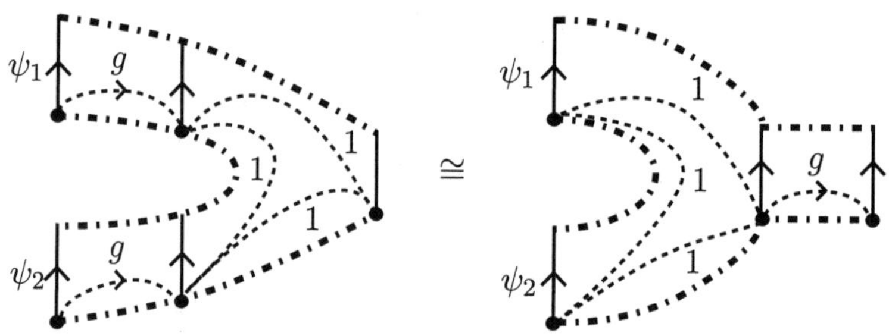

$$(\rho(g)\psi_1)\cdot(\rho(g)\psi_2) = \rho(g)(\psi_1\psi_2)$$

FIGURE 26. Showing that G acts on $\mathcal{O}$ as a group of automorphisms.

the "square" cobordism with the bundle whose holonomy is g along each of its "constrained" edges. There is also a composition law $\mathcal{O}_{ab} \times \mathcal{O}_{bc} \to \mathcal{O}_{ac}$, which is G-equivariant. These are illustrated in Figures 24 and 25.

In the open/closed case the conditions analogous to equations (2.2) to (2.12) are the following.

Focusing first on a single label a, we have a not necessarily commutative Frobenius algebra $(\mathcal{O}_{aa} = \mathcal{O}, \theta_{\mathcal{O}})$ together with a G-action $\rho : G \to \mathrm{Aut}(\mathcal{O})$:

$$\rho_g(\psi_1\psi_2) = \big(\rho_g\psi_1\big)\big(\rho_g\psi_2\big) \tag{2.98}$$

which preserves the trace $\theta_{\mathcal{O}}(\rho_g\psi) = \theta_{\mathcal{O}}(\psi)$. See Figure 26.

There are also G-twisted open/closed transition maps

$$\begin{aligned} \iota_{g,a} = \iota_g &: \mathcal{C}_g \to \mathcal{O}_{aa} = \mathcal{O} \\ \iota^{g,a} = \iota^g &: \mathcal{O}_{aa} = \mathcal{O} \to \mathcal{C}_g \end{aligned} \tag{2.99}$$

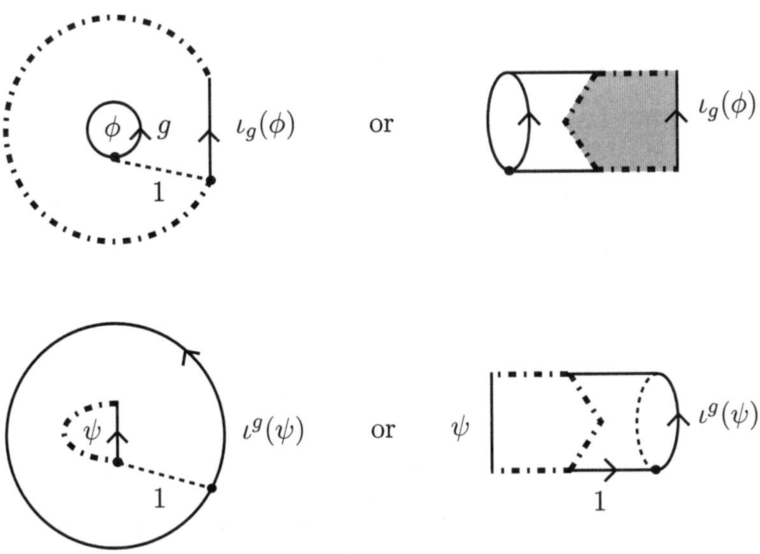

FIGURE 27. The open/closed transitions ι_g and ι^g.

which are equivariant:

(2.100)
$$\begin{array}{ccc} \mathcal{C}_{g_1} & \xrightarrow{\alpha_{g_2}} & \mathcal{C}_{g_2 g_1 g_2{}^{-1}} \\ \downarrow{\scriptstyle \iota_{g_1}} & & \downarrow{\scriptstyle \iota_{g_2 g_1 g_2{}^{-1}}} \\ \mathcal{O} & \xrightarrow[\rho_{g_2}]{} & \mathcal{O} \end{array}$$

(2.101)
$$\begin{array}{ccc} \mathcal{C}_{g_2{}^{-1} g_1 g_2} & \xrightarrow{\alpha_{g_2}} & \mathcal{C}_{g_1} \\ \uparrow{\scriptstyle \iota^{g_2{}^{-1} g_1 g_2}} & & \uparrow{\scriptstyle \iota^{g_1}} \\ \mathcal{O} & \xrightarrow[\rho_{g_2}]{} & \mathcal{O} \end{array}$$

These maps are illustrated in Figure 27. The open/closed maps must satisfy the G-twisted versions of conditions 3a-3e of §2.1.2. In particular, the map $\iota : \mathcal{C} \to \mathcal{O}$ obtained by putting the ι_g together is a ring homomorphism, i.e.,

(2.102) $$\iota_{g_1}(\phi_1)\iota_{g_2}(\phi_2) = \iota_{g_2 g_1}(\phi_2\phi_1) \qquad \forall \phi_1 \in \mathcal{C}_{g_1}, \phi_2 \in \mathcal{C}_{g_2}.$$

Since the identity is in $\mathcal{C}_1$ the condition (2.8) is unchanged. The G-twisted centrality condition is

(2.103) $$\iota_g(\phi)(\rho_g\psi) = \psi\iota_g(\phi) \qquad \forall \phi \in \mathcal{C}_g, \psi \in \mathcal{O},$$

and is illustrated in Figure 28.

The G-twisted adjoint condition is

(2.104) $$\theta_{\mathcal{O}}\big(\psi\iota_{g^{-1}}(\phi)\big) = \theta_{\mathcal{C}}\big(\iota^g(\psi)\phi\big) \qquad \forall \phi \in \mathcal{C}_{g^{-1}},$$

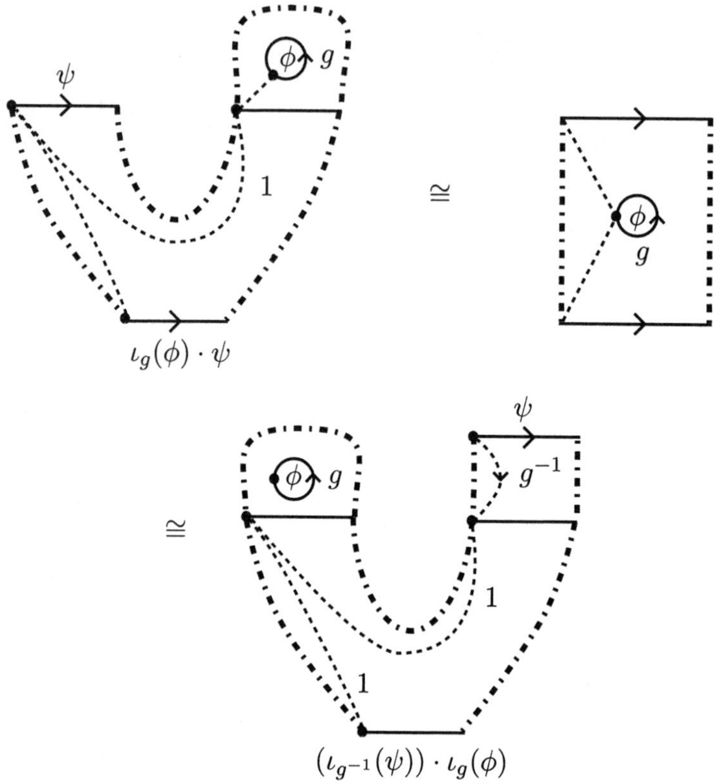

FIGURE 28. The G-twisted centrality axiom.

and is shown in Figure 29.

The G-twisted Cardy conditions place restrictions not only on the algebras $\mathcal{O}_{aa}$, but also on the spaces of morphisms $\mathcal{O}_{ab}$ for all a, b. For each $g \in G$ we must have

$$\pi^{\ a}_{g,b} = \iota_{g,b}\iota^{g,a}. \tag{2.105}$$

Here $\pi^{\ a}_{g,b}$ is defined by

$$\pi^{\ a}_{g,b}(\psi) = \sum_{\mu} \psi^{\mu}\psi(\rho_g\psi_{\mu}), \tag{2.106}$$

where we sum over a basis ψ_μ for $\mathcal{O}_{ab}$, and take ψ^μ to be the dual basis of $\mathcal{O}_{ba}$. See Figure 30.

We may now formulate

THEOREM 2.15. *The above conditions form a complete set of sewing conditions for G-equivariant open/closed 2d TFT.*

This will be proved in §2.7. Note that the above axioms are slightly redundant since (2.100) and (2.104) together imply (2.101).

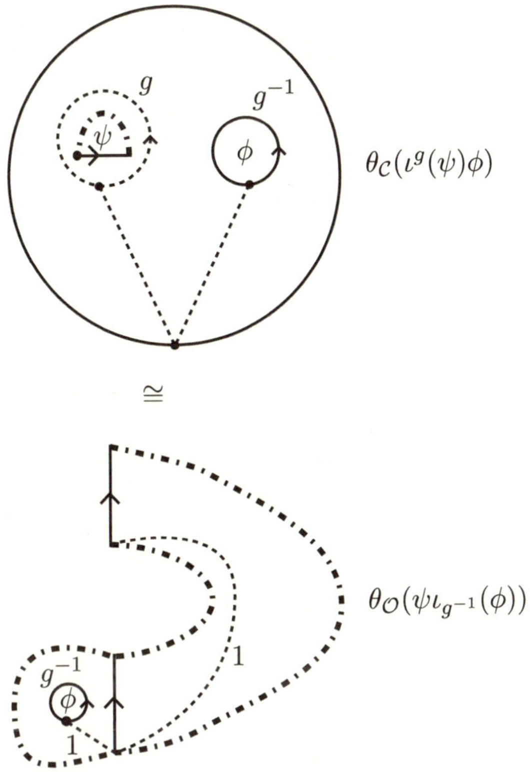

FIGURE 29. The G-twisted adjoint relation. The upper figure is a sphere with two disks removed — the outer circle is not meant to be a boundary.

2.6.5. Solution of the sewing conditions for semisimple $\mathcal{C}$. We now show that, when $\mathcal{C}$ is semisimple, the solutions of the above sewing conditions are provided by G-equivariant bundles on $X = \mathrm{Spec}(\mathcal{C}_1)$ twisted by the B-field defined by $\mathcal{C}$.

Let us first say a word about these bundles. To give a finite dimensional representation of the crossed-product algebra $A(X) \ltimes G$ is to give a representation of $A(X)$ — i.e., a vector bundle E on X — together with an intertwining action of G. Thus representations of $A(X) \ltimes G$ are precisely G-vector bundles on X. For a finite group G there are many equivalent ways of defining the notion of a *twisted* G-vector bundle on X, twisted by a B-field b representing an element of $H^2_G(X, \mathbb{C}^\times)$: the simplest for our purposes is to say that a twisted bundle is just a representation of $A(X) \ltimes_b G$. (Unfortunately this description does not work when G is not finite, and so it is not the one used in [**26**]. We shall explain the relationship with the description of [**26**] at the end of this subsection.)

The problem is easily reduced to consideration of a single G-orbit, so we may assume $X = G/H$ for some subgroup H of G. Accordingly, the closed

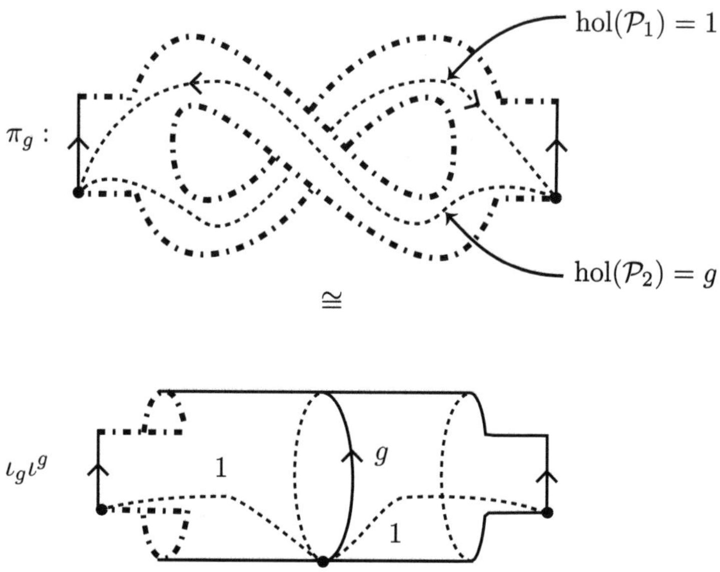

FIGURE 30. The G-twisted Cardy condition. In the double-twist diagram the holonomy around $\mathcal{P}_1$ is 1 and the holonomy around $\mathcal{P}_2$ is g.

string Frobenius data is specified by a 2-cocycle b and a single constant $\theta_c \in \mathbb{C}^\times$ defining the trace: $\theta(\ell_{g,x}) = \delta_{g,1}\theta_c$. As usual, the isomorphism class only depends on $[b] \in H^2(H, \mathbb{C}^\times)$.

THEOREM 2.16. *Let $\mathcal{C} = T(X, b, \theta_c)$ be a Turaev algebra with $\mathcal{C}_1$ semisimple and $X = G/H$. For a single label a the most general solution $\mathcal{O} = \mathcal{O}_{aa}$ of the sewing constraints is determined by a choice of square root $\theta_o = \sqrt{\theta_c}$ and a projective representation V of H with the cocycle b_o which is the restriction of b.*

The algebra $\mathcal{O}$ is the algebra of sections of the G-equivariant bundle of algebras over X:

$$\mathcal{O} := \Gamma\big(G \times_H (\mathrm{End}(V))\big) = \mathrm{Ind}_H^G\big(\mathrm{End}(V)\big), \tag{2.107}$$

and the trace is determined by θ_o:

$$\theta_{\mathcal{O}}(\Psi) = \theta_o \sum_{x \in G/H} \mathrm{Tr}_V(\Psi(x)). \tag{2.108}$$

PROOF. Let us suppose that we are given the Turaev algebra $\mathcal{C}$ with $\mathcal{C}_1$ semisimple, together with $\mathcal{O}, \theta_{\mathcal{O}}, \iota_g, \iota^g$ satisfying the sewing conditions. Let X be the G-space $\mathrm{Spec}(\mathcal{C}_1)$. Then, from our results in the non-equivariant case, we know that $\mathcal{O} = \mathrm{End}_{\mathcal{C}_1}(\Gamma(E)) = \Gamma(\mathrm{End}(E))$ for some vector bundle $E \to X$, unique up to tensoring with a line bundle $L \to X$. Thus $\mathcal{O} = \oplus \mathcal{O}_x$, where $\mathcal{O}_x = \mathrm{End}(E_x)$. We also know that the trace on $\mathcal{O}$ must be given by (2.35). The same square root θ_o of θ_x must be taken for each $x \in X$ to make $\theta : \mathcal{O} \to \mathbb{C}$ invariant under G. Now G acts compatibly on $\mathcal{C}_1$ and $\mathcal{O}$ by algebra isomorphisms, so $g \in G$ maps $\mathcal{O}_{x_0}$ to $\mathcal{O}_x$ by an algebra isomorphism. This proves (2.107), where $V = E_{x0}$. Finally, the Turaev algebra $\mathcal{C}$ is the product $\oplus \mathcal{C}_x$, where $\mathcal{C}_x$ is the twisted group-ring of G_x with the twisting b_x. The algebra homomorphism $\mathcal{C} \to \mathcal{O}$ makes $\mathcal{C}_x$ act on E_x, and so V is a projective representation of $H = G_{x_0}$ with the cocycle b_{x_0}.

This proves that $\mathcal{O}$ is of the form stated. One must still check that the definition (2.107) does provide a solution of the sewing conditions, but that presents no problems. □

REMARK 2.17. Although in the hypothesis of the theorem we were given a cocycle b representing an element of $H^2_G(X, \mathbb{C}^\times)$, the conclusion uses only its restriction b_{x_0}. This should not surprise us, as cohomologous cocycles b define isomorphic Turaev algebras, and $H^2_G(X, \mathbb{C}^\times)$ is canonically isomorphic to the group cohomology $H^2_H(\text{point}; \mathbb{C}^\times)$ when $X = G/H$.

We can now deduce a complete description of the category of boundary conditions, using exactly the same arguments by which we obtained Theorem 2.8 from Theorem 2.6.

THEOREM 2.18. *If $\mathcal{C}$ is a Turaev algebra with $\mathcal{C}_1$ semisimple, corresponding to a space-time X with a B-field b, then the category of boundary conditions for $\mathcal{C}$ is equivalent to the category of b-twisted G-vector bundles on X, uniquely up to tensoring with a G-line bundle on X. Its Frobenius structure is determined by a choice of the dilaton field θ.*

The meaning of this theorem needs to be explained. The linear category of equivariant boundary conditions for a given Turaev algebra is an example of what is called an "enriched" category: for each pair of objects a, b the vector space $\mathcal{O}_{ab}$ has an action of the group G. Now the category Vect_G of finite dimensional vector spaces with G-action is a symmetric tensor category, with the neutral object $\mathbb{C}$. To say that we have a category *enriched* in a tensor category such as Vect_G means that we have

(i) a set of objects,
(ii) for each pair a, b of objects an object $\mathcal{O}_{ab}$ of Vect_G, and
(iii) for each triple a, b, c of objects an associative "composition" morphism

$$\mathcal{O}_{ab} \otimes \mathcal{O}_{bc} \to \mathcal{O}_{ac}$$

of G-vector spaces.

The axioms are almost identical to the axioms for a category, but the space of morphisms has extra structure. In such a situation the category is said to be an enrichment of the ordinary linear category in which the morphisms from b to a are $F(\mathcal{O}_{ab})$, where $F : \mathrm{Vect}_G \to \mathrm{Vect}$ is the functor defined by $F(V) = \mathrm{Hom}_G(\mathbb{C}, V) = V^G$. There is, however, another ordinary category associated to the enriched category by simply forgetting the G-action, so that the morphisms from b to a are simply $\mathcal{O}_{ab}$ as a vector space.

An example of a category enriched in Vect_G is the category of finite dimensional representations of $\tilde{G}$, where $\tilde{G}$ is a central extension (with a fixed cocycle) of G by the circle, where the central circle acts by scalar multiplication. Indeed, given two such representations $V_1^* \otimes V_2$ is a representation of G.

Theorem 2.18 should really be expanded as follows. The category of b-twisted G-vector bundles on X has a natural enrichment in Vect_G, in which the G-vector space of morphisms consists of the homomorphisms of b-twisted vector bundles which are not necessarily equivariant with respect to the G-action. This enrichment is equivalent to the category of equivariant boundary conditions. The underlying ordinary category is the category of boundary conditions for the orbifold theory.

Theorem 2.18 has a converse, which is the G-equivariant extension of the discussion of §2.2.3.

THEOREM 2.19. *If $\mathcal{B}$ is a linear category enriched in* Vect_G*, with G-equivariant traces making it a Frobenius category, and the linear category obtained from $\mathcal{B}$ by forgetting the G-action is semisimple with finitely many irreducible objects, then $\mathcal{B}$ is equivalent to the category of equivariant boundary conditions for a canonical equivariant topological field theory. The Turaev algebra defining the theory is $\oplus_g \mathcal{C}_g$, where an element of $\mathcal{C}_g$ is a family $\phi_a \in \mathcal{O}_{aa}$, indexed by the objects a of $\mathcal{B}$, satisfying*

$$\phi_a \circ f = (g \cdot f) \circ \phi_b \tag{2.109}$$

for each $f \in \mathcal{O}_{ab}$.

To prove this, one must show that (2.109) really does define a Turaev algebra. The details are straightforward and we will omit them.

2.6.6. Equivariant boundary states. To conclude our discussion, let us consider the equivariant analogues of the "boundary states" discussed in §2.3. Our notion of the category of boundary conditions for a G-gauged theory is intrinsically G-invariant, and we have already pointed out that it gives us exactly the same category as we would obtain from the orbifold theory in which we have summed over the gauge fields. To reformulate this in terms of boundary states we begin with the definition.

In the gauged theory associated to a Turaev algebra $\mathcal{C} = T(X, b, \theta)$ the observables at any point of the world sheet are precisely the elements of $\mathcal{C}$. The boundary state $B_a \in \mathcal{C}$ associated to a boundary condition a is characterized by the property that the correlation function of observables $\phi_1, \ldots, \phi_k$ evaluated at points of a surface Σ with boundary S^1 and boundary condition a (with arbitrary holonomy around the boundary) is equal to that of the same observables on the closed surface obtained by capping-off the boundary, with the additional insertion of B_a at the center of the cap. It suffices (because of the factorization properties of a field theory) to check the case when Σ is a disc. The correlation function on the disc is obtained by propagating $\phi_1 \cdots \phi_k \in \mathcal{C}_g$ to $\mathcal{H}(\emptyset) = \mathbb{C}$ by the annulus whose non-incoming boundary circle is constrained by the condition a, along which the holonomy is necessarily g. Our rules tell us that the result is

$$\theta_{\mathcal{O}_{aa}}(\iota_{g,a}(\phi_1 \cdots \phi_k)).$$

Equating this to $\theta_{\mathcal{C}_1}(\phi_1 \cdots \phi_k B_a)$, we see that

$$B_a = \sum_g \iota_{g,a}(1).$$

The map $a \mapsto B_a$ evidently has its image in the G-invariant part — i.e., the center — of the Turaev algebra. It extends to a homomorphism

$$K_{G,h}(X) \to T(X, b, \theta)^G,$$

and we have

THEOREM 2.20. *The G-invariant boundary states generate a lattice in $T(X, b, \theta)^G$ related to the twisted equivariant K-theory via*

$$K_{G,h}(X) \otimes_{\mathbb{Z}} \mathbb{C} = T(X, b, \theta)^G. \tag{2.110}$$

REMARK 2.21. (1) Equation (2.110) is related to an old observation of [**121**]. If $X = G$, with G acting on itself by conjugation, then $T(X, 0)^G$ is the Verlinde algebra occurring in the conformal field theory of orbifolds for chiral algebras with one representation [**121**]. The different orbits are the conjugacy classes of G. Focusing on one conjugacy class $[g]$ we can compare with the above results. One basis of states is provided by a choice of a character of the centralizer of g. These are just the G-invariant boundary states found above.

(2) The translation of the above results to the language of branes at orbifolds is the following. The boundary states corresponding to the different b-irreps V_i are the "fractional branes" of [**141**]. The use of projective representations was proposed in [**141**], and argued to correspond to discrete torsion in [**132, 136**]. A different proof of

the fact that the cocycle for the open sector and that of the closed sector b are cohomologous can be found in [**15**].

To conclude this section, let us return to explain the relation between the definition of twisted equivariant K-theory by $A(X) \ltimes_b G$-modules and the definition given in [**26**].

In [**26**] the elements of the twisted equivariant theory are described as follows. First, the twisting class $b \in H^3_G(X, \mathbb{Z})$ is represented by a bundle P of projective Hilbert spaces on X equipped with a G-action covering the G-action on X. Then elements of $K_{G,P}(X)$ are represented by families $\{T_x\}_{x\in X}$ of fibrewise Fredholm operators in the bundle P. Let us show how to associate such a pair $(P, \{T_x\})$ to a finitely generated $A(X) \ltimes_b G$-module. Such a module is the same thing as a finitely generated $A(X)$-module equipped with a compatible action of the extended group $\tilde{G}$ associated to b which we have already described. Equivalently, it is a finite dimensional vector bundle E on X with an action of $\tilde{G}$ on the total space which covers the action of G on X. Let us choose a fixed infinite dimensional Hilbert space $\mathcal{H}$. Then $\hat{E} = E \otimes \mathcal{H}$ is a Hilbert bundle on X, and the associated bundle $P = \mathbb{P}(\hat{E})$ of projective spaces has a natural action of G, and it represents the class of b in $H^3_G(X, \mathbb{Z})$. (Cf. the proof of Proposition 6.3 in [**26**].) If $T : \mathcal{H} \to \mathcal{H}$ is a fixed surjective Fredholm operator with a one-dimensional kernel, then $\mathrm{id}_E \otimes T : P \to P$ represents an element of $K_{G,P}(X)$ according to the definition of [**26**].

If the cocycle b is a coboundary — or even if $b_x(g_1, g_2)$ is independent of $x \in X$ — it is plain that the two rival definitions of equivariant K coincide. A Mayer-Vietoris argument can then be used to show that they coincide for all b.

The essential point here is that, when X and G are finite, the twisting class b is of finite order, and that makes it possible to represent the K-classes by families of Fredholm operators of constant rank, and hence by finite dimensional vector bundles.

2.7. Appendix: Morse theory proof of the sewing theorems

In this appendix we shall use Morse theory to give uniform proofs of four theorems. The first is the very well-known result that a two-dimensional topological field theory is precisely encoded in a commutative Frobenius algebra. The second is the corresponding statement for open and closed theories: this is Theorem 2.1 of §2.1. The third and fourth are the equivariant analogues of the first two, i.e., Theorems 2.11 and 2.15 of §2.6.

2.7.1. The classical theorem. We wish to prove that when we have a commutative Frobenius algebra $\mathcal{C}$ we can assign to an oriented cobordism Σ from S_0 to S_1 a linear map

$$U_\Sigma : \mathcal{C}^{\otimes p} \to \mathcal{C}^{\otimes q},$$

where the oriented 1-manifolds S_0 and S_1 have p and q connected components respectively.

We can always choose a smooth function $f : \Sigma \to [0,1] \subset \mathbb{R}$ such that $f^{-1}(0) = S_0$ and $f^{-1}(1) = S_1$, and which has only "Morse" singularities, i.e., the gradient df vanishes at only finitely many points $x_1, \ldots, x_n \in \Sigma$, and

(i) the Hessian $d^2 f(x_i)$ is a non-degenerate quadratic form for each i, and

(ii) the critical values $c_1 = f(x_1), \ldots, c_n = f(x_n)$ are distinct, and not equal to 0 or 1.

Each critical point has an *index*, equal to 0, 1, or 2, which is the number of negative eigenvalues of the Hessian $d^2 f(x_i)$.

The choice of the function f gives us a decomposition of the cobordism into "elementary" cobordisms. If

$$0 = t_0 < c_1 < t_1 < c_2 < t_2 < \cdots < c_n < t_n = 1,$$

and $S_t = f^{-1}(t)$, then each S_{t_i} is a collection of, say, m_i disjoint circles, with $m_i = m_{i-1} \pm 1$, and $\Sigma_i = f^{-1}([t_{i-1}, t_i])$ is a cobordism from $S_{t_{i-1}}$ to S_{t_i} which is trivial (i.e., a union of cylinders) except for one connected component of one of the four forms of Figure 2.

For a given Frobenius algebra $\mathcal{C}$ we know how to define an operator

$$U_{\Sigma_i} : \mathcal{C}^{\otimes m_{i-1}} \to \mathcal{C}^{\otimes m_i}$$

in each case. (In the third case the map we assign is

$$\phi \mapsto \sum \phi\phi_i \otimes \phi^i,$$

where $\{\phi_i\}$ and $\{\phi^i\}$ are dual bases of $\mathcal{C}$ such that $\theta_{\mathcal{C}}(\phi^i \phi_j) = \delta_{ij}$.) We should notice two points. First, we need $\mathcal{C}$ to be commutative, for otherwise we would need to have an *order* on the two incoming circles of a pair of pants, and no such order is given. Secondly, the assignments we make have the property that reversing the direction of time in a cobordism replaces the operator by its adjoint with respect to the Frobenius inner product on the state spaces. This property will be a firm principle in all our constructions, and it reduces the number of cases we have to check in the tedious arguments below.

The important task now is to show that the composite operator $U_{\Sigma_n} \circ \cdots \circ U_{\Sigma_1}$ is independent of the chosen Morse function f.

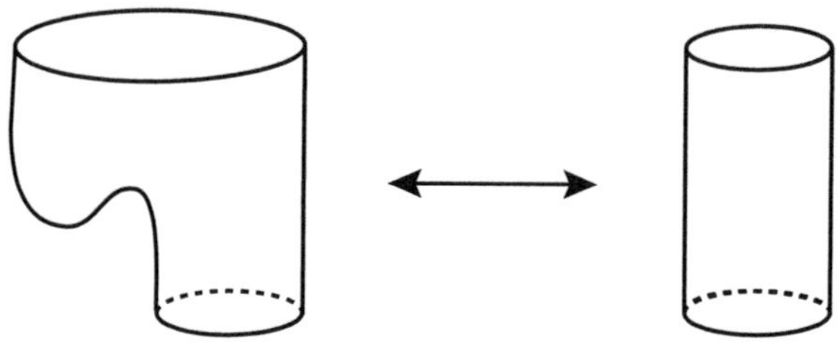

FIGURE 31

Two Morse functions f_0 and f_1 can always be connected by a smooth path $\{f_s\}_{0\le s\le 1}$ in which f_s is a Morse function except for a finite set of parameter values s at which one of the following two things happens:

(i) f_s has one degenerate critical point where in local coordinates (u,v) it has the form $f_s(u,v) = \pm u^2 + v^3$, or

(ii) two distinct critical points x_i, x_j of f_s have the same critical value $f_s(x_i) = f_s(x_j) = c$.

In the first case, two critical points of adjacent indices are created or annihilated as the parameter passes through the non-Morse value s, and the cobordism changes by Figure 31, or vice-versa, or by the time-reversal of these pictures. The well-definedness of U_Σ under this kind of change is ensured by the identity $1 \cdot a = a$ in the algebra $\mathcal{C}$.

Case (ii) is more problematical. Because operators of the form $U \otimes 1$ and $1 \otimes U'$ commute, we easily see that there is nothing to prove unless the two critical points x_i and x_j are connected in the "bad" critical contour S_c, in which case they must both have index 1.

Let us consider the resulting two-step cobordism which is factorized in different ways before and after the critical parameter value s. It will have just one non-trivial connected component, which, because an elementary cobordism changes the number of circles by 1, must be a cobordism from p circles to q circles, where $(p,q) = (1,1), (2,2), (1,3)$ or $(3,1)$. We need to check only one of (1,3) and (3,1), as they differ only by time-reversal. Because the Euler number of a cobordism is the number of critical points of its Morse function (counted with the sign $(-1)^{\text{index}}$), the non-trivial component has Euler number -2, so is a 2-holed torus when $(p,q) = (1,1)$ and a 4-holed sphere in the other cases.

In the case (1,1), depicted in Figure 32, a circle splits into two which then recombine. There is nothing to check, because, though a torus with two holes can be cut into two pairs of pants by many different isotopy classes of cuts, there is only one possible composite cobordism, and we have only one possible composite map $\mathcal{C} \to \mathcal{C} \otimes \mathcal{C} \to \mathcal{C}$.

In the case (3,1), two circles of the three combine, then the resulting circle combines with the third. The picture is Figure 33. Clearly this case is covered by the associative law in $\mathcal{C}$.

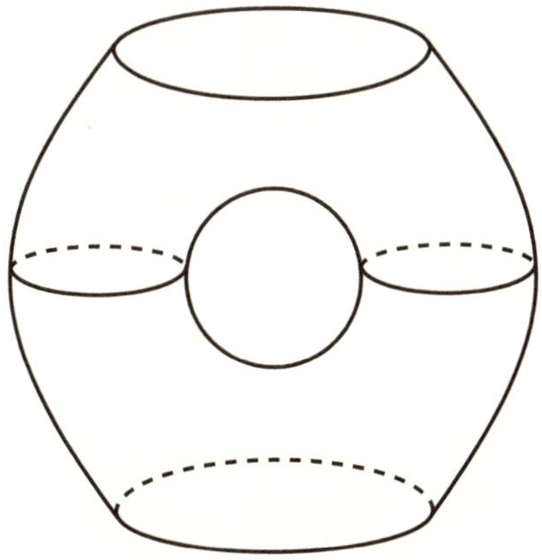

FIGURE 32. The diagram depicts a torus truncated by two horizontal planes which are level surfaces of the Morse function. The two critical points are at the top and bottom of the inner circle. If the torus is tilted— as a rigid body— then the two critical points of the height function can be made to lie in the same horizontal plane.

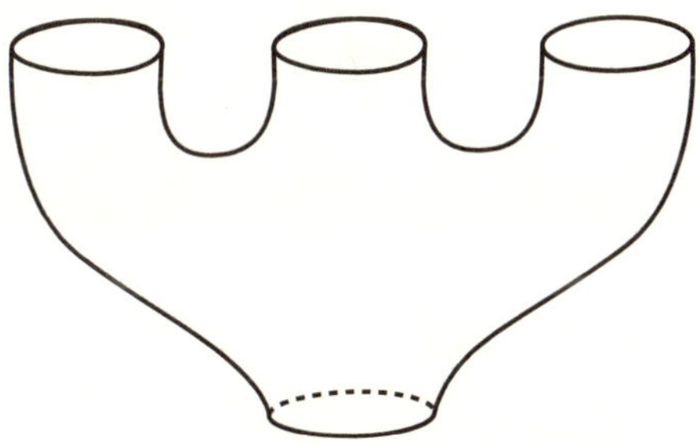

FIGURE 33

FIGURE 34. Ways of embedding a cobordism with two critical points in $\mathbb{R}^3$. The right-hand diagram depicts the situation whose contour lines are drawn in Figure 35, (i).

In the case (2,2) we are again factorizing a 4-holed sphere into two elementary cobordisms. This can be done in many ways, as we see from the pictures Figure 34. The best way of making sure we are not overlooking any possibility is to think of the contour just below the doubly-critical level,

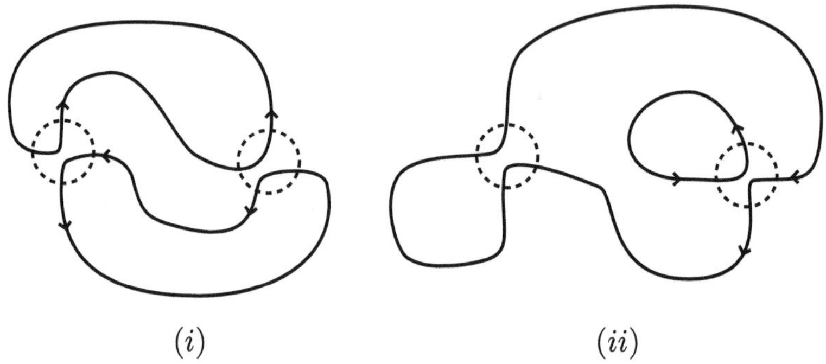

FIGURE 35

which, if it consists of two circles, must have one of the two forms (i) or (ii) in Figure 35. (Consider the possible ways of connecting the "terminals" inside the dotted circles.) But, whatever happens, the only algebraic maps the cobordism can lead to are

$$\mathcal{C} \otimes \mathcal{C} \to \mathcal{C} \to \mathcal{C} \otimes \mathcal{C}$$

and

$$\mathcal{C} \otimes \mathcal{C} \to \mathcal{C} \otimes \mathcal{C} \otimes \mathcal{C} \to \mathcal{C} \otimes \mathcal{C},$$

given by

$$\phi \otimes \phi' \mapsto \phi\phi' \mapsto \sum \phi\phi'\phi_i \otimes \phi^i$$

and

$$\phi \otimes \phi' \mapsto \sum \phi\phi_i \otimes \phi^i \otimes \phi' \mapsto \sum \phi\phi_i \otimes \phi^i\phi'$$

respectively, where $\{\phi_i\}$ and $\{\phi^i\}$ are dual bases of $\mathcal{C}$ such that $\theta_{\mathcal{C}}(\phi^i\phi_j) = \delta_{ij}$. These two maps are equal because of the identity

$$\sum \phi'\phi_i \otimes \phi^i = \sum \phi_i \otimes \phi^i\phi', \tag{2.111}$$

which holds in any Frobenius algebra because the inner product of each side with $\phi^j \otimes \phi_k$ is $\theta_{\mathcal{C}}(\phi^j\phi'\phi_k)$.

That completes the proof of the theorem. Notice that we have used all the axioms of a commutative Frobenius algebra.

2.7.2. Open and closed theories. As in the preceding argument we consider a cobordism Σ from S_0 to S_1, but now S_0 and S_1 are collections of circles and intervals, and the boundary $\partial\Sigma$ has a constrained part $\partial_{\text{constr}}\Sigma$, which we shall abbreviate to $\partial'\Sigma$, which is a cobordism from ∂S_0 to ∂S_1. We choose $f : \Sigma \to [0, 1]$ as before, but now there are two kinds of critical points of f: interior points of Σ at which the gradient df vanishes, and points of $\partial'\Sigma$ at which the gradient of the restriction of f to the boundary vanishes. For an internal critical point, "non-degenerate" has its usual meaning. A critical point x on the boundary is called non-degenerate if it is a non-degenerate critical point of the restriction of f to $\partial'\Sigma$, and in addition the derivative of f normal to the boundary does not vanish at x.

As before, we say f is a *Morse function* if all its critical points are non-degenerate, and all the critical values are distinct and $\neq 0, 1$. We can always choose such a function.

There are now four kind of boundary critical points, which we can denote $0\pm, 1\pm$, recording the index and the sign of the normal derivative. Six things can happen as we pass through one of them. At those of type $0+$ or $1-$, an open string is created or annihilated. At type $0-$ either two open strings join end-to-end, or else an open string becomes a closed string. Type $1+$ is the time-reverse of $0-$. If we have a Frobenius category $\mathcal{B}$, we know what to do in each of the six cases.

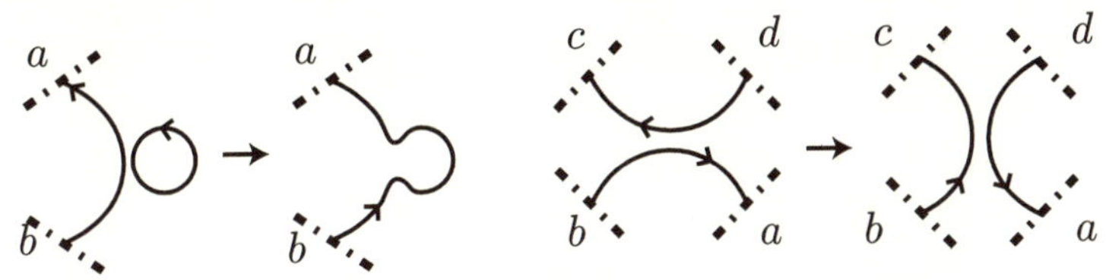

FIGURE 36

An internal critical point has index 0,1, or 2, as before. Only if the index is 1 can the corresponding cobordism involve an open string. Up to time reversal, there are three index 1 processes: two closed strings can become one, an open string can "absorb" a closed string, and two open strings can "reorganize themselves" to form two new open strings as in Figure 36.

For a given Frobenius category $\mathcal{B}$, we assign to (open)+(closed)→(open) the map

$$\mathcal{O}_{ab} \otimes \mathcal{C} \to \mathcal{O}_{ab}$$

given by $\phi \otimes \psi \mapsto \phi\psi$. Here, as we usually do, we are regarding $\mathcal{O}_{ab}$ as a $\mathcal{C}$-module, writing

$$\phi\psi = \iota_a(\phi)\psi = \iota_b(\phi)\psi.$$

To (open)+(open)→(open)+(open) we assign the map

$$\mathcal{O}_{ab} \otimes \mathcal{O}_{cd} \to \mathcal{O}_{ad} \otimes \mathcal{O}_{cb}$$

given by

$$\psi \otimes \psi' \mapsto \sum \psi\psi_i \otimes \psi'\psi^i,$$

where ψ_i and ψ^i are dual bases of $\mathcal{O}_{bd}$ and $\mathcal{O}_{db}$.

We must now consider what happens when we change the Morse function. As before, two Morse functions can be connected by a path $\{f_s\}$ in which each f_s is a Morse function except for finitely many values of s at which either one critical point is degenerate or else two critical values coincide. We begin with the degenerate case. There are now three kinds of degeneracy which we must allow, for besides internal degeneracies which are just as in the closed string case we can have two kinds of degeneracy on the boundary: either $f|\partial'\Sigma$ has a cubic inflexion, or else the normal derivative vanishes at a boundary critical point.

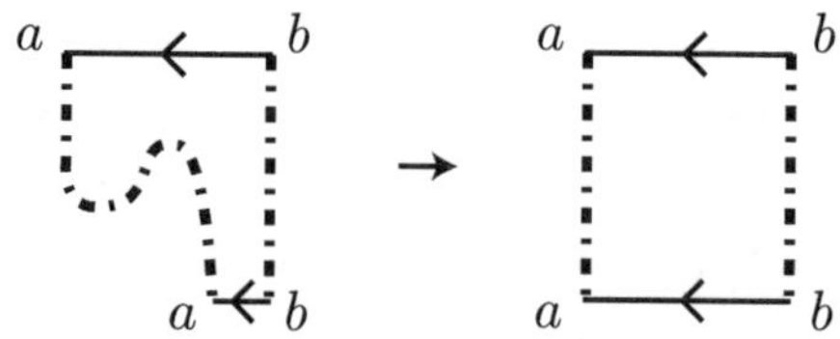

FIGURE 37

When s passes through a boundary inflexion, two non-degenerate boundary critical points of opposite index but with normal derivatives of the same sign are created or annihilated. This means that the cobordism changes between the two figures of Figure 37 (or the time-reversal). These changes are covered by the axiom that the category $\mathcal{B}$ has identity morphisms.

When the normal derivative vanishes at a boundary critical point what happens is that an internal critical point has moved "across the boundary of Σ, i.e. it moves into coincidence with a boundary critical point and changes the sign of the normal derivative there. There are four cases:

$$(0-) \ + \ (\text{index } 0) \ \to \ (0+),$$
$$(0+) \ + \ (\text{index } 1) \ \to \ (0-),$$

and the time-reversals of these. In the first case, the composite cobordism in which a small closed string is created and then breaks open is replaced by the elementary cobordism in which an open string is created. This corresponds to the axiom that $\mathcal{C} \to \mathcal{O}_{aa}$ takes $1_{\mathcal{C}}$ to 1_a. In the second case, in the composite cobordism, an open string is created, and then it either "absorbs" an existing closed string or else "rearranges" itself with an existing open string; these composites are to be equivalent, respectively, to the elementary breaking of a closed or open string. Putting $\psi = 1_a$ in the formulae above we see that this is allowed by the Frobenius category axioms.

When we have an internal degenerate critical point, what happens, up to time-reversal, is that a closed string is created and then joins an existing open or closed string, this should be the same as the trivial cobordism. Again, the unit axioms cover this.

Finally, we have to consider what happens when two critical values cross. They can be two boundary critical points, two internal ones, or one of each.

If two boundary critical points are linked by a critical contour, it has the form in Figure 38. These give us four cases to check, where the contour below the critical level is as in Figure 39.

Case $(i)_a$ is accounted for by the associativity of composition in the category $\mathcal{B}$; case $(i)_b$ by the open string analogue of the identity (2.111); case

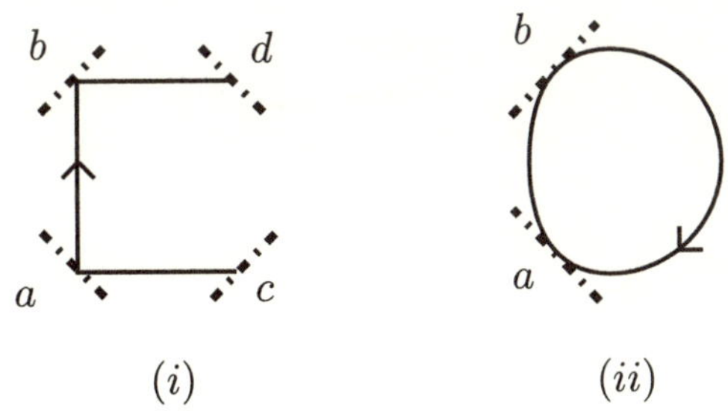

FIGURE 38

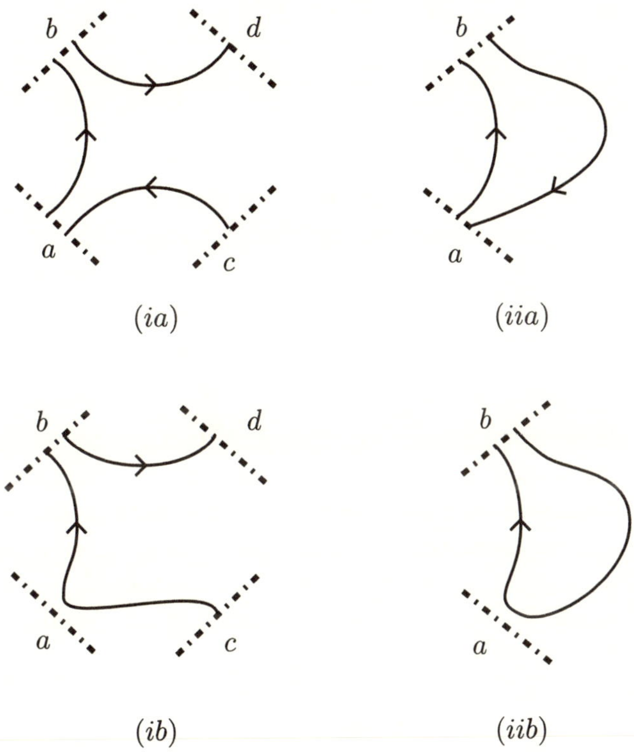

FIGURE 39

$(ii)_a$ by the trace axiom $\iota^a(\psi_1\psi_2) = \iota^b(\psi_2\psi_1)$, which follows by combining (2.5),(2.9),and (2.10); and case $(ii)_b$ by the Cardy identity.

When we have one boundary and one internal critical point at the same level we may as well assume the boundary point is of type $0-$ and the internal critical point is of index 1, and that they are joined in the critical contour,which must have one of the four forms in Figure 40.

At the boundary point either an open string becomes closed, or else two open strings join. We shall consider each possibility in turn. In the first case, if the boundary point is encountered first, then at the interior point

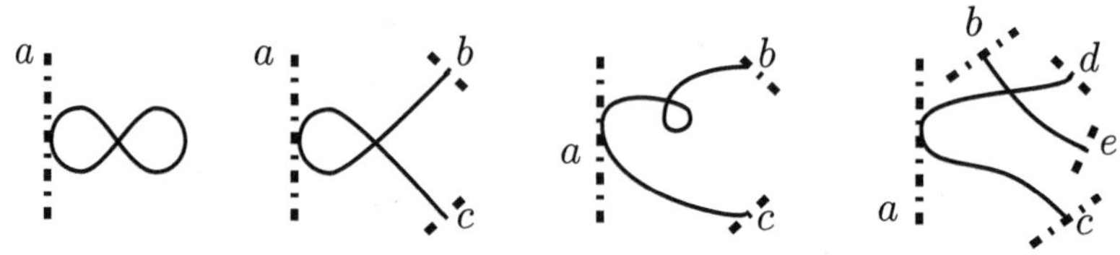

FIGURE 40

three things can happen: the closed string can split into two closed strings, or it can combine with another closed or open string. Thus the possibilities are

$$o \to c \to c + c$$
$$o + c \to c + c \to c$$
$$o + o \to c + o \to o.$$

When the internal point is encountered first there is only one possibility in each case, and the three sequences are replaced respectively by

$$o \to o + c \to c + c$$
$$o + c \to o \to c$$
$$o + o \to o + o \to o.$$

We have to check three identities. The first two reduce to the fact that $\iota^a : \mathcal{O}_{aa} \to \mathcal{C}$ is a map of modules over $\mathcal{C}$. The third is the Cardy condition.

Now let us consider the case where two open strings join at the boundary critical point. If we meet the boundary point first, there are again three things that can happen at the internal critical point: the open string can emit a closed string, or else it can interact with another closed or open string. The possibilities are

$$o + o \to o \to o + c$$
$$o + o + c \to o + c \to o$$
$$o + o + o \to o + o \to o + o.$$

In the second and third of these cases there is only one thing that can happen when the order of the critical points is reversed: they become

$$o + o + c \to o + o \to o$$
$$o + o + o \to o + o + o \to o + o.$$

The identities relating the corresponding algebraic maps $\mathcal{O}_{ab} \otimes \mathcal{O}_{bc} \otimes \mathcal{C} \to \mathcal{O}_{ac}$ and $\mathcal{O}_{ab} \otimes \mathcal{O}_{bc} \otimes \mathcal{O}_{de} \to \mathcal{O}_{ae} \otimes \mathcal{O}_{dc}$ are immediate.

The first sequence, however, can become either

$$o + o \to o + o + c \to o + c$$

or

$$o + o \to o + o \to o + c.$$

The first of these presents nothing of interest algebraically, but to deal with the second we need to check that

$$\sum \psi\psi'\phi_i \otimes \phi^i = \sum \psi\psi^k \otimes \iota^b(\psi'\psi_k)$$

for $\psi \in \mathcal{O}_{ab}$, $\psi' \in \mathcal{O}_{bc}$, and dual bases ϕ^i, ϕ_i of $\mathcal{C}$ and ψ^k, ψ_k of $\mathcal{O}_{bc}, \mathcal{O}_{cb}$. This relation holds because the inner product of the left-hand side with $\psi_m \otimes \phi_j$ is $\theta_b(\psi\psi'\phi_j\psi_m)$, while the inner product of the right-hand side with $\psi_m \otimes \phi_j$ is

$$\begin{aligned}\sum_k \theta_b(\psi\psi^k\psi_m)\theta(\iota^b(\psi'\psi_k)\phi_j) &= \sum_k \theta_b(\psi_m\psi\psi^k)\theta_b(\psi_k\phi_j\psi') \\ &= \theta_b(\psi_m\psi\phi_j\psi') = \theta_b(\psi\psi'\phi_j\psi_m). \qquad (2.112)\end{aligned}$$

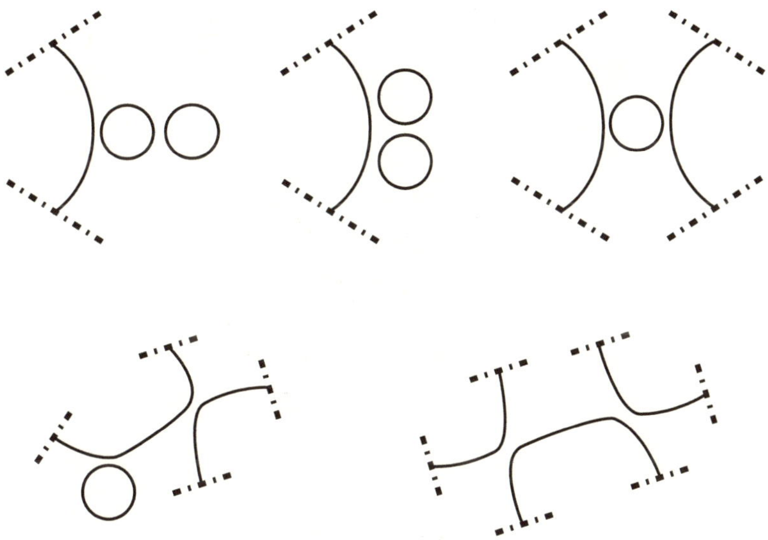

FIGURE 41

Finally, we must consider what happens when there are two internal critical points on the same level. Here we have the possibilities which we have already discussed in the closed case, but must also allow any or all of the strings involved to be open. We can analyse the situation according to the number of connected components of the part of the contour immediately below the doubly critical level which pass close to the critical points. There must be one, two, or three such components. If there are three they can form five configurations (apart from the case when all three are closed), as depicted in Figure 41. The well-definedness of the composite map in all these cases follows immediately from the associative law of composition in the Frobenius category.

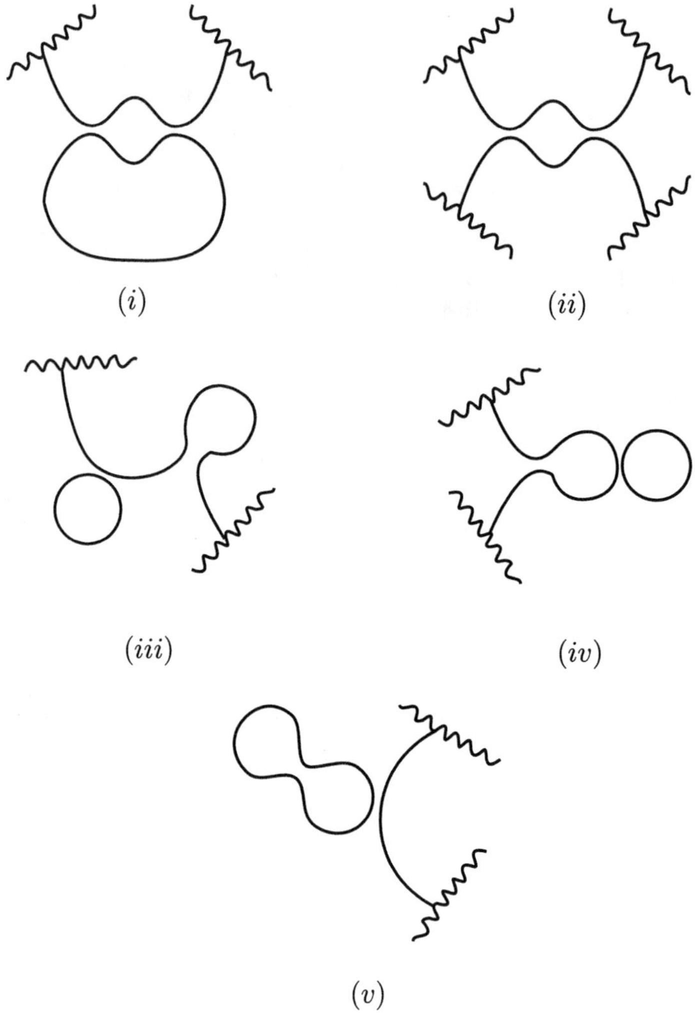

FIGURE 42

If there are two components below the critical level then they can again form five configurations (for either the two components meet twice, or else they meet once, and one of them has a self-interaction), depicted in Figure 42. But we have only three cases to check, as the second is the time-reversal of one from Figure 41, and the last two are time-reversals of each other. Figure 42, (i) corresponds to the fact that the composition

$$\mathcal{O}_{ab} \otimes \mathcal{C} \to \mathcal{O}_{ab} \to \mathcal{O}_{ab} \otimes \mathcal{C}$$

can be effected by cutting the composite cobordism in different ways, but there is nothing to check, as there is only one possible algebraic map.

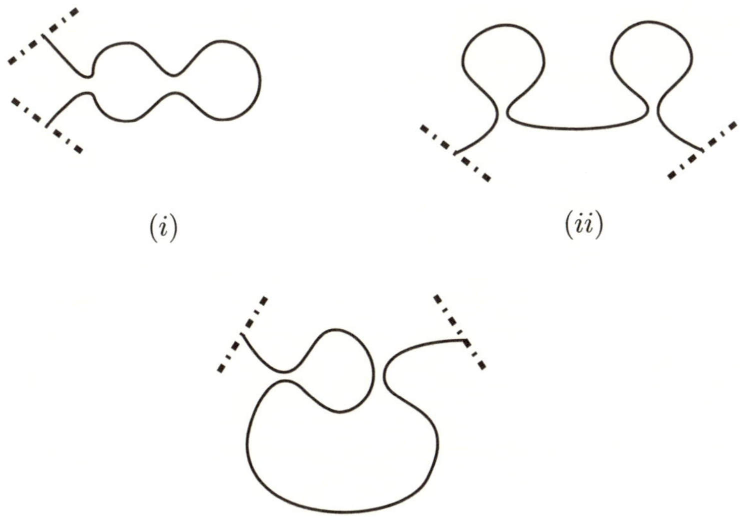

(i) (ii) (iii)

FIGURE 43

In Figure 42 case (iii), one order of the critical points gives us the same composition

$$\mathcal{O}_{ab} \otimes \mathcal{C} \to \mathcal{O}_{ab} \to \mathcal{O}_{ab} \otimes \mathcal{C}$$

as before, while the other order gives

$$\mathcal{O}_{ab} \otimes \mathcal{C} \to \mathcal{O}_{ab} \otimes \mathcal{C} \otimes \mathcal{C} \to \mathcal{O}_{ab} \otimes \mathcal{C};$$

but it is very easy to check that both maps take $\psi \otimes \phi$ to $\sum \psi\phi\phi_i \otimes \phi^i$ in the notation we have already used.

In Figure 42 case (iv), we must again compare compositions

$$\mathcal{O}_{ab} \otimes \mathcal{C} \to \mathcal{O}_{ab} \otimes \mathcal{C} \otimes \mathcal{C} \to \mathcal{O}_{ab} \otimes \mathcal{C}$$

and

$$\mathcal{O}_{ab} \otimes \mathcal{C} \to \mathcal{O}_{ab} \to \mathcal{O}_{ab} \otimes \mathcal{C}.$$

This time we must check that

$$\sum \psi\phi_i \otimes \phi^i\phi = \sum \psi\phi\phi_i \otimes \phi^i.$$

This is the same formula which we met at the end of our discussion of closed string theories.

Finally, suppose that the contour below the critical level has only one connected component. There are three possible configurations, corresponding to the three ways of pairing four points on an interval. They are in Figure 43. The first two of these are time-reversals of cases we have already treated. The last one leads — in either order — to a factorization

$$\mathcal{O}_{ab} \to \mathcal{O}_{ab} \otimes \mathcal{C} \to \mathcal{O}_{ab}.$$

There is only one possibility for this, so there is nothing to check.

That completes the proof of the theorem about open and closed theories.

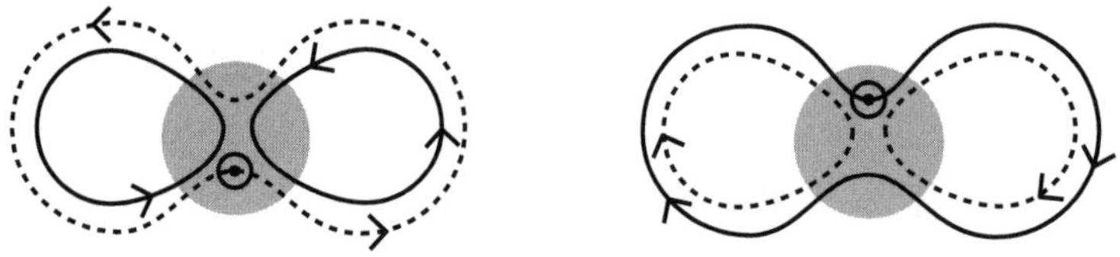

FIGURE 44

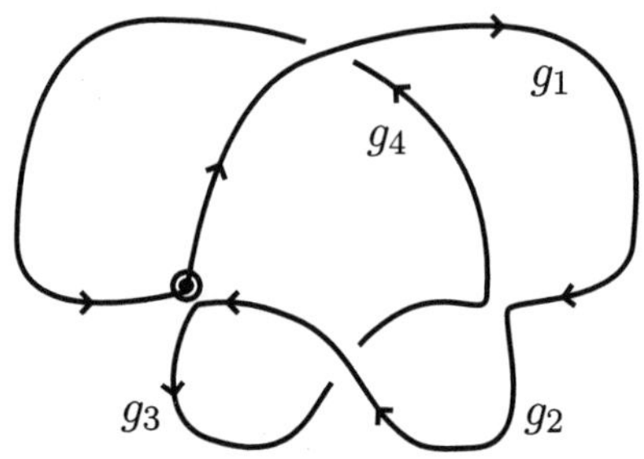

FIGURE 45

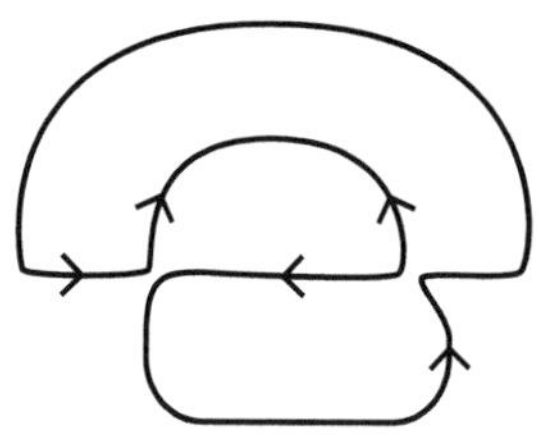

FIGURE 46

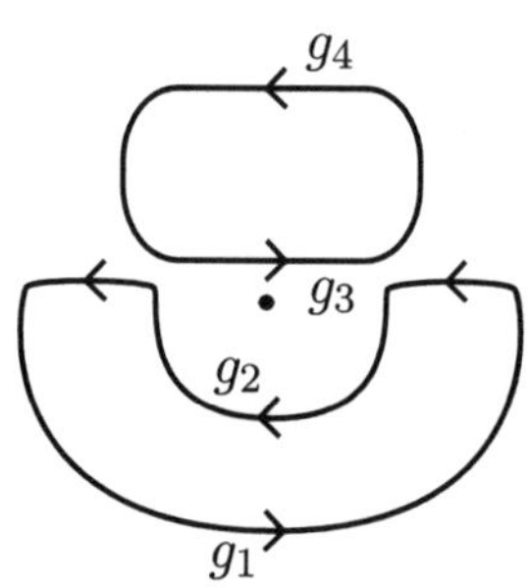

FIGURE 47

2.7.3. Equivariant closed theories. We must now redo the discussion in the first part of this appendix, but for surfaces and circles equipped with a principal G-bundle, where G is a given finite group.

The first observation is that any circle with a bundle is isomorphic to a standard bundle S_g with holonomy $g \in G$ on the standard circle S^1. Furthermore the set of morphisms from S_g to $S_{g'}$ is $\{h \in G : \ hgh^{-1} = g'\}$. In other words, the category of bundles on S^1 is equivalent to the category $G//G$ formed by the group G acting on itself by conjugation. An equivariant theory therefore gives us a vector space $\mathcal{C}_g$ for each g, and together the $\mathcal{C}_g$ form a G-vector bundle on G. Conversely, given the G-vector bundle $\{\mathcal{C}_g\}$ and a circle S with a bundle P on it, the theory gives us the vector space $\mathcal{H}(S,P)$ whose elements are rules which associate $\psi_{x,t} \in \mathcal{C}_{g_{x,t}}$ to each $x \in S$ and trivialization $t : P_x \to G$, where $g_{x,t}$ is the holonomy of P with base point (x,t), and we require that

$$\psi_{x',t'} = g\psi_{x,t}$$

if g is the holonomy of P along the positive path from (x,t) to (x',t'). For this to be well-defined we need the condition that $g_{x,t}$ acts trivially on $\mathcal{C}_{g_{x,t}}$, whose necessity we have already explained in §2.6.

Next we consider the trivial cobordism from S_g to $S_{g'}$. The possible extensions of the bundles on the ends over the cylinder correspond to the possible holonomies from the incoming base point to the outgoing base point, i.e., to the set of morphisms $\{h \in G : \ hgh^{-1} = g'\}$ in $G//G$. Clearly these cylinders induce the isomorphisms $\mathcal{C}_g \to \mathcal{C}_{g'}$ which we already know. But two such cobordisms are to be regarded as equivalent if there is a diffeomorphism from the cylinder (with its bundle) to itself which is the identity on the ends. The mapping class group of the cylinder is generated by the Dehn twist around it, so the morphism corresponding to h is equivalent to that for $hg = g'h$. This means that g must act trivially on $\mathcal{C}_g$, as we already know.

Now we come to the possible bundles on the four elementary cobordisms of Figure 2. The bundle on a cap must of course be trivial. The pair-of-pants cobordisms that are relevant to us arise as the regions between nearby level curves separated by a critical level. We can draw them as in Figure 44, where the solid contour is below the critical level, and the dashed one is above it. We can trivialize the G-bundle in the neighbourhood of the critical point (i.e., within the shaded area), and then the bundle on the cobordism is determined by giving the holonomies g_1, g_2 along the ribbons (i.e., the unshaded part of the surface), as indicated. The operator we associate to case (i) is the multiplication map

$$m_{g_1,g_2} : \mathcal{C}_{g_1} \otimes \mathcal{C}_{g_2} \to \mathcal{C}_{g_1g_2}$$

of (2.80). In writing it this way we are choosing an ordering of the ribbons, i.e., a base point on the outgoing loop. The two orderings are related by the conjugation

$$\alpha_{g_2} : \mathcal{C}_{g_1 g_2} \to \mathcal{C}_{g_2 g_1},$$

so the consistency condition for us to have a well-defined assignment is that

$$m_{g_2,g_1}(\psi_2 \otimes \psi_1) = \alpha_{g_2}(m_{g_1,g_2}(\psi_1 \otimes \psi_2)).$$

We see that this holds in any Turaev algebra by combining (2.82) with the facts that G acts on the algebra by algebra-automorphisms, and that α_{g_2} acts trivially on $\mathcal{C}_{g_2}$. As the mapping class group of the pair of pants is generated by the three Dehn twists parallel to its boundary circles, there are no new conditions needed to make the assignment of the operator to the pair of pants well-defined.

The homomorphism

$$c_{g_1,g_2} : \mathcal{C}_{g_1 g_2} \to \mathcal{C}_{g_1} \otimes \mathcal{C}_{g_2}$$

corresponding to the cobordism in Figure 15 is fixed by the requirement of adjunction, bearing in mind that the dual space to $\mathcal{C}_g$ is $\mathcal{C}_{g^{-1}}$. It is given by

$$c_{g_1,g_2}(\phi) = \sum \phi\phi^i \otimes \phi_i,$$

where $\{\phi_i\}$ is a basis for $\mathcal{C}_{g_2}$, and $\{\phi^i\}$ is the dual basis of $\mathcal{C}_{g_2^{-1}}$.

Any cobordism with a bundle can be factorized by Morse theory just as before; bundles are inherited by the elementary cobordisms. The difficult part of the discussion is considering what happens when we change the Morse function. But in fact the only step which presents anything significant is the consideration of the interchange of two critical points of index 1 on the same level, i.e., the cobordisms of Figures 32, 33, 34.

Let us consider the case in Figure 32, where a string divides and then rejoins — i.e., a torus with two holes, one incoming and one outgoing. We draw the picture in the form in Figure 45. (We do not draw it in the apparently more perspicuous form in Figure 46, as then the neighbourhoods of the two critical points would have opposite orientation in the plane.)

The cobordism corresponds to a map $\mathcal{C}_{4321} \to \mathcal{C}_{2341}$, where, as in the following, we have abbreviated $\mathcal{C}_{g_4 g_3 g_2 g_1}$ to $\mathcal{C}_{4321}$. If the left-hand critical point is encountered first, the map we obtain is

$$\mathcal{C}_{4321} \to \mathcal{C}_{43} \otimes \mathcal{C}_{21} \cong \mathcal{C}_{34} \otimes \mathcal{C}_{12} \to \mathcal{C}_{3412} \cong \mathcal{C}_{2341},$$

$$\begin{aligned} \phi \mapsto \sum \phi\phi^i \otimes \phi_i \mapsto \sum \alpha_3(\phi\phi^i) \otimes \alpha_1(\phi_i) &\mapsto \sum \alpha_3(\phi\phi^i)\alpha_1(\phi_i) \\ &\mapsto \sum \alpha_2(\alpha_3(\phi\phi^i)\alpha_1(\phi_i)), \end{aligned}$$

where ϕ_i runs through a basis for $\mathcal{C}_{21}$, and we write α_3 for α_{g_3}, and so on. (The maps indicated by $\cong$ in the previous line correspond to moving the choice of base point on the various strings.)

With the other order, we get

$$\mathcal{C}_{4321} \cong \mathcal{C}_{3214} \to \mathcal{C}_{32} \otimes \mathcal{C}_{14} \cong \mathcal{C}_{23} \otimes \mathcal{C}_{41} \to \mathcal{C}_{2341},$$

$$\begin{aligned} \phi \mapsto \alpha_4^{-1}(\phi) \mapsto \sum \alpha_4^{-1}(\phi)\psi^i \otimes \psi_i &\mapsto \sum \alpha_2(\alpha_4^{-1}(\phi)\psi^i) \otimes \alpha_4(\psi_i) \\ &\mapsto \sum \alpha_2(\alpha_4^{-1}(\phi)\psi^i)\alpha_4(\psi_i), \end{aligned}$$

where ψ_i runs through a basis of $\mathcal{C}_{14}$.

Thus we must prove that

$$\sum \alpha_{23}(\phi\phi^i)\phi_i = \sum \alpha_{24^{-1}}(\phi)\alpha_2(\psi^i)\alpha_4(\psi_i).$$

We can deduce this from the axiom (2.84) of §2.6, with $h = g_2 g_4^{-1} g_1^{-1} g_2^{-1}$ and $g = g_1^{-1} g_2^{-1}$, as follows. We rewrite the right-hand side of the equation as

$$\sum \alpha_{24^{-1}}(\phi)\eta_i \alpha_g(\eta^i),$$

where η_i is the basis $\alpha_2(\psi^i)$ of $\mathcal{C}_h$, so that $\eta^i = \alpha_2(\psi_i)$ and

$$\alpha_g(\eta^i) = \alpha_1^{-1}(\psi_i) = \alpha_4(\psi_i).$$

By the axiom this equals

$$\sum \alpha_{24^{-1}}(\phi)\alpha_h(\xi_i)\xi^i = \sum \alpha_{24^{-1}}(\phi)\alpha_h(\phi^i)\phi_i.$$

Finally,

$$\alpha_{24^{-1}}(\phi)\alpha_h(\phi^i) = \alpha_{24^{-1}}(\phi\phi^i) = \alpha_{23}(\phi\phi^i),$$

because $\phi\phi^i \in \mathcal{C}_{43}$, and so $\alpha_{24^{-1}}(\phi\phi^i) = \alpha_{24^{-1}}\alpha_{43}(\phi\phi^i) = \alpha_{23}(\phi\phi^i)$. Thus we have dealt with the case of Figure 32.

In fact this case is decidedly the most complicated of the set. We shall do one more, namely case (i) of Figure 35, in which two strings join and then split. We draw the diagram as in Figure 47, corresponding to the two compositions

$$\mathcal{C}_{43} \otimes \mathcal{C}_{21} \to \mathcal{C}_{4321} \cong \mathcal{C}_{1432} \to \mathcal{C}_{14} \otimes \mathcal{C}_{32} \cong \mathcal{C}_{41} \otimes \mathcal{C}_{23}$$

$$\mathcal{C}_{43} \otimes \mathcal{C}_{21} \cong \mathcal{C}_{34} \otimes \mathcal{C}_{12} \to \mathcal{C}_{3412} \cong \mathcal{C}_{4123} \to \mathcal{C}_{41} \otimes \mathcal{C}_{23}.$$

The first sequence gives us

$$\begin{aligned} \psi \otimes \psi' \mapsto \psi\psi' \mapsto \alpha_1(\psi\psi') \mapsto \sum \alpha_1(\psi\psi')\phi^i \otimes \phi_i \\ \mapsto \sum \alpha_4(\alpha_1(\psi\psi')\phi^i) \otimes \alpha_2(\phi_i), \end{aligned}$$

where ϕ_i is a basis for $\mathcal{C}_{32}$. The second sequence gives

$$\begin{aligned} \psi \otimes \psi' \mapsto \alpha_3(\psi) \otimes \alpha_1(\psi') \mapsto \alpha_3(\psi)\alpha_1(\psi') \\ \mapsto \psi\alpha_{3^{-1}1}(\psi') \mapsto \sum \psi\alpha_{3^{-1}1}(\psi')\psi^i \otimes \psi_i, \end{aligned}$$

where ψ_i is a basis for $\mathcal{C}_{23}$. But we can assume that $\psi_i = \alpha_2(\phi_i)$, and hence that $\psi^i = \alpha_2(\phi^i)$. So, noticing that $\alpha_1(\psi\psi')\phi^i \in \mathcal{C}_{14}$, and hence that

$$\alpha_4(\alpha_1(\psi\psi')\phi^i) = \alpha_{1^{-1}}(\alpha_1(\psi\psi')\phi^i),$$

what we need to prove is just that

$$\psi'\alpha_{1^{-1}}(\phi^i) = \alpha_{3^{-1}}(\alpha_1(\psi')\phi^i).$$

This is true because $\alpha_1(\psi')\phi^i \in \mathcal{C}_{13^{-1}}$, and so is fixed by $\alpha_{13^{-1}}$.

We shall leave the remaining verifications to the reader.

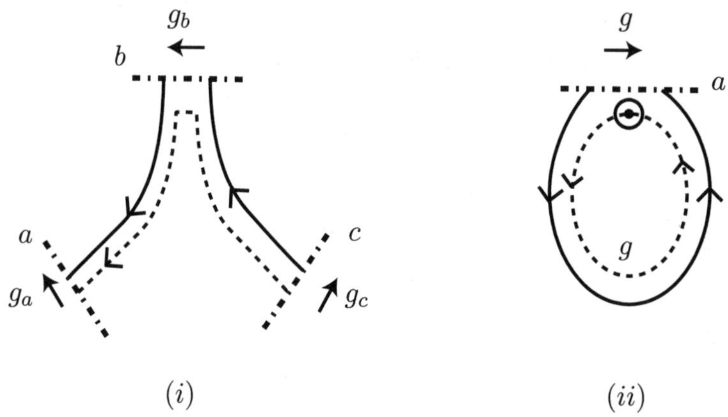

FIGURE 48

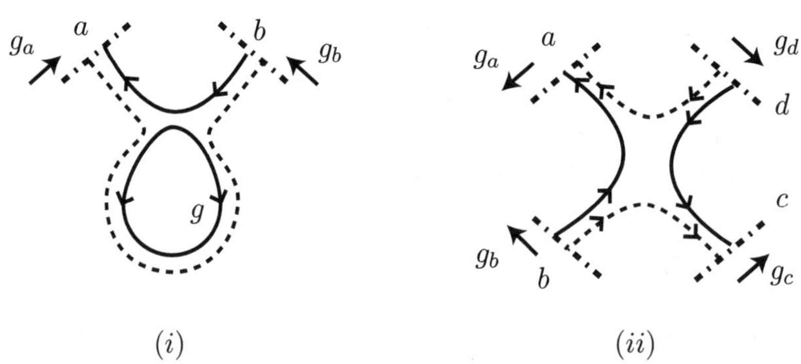

FIGURE 49

2.7.4. Equivariant open and closed theories. We now have to redo the open and closed case taking account of G-bundles on the cobordisms.

We assign the vector space $\mathcal{O}_{ab}$ to an open string from b to a equipped with a trivialization of the bundle on it. Changing the trivialization by an element $g \in G$ corresponds to the action ρ_g of g on $\mathcal{O}_{ab}$, which also corresponds to the map induced by a rectangular cobordism with holonomy g along its constrained edges.

We must consider the maps to be associated to the elementary cobordisms corresponding to the critical points of a Morse function. Up to time-reversal, two interesting things can happen at a boundary critical point: either two open strings join end-to-end or an open string becomes closed. We have the pictures of Figure 48. As before, the solid line is the contour below the critical point, and the dashed line that above it. In Figure 48 (i), g_a, g_b, g_c are the holonomies between nearby points on the respective D-branes, expressed in terms of the chosen trivializations on the strings. (They satisfy $g_c g_b = g_a$.) The map $\mathcal{O}_{ab} \otimes \mathcal{O}_{bc} \to \mathcal{O}_{ac}$ that we associate to this situation is

$$\psi \otimes \psi' \mapsto \rho_{g_a}(\psi)\rho_{g_b}(\psi').$$

The dual operation $\mathcal{O}_{ac} \to \mathcal{O}_{ab} \otimes \mathcal{O}_{bc}$ is

$$\psi \mapsto \sum \rho_{g_1}(\psi\xi^i) \otimes \rho_{g_2}(\xi_i),$$

where ξ_i and ξ^i are dual bases of $\mathcal{O}_{bc}$ and $\mathcal{O}_{cb}$.

In case (ii) of Figure 48, the open string becomes a closed string whose holonomy is g with respect to the indicated base point and the trivialization coming from the beginning of the open string. The corresponding map is ι^g, with adjoint ι_g.

There are also the two kinds of operation coming from internal critical points which involve open strings. They are illustrated in Figure 49. The map $\mathcal{C}_g \otimes \mathcal{O}_{ab} \to \mathcal{O}_{ab}$ corresponding to Figure 49 (i) is $\phi \otimes \psi \mapsto \rho_{g_a}(\iota_g(\phi)\psi))$, while the map $\mathcal{O}_{ab} \otimes \mathcal{O}_{cd} \to \mathcal{O}_{ad} \otimes \mathcal{O}_{cb}$ corresponding to Figure 49 (ii) is

$$\psi \otimes \psi' \mapsto \sum \rho_{g_a}(\psi)\psi_i \otimes \rho_{g_c}(\psi')\rho_{g_b g_a^{-1}}(\psi^i),$$

where $\{\psi_i\}$ is a basis of $\mathcal{O}_{bd}$.

We now have all the same verifications to make as in the non-equivariant case. They are very tedious, but are in 1-1 correspondence with what we have already done, and present nothing new. As an example of the modifications needed, let us point out that the very frequently used formula (2.111), which holds in any Frobenius category when $\phi' \in \mathcal{O}_{ab}$ and ϕ^i and ϕ_i are dual bases for $\mathcal{O}_{ab}$ and $\mathcal{O}_{ba}$, generalizes — with the same proof — when there is a G-action on the category to

$$\sum \phi'\phi_i \otimes \alpha_g(\phi^i) = \sum \phi_i \otimes \alpha_g(\phi^i\phi')$$

for any $g \in G$.

We shall say no more about the proof.

2.8. Notes

There is a rather large literature on 2d TFT and it is impossible to give comprehensive references. Here we just indicate some closely related works. The 2d closed sewing theorem is a very old result implicit in the earliest papers in string theory. The algebraic formulation was perhaps first stated by Friedan. Accounts have been given in [**117, 410, 401**] and in the Stanford lectures by Segal [**413**]. Sewing constraints in 2D open and closed string theory were first investigated in [**332**]. Extensions to nonorientable worldsheets were described in [**74, 7, 71, 244**].

The work in this chapter was first described at Strings 2000 [**361**] and summarized briefly in [**362**]. It was described more completely in lectures at the KITP in 2001 and at the 2002 Clay School [**360**]. In [**359**] one can find alternative (more computational) proofs and examples to those we give here, together with the original figures. Some of our results were independently obtained in the papers of C. Lazaroiu [**325, 323, 322, 324**] although the emphasis in these papers is on applications to disk instanton corrections in low energy supergravity. Regarding G-equivariant theories, there is a very large literature on D-branes and orbifolds not reflected in the above references. In the context of 2D TFT two relevant references are [**299, 341**]. Alternative discussions on the meaning of B-fields in orbifolds (in TFT) can be found in [**152, 155, 428, 427, 426**]. Our treatment of cochain-level theories and A_∞-algebras has been developed considerably further by Costello [**100**].

CHAPTER 3

Open strings and Dirichlet branes

In this chapter we will begin our treatment of Dirichlet branes from the point of view of two-dimensional conformal field theory (CFT), and take this as far as we can without calling on modern representation theory.

To warm up, in §3.1 we review the relations between quantum mechanics and various cohomology theories: de Rham, Dolbeault, their embeddings in Hodge theory, and so on. The structures we will get from CFT can for many purposes be considered to be deformations of these well understood theories.

The discussion of CFT begins in §3.2 with a brief overview of how CFT is defined physically, as a special case of two-dimensional quantum field theory. Most of the discussion is rather conceptual, but we discuss the case of free field theory in some detail, so that we can give the standard arguments for T-duality in §3.2.3.6.

We continue in §3.3 with a brief overview of superconformal field theory and its topological twistings, as discussed in MS1, Chapters 12 and 13. This includes the physics definitions of the "A-model" and "B-model," the chiral ring and the structure of the $N = 2$ superconformal algebra. Since these definitions are not based on target space geometry, physics allows making conjectures which go beyond standard mathematical frameworks such as algebraic or differential geometry. However, making contact with mathematics requires us to assume that these models also have geometric definitions, namely the nonlinear sigma model. We review this in detail in §3.4.

We then discuss boundary conditions and open strings. Again, we start with a general physical discussion in §3.5, and then restrict attention to boundary conditions in the topologically twisted A- and B- models in §3.6. We finally explain the relation to the calibrated submanifolds of Chapter 1, and develop just enough of the formalism (boundary conditions associated to holomorphic vector bundles and to the structure sheaf of a point) to support the more general discussion to come in Chapter 5.

Besides MS1, other standard references on $(2,2)$ SCFT include [**192, 119**] and Chapter 19 of [**394**]. A nice introduction to supersymmetry for mathematicians is [**153**].

3.1. Topological quantum mechanics and cohomology theories

While the standard physics definitions of local quantum field theory look rather different from the definition we used for TFT in Chapter 2, we can use that definition to motivate the physics definitions. Thus, let us again define QFT as a functor from a geometric category, namely a category of manifolds with boundary, to a category of complex vector spaces and linear maps. Now, however, we take our manifolds to carry a Riemannian metric. Thus, objects in the geometric category are closed $(d-1)$-manifolds with metric, while morphisms are d-manifolds with metric which provide cobordisms between the objects. Of course, the corresponding linear objects and morphisms will be parameterized by this metric information as well.

Let us briefly review the case of $d = 1$ before moving on to field theory. This is the well-known relation between supersymmetric quantum mechanics and Hodge theory. Now, an object is a zero-dimensional manifold; in other words a finite set of points. We denote the complex vector space corresponding to a point as $\mathcal{H}$. The simplest morphism is the interval $[0, t]$, which corresponds to a linear operator on $\mathcal{H}$ for each t. The consistency of gluing now follows from the requirement that these operators form a semigroup,

$$\exp(-tH) : \mathcal{H} \to \mathcal{H}. \tag{3.1}$$

Normally one can take $\mathcal{H}$ to be a Hilbert space and the semigroup action to be self-adjoint and bounded, and we do so. In this case, we can define the Hamiltonian H, a self-adjoint (and typically unbounded) operator H which generates the semigroup.

To get analogs of the interesting cobordisms of $d > 1$, we can allow arbitrary graphs as morphisms. The metric data consists of an assignment of positive real lengths to edges. Any such graph can be built by gluing together intervals using a cubic vertex

$$V : \mathcal{H} \times \mathcal{H} \to \mathcal{H} \tag{3.2}$$

which defines a commutative associative product. Thus we have an algebra with a semigroup action. Finally, to get a Frobenius algebra, we choose a trace $\theta : \mathcal{H} \to \mathbb{C}$, compatible with the inner product:

$$(a, b) = \theta(a^* b).$$

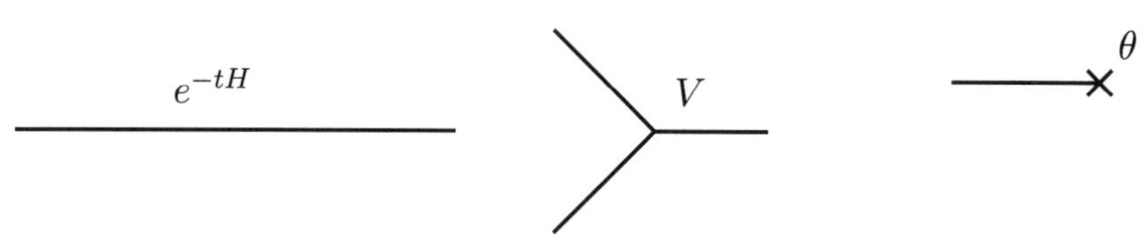

FIGURE 1. QM and graphs

The example of primary interest for us is the algebra of complex-valued functions on a manifold with metric (M, g). Thus, let $\mathcal{H}$ be $L^2(M, \mathbb{C})$, the product (3.2) be multiplication of functions, and the trace to be integration with the measure being the volume form Vol_g. Finally, we take (3.1) to be the evolution operator for the heat equation on M,

$$-\frac{\partial}{\partial t} \exp(-tH) \cdot f = \Delta \exp(-tH) \cdot f$$

with Δ the scalar Laplacian, in coordinates

$$\Delta = -\frac{1}{\sqrt{g}} \partial_i \sqrt{g} g^{ij} \partial_j.$$

Of course this is solved by $H = \Delta$. Since the Schrödinger equation is the same equation with t pure imaginary, this is often called "imaginary time quantum mechanics."[1] So far there is no restriction on M; in particular it need not be Ricci-flat. Such restrictions will appear when we consider $d \geq 2$.

The physical observables in this theory are the spectrum of the Laplacian, and matrix elements of other operators. The most important other operator is that of multiplication by a function; in other words given $f \in C^\infty(M, \mathbb{C})$ we define

$$O_f : \mathcal{H} \to \mathcal{H} : \qquad \psi \to f \cdot \psi.$$

These operators clearly form a commutative associative algebra, essentially the same as that defined by (3.2).

Another example is the derivative operator along a vector field $v \in TM$,

$$\Pi[v] = -iv^I \frac{\partial}{\partial X^I}.$$

Physically, the special case of v an isometry is called a "momentum operator." It gives rise to a "conserved charge," meaning that since

$$[H, \Pi(v)] = 0,$$

the two operators can be simultaneously diagonalized, so that momentum eigenvalues (or more simply, momenta) are independent of time.

While there are relations between this data and simpler invariants of (M, g), they are very subtle and intricate (e.g., see [**41**]). To obtain theories with more direct relations to topology, we consider supersymmetric quantum mechanics.

3.1.1. Supersymmetric quantum mechanics. By definition, this is a theory with a set of N linearly independent Hermitian operators Q_I satisfying

$$\{Q_I, Q_J\} = \delta_{IJ} H, \tag{3.3}$$

[1]Or "Euclidean time," by analogy to quantum field theory.

where H is the Hamiltonian. The case $N > 1$ is sometimes referred to as "extended supersymmetry."

The basic $N = 1$ supersymmetric quantum mechanics (SQM) is the theory of the Dirac equation. We take

$$\mathcal{H} \equiv \Gamma(M, S), \tag{3.4}$$

the sections of S a spinor bundle over M. Over each point $x \in M$, the fiber S_x admits the action of a Clifford algebra $\mathrm{Cl}(T_xM)$ with $\dim M$ real generators, called the "fermions", which span T_xM. The defining relations of $\mathrm{Cl}(T_xM)$ are

$$\psi^2 = (\psi, \psi) \cdot 1, \quad \forall \psi \in T_xM.$$

Choosing an explicit basis e_i in T_xM and denoting the corresponding linear coordinates on T_xM by ψ^i, the defining relations can be recognized as canonical anticommutation relations:

$$\{\psi^i, \psi^j\} = 2g^{ij}|_x. \tag{3.5}$$

We can take the supercharge Q to be the Dirac operator:

$$Q = \not{D} = -i\psi^k D_k,$$

where D_k is a covariant derivative. Then (3.3) is satisfied, with H the Laplacian on sections of the spinor bundle.

Another important operator, denoted $(-1)^F$, can be defined (up to an overall sign) by the property that it has eigenvalues ± 1 and anticommutes with all ψ^i:

$$(-1)^F \psi^i = -\psi^i (-1)^F. \tag{3.6}$$

If $\dim M$ is even, its ± 1 eigenspaces define the splitting of the spinor bundle S into the two irreducible spinor representations,

$$S \cong S_+ \oplus S_-, \tag{3.7}$$

which are intertwined by the Dirac operator Q. The irreducible spinor representations are referred to as "chiral" in physics, and the ± 1 eigenvalues as chirality.

3.1.1.1. $N = 2$ *supersymmetry*. A naive attempt to construct a theory with N-extended supersymmetry by taking N fermions ψ^i_I and letting them act on the tensor product of several spinor bundles fails, because $[D_i, D_j] \neq 0$, and as a result different supercharges do not anticommute. But in the special case $N = 2$ there is a modification of this construction that does work.

In the $N = 2$ case the supersymmetry algebra can be written in terms of a complex supercharge $Q = Q_1 + iQ_2$:

$$\{Q, Q^\dagger\} = 2H, \qquad \{Q, Q\} = \{Q^\dagger, Q^\dagger\} = 0. \tag{3.8}$$

The Clifford algebra is

$$\{\psi_1^i, \psi_1^j\} = \{\psi_2^i, \psi_2^j\} = 2g^{ij}, \qquad \{\psi_1^i, \psi_2^j\} = 0.$$

It is useful to define complex fermions

$$dx^i \equiv \frac{1}{2}\left(\psi_1^i + i\psi_2^i\right), \qquad \iota_j \equiv \frac{1}{2} g_{jk}\left(\psi_1^k - i\psi_2^k\right)$$

which satisfy the algebra of differential forms and their adjoints,

$$\{dx^i, \iota_j\} = \delta^i_j, \qquad \{dx^i, dx^j\} = \{\iota_i, \iota_j\} = 0. \tag{3.9}$$

Thus, in this case, we can take the vector space $\mathcal{H}$ to be the space $\Omega^*(M, \mathbb{C})$ of complex-valued differential forms on M. It has a $\mathbb{Z}$-grading by form degree, in physics called "fermion number." The grading is defined by the action of the operator

$$F = dx^i \iota_i \tag{3.10}$$

(note that this is consistent with (3.6)).

We now identify the operators Q and $Q^\dagger$ as follows:

$$Q = d \equiv dx^i \frac{\partial}{\partial x^i}$$

and

$$Q^\dagger = d^\dagger = (-1)^{mF+1} * d *$$

where $*$ is the Hodge star and $m = \dim M$. These satisfy (3.8) where H is the Laplacian acting on differential forms

$$2H = dd^\dagger + d^\dagger d. \tag{3.11}$$

In the physics literature, the operator Q is often called the "BRST operator", by analogy to other discussions (e.g., covariant quantization of the string). One sometimes sees $Q^\dagger$ referred to as the "B ghost,"

$$B \equiv Q^\dagger.$$

In any case, the key point is that

$$Q^2 = 0,$$

and thus we can use Q to define a cohomology theory;

$$H_Q^*(M, \mathbb{C}) \equiv \frac{\operatorname{Ker} Q}{\operatorname{Im} Q}.$$

In the case at hand, $Q = d$, and we see that the de Rham complex arises physically in $N = 2$ supersymmetric quantum mechanics.

One also has the standard argument that the algebra structure determines a graded product on the cohomology: if $Qa = Qb = 0$, then

$$(a + Qx)(b + Qy) = ab + Q(ay + bx + \frac{1}{2}(Qx)y + \frac{1}{2}xQy).$$

The theory obtained from $N = 2$ supersymmetric quantum mechanics by restricting to the Q-cohomology is sometimes referred to as "topological quantum mechanics" (or even the "topologically twisted theory," by analogy to the field theory case below). The terminology does **not** originate in the fact that this theory encodes de Rham cohomology, but rather because the QFT functor to the linear category does not depend on the metric we place on our graphs, and thus is "topological" in the sense of Chapter 2. This is because the Hamiltonian (3.11) is identically zero on the cohomology, so time evolution (3.1) is trivial.

While Q can be defined solely using the algebra of differential forms on M, defining B and H requires additional choices, in this case a metric. In this sense, supersymmetric quantum mechanics contains not just de Rham cohomology, but its realization using Hodge theory. As is standard there [**197**], for a compact manifold M one can choose a canonical representative for each Q-cohomology class, the harmonic form a satisfying

$$Qa = Ba = 0,$$

and this establishes an isomorphism between the zero eigenspace of H and Q-cohomology. This result is important from the physical point of view: it shows that in $N = 2$ SQM the space of ground states depends only on the topology of M, and not on its Riemannian metric. Similar results hold in higher-dimensional field theories [**464**].

3.1.2. Functional integral approach. So far we have simply cast the de Rham and Hodge theories into a more physical language. As first realized by Witten in the early 1980's, by borrowing more from physics, we get a powerful new approach to many mathematical problems.

The key ingredient in most of these developments is the functional integral definition of QFT. A heuristic introduction to this as well as many of its applications appears in MS1. Here we briefly review it for quantum mechanics. For a recent rigorous discussion, see [**89**].

The simplest problem we can treat in this way arises if we consider the morphism in the geometric category which is a closed loop. As this has no boundary, its image under the QFT functor is simply a complex number, called the "partition function" (or "one-loop partition function") and denoted $Z(t)$. Given an explicit representation of (3.1), it could be computed as

$$Z(t) = \mathrm{Tr}\ e^{-tH} \equiv \sum_{i \in I} \langle i | e^{-tH} | i \rangle \tag{3.12}$$

where the sum is taken over the index set I of an orthonormal basis.

Let us use the semigroup property to decompose the time evolution operator into $t \cdot k$ operators, each acting for time $1/k$:

$$Z(t) = \sum_{i_1,\ldots,i_{tk} \in I} \langle i_1 | e^{-H/k} | i_2 \rangle \langle i_2 | e^{-H/k} | i_3 \rangle \cdots \langle i_{tk} | e^{-H/k} | i_1 \rangle.$$

We then consider the large k limit. One expects that, in this limit, the kernels

$$\langle i|e^{-H/k}|j\rangle$$

will concentrate on the diagonal,

$$e^{-H/k} = 1 - \frac{1}{k}H + \text{negligible},$$

allowing us to write them explicitly. This can be proven, and thus we can write the partition function as a multiple integral with a known integrand. For the specific case of quantum mechanics with target M, this becomes

$$Z(t) = \int_{M^{t\cdot k}} \prod_{i=1}^{t\cdot k} dx_i \, e^{-S[x_i]} \tag{3.13}$$

where $x_i \in M$ and the integrand is written as an exponential of a function $S[x_i]$, the "action."

Finally, we can try to think of the $k \to \infty$ limit as resulting in a continuous form of this integral, an integral over continuous "paths" $S^1 \to M$. This is of course the tricky step mathematically, as the nature of the limiting measure is not entirely obvious. In general, this depends on the specific Hamiltonian and thus on the action functional S.

In the case $d = 1$, this step is well understood. In the case at hand in which H is a Laplacian on a Riemannian manifold, and in its supersymmetric generalizations, the limit leads to Wiener measure and its generalizations, supported on continuous but almost nowhere differentiable paths. Using this, one can draw rigorous conclusions from the path integral. We discuss the situation in $d \geq 2$ below.

Following these arguments, one finds that in ordinary quantum mechanics, the action is

$$S = \int_0^t ds \left| \frac{dx(s)}{ds} \right|^2,$$

where s parameterizes S^1, and the norm is defined using the metric on M. Thus, we have an integral representation of (3.12).

A similar formal argument can be made in any dimension, using a series of simplicial approximations which converges on a d-dimensional manifold with metric. This leads to functional integrals of the general form

$$Z = \int [D\phi] \; e^{-S[\phi]}, \tag{3.14}$$

in which the "field" $\phi : \Sigma \to M$ is a map from a d-dimensional manifold Σ to a target space M. The action takes the form

$$S = \int_\Sigma |d\phi|^2, \tag{3.15}$$

where ϕ^i are local coordinates on M, and the norm now depends on metrics on both Σ and M. We will be more explicit in the case of $d = 2$ below.

3.1.2.1. *Functional integral and sewing.* In principle, the functional integral offers a direct definition of the QFT functor, in which the sewing theorem of §2.1 is manifest. We now explain this very powerful point of view, keeping in mind that its mathematical utility is presently rather limited by the lack of any general and rigorous definition of the functional integral which would make the following claims precise.

The key property of the action functional S is that it is *local* – it is an integral of a function on Σ, constructed from the field $\phi(\sigma)$ and a finite number of its derivatives evaluated at a point $\sigma \in \Sigma$. The functional measure $[D\phi]$ has a similar independence property – joint expectations of products of fields at distinct points of Σ factorize.

To see why this implies the sewing theorem, let us briefly explain how the functional integral defines the QFT functor. In physics, this is called going to the canonical formulation. Recall that given a morphism Σ in the geometric category, namely a d-manifold with boundary, and an element v of the Hilbert space $\mathcal{H}$ associated to the boundary $\partial\Sigma$, the functor is supposed to give us a number. Now it is clear that an integral will (in principle) result in a number; what remains to be explained is how the choice of v is taken into account.

This is done through the choice of boundary conditions. While for a closed manifold Σ, we perform the functional integral (3.14) over "all" fields $\phi : \Sigma \to M$, for a manifold with boundary we need to specify the behavior of the fields on the boundary. We might do this by specifying a measure[2] $\mu[\phi]$ on the space of maps $\phi : \partial\Sigma \to M$. Now *a priori*, there are many possibilities, ranging from an atomic measure with support on a constant map to a single point on M, to a "free" measure with support on all maps. In any case, there is a choice here.

We then define the functional integral with boundary by defining a conditional path integral measure $[D\phi|\mu]$, which agrees with μ on the boundary:

$$\int [D\phi|\mu] \; F(\phi|_{\partial\Sigma}) = \int d\mu[\phi] \; F(\phi).$$

Then

$$U_\Sigma[\mu] = \int [D\phi|\mu] \; e^{-S[\phi]}. \tag{3.16}$$

Now, if we can identify the Hilbert space $\mathcal{H}$ with some linear space of measures μ, the functional integral with boundary (3.16) will provide the morphisms U_Σ of the QFT functor in a form in which gluing is manifest.

[2]More precisely, to make the constructions which follow, this should be the "square root" of a measure, or "half-density."

Consider a connected sum decomposition $\Sigma = \Sigma_1 \cup \Sigma_2$ along a submanifold $\Gamma \subset \partial\Sigma_i$, $i = 1, 2$. We want to show that the linear map corresponding to Σ produced by the integral (3.16) can be obtained by contracting the linear maps corresponding to the Σ_i along a component corresponding to their common boundary. Define a basis μ_i (with $i \in I$) for the space of measures $\mu[\phi]$, corresponding to the orthonormal basis $|i\rangle \in \mathcal{H}$ then

$$U_\Sigma = \sum_{i \in I} U_{\Sigma_1}(\ldots, \mu_i) U_{\Sigma_2}(\ldots, \mu_i). \tag{3.17}$$

Now, locality of the action functional S implies that the weight $\exp(-S)$ decomposes into a product of two terms, one the weight $\exp(-S)$ associated to Σ_1, the other the weight for Σ_2. Locality also implies that the functional measure on Σ will factorize as a product measure, up to its dependence on the common boundary Γ. The remaining step is to see that (3.17) holds for the measure; in other words to write the functional measure on Σ as a sum over the basis μ_i as

$$[D\phi]|_\Sigma = \sum_{i \in I} [D\phi|\mu_i]_{\Sigma_1} [D\phi|\mu_i]_{\Sigma_2}$$

If this is true for arbitrary decompositions $\Sigma = \Sigma_1 \cup \Sigma_2$, then the equality of different decompositions of the same Σ will imply the sewing theorem.

In the case of QM, making this precise is not hard. If we regard the kernel (3.1) as a bilinear functional on measures and take the $t \to 0$ limit, we get the standard inner product on $\mathcal{H} \cong L^2(M, \mathbb{C})$. However, at present the analog for $d \geq 2$ can only be done rigorously in very special cases (exactly solvable theories). Doing this in more generality would be a major advance.

3.1.3. Index theorem. An excellent example of Witten's point of view on supersymmetric quantum mechanics is provided by the following rederivation and proof of the Atiyah-Singer index theorem for the Dirac operator [**472, 43**].

It is a standard result in quantum mechanics that the evolution operator (3.1), and thus the partition function, has a small t expansion,

$$Z(t) = \frac{\mathrm{Vol}\, M}{t^{d/2}} \left(1 + \mathcal{O}(t)\right).$$

To see this from the functional integral, we observe that in the limit $t \to 0$, (3.13) heuristically reduces to an integral over M. Of course, one must first take the limit $k \to \infty$, so this is too naive, but one can show that the corrections from properly taking this limit reduce to the leading correction to a saddle point approximation for the functional integral, in other words a Gaussian integral over the tangent space to the space of loops, leading to the $1/t^{d/2}$ factor.

This is interesting, and becomes even more so if we can find a quantity which (unlike the partition function) is *independent* of t. The prototypical example is the index of the Dirac operator, which can be realized as a modified partition function in $N = 1$ supersymmetric quantum mechanics,

$$I(t) = \mathrm{Tr}\, e^{-tH}(-1)^F \tag{3.18}$$

where $(-1)^F$ is defined in (3.6). To see this, first note that the t dependence comes from the subspace of $\mathcal{H}$ on which $H > 0$. But, since $H = Q^2$, any eigenvector v of H in this subspace will be paired with another eigenvector Qv of equal eigenvalue but opposite $(-1)^F$, and will cancel out of (3.18) (see MS1 §10.2 for a detailed explanation). Finally, since $(-1)^F$ anticommutes with the Dirac operator, the two chiralities of spinor are weighted by opposite signs, leading to

$$I(t) = \dim\ker \not{D}_+ - \dim\ker \not{D}_- \equiv \mathrm{ind}\ \not{D}.$$

Because of this, $I(t)$ must equal the leading (t-independent) term in its small t asymptotics. By essentially the same arguments we outlined for $Z(t)$, this can be computed by Gaussian functional integration, leading to an expression for (3.18) as the integral of a local density constructed from the curvature of the metric and connection— in other words the Atiyah-Singer index formula.

Another path integral approach to the same result is the argument from localization. See [**47**] for an introduction, as well as MS1 Chapter 9. This starts from the supersymmetric QM path integral. There are several ways to write this, either in terms of maps from a $1|1$-dimensional superspace to M, or in components. We will follow the second approach, and introduce, along with the local coordinates X^i on M, fermionic maps ψ^i from Σ to $TM|_x$. The action is then

$$S = \int_\Sigma ||\partial X||^2 + g_{ij}\psi^j(D_t\psi)^i,$$

where D_t is a covariant derivative on TM, explicitly

$$(D_t\psi)^i = \partial_t\psi^i + (\partial_t X^k)\Gamma^i_{jk}\psi^j$$

in terms of the Levi-Civita connection Γ^i_{jk} compatible with the metric g.

Now, supersymmetry of the functional integral follows from the invariance of the action under the following infinitesimal change of variables,

$$\delta X^i = \psi^i, \qquad \delta\psi^i = \partial_t X^i.$$

Next, one can argue that integrals with such odd symmetries localize on the fixed points of the symmetry, the configurations which "preserve supersymmetry." Thus, the functional integral reduces to an integral over these supersymmetric configurations, each weighted by a Gaussian (or "one-loop") factor.

In the proof of the index theorem, the supersymmetric configurations are simply the constant loops $\partial_t X^i = 0$, and this argument leads very directly to the index formula. Even better, if the moduli spaces of supersymmetric configurations which appear are finite dimensional, then since Gaussian functional integrals are tractable, the argument generalizes fairly straightforwardly to $d > 1$. Again, we refer to MS1 for a discussion of applications of localization to mirror symmetry.

3.1.4. Dolbeault cohomology and extended supersymmetry. A different way to obtain extended $N = 2$ supersymmetry is to postulate a restricted geometry for the target manifold M. Let us start with $N = 1$ SQM and consider the ansatz

$$Q_2 = -iJ^j_k \psi^k D_j,$$

where J^j_k is a tensor field. If we ask what restrictions the relations (3.3) for $I = 1, 2$ place on J, we find two algebraic conditions

$$g^{ik} J^j_k + g^{jk} J^i_k = 0$$

and

$$J^k_i J^j_k = -\delta^j_i,$$

as well as a differential condition:

$$N^i_{jk}(J) \equiv J^l_j \partial_l J^i_k - J^l_k \partial_l J^i_j - J^i_l \partial_j J^l_k + J^i_l \partial_k J^l_j = 0.$$

The algebraic conditions say that J is an almost complex structure compatible with the metric. The differential condition is equivalent to the integrability of J. Thus we have an extra supersymmetry provided M is a Kähler manifold. The space of states of this SQM is still $\Gamma(M, S)$, but can also be written in terms of differential forms:

$$\mathcal{H} = \Omega^{0,*}(M, \mathcal{K}^{1/2}),$$

where $\mathcal{K} \equiv \Omega^{n,0}(M)$ is the canonical line bundle on M.

Another way to see what is going on is to use the decomposition of the complexified tangent bundle

$$T_{\mathbb{C}}M = TM \otimes \mathbb{C} = T^{1,0}M \oplus T^{0,1}M$$

to rewrite the Clifford algebra $\mathrm{Cl}(T_x M)$ as the complex analog of (3.9),

$$\psi^{\bar{j}} \to d\bar{z}^{\bar{j}}; \qquad g_{i\bar{j}}\psi^i \to \iota_{\bar{j}}.$$

Then in terms of differential forms the operator $Q = Q_1 + iQ_2$ becomes

$$Q = -id\bar{z}^{\bar{j}} D_{\bar{j}},$$

where $D_{\bar{j}}$ is the antiholomorphic covariant derivative on the holomorphic line bundle $\mathcal{K}^{1/2}$. From this formula and the fact that $[D_{\bar{j}}, D_{\bar{k}}] = 0$ it follows immediately that $Q^2 = 0$. The Q-cohomology will be the Dolbeault cohomology

$$H_Q(M) \cong H^{0,*}_{\bar{\partial}}(M, \mathcal{K}^{1/2}),$$

and $H = \{Q, Q^\dagger\}$ will again be a Laplacian. We can also define the $\mathbb{Z}$-grading

$$R = d\bar{z}^{\bar{\jmath}} \iota_{\bar{\jmath}}, \tag{3.19}$$

which is simply the (antiholomorphic) form degree.

One can replace the line bundle $\mathcal{K}^{1/2}$ with any holomorphic line bundle $\mathcal{L}$. For example, one can take $\mathcal{L}$ to be trivial; then Q becomes an ordinary Dolbeault operator on forms of type $(0,p)$, and H becomes an ordinary Laplacian on forms. One can go further and relax the condition that M be Kähler and only require that M be a complex manifold with a Hermitian metric. To realize the $N = 2$ supersymmetry algebra on forms of type $(0,p)$, possibly tensored with a holomorphic line bundle $\mathcal{L}$, one again takes $Q = \bar{\partial}$ and $B = Q^\dagger$. The corresponding Hamiltonian is the Dolbeault Laplacian, which even for trivial $\mathcal{L}$ is different from the standard de Rham Laplacian on forms (they agree when M is Kähler [**197, 249**]).

The above construction of $N = 2$ supersymmetry algebra generalizes: a target space with k linearly independent complex structures compatible with the metric has $k+1$ supersymmetries. The most interesting additional case is M hyperkähler, which gives rise to $N = 4$ supersymmetry. We refer to [**239**].

3.1.4.1. *Notations for extended supersymmetry.* We have discussed two different types of systems with $N = 2$ supersymmetry. The first type was based on an arbitrary Riemannian manifold M and made use of complex fermions taking values in $T_{\mathbb{C}}M$, while the second type was based on a Kähler manifold and made use of real fermions taking values in TM. To distinguish these two realizations of the $N = 2$ supersymmetry algebra, we will refer to them as $N = (1,1)$ and $N = (2,0)$ models. The terminology arises from 2d field theory, where supercharges can have either positive or negative spinor chirality: the $N = (1,1)$ model can be obtained by dimensional reduction from a 2d field theory with one real supercharge of each chirality, while the $N = (2,0)$ model is similarly obtained from a 2d field theory with two real supercharges of the same chirality. Similarly, quantum-mechanical models with $N = 4$ supersymmetry can be obtained either by reduction of 2d models with $N = (2,2)$ supersymmetry (in which case the manifold M must be Kähler and the fermions take values in $T_{\mathbb{C}}M$), or by reduction of 2d models with $N = (4,0)$ supersymmetry (in which case the manifold M must be hyperkähler and the fermions take values in TM).

3.1.4.2. *R symmetry.* The N-extended supersymmetry algebra (3.3) can be naturally combined with a linear action of the group $SO(N)$ on the supercharges Q_I, as a semidirect product. If this action can be lifted to an action on the Hilbert space $\mathcal{H}$ which preserves all the other structures of the quantum mechanics (such as the algebra (3.2)), then we speak of it as a "symmetry of the theory."

DEFINITION 3.1. The R *symmetry* group is the subgroup $G_R \subseteq SO(N)$ of symmetries of the theory.

For example, the $N = (1,1)$ model has a $U(1)$ R symmetry whose generator is the fermion number F. The $N = (2,0)$ model also has a $U(1)$ R symmetry whose generator R acts trivially on the bosonic fields ϕ^i and acts on the fermion ψ as the complex structure tensor:

$$R \cdot \psi = iJ\psi.$$

The $N = (2,2)$ model has $U(1) \times U(1)$ R-symmetry, while the $N = (4,0)$ model has $SU(2)$ R-symmetry.

3.1.5. Bundle-valued cohomology. We can generalize $N = 2$ models by choosing a vector bundle V with structure group G and connection ∇, taking

$$\mathcal{H} = \Omega^*(M, V),$$

and covariantizing all derivatives appropriately. Choosing a frame $e^I \in \Gamma(M, V)$, we can write the covariant differential D in terms of a connection one-form A taking values in $V \otimes V^*$:

$$D = d + A.$$

In general, the supersymmetry algebra has a curvature term,

$$Q^2 \sim F.$$

To get a cohomology theory (and the correct supersymmetry algebra), this must vanish. In the $N = (1,1)$ SQM, which could be defined for any Riemannian target space M and where $Q = D$, the only general way to achieve this is to take a flat connection on V. Q-cohomology in this case is isomorphic to the twisted de Rham cohomology of the flat vector bundle V. (See §6.2.2 for further discussion of this twisted cohomology.) In the $N = (2,0)$ SQM, where M is complex, the curvature term is proportional to $F^{0,2}$, and we can take a connection with $F^{0,2} = 0$. The resulting Q-cohomology is the bundle-valued Dolbeault cohomology

$$H^{0,*}_{\bar{\partial}}(M, V).$$

Again we can identify Q-cohomology with the space of zero-energy states, which corresponds to considering harmonic representatives of cohomology classes.

In physics terms, this SQM is the quantum mechanics of a particle with "color." Let us describe its functional integral definition, as the same formalism is used to define the coupling of the end of an open string to the connection on a Dirichlet brane. Consider the time evolution operator (3.1) as an element of $\mathrm{Hom}(\mathcal{H}, \mathcal{H}) \cong \mathcal{H} \otimes \mathcal{H}^*$. The corresponding functional integral will be a complex-valued function of an "initial" boundary condition $(\phi(0), e) \in (M, V_{\phi(0)})$ and a "final" boundary condition $(\phi(t), e) \in (M, V_{\phi(t)})$.

It is defined as an integral over paths similar to what we discussed in 3.1.2, but now weighted by the holonomy of ∇ along the path, a matrix-valued functional of the path. A standard way to write the holonomy is as a path-ordered exponential,

$$\int [d\phi] e^{-S_0[\phi]} \left(P e^{i \int \phi^* A} \right)^I_J$$

One may question this formula on the grounds that such a functional integral is not local. This can be remedied by introducing additional quantum mechanical degrees of freedom, Σ, and representing the holonomy as a local path integral in terms of these. To do this, we need a system whose quantization produces a finite dimensional Hilbert space, analogous to the quantization of fermions. In fact the simplest way to get this is to embed the general case in a fermionic system. Thus, we embed G in $SO(N_c)$ for some N_c, and introduce a new Grassmann algebra with N_c generators λ_I. The quantization of this system will produce fermionic operators λ_I which act in the spinor representation of $SO(N_c)$. One can then write a local action

$$S = i \int ds \; (\lambda_I \partial_s \lambda^I + \partial_s \phi^i A_i^{IJ}(\phi) \lambda_I \lambda_J).$$

The first term is a kinetic term, while the second is the infinitesimal form of the holonomy, acting in the spinor representation. One can then decompose this into G representations and restrict attention to the desired one; see [**334**] for details.

3.2. Two-dimensional QFT, CFT and TFT

All of the $d = 1$ theories we just reviewed have generalizations to $d = 2$. We again start with the categorical framework, and gradually shift over to the more standard physical approaches.

We first need to choose a simple set of objects and morphisms for the closed theory. As in the arguments leading to the functional integral, we would like to think of the morphisms as built up by concatenation of some elementary morphisms. For example, given a $(d-1)$-manifold Y, a natural morphism to consider is $Y \times [0, t]$ for $t \in \mathbb{R}^+$, with the product metric. This generates a semigroup action on a Hilbert space $\mathcal{H}_Y$ analogous to (3.1), which is again referred to as Euclidean time evolution. We can also write this as the exponential of a self-adjoint operator, the Hamiltonian H (which implicitly depends on Y).

We can now state several of the most important physical axioms of QFT. Strictly speaking these apply to "unitary QFT" as physicists do consider more general QFT's, in which the "Hilbert space" $\mathcal{H}$ carries an indefinite metric.[3] However, in this book we will only discuss unitary QFT.

Axiom 1: $\mathcal{H}_Y$ is a Hilbert space.

Axiom 2: The spectrum of H is bounded from below. H eigenvalues are usually called "energies."

Axiom 3: The eigenspace of H with the minimum energy is one-dimensional. It is called the "vacuum" or "ground state," and is often denoted $|0\rangle$.[4]

Axiom 4: For any $E \in \mathbb{R}$, the subspace of $\mathcal{H}_Y$ with $H \leq E$ is finite dimensional.

3.2.1. Two dimensions. We now take $Y \cong S^1$. Its metric is parameterized by a single real number, the circumference ℓ. We let $\mathcal{H}_\ell$ be the corresponding complex vector space.

The semigroup of time evolution morphisms is now given by the annuli carrying the product metric on $Y \times [0,t]$. Of course, these do not suffice to generate all cobordisms with metric.

We can get a sufficient generating set by considering two larger families. The first family of morphisms is topologically $Y \times [0,1]$, but with a more general metric g, such that the circumference of the "incoming" bounding S^1 is ℓ, and that of the "outgoing" bounding S^1 is ℓ'. We denote this morphism as D_g.

FIGURE 2. Two-dimensional QFT morphisms

The second family is obtained from the first by removing a small disc of radius r around a marked point pt. Denote this as O_g. We regard the new bounding S^1 as incoming.

One might think that we will need another family of morphisms to construct a general Σ, namely the disks with general metrics. We will argue shortly that these can be obtained by taking the limit of D_g in which one

[3]For example, in the standard covariant world-sheet quantization of string theory, the state space has an indefinite metric. This has two origins: first, the target space-time M has indefinite metric, and second, from gauge fixing of local symmetries.

[4]If the $(d-1)$-manifold is $\mathbb{R}^{d-1}$ with a flat metric, it is conventional to define the Hamiltonian H so that the vacuum has zero energy. In general, the vacuum energy depends nontrivially on the metric of the spatial slice. This is known as the Casimir effect.

boundary shrinks to a puncture, and placing a definite state (the vacuum) on that boundary.

Any morphism of the type O_g can be used to define a product,

$$V_g : \mathcal{H}_\ell \times \mathcal{H}_r \to \mathcal{H}_{\ell'}. \tag{3.20}$$

In a bit more detail, given states in $\mathcal{H}_\ell$ and $\mathcal{H}_r$, we use the first to determine the boundary conditions on the incoming S^1 of the annulus, and the second to determine boundary conditions on the S^1 obtained by excising the disk. The state on the outgoing boundary then lives in $\mathcal{H}_{\ell'}$.

Given a fixed $\phi \in \mathcal{H}_r$, one can also think of this as defining a linear operator,

$$O_g[\phi] : \mathcal{H}_\ell \to \mathcal{H}_{\ell'}. \tag{3.21}$$

The map from $\mathcal{H}_r$ to this space of linear operators is called the *state-operator correspondence*. Its image is the subspace of *local operators*.

Of course, there are many possible metrics g which could be used in O_g. Since one can attach morphisms $D_{g'}$ to "grow" the metric, one is tempted to use as the fundamental definition of local operator the operator obtained by taking the limit of "zero volume" g. More precisely, one can define $O_{\text{local}}[\phi]$ by taking (3.20) with $\ell = \ell'$, with O_g obtained by starting with a product metric on $Y \times [0,t]$, excising a disc centered at a fixed point pt, say $(x, t/2)$, and then taking the limits $t \to 0$ and $r \to 0$.

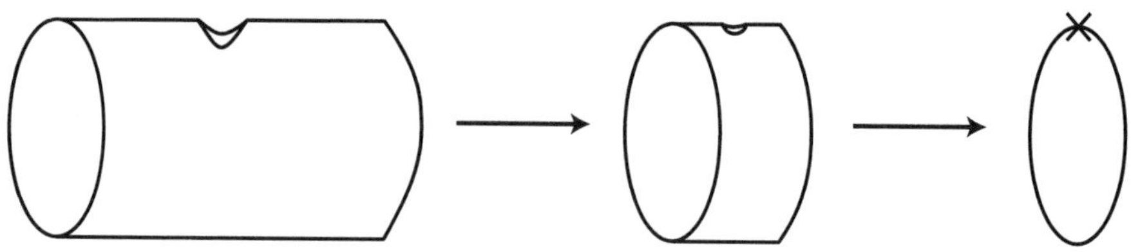

FIGURE 3. A local operator as a limit of a geometric morphism

3.2.1.1. *Comparison with QM.* Given a geometric category of this type, one might imagine that one could always extract a quantum mechanics as defined in §3.1. One would take $Y \cong S^1$ with circumference $\ell \to 0$, and consider morphisms associated to "approximately one-dimensional" surfaces Σ. For example, the time evolution morphisms are products $Y \times [0,t]$ with t held fixed as $\ell \to 0$.

Without going into details, one can show that in such a limit most of the Hilbert space $\mathcal{H}$ "is lifted to infinite energy;" in other words the time evolution operator becomes a projector on some subspace $\mathcal{H}' \subset \mathcal{H}$. The simplest conjecture for what remains is that it is an $L^2(M, \mathbb{C})$ for some finite dimensional manifold M.

If this limit really were a QM, the next step would be to identify M by using the cubic vertex (3.2) to define a commutative associative algebra. We would then define the spectrum of this algebra, and show that this is the manifold M. That this can be done is the core of the argument that quantum mechanics is not a "fundamentally new" mathematical structure, but rather a different way of thinking about known structures such as function spaces and metrics on manifolds.[5]

However, if we try to take the limit of (3.21) to obtain a cubic vertex, we will find that the resulting operators $O_{\text{local}}[\phi]$ are unbounded and their products are almost always singular. Thus, there is no natural commutative associative algebra in the problem.

This is not just a technical obstacle but is fundamental to the structure of QFT. It is the main reason that we have as yet nothing as simple as the relation $C^0(M) \leftrightarrow M$ which is the foundation for the QM discussion.

3.2.1.2. *Local operators and the OPE.* A good deal of formal development is needed to get past this difficulty. While we are not going to go deeply into this, let us at least define the operator product expansion (OPE). For some CFT's, this can be made rigorous using the formalism of vertex algebras.

If we grant the gluing axioms, then to work with local operators it is simpler not to cut up a surface Σ into D_g's and O_g's and take limits, but rather to think of each local operator as associated to a puncture on Σ. Thus a local operator is parameterized by a point $p \in \Sigma$. Furthermore, the definition of such an operator is usually made by choosing a coordinate x in the neighbourhood of p; thus we write the operator as $\phi(x)$.

We then take as the basic observables, the *correlation functions* of a product of local operators,

$$F(x_1, x_2, \ldots, x_n) \equiv \langle \phi_1(x_1)\ \phi_2(x_2)\ \cdots\ \phi_n(x_n)\rangle_\Sigma. \tag{3.22}$$

These can be defined in various ways. In the categorical language, if we can assemble Σ from the D_g's and O_g's, we have

$$\begin{aligned} F(x_1, x_2, \ldots, x_n) = \phi_1 \cdot D_{g^{1,2}}\ O_{\text{local}}[\phi_2]\ D_{g^{2,3}} \\ \cdots D_{g^{n-2,n-1}}\ O_{\text{local}}[\phi_{n-1}]\ D_{g^{n-1,n}} \cdot \phi_n. \end{aligned} \tag{3.23}$$

We will refer to such a quantity as an unnormalized correlation function. In the functional integral formalism, such functions are averages under the functional measure,

$$F(x_1, x_2, \ldots, x_n) = \int [d\phi] e^{-S[\phi]}\ \phi_1(x_1)\ \phi_2(x_2)\ \cdots\ \phi_n(x_n). \tag{3.24}$$

---

[5]Of course, this is a very ahistorical way of phrasing the relation. The mathematical development of these structures was strongly influenced by thinking about quantum mechanics.

To be precise, the term "correlation function" is more often used for *normalized* expectation values,

$$F(x_1, x_2, \ldots, x_n) = \frac{1}{Z} \int [d\phi] e^{-S[\phi]} \, \phi_1(x_1)\, \phi_2(x_2) \, \cdots \, \phi_n(x_n), \tag{3.25}$$

with Z as defined in (3.14) (so that $[d\phi]/Z$ is a probability measure). In any case, for a given set of operators, the correlation functions are well-defined on Σ^n minus the diagonal.

Given the complete set of correlation functions, one can reconstruct $\mathcal{H}$ and the other categorical data using arguments based on the Gel'fand-Naimark-Segal construction, so this is an equally valid presentation of the QFT.[6]

Let us now discuss the product of local operators. It was first postulated by Wilson that such a product can be written as an infinite series expansion in local operators, the operator product expansion or OPE. It takes the form

$$\phi_1(x_1)\, \phi_2(x_2) = \sum_i C_{12,i} |x_1 - x_2|^{\Delta_{12,i}} \phi_i(x_1) \tag{3.26}$$

with universal coefficients $C_{12,i}$ and $\Delta_{12,i}$ depending only on the choice of operators; all of the position dependence is explicit.

One can show from the QFT axioms that while in general some $\Delta_{12,i} < 0$, expressing the singular nature of the product, the number of divergent terms is finite. Furthermore, the most singular possible term is the one in which $\phi_i = \phi_0 \equiv |0\rangle$, the vacuum. Its image under the state-operator correspondence is (up to the overall coefficient) the identity operator, and thus one often writes $\phi_0 = 1$ as well. In this case, $\Delta_{12,0}$ is determined by the eigenvalues of the Hamiltonian H acting on ϕ_1, ϕ_2 (which must be equal).

In (3.24), all the operators appear symmetrically. To get a form of (3.23) with this property, we can replace the "caps" ϕ_1 and ϕ_n with the identity ϕ_0. Up to the overall coefficient, this is the same as using disk morphisms to close off each of the two ends.

Conversely, given a functional integral representation, the local functionals of the fields define a preferred set of local operators, defined as follows. Let Σ be a disk with boundary a circle of radius r and a standard metric. We then do the path integral weighted by the local functional F evaluated at the origin, to obtain an element of $\mathcal{H}_r$. This element depends linearly on F, so in a sense this is the inverse of the state-operator correspondence.

The gluing relations imply strong constraints on the OPE coefficients, usually called the "associativity of the OPE" (although since the positions enter, this is not standard associativity). In simple cases (the $c \leq 1$ CFT's we will discuss below), these can actually be solved and uniquely determine the theory.

---

[6] To do this, the Euclidean correlation functions we are discussing must satisfy the Osterwalder-Schrader axioms; see **[182]** for these axioms and the reconstruction theorem.

3.2.1.3. *The stress tensor.* Usually denoted T, this is the most important local operator and is present in all QFT's. It can be defined in terms of the functional integral as follows:

$$\langle T^{\mu\nu}(x)\cdots\rangle \equiv \frac{1}{\sqrt{g(x)}}\frac{\delta}{\delta g_{\mu\nu}(x)}\int [D\phi]e^{-S[\phi,g]}\cdots . \tag{3.27}$$

In words, the insertion of the stress tensor $T_{\mu\nu}(x)$ in a correlation function generates an infinitesimal metric deformation at the point x. On the other hand, using the definition (3.20) and the relation to the linear category, $T_{\mu\nu}(x)$ acts as a linear operator on $\mathcal{H}_d$. If we know this action, we can in principle compute the morphisms D_g for any g.

As an example, consider the Euclidean time evolution (3.1), defined by the morphism $Y \times [0,t]$ with the product metric. As we discussed, this is the exponential of a Hamiltonian H. One can show that this can be defined in terms of the stress tensor as

$$H = \int_Y T_{\mu\nu}v^\mu v^\nu,$$

where v^μ is the unit vector field in the time direction (the $[0,t]$ factor).

Thus, the Hamiltonian is determined by the stress tensor. On the other hand, since the latter is a local operator, it is far more constrained.

3.2.2. Two-dimensional conformal field theory. We can now state the definition of a CFT: it is a QFT in which conformal rescaling of the metric acts by conjugation. For the family of morphisms D_g, we can state this as

$$D_{[e^h g]} = e^{c\cdot\alpha[h]}\; L^{-1}[h|_{B_1}]\; D_g\; L[h|_{B_2}] \tag{3.28}$$

The analogous statement (conjugating the state on each boundary) is true for any Σ.

Here L is a linear operator depending only on the restriction of h to one of the boundaries of the annulus. All the dependence on the conformal rescaling away from the boundary is determined by a *universal* (independent of the particular CFT) functional $\alpha[h] \in \mathbb{R}$, which appears in an overall multiplicative factor $e^{c\cdot\alpha[h]}$. The quantity c, called "Virasoro central charge" (or, in this chapter, just central charge) will be defined more carefully shortly.

Let us first consider the special case of an overall rescaling, with h constant. As in the QM discussion, the corresponding operators $L[h]$ form a semigroup, with a self-adjoint generator H.[7] Then, since according to the axioms of QFT the spectrum of H is bounded below, we can promote this to a group action. This can be used to map any of the Hilbert spaces $\mathcal{H}_d$ to a single $\mathcal{H}_\ell$ for a fixed value of ℓ, say $\ell = 1$. We will now do this and use the simpler notation $\mathcal{H} \cong \mathcal{H}_1$, without further comment.

[7]This is related to the Virasoro generators introduced below as $H = L_0 + \bar{L}_0$.

How do we determine the $L[h]$? In outline, this is done as follows. First, we uniformize Σ – in other words, we find a complex diffeomorphism ϕ from our surface with boundary Σ to a constant curvature surface. We then consider the restriction of ϕ to each of the boundary components B_i, to get an element ϕ_i of Diff $S^1 \times \mathbb{R}^+$, where the $\mathbb{R}^+$ factor acts by an overall rescaling. We then express each ϕ_i as the exponential of an element l_i in the Lie algebra Diff S^1. Finally, we find an appropriate projective representation of this Lie algebra on $\mathcal{H}$.

Actually carrying this out, one discovers some very important subtleties, whose proper understanding leads to most of the exact results for these theories. The first of these is that the Lie algebra Diff S^1 which appears is actually a subalgebra of a direct sum of *two* commuting algebras, which act independently on "left moving" and "right moving" factors in $\mathcal{H}$.[8] Thus, we can write $\mathcal{H}$ as a direct sum of irreps of this direct sum algebra,

$$\mathcal{H} = \oplus_i \mathcal{H}_{L,i} \otimes \mathcal{H}_{R,i}. \tag{3.29}$$

Each of these two commuting algebras is a central extension of the Lie algebra Diff S^1, usually called the Virasoro algebra or Vir.

Before discussing the representation theory of this algebra, let us explain how conformal invariance implies that, in a given correlation function, the OPE (3.26) has a finite radius of convergence. Consider a correlation function containing two operators ϕ_1 and ϕ_2, with positions such that there is a circle surrounding them and no other operators. In any 2d QFT, we can define a Hilbert space $\mathcal{H}$ on this circle. But in CFT, we can rescale it to be arbitrarily small, so that a state in $\mathcal{H}$ is again a local operator.

Iterating, we find that the state produced by any finite product of local operators corresponds to a local operator. In this sense, the state-operator correspondence for CFT is an isomorphism.

3.2.2.1. *Constraints from Virasoro representation theory.* Consider the natural action of Diff S^1 on functions on an S^1 parameterized by $\theta \in [0, 2\pi)$. After complexification, we can take the following set of generators,

$$l_n = -ie^{in\theta}\frac{\partial}{\partial\theta} \qquad n \in \mathbb{Z}, \tag{3.30}$$

which satisfy the relations

$$[l_m, l_n] = (m-n)l_{m+n}. \tag{3.31}$$

The Virasoro algebra is the universal central extension of this, with generators L_n with $n \in \mathbb{Z}$, $c \in \mathbb{R}$, and the relations

$$[L_m, L_n] = (m-n)L_{m+n} + \frac{c}{12}n(n^2-1)\delta_{m+n,0}. \tag{3.32}$$

The parameter c is again the Virasoro central charge.

---

[8]One also says "holomorphic" and "antiholomorphic," or "chiral" and "antichiral."

It is easy to show that the central extension is required in any non-trivial unitary CFT. We outline the argument, not because we need it in detail, but as a warm-up for a similar argument in the $N = 2$ superconformal algebra which will be important for us. First unitarity and other QFT axioms require the Virasoro representation to act on a Hilbert space, so that $L_{-n} = L_n^\dagger$. In particular, L_0 is self-adjoint and can be diagonalized. Then, Axiom 2 above requires us to take a "highest weight representation," meaning one in which the spectrum of L_0 is bounded below. The L_0 eigenvector with the minimum eigenvalue, call this h, is by definition the "highest weight state."[9] Call this state $|h\rangle$, so that

$$L_0|h\rangle = h|h\rangle, \tag{3.33}$$

and normalize it so that $\langle h|h\rangle = 1$.

As a warm-up, consider

$$\langle h|[L_1, L_{-1}]|h\rangle = \langle h|2L_0|h\rangle = 2h.$$

Since $|h\rangle$ is a highest weight state, one can show that $L_1|h\rangle = 0$ (otherwise it would have a lower L_0 eigenvalue). Therefore, this also equals

$$\langle h|L_1 L_{-1}|h\rangle = ||L_{-1}|h\rangle||^2$$

since $L_1 = L_{-1}^\dagger$. Finally, since this is a norm in a Hilbert space, we conclude that $h \geq 0$, with equality only if $L_{-1}|h\rangle = 0$. Thus, we verify Axiom 2, and get some information on the vacuum with $h = 0$. In fact, $L_{-1}|0\rangle = 0$ can be related to the translation invariance of the vacuum, another axiom.

The argument that $c > 0$ runs the same way, by considering

$$\langle 0|[L_2, L_{-2}]|0\rangle = \frac{c}{2} \geq 0,$$

with equality only if $L_{-2}|0\rangle = 0$. One can also show that equality here implies all $L_n|0\rangle = 0$ and consequently complete triviality of the CFT. Continuing along these lines, the entire structure of a Virasoro representation is determined by the two numbers h and c.

It is useful to rephrase the above discussion in terms of local operators instead of states. We take Σ to be the infinite cylinder $\mathbb{R}\times S^1$, or equivalently the punctured complex plane $\mathbb{C}^*$ with the complex coordinate z. One can show that in a CFT the component T_{zz} of the stress tensor can be expressed in terms of the Virasoro generators:

$$T_{zz} \equiv T(z) = \sum_{n\in\mathbb{Z}} L_n z^{-n-2}.$$

The component $T_{\bar{z}\bar{z}}$ is antiholomorphic and can be similarly expressed in terms of the generators $\bar{L}_n$ of the second copy of the Virasoro algebra:

$$T_{\bar{z}\bar{z}} \equiv \bar{T}(\bar{z}) = \sum_{n\in\mathbb{Z}} \bar{L}_n \bar{z}^{-n-2}.$$

[9]The inversion of the nomenclature here is standard.

The mixed component $T_{z\bar{z}} = T_{\bar{z}z}$ is a c-number which vanishes for a flat metric. The state corresponding to $T(z)$ is $L_{-2}|0\rangle$.

3.2.2.2. *Primary fields.* The local operator corresponding to a highest weight state $|h\rangle$ is called a *conformal primary* and satisfies

$$T(z)\mathcal{O}_h(w,\bar{w}) = \frac{h}{(z-w)^2}\mathcal{O}_h(w,\bar{w}) + \frac{1}{z-w}\partial_w \mathcal{O}_h(w,\bar{w}) + \cdots,$$

where dots denote terms which stay finite in the limit $z \to w$. This is a special example of the operator product expansion and illustrates that the product of two local operators typically does not have a good limit when the insertion points approach each other. From this formula one can deduce that local operators corresponding to states $L_{-n}|h\rangle$ are of the form $\partial_w^n \mathcal{O}_h$.

The full symmetry algebra of the CFT is the direct sum of commuting "left" and "right" Virasoro algebras, which we denote L_L and L_R. A representation of these is determined by the values (h_L, c_L, h_R, c_R). In all theories we discuss, $c_L = c_R$. Furthermore, all operators which descend to the topological theory will have $h_L = h_R$.

3.2.2.3. *Sewing and factorization.* For Σ without boundary, the geometric functor assigns a number $Z[\Sigma]$, called the partition function. While $Z[\Sigma]$ depends on the metric g on Σ, its variation under a conformal transformation $g \to \alpha g$ (with $\alpha \in C(\Sigma, \mathbb{R}^+)$) is determined by (3.28), while it depends nontrivially on the complex structure of Σ. This is characterized by a finite number of parameters and thus partition functions are functions on a moduli space.

The set of all partition functions (for every genus surface) determines the CFT, as is demonstrated using "factorization." This is based on the fact that a boundary of complex structure moduli space (for a Riemann surface) is associated with a limit in which the surface degenerates to a lower genus surface with punctures.

We will consider the behavior of $Z[\Sigma]$ near a boundary in which Σ develops a long neck, ultimately breaking into two surfaces, each with a puncture. Such a degeneration can be parameterized by a "length-twist" parameter τ, whose real part is the length of the neck divided by its circumference, and whose imaginary part is an angle of rotation.

Using Axiom 4 and the sewing axioms, $Z[\Sigma]$ will have an expansion

$$Z = \sum_{h_i} C_i e^{-\tau h_{Li} - \bar{\tau} h_{Ri}}.$$

The coefficients C_i of individual terms can then be identified with a sum of correlation functions in which operators of dimensions (h_{Li}, h_{Ri}) are inserted at the punctures.

By taking multiple degeneration limits, the partition functions (in principle) determine all correlation functions. One then uses the state-operator correspondence above to reconstruct the geometric category.

3.2.2.4. *Classification by Virasoro central charge c.* This is the most important invariant of a CFT. It is analogous to the dimension d of a manifold, indeed for the sigma models we discuss below the two are proportional.

To explain this, let us consider the "density of states" $N(\lambda)$, defined as the number of eigenvalues of the Hamiltonian H which are less than a specified $\lambda \in \mathbb{R}$. Its asymptotic behavior for large λ is controlled by the dimension; for quantum mechanics this is (Weyl's theorem)

$$N(\lambda) \sim \lambda^{\dim M/2}. \tag{3.34}$$

In conformal field theory, taking $H = L_0$, this is (Cardy's formula)

$$N(h) \sim \exp\left(2\pi\sqrt{6ch}\right). \tag{3.35}$$

While this grows faster than (3.34) for any finite dimensional M, since it is subexponential, quantities like the partition function (3.12) are well-defined for any $\mathrm{Re}\, t > 0$. In fact, they have modular properties under an $SL(2,\mathbb{Z})$ group action. This arises physically because they come from functional integrals with Σ an elliptic curve.

The representation theory of the Virasoro algebra is well understood. How far does this help us with understanding CFT? One can understand the basic picture by considering the simplest picture of a highest weight representation, which is given by the "Verma module." This is simply obtained from the free action of the universal enveloping algebra on a highest weight state; using the algebra, all such elements can be written in the form

$$\prod_{i>0} L_{-i}^{N_i}|h\rangle. \tag{3.36}$$

Granting that all of these states are independent, the partition function (3.12) is simply an η function, and thus we can compare its asymptotic number of states with (3.35). If it is comparable, we can hope to decompose the full Hilbert space into a finite sum of the form (3.29), in which case the representation theory will be highly constraining. On the other hand, if (3.35) grows much faster, we cannot hope to do this.

In fact, the asymptotic number of states for (3.36) corresponds to $c = 1$ in (3.35), so the general theory divides into two cases. For $c \leq 1$, representation theory and physical arguments have led to a complete classification and complete solutions. These theories are the "minimal models" with $c < 1$, and the "free boson" and its orbifolds for $c = 1$. We refer to [**112**] for a complete discussion.

3.2.2.5. *Constructions of $c > 1$ CFT.* Here representation theory by itself does not give very strong results, and we need to appeal to other definitions. In the study of mirror symmetry, three definitions are commonly used. Our primary approach will be the nonlinear sigma model, which we will discuss in some detail in §3.2.6 and §3.3.2.

A second general approach, much used in MS1, is the linear sigma model (a special case of which is the "Landau-Ginzburg model"). This is more powerful physically than the nonlinear sigma model approach, and many of the first results along the lines we will discuss were obtained in this way. On the other hand, using it requires far more physical technique than we can fit into this book. Thus, except for a brief discussion in §3.2.6.3, we have decided not to rely on it.

Finally, one can remain within the algebraic approach, by considering tensor products and other combinations of $c \leq 1$ theories, to obtain the subclass of "rational" CFT's. Although a very special subclass, these provide independent confirmation of the physical arguments used in the other two approaches. We will discuss the case of orbifolds of flat space in some detail in §5.6, and briefly outline the construction of the "Gepner models" in §3.3.6.

3.2.3. Free bosonic CFT. The simplest example of a CFT and the first example in every textbook is the free boson. We now describe this theory, both as a concrete example and because many of these results will play an essential role in our discussion.

The "free bosonic field" is a random map $\varphi : \Sigma \to M$, where M is a Riemannian manifold with flat metric g_{ij}, which we take to be a constant real symmetric matrix. One can in this case make precise definitions of the path integral (3.14) and action (3.15). We refer to [**182**] for this, and describe it more informally here.

Let us choose a complex coordinate z on Σ, and write the action as

$$S = \frac{1}{2\pi} \int_\Sigma (\partial\varphi, \bar{\partial}\varphi)$$

where $\partial = \partial/\partial z$, $\bar{\partial} = \partial/\partial \bar{z}$, and $(,)$ is the inner product on TM.[10] Since the integrand is a $(1,1)$-form, we see that the action depends on the metric on Σ only through its complex structure, so a QFT based on it is a candidate for a CFT. Of course, by (3.28) and the previous arguments that $c > 0$, this property (independence of the conformal factor) must be violated by the quantization procedure.

3.2.3.1. *Functional integral formalism.* Let us begin with $M \cong \mathbb{R}^d$. Since the action is quadratic in φ, this functional integral is an infinite dimensional Gaussian integral. Its essential features can be understood by analogy to those of a finite dimensional Gaussian integral, say

$$Z[C] = \left(\frac{(2\pi)^N}{\det C}\right)^{1/2} = \int_{\mathbb{R}^N} d^N x \, \exp\left(-\frac{1}{2} x^t \cdot C \cdot x\right) \tag{3.37}$$

with $x \in \mathbb{R}^N$ and C a real symmetric matrix. By analogy, we would like to write $Z[\Delta] = (\det \Delta)^{-d/2}$, where $\Delta = \partial\bar{\partial}$ is the scalar Laplacian on Σ, and

[10] Equivalently, $S = (1/2)(1/2\pi) \int dxdy((\partial_x\varphi)^2 + (\partial_y\varphi)^2)$. The $1/2\pi$ prefactor is a convention, chosen to obtain a simple normalization for the Green's function (3.43).

the determinant is defined as a product over a complete set of eigenvalues

$$\det \Delta \equiv \prod_i \lambda_i; \qquad \Delta\varphi_i = \lambda_i \varphi_i.$$

Of course this is a divergent product, but a variety of suitable definitions have been developed, which realize (3.28). A good example is zeta-function regularization. As we will not need the details, we refer to [**144, 182**].

In addition to these "ultraviolet" or UV divergences, there is another problem. Since the scalar Laplacian has a zero eigenvalue (the constant function), its determinant is zero. To deal with this "zero mode," we decompose the field as

$$\varphi(z) = \varphi_0 + \tilde{\varphi}(z) \tag{3.38}$$

with $\varphi_0 \in M$ and $\int_\Sigma \tilde{\varphi} = 0$. The measure then decomposes as

$$[D\varphi] = \int_M d\varphi_0 \int [D\tilde{\varphi}], \tag{3.39}$$

and we can write the formal expression

$$Z = (\det{}'\Delta)^{-d/2} \times \int_M \sqrt{\det\ g},$$

where $\det{}'$ is a product over the non-zero eigenvalues of Δ,

$$\det{}'\Delta \equiv \prod_{\lambda_i \neq 0} \lambda_i.$$

After regularization, this expression defines a real-valued functional on metrics on Σ. It can be reduced to a function of the complex moduli of Σ by either using the ideas of §3.2.2, or simply restricting to a particular conformal class (say constant curvature metrics). This function can be written explicitly in terms of automorphic functions [**12, 11**].

Let us go on to discuss correlation functions, as defined in (3.24). These are an infinite dimensional analog of expectation values in the matrix Gaussian integral (3.37) such as (here $v_1, v_2 \in (\mathbb{R}^N)^*$)

$$\langle v_1 \cdot x\ v_2 \cdot x \rangle \equiv \frac{1}{Z[C]} \int_{\mathbb{R}^N} d^N x\ \exp\left(-\frac{1}{2} x^t \cdot C \cdot x\right)\ v_1 \cdot x\ v_2 \cdot x.$$

These are easily obtained by differentiating the following generating function:

$$Z[C,j] \equiv \int_{\mathbb{R}^N} d^N x\ \exp\left(-\frac{1}{2} x^t \cdot C \cdot x + j \cdot x\right) \tag{3.40}$$

$$= Z[C] \exp\left(\frac{1}{2} j^t \cdot C^{-1} \cdot j\right) \tag{3.41}$$

as

$$\langle v_1 \cdot x\ v_2 \cdot x \cdots\ v_n \cdot x \rangle = \frac{1}{Z[C,j]}\ v_1 \cdot \frac{\delta}{\delta j}\ v_2 \cdot \frac{\delta}{\delta j} \cdots\ v_n \cdot \frac{\delta}{\delta j} Z[C,j]. \tag{3.42}$$

Thus any correlation function can be expressed in terms of the "Green function" or formal inverse to the Laplacian,

$$\Delta_z G(z,z') = \delta^{(2)}(z-z')$$

where the Laplacian Δ_z acts on the first argument. In two dimensions, a simple calculation leads to

$$\langle \varphi(z_1)\varphi(z_2)\rangle = G(z,z') = -\log|z-z'|^2. \tag{3.43}$$

This formalism is the starting point for perturbative quantum field theory, and is developed in every textbook on the subject. As other examples, we have

$$\langle \partial\varphi(z_1)\partial\varphi(z_2)\rangle = \frac{1}{(z_1-z_2)^2}. \tag{3.44}$$

and

(3.45)

$$\langle \partial\varphi(z_1)\,\partial\varphi(z_2)\,\partial\varphi(z_3)\,\partial\varphi(z_4)\rangle$$
$$=\frac{1}{(z_1-z_2)^2(z_3-z_4)^2}+\frac{1}{(z_1-z_3)^2(z_2-z_4)^2}+\frac{1}{(z_1-z_4)^2(z_2-z_3)^2}.$$

3.2.3.2. *Vertex algebra formalism.* Using the standard physical framework of canonical quantization, one can derive the Hilbert space $\mathcal{H}$ and Hamiltonian H of §3.2. We refer to [**112**] or any other textbook on QFT for this approach. What we will do here is define the Heisenberg vertex algebra, the simplest non-trivial example, and compare it with the functional integral results we just derived.

The axioms of a vertex algebra are given in [**157, 158, 246**]. It can be shown that their general realization, consistent with (3.44), is the "$U(1)$ current algebra," sometimes called the Heisenberg algebra in the math literature.[11] We introduce generators α_n with $n \in \mathbb{Z}$, satisfying the relations

$$[\alpha_m, \alpha_n] = m\delta_{m+n,0}. \tag{3.46}$$

Note that the generator α_0 is central.

A highest weight representation of this algebra is determined by a single parameter p. We introduce a highest weight vector $|p\rangle$ defined by the conditions

$$\alpha_n|p\rangle = 0; \qquad n>0$$
$$\alpha_0|p\rangle = p|p\rangle.$$

The operators α_n with $n>0$ are called "annihilation operators," as they annihilate the highest weight vector. Those with $n<0$ are "creation operators," as they create new vectors.

We then define the representation as the linear space obtained by acting on $|p\rangle$ with the universal enveloping algebra, modulo the relations (this is the

[11] It is an infinite product of Heisenberg algebras as the term is used in physics. It is also called the algebra of canonical commutation relations or CCR in two dimensions.

"Fock space"). A basis for this space is labelled by a multi-index N which is an infinite sequence of nonnegative integers with finitely many nonzero entries,

$$\alpha_{-n}^{N_n}\alpha_{-n+1}^{N_{n-1}}\cdots\alpha_{-1}^{N_1}\,|p\rangle. \tag{3.47}$$

The integers N_i are usually called "occupation numbers."

The relation to the previous definition of free boson can be made by defining

$$\frac{\partial}{\partial z}\varphi(z) = \partial\varphi(z) = \sum_{n\in\mathbb{Z}} z^{-n-1}\alpha_n. \tag{3.48}$$

This could be integrated to obtain $\varphi(z)$, but we will not need this.

As an example, let us verify (3.44) algebraically,

$$\langle 0|\partial\varphi(z_1)\partial\varphi(z_2)|0\rangle = \sum_{n\geq 1} n(z_1)^{-n-1}(z_2)^{n-1} = \frac{1}{(z_1-z_2)^2}.$$

3.2.3.3. *Stress tensor and Virasoro algebra.* We now explain how one would use this formalism to derive (3.32) for the free boson, and determine the central charge c. Very similar but more lengthy computations would suffice to derive the $N=2$ superconformal algebra and justify the structure theorems used in §3.3.2.

Applying the definition (3.27) with S treated as a classical functional, one finds that each component of the stress tensor is quadratic in $\partial\varphi$ and $\bar\partial\varphi$. One component is purely holomorphic,

$$T_{zz}(z) = \frac{1}{2}(\partial\varphi(z))^2$$

However, from (3.44) we see that this expression does not really make sense in the quantum theory as such a product of local operators is divergent.

An obvious way to try to fix the problem is to subtract the divergence. The form of (3.44) suggests defining

$$T(z) = \frac{1}{2}\lim_{z'\to z}\left(\partial\varphi(z)\partial\varphi(z') - \frac{1}{(z-z')^2}\right). \tag{3.49}$$

A related algebraic operation is "normal ordering." It is denoted by colons, e.g.,

$$T(z) = \frac{1}{2} :\partial\varphi(z)\partial\varphi(z): .$$

It is defined by taking a product of operators, performing the mode expansion (3.48), and then reordering so that all annihilation operators appear to the left of all creation operators.

In the case at hand, the two definitions are equivalent. More generally, we can define the normal product of operators by taking the non-singular terms of the OPE in the coincidence limit,

$$(3.50)\quad \partial\varphi(z_1)\,\partial\varphi(z_2) \;\to \frac{1}{(z_1-z_2)^2} + \;:\partial\varphi(z_1)\partial\varphi(z_2): + \text{ nonsingular}$$
$$\to \text{(singular)} + :\partial\varphi(z)\partial\varphi(z):$$

While useful, the price we pay for dropping the singular terms is that this product satisfies no analog of associativity.

Using any of these prescriptions, we can compute the mode expansion of the stress-energy tensor,

$$T(z) = \sum_{n\in\mathbb{Z}} L_n z^{-n-2},$$

and verify (3.32) algebraically, determining the constant c. One finds $c = 1$ with

$$L_n = \frac{1}{2}\sum_{m\in\mathbb{Z}} :\alpha_m\alpha_{n-m}: .$$

In particular,

$$L_0 = \frac{1}{2}(\alpha_0)^2 + \sum_{m\geq 1}\alpha_m\alpha_{-m}$$

Using this, a trivial computation shows that, on the highest weight vector,

$$(3.51)\qquad L_0|p\rangle = \frac{p^2}{2}|p\rangle.$$

Furthermore, since α_0 is central and the creation operators can only raise h, we see that

$$(3.52)\qquad h = \frac{p^2}{2} + N \geq \frac{p^2}{2}; \qquad N \in \mathbb{Z}^{\geq 0}$$

for every eigenstate.

3.2.3.4. *Complete theory and gradings.* The complete theory of a free boson is defined by taking the tensor product of left and right moving copies of this algebra. A highest weight representation is then labelled by two "charges" (or "momenta") (p_L, p_R); call it

$$V_{p_L,p_R}.$$

The remaining choice in defining the theory is the particular direct sum of irreducibles,

$$(3.53)\qquad \mathcal{H} = \oplus_{p_L,p_R} N_{p_L,p_R} V_{p_L,p_R},$$

where the factors N_{p_L,p_R} are integer multiplicities.

The charges (p_L, p_R) define a bigrading on $\mathcal{H}$ which is preserved by the OPE (this is called "charge conservation"). Thus, the set of charges (p_L, p_R) appearing in (3.53), call this $Q \subset \mathbb{R}^2$, must be closed under addition.

As the simplest example, the theory with target space $M \cong \mathbb{R}$ is obtained by taking all $p_L = p_R \in \mathbb{R}$ with multiplicity one (more precisely, $\mathcal{H}$ contains a factor $L^2(M, \mathbb{C})$, whose Fourier decomposition realizes the direct sum in (3.53)).

A simple argument for charge conservation uses the relation

$$\alpha_0 = \oint dz\, \partial\varphi(z)$$

which follows from (3.48). Because of this, one can express the charge p_L of the state created by any set of local operators as an integral of the current $\partial\varphi(z)$ on a contour enclosing the set. Holomorphy and the residue formula then imply additivity. While we will not do it here, one can also define local operators (usually called "vertex operators") which intertwine representations with different (p_L, p_R).

3.2.3.5. *The compactified boson.* Let us now turn to the case $M \cong S^1$, with circumference $2\pi R$. To define the functional integral, we modify the decomposition (3.38), adding terms for the maps with nontrivial winding number. A general map to S^1 can be written uniquely as a sum of a harmonic map $f : \Sigma \to S^1$ and a map $\tilde{\varphi} : \Sigma \to \mathbb{R}$ satisfying $\int_\Sigma \tilde{\varphi} = 0$. The differential of a harmonic map to S^1, call it df, is then an element of $H^1(\Sigma, \mathbb{R})$ with quantized periods. Thus we can choose a finite dimensional integral basis f_i for the harmonic maps, and write

$$\varphi(z) = \varphi_0 + \tilde{\varphi}(z) + 2\pi \sum_i m^i f_i(z); \qquad m^i \in \mathbb{Z}^{b_1(\Sigma)}. \tag{3.54}$$

The functional measure now becomes

$$\int [D\varphi] e^{-S[\varphi]} \longrightarrow \int_M d\varphi_0 \int [D\tilde{\varphi}] \sum_{m \in \mathbb{Z}^{b_1(\Sigma)}} e^{-S[\varphi]}. \tag{3.55}$$

Evaluating the sum, one finds that the new term in the partition function is a theta function on Σ. We refer to [**112**] for the details.

The discussion of §3.2.3.2 applies without change, up to the point where we specify the sum over irreducibles, (3.53). This could be determined by comparing the partition function on a torus with modular parameter τ, computed as above, with the trace over the Hilbert space

$$Z(\tau) = \mathrm{Tr}_{\mathcal{H}} \exp\left(2\pi i \tau L_{0,L} - 2\pi i \bar{\tau} L_{0,R}\right). \tag{3.56}$$

There is also an algebraic approach to determining $Z(\tau)$, along the lines of Chapter 2. This is to express the sewing constraints in terms of the multiplicity data in (3.53), and then find the general solutions of these constraints. At present this can be carried out only for $c \leq 1$ theories.

An implicit consequence of the sewing constraints is the invariance of the partition function under the action of "large" diffeomorphisms of Σ (those which are not continuously connected to the identity). For T^2 these are classified by $SL(2, \mathbb{Z})$ and this invariance is usually called *modular invariance.*

The simplest example is that for (3.56) to be invariant under $\tau \to \tau + 1$, we need

$$h_L - h_R \in \mathbb{Z}$$

for all states. From (3.51), this requires

$$p_L^2 - p_R^2 \in 2\mathbb{Z}. \tag{3.57}$$

Another way to phrase this, which generalizes to d dimensions, is to regard the left hand side of (3.57) as giving the additive subgroup Q of charges (p_L, p_R) the structure of a signature $(1,1)$ lattice. Then (3.57) states that Q must be an even lattice. Given explicit results for (3.56), one can go on to show that for $Z(\tau)$ to be invariant under $\tau \to -1/\tau$, Q must be a self-dual lattice.

Thus, Q must be an even self-dual $(1,1)$ lattice. Any such lattice can be obtained by acting on $\mathbb{Z}^2$ by an automorphism preserving (3.57), in other words

$$(p_L, p_R) \to (p_L \cosh\alpha + p_R \sinh\alpha, p_L \sinh\alpha + p_R \cosh\alpha)$$

for $\alpha \in \mathbb{R}$. Thus the compact free bosonic theories are classified by a single parameter α, consistent with the single parameter R we introduced in defining the functional integral.[12]

3.2.3.6. *T-duality: functional formulation.* As we discussed in §1.3, the central new feature of CFT which will lead to mirror symmetry is T-duality. We now discuss this in some detail, both because of its fundamental role in mirror symmetry, and because it is a prototypical "duality" argument of the sort which has become very important in the broader study of quantum field theory and superstring theory.

Let us first discuss $M \cong S^1$. The claim is that

$$\mathrm{CFT}(S^1, R) \cong \mathrm{CFT}(S^1, 1/R), \tag{3.58}$$

in other words there is a unitary transformation from the Hilbert space of $\mathrm{CFT}(S^1, R)$ to that of $\mathrm{CFT}(S^1, 1/R)$ which takes the morphisms of the first theory into those of the second.

In particular, this requires equality of the partition functions. As in §3.2.2.3, equality of the partition functions for every Σ (*i.e.*, Riemann surfaces of every genus) implies the general claim. We now demonstrate this using the functional integral.

To make the R dependence explicit, we take $M \cong S^1$ with a fixed Euclidean metric with circumference 2π, and rewrite the action as

$$S = \frac{R^2}{4\pi} \int_\Sigma d\varphi \wedge *d\varphi. \tag{3.59}$$

---

[12]The complete classification of $c = 1$ theories is also known [**112**]. The others are obtained by orbifolding by discrete symmetries, along lines we discuss later for $(2,2)$ theories.

While the main point is to understand T-duality on S^1, the core of the argument can be understood by first considering the field φ to take values in $\mathbb{R}$. We will then come back and treat S^1.

First, the functional integral with action (3.59) can be shown to be equivalent to another functional integral, over φ and a second field $\Pi \in T^*\Sigma$, with action

$$S = \frac{\pi}{R^2} \int \Pi \wedge *\Pi + i \int \Pi \wedge d\varphi. \tag{3.60}$$

Because the new field Π appears only quadratically, its functional integral can be done exactly along the lines of (3.40) in §3.2.3.1. The result is obtained by evaluating Π at the saddle point $\delta S/\delta\Pi = 0$,[13] given by

$$\frac{2\pi i}{R^2}\Pi = *d\varphi. \tag{3.61}$$

and reproduces (3.59).

On the other hand, we can instead first integrate over φ to get a new functional integral over Π. We first integrate the $(\Pi, d\varphi)$ term by parts, obtaining

$$S = \frac{\pi}{R^2} \int \Pi \wedge *\Pi - i\varphi \, d\Pi.$$

Then, since φ appears only linearly in the action, we can do its functional integral. Formally, this is done by analogy with the finite dimensional integral

$$\int_{\mathbb{R}} d\lambda \; e^{i\lambda x} = \delta(x)$$

where $\delta(x)$ is the Dirac delta function. Thus, the result of the φ functional integral will be a measure with support on fields satisfying the constraint

$$d\Pi = 0. \tag{3.62}$$

We then solve this constraint in terms of a new scalar field, the "dual boson" $\hat{\varphi}$,

$$\Pi = \frac{1}{2\pi} d\hat{\varphi}. \tag{3.63}$$

We will justify this choice of normalization shortly.

Substituting back, one obtains

$$S = \frac{1}{4\pi R^2} \int d\hat{\varphi} \wedge *d\hat{\varphi}. \tag{3.64}$$

This is of the same form as (3.59) with the substitution $R \to 1/R$, supporting the claim (3.58).

[13] Because the action does not involve derivatives of Π, the usual one-loop integration around the saddle point in this case leads to a trivial constant factor.

To complete the argument, we must extend it to maps $\varphi : \Sigma \to S^1$. We express these using (3.54) as

$$\varphi(z) = \varphi_0 + \tilde{\varphi}(z) + 2\pi \sum_i m^i f_i(z); \qquad m^i \in \mathbb{Z}^{b_1(\Sigma)}.$$

This turns (3.60) into

$$S = \frac{\pi}{R^2} \int \Pi \wedge *\Pi + i \int \Pi \wedge d\tilde{\varphi} + i \sum_i m^i \Pi \wedge df_i. \tag{3.65}$$

Since $d\tilde{\varphi}$ is single-valued, the argument that doing the functional integral over Π reproduces (3.59) goes through unchanged.

However, repeating the arguments which led to (3.64), when we solve (3.62), we find that the general solution (3.63) is given by a map $\hat{\varphi} : \Sigma \to \mathbb{R}$, whereas what we want is a functional integral over maps to S^1. In other words, we appear to be integrating over too many maps. The constraint which reduces this to the correct integral arises because the functional measure (3.55) now includes an additional sum over m^i,

$$\sum_{m \in \mathbb{Z}^{b_1(\Sigma)}} \exp\left(i m^i \int d\hat{\varphi} \wedge df_i \right). \tag{3.66}$$

Note that the 2π in (3.54) was compensated by the $1/2\pi$ in (3.63).

Doing this sum, we get a measure with support on

$$\frac{1}{2\pi} \int d\hat{\varphi} \wedge df_i \in \mathbb{Z} \qquad \forall i.$$

This constraint is trivially satisfied by a single-valued function $\hat{\varphi}$. Using Poincaré duality for the basis f_i of harmonic one-forms, its general solution is

$$\hat{\varphi} = \hat{\varphi}_0 + \tilde{\hat{\varphi}} + 2\pi n^j f_j(z); \qquad n^j \in \mathbb{Z}^{b_1(\Sigma)}, \tag{3.67}$$

where the terms are defined as in (3.54). Thus the field $\hat{\varphi}$ is a map to S^1 in the same sense as φ.

To summarize, we found that by simple manipulations on the Gaussian functional integral (3.60) (linear changes of variable and evaluation at a saddle point), we could obtain both (3.59) and (3.64), which differ only by the substitution $R \to 1/R$. Thus these two functional integrals must be equal.

The same argument works for an arbitrary compact Σ (we will discuss the case with boundaries in §3.5.4). By considering degeneration of the complex structure of Σ and using the sewing axioms, this implies that the CFT's must be isomorphic. Thus we have proven (3.58).

3.2.3.7. *T-duality on vertex operators.* A simple relation between the vertex algebras in the T-dual theories is obtained by combining (3.61) with (3.63), to get

$$\frac{1}{R^2}d\hat{\varphi} = *d\varphi.$$

This can also be written as

$$J_L \to J_R; \quad J_R \to -J_L; \quad V_{p_L,p_R} \to V_{p_R,-p_L}. \tag{3.68}$$

Thus T-duality acts on the charge lattice Q as the non-trivial automorphism in $SO(1,1;\mathbb{Z})$. One can also compose this with the automorphism which exchanges left and right movers, to get

$$J_L \to J_L; \quad J_R \to -J_R; \quad V_{p_L,p_R} \to V_{p_L,-p_R}. \tag{3.69}$$

3.2.3.8. *Generalization to T^d.* This is fairly straightforward. We start with the action

$$S = \int d^2z \, (g_{ij} + B_{ij})\partial\varphi^i\bar{\partial}\varphi^j,$$

where in addition to the metric g_{ij} we can add an antisymmetric two-index tensor B_{ij}. We then proceed as before. The consistent theories are again those in which the charges (p_L, p_R) lie in a lattice of signature (d,d). The space of such lattices is the automorphism group $SO(d,d;\mathbb{R})$ modulo $SO(d,\mathbb{R}) \times SO(d,\mathbb{R})$. Locally, this is a homogeneous space of real dimension d^2, which matches the parameter counting of the matrix $g+B$. Finally, two theories are isomorphic if they are related by a change of basis in $SO(d,d;\mathbb{Z})$. This includes an $SL(d,\mathbb{Z})$ subgroup induced from change of basis on T^d. It also includes the T-duality transformations on any subset of the coordinates.

3.2.4. Factorization of $U(1)$ CFT.

As we discussed in §3.2.3.2, from the algebraic point of view, the defining feature of the free boson is the $U(1)$ current algebra (3.46). In fact one can prove that **all** occurences of $U(1)$ current algebra in CFT are described by the free boson. Since this is the foundation of the general classification of boundary conditions in the A and B-models, let us explain how this goes.

Consider a unitary CFT X with central charge c containing a $U(1)$ current J, i.e., an operator with OPE

$$J(z_1)\, J(z_2) \to \frac{1}{(z_1-z_2)^2} + \text{nonsingular}.$$

The basic example is the free boson with $J = \partial\varphi$, for which this is (3.50). But $U(1)$ currents are far more common. In fact, any $U(1)$ action on a CFT by automorphisms leads to at least one $U(1)$ current (and usually two, holomorphic and antiholomorphic). Later, the $N=2$ SCA will be our primary example.

Such a CFT can be "factorized" into two parts, a free boson with $c = 1$ and stress tensor

$$T_{U(1)} = \frac{1}{2} : J\ J :,$$

and a quotient CFT X' with $c' = c - 1$ and

$$T_{X'} = T_X - T_{U(1)}.$$

By factorization, one loosely means that the X Hilbert space is a tensor product of those of X' and the free boson, with independent OPE's. In particular, the operator J acts trivially on X'.

To be more precise, the decomposition (§3.2.3.4) of the free boson Hilbert space into highest weight representations V_{p_L,p_R} lifts to the Hilbert space of X as

$$\mathcal{H} = \oplus_{p_L,p_R} \mathcal{H}^{X'}_{p_L,p_R} \otimes V_{p_L,p_R}. \tag{3.70}$$

There are similar product relations for the OPE and correlation functions in the X theory, in terms of those for X' and the free boson.

This goes back to [**183**] and can be verified algebraically as follows. First, we have already checked that $T_{U(1)}$ defines a Virasoro algebra with $c = 1$, in §3.2.3.2. Second, the postulates we gave suffice to compute the OPE of $T_{X'}$, and show that it defines a Virasoro algebra with $c' = c - 1$.

Now, after a little algebra one sees that the two Virasoro algebras commute,

$$[L_m^{U(1)}, L_n^{X'}] = 0$$

Since the Virasoro action completely determines the position dependence of correlation functions, this implies that any correlation function in the X theory is a sum of products of correlation functions in the two factors, and this implies the rest.

One consequence of this is that automorphisms of the $U(1)$ factor lift to automorphisms of X. In particular, the action of T-duality on this factor lifts to X.

Another consequence is that the theory of boundary conditions, which we discuss later, factorizes in a similar way. Thus, the classification of boundary conditions for the free boson will provide part of the classification of boundary conditions for the theory X.

In the next section, we will introduce the $N = 2$ superconformal algebra, and show that it contains a $U(1)$ current algebra, so that these results apply. It will then turn out that T-duality on this subsector is the CFT definition of mirror symmetry.

3.2.5. Deformation theory. In principle, the deformation theory of local QFT and CFT is already determined by the QFT functor, or equivalent presentations of the same data (the algebra of operators, or a complete understanding of the functional integral). The simplest way to deform a

QFT is to add an operator to the action in the functional integral. Thus, we define the partition functions $Z[g]$ of a family of QFT's with parameters (usually called "coupling constants" or just couplings) g^i, as

$$Z[g] = \int [D\varphi] \; e^{-S_0[\varphi]+\delta S} \tag{3.71}$$

with

$$\delta S = \sum_i g^i \int_\Sigma d^d x \; O_i(x), \tag{3.72}$$

where the $O_i(x)$ are local operators.

The simplest way to think about (3.71) is as the generating function of correlation functions in the undeformed theory; in other words its derivatives at zero provide a shorthand description for the set of correlation functions of finitely many operators $\int O_i$. Of course, this is "formal" in the sense that we are making no requirement that this Taylor series converges. To prove that a deformation exists, this point would need to be addressed.[14]

There is another, equally important sense in which (3.71) is a "formal" expression, requiring more information for a precise definition. Products of operators at coincident points are usually divergent, as we saw in (3.49). To define correlation functions of integrated operators, we must subtract these divergences, in other words define *renormalized correlation functions*. In the physics literature, expressions such as (3.71) usually denote the generating function of renormalized correlation functions.

3.2.5.1. *Renormalization theory.* We outline only a few of the most important results of this theory. A sample computation appears in §3.2.6.

First, a simple argument involving behavior under conformal rescaling of the metric implies that a local operator which can be used to deform a $d = 2$ CFT to another CFT must have the scaling dimension $h = 1$, as defined in (3.33).[15] Thus by Axiom 4, the tangent space to the space of CFT's is always finite dimensional.

What if we try to make a deformation with $h \neq 1$? A proper discussion requires introducing the renormalization group (RG), which controls the relation between the original local operator deformations supported on points (and thus, determining the behavior on extremely short distance scales), and the behavior at general distance scales. We refer to MS1 chapter 14, and especially to Witten's lectures in [**109**], for a brief overview of the RG.

---

[14] While the physics discussion is complicated, there are cases in which it is clear that this series has finite radius of convergence, for example the deformation varying the radius R of the compactified boson of §3.2.3.5.

[15] More generally the total scaling dimension must equal the space-time dimension d, so that the operator transforms as a density. Recall that in $d = 2$ the total scaling dimension is the eigenvalue of $L_0 + \bar{L}_0 = h_L + h_R$. One also requires $L_0 - \bar{L}_0$, the generator of $SO(2)$ rotations, to act trivially, so $h_L = h_R \equiv h$.

The simplest consequence of these arguments is a trichotomy between the cases of $h < 1$, $h = 1$ and $h > 1$. Deformations with $h > 1$ do not lead to new QFT's, and are called "irrelevant." On the other hand, a deformation by an operator with $h < 1$, called a "relevant operator," spoils conformal invariance, and becomes more important at longer distances. As we explain further below, one can rephrase this violation of scale invariance, as a flow through a space of QFT's, called the RG flow induced by the operator.

One can show that Virasoro central charge always decreases under RG flow, and thus the RG flow has a limit which is a different CFT. Traditionally the start and end points of the flow are called the "UV" and "IR" CFT's, so

$$c_{UV} > c_{IR}. \tag{3.73}$$

For a generic flow, $c_{IR} = 0$ and the endpoint is trivial, but by tuning parameters (the choice of UV theory and the deformation), one can obtain a non-trivial IR CFT. A loose but useful analogy can be drawn to the process of rescaling the metric of a Riemannian manifold as $g \to \lambda g$. In the $\lambda \to 0$ limit, one generically obtains a point (this could be made precise using the Gromov-Hausdorff topology, see §7.3.6), but one can find examples (say $S^1 \times \mathbb{R}$ with the flat metric) which "collapse" to a non-trivial lower dimensional manifold.

An operator with $h = 1$, called "marginal," generates a deformation which, at least infinitesimally, produces a new CFT with the same central charge c. However, the scaling argument we cited above only works to linear order; renormalizing products of operators leads to corrections which are nonlinear in the couplings. If these are non-zero and cannot be absorbed into redefinitions, scale invariance and thus conformal invariance are broken. In this case, the deformation again leads to a non-trivial RG flow (decreasing c), and is called "marginally relevant."

An explicit description of the RG flow is given by the *beta function*, which expresses the variation of the couplings with the choice of renormalization scale. Denoting this scale as Λ, we have

$$\Lambda \frac{\partial}{\partial \Lambda} g_i = \beta^i. \tag{3.74}$$

Thus β^i is a vector field on the space of QFT's, parameterized by an explicit choice of couplings as in (3.72) A scale-invariant QFT (a CFT) is an RG fixed point and thus a zero of the beta function. If the beta function is positive, the coupling is irrelevant and does not lead to a deformation, while if it is negative it leads to an RG flow whose endpoint satisfies (3.73).

In many cases CFT deformation theory has a geometric counterpart. For example, there is a general argument [**192**] that closed string deformations in $(2,2)$ SCFT's are unobstructed. The sigma model construction we review next shows that this includes as a limiting case the mathematical result that Calabi-Yau complex structure moduli spaces are unobstructed. The analogous statement for open string deformations is false, corresponding to the mathematical statement that vector bundle deformations on a CY can be obstructed. We will discuss this relation further in §3.6.2.3 and §3.6.2.4.

3.2.6. Nonlinear sigma models. A very general source of QFT's and CFT's is the nonlinear sigma model. This is the $d \geq 2$ generalization of the theory of a quantum mechanical particle moving on a target space M which was our basic example in §3.1.

For $d = 2$, we can think of a map $X : \Sigma \to M$ as tracing out the space-time history of a string in M, and thus we will often refer to Σ as the "string world-sheet." Anticipating the role of conformal invariance, we generally think of Σ as one-complex dimensional, rather than two-real dimensional.

Now, the nonlinear sigma model is a field theory defined as a functional integral over all maps $\varphi : \Sigma \to X$, where X is the target manifold, and the action is

$$S[\varphi] = \frac{1}{8\pi} \int_\Sigma d^2z \, \hat{G}_{ij} \frac{\partial \varphi^i}{\partial z} \frac{\partial \varphi^j}{\partial \bar{z}}. \tag{3.75}$$

Here z is a complex coordinate on Σ, φ^i are local real coordinates for X, and $\hat{G}_{ij}$ is a tensor in the square of the cotangent bundle over X, which need not be symmetric; one can also write it as a sum

$$\hat{G} = g + B,$$

of a metric g_{ij} and an antisymmetric tensor (or two-form) B_{ij}.

By definition, the Euler-Lagrange equations, or "equations of motion," are the condition for the action to remain stationary (to first order) under an infinitesimal variation of the fields. Here, they are

$$0 = \frac{\delta S}{\delta \varphi^k} \propto \frac{\partial}{\partial z}\left(\hat{G}_{kj}(\varphi)\frac{\partial \varphi^j}{\partial \bar{z}}\right) + \frac{\partial}{\partial \bar{z}}\left(\hat{G}_{ik}(\varphi)\frac{\partial \varphi^i}{\partial z}\right) - g_{ij,k}\frac{\partial \varphi^i}{\partial z}\frac{\partial \varphi^j}{\partial \bar{z}}. \tag{3.76}$$

A "classical solution of the theory" is a solution of these equations. Clearly their simplest solution is φ constant on Σ. Another simple family of solutions satisfies the ansatz $\partial\varphi/\partial \operatorname{Im} z = 0$; in this case B drops out and the equations reduce to those for $d = 1$, whose solutions are geodesics in the metric g_{ij}.

More generally, but restricting attention to $dB = 0$, these are the equations defining a harmonic map,

$$\Delta_{\varphi^* g}\varphi = 0.$$

Thus, another solution is to take $\varphi(\Sigma)$ to be a smooth volume minimizing two-cycle.

3.2.6.1. *Perturbation theory.* These classical solutions are also important in the quantum field theory, as the simplest way to make sense of the functional integral (3.14) is the semiclassical method. We assume $\hbar$ is small, and write Z as a formal sum over the "saddle points" φ_i of the integral,

$$Z \sim \sum_{\delta S[\varphi_i]=0} \frac{1}{(\det{}'\Delta)^{1/2}} e^{-S[\varphi_i]/\hbar}. \tag{3.77}$$

The prefactor is obtained by expanding $S[\varphi_i + \delta\varphi_i]$ to quadratic order in a small perturbation $\delta\varphi$, and doing the Gaussian integral over $\delta\varphi$, as in §3.2.3. This computation is facilitated by working in an appropriate coordinate system; given a general Riemannian metric g_{ij} and a distinguished point φ, the natural choice is Riemann normal coordinates around φ, in which

$$g_{ij}(\varphi + \delta\varphi) = g_{ij}(\varphi) - \frac{1}{4} R_{ikjl}(\varphi)\delta\varphi^k \delta\varphi^l + \cdots,$$

where R_{ikjl} is the Riemann tensor.

Using this expansion and taking $B = 0$ for simplicity, we can rewrite the terms in the action (3.75) which are quadratic in $\delta\varphi$ as

$$\delta^2 S[\varphi] = \frac{1}{8\pi} \int_\Sigma d^2z\, g_{ij}(\varphi)\, \partial\delta\varphi^i \bar{\partial}\delta\varphi^j - \frac{1}{4} R_{ikjl}(\varphi)\, \partial\varphi^i \bar{\partial}\varphi^j \delta\varphi^k \delta\varphi^l, \tag{3.78}$$

thus defining the operator Δ appearing in (3.77).

Again, the divergence of $\det{}'\Delta$ must be dealt with by renormalization. In the original discussions, this was defined as a formal procedure in which divergent "counterterms" were added to the action (3.78), to cancel divergences arising from the functional integral. From this point of view, the main goal of renormalization theory is to show that the divergences are the sum of a finite number of local functionals of the fields, and thus can be cancelled by an action with a finite number of terms. A renormalizable action is one in which all of the necessary terms appear.

In the more modern RG language, one phrases this differently. One defines a family of "cutoff" or regulated theories depending on an additional parameter Λ (usually taken of dimensions inverse length), in which all fluctuations on length scales shorter than $1/\Lambda$ are removed from the functional measure. One then computes the variation of the cutoff functional integral with respect to Λ. Finally, one postulates a variation of the action S which compensates the previous variation, to enforce the principle that the final results should be independent of Λ.

The result of these computations can be summarized in the beta function (3.74). In broad terms, this is determined by the scale dependence of the partition function,

$$\Lambda \frac{\partial}{\partial \Lambda} \log Z = -\sum \beta^i O_i,$$

where the O_i are the local operators defined in (3.72). The precise definition (in particular, whether one restricts the sum to the finite number of operators needed to cancel divergences, or allows a larger sum) depends on the formalism being used.

The general behavior of the RG depends strongly on the dimension d of space-time. Typically the scale dimension of a given operator will have a classical (or "engineering") dimension, and quantum corrections which (in perturbation theory) are given by a Taylor series in $\hbar$. The *critical dimension* is the choice of d for which the classical scale dimension equals d, and thus the RG is controlled by quantum corrections.

The nonlinear sigma model is most interesting if a general action (3.75) is in its critical dimension. This will be the case if the required scaling dimension d is entirely made up by the scaling of the derivatives, so that we can take the classical scale dimension of the field φ to be zero. Since (3.75) has two derivatives, this forces $d = 2$. For $d > 2$ one can check that any nonlinear term in $\hat{G}$ is irrelevant, so RG flow drives the metric to be Euclidean.

In $d = 2$, we can regard the metric tensor $\hat{G}_{ij}$ as a formal generating function for an infinite series of coupling constants, or (better) regard the space of QFT's as parameterized by a space of metrics. Thus the beta function is a vector field on the space of metrics, which can be computed using the formalism of §3.2.3.1, leading to (for the special case $dB = 0$)

$$\Lambda\frac{\partial}{\partial\Lambda}g_{ij} = \beta_{ij} = -R_{ij}[g] \tag{3.79}$$

(the Ricci flow) at leading (one-loop or $\hbar^0$) order.

One can continue to expand the renormalized action, the partition function and correlation functions to higher orders in $\hbar$, to obtain the standard (weak coupling) *perturbative expansion.* In the sigma model, this turns out to be equivalent to a derivative expansion; for example the next term in (3.79) is quadratic in the Riemann tensor, with two additional derivatives compared to the Ricci tensor. Thus there is a limit (the "large volume," "large structure" or "supergravity" limit) in which the corrections go to zero.

The RG flow (3.79) was discovered by Friedan [**160**]. He went on to show that, to all orders in $\hbar$, the renormalization procedure is covariant under change of coordinates on X, so that the QFT depends only on the diffeomorphism class of the metric and B-field. In this sense, the sigma model can be regarded as a functor from Riemannian geometry to QFT.

The renormalization theory of the supersymmetric sigma model (§3.3.2) is similar and one again finds (3.79) at leading order. In the $(2,2)$ models we will discuss, the first correction appears at order $(\text{Riemann})^4$ [**198**].

Unfortunately, the perturbative expansion (for both the bosonic and supersymmetric sigma models) is an *asymptotic* series in $\hbar$, with factorial growth, and thus cannot be regarded as a satisfactory definition of the QFT, either mathematically or physically. This problem (which is entirely different from that of renormalization) is of such a long-standing and fundamental nature that it tends to be swept under the rug in general discussions such as this.

By now there are many arguments that in the case at hand, and in analogous cases such as four-dimensional Yang-Mills theory, this is essentially a technical problem, in the sense that there do exist QFT and CFT partition functions for which perturbation theory provides a good asymptotic expansion. Furthermore, there is a fairly good sense for which points in the physics arguments depend in an essential way on perturbation theory, and which points can be proven (or at least expected to hold) without relying on the perturbative expansion. It is this understanding which we implicitly rely on when we say that a physics argument is "heuristic," yet consider it to be convincing.

In simpler cases, such as the Landau-Ginzburg theory we discuss in §3.6.8, there even exist explicit (albeit extremely complicated) "constructions" of the QFT, in terms of convergent series expansions [**182**]. Perhaps someday similar (or simpler) arguments will provide a solid mathematical basis for all of the physics we discuss in this book.

3.2.6.2. *CFT and the renormalization group.* For a sigma model to be a CFT, the beta function must vanish. From (3.79), at leading order (and if $dB = 0$) the metric must be Ricci-flat. Thus in the large volume limit, any Ricci-flat target space X can be used to define a CFT.

At finite but large volume, one must study the corrections to (3.79), and check that there is no obstruction to extending the solution to higher orders. While in no case do we have exact results for the beta function, for the $N = 2$ supersymmetric sigma models to be discussed below, it is not hard to see using a superspace formalism that the beta function takes the form

$$\beta_{ij} = R_{ij} + \sum_{k\geq 1} \hbar^k \partial_i \bar{\partial}_j F_k$$

in terms of globally defined functions F_k on X. As argued in [**374**], the resulting $\beta = 0$ condition can always be solved to all orders in $\hbar$.

On the other hand, no non-trivial bosonic sigma models with $R \neq 0$, $dB = 0$ and constant dilaton are known. There are non-trivial models with $dB \neq 0$ (the Wess-Zumino-Witten models), but since our primary interest is in supersymmetric sigma models, we now leave this topic.

The cases with $\beta \neq 0$ are best understood in the language of the RG. It is natural to think of (3.74) as defining a vector field or flow on a space of coupling constants, which we think of as defining a space of QFT's. A zero $\beta = 0$ is then a fixed point of the flow, but one can also make sense of the flow, as the underlying definition of cutoff QFT gives a definition (in principle) of all observables (correlation functions) as functions of Λ.

Consider a flow which asymptotes as $\Lambda \to \infty$ to a fixed point g_{UV}, and asymptotes as $\Lambda \to 0$ to a fixed point g_{IR}. We can physically take the limit to the fixed point by taking the limit $\Lambda \to 0$ in correlation functions, equivalently by considering the limit of very long distance scales (usually called the "infrared" or IR). Similarly the $\Lambda \to \infty$ limit defines the "ultraviolet" or UV limit.

Physically, the question of whether there exists a true definition of a given functional integral, respecting the axioms of QFT, is believed to be equivalent to the question of whether the RG flow can be defined as asymptoting from a well-defined UV fixed point. This leads to an important distinction between $\beta > 0$ (IR free) and $\beta < 0$ (UV free). Only the second (UV free) theories are believed to exist as stand-alone QFT's; the first class can only exist as subtheories of CFT's or UV free QFT's with more degrees of freedom (e.g., sigma models on higher-dimensional X). In the terminology of §3.2.5, $\beta < 0$ is a marginally irrelevant deformation, while $\beta > 0$ is marginally relevant.

The sign in (3.79) is such that the UV free sigma models are those defined on manifolds of positive Ricci curvature such as the sphere, or complex manifolds with $c_1(X) > 0$. These are also called "massive models" as another physical consequence of $\beta < 0$ is the formation of a "mass gap" and a rather different physical interpretation than the one we will discuss below. We should say that many of these models can be topologically twisted and lead to a theory which, while missing an ingredient (the boundary $U(1)$ charge) which will be crucial for us below, in other ways is similar to what we will discuss. Perhaps the simplest way to bring them into our framework is to instead consider the related sigma models, with target the total space of the canonical bundle K_X or the cotangent bundle T^*X. We refer to MS1 for a direct discussion of these models.

3.2.6.3. *Linear sigma models.* In the RG approach, one can define a CFT using a weaker condition than $\beta = 0$. Instead of insisting that our QFT sits at a fixed point, one can consider a flow which asymptotes to a fixed point. While this is not literally a CFT, by taking the IR limit one can get a CFT (the IR fixed point).

A primary example, much used in the physics literature, is the linear sigma model. This is defined as a flow whose UV limit is a free boson theory, obtained by adding a potential $V(\varphi)$ to the action,

$$S_{LSM} = \int d^2z \left(|\partial\varphi|^2 + V(\varphi) \right).$$

Classically, one would expect a particle described by this action to try to minimize its energy, in other words to sit at a minimum of the function

$V(\varphi)$. Taking for example

$$V(\varphi) = f(\varphi)^2,$$

this condition would define a non-trivial target space X (the "constraint surface"), as the real hypersurface $f = 0$ in $\mathbb{R}^n$.

Of course, the discussion in QFT requires discussing renormalization. The starting point is the observation that any potential $V(\varphi)$ is a relevant operator, from the point of view of the original free boson theory. Thus, the RG will tend to eliminate the fields which parameterize fluctuations normal to the constraint surface, leading to a nonlinear sigma model with target X. This description is particularly valuable for the supersymmetric case as one can argue that various observables (in particular, those of the topologically twisted B-model) are preserved by the flow. We refer to MS1 for further detailed discussion of these models.

3.3. Supersymmetric and topological field theories

As we saw in the previous section, while the framework of CFT might be thought of as a natural variation on the theme of spectral geometry, mathematically it is still in its early stages of development. Most of the contact with mathematics has been through the related framework of two-dimensional topological field theory. As we saw in Chapter 2, this can be defined precisely. And as we will explain shortly, it is directly related to CFT with extended supersymmetry.

We then discuss our primary examples, the supersymmetric nonlinear sigma models, and their A- and B-twistings to produce topological models, in detail.

3.3.1. Twisting and topological field theory. As in our discussion of quantum mechanics, we now want to isolate a simpler, "topological" sector within our QFT, following the general approach of cohomological TFT [**467, 468, 109**].

Thus, we start with a standard QFT with metric dependence, and then look for an operator Q such that

- $Q^2 = 0$, so we can define its cohomology.
- The stress tensor T is trivial in Q-cohomology, i.e.,

$$T = \{Q, b\} \tag{3.80}$$

  for some local operator b.

Since according to (3.29) all metric dependence is determined by the stress tensor, passing to Q-cohomology eliminates all dependence on the metric. Thus we obtain a TFT satisfying the axioms of Chapter 2.

Physically, the equation $Q^2 = \frac{1}{2}\{Q, Q\} = 0$ arises naturally if Q is fermionic, and with (3.80) this suggests that Q might be obtained as a supercharge in a supersymmetric quantum field theory. Just as for SQM, the basic defining property of a supersymmetric QFT is the existence of supercharges, a set of N linearly independent Hermitian operators Q_I satisfying

$$\{Q_I, Q_J\} = \delta_{IJ} H \tag{3.81}$$

where H is the Hamiltonian. One can show that with respect to rotations Q_I must transform as a sum of several copies of spinor representations. Recall that in 2d there are two inequivalent one-dimensional spinor representations which are exchanged by orientation reversal. In a parity-invariant theory we can work with Dirac spinors which take values in the sum of the two spinor representations. If there are p such spinor supercharges, one says that the theory has p-extended supersymmetry. Equivalently, one can say that the theory has $N = (p,p)$ supersymmetry, indicating that there are p left-handed and p right-handed supercharges. We will be mostly dealing with $N = (2,2)$ QFT's and in particular with sigma models with $N = (2,2)$ supersymmetry.[16]

As usual, we assume that supersymmetry is locally realized, meaning that there are fermionic local operators $G_I^\mu(x)$ (the supercurrents), which satisfy $\partial_\mu G_I^\mu = 0$ on a flat space-time, so that

$$Q_I = \int_Y G_I^\mu v_\mu.$$

Here v^μ is a unit time-like vector, as before.

If $N \geq 2$, we can try to get a TFT by taking $Q = Q_1 + iQ_2$; then $Q^2 = 0$ follows from the supersymmetry algebra, as does the fact that H is Q-exact. However, the stress tensor T is typically not Q-exact or even Q-closed. In general one needs more structure than $N = 2$ supersymmetry to get a TFT.

What one needs is a conserved bosonic current J_μ, $\partial_\mu J^\mu = 0$, satisfying[17]

$$\{Q, b_{\mu\nu}\} = T_{\mu\nu} - \frac{1}{4}\varepsilon_{\mu\alpha}\partial^\alpha J_\nu - \frac{1}{4}\varepsilon_{\nu\alpha}\partial^\alpha J_\mu \tag{3.82}$$

for some fermionic tensor $b_{\mu\nu}$. Then one can define a new stress tensor which is conserved, symmetric, and Q-exact:

$$T'_{\mu\nu} = T_{\mu\nu} - \frac{1}{4}\varepsilon_{\mu\alpha}\partial^\alpha J_\nu - \frac{1}{4}\varepsilon_{\nu\alpha}\partial^\alpha J_\mu. \tag{3.83}$$

The Hamiltonian corresponding to T' is the same as for T, therefore the integrated version of (3.82) reads

$$\{Q, G\} = H,$$

[16] In two dimensions it is also possible to have unitary theories with unequal numbers of left-handed and right-handed supercharges, i.e., $N = (p,q)$ supersymmetry with $p \neq q$. In such theories the basic commutation relations (3.81) are modified to

$$\{Q_{I\pm}, Q_{J\pm}\} = \frac{1}{2}\delta_{IJ}(H \pm P), \qquad \{Q_{I\pm}, Q_{J\mp}\} = 0,$$

where Q_+ and Q_- are right-handed and left-handed supercharges, respectively, and P is the generator of spatial translations, i.e., the spatial momentum operator.

[17] Here $\varepsilon_{\nu\alpha}$ is the two-dimensional antisymmetric tensor with $\varepsilon_{12} = 1$.

where

$$G = \int_Y b_{\mu\nu} v^\mu v^\nu .$$

This suggests that the b-ghost $b_{\mu\nu}$ is some linear combination of supercurrents G^μ_I.

An additional constraint comes from requiring that Q be a scalar with respect to Lorentz transformations.[18] In a supersymmetric field theory, Q is a component of a spinor, but since we modified the stress tensor, the generator of Lorentz transformations is also modified:

$$M_{\mu\nu} = \int_Y \left(x_\mu T_{\rho\nu} - x_\nu T_{\rho\mu} \right) v^\rho \to M'_{\mu\nu} = \int_Y \left(x_\mu T'_{\rho\nu} - x_\nu T'_{\rho\mu} \right) v^\rho$$

It turns out that Q indeed commutes with $M'_{\mu\nu}$ if

$$[R, Q] = Q, \quad [R, G] = -G,$$

where

$$R = \int_Y J_\mu v^\mu .$$

In other words, J_μ corresponds to $U(1)$ R-symmetry, and Q has R-charge 1. This implies that Q-cohomology is graded by the R-charge. With the normalization of the R-charge adopted above, this grading is often rational rather than integral.

To summarize, we will seek $N = 2$ SQFT's with $U(1)$ R-symmetry, as candidates to construct topological field theories. In fact the combination of this requirement with conformal symmetry in $d = 2$, which defines the $N = 2$ superconformal algebra (SCA), is highly constraining. However we postpone working out its general consequences until after we introduce our basic example.

3.3.2. Supersymmetric sigma models. The most important class of supersymmetric QFTs for us are the supersymmetric sigma models. We can supersymmetrize the sigma model by following the same procedure as before. In functional integral terms, one introduces a "partner fermion" ψ^i for each of the bosons X^i (independent local coordinates on M). One then postulates an action which respects a supersymmetry of the general form

$$\delta X^i = \epsilon \psi^i; \qquad \delta \psi^i = \epsilon \partial X^i . \tag{3.84}$$

Here ϵ is a fermionic parameter. Now of course the partial derivative ∂ on Σ has two components, so we have more choices to make. The most symmetric theories allow using either component, and thus have one supersymmetry of each chirality. Thus, introducing fermions ψ^i, starting from any manifold with metric M, we can get a sigma model with $(1, 1)$ supersymmetry. Its general theory is broadly similar to that of the bosonic sigma model. In

[18]If Q is not a scalar, it will not be conserved when the theory is considered on a curved manifold.

particular, it will give rise to an SCFT if the beta function vanishes, and at leading order this condition again requires M to be Ricci-flat. However, the higher order corrections are different.

Although these models have two supersymmetries, there is no analog of the current J which as we saw in §3.3.1 was required to make a topological theory. Thus we need extended supersymmetry. Just as in supersymmetric quantum mechanics, to realize the algebra (3.8) in a supersymmetric sigma model, the target space X must be complex. In this case, the standard supersymmetric sigma model will have $(2,2)$ supersymmetry.

More generally, if X has k linearly independent complex structures, the sigma model will have $(k+1, k+1)$ world-sheet supersymmetry. Essentially the only non-trivial case is hyperkähler geometry and $(4,4)$ supersymmetry. It is also possible to have (p,q) supersymmetry with $p \neq q$ by adding non-parity invariant couplings (including special cases for the B field). The case of $(0,2)$ supersymmetry has been recently shown in [**285, 474**] to be related to the sheaves of chiral algebras on a complex manifold X constructed in [**345, 187, 189, 188**].

In the following we restrict attention to $(2,2)$ supersymmetry. The assumption of a covariantly constant complex structure is equivalent to assuming that X is a Kähler manifold. We now switch to complex coordinates denoted by ϕ^i and its complex conjugate $\phi^{\bar{\imath}}$.

To define the fermions, we introduce two spin^c structures as discussed in Chapter 2. Recall that a spin^c structure is a pair of holomorphic line bundles L_1 and L_2, with an isomorphism $L_1 \otimes L_2 \cong K \equiv T^*\Sigma$. The basic example is $L_1 \cong L_2 = K^{1/2}$, a square root of K, but we will shortly want more generality. The second spin^c structure will be $L_3 \otimes L_4 \cong T^*\Sigma$.

The fermions are defined as sections of bundles on Σ as follows:

$$
\begin{aligned}
\psi^i_+ &\in \Gamma(L_1 \otimes \phi^* T_X) \\
\psi^{\bar{\jmath}}_+ &\in \Gamma(L_2 \otimes \phi^* \bar{T}_X) \\
\psi^i_- &\in \Gamma(\bar{L}_3 \otimes \phi^* T_X) \\
\psi^{\bar{\jmath}}_- &\in \Gamma(\bar{L}_4 \otimes \phi^* \bar{T}_X),
\end{aligned}
\tag{3.85}
$$

where T_X is the holomorphic tangent bundle on X and bar denotes the corresponding antiholomorphic bundle. D represents the covariant derivative $D\psi^i_- = \partial\psi^i_- + \partial\phi^j \Gamma^i_{jk}\psi^k_-$, where ∂ is the holomorphic part of the de Rham differential as usual.[19]

[19]Note that many other conventions for writing $N = (2,2)$ theories can be found in the literature. For example, the $\pm$ notation is sometimes used for the sign of the U(1) charge.

The action is then

$$\frac{1}{4\pi}\int_\Sigma d^2z\Bigg\{g_{i\bar{\jmath}}\left(\frac{\partial\phi^i}{\partial z}\frac{\partial\phi^{\bar{\jmath}}}{\partial\bar{z}}+\frac{\partial\phi^i}{\partial\bar{z}}\frac{\partial\phi^{\bar{\jmath}}}{\partial z}\right)+B_{i\bar{\jmath}}\left(\frac{\partial\phi^i}{\partial z}\frac{\partial\phi^{\bar{\jmath}}}{\partial\bar{z}}-\frac{\partial\phi^i}{\partial\bar{z}}\frac{\partial\phi^{\bar{\jmath}}}{\partial z}\right)$$
$$+ig_{i\bar{\jmath}}\psi_-^{\bar{\jmath}}D\psi_-^i+ig_{i\bar{\jmath}}\psi_+^{\bar{\jmath}}\bar{D}\psi_+^i+R_{i\bar{\imath}j\bar{\jmath}}\psi_+^i\psi_+^{\bar{\imath}}\psi_-^j\psi_-^{\bar{\jmath}}\Bigg\}, \tag{3.86}$$

where $g_{i\bar{\jmath}}$ is the Kähler metric.

The tensor $B_{i\bar{\jmath}}$ is called the *B-field*. In general, the B-dependent part of the action is simply the integral of the pullback of the 2-form $B\in\Omega^2(X)$ to the world-sheet Σ. For our purposes, it will suffice to assume that $dB=0$. We will show in §3.4.1 that in this case, the sigma model only depends on the class of B in $H^2(X,\mathbb{R})/H^2(X,\mathbb{Z})$. Furthermore, in most of our examples $H^{2,0}(X)$ is trivial and thus B can be taken to be a real (1,1)-form; in this case we let $B=\frac{1}{2}B_{i\bar{\jmath}}d\phi^i\wedge d\phi^{\bar{\jmath}}$.

The supersymmetries are given by the following transformations:

$$\begin{aligned}
\delta\phi^i &= i\alpha_-\psi_+^i+i\alpha_+\psi_-^i\\
\delta\phi^{\bar{\imath}} &= i\tilde{\alpha}_-\psi_+^{\bar{\imath}}+i\tilde{\alpha}_+\psi_-^{\bar{\imath}}\\
\delta\psi_+^i &= -\tilde{\alpha}_-\partial\phi^i-i\alpha_+\psi_-^j\Gamma^i_{jk}\psi_+^k\\
\delta\psi_+^{\bar{\imath}} &= -\alpha_-\partial\phi^{\bar{\imath}}-i\tilde{\alpha}_+\psi_-^{\bar{\jmath}}\Gamma^{\bar{\imath}}_{\bar{\jmath}\bar{k}}\psi_+^{\bar{k}}\\
\delta\psi_-^i &= -\tilde{\alpha}_+\bar{\partial}\phi^i-i\alpha_-\psi_+^j\Gamma^i_{jk}\psi_-^k\\
\delta\psi_-^{\bar{\imath}} &= -\alpha_+\bar{\partial}\phi^{\bar{\imath}}-i\tilde{\alpha}_-\psi_+^{\bar{\jmath}}\Gamma^{\bar{\imath}}_{\bar{\jmath}\bar{k}}\psi_-^{\bar{k}}
\end{aligned}\tag{3.87}$$

with fermionic parameters α_-, α_+, $\tilde{\alpha}_-$, and $\tilde{\alpha}_+$ which are sections of L_1^{-1}, $\bar{L}_3^{-1}$, L_2^{-1} and $\bar{L}_4^{-1}$, respectively.

In the special case $L_1=K^{1/2}$, the fermionic fields can be assembled into a Dirac spinor taking values in the real tangent bundle of X. The fermionic parameters $\alpha_+,\tilde{\alpha}_+$ and $\alpha_-,\tilde{\alpha}_-$ are right-handed and left-handed spinors, respectively. This case corresponds to the ordinary (untwisted) sigma model with $N=(2,2)$ supersymmetry.

We can define conserved supercurrents G which generate the symmetry transformations (3.87) in the sense of

$$\delta W=-i\{Q(\alpha),W\},\qquad Q(\alpha)=\int d\sigma_2\;G\alpha \tag{3.88}$$

for any operator W. Since there are four supersymmetries, there are four independent supercurrents, which we denote $G_+,\tilde{G}_+,G_-,\tilde{G}_-$. It turns out that there is an additional $U(1)$ current J_μ, whose holomorphic and anti-holomorphic components will be denoted J and $\bar{J}$. At the classical level the stress tensor, the supercurrents, and the R-current are given by

$$
\begin{aligned}
T(z) &= -g_{i\bar{\jmath}}\frac{\partial\phi^i}{\partial z}\frac{\partial\phi^{\bar{\jmath}}}{\partial z} + \tfrac{1}{2}g_{i\bar{\jmath}}\psi_+^i\frac{\partial\psi_+^{\bar{\jmath}}}{\partial z} + \tfrac{1}{2}g_{i\bar{\jmath}}\psi_+^{\bar{\jmath}}\frac{\partial\psi_+^i}{\partial z} \\
G_+(z) &= \tfrac{1}{2}g_{i\bar{\jmath}}\psi_+^i\frac{\partial\phi^{\bar{\jmath}}}{\partial z} \\
\tilde{G}_+(z) &= \tfrac{1}{2}g_{i\bar{\jmath}}\psi_+^{\bar{\jmath}}\frac{\partial\phi^i}{\partial z} \\
J(z) &= \tfrac{1}{4}g_{i\bar{\jmath}}\psi_+^i\psi_+^{\bar{\jmath}}
\end{aligned}
\tag{3.89}
$$

with similar expressions for the anti-holomorphic $\bar{T}(\bar{z})$, $G_-(\bar{z})$, $\tilde{G}_-(\bar{z})$ and $\bar{J}(\bar{z})$. Note that left-handed (resp. right-handed) supercurrents are holomorphic (resp. antiholomorphic). This is a consequence of superconformal symmetry which the sigma model has at the classical level. As discussed above, it is present in the quantized theory if the beta-functions for the metric and the B-field vanish.

Upon quantization, one must check that these operators can be renormalized so as to preserve the $(2,2)$ algebra. This can be shown to all orders in the α' expansion, a task which is made relatively easy by the existence of a superfield formalism [**10**].

3.3.3. The $(2,2)$ superconformal algebra. String states live in representations of the superconformal algebra. Classically, this is the symmetry algebra generated by the transformations (3.87), where the parameters α_-, $\tilde{\alpha}_-$ are taken to be holomorphic sections, and α_+, $\tilde{\alpha}_+$ are taken to be antiholomorphic sections. These supersymmetry transformations anticommute to generate the conformal algebra and $U(1)\times U(1)$ R symmetry transformations.

Let us focus on the holomorphic modes for definiteness, and drop the subscript $-$ on the supercurrents G_- and $\tilde{G}_-$ until further notice. Before writing down the commutation relations of $N=2$ SCA, recall that in $N=2$ superconformal field theory the Hilbert space depends on a choice of $\lambda\in\mathbb{C}^*$ which labels the isomorphism class of the line bundle L_1. We also write

$$\lambda = e^{2\pi i a}. \tag{3.90}$$

The mode expansions for currents in such a sector are

$$
\begin{aligned}
T(z) &= \sum_{n\in\mathbb{Z}} L_n z^{-n-2}, \\
G(z) &= \sum_{n\in\mathbb{Z}} G_{n-a} z^{-n+a-3/2}, \\
J(z) &= \sum_{n\in\mathbb{Z}} J_n z^{-n-1}.
\end{aligned}
$$

Two special cases are particularly important; they are called the

$$\textit{Ramond}\ (\lambda = 1 \text{ or } a = 0)$$

and

$$\textit{Neveu-Schwarz}\ (\lambda = -1 \text{ or } a = 1/2)$$

sectors, usually abbreviated R and NS respectively.

In terms of these modes, the $N = 2$ SCA commutation relations are

(3.91)

$$\begin{aligned}
[L_m, L_n] &= (m-n)L_{m+n} + \frac{\hat{c}}{4}n(n^2-1)\delta_{m+n,0} \\
[J_m, J_n] &= \hat{c}m\delta_{m+n,0} \\
[L_n, J_m] &= -mJ_{m+n} \\
[L_n, G_{m-a}] &= (n/2 - m + a)G_{m+n-a} \\
[J_n, G_{m-a}] &= G_{m+n-a} \\
[J_n, \tilde{G}_{m-a}] &= -\tilde{G}_{m+n-a} \\
\{G_{n+a}, \tilde{G}_{m-a}\} &= 2L_{m+n} + (n-m+2a)J_{n+m} + \hat{c}\left((n+a)^2 - \frac{1}{4}\right)\delta_{m+n,0}.
\end{aligned}$$

As for the Virasoro algebra, this presentation can be used to define highest weight representations, which can be completely classified [**54**]. Let us state the few results which we will need.

First, the Cartan subalgebra of the SCA is generated by three elements, L_0, $\hat{c}$ and J_0. Two of these are the same as in our discussion of the Virasoro algebra, in particular the L_0 eigenvalue h is the same. One usually uses a different normalization for the central charge, $\hat{c} = c/3$, so that for the sigma model $\hat{c}$ will equal the complex dimension of X. The new Cartan element (or conserved charge) is J_0. Its eigenvalue q is the R-charge, sometimes also called "the $U(1)$ charge."

The closed string has both a left-moving and a right-moving $N = 2$ algebra, so a closed string state will have both left-moving weight and charge which we denote h_L and q_L, and right-moving weight and charge which we denote h_R and q_R. Finally, one can take a_L and a_R independent. This is used in superstring theory, where one speaks of NS-NS, NS-R, R-NS and R-R sectors, but will not be essential for us.

As we discussed in §3.2.3.4, the charges (q_L, q_R) define a bigrading on CFT. The decomposition into a sectors also leads to an $\mathbb{R}/\mathbb{Z}$ grading (or bigrading), however it is $a + 1/2$ which is additively conserved in this case. In other words, the action of an NS operator preserves the a value of a state, whereas R (and other) operators change it. This is because L_1 and L_2 are being used to generalize a spin structure, so the natural choice is $L_1 \cong L_2 \cong K^{1/2}$, not trivial. In a bit more detail, we need to find the condition under which the cobordism from $S^1 \times S^1 \to S^1$ as defined in §2.1.6 exists, respecting $L_1 \otimes L_2 \cong T\Sigma$. Clearly the simplest case in which this is always possible is $L_1 \cong L_2 \cong K^{1/2}$.

3.3.4. Topological twisting and chiral algebra. We now look at how the discussion of §3.3.1 is realized in the $N = 2$ SCA. Rewriting (3.83) in complex coordinates on Σ, we get

$$T' = T - \frac{1}{2}\partial J, \qquad \bar{T}' = \bar{T} + \frac{1}{2}\bar{\partial}\bar{J}.$$

This motivates defining twisted Virasoro generators

$$L'_m = L_m - \frac{n}{2}J_m$$

which satisfy

$$[L'_m, G_n] = -nG_{m+n}. \tag{3.92}$$

Thus, the zero mode G_0 is conserved under conformal transformations generated by the twisted stress-energy tensor, motivating the choice

$$Q_L = G_{L,0}. \tag{3.93}$$

We can argue similarly for Q_R, leading us to choose either $G_{R,0}$ or $\tilde{G}_{R,0}$. Let us postpone combining left and right to §3.3.4.4, and for now write $Q = G_0$.

Note that G_0 lives in the Ramond sector. Another way to see why this is natural is to rewrite the definition as

$$Q_L = \frac{1}{2\pi i}\oint dz\, s(z)G(z)$$

Here $s(z)$ is a section of L_1^{-1}, which is needed since $G(z)$ is a holomorphic 1-form taking values in L_1. In general, this will be globally defined only when L_1 is trivial, i.e., in the Ramond sector.

3.3.4.1. *Ramond ground states.* We now define the Hilbert space of the topological theory as the Q-cohomology, and try to follow the other definitions in §3.1.1.1. For example, the relation to Hodge theory can be seen by considering the following relation in (3.91),

$$\{G_0, \tilde{G}_0\} = 2L_0 - \frac{c}{12}. \tag{3.94}$$

By the same argument used there, one can find a canonical representative of a Q-cohomology class s, the state $|s\rangle$ satisfying $\tilde{G}_0|s\rangle = 0$. This state will have $h = c/24$. Furthermore, since $\tilde{G}_0 = G_0^\dagger$, the argument of §3.2.2.1 can be used to show that

$$h \geq \frac{c}{24}$$

in the Ramond sector. Thus the Hilbert space of the topological theory consists of the states of minimal energy, the *Ramond ground states.*

3.3.4.2. *Chiral operators.* We would next like to define the analog of cohomology algebra, but, as we discussed, Ramond operators do not form an algebra; the product of two Ramond operators is in fact an NS operator.

Since the NS operators form a closed subalgebra, they are more promising in this regard. However, we might worry about the usual singularities in operator products.

In addition, the restriction to cohomology is not as simple. Now every mode of $G(z)$ is nilpotent, so we could try to define

$$Q_{NS} = G_{-1/2}$$

and take its cohomology. However the choice of mode number 1/2 may seem arbitrary, and indeed by (3.92) it is not invariant, even under twisted conformal transformations.

Nevertheless, it is useful to do this. Thus, we define a *chiral primary* state to be a conformal primary satisfying

$$G_{-1/2}|h,q\rangle = 0.$$

Since the conformal primary condition (§3.2.2.2) implies that $\tilde{G}_{1/2}|h,q\rangle = 0$, these are in fact the canonical representatives in the sense of Hodge theory.

The analog of (3.94) is

$$\{\tilde{G}_{1/2}, G_{-1/2}\} = 2L_0 - J_0.$$

Taking expectation values in a highest weight state $|h,q\rangle$, we find

$$\langle h,q|\tilde{G}_{1/2}\ G_{-1/2}|h,q\rangle = 2h - q.$$

Since $(\tilde{G}_m)^\dagger = G_{-m}$, this must be non-negative, and we find that

$$h \geq \frac{q}{2}, \tag{3.95}$$

with equality

$$h = \frac{q}{2} \tag{3.96}$$

precisely for the chiral states. Since $h \geq 0$, we also infer that chiral states have

$$q \geq 0. \tag{3.97}$$

If one instead takes

$$\tilde{G}_{-1/2}|h,q\rangle = 0,$$

one defines the *antichiral* primaries. Now by considering

$$\{G_{1/2}, \tilde{G}_{-1/2}\} = 2L_0 + J_0,$$

one can derive that in addition

$$h \geq -\frac{q}{2}$$

with equality for antichiral primaries.

Finally, by considering

$$\{\tilde{G}_{3/2}, G_{-3/2}\} = 2L_0 - 3J_0 + 2\hat{c},$$

one can find an upper bound on the R-charge of a chiral operator,

$$q \leq \hat{c}. \tag{3.98}$$

3.3.4.3. *Chiral algebra.* The key feature of the chiral operators is that they close nicely under the operator product to form the *chiral algebra* (or, less precisely, the "chiral ring"). It follows from (3.95) that the operator product of two chiral operators takes the form

$$O_1(z_1)\, O_2(z_2) \to O_3(z_1) + \mathcal{O}((z_1 - z_2)^{\epsilon}) \tag{3.99}$$

with $\epsilon > 0$ and $h_3 = h_1 + h_2$.

Notice that the leading term in this expansion is independent of the positions of the operators. Furthermore, since the subleading terms in this expansion do not satisfy (3.96), they cannot be chiral primary operators, and therefore must be Q-exact. Thus, if we restrict attention to the cohomology of Q, the operator product algebra is *independent* of the positions of the operators, as it must be in a TFT.

One can of course make the analogous construction with the antichiral operators, now taking $Q = \tilde{G}$ instead of $Q = G$. Physical consistency conditions related to the positivity of the norm (CPT invariance) imply that the resulting "antichiral ring" is the complex conjugate of the chiral ring, and thus provides no additional information.

To summarize, we found that the Q-cohomology in an $N = 2$ SCFT has a supercommutative algebra structure. This is a particular case of

DEFINITION 3.2. The *Q-chiral algebra* associated to a QFT with a supersymmetry Q satisfying $Q^2 = 0$ is the cohomology of Q with the product law induced from the operator product expansion.

For historical reasons Q is often called the "BRST charge."

For our purposes, it will suffice to use this definition with $Q = Q_{NS}$, however this begs the question of how to make contact with the Ramond sector and the correct definition (3.93). We will discuss this further in §3.3.5.1.

3.3.4.4. *A- and B-models.* Let us now consider a $(2,2)$ SCFT with left and right sectors. An operator product will have an expansion similar to (3.99), but now with subleading terms which have both holomorphic and antiholomorphic dependence on the coordinates. To reduce to a TFT, we need to eliminate this.

This can again be done by taking cohomology, but now there is a choice involved. Suppose we eliminate holomorphic dependence by taking the cohomology with respect to a left-moving supercharge $Q_L = G_L$. We can now eliminate the antiholomorphic dependence by taking cohomology with respect to a Q_R, but now this can be either G_R or $\tilde{G}_R$. Since the choice of G_L already broke the CPT invariance, these will lead to different, in general non-isomorphic algebras.

Since the left and right $N = 2$ SCA's (graded) commute with each other, this double cohomology is the same as the cohomology with respect to the sum $Q_L + Q_R$. Thus, we could use any of the operators

$$\begin{aligned} Q_A &= G_{+,0} + \tilde{G}_{-,0} \\ \tilde{Q}_A &= \tilde{G}_{+,0} + G_{-,0} \\ Q_B &= \tilde{G}_{+,0} + \tilde{G}_{-,0} \\ \tilde{Q}_B &= G_{+,0} + G_{-,0} \end{aligned} \tag{3.100}$$

as the BRST charge. Since the sigma model is left-right symmetric, in fact the Q and $\tilde{Q}$ chiral algebras are isomorphic, but the Q_A and Q_B chiral algebras are non-isomorphic, in general. They will define the A- and B-topological models respectively.

The relation (3.96) can be adapted to this case by defining q to be the charge (eigenvalue) with respect to

$$\begin{aligned} J_A &= J_{L0} - J_{R0}; \qquad \text{A-model} \\ J_B &= -J_{L0} - J_{R0}; \qquad \text{B-model} \end{aligned} \tag{3.101}$$

With these definitions, we have

$$[J_X, Q_X] = Q_X; \qquad X \in \{A, B\}$$

in both cases.

It follows that the Hilbert space of the A-model is isomorphic to the space of superconformal primaries satisfying $h = q/2$, $\bar{h} = -\bar{q}/2$, i.e., to the space of states which are right-antichiral and left-chiral. Similarly, the Hilbert space of the B-model is isomorphic to the space of primaries which are both right and left anti-chiral.

Thus, the twisted theory has a $U(1)$ symmetry, or grading. The corresponding charge is usually called the *ghost number* and is defined so that the BRST charge has ghost number one.[20]

3.3.4.5. *Comparison with general discussion of twisting.* In this subsection we used special features of SCFT to simplify the discussion. Let us now make contact with the more general discussion of §3.3.1, in part to explain what part of this can be applied for more general 2d supersymmetric theories.

In fact, the construction we just discussed is equivalent to our previous one; however the $U(1)$ current used in (3.83) to modify the stress tensor is different for the A- and B-models. Let us provisionally denote it $\mathcal{J}_\mu$ to

[20]Since this is essentially the same fermion number and $\mathbb{Z}$ grading which appeared in the QM discussion, one might wonder where this strange intrusion of the occult into the physics terminology comes from. We will reveal this in §3.3.5.4.

distinguish it from the *a priori* different R-current J_μ. Rewriting (3.83) in complex coordinates on Σ, we get

$$T' = T - \frac{1}{2}\partial \mathcal{J}, \qquad \bar{T}' = \bar{T} + \frac{1}{2}\bar{\partial}\bar{\mathcal{J}}.$$

Repeating the discussion below §3.3.4, we see that the A-model BRST operator $Q_A = G_{+,0} + \tilde{G}_{-,0}$ is invariant under arbitrary conformal transformations. It is also easy to check that with such a choice of $\mathcal{J}$ none of the other three BRST operators is conformally invariant. To make $\tilde{Q}_A$ conformally invariant, one has to flip the sign of $\mathcal{J}_\mu$, while to make Q_B conformally invariant one has to let $\mathcal{J} = -J$, $\bar{\mathcal{J}} = \bar{J}$. In real orthogonal coordinates, this is equivalent to $\mathcal{J}_\mu = -\varepsilon_{\mu\nu}J^\nu$. This current is usually called the axial R-current, while the usual R-current in the sigma model is called the vector R-current. $N = 2$ superconformal invariance requires both the vector and axial R-currents to be conserved, so a conformally invariant $N = 2$ sigma model can be twisted into either the A- or B-model. But if we consider more general $N = 2$ sigma models with a nonvanishing beta-function (e.g., if we take the target X to be a Kähler manifold with $c_1(X) \neq 0$), then only the vector R-current is conserved on the quantum level, and only the A-model can be defined.

Note that in the absence of conformal invariance the axial R-charge is not conserved, and therefore the A-model does not have a $\mathbb{Z}$-grading if the target space is not a Calabi-Yau manifold. However it still has a $\mathbb{Z}_2$ grading, since the distinction between fermions and bosons is maintained.

3.3.4.6. *The $U(1)$ subalgebra and spectral flow.* Note that, up to an overall constant, the relation

$$[J_m, J_n] = \hat{c} m \delta_{m+n,0}$$

is the $U(1)$ current algebra (3.46). Thus, we can use the factorization discussed in §3.2.4 to describe the "$U(1)$ sector" of any $(2,2)$ SCFT in terms of a free boson.

To deal with the constant, we define $J_n = \sqrt{\hat{c}}\alpha_n$ or equivalently

$$J = \sqrt{\hat{c}}\partial\varphi. \tag{3.102}$$

This implies the relation

$$q = \sqrt{\hat{c}} \cdot p \tag{3.103}$$

between the "$U(1)$ charge" p defined in our earlier discussion, and our present definition in which the charges of $G^\pm$ are ± 1.

Thus, we can write the partition function of a $(2,2)$ SCFT X as the graded direct sum (3.70),

$$\mathcal{H}^X = \oplus_{q_l, q_R} \mathcal{H}^{X'}_{q_L, q_R} \otimes V_{q_L, q_R}.$$

A general chiral state in X will decompose into non-trivial states in both factors. Of course, the vacuum is trivial in both factors.

Are there any chiral operators which are the identity in X' but non-trivial in the $U(1)$ sector? Using $h_X = h_{X'} + h_{U(1)}$ and Axiom 3 of §3.2, all we need to check is that the dimension h_X is equal to $h_{U(1)}$. Such an operator would have to satisfy both $h = q/2$ and (3.52), which keeping in mind (3.103) becomes

$$h = \frac{q^2}{2\hat{c}} + N; \qquad N \in \mathbb{Z}^{\geq 0}$$

which is satisfied only by $q = \hat{c}$.

Let Ω_L denote such a local operator with $q_L = \hat{c}$ and $q_R = 0$. In the sigma model, we can identify this local operator with the holomorphic $(d,0)$-form, explicitly

$$\Omega_L(z) = \Omega_{i_1 \dots i_d} \psi_+^{i_1} \cdots \psi_+^{i_d}. \tag{3.104}$$

Similarly, we can define an analog Ω_R of the $(0,d)$-form. This identification is clear in the large volume (semiclassical) limit as there are no other candidate operators of this charge and dimension. That it always holds in the quantum theory is a sort of "nonrenormalization theorem" which is a nontrivial consequence of the factorization §3.2.4.

In the SCFTs used in string theory compactification, we can even find an operator analog of the relation between the $(d,0)$-form and the covariantly constant spinor,

$$\Omega_{i_1 \dots i_d} = \epsilon^\dagger \Gamma_{i_1 \dots i_d} \epsilon,$$

in the existence of an operator Υ_L satisfying

$$\Upsilon_L^2 = \Omega_L. \tag{3.105}$$

and with $U(1)$ charge $q_L = \hat{c}/2$. This local operator is called the *spectral flow operator*. It intertwines the Ramond and Neveu-Schwarz sectors and is responsible for space-time supersymmetry in superstring theory.

Another application of this construction is to solve the problem with the state-operator correspondence we mentioned above, by giving us an explicit isomorphism between R and NS sectors. The simplest way to phrase this does not use Υ, but rather the following definition of spectral flow in terms of the decomposition (3.70): it is the intertwining operator

$$S_{\delta q_L} : V_{q_L, q_R} \to V_{q_L + \delta q_L, q_R}. \tag{3.106}$$

which in the basis (3.47) simply shifts the highest weight q, keeping all occupation numbers N_n (as in (3.47)) fixed. The NS-R correspondence is then the spectral flow with $\delta q_L = \hat{c}/2$.

Repeating this operation leads to a spectral flow with $\delta q = \hat{c}$ from the NS sector to itself (resp. R to itself), corresponding to the action of Ω as in (3.105). Let us return to this momentarily.

3.3.5. Frobenius structure and twisting. Recall from Chapter 2 that a Frobenius structure on an algebra is a linear functional (trace) which, when composed with the algebra product, induces a nondegenerate inner product.

In CFT, we can use the state-operator correspondence to get a natural trace,

$$\mathrm{tr}\, O \equiv \langle 0|O\rangle$$

where $|0\rangle$ is the vacuum. This is the same as the one-point function of O on the sphere. Using the axioms of unitary CFT, one can show that composition with the OPE reproduces the Hilbert space inner product.

While this trace can be restricted to the algebra of chiral primary operators in SCFT, the resulting inner product is highly degenerate, and does not define a Frobenius structure. Rather, in TFT, one replaces $|0\rangle$ with the canonical state $\Omega = \Omega_L \otimes \Omega_R$ defined in §3.3.4.6. Given Ω (this involves a choice of normalization), we define

$$\langle O\rangle_{TFT} = \langle \Omega|O|0\rangle_{SCFT}. \tag{3.107}$$

This then defines a pairing

$$\langle O_i O_j\rangle_{TFT}. \tag{3.108}$$

Using factorization of the $U(1)$ subalgebra, this is proportional to an inner product in the X' theory (in the notation of §3.2.4), and is thus nondegenerate.

The pairing (3.108) has degree $\hat{c}$ with respect to both left-handed and right-handed R-charges and leads to the SCFT generalization of the Kodaira-Serre duality relation

$$H^{0,p}_{\bar{\partial}}(\Omega^q) \cong H^{0,d-p}_{\bar{\partial}}(\Omega^{d-q})^*. \tag{3.109}$$

Clearly this is related to the spectral flow $\delta q = \hat{c}$ we mentioned at the end of the last subsection. Note that because of the sign flip of the $U(1)$ charge in (3.109), this is not a symmetry of the SCFT. We can obtain a symmetry either by combining it with $J \to -J$, or by iterating it again, leading to the result that an overall spectral flow with $\delta q = 2\hat{c}$ is a symmetry of the SCFT. We will use this in §5.3.4.

3.3.5.1. *Twisting in the functional integral.* Although (3.107) may seem a bit *ad hoc*, there is a more physical derivation from anomaly considerations, which also shows how spectral flow appears naturally within the twisted theory. The point is to carefully consider the effect of the twisting (3.83) on the functional integral. Recall that in operator language, this is a modification of the stress tensor by terms depending on the R-current $\mathcal{J}_\mu$. Since the stress tensor determines the coupling of the theory to the world-sheet curvature, the twisted theory on a sphere is different from the untwisted one. In the functional integral approach this is reflected in the fact that fermions take values in bundles twisted by suitable spin^c structures, as described in §3.3.2.

In the twisted theory, the R-current has a "gravitational anomaly", i.e., the R-charge is not conserved on a curved world-sheet. For example, one can show that the one-point function on the sphere is nonvanishing precisely if the total R-charge of the operator is $2\hat{c}$, in which case it is proportional to (3.107). More generally, a region $R \subset \Sigma$ contains a "background R-charge" $\delta q(R)$ proportional to the integrated curvature two-form $\mathcal{R}$ of the metric, as

$$\delta q(R) = -\frac{\hat{c}}{2\pi} \int_R \mathcal{R}. \tag{3.110}$$

For example, a correlation function on a closed surface with Euler characteristic χ is non-zero only if the total $U(1)$ charge of the operators is $\chi\hat{c}$. Furthermore, taking into account the effects of this background R-charge on the decomposition of Σ into geometric morphisms, one finds that various spectral flows are induced on the intermediate Hilbert spaces, which intertwine the NS and R sector pictures we gave into a single TFT.

3.3.5.2. *$N = 2$ supersymmetric deformations.* These correspond to 2-form local operators of charge zero whose integral over Σ is Q-closed. In fact, all such 2-forms can be related to the scalar operators we have been discussing as follows.

Suppose W_a is a local operator of charge p. The operator dW_a (where d is the *world-sheet* de Rham operator) will have trivial correlation functions with other operators since the location of the operator insertions is unimportant in a TFT. It follows that it must be Q-exact, i.e.,

$$dW_a = \{Q, W_A^{(1)}\}, \tag{3.111}$$

for some operator 1-form $W_A^{(1)}$ with ghost number $p-1$. We may repeat this process again by setting

$$dW_a^{(1)} = \{Q, W_A^{(2)}\}, \tag{3.112}$$

for some operator $W_A^{(2)}$ with ghost number $p-2$. But $W_A^{(2)}$ is a 2-form and so we can naturally integrate it over Σ. We therefore find a deformation of the theory given by

$$S \mapsto S + t \int_\Sigma W_A^{(2)} \, d^2 z. \tag{3.113}$$

In order to preserve the grading, we must take $p = 2$.

3.3.5.3. *Correlation functions and prepotential.* As in §3.2.5, we would now like to define a generating function for correlation functions in TFT. For closed string correlation functions this is usually called the *prepotential.* The standard definition combines the previous ingredients as follows. Naively, we would like to write

$$\mathcal{F}[t] = \langle \exp\left(\sum_i t_i \int O^i \right) \rangle, \tag{3.114}$$

where the integrated operators are the $W_i^{(2)}$ defined in (3.113). However, since the $W_i^{(2)}$ have zero R-charge, this will be trivial.

To get a nontrivial expectation value, we must instead use an operator with total R-charge $q_L + q_R = 2\hat{c}$. One can do this by the seemingly *ad hoc* prescription of taking $n = \hat{c}$ of the operators in the exponential in (3.114) to be the original W_i, as this will provide a total ghost number $2\hat{c}$.

Of course, one must then ask how the result depends on the choice of $n = \hat{c}$ distinguished operators. The somewhat nontrivial claim, which is proven (for example) in [**119**, §8], is that it is *independent* of this choice. This can be seen using properties of the $N = 2$ SCA, or more geometrically by developing the relation to twisting described in §3.3.5.1. Thus, one can regard (3.114) as the formal definition of the prepotential, leaving implicit the step of distributing the ghost number anomaly $2\hat{c}$ among the various operators O_i to get a nonzero result.

3.3.5.4. *A historical digression.* Why is the $U(1)_R$ charge so often referred to as "ghost number"? The history of this terminology [**438**] starts with Feynman's 1963 work on the quantization of Yang-Mills theory. Having noticed that his original diagrammatic approach to computation in QFT did not work for these theories, he introduced additional fictitious particles to solve the problem. These were physically unobservable and eventually received the name "ghost particles." Meanwhile, work by Faddeev and Popov, DeWitt, and others, showed that the ghost particles were an unavoidable feature of a diagrammatic expansion of the functional integral in any theory with a nonabelian gauge symmetry, expressing a Jacobian arising from a change of variables from a gauge fixing term to the gauge parameters.

Another important example is the world-sheet theory of string theory, in which the gauge symmetry is two-dimensional general coordinate invariance. The Faddeev-Popov approach was used in string quantization starting with Friedan and Alvarez in the early 1980's [**159, 9**]. The resulting "world-sheet reparameterization ghosts" have an anomalous $U(1)$ symmetry as in (3.110), but with the coefficient $\hat{c}$ replaced by 3. This constant can be tied to $3 = \dim_{\mathbb{C}} SL(2,\mathbb{C})$, the conformal symmetry group of $\mathbb{CP}^1$, and to the coefficient in the formula $\dim_{\mathbb{C}} \mathcal{M}_g = 3g-3$ for the dimension of the moduli space of genus g Riemann surfaces as well.

The same structure appears in superstring theory, and most of the structure we discussed was discovered physically in this case. For example, it turns out both mathematically and physically that, instead of $\mathbb{CP}^1$, it is more natural to define correlation functions on a three-punctured sphere, justifying the prescription of §3.3.5.3 for the case $\hat{c} = 3$. More generally, starting with a TFT with $\hat{c} = 3$, one can define topological string theory at arbitrary genus by systematically using R-charge in the role of ghost number; see [**119, 242**] for details. A similar but considerably more complicated discussion applies to the full superstring theory [**395**].

3.3.6. $N = 2$ minimal models and Gepner models. In direct analogy to our discussion of Virasoro representation theory, there is a critical value $\hat{c} = 1$ of the $N = 2$ SCA central charge, below which the representation theory becomes simple, which allows classifying and exactly solving all such unitary SCFTs, the "$N = 2$ minimal models." Let us briefly discuss these results here, going into more detail in §3.6.8.

First, there is a discrete set of allowed $\hat{c}$ values less than one,

$$\hat{c} = 1 - \frac{2}{k+2}, \qquad k \in \mathbb{N}.$$

For each $\hat{c}$, there is a discrete set of allowed R-charges,

$$q = \frac{n}{k+2}, \qquad n = 0, 1, \ldots, k.$$

These $N = 2$ SCA representations can be combined into complete SCFTs which must satisfy the constraint of modular invariance. This leads to a very simple result: the $\hat{c} < 1$ $N = 2$ SCFTs are in one-to-one correspondence with the simply laced (ADE) Dynkin diagrams, with k the dual Coxeter number, and chiral operators associated to nodes.

This result seemed rather mystical at first, but it was soon realized that the reason for the ADE classification was that $N = 2$ minimal models are in one-to-one correspondence with rigid complex singularities, which are also classified by ADE. The simplest way to see this is to consider the "Landau-Ginzburg" construction of the $N = 2$ minimal models, discussed in MS1 Chapters 13 and 16. This construction starts with a free sigma model (the theory (3.86) with $X \cong \mathbb{C}^n$ with the Euclidean metric), and then modifies the action by means of a *superpotential,* a holomorphic function $W : X \to \mathbb{C}$.

This has two main physical effects. First, in the terminology of 3.2.5, the superpotential is a relevant operator, so this deformation can in principle lead to a new CFT, with $\hat{c} < n$. However, characterizing this just using the RG is not so easy.

Since we have $N = 2$ supersymmetry, however, we can proceed as follows. One can show that the chiral ring in the resulting theory is a quotient ring (sometimes called the Jacobian ring of W),

$$R \cong \frac{\mathbb{C}[\phi^1, \ldots, \phi^n]}{\{\partial W/\partial\phi^1, \ldots, \partial W/\partial\phi^n\}}.$$

Since in CFT the chiral ring is graded, we conclude that W must be quasi-homogeneous. The QFT definition in fact implies that the $U(1)$ charge of W is 1,[21] so this determines the gradings. Finally, we know that charge 1 operators are marginal, so we must choose W to avoid this; this leads to the rigid singularities. The identification can be confirmed by comparing

[21] To be more precise, from the superfield form of the superpotential term $\int d^2\theta\, W$, the axial R-charge of W is 2. In an SCFT, this will be the sum of equal left and right $U(1)$ charges.

these chiral rings with the exact solutions derived from SCFT representation theory.

The simplest case is $W = \phi^{k+2}$, which leads to the A_k series of minimal models. It is easy to check that these results are consistent with our previous claims: the chiral ring is generated by the operator ϕ with R-charge $1/(k+2)$, while the maximal R-charge $\hat{c} = k/(k+2)$ is realized by ϕ^k.

Since, as we discussed earlier, $\hat{c}$ can be identified with the dimension of the Calabi-Yau, there is a simple physical definition which in a sense leads to SCFT analogs of Calabi-Yau manifolds with certain non-integral dimensions.

3.3.6.1. *Gepner models and linear sigma models.* Before proceeding, we should say that introducing Gepner models here is a bit out of the logical order, as their modern understanding requires many topics we have not yet discussed. Nevertheless we do it, in part to give some sense of the history – Gepner's work [**175**], done in 1988, predates almost everything else we discuss. It also remains the high-watermark of the results which can be obtained by purely algebraic methods of SCFT, and which are thus independent of the heuristic physical arguments which back up the rest of our discussion. Indeed, the mathematical understanding of SCFT [**31, 247, 248**] is reaching the point where these results could be made rigorous.

Let us consider the A_3 model defined by $W = \phi^5$, with central charge $\hat{c} = 3/5$. To compare it with an SCFT obtained from a sigma model with a conventional Calabi-Yau target space, we might try taking a tensor product of five copies of this model. From (3.91), the diagonal $N = 2$ SCA generated by the sum of the generators from each model,

$$T = \sum_{i=1}^{5} T_i$$

(resp. $G^{\pm}$ and J), is an SCA with $\hat{c} = 3$. Is there a sense in which the A_3 minimal model is the "fifth root of the Fermat quintic" ?

Yes, but there is a subtlety in making this precise. Strictly speaking, the A_k minimal model *does not* contain the operator $\phi^k = \Upsilon^2$ of (3.105), nor any of the operators ϕ^n. Rather, it contains some subset of the operators $\phi^n\bar{\phi}^m$, with $U(1)$ left and right charges (m, n), which satisfies the constraints of modular invariance. No such subset contains $(k, 0)$, and in fact the only such subset for a single A_k is the set of operators with $m = n$. Thus the literal tensor product $A_3^{\otimes 5}$ cannot contain the Υ^2 of the diagonal SCA either.[22]

However, Gepner showed that a $\mathbb{Z}_5$ orbifold of this product of minimal models does contain the desired Υ^2 operator, so it could be identified with a sigma model with target a quintic Calabi-Yau. He went on to give various

[22]This also means that minimal models or their products cannot be twisted into a 2d TFT.

pieces of evidence that this SCFT was isomorphic to the sigma model with target the Fermat quintic, at some "stringy" value of the Kähler moduli at which classical geometric considerations might break down.

A better understanding of this construction was later obtained from the linear sigma model, as discussed in MS1 Chapter 15. The linear sigma model introduces a complexified Kähler parameter, which can be used to interpolate between the Gepner model (or LG orbifold), and the nonlinear sigma model. As such, it exhibits all of the structure which is visible in TFT, and is a more flexible basis for subsequent physical developments.

Despite these advantages, there is also much to be said for having exact results. Later we will outline results on boundary states in Gepner models [**402, 69**], which were very important in coming to the general picture we will discuss. As another example, there are by now many exact results for orientifolds of Gepner models [**73**], a related subject (which we will not discuss in this book) for which other formulations remain work in progress.

3.4. Topological sigma models of closed strings

We resume the discussion based on the $(2,2)$ nonlinear sigma model. Rather than construct this SCFT and then reduce to the topological subsector, one can write twisted sigma model actions directly, which slightly simplifies the discussion.

3.4.1. The A-model. In the sigma model, twisting can be done by modifying the bundles in which the fermions take values. Let us begin with the A-model. From now on, we will generally denote the A-model target space by Y, and use X as the target space of the B-model.[23]

We "twist" the superconformal field theory by modifying the bundles in which the fermions take values. Instead of (3.85), we now take

$$\begin{aligned} \chi^i &= \psi^i_+ \in \Gamma(\phi^* T_Y) \\ \chi^{\bar{\imath}} &= \psi^{\bar{\imath}}_- \in \Gamma(\phi^* \bar{T}_Y) \\ \psi^{\bar{\imath}}_z &= \psi^{\bar{\imath}}_+ \in \Gamma(K \otimes \phi^* \bar{T}_Y) \\ \psi^i_{\bar{z}} &= \psi^i_- \in \Gamma(\bar{K} \otimes \phi^* T_Y). \end{aligned} \tag{3.115}$$

Note that the action (3.86) still makes sense (i.e., it is invariant under rotations of the world-sheet) with this assignment.

Now, the symmetry defined by setting $\alpha = \alpha_- = \tilde{\alpha}_+$ and $\alpha_+ = \tilde{\alpha}_- = 0$ in (3.87) satisfies the conditions for a BRST symmetry, and the operator Q which generates this symmetry satisfies $Q^2 = 0$.

The action can now be written as

$$S = i \int_\Sigma \{Q, \mathscr{D}\} - 2\pi i \int_\Sigma \phi^*(B + i\omega), \tag{3.116}$$

[23]This apparently backwards convention is used since it renders the notation in some of the sections on algebraic geometry more standard.

where

$$\mathscr{D} = 2\pi g_{i\bar{j}}(\psi_z^{\bar{j}}\,\bar{\partial}\phi^i + \partial\phi^{\bar{j}}\,\psi_{\bar{z}}^i), \tag{3.117}$$

$\omega = ig_{i\bar{j}}dz^i \wedge d\bar{z}^j$ is the Kähler form, and $B + i\omega \in H^2(Y, \mathbb{C})$ is the *complexified Kähler form.*

Since the complex structure only appears in $\mathscr{D}$, varying it leads to a Q-exact variation of the action, which in the topological theory is trivial. Thus the correlation functions in this topological A-model depend only on the complexified Kähler form $B+i\omega$. Indeed, from (3.116) they depend only on its cohomology class.

Furthermore, since the functional integral (3.14) depends only on e^{-S}, it is invariant under a shift of B by an element of integral cohomology. Thus the naive moduli space of A-models[24] is the *complexified Kähler moduli space* of Y, defined to be the cone in $H^2(Y, \mathbb{C})/H^2(Y, \mathbb{Z})$ in which ω satisfies the positivity condition of Kähler geometry.[25]

The local operators are general functions of the fields ϕ and ψ. As we discussed, the Q-cohomology is generated by the scalar operators, which cannot include $\psi_z^{\bar{i}}$ or $\psi_{\bar{z}}^i$. Nor can they include the derivatives of the fields $\partial^n\phi$ or $\partial^n\chi$ (one can use the SCA to rewrite these as descendants).

Thus, the chiral operators can all be written as

$$W[a] = a_{I_1 I_2 \ldots I_p}\chi^{I_1}\chi^{I_2}\cdots\chi^{I_p}, \tag{3.118}$$

where $a = a_{I_1 I_2 \ldots I_p} d\phi^{I_1} d\phi^{I_2}\cdots d\phi^{I_p}$ is a p-form on Y. The I_n's are real indices — in other words they may be holomorphic or antiholomorphic. One can then compute

$$\{Q, W[a]\} = -W[da]. \tag{3.119}$$

That is, for the A-model, **Q-cohomology is de Rham cohomology** and the space of operators is given by $H^*(Y, \mathbb{C})$.

3.4.1.1. *Correlation functions.* We want to compute

$$\langle W_a W_b W_c \cdots \rangle = \int [d\phi d\chi d\psi]\, e^{-S}\, W_a W_b W_c \cdots, \tag{3.120}$$

by integrating over maps $\phi : \Sigma \to Y$. We begin by noting that the second term of (3.116) depends only on the class $\phi_*(\Sigma) \in H_2(Y, \mathbb{Z})$, and thus we can pull it out,

$$Z = \sum_{\phi_*(\Sigma)} e^{-2\pi i \int_\Sigma \phi^*(B+i\omega)} \int_{\phi_*(\Sigma)\text{ fixed}} [d\phi \cdots]\, e^{-i\int\{Q,\mathscr{D}\}}.$$

The term in the action that remains is Q-exact and is therefore trivial. Although one's first temptation might be to replace something that is trivial

[24]After adding world-sheet instanton corrections, it turns out that one can analytically continue beyond the boundary of the Kähler cone.

[25]This is that $\int_C \omega > 0$ for any holomorphic curve $C \subset Y$.

by zero, we do the opposite and rescale it by a factor that tends to infinity! Since the integrand is positive semi-definite, this effectively restricts the path integral to the maps ϕ where this part of the action is zero, the *world-sheet instantons.* In other words, the saddle-point approximation of instantons is *exact* for topological field theories. The world-sheet instantons are given by $\mathscr{D} = 0$ in (3.117). These are the *holomorphic maps* satisfying $\bar{\partial}\phi^i = 0$.

The infinite-dimensional space of all maps $\phi : \Sigma \to Y$ is therefore replaced by the finite-dimensional space of holomorphic maps when we perform the path integral. Supersymmetry then cancels the Pfaffians associated with the fermionic path integral and the remaining determinants from the ϕ integrals, to produce a natural measure on the moduli space of holomorphic maps. We refer to [**468**] and MS1 for the details of this.

The result for the 3-point function is that

$$\langle W_a W_b W_c \rangle = \int_Y a \wedge b \wedge c + \sum_{\alpha \in I} N^{\alpha}_{abc} e^{2\pi i \int_\Sigma \phi^*(B+i\omega)}, \tag{3.121}$$

where I is the set of instantons and N^{α}_{abc} are numbers given by the intersection theory on the moduli space of rational curves (i.e., holomorphic embeddings of Σ) in Y, including the possibility of multiple covers [**85, 24**]. Here it is assumed that the degrees of forms a, b, c sum up to d (otherwise the 3-point function vanishes). The 1-point function is

$$\langle W_a \rangle = \int_Y a. \tag{3.122}$$

It is nonvanishing only if a is the top form on Y. These two correlators together completely determine the Frobenius algebra structure on $H^*(Y, \mathbb{C})$ and therefore arbitrary correlators.

In the large volume limit one can neglect instanton corrections in (3.121), and then the algebra structure is simply given by the wedge product of forms. At finite volume the deformed ring is called the "quantum cohomology ring" of Y. We have impinged on a vast subject here which we do not have space to explore more fully. We refer to [**101**] and references therein for a detailed account.

For string theory, the most important case is when Y is a Calabi-Yau 3-fold. Then the structure of the quantum cohomology ring is completely encoded in the 3-point functions of degree-2 forms. Degree-2 forms are special from the point of view of 2d TFT, since they give rise to deformations of the A-model which preserve the ghost number. One can show that these deformations simply deform the symplectic form ω and the B-field B (which are both closed 2-forms).

We emphasize again that the structure of the operator algebra depends only upon $B + i\omega$ and not the complex structure of Y. In fact, as explored in [**468**], we do not need any complex structure on Y, nor do we require the Calabi-Yau condition. Y can be any symplectic manifold with a compatible almost complex structure.[26] Instantons then correspond to pseudo-holomorphic curves. Since the topological A-model knows about only a small subset of the data of the untwisted theory, it should not come as a surprise that it can be applied to a wider class of target spaces.

3.4.2. The B-model. To obtain this, we make a different redefinition. Let $\psi_{\pm}^{\bar{j}}$ be sections of $\phi^*(\bar{T}_X)$, while ψ_+^j is a section of $K \otimes \phi^*(T_X)$ and ψ_-^j is a section of $\bar{K} \otimes \phi^*(T_X)$. Define world-sheet scalars

$$\begin{aligned} \eta^{\bar{j}} &= \psi_+^{\bar{j}} + \psi_-^{\bar{j}} \\ \theta_j &= g_{j\bar{k}}(\psi_+^{\bar{k}} - \psi_-^{\bar{k}}), \end{aligned} \tag{3.123}$$

and define a 1-form ρ^j with $(1,0)$-form part given by ψ_+^j and $(0,1)$-form part given by ψ_-^j.

Now consider a variation corresponding to the original supersymmetric variation with $\alpha_{\pm} = 0$ and $\tilde{\alpha}_{\pm} = \alpha$. As in the A-model, this produces a BRST charge Q satisfying $Q^2 = 0$ (up to equations of motion).

Now we may rewrite the action in the form

$$S = i\int \{Q, \mathscr{D}\} + U, \tag{3.124}$$

where

$$\begin{aligned} \mathscr{D} &= g_{j\bar{k}}\left(\rho_z^j \bar{\partial}\phi^{\bar{k}} + \rho_{\bar{z}}^j \partial\phi^{\bar{k}}\right) \\ U &= \int_{\Sigma}\left(-\theta_j D\rho^j - \tfrac{i}{2}R_{j\bar{j}k\bar{k}}\rho^j \wedge \rho^k \eta^{\bar{j}}\theta_l g^{l\bar{k}}\right). \end{aligned} \tag{3.125}$$

An additional complication arises in the B-model because the fermions are twisted in a more asymmetric fashion than in the A-model. For a general target space X, there is a chiral anomaly associated with an ambiguity in defining the phase of the Pfaffian associated to the fermionic path integrals. This anomaly is zero if we require $c_1(T_X) = 0$, i.e., if X is a Calabi-Yau manifold.[27]

It is obvious from (3.125) that variations of the metric on Σ change the action only by Q-exact terms. It is less obvious but true that varying the Kähler form ω and the $(1,1)$ component of the B-field also change the action by BRST-exact terms, and thus correlation functions in the B-model are independent of these parameters. In SCFT terms it follows because these operators are trivial in Q_B-cohomology. It can also be shown in the sigma model by adding auxiliary fields to get an equivalent formulation in which these variations are Q-exact [**318**].

---

[26] If Y is not a Calabi-Yau manifold, the A-model only has $\mathbb{Z}_2$ grading, as explained earlier.

[27] More precisely, in the case when Σ has empty boundary, it is sufficient to require $2c_1(T_X) = 0$. But one needs the stronger condition $c_1(T_X) = 0$ if one considers Σ with a nonempty boundary, as we will do later.

Local observables are now written

$$W[A] = \eta^{\bar{k}_1} \cdots \eta^{\bar{k}_q} A^{j_1 \ldots j_p}_{\bar{k}_1 \ldots \bar{k}_q} \theta_{j_1} \cdots \theta_{j_p}, \tag{3.126}$$

where

$$A = d\bar{z}^{\bar{k}_1} \cdots d\bar{z}^{\bar{k}_q} A^{j_1 \ldots j_p}_{\bar{k}_1 \ldots \bar{k}_q} \frac{\partial}{\partial z_{j_1}} \cdots \frac{\partial}{\partial z_{j_p}}, \tag{3.127}$$

is a $(0, q)$-form on X valued in $\bigwedge^p T_X$. One might call A a "$(-p, q)$-form". Note that we can use contraction with the holomorphic d-form Ω to give an isomorphism between the spaces of $(-p, q)$-forms and $(d - p, q)$-forms. This isomorphism is often used implicitly and explicitly in discussions of mirror symmetry, as we will see in §3.4.3.

Now,

$$\{Q, W[A]\} = -W[\bar{\partial} A], \tag{3.128}$$

and so, for the B-model, **Q-cohomology is Dolbeault cohomology** on forms valued in exterior powers of the holomorphic tangent bundle.

The instantons in the B-model are trivial. Setting $\mathscr{D} = 0$ in (3.125) requires $\bar{\partial}\phi^{\bar{k}} = \partial\phi^{\bar{k}} = 0$, i.e., ϕ is a constant map mapping Σ to a point in X. In fact the correlation functions receive no quantum corrections at all. This follows from the structure of supersymmetric perturbation theory, and can also be seen by the following argument. Since the action is Q-exact, correlation functions must be unaffected by rescaling the action by an arbitrary constant. This is the loop counting parameter $\hbar$ and thus correlation functions must be independent of $\hbar$, in other words equal to their classical limit $\hbar \to 0$.

The correlator in the B-model on a sphere can be shown to be

$$\langle W_A W_B \cdots \rangle = \int_X \Omega \wedge \iota_{AB\ldots} \Omega$$

Here we assume that both form degrees and polyvector degrees of $A, B, \ldots,$ sum up to d; thus $AB\ldots$ is a $(0, d)$-form with values in $\Lambda^d T_X$, and its contraction with the $(d, 0)$-form Ω is an ordinary $(0, d)$-form.

We can make this more explicit in the most interesting case when X is a Calabi-Yau 3-fold, and A, B, C are $(0, 1)$-forms with values in T_X. Then the 3-point function is

$$\langle W_A W_B W_C \rangle = \int_X \Omega^{jkl} A_j \wedge B_k \wedge C_l \wedge \Omega, \tag{3.129}$$

where $A = A^j \frac{\partial}{\partial \phi^j} \in H^1_{\bar{\partial}}(X, T_X)$, etc. The object Ω^{jkl} can be obtained from the antiholomorphic 3-form $\bar{\Omega}$ using the Kähler metric to raise indices: while naively this introduces dependence on the Kähler metric, this dependence is trivial in cohomology. Deformations of the B-model corresponding to such $(-1, 1)$-forms are deformations of the complex structure of X (recall that infinitesimal deformations of the complex structure are classified by elements of $H^1(X, T_X)$). If the cohomology groups $H^0(X, \Lambda^2 T_X)$ or $H^2(X, \mathcal{O})$ are nontrivial, one also has less obvious deformations of the B-model associated to $(-2, 0)$-forms and $(0, 2)$-forms. Such deformations are absent for manifolds of $SU(3)$ holonomy, but are present when X is a complex torus, or a K3 surface, or their product. Their geometric significance can be explained using the notion of generalized complex structure introduced by Hitchin [**236**], see §6.2.5.

Note that the B-model does require that X have a complex structure and that it be Calabi-Yau. However, it does not require any mention of the Kähler form ω. This means that the B-model depends only on the complex structure of X. In analogy with the A-model, one might suspect that the B-model can be defined for arbitrary complex manifolds with $c_1(T_X) = 0$, which need not admit a Kähler form. This is indeed so [**293**].

3.4.3. Closed string mirror symmetry. We only give the briefest review of aspects we will use below, otherwise referring to MS1 and [**101**]. There are several definitions of mirror symmetry varying in strength. We require only a fairly weak definition which asserts that two Calabi-Yau manifolds X and Y are mirror if the operator algebra of the A-model with target space Y is isomorphic to the operator algebra of the B-model with target space X.

The original definition is stronger and is a statement concerning conformal field theories. From the string theory viewpoint the most interesting case is when X and Y are Calabi-Yau 3-folds; in this case the strongest definition would be that the type IIA string compactified on Y and Type IIB string theory compactified on X give equivalent physics in four dimensions.

A simple analysis of the dimensions of the vector spaces of the operator algebra yields the simple statement that $h^{p,q}(Y) = h^{d-p,q}(X)$ and thus $\chi(Y) = (-1)^d \chi(X)$.

3.4.3.1. *The mirror map.* The operator algebra for the A-model on Y depends on a choice of $B + i\omega$ on Y and the operator algebra for the B-model on X depends on a choice of complex structure for X. Thus a precise statement of mirror symmetry must map the moduli space of $B + i\omega$ of Y to the moduli space of complex structures of X. This mapping is called the "mirror map." In §5.6.1 we will work this out in some detail in one of the simplest examples, the noncompact Calabi-Yau which is the total space of the line bundle $\mathcal{O}_{\mathbb{P}^2}(-3)$.

Let us here review this for the most-studied example of a mirror pair of Calabi-Yau threefolds, following [**193, 85**]. We take Y to be the "quintic threefold," i.e., a hypersurface in $\mathbb{P}^4$ defined by the vanishing of a homogeneous polynomial of degree 5. Since $h^{1,1}(Y) = 1$, the moduli space of complexified Kähler classes is one dimensional. Let e denote the positive[28]

[28]That is, $\int_Y e^3 > 0$.

generator of $H^2(Y,\mathbb{Z})$. Then (by an abuse of notation) we refer to the cohomology class of the complexified Kähler form as $(B+i\omega)e$, i.e., B and ω are real numbers in the context of the quintic. Thus we can think of the size of Y (i.e., ω) as determined by the size of the ambient $\mathbb{P}^4$. Y also has $h^{2,1}=101$ and thus 101 deformations of complex structure, but this is of no interest to us here.

Its mirror X can be constructed by dividing Y by a $(\mathbb{Z}_5)^3$ orbifold action. The result has orbifold singularities, and to get a smooth Calabi-Yau these must be resolved, yielding 100 new degrees of freedom for $B+i\omega$. However, all we care about is the complex structure of X, which is determined by the choice of quintic polynomial. The most general quintic compatible with the $(\mathbb{Z}_5)^3$ orbifold action is given by

$$x_0^5+x_1^5+x_2^5+x_3^5+x_4^5-5\psi x_0x_1x_2x_3x_4. \tag{3.130}$$

Thus the complex structure is determined by the single complex parameter ψ. The mirror map will then be a mapping between $B+i\omega$ on the A-model side and ψ on the B-model side.

The mirror map turns out to be quite complicated and is actually a many-to-many mapping. Because of this, one generally starts with a basepoint, which is usually the large radius limit on the A-model side, and finds the mirror map in some neighbourhood of this basepoint. One can then try to analytically continue the mirror map to a larger region.

One may analyze the moduli space intrinsically without any reference to a specific compactification by studying the general features of scalar fields in $N=2$ theories of supergravity in four dimensions. The result is that the moduli space is a so-called "special Kähler manifold" [**108, 431, 85**]. For a nice mathematical treatment of this subject see [**154**].

The special Kähler structure of the moduli space leads to the existence of favored (but not uniquely defined) coordinates, the "special coordinates" which obey certain flatness constraints. On the A-model side, the components of $B+i\omega$ are special coordinates. However, on the B-model side, complex parameters such as ψ in (3.130) do *not* form special coordinates.

A more natural way to characterize the complex structure is to use Hodge structures, and more specifically the class of the holomorphic three-form $\Omega\in\mathbb{P}(H^3(X,\mathbb{C}))$. In the case of the mirror quintic, the class of the holomorphic three-form uniquely characterizes a complex structure, but of course not all points in $\mathbb{P}(H^3(X,\mathbb{C}))$ correspond to complex structures of Calabi-Yau threefolds, as the dimension is too large. Locally, the submanifolds of $\mathbb{P}(H^3(X,\mathbb{C}))$ which parameterize Calabi-Yau threefolds are *complex Lagrangian* submanifolds, determined by a holomorphic *prepotential.* To define this, we can choose a symplectic basis of $H_3(X,\mathbb{Z})$. This means a basis α_m,β^m for $m=0,\ldots,h^{2,1}(Y)$ with the intersection numbers

$$\alpha_m\cap\alpha_n=0,\quad \alpha_m\cap\beta^n=\delta_m^n,\quad \beta^m\cap\beta^n=0. \tag{3.131}$$

The A-cycle periods

$$\varpi_m = \int_{\alpha_m} \Omega \tag{3.132}$$

then form a set of homogeneous special coordinates. One can also show that the B-cycle periods are determined locally by a holomorphic function $\mathcal{F}$ as

$$\frac{\partial \mathcal{F}}{\partial \varpi_m} = \int_{\beta_m} \Omega. \tag{3.133}$$

The same structure is present in the A-model, now embedding the complexified Kähler moduli space as a complex Lagrangian submanifold

$$H^2(Y,\mathbb{C}) \subset \mathbb{P}\left(\oplus_{k=0,2,4,6} H^k(Y,\mathbb{C})\right).$$

Special geometry then requires that the mirror map is a *projective linear symplectic* map between the two ambient spaces.

Since the mirror map is linear, it is largely determined by matching the monodromy induced by the paths $B \mapsto B+1$ around large radius limit points. A systematic method for doing this was analyzed in [**366**]. To nail down the last constants one really needs to explicitly count some rational curves on Y and map the correlation functions of the A-model to that of the B-model directly. Having said that, there is a conjectured form of the mirror map (which was implicitly used in [**85**]) which appears to work in all known cases. We refer to [**101**] for more details.

3.4.3.2. *SCFT arguments for mirror symmetry.* The simplest observation which suggests this is based on the $N=2$ SCA, and the factorization of the $U(1)$ subalgebra discussed in §3.2.4. Factorization implies that any automorphism of the $U(1)$ (free boson) theory lifts to an automorphism of the full $N=2$ SCFT.

We now consider the T-duality automorphism in this context. From (3.69), this acts as

$$J \to J; \qquad \bar{J} \to -\bar{J}. \tag{3.134}$$

The corresponding automorphism of (3.91) takes

$$G_- \leftrightarrow \tilde{G}_-$$

while preserving the other generators. Then, from (3.100) we see that this exchanges the A- and B-models.

Thus, given an A-model with target space Y, we know that there is an SCFT whose B-model is isomorphic, and vice versa. Of course this does not yet tell us that there is any Calabi-Yau X which leads to this B-model. Indeed, there are evident counterexamples to the reverse claim, for example if we start with the B-model of a rigid Calabi-Yau X (i.e., with $h^{2,1}=0$).

However, in an explicit construction, there can be a natural guess for how this T-duality automorphism relates X and Y. The original example was the Gepner model construction of the quintic Calabi-Yau discussed in

§3.3.6, as a $\mathbb{Z}_5$ quotient of a product of five A_3 minimal models. One can explicitly apply the SCFT automorphism to each of the minimal models, and thus construct the mirror SCFT. Upon examination, it turns out to be the same as a $\mathbb{Z}_5^4$ quotient of the product of minimal models, which is plausibly the $\mathbb{Z}_5^3$ quotient we cited above. We refer to [**192**] for the details.

This type of reasoning was eventually understood more geometrically and led to an explicit construction of mirror pairs within the class of toric hypersurfaces [**32**]. This class is large but by no means exhausts the Calabi-Yau threefolds, which is one motivation to look for other pictures of mirror symmetry. Still, the great virtue of this argument is that it is formulated in the SCFT, so that once we grant that X and Y are mirror pairs in this sense, we can draw exact conclusions, which implicitly take into account any and all differences between string theory and conventional geometry.

3.5. Boundary CFT

In this section we consider a world-sheet Σ with boundaries. A careful analysis of this is rather technical, and will not be needed for this book. For more details, one can consult [**241**], much of whose analysis appears in MS1, and [**6, 336**] for an even more thorough treatment.

3.5.1. General discussion. Our goal is to define a category of open and closed strings, satisfying the axioms of TFT given in Chapter 2. The simplest discussion would be to start with a closed string TFT, i.e., a Frobenius algebra $\mathcal{C}$, and define the open string structure in terms of this, perhaps in analogy to the construction of sheaves on complex manifolds. Although some ideas for how to incorporate more geometry of an underlying complex manifold M were presented in §2.5, so far the direct approach of Chapter 2 has only been made to work for simple algebras $\mathcal{C}$, i.e., zero-dimensional M.

The original discussion of open string TFT on a Calabi-Yau was due to Witten [**463**]. This started from the definition of the A and B topologically twisted sigma models, which we gave in §3.4, and identified the geometric boundary conditions which were compatible with the twisting. We will refer to these as "topological Dirichlet branes" in the following.

The results, as we will shortly demonstrate, are that the topological D-branes in the A-model are Lagrangian submanifolds with bundles carrying flat connections, while in the B-model they are holomorphic submanifolds carrying holomorphic vector bundles. One can also see the origin of the world-sheet disk instanton corrections which lead to the A_∞ structure of the Fukaya category.

While this gives us the basic structure of the problem, understanding mirror symmetry physically requires more input from CFT. Thus, we continue with some of the physical theory of D-branes, in particular explaining the action of T-duality.

3.5.2. Generalities on open string CFT. Before we begin, let us discuss the general features of the problem at the level of §3.2, following [**86**].

While the previous discussion already involved boundary conditions in defining the Hilbert space $\mathcal{H}$ and the QFT functor, we now want to define a *second*, essentially different type of boundary. In the language of Chapter 2, the Hilbert space $\mathcal{H}$ is the CFT analog of $\mathcal{C}$, while the new boundaries corresponding to D-branes will come out of a set $\mathcal{B}_0$ of "simple" boundary conditions.

The primary physical constraint on a CFT boundary condition is that it respects the conservation of energy and momentum on the world-sheet. This is a local condition on the stress tensor T defined in (3.27),

$$0 = t^\mu n^\nu T_{\mu\nu}|_{\partial\Sigma}, \tag{3.135}$$

where t^μ and n^ν are the tangent and normal vectors (respectively) to the boundary.

Suppose we consider a component $S \cong S^1$ of $\partial\Sigma$; then, as in §3.2.2.1, we can think of the boundary condition as a "state" $|B\rangle$ in the Hilbert space $\mathcal{H}$, carrying an action of two commuting Virasoro algebras L_n and $\tilde{L}_n$. Expanding the stress tensor in these, the condition (3.135) becomes

$$0 = \left(L_n - \tilde{L}_{-n}\right)|B\rangle. \tag{3.136}$$

One can show [**255**] that solutions to this equation are in one-to-one correspondence with primary fields. The simplest argument for this [**36**] uses the decomposition (3.29) and the existence of a unique inner product on each $\mathcal{H}_{R,i}$ such that $L_n = L^\dagger_{-n}$. This allows us to reinterpret $|B\rangle$ as an operator

$$O_B \in \oplus_i(\mathcal{H}_{L,i} \otimes \mathcal{H}^*_{R,i}) \tag{3.137}$$

satisfying

$$L_n O_B = O_B L_n.$$

By Schur's lemma, such an operator O_B is a sum of projectors on irreducible representations of the Virasoro algebra. The boundary state corresponding to such a projector is called an *Ishibashi state.*[29] The standard boundary conditions constructed in physics, the "Cardy states" (for $c \leq 1$) or "boundary states," are then linear combinations of Ishibashi states.

As in the closed string discussion, the further discussion divides into two cases. For $c \leq 1$, we can use the Virasoro algebra representation theory to make a complete classification of boundary conditions. An important role is played by the "Cardy conditions" which formed part of the axioms in Chapter 2, expressing the fact that a world-sheet as depicted in Figure 12

[29]Note that these are not elements of $\mathcal{H}$, as they are nonnormalizable. Rather, since physics requires that there be a sensible pairing $\langle B|\phi\rangle$, they are unbounded linear functionals on $\mathcal{H}$.

of Chapter 2 must have consistent open and closed string interpretations. The result is a finite set $\mathcal{B}_0$, which in the simplest (diagonal) models has cardinality equal to the number of Virasoro irreps.

For $c > 1$, there is no comparable general theory. Indeed, it is not even known whether (3.136) suffices to define the problem or whether additional constraints are required. One can however get partial results in any of the three frameworks we discussed. For boundary conditions in Gepner models, see [**402, 69**], while for the linear sigma model see [**241**] and MS1.

3.5.2.1. *Dirichlet branes in sigma models.* In the nonlinear sigma model, it is natural to define boundary conditions geometrically, which is how we will proceed. We recall the classical equations of motion for the sigma model (3.76), which for $B = 0$ defined harmonic maps $\phi : \Sigma \to M$. Thus we seek natural boundary conditions for the harmonic map equations. At least to start with, they should also be local (a condition at each point on the fields and finitely many derivatives), to describe D-branes without additional physical degrees of freedom.

Mathematically, one might require that the boundary conditions preserve ellipticity. In the simplest case of target space $\mathbb{R}$, we are asking for elliptic boundary conditions for the scalar Laplacian. As is well known, there are two such local boundary conditions,

- Dirichlet boundary conditions: $\phi|_{\partial\Sigma} = \phi_0$ with $\phi_0 \in \mathbb{R}$.
- Neumann boundary conditions: $n^\mu \partial_\mu \phi|_{\partial\Sigma} = 0$.

One could also phrase the Dirichlet condition as $t^\mu \partial_\mu \phi|_{\partial\Sigma} = 0$, which superficially makes the two cases look more parallel. However, as the Dirichlet boundary condition does in fact depend on the parameter ϕ_0, making the correct parallel requires identifying a comparable parameter for the Neumann case (§3.5.2.2).

We can also rephrase these boundary conditions as a relation between left and right movers, by noting that

$$J_L = (t^\mu + n^\mu)\partial_\mu \phi \,, \qquad J_R = (t^\mu - n^\mu)\partial_\mu \phi \,, \tag{3.138}$$

so that

$$\begin{aligned} J_L &= -J_R \qquad \text{Dirichlet;} \\ J_L &= J_R \qquad \text{Neumann.} \end{aligned} \tag{3.139}$$

If we generalize to the target space $\mathbb{R}^d$, besides the evident generalizations of the above, there are also "mixed" boundary conditions, in which the boundary $\phi|_{\partial\Sigma}$ must map into a fixed submanifold $L \subset M$. One can show that these are ellipic by going to a coordinate system made up of tangential and normal coordinates to L; the leading term in the harmonic map equation is again a Laplacian, and the mixed boundary conditions in these coordinates reduce to Neumann and Dirichlet respectively.

One refers to a boundary condition with p Neumann components and $d-p$ Dirichlet components, so that $\dim_{\mathbb{R}} L = p$, as a **Dirichlet p-brane.** For example, a 0-brane is associated to a point, a 1-brane to a real line (or string), and so forth.

A more systematic way to find appropriate boundary conditions for an equation of motion $\delta S = 0$ is to require that the boundary term in the variation of the action vanishes,

$$0 = \delta S|_{\partial\Sigma}.$$

This eliminates a boundary term which would arise from the integration by parts performed in the usual Euler-Lagrange derivation. Because of this, the standard physical arguments for conservation of energy and momentum go through without change, satisfying our primary condition (3.135). In addition, one has to require that the boundary conditions together with the equations of motion would not lead to an overdetermined initial-value problem.

Let us illustrate how this works for the nonlinear sigma model with target M on an infinite "strip" Σ with Euclidean metric, parameterized by $\sigma_1 \in \mathbb{R}$ and $\sigma_2 \in [0,1]$, as shown in Figure 4. The boundary of Σ on which we are imposing boundary conditions consists of two components given by $\sigma_2 = 0$ and $\sigma_2 = 1$. We are considering oriented open strings and thus we distinguish the "start" $\sigma_2 = 0$ and "end" $\sigma_2 = 1$ of the string.

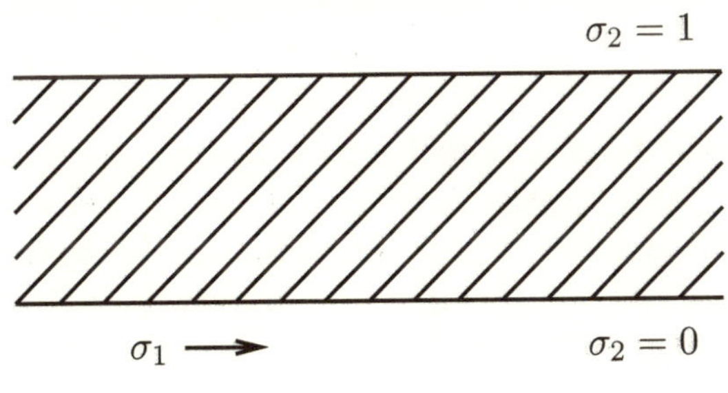

FIGURE 4. Open string world-sheet

The action of the bosonic sigma model is

$$S = \frac{1}{2}\int_{\Sigma} d^2\sigma\ G_{IJ}(\phi) \sum_{i=1,2} \partial_i\phi^I \partial_i\phi^J,$$

where ϕ is a map from Σ to M and $\partial_i = \frac{\partial}{\partial\sigma_i}$. Varying the action and integrating by parts, we find a boundary term

$$\int_{\partial\Sigma} d\sigma_1 G_{IJ}(\phi)\delta\phi^I \partial_2\phi^J,$$

which must be set to zero. One way to do this is to constrain ϕ on the boundary to map to a submanifold $L \subset M$. This implies that the tangential derivative $\partial_1\phi^I$ lies in $TL \subset TM$. To ensure that the boundary term in the variation of the action vanishes, one may complement this with the requirement that the normal derivative $\partial_2\phi^I$ belongs to the normal bundle NL. The resulting boundary condition is of Dirichlet type in the directions normal to L and of Neumann type in the directions tangent to L. Note that one cannot impose both Dirichlet and Neumann conditions on a particular component ϕ^I, since when taken together with the equations of motion this would be an overdetermined system of equations.

3.5.2.2. *Boundary conditions involving a gauge field.* If L has dimension larger than one, these boundary conditions can be generalized. Let us keep the condition that the NL component of $\partial_1\phi^I$ vanishes, but subject the TL component of $\partial_2\phi^I$ to a more general condition

$$G_{IJ}(\phi)\partial_2\phi^J = F_{IJ}(\phi)\partial_1\phi^J.$$

For consistency, all indices here must be in the TL subspace. We will refer to this as a deformed Neumann boundary condition. If the matrix function F_{IJ} is of the form $\partial_I A_J(\phi) - \partial_J A_I(\phi)$ for some vector function $A_I(\phi)$, then the boundary part of the variation of the action becomes, after further integration by parts:

$$\delta \int_{\partial\Sigma} d\sigma_1 A_I(\phi)\partial_1\phi^I.$$

Therefore if we add to the action a boundary term

$$S_b = -\int_{\partial\Sigma} d\sigma_1 A_I(\phi)\partial_1\phi^I, \tag{3.140}$$

the variation of the total action $S + S_b$ will have no boundary part.

To make the geometric significance of A_I and F_{IJ} clearer, let us introduce a 1-form $A = A_I(\phi)d\phi^I \in \Omega^1(L)$. Then F_{IJ} are components of the 2-form $F = dA \in \Omega^2(L)$, and the boundary action can be written in a manifestly diffeomorphism invariant form:

$$S_b = -\int_{\partial\Sigma} \phi^* A.$$

Unitarity requires A to be purely imaginary, so we redefine A by a factor $-i$ and write the boundary action as

$$S_b = i\int_{\partial\Sigma} \phi^* A.$$

After such a redefinition, the theory is unitary if A is a real 1-form on L.

It is clear that the action is not changed if we replace A by $A' = A + df$ for any function f on L. This suggests that A need not be a globally defined 1-form, but only a connection on a principal $U(1)$ bundle over L. To show that this is indeed the case, let us pick a particular connected component $\partial_0\Sigma$ of the boundary. We can assume that $\partial_0\Sigma$ is closed, i.e., diffeomorphic to a circle. Let us extend the map $\phi|_{\partial_0\Sigma} : S^1 \to L$ to a map $\tilde{\phi} : D^2 \to L$, where D^2 is a disc. Then the boundary action can be written as

$$i \int_{D^2} \tilde{\phi}^* F.$$

This depends only on F and not on A. It does depend on the way ϕ is extended to $\tilde{\phi}$, but the difference between any two extensions is

$$i \int_{S^2} \tilde{\phi}^* F = 2\pi i n$$

for some $n \in \mathbb{Z}$. Since the boundary action enters the path integral only through $\exp(-S_b)$, this ambiguity is immaterial.

Even if $F_{IJ} = 0$, i.e. when the boundary condition is Neumann in the directions tangent to L, the $U(1)$ connection need not be gauge-equivalent to the trivial one, if L is not simply-connected. This observation goes a long way towards remedying the asymmetry between Dirichlet and Neumann boundary conditions mentioned in §3.5.2.1, as now the Neumann boundary condition also admits a choice of parameter, namely the flat connection. In the particular case of branes on T^d, since flat connections are parameterized by a dual torus, there is a complete symmetry between the two. We will soon show that this is how T-duality acts on Dirichlet branes.

3.5.2.3. *Quantizing the world-sheet.* The choice of L (or, more generally, of the triple (L, E, ∇), where E is a $U(1)$ bundle over L and ∇ is a connection on E) is the basic datum defining a boundary condition or Dirichlet brane.

Now from the point of view of QFT, the choice of L is the analog of postulating a general metric in the original sigma model. Just as not all metrics lead to CFTs, only those with vanishing beta function, so too not all submanifolds L lead to consistent boundary conditions in the CFT. By studying the renormalization of the boundary theory, one can derive a boundary beta function, which plays the same role in this context.

In the leading approximation (all length scales large compared to the string length l_s), the vanishing of the boundary beta function corresponds to the condition that L be a *volume extremizing* submanifold. In equations, defining the volume functional as the integral of the volume element associated to the pullback of g,

$$S = \mathrm{Vol}(L) = \int_L \sqrt{\det \phi^*(g)}, \tag{3.141}$$

we have

$$\beta \sim \delta \mathrm{Vol}(L) = 0.$$

The simplest case is a minimal volume submanifold. In CFT terms, these would be distinguished by the absence of relevant boundary operators.

Now, finding minimal volume submanifolds is generally difficult. However, there are some easy cases, such as a holomorphic submanifold of a

Kähler manifold. As we discussed in Chapter 1, this generalizes to the concept of calibrated submanifold, which although more difficult to study than holomorphic submanifolds, is at least a linear condition.

There is a close analogy between these problems and the corresponding closed string problems, in that the general problem of finding Ricci-flat manifolds includes the simpler special case of finding Kähler-Einstein metrics. Both simplifications can be understood physically as consequences of supersymmetry, and a general argument that solutions preserving supersymmetry must satisfy first order equations (sometimes called BPS equations). We will not develop this further here, but refer to [**382, 172**] and MS1.

3.5.2.4. *The Born-Infeld action.* More generally, if we wish to include the gauge field A on L, both A and L are subject to RG flow, and conformal invariance requires the corresponding boundary beta functions to vanish. To leading order in a power series expansion in the curvature F, this requires $d * F = 0$, in other words F must satisfy Maxwell's equation.

If F is not small, Maxwell's equations can get nonlinear corrections. Now, in physics terms, the curvature F is naturally regarded as having dimensions of inverse length squared. By way of illustration, a related dimensionless quantity is the holonomy around a closed loop; for a small loop this will be linear in the area enclosed by the loop. Thus, by "small," one means $|F| \ll 1/l_s^2$, where l_s is the string length.

It is possible to determine the condition for conformal invariance in a closed form in the case when F is not necessarily small, but its derivatives are, i.e., assuming $l_s||\partial F|| \ll ||F||$. It is that the Dirichlet brane world-volume fields (the embedding $\iota : L \to M$ and the connection A on L) satisfy the equations of motion following from the *Born-Infeld action*,

$$S = \int_L d^n x \sqrt{\det(g|_L + F)}. \tag{3.142}$$

Here $g|_L$ is the pullback of the space-time metric to L regarded as a symmetric $n \times n$ matrix, $F = dA$ is the curvature two-form regarded as an antisymmetric $n \times n$ matrix, and det is the determinant of the $n \times n$ matrix obtained by adding the two. We will derive this action using T-duality in §3.5.4.

3.5.2.5. *Open string sectors.* To define an open string, two boundary conditions are required, one for each end of the string. Carrying through the quantization procedure on the field theory with boundary conditions a and b, we will derive a Hilbert space $\mathcal{H}_{ab}$ of open string states.

A simple world-sheet geometry in which to study the consequences of conformal invariance on the open string is the punctured upper half plane $\mathrm{Im}\, z \geq 0$, $z \neq 0$, depicted in Figure 5. This is a Riemann surface with two boundaries, the half line $\mathrm{Re}\, z > 0$ and the half line $\mathrm{Re}\, z < 0$. We choose the boundary condition a on the first and b on the second.

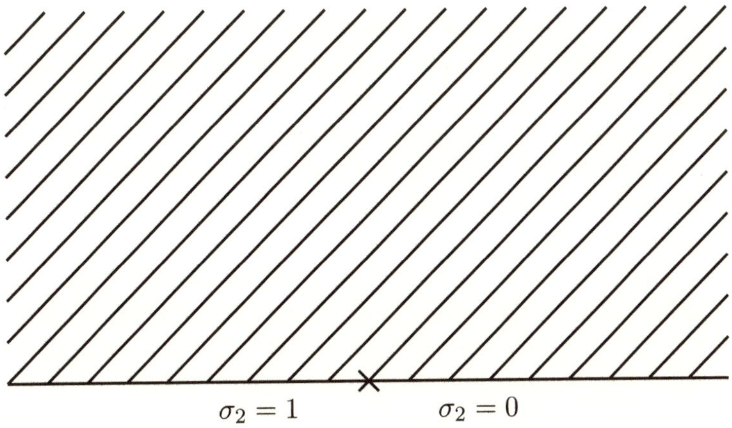

FIGURE 5. Open string world-sheet as punctured upper half plane.

Another world-sheet geometry, leading to pictures more similar to those in Chapter 2, is the strip $0 \leq x \leq \pi$, $t \in \mathbb{R}$. The two are related by the conformal transformation $z = \exp(x + it)$.

Because of the boundary condition (3.135), the operators L_R and L_L are equated on the boundary. A convenient way to think about this is to treat left movers (operators with antiholomorphic dependence, in particular the stress tensor T_L) as the complex conjugates of holomorphic operators defined in the lower half plane. This allows us to use the same formalism as for the closed string, but now the open string Hilbert space admits the action of a single Virasoro algebra.

In superconformal theory, this generalizes to a single $N = 2$ SCA. For example, whereas closed string states are bigraded by (J_L, J_R) eigenvalues, the open strings carry a single $\mathbb{Z}$-grading.

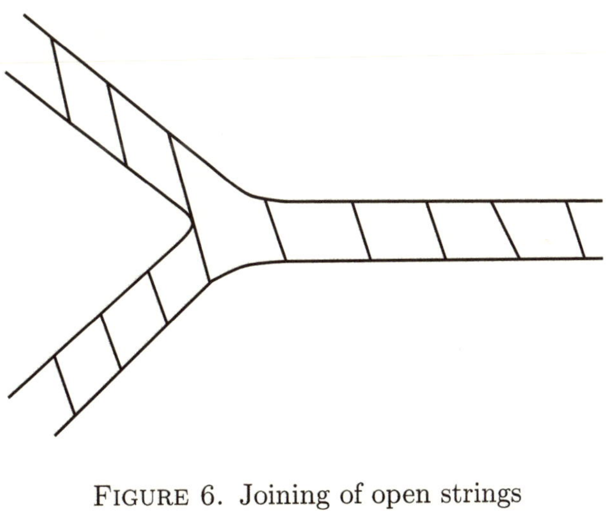

FIGURE 6. Joining of open strings

Since the interactions in open string CFT are associated to the same world-sheets pictured in Chapter 2, the interactions satisfy many of the same formal properties. In particular, it is natural to identify boundary conditions as objects in a category, and elements of an open string Hilbert space $\mathcal{H}_{ab}$ as morphisms. Of course, once we go to CFT, we will no longer have an associative multiplication; rather this will be replaced with a CFT operator product as in the closed string discussion.

3.5.2.6. *Deformations and world-volume gauge fields.* Once we have a sensible conformal boundary condition, we can consider its deformation theory. As in §3.2.5, this can be defined by adding new terms to the action, but now these terms should be restricted to the boundary,

$$Z[g] = \int [D\phi]\ e^{-S_0[\phi]+\sum_i g^i \int_{\partial\Sigma} dt\ O_i(t)}. \tag{3.143}$$

Scale invariance now tells us that O_i should be an operator with $h = 1$.

In the nonlinear sigma model (at leading order in l_s), the natural boundary operators with $h = 1$ are obtained by multiplying the derivatives of the boson $\partial_i\phi$, by a function $f(\phi)$. Although the function can be general, only its restriction to L survives the RG; furthermore, upon computing the leading order beta function we will find that f must be harmonic (in a suitable sense). Thus we will identify f physically as a field on L, which (in the classical limit) satisfies an equation of motion, the exact beta function.

The simplest example appears for Neumann boundary conditions, so that the tangential derivative $\partial_t\phi$ is nonvanishing. In this case the deformations are

$$O(t) = A(\phi(t))\partial_t\phi(t), \tag{3.144}$$

where $A(\phi) \in T^*L$. As we discussed in §3.5.2.2, this deformation corresponds to changing the gauge field A on L.

On the other hand, for a coordinate with a Dirichlet boundary condition, the normal derivative $\partial_n\phi$ is nonvanishing, and we can write

$$O(t) = v(\phi(t))\partial_n\phi(t) \tag{3.145}$$

for a general function $v(\phi)$. Generalizing this to several Dirichlet bosons, the parameter $v(\phi)$ lives in N^*L, the conormal bundle of L.

Since we have a metric, N^*L is isomorphic to NL, so it is natural to expect these operators to correspond to deformations of the submanifold L. This can be verified by physics arguments.[30] This is another example of the sense in which a CFT "contains its own deformation theory," and we conclude that the corresponding moduli were already taken into account in the original choice of embedding.

[30]The identification of v with a normal deformation is clearest in the closed string channel, i.e., taking the coordinate x normal to the boundary as Euclidean time, so that $\int dt\, \partial_x\phi$ is the momentum operator.

3.5.2.7. *Dependence on the B-field.* There is one further generalization we will need. Consider the term

$$S_B = \frac{1}{8\pi} \int_\Sigma d^2z\, B_{IJ}(\phi) \frac{\partial \phi^I}{\partial z} \frac{\partial \phi^J}{\partial \bar{z}}. \tag{3.146}$$

in the original sigma model action (3.75), where B is a two-form on M. This action has a gauge symmetry analogous to, but different from, the one for A: under

$$\delta B = d\Lambda, \quad \Lambda \in \Omega^1(M), \tag{3.147}$$

we have

$$\delta S_B = \frac{1}{8\pi} \int_\Sigma d^2z \frac{\partial \Lambda_J(\phi)}{\partial z} \frac{\partial \phi^J}{\partial \bar{z}}. \tag{3.148}$$

On a closed world-sheet this is a total derivative, so the variation (3.147) is a local symmetry of the theory. Thus B is a "two-form gauge connection," in a sense which can be formalized using the language of gerbes [**77, 234**].

On an open world-sheet, the resulting boundary term is

$$\delta S_B = \frac{1}{8\pi} \int_{\partial\Sigma} d\sigma_1\, \Lambda_J(\phi) \partial_1 \phi^J \tag{3.149}$$

and takes precisely the same form as (3.140). This implies that (3.147) by itself is not a symmetry. Rather, there is a symmetry under the simultaneous transformation

$$\delta B = d\Lambda, \qquad \delta A = -\Lambda|_L \tag{3.150}$$

which preserves the combination $B|_L + F$.

This has various implications to be discussed. The simplest is that (3.142), in the presence of $B \neq 0$, becomes

$$S = \int_L d^n x \sqrt{\det(g|_L + B|_L + F)}. \tag{3.151}$$

This also follows from the fact that g and B derive from the combination $G + B$ in (3.75).

In general, of course, a connection on a $U(1)$ bundle is not described by a globally defined one-form A. In this case, it is not right to think of B as a globally defined 2-form, rather it is a *connection on a* $U(1)$ *gerbe*. For a more detailed discussion, see [**283**].

3.5.2.8. *Chan-Paton factors and Yang-Mills fields.* In the physics discussion, a Dirichlet brane carrying a nonabelian connection, and more general systems of several Dirichlet branes, are defined by introducing "Chan-Paton factors." These are labels which are taken out of some index set I. One then defines a configuration of Dirichlet branes as a map from this into the set of simple boundary conditions $B : I \to \mathcal{B}_0$. The open string Hilbert space is then

$$\mathcal{H} = \oplus_{I,J} \mathcal{H}_{IJ}$$

where $\mathcal{H}_{IJ} \cong \mathcal{H}_{B(I),B(J)}$. Operator products of the corresponding boundary operators must respect this structure, i.e., it defines a linear map

$$\mathcal{H}_{IJ} \otimes \mathcal{H}_{JK} \to \mathcal{H}_{IK}.$$

Of course, all this is subsumed by the statement that Dirichlet branes form an additive category. However there are some interesting physics subtleties and ramifications of this.

First, consider N identical Dirichlet branes, in other words a subset $I' \subset I$ with cardinality $|I'| = N$, all of whose elements map to the same $B \in \mathcal{B}_0$. For definiteness take $I' = \{1, 2, \ldots, N\}$. The boundary operators we just discussed now come with multiplicity, and the resulting structure can be summarized by treating their coefficients (couplings) as $N \times N$ matrices. For example, we can generalize (3.143) to a boundary deformation

$$\delta S = \int_{\partial\Sigma} dt\; A_I^J(\phi(t))\partial_t\phi(t), \tag{3.152}$$

where the matrix structure of A_I^J is treated in the same way as for the "colored particle" of §3.1.5.

For (3.144), the interpretation of this is evident: the coupling A is a matrix of one-forms, which geometrically is a connection on a rank N vector bundle V. This suggests, and it is correct, that the claim in §3.5.2.6 that conformal invariance requires the $U(1)$ connection on a single brane to satisfy Maxwell's equations, generalizes to the claim that the $U(N)$ connection must satisfy the Yang-Mills equations, $D * F = 0$.

To verify this, we would need to generalize the arguments in §3.5.2.2, and show that now they lead to nonabelian gauge invariance,

$$A \to A + df + [A, f], \qquad f \in \mathrm{Mat}_N(C(L)). \tag{3.153}$$

If we can show this, since there are physical consistency arguments which require that the equations of motion are gauge invariant, these must be the Yang-Mills equations (at leading order in l_s).

The standard discussion of this generalization [**395**] is somewhat intricate. The starting point is to write the action (3.143) for a world-line with nonabelian connection using the path-ordered exponential as in §3.1.5. However, in string theory this requires additional renormalization. Consider the quadratic term in the expansion of this exponential,

$$\sum_{i,j} g_i g_j \int dt_1 \int dt_2\; O_i(t_1)\; O_j(t_2)\;.$$

In the coincidence limit $t_1 = t_2$, this will typically diverge. Renormalizing this divergence leads to a finite ambiguity in defining the couplings. For example, we could choose to include an additional contribution

$$\sum_k C_k^{ij} g_i g_j \int dt_1\; O_k(t_1)\;,$$

called a "contact term."

While this would seem to add a great deal of ambiguity to the definition of the functional integral, the primary source of this ambiguity is simply the fact that the coupling constants g_i in (3.143), in the various explicit forms we discussed, are a system of local coordinates on the space of boundary conditions. Of course, without additional structure, there is no preferred coordinate system. One can check that general nonlinear redefinitions $g_i \to g'_i(g)$ introduce such contact terms. Nor is this particular to boundary couplings; a similar discussion applies for spaces of CFTs or QFTs.

To fix the contact terms, the standard physical analysis appeals to physical consistency conditions (which we will not explain) which for a theory of "spin one" fields such as $A(\phi)$ can be satisfied if there is gauge invariance, thus favoring the choice of contact terms leading to (3.153). Such an argument is hardly conclusive and indeed it has been argued more recently that different choices are possible, for example leading to noncommutative gauge theory [**418**].

A satisfactory treatment of these issues requires string field theory, as in [**168, 169**]. As we discussed in Chapter 2, this framework is closely related to A_∞ structure.

3.5.2.9. *Dirichlet matrix boundary conditions.* Let us now turn to the expression (3.145). Formally, the discussion is entirely parallel, and leads to the statement that if we consider N identical Dirichlet branes, each embedded into the submanifold $L \subset M$, then the infinitesimal variations of their embeddings, which naively live in $NL \otimes \mathbb{R}^N$, actually live in $NL \otimes \mathrm{Mat}_N(\mathbb{R})$.

If the reader has not seen this before, it should be quite a surprise. It is a clear sign that noncommutative geometry has an essential role to play in the theory of Dirichlet branes. While subsequent developments we discuss and many others have certainly demonstrated this, we suspect there is far more to say here.

To start with a more elementary question, one may wonder how these $N^2 \times \dim NL$ functions are related to the number $N \times \dim NL$ of functions which describe deformations of a set of N embedded submanifolds. The answer is that the extra $(N^2 - N) \times \dim NL$ functions do not correspond to marginal deformations, because of a combination of gauge invariance and obstructions. We will explain this further in §3.5.4, §3.6.2.3 and §5.21.

3.5.3. Superstring, GSO projection and orientation. So far, we have only discussed the world-sheet bosons, a discussion which applies equally to the bosonic and superstring theories.

The treatment of the fermions in the superstring is largely dictated by the requirement that $(1, 1)$ world-sheet supersymmetry is preserved. However there are discrete choices which enter. In particular, the choice of the orientation of the Dirichlet brane world-volume is made at this point. This is often omitted in discussions of topological string theory, as one can make all of the basic definitions and even get all the way to the Gromov-Witten invariants without making it explicit. However we will need it.

A full discussion requires explaining the so-called GSO (Gliozzi-Scherk-Olive) projection. Unfortunately this gets into technicalities of superstring theory which we would prefer to avoid, so we simplify this discussion for our purposes. A full discussion can be found in the textbooks [**191, 394**].

3.5.3.1. *Boundary conditions in the supersymmetric sigma model.* Let L be a submanifold of the target X. As before, we require $\phi|_{\partial\Sigma} \subset L$ and impose Dirichlet conditions in the directions normal to L and modified Neumann conditions in the directions tangent to L.

These conditions can be compactly written as

$$\frac{\partial\phi^I}{\partial z} = R^I_J(\phi)\frac{\partial\phi^J}{\partial\bar{z}} + \text{fermions}, \tag{3.154}$$

where R is an orthogonal matrix with respect to the metric g_{IJ}. Eigenvectors of R with eigenvalue -1 give Dirichlet conditions and thus span the directions normal to L. We parameterize the action of R on the vectors tangent to L in terms of a 2-form $F \in \Lambda^2 T^*L$:

$$R|_{TL} = (g|_L - F)^{-1}(g|_L + F).$$

F should be closed and represents the curvature of the gauge field on L. This becomes clearer if we rewrite the boundary condition in the following form:

$$g_{IJ}\partial_2\phi^J = F_{IJ}\partial_1\phi^J + \text{fermions}. \tag{3.155}$$

If we set terms containing fermions to zero, this is exactly the modified Neumann condition with gauge-field curvature F.

Now, the boundary conditions for fermions in the superstring are fixed by requiring that they preserve $N = 1$ world-sheet supersymmetry. Given the transformation laws (3.84), it should be plausible that this requires the left and right movers to be related by the same matrix R,

$$\psi^I_+ = R^I_J\psi^J_-, \tag{3.156}$$

as is easily verified. With more analysis one can extend this to a full non-linear treatment. This requires fermionic terms in (3.154), and in the case of multiple branes, explicit boundary degrees of freedom as in §3.1.5. We refer the reader to the literature, e.g., [**334**].

These boundary conditions are conformally invariant at the classical level. Quantum mechanically, conformal invariance is equivalent to spacetime equations of motion for the gauge field, which (as for the bosonic string) follow from the Born-Infeld action (3.151).

3.5.3.2. *GSO projection.* Note that we have not yet needed to specify the orientation or a spin structure on L. This enters upon quantizing the theory, as can be seen in various ways.

Let us explain how this enters into the operator formulation. First, we want to define an operator $(-1)^F$ acting on $\mathcal{H}$, defined by the property that it anticommutes with the supercharges,

$$0 = (-1)^F G + G(-1)^F$$

for all of $G_+, G_-, \tilde{G}_+$ and $\tilde{G}_-$.

It is evident how to do this in the sigma model – we just take $(-1)^F$ to anticommute with the world-sheet fermions as in (3.6). More generally, since G has R-charge $q = +1$, we can take

$$(-1)^F = Ce^{i\pi(q_L+q_R)}. \tag{3.157}$$

Here $C = \pm 1$ is a constant which is not yet fixed.

Now, the GSO projection is essentially the projection on the sector with

$$(-1)^F = +1. \tag{3.158}$$

Since $(-1)^F$ is a multiplicatively conserved quantum number (i.e., a $\mathbb{Z}_2$ grading), the projection is consistent with the interactions.

This is somewhat oversimplified. But before going into more detail, let us say what this has to do with the orientation of L, or of X. Actually, the definitions involve the full ten-dimensional space-time $\mathbb{R}^{3,1} \times X$, or whatever it might be; call it M. Now, just as the bosonic string Hilbert space is an enlargement of the space of functions on M, the fermions of the superstring further enlarge this to the Hilbert space of the $N = 1$ supersymmetric quantum mechanics of §3.1.1, which was defined using a spinor bundle S over M. Thus, we might expect the projection (3.158) to implement the reduction to a single chirality S_+ or S_-, depending on the choice of C in (3.157). This choice implicitly chooses an orientation for M.

Strictly speaking, the preceding statement is true only in the Ramond sector, as only there do the fermions have zero modes which form a Clifford algebra. However, since in topological string theory the Neveu-Schwarz and Ramond sectors are essentially interchangeable (§3.3.3), we can ignore this detail for our discussion. Similarly, after mentioning that in the closed string one should really define left and right-moving operators $(-1)^{F_L}$ and $(-1)^{F_R}$ and project on both, we move on.

We thus see how the orientation of M enters the theory. What about the orientation of L? When we quantize the open string, we will need to choose a splitting of the spinors on M (take $S(M)_+$ for definiteness),

$$S(M)_+ \cong \bigoplus (S(TL)_\pm \otimes S(NL)_\pm).$$

where TL and NL in this expression refer to the tangent and normal bundles of L as a submanifold of M, say $\mathbb{R}^{3,1} \times X$ in our discussion. The point is that the choice of chirality $\pm$ in the two factors is correlated.

Now, since the relative chirality of two spinors in $\mathbb{R}^{3,1}$ is physically observable (it enters in defining discrete symmetries such as parity and charge conjugation), we can use this correlation to distinguish the two chiralities of spinor on L, and thus define the orientation of L.

Developing this argument in detail requires more physics than we can cover here. But one simple observation we can make is that string theory sees the chirality of spinors in the normal bundle, as well as those of the tangent bundle. Thus orientation always enters, even in seemingly degenerate cases such as L a point in M. It is referred to as the distinction between a "brane" and "antibrane," as we discuss in detail in §5.1.

There is another important observation to be made, which helps answer the following question: if we need to talk about orientation in our discussion, how do we define this in SCFT, without bringing in explicit geometric concepts? In particular, since mirror symmetry relates two very different geometric pictures, it would be helpful to have some non-geometric definition underlying the two geometric definitions.

The key observation is that all of the orientation dependence can be traced back to (3.157), an operator which is defined purely in terms of R-charges. Thus, at every point in the discussion where we need to specify an orientation, we can replace this with some condition depending only on the R-charges. We will use this observation several times in Chapter 5.

3.5.4. T-duality for Dirichlet branes. Let us now make precise the idea introduced in §1.3.1, that the action of T-duality on a Dirichlet brane is to convert it into a Dirichlet brane of a different dimensionality.

We take $M \cong T^d$, and follow the functional integral approach of §3.2.3.6. These arguments generalize straightforwardly to the case with boundary.

In fact, one can see the main point without any further details. We can summarize the central point of the previous discussion by combining (3.61) with (3.63), to obtain

$$d\hat{\varphi}_i = g_{ij} * d\varphi^j .$$

We might say that T-duality (a symmetry of CFT) follows from Hodge duality (on the world-sheet field φ).

On the boundary, this becomes

$$\partial_t \hat{\varphi}_i = g_{ij} \partial_n \varphi \tag{3.159}$$

$$\partial_n \hat{\varphi}_i = -g_{ij} \partial_t \varphi \tag{3.160}$$

Looking back at §3.5.2.1, we see that the operation of T-duality on the world-sheet directly exchanges the role of Dirichlet and Neumann boundary conditions. For example, a D0-brane on S^1 (Dirichlet in φ), will become a

D1-brane on the dual S^1 (Neumann in $\hat{\varphi}$). This generalizes to T^d directly: if we T-dualize all d coordinates, a Dp-brane will become a D$(d-p)$-brane on the dual T^d.

Proceeding to §3.5.2.6, we also see that T-duality exchanges the deformation operators:

$$v_i \int dt\, \partial_n \varphi^i \text{ (Dirichlet)} \qquad \leftrightarrow \qquad A_i \int dt\, \partial_t \hat{\varphi}^i \text{ (Neumann)}.$$

The first of these generates motion in T^d, and since the CFT is free it can be integrated to show that the resulting boundary condition is just $\varphi = \varphi_0 + v$.

While naively the second of these operators is a total derivative on the boundary, since $\hat{\varphi}$ is a map to S^1 we must be more careful. Using the expansion (3.67), we see that this term in the action is proportional to $A_j n^j$, leading to a factor $\exp(iA_j n^j)$ in the functional integral. This is precisely the holonomy in the flat connection $A_j dx^j$, for a path represented by n^j in $\pi_1(T^d)$. Thus we conclude that the naive interpretation

$$\phi^i \leftrightarrow A_i \tag{3.161}$$

between the position of the D0-brane, and the parameter of the flat connection on the Dd-brane, is correct.

One puzzling feature of this argument is that the space of flat $U(1)$ connections on a torus has a marked point (the trivial flat connection), while the moduli space of a D0-brane on the dual torus does not have a marked point. To resolve this puzzle, recall that in conformal field theory there is a gauge symmetry (3.150). If we choose Λ to be a closed 1-form, then the B-field is unchanged, while the gauge field A is shifted by an arbitrary closed 1-form. The fact that the trivial flat connection is not invariant under such transformations means that the space of flat gauge fields A does not have a natural origin. For some further discussion of this point, see §6.3.

3.5.5. The Born-Infeld action from T-duality. We now want to use T-duality to deduce the Born-Infeld action

$$S = \int_L d^n x \sqrt{\det(g+F)}. \tag{3.162}$$

Here g is the pullback of the space-time metric to L regarded as a symmetric $n \times n$ matrix, $F = dA$ is the curvature two-form regarded as an antisymmetric $n \times n$ matrix, and det is the determinant of the $n \times n$ matrix obtained by adding the two. As mentioned above, its critical points correspond to zeroes of the boundary beta-functions, provided F is slowly varying on the string scale l_s. The idea is to use T-duality to relate a brane with an almost constant F with a brane with $F = 0$. For the latter brane, the action is assumed to be known and equal to its volume.

Let us consider a Dp-brane with $F = 0$ which is a section L of a trivial T^d fibration $Y \to \mathbb{R}^p$, and apply T-duality along the fibers. Let $\phi : \mathbb{R}^p \to T^d$

define the section; then the pullback metric becomes

$$\phi^*(g) = g_{ij}dx^i dx^j + g_{ab}d\phi^a d\phi^b$$

where $1 \leq i,j \leq p$ and $1 \leq a,b \leq d$. We assume $g_{ia} = 0$, but allow g_{ab} to vary on the base.

Substituting this into the volume (3.141), we have

$$S = \int_L d^p x \sqrt{\det_{1\leq i,j\leq p} (g_{ij} + \partial_i \phi^a \partial_j \phi^b g_{ab})}.$$

Now, applying the T-duality relation (3.161), this becomes

$$S = \int_L d^p x \sqrt{\det_{1\leq i,j\leq p} (g_{ij} + g^{ab} F_{ia} F_{jb})}. \tag{3.163}$$

where $F_{ia} = \partial_i A_a$, the field strength.

This is almost equal to (3.142) in the metric

$$g = g_{ij}dx^i dx^j + (g^{-1})_{ab} dy^a dy^b,$$

up to an overall multiplicative factor of the inverse volume of the fiber $(\det g_{ab})^{-1/2}$. To check this, one needs to relate the determinant of the $p\times p$ matrix appearing in (3.163) to the determinant of the $(p+d)\times(p+d)$ matrix appearing in (3.142), which follows from standard identities.[31]

Thus, we find that the D$(p+d)$-brane in the T-dual theory is described by a slightly modified version of the Born-Infeld action,

$$S = \int_L d^{p+d}x\, e^D \sqrt{\det(g+F)} \tag{3.164}$$

with

$$D = -\frac{1}{2}\log\det g.$$

This is correct and the additional factor e^D is indeed an additional possible choice in the closed string theory, known as the *dilaton*. While it will not be important in our subsequent considerations, it is generally quite important in string theory as it controls the strength of space-time quantum corrections (the string coupling constant). But its presence here is one signal that the considerations we are discussing now, based on a series expansion in α', would be expected to run into difficulties at singular fibers.

The analogous description for N Dirichlet branes is only partially understood. The Born-Infeld action (3.142) is valid in the limit where F changes slowly on the string scale, while its magnitude can be large. This makes sense in the abelian case, where F is gauge invariant, but not in general.

[31] One way to do this is to first make a similarity transformation, which turns the $p\times p$ determinant into $\det(1+g^{-1/2}Fg^{-1}Fg^{-1/2})$, and the $(p+d)\times(p+d)$ determinant into $\det\begin{pmatrix} \mathbf{1}_p & g^{-1/2}Fg^{-1/2} \\ g^{-1/2}Fg^{-1/2} & \mathbf{1}_d \end{pmatrix}$. One then applies the expansion $\det(1+M) = \exp\left[-\left(\sum_{k\geq 1} \frac{\mathrm{tr}(-M)^k}{k}\right)\right]$ and matches terms.

One simple recipe for defining the nonabelian Born-Infeld action is the "symmetrized trace prescription" [**450**], but (as stressed there) this is not exact. A recent discussion appears in [**376**].

3.6. Supersymmetric and topological boundary conditions

We are now ready to make contact with the discussion in Chapter 1. There, we argued that the condition of space-time supersymmetry required the D-branes to lie in calibrated submanifolds.

Once Σ has a boundary, it is impossible to preserve the entire $N = (2, 2)$ SCA, because the reflection condition (3.156) relates the left-moving and right-moving fermions. On the other hand, to be able to topologically twist, we must be able to consistently define one $N = 2$ SCA. Furthermore, this must be compatible with the existence of the world-sheet operator Ω of §3.3.4.6.

This can be done in two ways, corresponding to the two twistings. As one would expect, the primary condition is that the appropriate (A- or B-model) Q operator defined in (3.100) must be well defined in the open string sector. Using the SCA, this determines the boundary conditions on the $U(1)_R$ current J, as

$$J_L = -J_R \qquad \text{A-model;} \tag{3.165}$$

$$J_L = J_R \qquad \text{B-model.} \tag{3.166}$$

Note from (3.139) that, if we think of (J_L, J_R) as arising from a free boson, then these would correspond to Dirichlet and Neumann respectively.

Either of these boundary conditions reduces the bigrading of the $(2, 2)$ SCA to a single grading, defined by the complementary linear combination of currents. The conserved charge in the two models is thus

$$\begin{aligned} J = J_L - J_R &= \int d\sigma\ \partial_\tau \varphi \qquad \text{A-model;} \\ J = J_L + J_R &= \int d\sigma\ \partial_\sigma \varphi \qquad \text{B-model,} \end{aligned} \tag{3.167}$$

where the integral expressions use the realization of the $U(1)$ algebra as a free boson discussed in §3.2.4 and §3.3.4.6, to make contact with the explicit treatment of §3.2.3.

3.6.1. The A-model. Consistency of the A twist (3.115) with the boundary conditions (3.156) requires that $R^i_j = R^{\bar\imath}_{\bar\jmath} = 0$ (we use holomorphic coordinates). That is, only the off-diagonal terms $R^{\bar\imath}_j$ and $R^i_{\bar\jmath}$ are nonzero.

Let us set $F = 0$ for simplicity and choose a vector v which has eigenvalue $+1$ with respect to R, i.e., a tangent vector to $L \subset Y$. Let us introduce the almost complex structure J, which in holomorphic coordinates is of the form

$$J^m_n = i\delta^m_n, \quad J^{\bar m}_{\bar n} = -i\delta^{\bar m}_{\bar n}, \tag{3.168}$$

with off-diagonal entries equal to zero. It is then easy to see that the vector Jv has eigenvalue -1 with respect to R, i.e., it is normal to L. Furthermore, $J^2v = -v$, so a further application of J restores us to the tangent direction. Thus J exchanges the directions tangent and normal to the D-brane L. Clearly then L must be of middle dimension.

Note that if v and w are two tangent vectors to L with eigenvalue $+1$ under R, then w is orthogonal to Jv with respect to the metric g_{IJ}. Since, by definition, the Kähler form on Y is $\frac{1}{2}g_{IK}J^K_M d\phi^I d\phi^M$, we see that the Kähler form restricted to L is zero. Thus L is a Lagrangian submanifold of Y, and we have made contact with the A-type branes of Chapter 1.

There are additional, quantum constraints on A-branes. For the case at hand, these are known, so we make the following

DEFINITION 3.3. A *Lagrangian A-brane* is an equivalence class of Lagrangian 3-manifolds in Y equipped with a flat connection, modulo Hamiltonian deformations, which has trivial Maslov class (3.170) and satisfies the quantum obstruction condition (3.181).

But before explaining these constraints in detail, let us introduce the other known A-branes.

3.6.1.1. *Coisotropic branes.* A more careful analysis allowing for $F \neq 0$ [**382, 291**] shows that a Calabi-Yau n-fold may have A-branes of real dimension $n + 2p$ for non-negative integer p. The precise conditions on L and F turn out to be fairly intricate. First, the submanifold L must be coisotropic. This means that for any point $p \in L$ the skew-complement of TL_p with respect to the symplectic form ω is contained in TL_p. Equivalently, L is defined locally by the vanishing of the functions f_i, $i = 1, \ldots, \mathrm{codim}_{\mathbb{R}} L$, such that all the pairwise Poisson brackets of f_i vanish on L. In Dirac's terminology, this means that L is the first-class constraint surface. Consequently, the functions f_i can be regarded as generators of gauge transformations, or Hamiltonians. The corresponding vector fields are tangent to L and span an integrable distribution whose dimension is equal to $\mathrm{codim}_{\mathbb{R}} L$. Integrating this distribution, we get a foliation of L, whose leaves can be thought of as "gauge orbits." Identifying all the points in a leaf, we get a topological space which may be called the space of leaves, or in Dirac's terminology a reduced phase space. In general, it is not a manifold, or even a Hausdorff topological space.

The second condition on the pair (L, F) says that the contraction of the 2-form F with any vector tangent to the leaves of the foliation vanishes. Since F is closed, this also implies that F is a basic 2-form on the foliated manifold L, and descends to a closed 2-form f on the reduced phase space, provided that the latter makes sense as a manifold. Note that although the restriction of the symplectic form ω to L is degenerate, it descends to a nondegenerate 2-form in the directions transverse to the leaves of the foliation. Thus ω gives rise to a symplectic form σ on the reduced phase space (this is one of the key observations in Dirac's theory of systems with first-class constraints).

Finally, the third condition says that the $(1,1)$ tensor $J = \sigma^{-1} f$ on the reduced phase space squares to -1, i.e., it is an almost complex structure. Using the fact that σ and f are closed 2-forms, one can show that J is automatically integrable [**290**]. Thus the reduced phase space is a complex manifold. Furthermore, it is easy to see that $\sigma + if$ is a holomorphic symplectic form with respect to J, and therefore the complex dimension of the reduced phase space is even. We will denote it by $2p$.[32] One can easily see that the real dimension of L is equal to $n + 2p$.

The above three conditions on (L, F) can be more elegantly formulated in the language of *generalized complex geometry* [**236, 212, 284**] (see §6.2.5). In the case $F = 0$ they simplify to the requirement that L be Lagrangian; in this special case $p = 0$ and the reduced phase space is a single point. For Calabi-Yau one-folds (i.e., elliptic curves), this is the only possibility. In the case of a Calabi-Yau 3-fold Y, the only other possibility is $p = 1$ and L a five-dimensional real submanifold of Y. If Y is a strict Calabi-Yau threefold (the holonomy is not a proper subgroup of SU(3)), then $b_5(Y) = 0$, and any such L will be homologically trivial. Nevertheless, such a brane can be stable because the gauge field is not flat.[33]

3.6.1.2. *Ghost number anomaly and Maslov class.* Given a Lagrangian submanifold with a flat bundle, quantum considerations impose two further constraints. Here we discuss the first, that the A-brane must preserve the $U(1)_R$ grading (ghost number) of the operator product algebra. This is non-trivial in the quantum theory as this symmetry can be anomalous.

We refer to MS1 Chapter 40 for a computation of the $U(1)_R$ anomaly, and here explain its geometric origin. This is related to the problem of grading Floer cohomology [**377, 378**].

Let us fix a particular choice of a holomorphic 3-form Ω on Y. At any point p on a Lagrangian submanifold L the volume form of L may be written as a restriction

$$dV_L = w \cdot e^{-i\pi\xi(p)} \Omega|_L, \tag{3.169}$$

where w is a positive real number. ξ gives a map from L to a circle, $\xi : L \to S^1$. This in turn induces a map on the fundamental group

$$\xi_* : \pi_1(L) \to \pi_1(S^1) \cong \mathbb{Z}. \tag{3.170}$$

---

[32]We stress that the complex structure J on the reduced phase space is unrelated to the complex structure on the Calabi-Yau n-fold Y.

[33]Strictly speaking, the analysis of [**291**] applies only to the case of an abelian gauge field on the A-brane, so one can draw this conclusion only for such A-branes. But it is very plausible that one can always regard branes equipped with a vector bundle of rank N as a deformation of N coinciding branes of rank 1.

It can be thought of as an element of $H^1(L,\mathbb{Z})$ and is called the *Maslov class* of L.

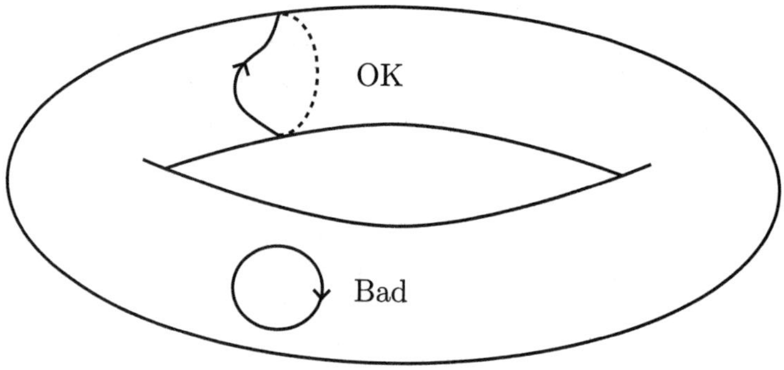

FIGURE 7. Loops which do and do not give an anomaly.

The anomaly is absent precisely when the Maslov class of L is zero. Clearly this is always the case when $\pi_1(L) = 0$. As a nontrivial example, consider Y a one-complex dimensional torus. Any line of Y is trivially Lagrangian. But as shown in figure 7, a contractible loop has a nontrivial Maslov class and so is ruled out as an A-brane.

For coisotropic A-branes, the ghost number anomaly has been computed in [**335**]. The analogue of the Maslov class turns out to be the following. Let $\dim_{\mathbb{R}} L = n+2p$. Consider a top form on L given by $\Omega|_L \cdot F^{\wedge p}$. One can show that this form is nonzero everywhere on L and therefore is proportional to the volume form of L. The proportionality coefficient is a nowhere vanishing function whose phase defines a map $\pi_1(L) \to \mathbb{Z}$. The vanishing of the map is equivalent to the absence of the ghost number anomaly.

3.6.2. Open strings for A-branes. We now discuss the open string spectrum between a pair of A-branes (L_1, E_1, ∇_1) and (L_2, E_2, ∇_2).

We begin with the case $L_1 \cong L_2 \cong L$ a Lagrangian submanifold, as analyzed by Witten in [**463**]. The general case will be treated in §3.6.2.5.

3.6.2.1. *Open strings from an A-brane to itself.* Let $L_1 \cong L_2 \cong L$, etc. Suppose we have a Lagrangian cycle L with a $U(N)$ vector bundle $E \to L$. As discussed above, open string states will be sections of $E^* \otimes E$ (perhaps tensored with other bundles), i.e., endomorphisms of E. If E is a line bundle, then sections of $\mathrm{End}(E)$ are simply functions on L.

The basic string theory analysis of the open string spectrum is parallel to that of §3.4.1, leading to the conclusion that all of the Q-cohomology can be represented by operators of the form

$$a_{I_1 I_2 \ldots}(\phi)\chi^{I_1}\chi^{I_2}\ldots, \tag{3.171}$$

where $a_{I_1 I_2 \ldots}(\phi) \in \mathrm{End}(E)_{\mathbb{C}}$, $\phi \in L$, and χ^{I_k} lies in the tangent bundle of L. Note that the functions $a_{I_1 I_2 \ldots}$ are allowed to be complex, or more precisely they are sections of the complexification of $\mathrm{End}(E)$. The BRST operator Q acts similarly to §3.4.1, and so the Hilbert space of open string states is given by the total de Rham cohomology group

$$\mathcal{H} \cong \bigoplus_k H^k(L, \mathrm{End}(E)_{\mathbb{C}}), \tag{3.172}$$

where the ghost number is given by k. The same discussion applies in the special case of open strings between two A-branes wrapping the same cycle L, but carrying different vector bundles E_1 and E_2 with different connections, to obtain

$$\mathcal{H} \cong \bigoplus_k H^k(L, \mathrm{Hom}(E_1, E_2)_{\mathbb{C}}). \tag{3.173}$$

The discussion of deformations of the action from §3.4.1 also generalizes to the boundary case. Now, a deforming operator is integrated along the one-dimensional manifold $\partial\Sigma$ rather than the two-dimensional Σ. Thus we look for ghost number *one* boundary operators whose ghost number zero descendants can be added to the boundary action. These correspond to elements of $H^1(L, \mathrm{End}(E)_{\mathbb{C}})$.

Let us see this geometrically in the simplest case of E a line bundle. Half of these deformations can be identified with the deformations of the flat unitary connection on E, while the other half are deformations of the isotopy class of L as a Lagrangian submanifold, as we will see in §6.1.1. The basic point, which may be familiar from classical mechanics, is that a deformation of a Lagrangian submanifold $v \in NL$ corresponds to a closed 1-form θ on L, which can be written locally in terms of a Hamiltonian function H,

$$v^j = \omega^{ij}\theta_i \sim \omega^{ij}\partial_i H.$$

Deformations which are globally of this form are Hamiltonian isotopies. Therefore elements of $H^1(L, \mathbb{R})$ correspond to Lagrangian deformations modulo Hamiltonian isotopy. Thus we see that A-branes in the topological theory are defined only up to isotopy equivalence.

The open string spectrum for a single (rank one) coisotropic A-brane has been computed in the classical approximation [**289**]. One way to formulate the answer goes as follows. As discussed in §3.6.1.1, a coisotropic A-brane has a natural foliation and a natural complex structure in the directions transverse to the leaves. Thus it makes sense to consider the sheaf of functions locally constant along the leaves and holomorphic in transverse directions. The open string Hilbert space is then the cohomology of this sheaf. For Lagrangian A-branes, this is the sheaf of locally constant $\mathbb{C}$-valued functions on L, so this reduces to (3.172).

3.6.2.2. *Correlation functions and superpotential.* Given the open string operators, we next want to compute their correlation functions. We can again write

$$\langle W_{a_1} \cdots W_{a_k} \rangle = \int [d\phi d\chi d\psi]\, e^{-S}\, W_{a_1} \cdots W_{a_k}, \tag{3.174}$$

where now the world-sheet Σ has boundaries, on which we insert the operators W_{a_i}. The primary case (sometimes called "tree level") is Σ, the disk with a single boundary.

In general terms, the computation [**463**] is very parallel to that in §3.4.1.1 for the closed string A-model. The same localization argument will tell us that the correlation function will be a sum over holomorphic maps $\phi : \Sigma \to Y$, now compatible with the boundary conditions. This will include the constant map, which will lead to a classical term. It will also include nontrivial maps, which lead to world-sheet instanton contributions.

Consider the three-point correlation function of operators

$$\begin{aligned} a &\in H^1(L, \mathrm{Hom}(E_1, E_2)), \\ b &\in H^1(L, \mathrm{Hom}(E_2, E_3)), \\ c &\in H^1(L, \mathrm{Hom}(E_3, E_1)). \end{aligned} \tag{3.175}$$

The classical contribution to the 3-point function is

$$\langle W_a W_b W_c \rangle = \int_L \mathrm{Tr}(a \wedge b \wedge c). \tag{3.176}$$

At this point we have crucially used the fact that Y is a Calabi-Yau threefold.

The general form of the instanton corrections is

$$\sum_{D_\alpha} \pm \exp\left(i \int_{D_\alpha} (B + i\omega) + i \oint_{\partial D_\alpha} A_i \right) N_{abc} \tag{3.177}$$

where the sum is over all holomorphic disks D_α with $\partial D_\alpha \subset L$ (including multiple covers).

Here A_i is the connection on the bundle E_i. To be a bit more precise, we divide the boundary ∂D_α into three segments labelled by $i \in \{1, 2, 3\}$; use A_i on the ith segment. The open string vertex operators sit at the boundaries between segments, for example W_a between segments 1 and 2. Of course, this prescription sees the ordering of the operators, and will lead to an associative but not necessarily commutative open string algebra. More details will follow below and especially in Chapter 8.

While a naive generalization of (3.175) to four- and higher-point correlation functions does not work, there is another definition of these correlation functions which reflects the mathematical statement that elements of H^1 correspond to (linearized) deformations. This uses the fact that given an operator W with R-charge 1, there exists a unique operator $W^{(1)}$ which has R-charge zero and is a one-form on Σ, satisfying

$$dW = \{Q, W^{(1)}\}. \tag{3.178}$$

One can then modify the boundary action by adding

$$\delta S_b = \int_{\partial\Sigma} W^{(1)}. \tag{3.179}$$

Since $W^{(1)}$ has R-charge zero, this preserves R-charge conservation and conformal invariance, at least at the linearized level.

By inserting (3.179) into (3.174), one can define a 4-point correlation function, and continuing in this way one defines n-point functions.

Given an explicit SLag, one can check that this agrees with our previous definition (3.140) of a variation of the $U(1)$ connection. However the present discussion applies to a general boundary SCFT.

The generating function for open string correlation functions on the disk, usually called the *superpotential*, can now be defined in direct analogy to §3.3.5.3, as

$$W[t] = \langle \exp\left(\sum_i t_i \int O^i\right)\rangle, \tag{3.180}$$

where integrated operators are defined using (3.178), but keeping in mind that $n = \hat{c}$ of the operators O^i are not integrated. One can check using properties of the $N = 2$ SCA that the choice of which operators are integrated does not affect the final result.

One difference from the closed string case is that, since the operators O_i are integrated over the boundary of the disk, we can keep track of their relative ordering. This allows us to promote the couplings t_i to matrices, as discussed in §3.5.2.8. While important, there is a more important difference from the closed string case, to which we turn.

3.6.2.3. *Tadpoles and obstructions.* Despite the formal similarity between the closed and open string computations, there is a key difference which will have major consequences, both mathematical and physical. In both cases, we can think of the open string states (which after all arise from a linearization of the full problem of string dynamics) as tangent vectors to a space of deformed configurations. In our A-models, elements of $H^{1,1}(Y,\mathbb{C})$ are deformations of the complexified Kähler class, while as discussed above elements of $H^1(L,\mathbb{C})$ correspond to deformations of the Lagrangian L and flat connection on E.

The key difference between these otherwise rather similar deformations is that while closed string deformations in $(2,2)$ SCFT can never be obstructed, open string deformations can be, and often are. In physics terms, while closed string states in the first instance only correspond to solutions of a linearization of the equation of motion, it can be proven that these solutions are actually tangent vectors to a family of solutions of the full equation.[34] This need not be the case for the open string deformations.

Let us see this explicitly, first in the large structure ($l_s \to 0$) limit. The "closed string equation of motion" for Calabi-Yau geometry is the Ricci-flatness condition. By Yau's theorem, its solutions are in one-to-one correspondence with points in the Kähler cone of Y. This is an open space and thus the claim is manifest (one must also consider the B-field equation of motion, but this is linear). The corresponding claim for the B-model is that deformation of complex structure is unobstructed; this was shown for Calabi-Yau threefolds in [**448, 449**].

At finite l_s, there is an argument based on $N = (2,2)$ SCFT that closed string deformations are never obstructed [**124**]. A different but related argument uses the fact that compactification of type II string theory on a Calabi-Yau threefold leads to a $d = 4, N = 2$ supergravity theory. The result then follows from space-time arguments of a type we will discuss in §5.4 using a fact about $d = 4, N = 2$ supergravity, namely that the superpotential for uncharged fields vanishes.

Let us now turn to the open string equations of motion. For an A-brane on a Lagrangian L, these include the condition that the connection on E be flat. For a rank N bundle E, the moduli space of flat connections has real dimension $N \cdot b_1(L)$. On the other hand, there exist special connections with larger endomorphism groups. In the extreme case of N copies of the same line bundle $E \cong \mathbb{C}^N \otimes \mathcal{L}$, we have

$$\dim_{\mathbb{R}} H^1(L, \operatorname{End} E) = N^2 \cdot b_1(L)$$

which for $N > 1$ is larger. This is the puzzle raised in §3.5.2.9.

Taking into account $U(N)$ gauge invariance removes $N^2 - N$ of these open string deformations. However for $b_1(L) > 1$ this is not enough; $(b_1(L) - 1)(N^2 - N)$ of the remaining deformations must be obstructed.

Having stated the puzzle, in general terms its resolution is no mystery. It is that the condition that the connection on E be flat is nonlinear, leading to obstructions. This is easy to see in the simplest case of $L \cong T^3$, because we can choose a frame on E in which a flat connection is simply a vector of constant matrices. Then, the flatness condition takes the form

$$[A_i, A_j] = 0 \qquad \forall i, j.$$

As promised, a generic deformation $A_i \neq 0$ is obstructed. In this example, the obstruction theory is entirely given by these quadratic constraints.

In general, obstruction theory is not so simple, even at the classical level. And once we see that the open string correlation functions get quantum

[34] We should emphasize that this statement depends critically on the assumed type of supersymmetry. More generally, closed string deformations can be obstructed as well.

corrections as in (3.177), we must realize that these corrections will modify the obstruction theory.

While physics motivates the arguments we just gave, and further characterizes the obstruction theory in ways we will describe, in fact the general form of the quantum obstruction has not yet been worked out by physicists.[35] However, its general form is clear. For a Lagrangian A-brane carrying a line bundle, it is that

$$\sum_{\alpha\in I} \pm \exp\left(i \int_{D_\alpha} (B + i\omega) + i \oint_{\partial D_\alpha} A\right) [\partial D_\alpha] = 0 \quad \text{in } H_1(L), \tag{3.181}$$

where the sum is over all holomorphic disks D_α with $\partial D_\alpha \subset L$ (including multiple covers), and the notation $[\partial D_\alpha]$ refers to the homology class of ∂D_α.[36]

Only Lagrangians satisfying (3.181) are valid A-branes. There clearly are solutions; for example $L \cong S^3$ always trivially satisfies this constraint. More general explorations appear in [**276, 277**], but a general geometric understanding appears to be missing. For example, it is not known if the 3-torus fibrations of SYZ [**433**] satisfy this condition. Note also that the condition (3.181) depends on $B + i\omega$ and the value of the connection A. Thus there can be A-branes which are good for a specific value of these parameters, but not in general.

The condition (3.181) is often called "tadpole cancellation" for somewhat obscure physics reasons. Actually, the terms "vanishing tadpole" and "tadpole cancellation" have so many different meanings in related physics contexts that they are probably best avoided.

3.6.2.4. *Superpotential and A-infinity structure.* Physically, the simplest way to understand these obstructions is in terms of the world-volume approach which we will discuss in §5.4. From this point of view, it is manifest that the obstruction conditions can all be written as components of the gradient of the superpotential (3.180). We will see some examples of this in Chapter 5, such as (5.86) and (5.82).

There are some very interesting further constraints which can be best understood from the world-sheet point of view. Let us briefly discuss the analog of the chiral algebra structure discussed for the closed string in §3.3.4.3.

Open string chiral operators form an algebra, by the same argument based on the OPE (3.99). This algebra is not necessarily commutative, but will still be associative. Thus, one is tempted to generalize the WDVV discussion to the noncommutative setting.

However, the fact of obstructed deformations again leads to an essential difference from the closed string discussion, going beyond noncommutativity.

---

[35]Though many particular examples are known, beginning with [**275**].

[36]It is an interesting exercise to show that (3.181) is invariant under Hamiltonian deformations of L.

One can think of this as following from the fact that *deformations modify the Q-cohomology.* This can again be seen in the simplest example of flat bundles on $L \cong T^3$, as the dimensions of $H^p(L, \operatorname{End} E)$ vary discontinuously depending on the endomorphism group. In more physical terms, these deformations modify operator dimensions (as in the example of §3.2.3.5) in a way which violates (3.96).

Given such a situation, if we only consider correlation functions of operators in the BRST cohomology, we will inevitably lose information. On the other hand, the world-sheet arguments as we have phrased them do not easily extend to more general operators.

The resolution of this difficulty is found by enlarging the concept of associative algebra to that of an A_∞-algebra. We will discuss the definition and properties of this structure in Chapter 8. In the terms defined there, the BRST operator Q is identified with the operator m_1, and the chiral ring product with the product m_2. Thus, we can incorporate operators which lie outside the BRST cohomology, by simply allowing the possibility that $m_1 \neq 0$.

We then regard the higher point correlation functions defined using (3.179) (in other words, those which are encoded by the superpotential (3.180), and appear in the deformation theory) as defining a series of higher order products

$$m_k(a_1, a_2, \ldots, a_k); \qquad k = 3, 4, \ldots. \tag{3.182}$$

One can show using world-sheet arguments that these products satisfy the axioms of A_∞-algebra [**230**], along with certain additional "higher genus" relations. The combined system of relations should be regarded as the open string analog of the WDVV equations.

The A_∞ structure on a category of branes plays an essential role in Kontsevich's original formulation of homological mirror symmetry, as we will discuss at length in Chapter 8.

3.6.2.5. *Open strings for many A-branes.* Suppose we have a set of A-branes L_a. For simplicity of exposition, let us initially assume that each bundle E_a is a line bundle. Given a pair of A-branes L_a and L_b we will have a Hilbert space of open strings beginning on L_a and ending on L_b. This Hilbert space has a grading, which, up to an additive constant, is the ghost number. This grading shift will turn out to be very important and we will discuss it extensively below. We use the following notation for this graded Hilbert space:

$$\operatorname{Hom}^*(L_a, L_b) = \bigoplus_{m \in \mathbb{Z}} \operatorname{Hom}^m(L_a, L_b). \tag{3.183}$$

We will also denote $\operatorname{Hom}^0(L_a, L_b)$ simply by $\operatorname{Hom}(L_a, L_b)$.

By the general arguments we just discussed, a state in such a Hilbert space of open strings corresponds to an open string chiral vertex operator.

Correlation functions of these operators define a category of A-branes, and ultimately an A_∞-category.

As we mentioned earlier, it is generally believed that the resulting structure will be the same as the category introduced and named after Fukaya [**165**]. We discuss this in more depth in Chapter 8. Here let us discuss the physical arguments behind this belief, namely the considerations which determine the dimensions of the Hilbert spaces of open strings, and their correlation functions.

To compute the Hilbert space of open strings stretched between two different Lagrangian A-branes L_1 and L_2, it is easiest to assume that L_1 and L_2 intersect transversely at a finite number of points. As in §3.6.2, the Q-invariance of the topological field theory can be used to argue that open strings can only arise from constant maps $\phi : \Sigma \to Y$. This means that an open string state is associated to a point of $L_1 \cap L_2$.

The previous section suggests that locally the Hilbert space should be given by the de Rham cohomology of this intersection, i.e., the cohomology of a point. Therefore the first guess is that there is a one-dimensional Hilbert space associated with each point of intersection. Then the dimension of $\mathrm{Hom}^*(L_1, L_2)$ would be given by the number of points of $L_1 \cap L_2$. But this cannot be right. We know the A-model is invariant under Hamiltonian deformations of L_1 or L_2, but the number of intersection points is not invariant.

The simplest invariant is the *oriented* intersection number $\#(L_1 \cap L_2)$, as it depends only on homology classes. To define it, we need to prescribe the relative orientation of L_1 and L_2; note that this is the first point at which orientations have entered our discussion. Of course, this will only give us topological information about the branes.

The A-model physics which leads to the correct Hilbert spaces is holomorphic world-sheet instantons with disk topology. These produce corrections to the BRST operator Q which pair the open strings at different points of intersection, removing them from the cohomology.

Let us introduce some notation. Label the intersection points in $L_1 \cap L_2$ as p_a, $a = 1, \ldots, M$. Thus we have open string vertex operators W_{p_a} that create an open string at the point p_a. Our putative Hilbert space will be denoted $V \simeq \mathbb{C}^M$. Each vertex operator W_{p_a} has a ghost number that we denote $\mu(p_a)$. This leads to a grading of V by ghost number

$$V_i = \bigoplus_{\mu(p_a)=i} \mathbb{C}$$
$$V = \bigoplus_i V_i. \tag{3.184}$$

Now, the world-sheet instanton corrections can be parameterized as

$$\{Q, W_{p_a}\} = \sum_b n_{ab} W_{p_b}, \tag{3.185}$$

for some coefficients n_{ab} to be determined. Thus the true Hilbert space will be determined as the Q-cohomology of some complex based on the vector space V. Since Q has ghost number one, the complex looks like

$$\cdots \xrightarrow{Q} V_{-1} \xrightarrow{Q} V_0 \xrightarrow{Q} V_1 \xrightarrow{Q} \cdots \tag{3.186}$$

We define $\mathrm{Hom}^i(L_a, L_b)$ as the cohomology of this complex at position i.

To compute n_{ab} we must perform an integral over the moduli space of instantons. This integral must be performed over the fermionic parameters as well as the obvious bosonic maps ϕ. By the usual rules of fermionic integration such an integral vanishes unless the fermionic parameters cancel in some way, i.e., we have no net fermionic zero modes. To be more precise, we require that the index of the Dirac operator for the instanton be equal to the ghost number of Q, i.e., one [**466**].

The index of the Dirac operator also measures the generic (or, to be precise, virtual) dimension of the moduli space of holomorphic maps. We refer to [**469**] for a nice account of what happens in the non-generic situation. In the generic case, we thus compute n_{ab} simply by counting the number of points in the zero-dimensional instanton moduli space.

For an instanton connecting p_a to p_b, the index of the Dirac operator is given by the difference in ghost numbers $\mu(p_b) - \mu(p_a)$. Thus we expect that the generic dimension of the moduli space of instantons is given by

$$\dim \mathscr{M} = \mu(p_b) - \mu(p_a) - 1. \tag{3.187}$$

We refer the reader to [**150**] for further information on this point.

Thus, the physical definitions determine the coefficients n_{ab} as well-defined moduli space integrals – in principle. In fact these computations have not been pushed through in the physics literature. Rather, the similarity between the formalism we just described and the structure of the Fukaya category has led to the general belief that a full physical treatment would agree with the definitions of the Fukaya category.

3.6.2.6. *Ghost numbers.* An astute reader should have noticed that we have nowhere specified a way that one can actually compute $\mu(p_a)$. Given the dimensions of moduli spaces of instantons, the relation (3.187) only gives enough information to compute the *relative* ghost number of two points of intersection of L_a and L_b. Indeed, we have the following fact, which will turn out to be quite important in Chapter 5:

CLAIM 3.4. The topological A-model does not contain enough information to determine the absolute ghost number of an open string associated to a point of intersection of two A-branes.

Just how much ambiguity in the ghost number do we actually have? Given a pair of A-branes L_1 and L_2 we are free to shift the ghost numbers of the open strings from L_1 to L_2 by some fixed integer. We also saw in §3.6.2 that if $L_1 = L_2$ then the ghost number was given by the degree of de Rham cohomology, which is perfectly well-defined. Furthermore, we would like to preserve ghost number in the operator product

$$\mathrm{Hom}^i(L_1, L_2) \otimes \mathrm{Hom}^j(L_2, L_3) \to \mathrm{Hom}^{i+j}(L_1, L_3). \tag{3.188}$$

The ambiguity in the ghost number can then be accounted for by assigning a ghost number $\mu(L)$ to each A-brane itself. One then *defines* the ghost number of an element of $\mathrm{Hom}^i(L_a, L_b)$ as

$$i + \mu(L_b) - \mu(L_a). \tag{3.189}$$

It is easy to see that this definition has all the properties we desire.

We may restate the above as follows. The topological A-model has a symmetry which allows us to shift the ghost numbers of the open string states by assigning arbitrary ghost numbers to the A-branes and defining the ghost number as in (3.189). Note that this idea of assigning integers to Lagrangian submanifolds to fix this ambiguity was studied carefully in [**419**].

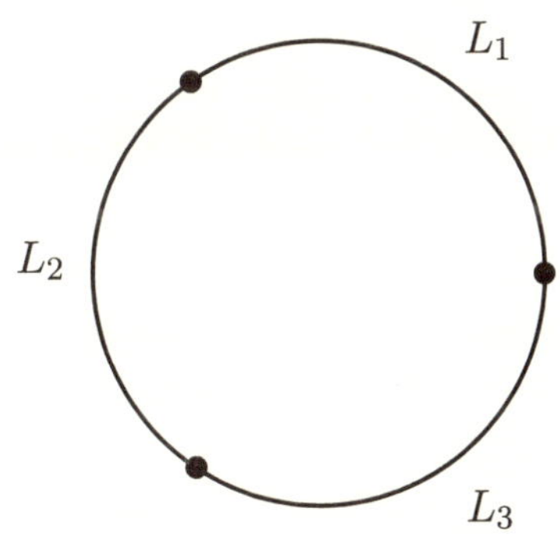

FIGURE 8. Disk instanton associated to three-point functions.

We will not give details on how to compute the correlation functions. It should be clear, however, that there will be instanton corrections involved. For example, to compute a three-point function at tree-level we will consider holomorphic disks in Y with boundary conditions shown in Figure 8. The cancellation of fermion zero modes will enforce ghost number conservation as usual.

In this section we have outlined the definition of the category of A-branes in the case that the objects L_a and L_b intersect transversely. Actually one may always use Hamiltonian deformations to deform *any* pair of Lagrangian into this case. Thus we actually have a complete definition of the category of Lagrangian A-branes.

The generalization of the Fukaya category to the case of higher rank bundles over each A-brane should be clear. Rather than associating $\mathbb{C}$ with each point of intersection, we have a matrix representing a linear map from the fiber of one bundle to the fiber of the other over the point of intersection. Generalizing to coisotropic A-branes is much more difficult and remains a problem for the future.

We emphasize that nothing in the A-model depends on the complex structure of Y. This is not at all obvious from the above definition of the Fukaya category, since the recipe for computing morphisms involves counting holomorphic maps to Y, which depend on a choice of a complex structure. On the other hand, the Fukaya category depends on $B + i\omega$ for both its objects and its composition of morphisms. The tadpole condition (3.181) has a $B + i\omega$ dependence and so certain objects might only exist for particular values of this parameter. The correlation functions depend on $B + i\omega$ through instanton corrections and so the compositions of morphisms are similarly dependent.

Finally we should point out that world-sheet instantons are generally expected to adversely affect notions based on the concept of a space-time metric. Thus it would be reasonable to expect that the concept of a Lagrangian submanifold is only really valid at large radius limit. The composition rules in the Fukaya category are based on power series associated to instanton effects. Beyond the radius of convergence of these power series it is reasonable to think that the Lagrangian submanifold description of A-branes has broken down.

3.6.3. B-type branes. We may repeat the analysis of the beginning of §3.6.1. The difference for the case of B-branes is that the B-model twist implies that we should impose $R^{\bar{\imath}}_{j} = R^{i}_{\bar{\jmath}} = 0$ for the reflection matrix in (3.154) and (3.156). That is, only the diagonal terms R^{i}_{j} and $R^{\bar{\imath}}_{\bar{\jmath}}$ are nonzero.

This means that the almost complex structure now *preserves* the tangent and normal directions to the D-brane, rather than exchanging them. It follows that the D-brane is a complex submanifold of X. Clearly this forces the real dimension of the D-brane to be even.

Note that, in contrast to the A-brane discussion, we have already used the orientation of X. If we keep track of this, there is another set of B-branes, the complex submanifolds with respect to the conjugate complex structure (equivalently, the antiholomorphic submanifolds and bundles). Of course a parallel discussion can be made for these; however, we should keep in mind that as BPS branes in the full string theory they are different.

Although B-branes of arbitrary even dimension exist, we will at first restrict attention to the case in which the B-brane fills X. That is, we impose either Neumann or mixed Dirichlet-Neumann boundary conditions. Eventually, we will be able to deduce the properties of all of the B-branes by combining an understanding of the maximal dimension B-branes with the homological algebra of Chapter 4.

As for A-branes, consideration of the B-field forces us to consider the possibility of a bundle over the B-brane, i.e., a bundle $E \to X$. Setting the B-field equal to zero, we may consider a constraint on this bundle from the requirement that the Q-variation of the action from the boundary term is zero. In this case, we find that the curvature F of the bundle is a 2-form of type (1,1) with values in End(E) [**463, 241, 242**]. In other words, $E \to X$ is a *holomorphic vector bundle.* The associated sheaf of holomorphic sections is a locally free coherent sheaf on X.

More generally, we find that the $(0,2)$ part of F must be equal to the negative of the $(0,2)$ part of B (times the identity endomorphism). For a Calabi-Yau 3-fold the $(0,2)$ part of B is always homologically trivial and can be made zero by a BRST transformation. Thus in this case the B-field has no effect on the category of B-branes.

If $B^{0,2}$ is homologically nontrivial (as can happen when X is a K3 surface or a complex torus), the situation is very different. In this case, one cannot attach to a space-filling B-brane a coherent sheaf on X, since the notion of a holomorphic section of E does not make sense. Instead, one can attach to it a coherent sheaf on a gerbe over X. For a brief discussion of how this works see, e.g., [**284**].

3.6.4. Open strings for B-branes. Using (3.156), the fermion θ_j on the boundary can be expressed through $\eta^{\bar{k}}$:

$$\theta_j = g_{j\bar{k}}(\psi_+^{\bar{k}} - \psi_-^{\bar{k}}) = F_{j\bar{k}}\eta^{\bar{k}}. \tag{3.190}$$

Thus a local operator will depend only on ϕ and $\eta^{\bar{\jmath}}$. It follows that local boundary operators are parameterized by $(0,q)$-forms with values in End(E).

Suppose we have two B-branes in the form of two vector bundles $E_1 \to X$ and $E_2 \to X$. The Chan-Paton degrees of freedom are associated with maps from E_1 to E_2. We denote the bundle of such maps by Hom(E_1, E_2).

We saw in §3.4.2 that in the B-model the BRST operator Q reduces to the Dolbeault operator in the large tension limit. Adding all these ingredients together, we see that an open string vertex operator for a string stretching from E_1 to E_2 is an element of the cohomology group

$$\oplus_q H_{\bar{\partial}}^{0,q}(X, \mathrm{Hom}(E_1, E_2)). \tag{3.191}$$

In contrast to the A-brane case, we can choose to declare the ghost number of an operator in (3.191) to be q without ambiguity.

As always, the B-model has no instanton corrections. If

$$\begin{aligned} a &\in H_{\bar{\partial}}^{0,p}(X, \mathrm{Hom}(E_1, E_2)), \\ b &\in H_{\bar{\partial}}^{0,q}(X, \mathrm{Hom}(E_2, E_3)), \\ c &\in H_{\bar{\partial}}^{0,r}(X, \mathrm{Hom}(E_3, E_1)), \end{aligned} \tag{3.192}$$

so that $p+q+r = \dim_{\mathbb{C}} X$, then the 3-point function is given by the classical expression

$$\langle W_a W_b W_c \rangle = \int_X \mathrm{Tr}(a \wedge b \wedge c) \wedge \Omega. \tag{3.193}$$

The corresponding operator product algebra is given by the ordinary wedge product of Hom-valued forms.

Higher point correlation functions are defined in the same way as discussed for A-branes in §3.6.2.2. The corresponding modification to the boundary action corresponds to deforming the Dolbeault operator by an element $\delta A^{(0,1)} \in H^1(X, \mathrm{End}(E))$ as

$$\bar{\partial} + \delta A^{(0,1)}.$$

From the sheaf-theoretic viewpoint, infinitesimal deformations take values in the global Ext group $\mathrm{Ext}^1(\mathscr{E}, \mathscr{E})$.

In [**463**] Witten showed that these correlation functions could be deduced from holomorphic Chern–Simons theory (§5.4.3.4). They also form part of an A_∞ structure, as we discuss in Chapter 8.

3.6.5. The D0-brane. Another particularly important B-type brane is the D0-brane, i.e., a brane with pure Dirichlet (constant) boundary conditions for all bosonic fields ϕ^i. This condition forces the boundary of the world-sheet to be mapped to a point $p \in X$.

Let us analyze its BRST-invariant boundary operators. BRST-invariance requires the fermionic fields $\eta^{\bar{i}}$ to vanish on the boundary, while the fields θ_i are arbitrary. Thus the most general BRST-invariant vertex operator of ghost number k has the form

$$a^{i_1 \ldots i_k} \theta_{i_1} \cdots \theta_{i_k}, \quad a^{i_1 \ldots i_k} \in \mathbb{C}. \tag{3.194}$$

Since the θ_i span $T^* X_p$ and are anticommuting, the tensor with components $a^{i_1 \ldots i_k}$ is completely skew-symmetric and should be regarded as an element of $\Lambda^k T X_p$.

Let us compute the space of morphisms from a B-brane corresponding to a holomorphic vector bundle E to the D0-brane at p. In this case, the boundary conditions force all fermions to vanish, so that boundary operators can be identified with elements of the vector space E_p^*.

One could go on to compute the space of morphisms from a B-brane supported on an arbitrary holomorphic submanifold Z to the D0-brane at p. While we will not do this, an easy result important for §5.3.3.3 is that nonzero morphisms exist only if $p \in Z$. This follows because morphisms correspond to Ramond ground states, but any non-zero winding contribution to the L_0 eigenvalue h of the string will lift it above the ground state. But the winding contribution is bounded below by the minimal distance,

$$h \geq \min_{x \in Z} d^2(p, x).$$

3.6.5.1. *Skyscraper sheaf.* In Chapter 4 we will discuss coherent sheaves, which loosely speaking generalize holomorphic bundles. It is natural to expect that the corresponding object in $D^b(X)$ is $\mathcal{O}_p$, the so-called skyscraper sheaf supported at p. The endomorphism algebra of the skyscraper sheaf in $D^b(X)$ is the exterior algebra of TX_p, which agrees with the space of boundary operators $\Lambda^k TX_p$. Its morphisms to other objects agree with those we discussed as well.

3.6.6. Supersymmetric branes and generalized calibrations. Finally, we are ready to demonstrate the relationship between superconformal boundary conditions and calibrated submanifolds as defined in §1.1.4. A-branes and B-branes as defined above preserve $N = 2$ supersymmetry, but not necessarily conformal or superconformal invariance. Given $N = 2$ supersymmetry, conformal invariance implies superconformal invariance, so it is sufficient to require the former. As explained above, this condition is equivalent to the vanishing of the boundary beta-function, which in the case $F = 0$ means that the submanifold L must be volume-minimizing.

For Lagrangian A-branes, minimality is equivalent to the condition

$$\Omega|_L = c \cdot \mathrm{vol}_L, \tag{3.195}$$

where Ω is the holomorphic top form on X and c is a constant. Such submanifolds are called special Lagrangian.

B-branes with $F = 0$ are simply complex submanifolds. A complex submanifold of dimension $2p$ in a Kähler manifold is automatically minimal and satisfies

$$\frac{\omega^p|_L}{p!} = \mathrm{vol}_L,$$

i.e., it is calibrated by the $2p$-form $\omega^p/p!$.

There is a more direct way to see the relation with calibrated geometry. Its advantage is that it works in exactly the same way whether $F = 0$ or not. The idea is to require the boundary condition to preserve Υ^2, the square of the spectral flow operator. To see how this works, let us begin with the case of a Lagrangian A-brane. We require that on the boundary of the world-sheet the following matching condition is satisfied:

$$\Omega_{i_1 \ldots i_n} \psi_+^{i_1} \cdots \psi_+^{i_n} = \bar{\Omega}_{\bar{i}_1 \ldots \bar{i}_n} \psi_-^{\bar{i}_1} \cdots \psi_-^{\bar{i}_n}. \tag{3.196}$$

Using the boundary condition $\psi_+ = R\psi_-$, one can show that this is equivalent to

$$\Omega|_L \sim \mathrm{vol_L}.$$

Similarly, for a coisotropic A-brane of real dimension $n+2p$, the matching of the spectral flow operators on the boundary implies [**287**]

$$\Omega|_L \wedge F^p = c \cdot \mathrm{vol}_L.$$

This generalizes the special Lagrangian condition for middle-dimensional A-branes.

For a rank-one B-brane of complex dimension p the spectral flow matching condition on the boundary is

$$\Omega_{i_1\ldots i_n}\psi_+^{i_1}\cdots\psi_+^{i_n} = \Omega_{i_1\ldots i_n}\psi_-^{i_1}\cdots\psi_-^{i_n} \tag{3.197}$$

Using $\psi_+ = R\psi_-$ together with the explicit form of the reflection matrix R, one finds the following condition [**287**]:

$$(\omega + iF)^{\wedge p} = c \cdot \mathrm{vol}_L. \tag{3.198}$$

Note that for small F it reduces to

$$\omega^{p-1} \wedge F = c' \cdot \mathrm{vol}_L$$

for some constant c'. This linear equation for F is the Hermitian-Yang-Mills equation (in the rank one case) and is a differential geometric counterpart of Mumford stability, as we discuss in §5.2.2. On the other hand, the nonlinear equation (3.198) is a differential-geometric counterpart of the Gieseker stability condition [**330**]. By analogy, the "special" condition (3.195) might be called the stability condition for Lagrangian A-branes. In Chapter 5 we will develop this analogy at length.

3.6.7. A failure of mirror symmetry. Given the complexities of the definition of the Fukaya category (most of which we omitted), it is rather surprising that the B-branes are so easy to analyze.

It would be remarkable if one could now invoke mirror symmetry and say that the category of A-branes on Y is equivalent to the category of B-branes on X at this point. However, this claim would be wrong with our current definition of B-branes.

The most obvious problem is that we have left out the B-branes on lower-dimensional submanifolds of X. These are somewhat more difficult to analyze, and we will do this by indirect means in Chapter 5. However, even adding these, we fall far short of the number of objects in the Fukaya category.

Another very significant point was the role of orientation. Whereas the A-branes necessarily include both orientations, in the B-brane discussion each brane comes in two orientations, with no obvious distinction in the formalism discussed so far. In the physics, these are "branes" and "antibranes," which are very different, as we will explain in Chapter 5. A correct definition will have to incorporate both.

More subtle clues come from the comparison of different CFT's which are supposed to lead to the same TFT, as discussed in [**135**]. For example, two nonlinear sigma models whose target spaces are birationally equivalent CY's, X and X', are in fact connected by deformation of stringy Kähler moduli. Both sit in a larger moduli space, most simply defined as the complex structure moduli space of their common mirror Y.

Let us consider such an example from the point of view of the twisted theories. Since we varied Kähler moduli, it is no surprise if the two A-models are different. But if we fix the complex structure moduli and make this variation, the B-models for X and X' should be the same.

With our current definitions, this leads to the prediction that the categories of coherent sheaves on X and X' are equivalent, but this is false, as we will see in the next chapter. To restore invariance of the TFT on Kähler moduli, we must find a larger category of B-branes which includes both categories of coherent sheaves as subcategories.

This type of argument can be carried even further by considering nontrivial loops in the stringy Kähler moduli space. As we discussed above in the context of the quintic, these loops induce an $\mathrm{Sp}(n, \mathbb{Z})$ monodromy action on the even homology lattice. This action includes elements which change the dimension of a brane, and even elements which can turn branes into antibranes. Thus the correct category of B-branes will have to treat branes and antibranes on an equal footing.

A simple and physically well motivated way to do this is to use K-theory. However, this throws away almost all of the information we have discussed. But the (late 1990's) physics point of view did not really suggest any other satisfactory way to proceed. To go further, we will need the tools of modern representation theory, as we discuss in the next chapter.

3.6.8. B-branes in Landau-Ginzburg models. While the focus of this book is 2d field theories and boundary conditions which have superconformal invariance, in this section (which assumes more physics background) we briefly describe another class of supersymmetric 2d theories which are not conformally-invariant, first discussed in §2.4. These are $N = (2,2)$ Landau-Ginzburg models, with or without boundaries.

A Landau-Ginzburg model depends on a noncompact Kähler manifold X (the target) and a holomorphic function W on X (the superpotential). Loosely speaking, one can regard such a theory as a deformation of the $N = (2,2)$ sigma model with target X by a holomorphic function W on X. However, this viewpoint is not very useful, since W necessarily becomes large in some regions of X.

As in most of the literature, we will confine ourselves to the case $X = \mathbb{C}^n$ with a standard flat metric. The action takes a rather simple form

$$\frac{1}{4\pi} \int_\Sigma d^2 z \left\{ \frac{\partial \phi^j}{\partial z} \frac{\partial \phi^{\bar{\jmath}}}{\partial \bar{z}} + \frac{\partial \phi^j}{\partial \bar{z}} \frac{\partial \phi^{\bar{\jmath}}}{\partial z} \right. \\ \left. + i\psi_-^{\bar{\jmath}} D \psi_-^j + i\psi_+^{\bar{\jmath}} \bar{D} \psi_+^j - \frac{1}{4} \partial_j W \partial_{\bar{\jmath}} \bar{W} - \frac{1}{2} \partial_i \partial_j W \psi_+^i \psi_-^j - \frac{1}{2} \partial_{\bar{\imath}} \partial_{\bar{\jmath}} \bar{W} \psi_-^{\bar{\imath}} \psi_+^{\bar{\jmath}} \right\}. \tag{3.199}$$

Note that the "potential energy" for the bosonic field ϕ has the form

$$V(\phi) = \frac{1}{4}|dW|^2.$$

Thus zero-energy field configurations must have $\phi = p$, where p is one of the critical points of W. The case of most interest to us is when the critical points of W are isolated, because then each critical point corresponds to a vacuum state in the quantized theory. Furthermore, the infrared behavior of the theory in any particular vacuum $\phi = p$ depends only on the germ of W at p. This is because taking the infrared limit is equivalent to rescaling $W \to \lambda W$ and taking the limit $\lambda \to \infty$. In this limit, the energy barriers between different vacua become infinitely high, and the wave-function for ϕ is concentrated near $\phi = p$.

As a direct consequence of this, singularity theory can be used to classify the infrared behavior of LG models. The simplest singularities (singularities of modality zero) have ADE classification [**14**], and it turns out that the corresponding infrared theories are precisely the ADE minimal models. More complicated singularities correspond to irrational superconformal field theories.

For any W the LG model has a $U(1)_A$ R-symmetry which multiplies ψ_+^i and $\psi_-^{\bar{\imath}}$ by $e^{i\alpha}$ and ψ_-^i and $\psi_+^{\bar{\imath}}$ by $e^{-i\alpha}$. Therefore it admits a topological twist of type B. There is no A-twist, in general, because the superpotential breaks the $U(1)_V$ R-symmetry of the sigma model down to $\mathbb{Z}_2$. If W is homogeneous, one can redefine the action of $U(1)_V$ so that the action is invariant. However, unless W is quadratic, the $U(1)_V$-charges of the fields are fractional. To avoid this, one can orbifold the theory in such a way that operators with fractional R-charge are projected out. The resulting Landau-Ginzburg orbifold theory admits an A-twist. There is an extensive literature both on A-twisted and B-twisted LG models, see e.g., [**241, 240, 336, 286, 70, 288, 72, 326, 229**]. We will only discuss the B-twist and the corresponding topological boundary conditions (B-branes), always assuming that $X = \mathbb{C}^n$.

Let us begin with the closed sector of the B-twisted LG model. On general grounds (see Chapter 2), the algebra of observables must be a supercommutative Frobenius algebra. According to [**454**], it is the Jacobian algebra

$$\mathbb{C}[x_1, \ldots, x_n]/I_{dW},$$

where I_{dW} is the ideal generated by the functions $\partial_i W$, $i = 1, \ldots, n$. One can motivate this answer by regarding the Jacobian algebra as the hypercohomology of the complex of sheaves

$$\Lambda^n T_X \xrightarrow{\iota_{dW}} \Lambda^{n-1} T_X \xrightarrow{\iota_{dW}} \cdots \xrightarrow{\iota_{dW}} \mathcal{O}_X.$$

For $X = \mathbb{C}^n$, it is equal to the cohomology of the complex of vector spaces

$$H^0(\Lambda^n T_X) \xrightarrow{\iota_{dW}} H^0(\Lambda^{n-1} T_X) \xrightarrow{\iota_{dW}} \cdots \xrightarrow{\iota_{dW}} H^0(\mathcal{O}_X),$$

which is nonvanishing only in top degree and isomorphic to the Jacobian algebra. On the other hand, for compact X the function W is constant, and the hypercohomology reduces to the sheaf cohomology of the vector bundle

$$\oplus_p \Lambda^p T_X,$$

as expected for the B-model with target X.

The Frobenius trace function is less obvious:

$$\mathrm{tr} : f \mapsto \sum_p \mathrm{Res}_p \frac{f\, dx_1 \wedge \cdots \wedge dx_n}{\partial_1 W \cdots \partial_n W}.$$

Here the multidimensional residue is defined as in **[197]** by the integral

$$(2\pi i)^{-n} \int_\Gamma \frac{f\, dx_1 \wedge \cdots \wedge dx_n}{\partial_1 W \cdots \partial_n W},$$

where Γ is a real n-cycle in a small neighbourhood of the critical point p defined by the equations $|\partial_i W| = \varepsilon > 0$, $i = 1, \ldots, n$. One can show **[197]** that the scalar product $(f, g) = \mathrm{tr}(fg)$ is nondegenerate, which is one of the axioms of 2d TFT.

Now let us turn to the open sector of the LG model. M. Kontsevich proposed that the category of B-branes is equivalent to the category of matrix factorizations of the superpotential W. Later this proposal was motivated by physical considerations **[286, 70, 288, 326, 229]**. We will confine ourselves to defining the category of matrix factorizations and refer an interested reader to the literature cited above for a physical interpretation and applications to string theory.

An object of the category of matrix factorizations is a pair of $k \times k$ matrices D_0, D_1 polynomially depending on $x_1, \ldots, x_n$ and satisfying the equation

$$D_0 D_1 = D_1 D_0 = (W - c) \cdot 1_{k \times k}.$$

The constant c is arbitrary, but it turns out that the category of matrix factorizations is nontrivial only if c is one of the critical values of W, and that there are no nontrivial morphisms between objects corresponding to different c. For this reason, it suffices to let c to be some particular critical value of W, which we may take to be 0 without loss of generality.

To define morphisms, it is convenient to think of the pair D_0, D_1 as an odd operator

$$D = \begin{pmatrix} 0 & D_1 \\ D_0 & 0 \end{pmatrix}$$

acting on a $\mathbb{Z}_2$-graded $\mathcal{O}$-module $V \simeq \mathbb{C}^k[x_1, \ldots, x_n] \oplus \mathbb{C}^k[x_1, \ldots, x_n]$ for some k. Here and below we denote $\mathcal{O} = \mathbb{C}[x_1, \ldots, x_n]$. Then D is an odd operator satisfying $D^2 = W$. One says that D is a twisted differential on V (an ordinary differential is an odd endomorphism D satisfying $D^2 = 0$). One refers to a pair (D, V) as a matrix factorization of W. Given two such pairs

(D, V) and (D', V') we define the space of morphisms to be the cohomology of the $\mathbb{Z}_2$-graded complex

$$\mathrm{Hom}_{\mathcal{O}}(V, V')$$

with the differential Q given by

$$Q\phi = D'\phi \pm \phi D, \quad \phi \in \mathrm{Hom}_{\mathcal{O}}(V, V'),$$

where the sign is plus or minus depending on whether ϕ is an odd or even element of the $\mathbb{Z}_2$-graded vector space $\mathrm{Hom}_{\mathcal{O}}(V, V')$. The composition of morphisms is defined in an obvious way. Note that the resulting category is $\mathbb{Z}_2$-graded, rather than $\mathbb{Z}$-graded as for the topological sigma model of type B. This is because the superpotential breaks the $U(1)_V$ symmetry down to $\mathbb{Z}_2$.

As explained in Chapter 2, in order to define an open-closed TFT, we need to specify an open trace function θ on the endomorphism algebra of each matrix factorization, as well as an open-closed map ι from the endomorphism algebra to the Jacobian algebra. These have been proposed in [**288**] and later derived more carefully from the Landau-Ginzburg path-integral in [**229**]. The open trace is

$$\vartheta(\phi) = \sum_p \frac{1}{n!}\mathrm{Res}_p \frac{\mathrm{str}\,(\phi\,(dD)^{\wedge n})}{\partial_1 W \cdots \partial_n W}, \quad \phi \in \mathrm{Hom}_{\mathcal{O}}(V, V),\ Q(\phi) = 0.$$

Here str denotes matrix supertrace.

The open-closed map is

$$\iota(\phi) = \frac{1}{n!}\mathrm{str}\left(\phi\,(dD)^{\wedge n}\right).$$

The dual closed-open map has been explicitly constructed in [**292**].

These data are expected to satisfy the axioms of Chapter 2. At the time of writing, a rigorous proof of this is lacking.

As a simple example, let us consider the LG model of type A_{m-2} in ADE classification. The corresponding superpotential is $W = x^m$. The most obvious matrix factorization is given by $D_0 = z^p, D_1 = z^{m-p}$. Let us denote this matrix factorization M_p, $p = 0, \ldots, m$. It turns out that any object in the category of matrix factorizations is equivalent to a direct sum of several copies of M_p with $p \in \{1, \ldots, m-1\}$ (the objects M_0 and M_m turn out to be isomorphic to the zero object), so it is sufficient to study the matrix factorizations M_p. For example, the endomorphism algebra of M_p is [**70, 288**] a $\mathbb{Z}_2$-graded algebra generated by an even element a and an odd element η with relations

$$\eta a = a\eta, \quad a^q = 0, \quad \eta^2 = -a^{p-2q}, \quad q = \min\{p, m-p\}$$

This shows that M_p is nontrivial (not isomorphic to the zero object) if $p = 1, \ldots, m-1$. One can also show that M_p and $M_{p'}$ are not isomorphic if $p \neq p'$.

Another simple example is $W = x_1^2 + \cdots + x_n^2$. In this case the category of B-branes is equivalent to the category of representations of the Clifford algebra with n generators [**240, 286**]. This example shows that the closed TFT (which is the same for all n) does not uniquely determine the open-closed TFT (which depends on whether n is even or odd).

CHAPTER 4

Representation theory, homological algebra and geometry

Let us now take a more mathematical approach, and switch our study to algebraic properties of categories of holomorphic vector bundles, and more generally, of coherent sheaves.

Although these techniques have been part of algebraic geometry for a long time, they have acquired new impetus recently, at least in part because they fit so well with ideas coming from string theory. In particular, the only known mathematical descriptions of branes in non-perturbative string theory involve constructions from homological algebra in a crucial way; the more intuitive geometrical approaches can only give valid descriptions of branes in a neighbourhood of a large volume limit point. Moreover, many global symmetries (or dualities) of string theories can so far only be understood mathematically as equivalences of derived categories (Fourier-Mukai transforms).

It is often assumed that this use of homological algebra in the description of non-perturbative string theories will eventually be replaced by a yet-to-be-discovered "stringy geometry". This is certainly an attractive prospect. However, any such geometry will have to be an extremely radical extension of traditional geometry; it seems likely that some sort of non-commutativity will be an essential feature.

Mathematically speaking, the ideas of this chapter fit into a general framework which can be called representation theory. Put simply, the basic idea is to try to understand the structure of a nonlinear object $\mathfrak{X}$ by studying spaces on which $\mathfrak{X}$ acts linearly. These spaces should be considered as the objects of a category in which the morphisms are linear maps commuting with the action of $\mathfrak{X}$.

A good example is the theory of group representations. Studying the set of all representations of a group on vector spaces, as well as the maps between them, allows one to understand the structure of the group better. In more technical language, this amounts to studying the group G via its category of (finite-dimensional) representations $\mathbf{Rep}(G)$. More generally, if A is an algebra, one can study the category $\mathbf{Mod}(A)$ of finitely generated modules over A. Again, this has proved to be an important tool in the study of algebras. Representations of finite groups are the special case when

$A = \mathbb{C}[G]$ is a group ring. Other easily-computable examples are provided by path algebras of quivers. Modules over these algebras are easily understood in terms of diagrams of vector spaces and maps between them. We give a basic introduction to quivers in §4.2.

The case of most importance for us is the category $\mathrm{Coh}(X)$ of coherent sheaves on a complex projective variety X. We can think of such a variety as a collection of affine varieties glued together. Each affine variety corresponds to a (commutative) algebra. Taking finitely generated modules over each of these algebras and glueing leads to the concept of a coherent sheaf. Thus, in the same way as a vector bundle is a global version of a vector space, a coherent sheaf is a global version of a module. Coherent sheaves are more general than vector bundles; in particular they can be supported on subvarieties. This will be discussed in more detail in §4.3.

Category theory is the basic language that mathematicians have developed to describe these different situations, but the general notion of a category is far too general for our purposes. The categories which arise in the contexts described above all share many of the properties of the category of modules over an algebra. Thus, one can define pointwise addition for maps between modules, any such map has a kernel and cokernel, and so on. Abstracting these properties leads to the notions of additive and abelian categories which are described in §4.1.

A useful idea when studying categories is to study generating sets. Thus, any representation of a finite group is a direct sum of irreducible (or simple) representations[1]. In more technical language one says that $\mathbf{Rep}(G)$ is semisimple. This means that everything about the category can be determined from knowing the set of irreducible representations. A slightly more complicated example is the category of representations of a quiver. Here again every object can be made by forming "bound states" of a finite number of simple objects. Such a category is said to have finite length. The difference from the semisimple case is that rather than everything being a direct sum of simple objects, there is some nontrivial glueing data which describes how the "bound states" are formed. This glueing data is described by the arrows of the quiver. Thus the category becomes more complicated. The categories of coherent sheaves occurring in algebraic geometry are even more complicated and are not of finite length. Nonetheless, we will see that a recurring theme is the idea of identifying certain "basic" or "irreducible" objects, and considering other objects to be "bound states" of these.

Representation theory only really becomes interesting when one introduces complexes of representations. The study of operations on complexes is known as homological algebra. A crucial tool in modern homological algebra is the derived category $\mathrm{D}(\mathcal{A})$ of an abelian category $\mathcal{A}$. The objects of $\mathrm{D}(\mathcal{A})$ are complexes of objects of $\mathcal{A}$ considered up to an equivalence relation

[1]We always work over the complex numbers.

called quasi-isomorphism. The details of this construction will be described in §4.4.

If X is a complex projective variety, we may consider the derived category of the abelian category of coherent sheaves on X. This category is usually denoted $\mathrm{D}(X)$ and referred to as the derived category of X. It is of crucial importance in string theory where it arises as the category of D-branes in the B-type topologically twisted theory. Many dualities occurring in the physics literature can be interpreted mathematically as equivalences of derived categories. In fact, Kontsevich's homological mirror conjecture states that mirror symmetry itself can be understood as a type of derived equivalence. Equivalences between derived categories of sheaves are known in the mathematics literature as Fourier-Mukai transforms. We present many examples in §4.6. A particularly interesting example is the categorical McKay correspondence, discussed in §4.7.

The existence of derived equivalences has the striking consequence that a fixed derived category D can be associated to various seemingly unrelated algebraic and geometric objects. For example, D can simultaneously be the derived category of sheaves on a projective space, and the derived category of modules over a finite-dimensional algebra. These various incarnations of D can be thought of as various geometric (or algebraic) phases of the same underlying theory. The tool which allows one to treat these different phases at the same time is the idea of a t-structure. We will introduce this concept in §4.4.6; an example corresponding to strings moving on the non-compact Calabi-Yau threefold $\mathcal{O}_{\mathbb{P}^2}(-3)$, the total space of the canonical line bundle over the projective plane, will be discussed in §5.8.3.2 in the next chapter.

One point which physicists often do not appreciate is that there are actually very few general facts known about representation theory in the sense described above. Much is now known about representation theory in low dimensions; the theory for finite groups (over $\mathbb{C}$) typifies the zero-dimensional case, and the theory of representations of a quiver is a good example of the one-dimensional case. A certain amount is also known about coherent sheaves on varieties of dimension two, particularly in the presence of the Calabi-Yau condition. But almost nothing is known about coherent sheaves or vector bundles on varieties of dimension three or more. Thus if one were to take a quintic threefold and fix a set of Chern classes, there would be no known method for saying anything non-trivial about moduli spaces of coherent sheaves with those Chern classes. In fact, it is not unreasonable to expect string theory to provide new methods for understanding such problems.

References: Some words about recommended books. The main reference for homological algebra and the derived category is Gelfand and Manin [**174**]. This is an excellent book, although the reader should keep an eye out for misprints and minor errors in the first edition. A standard reference for coherent sheaves is Hartshorne's justly famous textbook [**222**]. For basic category theory Blyth [**48**] is to be recommended, although even this small book contains more information than most geometers or algebraists will ever want to know. For derived categories, in addition to [**174**], there is the original treatment by Verdier [**456**] or Hartshorne's write-up of Grothendieck's notes [**221**]. For the less committed, Thomas's thought-provoking [**445**] provides a gentle introduction. Finally, those wanting to know more about representations of quivers might start by looking in Auslander and Reiten [**27**].

4.1. Categories (additive and abelian) and functors

In the introduction to this chapter we described various contexts in which it is useful to replace a nonlinear object by the category of its representations. The aim of this section is to introduce a general language which can be applied in all of these cases. The basic idea is that categories of representations are abelian, and that much of the theory developed for modules over rings extends without change to more general representation theoretic contexts such as categories of sheaves.

The material of this section is necessarily rather dry and content-free, since as with any mathematical language, the basic vocabulary consists of a rather large number of definitions which need to be fully absorbed before one can move on to consider more interesting statements. The reader is advised to keep a concrete example in mind, and to read this section in parallel with §§4.2 and 4.3, devoted to quiver representations and coherent sheaves respectively, where many relevant examples are given.

4.1.1. Categories. Categories appear all over mathematics. The basic definition is as follows.

DEFINITION 4.1. A category $\mathcal{C}$ consists of a collection of objects $\mathrm{Ob}(\mathcal{C})$ together with sets of morphisms

$$\mathrm{Hom}_{\mathcal{C}}(A,B)$$

for each pair of objects $A,B\in\mathrm{Ob}(\mathcal{C})$, and composition laws

$$\circ\colon \mathrm{Hom}_{\mathcal{C}}(A,B)\times\mathrm{Hom}_{\mathcal{C}}(B,C)\longrightarrow\mathrm{Hom}_{\mathcal{C}}(A,C)$$

for each triple $A,B,C\in\mathrm{Ob}(\mathcal{C})$, such that

(a) the composition law is associative, that is

$$f\circ(g\circ h)=(f\circ g)\circ h;$$

(b) for each object $A\in\mathrm{Ob}(\mathcal{C})$ there is a morphism $\mathrm{id}_A\in\mathrm{Hom}_{\mathcal{C}}(A,A)$ which is a left and right identity for the composition law.

In practice, it is common to confuse a category $\mathcal{C}$ with its objects, so that one writes $A\in\mathcal{C}$ rather than $A\in\mathrm{Ob}(\mathcal{C})$.

EXAMPLES 4.2. Almost all mathematical structures can be viewed naturally as objects of a category. For example, there is a category **Sets** whose objects are sets and whose morphisms are functions, a category **Top** whose objects are topological spaces and whose morphisms are continuous maps, and so on. In all such cases the composition law is composition of maps. In the case $\mathcal{C} = \mathbf{Rep}(G)$ mentioned in the introduction, the objects are finite-dimensional complex representations of a group G, and for any two such representations $U, V \in \mathbf{Rep}(G)$, the set of morphisms $\mathrm{Hom}_{\mathcal{C}}(U, V)$ is the set of G-equivariant linear maps $U \to V$, also called G-maps for short.

EXAMPLE 4.3. Given a ring R, there is a category $\mathbf{Mod}(R)$ in which the objects are the finitely generated left R-modules, and the morphisms are morphisms of left R-modules. In particular we use the notation $\mathbf{Mod}(\mathbb{Z})$ for the category of finitely generated abelian groups, and $\mathbf{Mod}(\mathbb{C})$ for the category of finite-dimensional complex vector spaces.

EXAMPLE 4.4. Any group G defines a category with one object X satisfying $\mathrm{Hom}(X, X) = G$. Composition of morphisms is given by the multiplication in G. Of course, one does not need the existence of inverses in G for this construction, so more generally, one can associate a one-object category to any monoid (i.e., a set with an associative product and an identity element).

EXAMPLE 4.5. Categories arise in topological field theory. In these examples the objects are boundary conditions in the theory, and the morphism sets $\mathrm{Hom}(P, Q)$ are spaces of states of topological strings stretching from P to Q. See Chapter 3 for examples.

EXAMPLE 4.6. Suppose Q is a quiver (directed graph, see the beginning of §4.2 for the formal definition). The path category $\mathcal{C}(Q)$ of Q is defined as follows. The objects of $\mathcal{C}(Q)$ are the vertices of Q. Given two vertices $i, j \in Q$ the set $\mathrm{Hom}_{\mathcal{C}(Q)}(i, j)$ is defined to be the set of finite length directed paths in Q beginning at i and ending at j. Composition of morphisms is defined by concatenation of paths, and the identity morphism corresponding to a vertex $i \in Q$ is the zero length path starting and finishing at i.

A morphism $f\colon M \to N$ in a category $\mathcal{C}$ is an isomorphism if there is a morphism $g\colon N \to M$ in $\mathcal{C}$ such that $g \circ f = \mathrm{id}_M$ and $f \circ g = \mathrm{id}_N$. One then says that the objects M and N are isomorphic in $\mathcal{C}$; this is usually written $M \cong N$.

As suggested by Example 4.6, one often visualises a category by thinking of the objects as vertices of a graph and the morphisms as arrows. Thus a diagram such as

$$\begin{array}{ccc} A & \xrightarrow{f} & B \\ {\scriptstyle h}\downarrow & & \downarrow{\scriptstyle g} \\ C & \xrightarrow{i} & D \end{array}$$

is supposed to represent four objects A, B, C, D of a category $\mathcal{C}$ together with morphisms $f \in \mathrm{Hom}_{\mathcal{C}}(A, B)$, $g \in \mathrm{Hom}_{\mathcal{C}}(B, D)$, etc. satisfying the relation $g \circ f = i \circ h$. In this situation one says that the diagram "commutes".

4.1.2. Functors. A functor is a map between categories which preserves the relevant structure. The definition is as follows.

DEFINITION 4.7. If $\mathcal{C}_1$ and $\mathcal{C}_2$ are categories then a (covariant) functor $F: \mathcal{C}_1 \to \mathcal{C}_2$ consists of a map $F\colon \mathrm{Ob}(\mathcal{C}_1) \to \mathrm{Ob}(\mathcal{C}_2)$ together with maps

$$F\colon \mathrm{Hom}_{\mathcal{C}_1}(A, B) \longrightarrow \mathrm{Hom}_{\mathcal{C}_2}(F(A), F(B))$$

for every pair of objects $A, B \in \mathrm{Ob}(\mathcal{C}_1)$. These maps must satisfy

$$F(f \circ g) = F(f) \circ F(g)$$

for all composable morphisms f, g in $\mathcal{C}_1$, and $F(\mathrm{id}_A) = \mathrm{id}_{F(A)}$ for all objects $A \in \mathrm{Ob}(\mathcal{C}_1)$.

EXAMPLE 4.8. For each integer i there is a functor $H_i\colon \mathbf{Top} \to \mathbf{Mod}(\mathbb{Z})$ which assigns to a topological space X the singular homology group $H_i(X)$ with coefficients in $\mathbb{Z}$. A continuous map of topological spaces $f\colon X \to Y$ induces a group homomorphism $H_i(f)\colon H_i(X) \to H_i(Y)$.

EXAMPLES 4.9. Suppose G is a group and $\mathcal{C}(G)$ is the corresponding one-object category (see Example 4.4). If H is another group then a functor $\mathcal{C}(G) \to \mathcal{C}(H)$ is just a group homomorphism $G \to H$.

It is also important to have a notion of a morphism between two functors.

DEFINITION 4.10. Given two functors $F, G\colon \mathcal{C}_1 \longrightarrow \mathcal{C}_2$, a morphism of functors $\eta\colon F \longrightarrow G$, also called a natural transformation, consists of morphisms

$$\eta(A)\colon F(A) \longrightarrow G(A)$$

for each object $A \in \mathcal{C}_1$, such that for each morphism $f\colon A \to B$ in $\mathcal{C}_1$, the diagram

$$\begin{array}{ccc} F(A) & \xrightarrow{\eta(A)} & G(A) \\ {\scriptstyle F(f)}\downarrow & & \downarrow{\scriptstyle G(f)} \\ F(B) & \xrightarrow{\eta(B)} & G(B) \end{array}$$

commutes. An isomorphism of functors is a morphism of functors in which each morphism $\eta(A)$ is an isomorphism.

EXAMPLES 4.11. Let G be a group and $\mathcal{C}(G)$ the corresponding one-object category (see Example 4.4). A functor $F\colon \mathcal{C}(G) \longrightarrow \mathbf{Mod}(\mathbb{C})$ is a finite-dimensional representation of G. If $F, G\colon \mathcal{C}(G) \longrightarrow \mathbf{Mod}(\mathbb{C})$ are two such functors then a morphism of functors $\eta\colon F \to G$ is a G-map between the corresponding representations.

Similarly, if Q is a quiver with path-category $\mathcal{C}(Q)$ (see Example 4.6), then a functor $F\colon \mathcal{C}(Q) \longrightarrow \mathbf{Mod}(\mathbb{C})$ is a finite-dimensional representation of Q (see §4.2). If $F, G\colon \mathcal{C}(Q) \longrightarrow \mathbf{Mod}(\mathbb{C})$ are two such functors then a morphism of functors $\eta\colon F \to G$ is a morphism between the corresponding representations.

DEFINITION 4.12. A functor $F\colon \mathcal{A} \longrightarrow \mathcal{B}$ is

(a) fully faithful if for each pair of objects $A_1, A_2 \in \mathcal{A}$ the induced map

$$F\colon \mathrm{Hom}_{\mathcal{A}}(A_1, A_2) \longrightarrow \mathrm{Hom}_{\mathcal{B}}(F(A_1), F(A_2))$$

is a bijection;

(b) an equivalence of categories if it is fully faithful, and is "surjective up to isomorphism", i.e., every object $B \in \mathcal{B}$ is isomorphic to an object $F(A)$ for some $A \in \mathcal{A}$.

It follows immediately from the definition that fully faithful functors are "injective on objects": $F(A) \cong F(B) \implies A \cong B$. It is also easy to show that the statement that $F\colon \mathcal{A} \longrightarrow \mathcal{B}$ is an equivalence of categories is equivalent to the existence of a functor $G\colon \mathcal{B} \longrightarrow \mathcal{A}$ such that there are isomorphisms of functors

$$G \circ F \longrightarrow \mathrm{id}_{\mathcal{A}} \text{ and } F \circ G \longrightarrow \mathrm{id}_{\mathcal{B}}.$$

Such a functor G is called a quasi-inverse (or just inverse) for F. The following old chestnut is a good example of an equivalence of categories.

EXAMPLE 4.13. Define a category $\mathbb{N}(\mathbb{C})$ whose objects are the non-negative integers, such that the set of morphisms from m to n is the set of complex-valued $n \times m$ matrices, with composition given by multiplication of matrices. There is a functor

$$F\colon \mathbb{N}(\mathbb{C}) \longrightarrow \mathbf{Mod}(\mathbb{C})$$

sending the object n to the vector space $\mathbb{C}^n$, and a matrix

$$M \in \mathrm{Hom}_{\mathbb{N}(\mathbb{C})}(m, n)$$

to the corresponding linear map $M\colon \mathbb{C}^m \to \mathbb{C}^n$. This functor F is an equivalence.

It tends to be rather natural to identify equivalent categories, although one should be a little careful, since as Example 4.13 shows, equivalent categories can have very different sizes.

Given a collection S of objects of a category $\mathcal{A}$, we can define a new category $\mathcal{B}$ whose objects are the elements of S, with morphisms

$$\mathrm{Hom}_{\mathcal{B}}(A_1, A_2) = \mathrm{Hom}_{\mathcal{A}}(A_1, A_2) \text{ for } A_1, A_2 \in S.$$

We call $\mathcal{B}$ a full subcategory of $\mathcal{A}$ and write $\mathcal{B} \subset \mathcal{A}$. There is an obvious inclusion functor $F\colon \mathcal{B} \to \mathcal{A}$ and this is clearly fully faithful. More general subcategories are obtained by throwing away morphisms as well as objects, but these won't be important in what follows.

If $F\colon \mathcal{B} \to \mathcal{A}$ is a fully faithful functor, then F defines an equivalence of $\mathcal{B}$ with the full subcategory of $\mathcal{A}$ consisting of those objects of $A \in \mathcal{A}$ such that $A \cong F(B)$ for some $B \in \mathcal{B}$. We can thus think of fully faithful functors as embeddings of categories.

4.1.3. Additive categories. One can easily dream up as many examples of categories as one wishes. However, the categories which appear in representation theory have some special properties. In particular, the fact that the objects are in some sense linear, and the maps between them are linear maps, gives the morphism sets some extra structure.

For example, if M, N are left modules over a ring A and

$$f, g \in \mathrm{Hom}_A(M, N)$$

are module maps, then one can define a module map $f + g\colon M \to N$ by

$$(f+g)(m) = f(m) + g(m) \quad \text{for } m \in M,$$

and if $h\colon L \to M$ and $j\colon N \to P$ are module maps then

$$(f+g) \circ h = f \circ h + g \circ h \qquad\qquad j \circ (f+g) = j \circ f + j \circ g.$$

In this way, each morphism set $\mathrm{Hom}_A(M, N)$ becomes an abelian group, and the composition law becomes biadditive: it defines a group homomorphism

$$\mathrm{Hom}_A(M, N) \times \mathrm{Hom}_A(N, P) \longrightarrow \mathrm{Hom}_A(M, P).$$

DEFINITION 4.14. A pre-additive category is a category in which the morphism sets have the structure of abelian groups and in which the composition law is biadditive.

To get the notion of an additive category it is convenient to throw in a couple of other properties which module categories always have, namely zero objects and finite direct sums.

A zero object in a preadditive category $\mathcal{C}$ is an object (usually denoted 0) such that for any object $A \in \mathcal{C}$ the morphism sets $\mathrm{Hom}_{\mathcal{C}}(A, 0)$ and $\mathrm{Hom}_{\mathcal{C}}(0, A)$ are the one-element (trivial) group 0. For any ring R the trivial module 0 is a zero object in $\mathbf{Mod}(R)$.

A direct sum of two objects M and N of a pre-additive category $\mathcal{C}$ is an object of $\mathcal{C}$, usually written $M \oplus N$, which has chosen morphisms

$s\colon M \to M \oplus N$ and $t\colon N \to M \oplus N$ such that for any object $P \in \mathcal{C}$ there is an isomorphism of groups

$$\mathrm{Hom}_{\mathcal{C}}(P, M) \times \mathrm{Hom}_{\mathcal{C}}(P, N) \cong \mathrm{Hom}_{\mathcal{C}}(P, M \oplus N),$$

induced by the map $(f, g) \mapsto s \circ f + t \circ g$. If M and N are modules over a ring R then it is easy to check that the usual module direct sum $M \oplus N$ is a direct sum in the category $\mathbf{Mod}(R)$ in the above sense.

DEFINITION 4.15. An additive category is a preadditive category $\mathcal{A}$ with a zero object $0 \in \mathcal{A}$ such that any two objects $M, N \in \mathcal{A}$ have a direct sum $M \oplus N \in \mathcal{A}$.

Often, one considers additive categories in which the morphism sets are not only abelian groups, but are in fact complex vector spaces, in such a way that the composition law is a bilinear map

$$\mathrm{Hom}_A(M, N) \times \mathrm{Hom}_A(N, P) \longrightarrow \mathrm{Hom}_A(M, P).$$

Such categories are called $\mathbb{C}$-linear. For example if A is an algebra over $\mathbb{C}$ (which is to say a ring containing $\mathbb{C}$ as a subring), then as well as adding module maps pointwise as above, one can also multiply maps pointwise by scalars

$$(\lambda f)(m) = \lambda \cdot f(m) \text{ for } \lambda \in \mathbb{C}.$$

EXAMPLES 4.16. For any ring R, the category $\mathbf{Mod}(R)$ is an additive category. If furthermore R contains $\mathbb{C}$ then $\mathbf{Mod}(R)$ is a $\mathbb{C}$-linear additive category. In particular, if G is a finite group, then the representation category $\mathbf{Rep}(G)$ is $\mathbb{C}$-linear. If X is a topological space, there is a $\mathbb{C}$-linear additive category whose objects are complex vector bundles on X and whose morphisms are morphisms of vector bundles.

A good way to think of an additive (or pre-additive) category $\mathcal{C}$ is as a ring with many identities. Each morphism set $\mathrm{Hom}_{\mathcal{C}}(A, B)$ is an abelian group. Consider the direct sum of all these groups

$$A(\mathcal{C}) = \bigoplus_{A, B \in \mathcal{C}} \mathrm{Hom}_{\mathcal{C}}(A, B).$$

The composition law on $\mathcal{C}$ induces a product on this abelian group $A(\mathcal{C})$, where for $f \in \mathrm{Hom}(A, B)$ and $g \in \mathrm{Hom}(C, D)$ one sets $g \cdot f = g \circ f$ if $B = C$ and $f \cdot g = 0$ otherwise. Thus one shouldn't think of an additive category as being something horrendously abstract, but rather as a slight weakening of the concept of a ring.

4.1.4. Abelian categories. An abelian category is an abstract category in which one can define kernels, cokernels and short exact sequences in such a way that these notions behave in the same way as those in the category of modules over a ring. Unfortunately the actual definition is rather abstract and, at least on first acquaintance, not particularly easy to work with.

In practice, the vast majority of mathematicians when faced with a statement about abelian categories will first think of some particular class of examples (e.g. modules over a ring, sheaves on a variety, ...), and only check afterwards that his or her arguments go through for a general abelian category. The reader is thus strongly advised to skim lightly over the definition of an abelian category and proceed to read the rest of this section concentrating on some concrete case such as the category $\mathbf{Rep}(G)$ of finite-dimensional representations a finite group G, or more generally the category $\mathbf{Mod}(R)$ of finitely generated modules over a ring.

Let $\mathcal{C}$ be an additive category. A morphism $f\colon C \to D$ in $\mathcal{C}$ is said to be injective (more properly mono) if the homomorphism of groups

$$f_*\colon \operatorname{Hom}_{\mathcal{C}}(X, C) \longrightarrow \operatorname{Hom}_{\mathcal{C}}(X, D)$$

given by post-composition with f is injective for all objects $X \in \mathcal{C}$. Similarly, a morphism $f\colon C \to D$ is said to be surjective (more properly epi) if the homomorphism

$$f^*\colon \operatorname{Hom}_{\mathcal{C}}(D, X) \longrightarrow \operatorname{Hom}_{\mathcal{C}}(C, X)$$

given by pre-composition with f is injective for all objects $X \in \mathcal{C}$.

Let $f\colon C \to D$ be a morphism in $\mathcal{C}$. A morphism $s\colon B \to C$ is said to be a kernel of f if the following sequence of abelian groups is exact:

$$0 \longrightarrow \operatorname{Hom}_{\mathcal{C}}(X, B) \xrightarrow{\ s_*\ } \operatorname{Hom}_{\mathcal{C}}(X, C) \xrightarrow{\ f_*\ } \operatorname{Hom}_{\mathcal{C}}(X, D)$$

for all $X \in \mathcal{C}$. In other words, any morphism of $\mathcal{C}$ whose composition with f is zero can be factored uniquely via s. Note that by definition s is injective.

Dually, a morphism $q\colon D \to E$ is a cokernel of f if the following sequence of abelian groups is exact:

$$\operatorname{Hom}_{\mathcal{C}}(C, X) \xleftarrow{\ f^*\ } \operatorname{Hom}_{\mathcal{C}}(D, X) \xleftarrow{\ q^*\ } \operatorname{Hom}_{\mathcal{C}}(E, X) \longleftarrow 0$$

for all $X \in \mathcal{C}$. By definition q is surjective.

It follows from the definitions that kernels and cokernels, when they exist, are unique up to isomorphism. More precisely, if $s_i\colon B_i \to C$ are both kernels of a morphism $f\colon C \to D$, then there is an isomorphism $t\colon B_1 \to B_2$ such that $s_2 \circ t = s_1$. Similarly for cokernels.

DEFINITION 4.17. An abelian category is an additive category $\mathcal{C}$ with the following two properties:

(a) every morphism has a kernel and a cokernel,
(b) every injective morphism is a kernel and every surjective morphism is a cokernel.

If $f\colon M \longrightarrow N$ is a morphism in an abelian category $\mathcal{A}$, the kernel and cokernel of f are denoted by $\ker(f)$ and $\operatorname{coker}(f)$. We also define the image of f to be

$$\operatorname{im}(f) = \ker(\operatorname{coker}(f)).$$

Note that strictly speaking $\ker(f)$ is an injective morphism $f\colon L \to M$. In practice one often describes the object L as the kernel of f and writes $L = \ker(f)$. This is a tricky point which can cause confusion. Note also that $\ker(f)\colon L \to M$ is only defined up to the notion of isomorphism defined above. Similar remarks apply to the cokernel and the image.

EXAMPLE 4.18. Let R be a ring. If $f\colon M \to N$ is a map of R-modules then the kernel of f is strictly speaking the inclusion morphism

$$\{m \in M : f(m) = 0\} \longrightarrow M.$$

Similarly, the image of f is the inclusion morphism

$$\{n \in N : \exists m \in M \text{ with } f(m) = n\} \longrightarrow N$$

in N. Finally, the cokernel of f is the quotient morphism $N \longrightarrow N/\operatorname{im}(f)$.

A complex in $\mathcal{A}$ is a sequence of morphisms

$$\cdots \longrightarrow M^{i-1} \xrightarrow{f^{i-1}} M^i \xrightarrow{f^i} M^{i+1} \longrightarrow \cdots$$

such that $f^i \circ f^{i-1} = 0$ for all i. Such a complex is said to be exact at M^i if $\ker(f^i)$ and $\operatorname{im}(f^{i-1})$ are isomorphic (strictly speaking as injective morphisms to M^i). An easy consequence of the definition is that every injective morphism $f\colon L \to M$ can be completed to a short exact sequence

$$0 \longrightarrow L \xrightarrow{f} M \xrightarrow{g} N \longrightarrow 0$$

where $g\colon M \to N$ is the cokernel of f. Similarly, every surjective morphism $g\colon M \to N$ fits into such a short exact sequence with $f\colon L \to M$ the kernel of g. In this situation we often loosely refer to L as a subobject of M and $N = M/L$ as the corresponding quotient object.

A special case of a complex is a resolution. A (right) resolution of an object E of $\mathcal{A}$ by a (not necessarily finite) set of other objects $\{M^i\}_{i\geq 0}$ is a complex

$$0 \longrightarrow E \longrightarrow M^0 \longrightarrow M^1 \longrightarrow M^2 \longrightarrow \cdots$$

which is exact at each place. A left resolution is defined dually in the obvious way. This is a really useful notion if we can guarantee that the objects $\{M^i\}_{i\geq 0}$ have some special properties. Right resolutions by injective objects, and left resolutions by projectives, are going to play a role presently.

DEFINITION 4.19. An object M of an abelian category $\mathcal{A}$ is called injective if, for every injective morphism (mono) $f\colon E \to F$ and for any morphism

$g\colon E \to M$ in $\mathcal{A}$, there is a morphism $h\colon F \to M$ making the following diagram commute:

$$\begin{array}{ccc} E & \xrightarrow{f} & F \\ {\scriptstyle g}\downarrow & & {\scriptstyle h}\downarrow \\ M & = & M. \end{array}$$

Dually, an object M of an abelian category $\mathcal{A}$ is called projective if, for every surjective morphism (epi) $f\colon E \to F$ and for any morphism $g\colon M \to F$ in $\mathcal{A}$, there is a morphism $h\colon M \to E$ making the following diagram commute:

$$\begin{array}{ccc} M & = & M \\ {\scriptstyle h}\downarrow & & {\scriptstyle g}\downarrow \\ E & \xrightarrow{f} & F. \end{array}$$

An abelian category $\mathcal{A}$ is said to have enough injectives if every object in $\mathcal{A}$ has a right resolution by injective objects. Dually, $\mathcal{A}$ has enough projectives if every object has a left resolution by projectives.

For example, given a ring R, it is an easy exercise to show that free R-modules are projective. Hence every R-module has a left resolution by projectives, and thus $\mathbf{Mod}(R)$ has enough projectives. On the other hand, it is well known that, for a finite group G, every G-submodule of a G-module is a direct summand, and that every G-module is a direct summand of some power of the regular representation $\mathbb{C}G$. It follows that every object in $\mathbf{Rep}(G)$, in particular every object in the category $\mathbf{Mod}(\mathbb{C})$ of finite-dimensional vector spaces, is both injective and projective, and thus injective and projective resolutions are trivial in these categories.

DEFINITION 4.20. An object E of an abelian category $\mathcal{A}$ is called simple if any subobject is either a zero object or is isomorphic to E.

EXAMPLE 4.21. If G is a finite group, then the simple objects in the representation category $\mathbf{Rep}(G)$ are exactly the irreducible representations V_χ of G.

Short exact sequences in an abelian category $\mathcal{A}$ are important because they provide a way of building up objects by glueing other objects together. Given a short exact sequence

$$0 \longrightarrow L \xrightarrow{f} M \xrightarrow{g} N \longrightarrow 0$$

we think of the object M as being a "bound state" of the objects L and N. The mathematical jargon is that M is an extension of L by N. Note that the simple objects are precisely those objects which can never be obtained by taking extensions of other objects in this way.

Given a short exact sequence as above, the object M will not usually be uniquely defined by L and N; in fact there is an abelian group $\mathrm{Ext}^1_{\mathcal{A}}(N, L)$

which classifies such extensions. More precisely, each short exact sequence as above defines an element of $\mathrm{Ext}^1_{\mathcal{A}}(N,L)$, and another such sequence

$$0 \longrightarrow L \xrightarrow{f'} M' \xrightarrow{g'} N \longrightarrow 0$$

defines the same element precisely if there is an isomorphism $s\colon M \to M'$ such that the following diagram commutes:

$$\begin{array}{ccccccccc} 0 & \longrightarrow & L & \xrightarrow{f} & M & \xrightarrow{g} & N & \longrightarrow & 0 \\ & & {\scriptstyle \mathrm{id}}\downarrow & & {\scriptstyle s}\downarrow & & {\scriptstyle \mathrm{id}}\downarrow & & \\ 0 & \longrightarrow & L & \xrightarrow{f'} & M' & \xrightarrow{g'} & N & \longrightarrow & 0 \end{array}$$

The fact that $\mathrm{Ext}^1_{\mathcal{A}}(N,L)$ is a group is not obvious from this description. See Example 4.52 for a full explanation.

Extensions or "bound states" of two simple objects are defined by a short exact sequence. Bound states of more than two objects are encoded in the notion of a filtration. A *Jordan-Hölder filtration* of an object E in an abelian category $\mathcal{A}$ is a finite filtration

$$0 = E_0 \subset E_1 \subset \cdots \subset E_{n-1} \subset E_n = E$$

such that each factor object $F_i = E_i/E_{i-1}$ is simple. If such a filtration exists, it will not in general be unique, but one can easily check that the simple factors F_i are uniquely determined up to isomorphism and reordering. Note that we obtain E by repeatedly glueing simple objects using the short exact sequences

$$0 \longrightarrow E_{i-1} \longrightarrow E_i \longrightarrow F_i \longrightarrow 0.$$

The category $\mathcal{A}$ is said to be of finite length if every object $E \in \mathcal{A}$ has a Jordan-Hölder filtration. In a finite length abelian category the simple objects can be thought of as the basic building blocks; all other objects can be made by repeatedly glueing simple objects together by extensions.

EXAMPLE 4.22. The category of finite-dimensional representations of a quiver has finite length, see Proposition 4.30.

EXAMPLE 4.23. The category of coherent sheaves on a variety of positive dimension (see §4.3) is never of finite length. Indeed, the only simple objects of $\mathrm{Coh}(X)$ are the skyscraper sheaves of points of X, and only sheaves supported in dimension zero can have a filtration by finitely many such sheaves.

4.1.5. Additive and exact functors. A functor $F\colon \mathcal{A} \longrightarrow \mathcal{B}$ between additive categories is called additive if for each pair of objects $A, B \in \mathcal{A}$ the map

$$F\colon \mathrm{Hom}_{\mathcal{A}}(A,B) \longrightarrow \mathrm{Hom}_{\mathcal{B}}(F(A),F(B))$$

is a homomorphism of abelian groups. Similarly, if $\mathcal{A}$ and $\mathcal{B}$ are $\mathbb{C}$-linear, then a functor $F\colon \mathcal{A} \to \mathcal{B}$ is called linear if the induced maps on Hom spaces are linear maps of vector spaces.

Suppose that $\mathcal{A}$ and $\mathcal{B}$ are abelian categories and $F\colon \mathcal{A} \longrightarrow \mathcal{B}$ is an additive functor. Given a short exact sequence

$$0 \longrightarrow M_1 \xrightarrow{\ f\ } M_2 \xrightarrow{\ g\ } M_3 \longrightarrow 0,$$

we can apply the functor F to get a complex

$$0 \longrightarrow F(M_1) \xrightarrow{F(f)} F(M_2) \xrightarrow{F(g)} F(M_3) \longrightarrow 0.$$

If this resulting complex is always exact, the functor F is said to be exact. If the complex is always exact at $F(M_1)$ and $F(M_2)$ but not necessarily at $F(M_3)$, then we call F left exact. Similarly if the complex is always exact at $F(M_2)$ and $F(M_3)$ but not necessarily at $F(M_1)$, then we call F right exact. Many functors occurring in nature are not exact but only left or right exact.

EXAMPLE 4.24. If $A \to B$ is a ring homomorphism, then there is an additive functor

$$- \otimes_A B\colon \mathbf{Mod}(A) \longrightarrow \mathbf{Mod}(B).$$

which sends an A-module M to the B-module $M \otimes_A B$, and a morphism of A-modules $f\colon M \to N$ to the morphism of B-modules

$$f \otimes_A B\colon M \otimes_A B \longrightarrow N \otimes_A B.$$

This functor is always right exact (as the reader can easily check), but in general not exact. When it is, one says that the ring B is flat over A. Similarly, if P is a fixed A-module then there is a right exact functor

$$- \otimes_A P\colon \mathbf{Mod}(A) \longrightarrow \mathbf{Mod}(A),$$

which sends a module M to $M \otimes_A P$. The module P is said to be flat over A if this functor is exact.

EXAMPLE 4.25. Let $\mathcal{A}$ be an abelian category and M a fixed object. There is a functor

$$\operatorname{Hom}_{\mathcal{A}}(M, -)\colon \mathcal{A} \longrightarrow \mathbf{Mod}(\mathbb{Z})$$

which sends an object $N \in \mathcal{A}$ to the abelian group $\operatorname{Hom}_{\mathcal{A}}(M, N)$, and a morphism $f\colon N_1 \to N_2$ in $\mathcal{A}$ to the homomorphism of abelian groups

$$\operatorname{Hom}_{\mathcal{A}}(M, N_1) \longrightarrow \operatorname{Hom}_{\mathcal{A}}(M, N_2)$$

given by post-composition with f. This functor is always left exact; it is exact if and only if the object M is a projective object of $\mathcal{A}$.

4.2. Representations of quivers

This section is essentially an extended example, introducing the category of representations of a quiver. We illustrate several of the above discussed general categorical notions, and end by discussing moduli spaces of quiver representations.

4.2.1. Quivers and their representations. A quiver Q is a directed graph, specified by a set of vertices Q_0, a set of arrows Q_1, and head and tail maps

$$h, t\colon Q_1 \longrightarrow Q_0.$$

We always assume that Q is finite, i.e., the sets Q_0 and Q_1 are finite. Here are two examples:

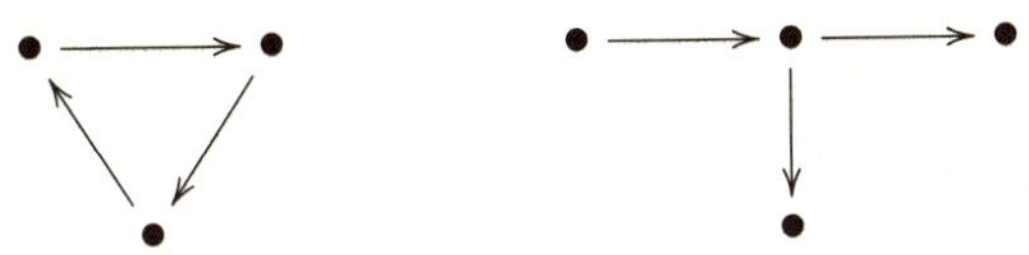

A (complex) representation of a quiver Q consists of complex vector spaces V_i for $i \in Q_0$ and linear maps

$$\phi_a\colon V_{t(a)} \longrightarrow V_{h(a)}$$

for $a \in Q_1$. A morphism between such representations (V, ϕ) and (W, ψ) is a collection of linear maps $f_i\colon V_i \longrightarrow W_i$ for $i \in Q_0$ such that the diagrams

$$\begin{array}{ccc} V_{t(a)} & \xrightarrow{\phi_a} & V_{h(a)} \\ \downarrow{\scriptstyle f_{t(a)}} & & \downarrow{\scriptstyle f_{h(a)}} \\ W_{t(a)} & \xrightarrow{\psi_a} & W_{h(a)} \end{array}$$

commute for all $a \in Q_1$. A representation of Q is finite-dimensional if each vector space V_i is. The dimension vector of such a representation is just the tuple of non-negative integers $(\dim V_i)_{i \in Q_0}$.

We write $\mathbf{Rep}(Q)$ for the category of finite-dimensional representations of Q. This category is obviously additive; we can add morphisms by adding the corresponding linear maps f_i, the trivial representation in which each $V_i = 0$ is a zero object, and the direct sum of two representations is obtained by taking the direct sums of the vector spaces associated to each vertex in the obvious way.

EXAMPLE 4.26. Take Q to be the one-arrow quiver

$$\bullet \longrightarrow \bullet$$

and let us classify the indecomposable objects of $\mathbf{Rep}(Q)$, that is, the objects $E \in \mathbf{Rep}(Q)$ which do not have a non-trivial direct sum decomposition $E = A \oplus B$.

By definition, an object of $\mathbf{Rep}(Q)$ is just a linear map of finite-dimensional vector spaces $f\colon V_1 \to V_2$. If $W = \operatorname{im}(f)$ is a nonzero proper subspace of V_2 then we can take a splitting $V_2 = U \oplus W$, and the corresponding object of $\mathbf{Rep}(Q)$ then splits as a direct sum of the two representations

$$V_1 \xrightarrow{f} W \quad \text{and} \quad 0 \longrightarrow U.$$

Thus if an object $f\colon V_1 \to V_2$ of $\mathbf{Rep}(Q)$ is indecomposable, the map f must be surjective. Similarly, if f is nonzero, then it must also be injective. Continuing in this way, one sees that $\mathbf{Rep}(Q)$ has exactly three indecomposable objects up to isomorphism:

$$\mathbb{C} \longrightarrow 0, \qquad 0 \longrightarrow \mathbb{C}, \qquad \mathbb{C} \xrightarrow{\text{id}} \mathbb{C}.$$

Every other object of $\mathbf{Rep}(Q)$ is a direct sum of copies of these basic representations.

4.2.2. The path algebra. Representations of a quiver can be interpreted as modules over a non-commutative algebra $A(Q)$ whose elements are linear combinations of paths in Q. We now describe this algebra $A(Q)$.

Let Q be a quiver. A non-trivial path in Q is a sequence of arrows $a_m \cdots a_0$ such that $h(a_{i-1}) = t(a_i)$ for $i = 1, \dots, m$:

$$\bullet \xrightarrow{a_0} \bullet \xrightarrow{a_1} \bullet \longrightarrow \cdots \longrightarrow \bullet \xrightarrow{a_m} \bullet\,.$$

We denote this path by $p = a_m \cdots a_0$. We write $t(p) = t(a_0)$ and say that p starts at $t(a_0)$ and, similarly, we write $h(p) = h(a_m)$ and say that p finishes at $h(a_m)$. For each vertex $i \in Q_0$, we denote by e_i the trivial path which starts and finishes at i. Two paths p and q are compatible if $t(p) = h(q)$ and, in this case, the composition pq can defined by juxtaposition of p and q in the obvious way. The length $l(p)$ of a path is the number of arrows it contains; in particular, a trivial path has length zero.

DEFINITION 4.27. The path algebra $A(Q)$ of a quiver Q is the complex vector space with basis consisting of all paths in Q, equipped with the multiplication in which the product pq of paths p and q is defined to be the composition pq if $t(p) = h(q)$, and 0 otherwise.

Notice that composition of paths is non-commutative; in most cases, if p and q can be composed one way, then they cannot be composed the other way, and even if they can, usually $pq \neq qp$. Hence the path algebra is indeed non-commutative.

Let us define $A_l \subset A$ to be the subspace spanned by paths of length l. Then $A = \bigoplus_{l \geq 0} A_l$ is a graded $\mathbb{C}$-algebra. The subring $A_0 \subset A$ spanned by the trivial paths e_i is a semisimple ring in which the elements e_i are orthogonal idempotents, in other words $e_i e_j = e_i$ when $i = j$, and 0 otherwise. Note also that the algebra A is finite-dimensional precisely if Q has no directed cycles.

PROPOSITION 4.28. *The category of finite-dimensional representations of a quiver Q is isomorphic to the category of finitely generated left $A(Q)$-modules.*

PROOF. Let (V, ϕ) be a representation of Q. We can then define a left module V over the algebra $A = A(Q)$ as follows: as a vector space it is

$$V = \bigoplus_{i \in Q_0} V_i,$$

and the A-module structure is extended linearly from

$$e_i v = \begin{cases} v, & v \in M_i, \\ 0, & v \in M_j \quad \text{for } j \neq i, \end{cases}$$

for $i \in Q_0$ and

$$av = \begin{cases} \phi_a(v_{t(a)}), & v \in V_{t(a)}, \\ 0, & v \in V_j \quad \text{for } j \neq t(a), \end{cases}$$

for $a \in Q_1$. This construction can be inverted as follows: given a left A-module V we set $V_i = e_i V$ for $i \in Q_0$ and define the map $\phi_a \colon V_{t(a)} \longrightarrow V_{h(a)}$ by $v \longmapsto a(v)$. One easily checks that morphisms of representations of (Q, V) correspond to A-module homomorphisms. □

4.2.3. The category of quiver representations. For a quiver Q, the category $\mathbf{Rep}(Q)$ of finite-dimensional representations of Q is abelian. This follows immediately from Proposition 4.28, but it is a good exercise to check the axioms directly. Note that a morphism $f \colon V \to W$ in the category $\mathbf{Rep}(Q)$ defined by a collection of morphisms $f_i \colon V_i \to W_i$ as above is injective (respectively surjective, an isomorphism) precisely if each of the linear maps f_i is.

There is an obvious collection of simple objects in $\mathbf{Rep}(Q)$. Indeed, each vertex $i \in Q_0$ determines a simple object S_i of $\mathbf{Rep}(Q)$, the unique representation of Q up to isomorphism for which $\dim(V_j) = \delta_{ij}$. If Q has no directed cycles, then these so-called vertex simples are the only simple objects of $\mathbf{Rep}(Q)$, but this is not the case in general.

EXAMPLE 4.29. Take the quiver

$$\bullet \leftrightarrows \bullet$$

and consider representations of dimension vector $(1, 1)$ in which both maps

$$x_a \colon V_0 \to V_1, \quad x_b \colon V_1 \to V_0,$$

are isomorphisms. We leave it to the reader to check that the isomorphism classes of such representations are parameterized by $\mathbb{C}^*$, and that all these representations are simple.

PROPOSITION 4.30. *If Q is a quiver, then the category* $\mathbf{Rep}(Q)$ *has finite length.*

PROOF. Given a representation E of a quiver Q, then either E is simple, or there is a nontrivial short exact sequence

$$0 \longrightarrow A \longrightarrow E \longrightarrow B \longrightarrow 0.$$

Now if B is not simple, then we can break it up into pieces in the same way. This process must stop, as every representation of Q consists of finite-dimensional vector spaces. In the end, we will have found a simple object S and a surjection $f\colon E \to S$. Take $E^1 \subset E$ to be the kernel of f and repeat the argument with E^1. In this way we get a filtration

$$\cdots \subset E^3 \subset E^2 \subset E^1 \subset E$$

with each quotient object E^{i-1}/E^i simple. Once again, this filtration cannot continue indefinitely, so after a finite number of steps we get $E^n = 0$. Renumbering by setting $E_i := E^{n-i}$ for $1 \leq i \leq n$ gives a Jordan-Hölder filtration for E. □

The basic reason for finiteness is the assumption that all representations of Q are finite-dimensional. This means that there can be no infinite descending chains of subrepresentations or quotient representations, since a proper subrepresentation or quotient representation has strictly smaller dimension.

EXAMPLE 4.31. Let Q be the Kronecker quiver

$$\bullet \rightrightarrows \bullet$$

and consider representations V of Q with dimension vector $(1,1)$. This means that V consists of two one-dimensional vector spaces (V_1, V_2) together with two maps $f_1, f_2\colon V_1 \to V_2$. It is easy to see that any such representation fits into a short exact sequence

$$0 \longrightarrow S_2 \longrightarrow V \longrightarrow S_1 \longrightarrow 0,$$

where S_i are the vertex simples at the two vertices. To determine the isomorphism classes of such representations we can first choose bases and thus identify V_1 and V_2 with $\mathbb{C}$; the maps (f_1, f_2) then determine an element of $\mathbb{C}^2$. Rescaling the bases just gives the scaling action of $\mathbb{C}^*$ on $\mathbb{C}^2$. If both the maps f_1 and f_2 are zero then $V = S_1 \oplus S_2$. For all other points of $\mathbb{C}^2$ the corresponding representation of Q is indecomposable, and the isomorphism classes of these representations are parameterized by the orbits of $\mathbb{C}^*$ in $\mathbb{C}^2 \setminus \{0\}$, which is to say by the points of $\mathbb{P}^1$. In terms of extension groups, one has $\operatorname{Ext}^1(S_1, S_2) = \mathbb{C}^2$.

If one instead took a quiver Q

$$\bullet \xrightarrow{\;n\;} \bullet$$

with n arrows from vertex 1 to vertex 2, then $\mathrm{Ext}^1(S_1, S_2) = \mathbb{C}^n$ and the isomorphism classes of indecomposable representations of Q with dimension vector $(1, 1)$ are parameterized by $\mathbb{P}^{n-1}$.

In this example, we see how representations of a quiver Q with two vertices and no oriented loops can be thought of as being "bound states" or extensions of the corresponding simple representations S_1 and S_2, and that the arrows in the quiver determine the dimension of the space $\mathrm{Ext}^1(S_1, S_2)$.

4.2.4. Quivers with relations.

In many geometric and algebraic contexts, we are interested in representations of a quiver Q where the morphisms associated to the arrows satisfy certain relations. Commutation relations, such as in Example 4.32 below, form possibly the simplest sort of examples, but more complicated relations also arise naturally.

Formally, a *quiver with relations* (Q, R) is a quiver Q together with a set $R = \{r_i\}$ of elements of its path algebra, where each r_i is contained in the subspace $A(Q)_{a_i b_i}$ of $A(Q)$ spanned by all paths p starting at vertex a_i and finishing at vertex b_i. Elements of R are called relations. A *representation* of (Q, R) is a representation of Q, where additionally each relation r_i is satisfied in the sense that the corresponding linear combination of homomorphisms from V_{a_i} to V_{b_i} is zero. Representations of (Q, R) form an abelian category $\mathbf{Rep}(Q, R)$. There is an analogue of Proposition 4.28, stating that $\mathbf{Rep}(Q, R)$ is equivalent to the category of finitely generated left modules over the non-commutative algebra

$$A(Q, R) = A(Q)/\langle\langle r_i \rangle\rangle,$$

where $\langle\langle r_i \rangle\rangle$ denotes the two-sided ideal of $A(Q)$ generated by the relations R.

EXAMPLE 4.32. Consider the problem of classifying commuting endomorphisms of a vector space V. This problem can be equivalently formulated as the problem of representing a quiver with relation (Q, R), where Q has one vertex with two loop arrows a_1, a_2 and $R = \{a_1 a_2 - a_2 a_1\}$. The corresponding algebra $A(R, Q)$ is the (commutative) polynomial algebra $\mathbb{C}[a_1, a_2]$ in two variables.

4.2.5. Quivers with superpotentials.

A special class of relations on quivers comes from the following construction, inspired by the physics of supersymmetric gauge theories to be discussed in the next chapter. Given a quiver Q, recall that the path algebra $A(Q)$ is non-commutative in all but the simplest examples, and hence the sub-vector space $[A(Q), A(Q)]$ generated by all commutators is non-trivial. The vector space quotient $A(Q)/[A(Q), A(Q)]$ is easily seen to have a basis consisting of the cyclic paths $a_n a_{n-1} \cdots a_1$ of Q, formed by composable arrows a_i of Q with $h(a_n) = t(a_1)$, up to cyclic permutation of such paths. By definition, a superpotential for the quiver Q is an element $W \in A(Q)/[A(Q), A(Q)]$ of this vector space, a linear combination of cyclic paths up to cyclic permutation.

Given a superpotential, define a set of relations of W by "formal differentiation of W by all arrows" as follows. Given an arrow $a \in Q_1$ of Q, define $\partial_a W$ to be the element of $A(Q)$ obtained by "opening up the cycles of W at a": consider each cycle making up W in which a appears, permute it cyclically so that a is the first arrow, and delete a from the cycle; then take the linear combination of these elements of $A(Q)$ with the corresponding coefficients. Now define a two-sided ideal of $A(Q)$, the ideal of relations defined by W, as

$$R_W = \langle\langle \partial_a W : a \in Q_1 \rangle\rangle.$$

Let the quiver algebra defined by the superpotential be the quotient

$$A_W = A(Q)/R_W.$$

Algebras obtained in this way frequently have many pleasant homological properties, and the corresponding quiver representations are often closely related to (three-dimensional) geometric constructions. More mathematical details and results can be found in [**49, 181**]; we consider here perhaps the simplest example with geometric content.

EXAMPLE 4.33. In analogy with Example 4.32, consider the quiver with one vertex and three loop arrows a_1, a_2, a_3. Define a superpotential W on this quiver by $W = a_1a_2a_3 - a_1a_3a_2$. Notice that the permutations (123) and (132) are not cyclic rotations of each other, and hence W is a nonzero element of $A(Q)/[A(Q), A(Q)]$. It is easy to check that the process described above leads to the ideal of relations

$$R_W = \langle\langle a_1a_2 - a_2a_1, a_2a_3 - a_3a_2, a_3a_1 - a_1a_3 \rangle\rangle,$$

and hence the superpotential algebra in this case is just the (commutative) polynomial algebra $\mathbb{C}[a_1, a_2, a_3]$, the ring of functions on affine 3-space. Compare also with Example 5.21.

As an exercise, the reader can check that the commutation relations between a_i, a_j on a quiver with one vertex and n loop arrows, leading to the commutative algebra $A(Q, R) = \mathbf{C}[a_1, \dots, a_n]$, can be written in superpotential form if and only if $n = 3$.

More substantial examples, where the algebra A_W is genuinely noncommutative, can be found below in Example 4.39, and in the next chapter.

4.2.6. The McKay quiver of a finite linear group. A class of quivers of great geometric interest, as well as a natural set of relations, arise from a simple but far-reaching definition of McKay [**355**].

Let $G \subset \mathrm{GL}(n, \mathbb{C})$ be a finite subgroup and write W for the n-dimensional representation defined by the embedding of G. Given an irreducible representation $\rho' \in \mathrm{Irr}(G)$, decompose the product

$$W \otimes \rho' = \bigoplus_{\rho \in \mathrm{Irr}(G)} \mathrm{Hom}^G(\rho, W \otimes \rho') \otimes \rho \tag{4.1}$$

into irreducible representations. The *McKay quiver* Q of $G \subset \mathrm{GL}(n, \mathbb{C})$ has vertex set equal to the set of irreducible representations $\{\rho \in \mathrm{Irr}(G)\}$, and has arrows denoted $\rho\rho'$ starting from ρ and finishing at ρ', marked by the vector space

$$M_{\rho\rho'} := \mathrm{Hom}^G(\rho, W \otimes \rho').$$

In practice, we often draw $\dim_{\mathbb{C}} \mathrm{Hom}^G(\rho, W \otimes \rho')$ arrows between the vertices corresponding to the dimensions of these vector spaces.

EXAMPLE 4.34. One of the simplest quivers of geometric significance arises from choosing G to be the cyclic group of order k, embedding a generator in $\mathrm{SL}(2, \mathbb{C})$ by the diagonal matrix $\mathrm{diag}(\omega, \omega^{-1})$, where ω is a fixed primitive k-th root of unity. The vertex set of the McKay quiver is $\rho_0, \ldots, \rho_{k-1}$, where ρ_j is the one-dimensional character mapping the generator ω to ω^j. The arrows are $\rho_j\rho_{j+1}$ and $\rho_{j+1}\rho_j$, with addition mod k. The resulting quiver has $2k$ arrows; the case $k = 3$ is shown below:

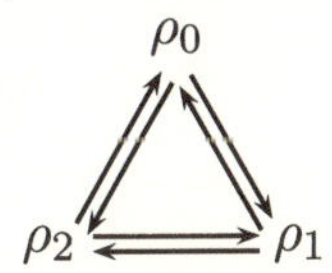

EXAMPLE 4.35. The following example will be used repeatedly in this chapter as well as the next one to illustrate general features of the theories we consider. Let G be the cyclic group of order three embedded in $\mathrm{SL}(3, \mathbb{C})$ by sending the generator to the diagonal matrix $\mathrm{diag}(\omega, \omega, \omega)$, where ω is a fixed primitive cube root of unity.

The given three-dimensional representation of G decomposes into one-dimensional representations as $W = \rho_1 \oplus \rho_1 \oplus \rho_1$, where ρ_1 is the one-dimensional character mapping the generator to ω. This implies that $W \otimes \rho_j = \rho_{j+1} \oplus \rho_{j+1} \oplus \rho_{j+1}$ for $j = 0, 1, 2$, where addition is mod 3. So the McKay quiver Q has vertices $Q_0 = \{\rho_0, \rho_1, \rho_2\}$, and three arrows each from ρ_{j+1} to ρ_j as shown below:

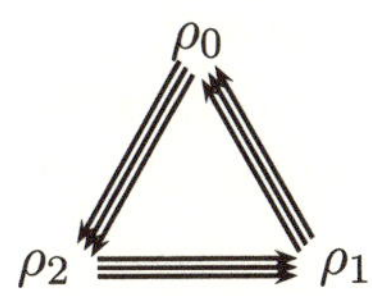

PROPOSITION 4.36. *There is a one-to-one correspondence between representations $\{V_\rho, f_{\rho\rho'}\}$ of the McKay quiver Q, and pairs (V, f), where V is a finite-dimensional G-module and $f \in \mathrm{Hom}^G(V, V \otimes W)$ is an equivariant map.*

PROOF. This is basically a tautology. Given a quiver representation $\{V_\rho, f_{\rho\rho'}\}$, set

$$V = \bigoplus_{\rho \in \mathrm{Irr}(G)} V_\rho \otimes \rho$$

to be the corresponding G-module; by the definition of the McKay quiver, the maps $f_{\rho\rho'}$ fit together to define the equivariant map f. Conversely, given (V, f), decompose into irreducible components. □

We now introduce a set of natural relations on Q. Choose a basis e_i of W consisting of G-eigenvectors; in the examples above, the actions have been defined so that the standard coordinate axes will do. Given a G-module V, an equivariant map $f : V \to V \otimes W$ can be decomposed into components $f_i : V \to V \otimes \langle e_i \rangle$; these maps can be thought of as multiplication operations on V by the dual coordinates x_i of the space W. If we now impose the condition that these operations of V commute, just as coordinates do under multiplication, then the equations $[f_i, f_j] = 0$ can be written out in the components $f_{\rho\rho'}$, and lead to a set of relations $R = \{r_i\}$ on the quiver Q. Thus, we obtain the following analogue of Proposition 4.36.

PROPOSITION 4.37. *There is a one-to-one correspondence between*

(1) finite-dimensional representations of the McKay quiver Q satisfying the relations R;

(2) finitely generated G-equivariant $\mathbb{C}[x_1, \ldots, x_n]$-modules.

PROOF. It suffices to note that the equivariant map $f \in \mathrm{Hom}^G(V, V \otimes W)$ associated to the pair $\{V_\rho, f_{\rho\rho'}\}$ by Proposition 4.36 decomposes into components as described above, and the commutativity of these operations endows the G-module V with a G-equivariant $\mathbb{C}[x_1, \ldots, x_n]$-module structure. □

REMARK 4.38. In §4.3.6, we will establish a third, geometric characterization of representations of (Q, R): the objects in (1)-(2) of Proposition 4.37 are also in one-to-one correspondence with

(3) G-equivariant coherent sheaves on affine space $\mathbb{A}^n$.

EXAMPLE 4.39. Recall Example 4.35, with the cyclic group of order three embedded in $\mathrm{SL}(3, \mathbb{C})$ diagonally using cube roots of unity. The group acts on affine space $\mathbb{C}^3$ via its embedding into $\mathrm{SL}(3, \mathbb{C})$. The coordinates x, y, z of $\mathbb{C}^3$ are eigen-coordinates for the given action of G. The corresponding quiver has nine edges which we can call $x_{(j+1)j}, y_{(j+1)j}, z_{(j+1)j}$, for $j \in Q_0 = \{0, 1, 2\}$ (simplifying the notation for the vertex set Q_0); addition is to be interpreted modulo 3. The ideal of relations introduced above is generated by

$$x_{(j+1)j}y_{j(j-1)} - y_{(j+1)j}x_{j(j-1)},$$

as well as the analogous expressions for the corresponding (y, z) and (z, x) pairs. It is immediately checked that these relations come from the superpotential

$$W = \sum_{j=0}^{2} \left(x_{(j+1)j} y_{j(j-1)} z_{(j-1)(j+1)} - y_{(j+1)j} x_{j(j-1)} z_{(j-1)(j+1)} \right)$$

on the McKay quiver Q, by the procedure described in §4.2.5 above. The appearance of superpotentials is a special feature of McKay quivers of finite groups embedded in $\mathrm{SL}(3, \mathbb{C})$ (compare [**181**]).

4.3. Coherent sheaves

We now turn to the study of our constructions in a geometric context, that of the category of coherent sheaves on an algebraic variety. A coherent sheaf is a generalization of, on the one hand, a module over a ring, and on the other hand, a vector bundle over a manifold. Indeed, in a suitable sense, the category of coherent sheaves is the "abelian closure" of the category of vector bundles on a variety.

4.3.1. Recollections on algebraic varieties. Recall that, given a field which we always take to be the field of complex numbers $\mathbb{C}$, an affine algebraic variety X is the vanishing locus

$$X = \{(x_1, \ldots, x_n) : f_i(x_1, \ldots, x_n) = 0\} \subset \mathbb{A}^n$$

of a set of polynomials $f_i(x_1, \ldots, x_n)$ in affine space $\mathbb{A}^n$ with coordinates $x_1, \ldots, x_n$. Associated to an affine variety is the ring $A = \mathbb{C}[X]$ of its regular functions, which is simply the ring $\mathbb{C}[x_1, \ldots, x_n]$ modulo the ideal $\langle f_i \rangle$ of the defining polynomials. Closed subvarieties Z of X are defined by the vanishing of further polynomials and open subvarieties $U = X \setminus Z$ are the complements of closed ones; this defines the Zariski topology on X. The Zariski topology is not to be confused with the complex topology, which comes from the classical (Euclidean) topology of $\mathbb{C}^n$ defined using complex balls; every Zariski open set is also open in the complex topology, but the converse is very far from being true. For example, the complex topology of $\mathbb{A}^1$ is simply that of $\mathbb{C}$, whereas in the Zariski topology, the only closed sets are $\mathbb{A}^1$ itself and finite point sets.

Projective varieties $X \subset \mathbb{P}^n$ are defined similarly. Recall that projective space $\mathbb{P}^n$ is the set of lines in $\mathbb{A}^{n+1}$ through the origin; an explicit coordinatization is by $(n+1)$-tuples

$$(x_0, \ldots, x_n) \in \mathbb{C}^{n+1} \setminus \{0, \ldots, 0\},$$

identified under the equivalence relation

$$(x_0, \ldots, x_n) \sim (\lambda x_0, \ldots, \lambda x_n) \text{ for } \lambda \in \mathbb{C}^*.$$

Projective space can be decomposed into a union of $(n+1)$ affine pieces

$$(\mathbb{A}^n)_i = \{[x_0, \dots, x_n] : x_i \neq 0\}$$

with n affine coordinates $y_j = x_j/x_i$ (omitting the ith). A projective variety X is the locus of common zeros of a set $\{f_i(x_0, \dots, x_n)\}$ of *homogeneous* polynomials. The Zariski topology is again defined by choosing for closed sets the loci of vanishing of further homogeneous polynomials in the coordinates $\{x_i\}$. The variety X is covered by the *standard open sets* $X_i = X \cap (\mathbb{A}^n)_i \subset X$, which are themselves affine varieties.

The notion of a general variety is a further generalization, and this is not the right place to expand on the definitions, which in any event are somewhat non-trivial; consult [**222**, Chapter 2] for the full story. Suffice it to say that for our purposes, a variety X is understood as a topological space with a finite open covering $X = \bigcup_i U_i$, where every open piece $U_i \subset \mathbb{A}^n$ is an affine variety with ring of global functions $A_i = \mathbb{C}[U_i]$; further, the pieces U_i are glued together by regular functions defined on open subsets. The topology on X is still referred to as the Zariski topology. Under our conventions, X also carries the complex topology, which again has many more open sets.

Given affine varieties $X \subset \mathbb{A}^n$, $Y \subset \mathbb{A}^m$, a morphism $f \colon X \to Y$ is given by an m-tuple of polynomials $\{f_1(x_1, \dots, x_n), \dots, f_m(x_1, \dots, x_n)\}$ satisfying the defining relations of Y. Morphisms on projective varieties are defined similarly, using homogeneous polynomials of the same degree. Morphisms on general varieties are defined as morphisms on their affine pieces, which glue together in a compatible way.

If X is a variety, points $P \in X$ are either singular or nonsingular; this is a local notion, so under our definition, it suffices to define a nonsingular point on an affine piece $U_i \subset \mathbb{A}^n$. A point $P \in U_i$ is nonsingular if, locally in the *complex* topology, a neighbourhood of $P \in U_i$ is a complex submanifold of $\mathbb{C}^n$; this is independent of the chart U_i chosen. For an equivalent algebraic definition, in terms of the equations $\{f_{ij}\}$ defining U_i in $\mathbb{A}^n$, consult [**222**, Chapter 1].

4.3.2. Coherent sheaves of modules. The motivating example of a coherent sheaf of modules on an algebraic variety X is the *structure sheaf* or *sheaf of regular functions* $\mathcal{O}_X$. This is a gadget with the following properties:

(1) On every open set $U \subset X$, we are given an abelian group (in this case, in fact also a commutative ring) denoted $\mathcal{O}_X(U)$, also written $\Gamma(U, \mathcal{O}_X)$, the *ring of regular functions on* U.
(2) *Restriction*: if $V \subset U$ is an open subset, a restriction map

$$\mathrm{res}_{UV} \colon \mathcal{O}_X(U) \to \mathcal{O}_X(V)$$

is defined, which simply associates to every regular function f defined over U, the restriction of this function to V. If $W \subset V \subset U$

are open sets, then the restriction maps clearly satisfy

$$\mathrm{res}_{UW} = \mathrm{res}_{VW} \circ \mathrm{res}_{UV}.$$

(3) *Sheaf Property*: suppose that an open subset $U \subset X$ is covered by a collection of open subsets $\{U_i\}$, and suppose that a set of regular functions $f_i \in \mathcal{O}_X(U_i)$ is given such that whenever U_i and U_j intersect, then the restrictions of f_i and f_j to $U_i \cap U_j$ agree. Then there is a unique function $f \in \mathcal{O}_X(U)$ whose restriction to U_i is f_i.

In other words, the sheaf of regular functions consists of the collection of regular functions on open sets, together with the obvious restriction maps for open subsets; moreover, this data satisfies the Sheaf Property, which says that local functions, agreeing on overlaps, glue in a unique way to a global function on U.

A *sheaf* $\mathcal{F}$ on the algebraic variety X is a gadget satisfying the same formal properties; namely, it is defined by a collection $\{\mathcal{F}(U)\}$ of abelian groups on open sets, called *sections of* $\mathcal{F}$ *over* U, together with a compatible system of restriction maps on sections $\mathrm{res}_{UV}\colon \mathcal{F}(U) \to \mathcal{F}(V)$ for $V \subset U$, so that the Sheaf Property is satisfied: sections are locally defined just as regular functions are. Of most interest in the present context are *sheaves of* $\mathcal{O}_X$*-modules*; the extra requirement is that the sections $\mathcal{F}(U)$ over an open set U form a module over the ring of regular functions $\mathcal{O}_X(U)$, and all restriction maps are compatible with the module structures. In other words, we are told how to multiply local sections by local functions, so that multiplication respects restriction. Said slightly differently, a sheaf of $\mathcal{O}_X$-modules is defined by the data of an A-module for every open subset $U \subset X$ with ring of functions $A = \mathcal{O}_X(U)$, so that these modules are glued together compatibly with the way the open sets glue. Hence, as discussed in the introduction to this chapter, a sheaf of modules is indeed a geometric generalization of a module over a ring (Example 4.3).

EXAMPLES 4.40.

(1) The simplest example of a sheaf of $\mathcal{O}_X$-modules is $\mathcal{O}_X$ itself; after all, local regular functions $\mathcal{O}_X(U)$ form a ring, which is a module over itself.
(2) The first non-trivial example is a "twisted" form of $\mathcal{O}_X$: a sheaf $\mathcal{L}$ of $\mathcal{O}_X$-modules on X is called a *line bundle* if for every sufficiently small open set $U \subset X$, $\mathcal{F}(U)$ is isomorphic to (not necessarily equal to!) $\mathcal{O}_X(U)$ as $\mathcal{O}_X(U)$-modules; however, these modules are glued together in a nontrivial way, so that globally one does not have an isomorphism $\mathcal{F}(X) \cong \mathcal{O}_X(X)$ between global sections of $\mathcal{F}$ and regular functions.

(3) More generally, a *locally free sheaf* $\mathcal{F}$ is a sheaf of $\mathcal{O}_X$-modules which satisfies $\mathcal{F}(U) \cong \mathcal{O}_X(U)^{\oplus n}$, the free $\mathcal{O}_X(U)$-module of rank n, for every sufficiently small open set $U \subset X$.

(4) Suppose that $Z \subset X$ is a closed subvariety. Then, more or less by definition, the sheaf of regular functions $\mathcal{O}_Z$ is an $\mathcal{O}_X$-module: if $U \subset X$ is an open set, then restriction of functions from U to the closed subset $U \cap Z$ defines a map of rings $\mathcal{O}_X(U) \to \mathcal{O}_Z(U)$, which allows us to turn $\mathcal{O}_Z(U)$ into an $\mathcal{O}_X(U)$-module.

(5) One can obviously combine the two previous constructions: if $\mathcal{F}$ is a locally free sheaf of $\mathcal{O}_Z$-modules on a closed subvariety, then it has an $\mathcal{O}_X$-module structure via the local restriction maps on regular functions, and hence $\mathcal{F}$ becomes a sheaf of $\mathcal{O}_X$-modules.

These constructions correspond, respectively, to the following local situations on a suitably small affine subset $U \subset X$ with ring of global functions $A = \mathcal{O}_X(U)$:

(1)-(2): the free rank-1 A-module, A itself;

(3): the free rank-n A-module $A^{\oplus n}$;

(4): given a surjective homomorphism $\phi\colon A \to B$, consider B as an A-module via the homomorphism ϕ;

(5): consider the free B-module $B^{\oplus n}$ as an A-module via the homomorphism ϕ.

As these examples suggest, as a first approximation, sheaves of modules are indeed something like "vector bundles on submanifolds". However, this statement is not precisely true: a general sheaf of $\mathcal{O}_X$-modules is definitely not just a vector bundle on a submanifold, as examples will presently demonstrate.

The next definition introduces a finiteness condition, which allows for more general sheaves than just locally free sheaves (3) of rank n. To wit, a sheaf $\mathcal{F}$ of $\mathcal{O}_X$-modules is called *coherent* if, for every sufficiently small open set $U \subset X$ with ring of functions $\mathcal{O}_X(U)$, there is an exact sequence of A-modules

$$\mathcal{O}_X(U)^{\oplus m} \to \mathcal{O}_X(U)^{\oplus n} \to \mathcal{F}(U) \to 0$$

that is compatible with restrictions. In other words, we require that the space of local sections of $\mathcal{F}$ should be the cokernel of a morphism between free $\mathcal{O}_X(U)$-modules of finite rank. In particular, this condition implies that $\mathcal{F}(U)$ is a finite-rank module over the ring of local functions $\mathcal{O}_X(U)$.

One important notion related to that of a sheaf is the notion of the *stalk of a sheaf* $\mathcal{F}_P$ at a point $P \in X$. This is the algebraic replacement of "fibre of a vector bundle"; its definition is a little subtle:

$$\mathcal{F}_P = \varinjlim_{P \in U} \mathcal{F}(U)$$

where the limit runs over the open sets $U \subseteq X$ containing the point P, and the sections $\mathcal{F}(U)$ are connected of course by restriction maps. The definition means that we want to concentrate on sections defined near the point $P \in X$, retaining "infinitesimal" information about them. Note that $\mathcal{F}_P$ is certainly an abelian group, but it is also a module over the ring

$$\mathcal{O}_{X,P} = \varinjlim_{P \in U} \mathcal{O}_X(U),$$

the *ring of local regular functions at* $P \in X$. The latter is, in turn, the algebraic replacement of the ring of "germs of functions" or "(convergent) Taylor series" at $P \in X$, well known from complex analysis. The ring $\mathcal{O}_{X,P}$ has a maximal ideal m_P, the regular functions vanishing at P, so that the quotient $\mathcal{O}_{X,P}/m_P$ is simply the base field $\mathbb{C}$, corresponding to the "value of the germ" or "constant Taylor coefficient". For a general sheaf $\mathcal{F}$ and a point $P \in X$, one can form its *fibre at* P, the $\mathbb{C}$-vector space $\mathcal{F}_P/m_p\mathcal{F}_P$, which conforms better to our intuition of what a "fibre of a vector bundle" should look like, though it carries a lot less information about the sheaf than its stalks.

If $\mathcal{F}$ is locally free of rank n, then its fibres are all of constant dimension n (but not conversely!). For a general coherent sheaf $\mathcal{F}$, the dimensions of these vector spaces jump as P varies. In particular, one defines the *support* of a sheaf $\mathcal{F}$ to be the set

$$\operatorname{supp}(\mathcal{F}) = \{P \in X : \mathcal{F}_P \neq 0\}$$

which is a closed subvariety $Z = \operatorname{supp}(\mathcal{F}) \subset X$. Note however that a sheaf on X, supported on Z, is *not* the same thing as a sheaf on Z considered as a sheaf on X as in Example 4.40 (4) above.

4.3.3. Homomorphisms of sheaves. Given sheaves $\mathcal{F}$, $\mathcal{G}$ of $\mathcal{O}_X$-modules, a *homomorphism* $\phi\colon \mathcal{F} \to \mathcal{G}$ between them is just a collection of maps

$$\phi_U\colon \mathcal{F}(U) \to \mathcal{G}(U)$$

defined on sections, which commute with restriction maps, and also respect the $\mathcal{O}_X(U)$-module structure. It is immediate from the definition that a homomorphism between sheaves defines an $\mathcal{O}_{X,P}$-module homomorphism

$$\phi_P\colon \mathcal{F}_P \to \mathcal{G}_P$$

between stalks. The set of homomorphisms is denoted $\operatorname{Hom}_{\mathcal{O}_X}(\mathcal{F}, \mathcal{G})$, which is easily seen to be an abelian group. Homomorphisms of sheaves can be composed by locally composing the maps between sections, and indeed, it is almost immediate from the definitions that sheaves of $\mathcal{O}_X$-modules form an additive category as defined in Definition 4.15. The next theorem is a bit more subtle, and for a proof we refer the reader to [**222**]:

THEOREM 4.41. *Given a sheaf homomorphism* $\phi\colon \mathcal{F} \to \mathcal{G}$,

- *the collection of kernels* $\ker(\phi_U)$ *for open sets* $U \subset X$ *define a sheaf* $\ker(\phi)$*, the kernel of* f*;*
- *the cokernels* $\operatorname{coker}(\phi_U)$ *for open sets* $U \subset X$ *can be modified in a canonical way to define a sheaf* $\operatorname{coker}(\phi)$*, the cokernel of* f*.*

With these definitions, the category of sheaves of $\mathcal{O}_X$*-modules is an abelian category as defined in Definition 4.17. The morphism* f *is*

- *injective if and only if*
  (1) $\phi_U \colon \mathcal{F}(U) \to \mathcal{G}(U)$ *is injective for all* $U \subset X$ *open, or*
  (2) $\phi_P \colon \mathcal{F}_P \to \mathcal{G}_P$ *is injective for all* $P \in X$*;*
- *surjective if and only if* $\phi_P \colon \mathcal{F}_P \to \mathcal{G}_P$ *is surjective for all* $P \in X$*.*

The main point of this theorem is that, as opposed to kernels, local cokernels do not glue to form a sheaf: the very important Sheaf Property is not necessarily satisfied, and the collection of cokernels needs to be modified ("sheafified") to obtain an honest sheaf. Correspondingly, a surjective morphism on sheaves does not need to be surjective on local sections, a fact with serious consequences to be discussed below (see Remark 4.63).

It is not too hard to prove that kernels and cokernels of morphisms between coherent sheaves are themselves coherent. Hence we can define the *category of coherent sheaves of* $\mathcal{O}_X$*-modules*, denoted $\operatorname{Coh}(X)$, and this category is abelian. Indeed, one often builds interesting sheaves as kernels or cokernels of homomorphisms between, or extensions of, already known sheaves.

EXAMPLE 4.42. Suppose that $Z \subset X$ is a closed subvariety. As discussed before, the structure sheaf $\mathcal{O}_Z$ can be thought of as an $\mathcal{O}_X$-module; in fact it is immediately seen that there is also a canonical surjective homomorphism of $\mathcal{O}_X$-modules $\mathcal{O}_X \to \mathcal{O}_Z$ fitting into a short exact sequence in $\operatorname{Coh}(X)$:

$$0 \longrightarrow \mathcal{I}_Z \longrightarrow \mathcal{O}_X \longrightarrow \mathcal{O}_Z \longrightarrow 0.$$

The kernel here is the *ideal sheaf* $\mathcal{I}_Z$, the sheaf of local regular functions on X which vanish along Z.

If Z is a proper subvariety, then $\operatorname{supp}(\mathcal{I}_Z) = X$. Let us investigate when is $\mathcal{I}_Z$ locally free. Away from Z, it is isomorphic to the structure sheaf $\mathcal{O}_{X\setminus Z}$, so if it is locally free, then it is of rank 1. But what this means is that, everywhere locally, the ideal of regular functions vanishing on Z must be generated by a single local function on X. This is exactly the definition of a *(Cartier) divisor*: a (necessarily codimension-1) subvariety $Z \subset X$ everywhere locally defined by a single equation. Conversely, if Z is not a Cartier divisor, then $\mathcal{I}_Z$ cannot be locally free; for example, ideal sheaves of points on surfaces or threefolds, though supported on the whole of X, are not locally free.

4.3.4. Operations and functors on the category of sheaves. If $\mathcal{F}$, $\mathcal{G}$ are sheaves of $\mathcal{O}_X$-modules, then one can define their direct sum $\mathcal{F}\oplus\mathcal{G}$ as the collection of direct sums of local sections, fitting into a trivial (split) extension of $\mathcal{O}_X$-modules

$$0 \longrightarrow \mathcal{F} \longrightarrow \mathcal{F}\oplus\mathcal{G} \longrightarrow \mathcal{G} \longrightarrow 0.$$

More interestingly, one can define the tensor product $\mathcal{F}\otimes_{\mathcal{O}_X}\mathcal{G}$ of $\mathcal{F}$ and $\mathcal{G}$, usually denoted simply by $\mathcal{F}\otimes\mathcal{G}$. This is a less trivial operation: the tensor products of local sections $\mathcal{F}(U)\otimes_{\mathcal{O}_X(U)}\mathcal{G}(U)$ do not necessarily form a sheaf, so one has to sheafify again. Further, the category of sheaves has internal Hom-sheaves: since the definition of sheaf homomorphisms is local, it makes sense to ask what is the sheaf of local homomorphisms. Since local rings of functions are commutative, the rule $U\mapsto \operatorname{Hom}_{\mathcal{O}_U}(\mathcal{F}|_U,\mathcal{G}|_U)$ defines the *local* Hom-*sheaf* $\mathscr{H}om(\mathcal{F},\mathcal{G})$, which is a sheaf of $\mathcal{O}_X$-modules on its own right; here $\mathcal{F}|_U,\mathcal{G}|_U$ are the restrictions of the sheaves $\mathcal{F},\mathcal{G}$ to the open set U, defined in the obvious way. One recovers the vector space of global Homs by taking sections over the whole of X:

$$\operatorname{Hom}_{\mathcal{O}_X}(\mathcal{F},\mathcal{G}) = \mathscr{H}om(\mathcal{F},\mathcal{G})(X)$$

On the other hand, using this construction we can define the dual of a locally free sheaf $\mathcal{F}$ as

$$\mathcal{F}^\vee = \mathscr{H}om(\mathcal{F},\mathcal{O}_X).$$

Evaluating local homomorphisms on local sections then gives a canonical map of sheaves

$$\mathcal{F}^\vee\otimes\mathcal{F}\to\mathcal{O}_X.$$

Interesting functors on the categories of sheaves of modules come from morphisms $f\colon X\to Y$ between algebraic varieties. Recall that a morphism of varieties, more or less by definition, is a system of compatible ring homomorphisms between regular functions on affine open sets; in particular, it is always continuous with respect to the Zariski topology. Thus, given a sheaf $\mathcal{E}$ on X, we can define its pushforward $f_*(\mathcal{E})$ by the rule

$$f_*(\mathcal{E})(U) = \mathcal{E}(f^{-1}(U)).$$

If $\mathcal{E}$ is an $\mathcal{O}_X$-module, then its pushforward is automatically an $\mathcal{O}_Y$-module: the multiplication of a local regular function on Y is defined by pulling back the function to X and multiplying there.

PROPOSITION 4.43. *If $f\colon X\to Y$ is a morphism between projective varieties (or more generally if f is proper), then f_* maps coherent $\mathcal{O}_X$-modules to coherent $\mathcal{O}_Y$-modules.*

The pullback functor is a little trickier to define (though frequently easier to compute). Given a sheaf $\mathcal{F}$ on Y, define $f^{-1}(\mathcal{F})$ as the sheaf obtained by sheafifying local section spaces defined as

$$U \mapsto \varinjlim_{f(U)\subset V} \mathcal{F}(V).$$

If $\mathcal{F}$ is a sheaf of $\mathcal{O}_Y$-modules, then this becomes a sheaf of modules over $f^{-1}(\mathcal{O}_Y)$. On the other hand, there is a canonical homomorphism of sheaves on X (check!) $f^{-1}(\mathcal{O}_Y) \to \mathcal{O}_X$, and thus we can set

$$f^*(\mathcal{F}) = f^{-1}(\mathcal{F}) \otimes_{f^{-1}(\mathcal{O}_Y)} \mathcal{O}_X.$$

This is by definition a sheaf of $\mathcal{O}_X$-modules, and it is coherent if $\mathcal{F}$ is. For example, we always have $f^*\mathcal{O}_Y \cong \mathcal{O}_X$.

All these operations and functors satisfy a plethora of compatibility relations, including the *projection formula* relating pullback, pushforward and the tensor product. In our treatment, these will be more naturally formulated in the context of derived functors, so we defer them to §4.5.

4.3.5. Line bundles and the Picard group. Before we press on with the general theory, let us take a detour to discuss an important special case, returning to the canonical map of sheaves

$$\mathcal{F}^\vee \otimes \mathcal{F} \to \mathcal{O}_X$$

discussed in the previous section. Notice that when $\mathcal{F}$ is locally free of rank 1, in other words when $\mathcal{F}$ is a line bundle, this map is an *isomorphism* of sheaves. Thus $\mathcal{F}^\vee$ behaves like a multiplicative inverse of the sheaf $\mathcal{F}$ under tensor product. This observation, together with the obvious remark that the tensor product of line bundles is again a line bundle, allows us to define the *Picard group* $\mathrm{Pic}(X)$ of a variety X as the set of all line bundles on X modulo isomorphism, with the tensor product operation and inverse $\mathcal{F} \mapsto \mathcal{F}^\vee$.

If X is smooth and projective, then one has a map

$$c_1 \colon \mathrm{Pic}(X) \to H^2(X, \mathbb{Z})$$

from the Picard group to the topological cohomology group $H^2(X, \mathbb{Z})$, where this latter cohomology is defined using the classical (complex) topology on X. There are several ways to define this map. The complex analytic method consists of choosing a connection in a line bundle $\mathcal{L}$, and considering an appropriate constant multiple of the trace of the curvature operator associated to the connection (Chern–Weil theory). A second definition based on sheaf cohomology is given below in Example 4.65.

EXAMPLES 4.44.

(1) Let $X = \mathbb{A}^n$ be affine n-space. Then it is easy to show that every line bundle is isomorphic to the trivial bundle $\mathcal{O}_{\mathbb{A}^n}$, and hence the Picard group $\mathrm{Pic}(\mathbb{A}^n)$ is trivial.

(2) Let $X = \mathbb{P}^n$ be projective n-space with homogeneous coordinates $x_0, \ldots, x_n$. Let us construct a nontrivial line bundle on X. Recall that we have an open cover $\mathbb{P}^n = \bigcup_{i=0}^n X_i$, where each piece satisfies $X_i \cong \mathbb{A}^n$. Note that on the i-th copy X_i, the ring of regular functions is the polynomial ring $A_i = \mathbb{C}[x_0/x_i, \ldots, x_n/x_i]$. Now on the intersection of open sets $X_{ij} = X_i \cap X_j$, we can glue these rings using multiplication by the (nonzero, nonvanishing) function $g_{ij} = x_i/x_j$. On triple overlaps $X_{ijk} = X_i \cap X_j \cap X_k$, this gives a well-defined glueing, since $x_i/x_j \cdot x_j/x_k \cdot x_k/x_i = 1$, and hence this produces a line bundle on $\mathbb{P}^n$ that is denoted by $\mathcal{O}_{\mathbb{P}^n}(1)$. Similarly, we can use the glueing function $(x_i/x_j)^k$ between open sets, for all integers $k \in \mathbb{Z}$, leading to the line bundle $\mathcal{O}_{\mathbb{P}^n}(k)$. It is more or less obvious from the definitions that $\mathcal{O}_{\mathbb{P}^n}(k) \otimes \mathcal{O}_{\mathbb{P}^n}(m) \cong \mathcal{O}_{\mathbb{P}^n}(k+m)$, and in particular $\mathcal{O}_{\mathbb{P}^n}(-k)$ is the inverse of $\mathcal{O}_{\mathbb{P}^n}(k)$. Slightly less trivially, every line bundle on $\mathbb{P}^n$ is isomorphic to $\mathcal{O}_{\mathbb{P}^n}(k)$ for some k; hence $\mathrm{Pic}(\mathbb{P}^n) \cong \mathbb{Z}$, the group of integers. Note incidentally, that $H^2(\mathbb{P}^n, \mathbb{Z}) \cong \mathbb{Z}$ (for example, by Mayer-Vietoris); the map $\mathrm{Pic}(\mathbb{P}^n) \xrightarrow{\sim} \mathbb{Z}$ is exactly the first Chern class map. Finally, now that we're here, a piece of notation: for any sheaf $\mathcal{F}$ on $\mathbb{P}^n$, denote by $\mathcal{F}(k) = \mathcal{F} \otimes \mathcal{O}_{\mathbb{P}^n}(k)$ its *(Serre) twist.*

For a smooth projective variety X, let $\mathrm{Pic}^0(X)$ denote the set of line bundles $\mathcal{L}$ with $c_1(\mathcal{L}) = 0$. A basic result due to Weil is that this set can be endowed with the structure of a (finite-dimensional) smooth projective variety, thus called the *Picard variety* of X. As $\mathrm{Pic}^0(X)$ also has a group structure, it has to be an abelian variety. We will not prove these statements, but we will construct $\mathrm{Pic}^0(X)$ as a complex torus in Example 4.65 below.

Points of $\mathrm{Pic}^0(X)$ correspond by definition to line bundles on X with $c_1 = 0$, but in fact more is true: $\mathrm{Pic}^0(X)$ is a (fine) moduli space: there is a line bundle $\mathcal{P}$ on the product $\mathrm{Pic}^0(X) \times X$ such that, for $z \in \mathrm{Pic}^0(X)$, its restriction $\mathcal{P}_z$ to $\{z\} \times X$ is isomorphic to the line bundle $\mathcal{L}_z$ corresponding to the point z. The line bundle $\mathcal{P}$ is not unique, since we can always tensor it with the pullback of a line bundle on $\mathrm{Pic}^0(X)$ without changing its defining property; any such $\mathcal{P}$ will be called a *Poincaré bundle*, soon to make a glorious return.

4.3.6. Equivariant sheaves. A mild generalization of the ideas developed so far will be useful for what follows. Suppose that X is a variety together with an action of a finite group G, meaning that we are given automorphisms $\phi_g \colon X \to X$ for all $g \in G$ which compose compatibly with multiplication in G. Given such an action, a coherent $\mathcal{O}_X$-module $\mathcal{F}$ is G-*equivariant*, or simply a *G-sheaf*, if there is a lift of the G-action to $\mathcal{F}$, in other words sheaf isomorphisms

$$\lambda_g^{\mathcal{F}} : \mathcal{F} \to \phi_g^*(\mathcal{F})$$

satisfying the cocycle condition

$$\lambda_{hg}^{\mathcal{F}} = \phi_g^*(\lambda_h^{\mathcal{F}}) \circ \lambda_g^{\mathcal{F}}.$$

Given G-sheaves $\mathcal{E}, \mathcal{F}$, the space of G-homomorphisms is defined to be the space of G-invariants in $\mathrm{Hom}_{\mathcal{O}_X}(\mathcal{E}, \mathcal{F})$. This defines the abelian category $\mathrm{Coh}^G(X)$ of coherent G-sheaves on X, to which all our previous constructions apply; the reader can find a thorough treatment in [**67**, Section 4].

A special case is worth spelling out. If $X = \mathbb{A}^n$ is affine space, then being a G-sheaf simply means that the space of global sections $\Gamma(\mathcal{F})$ is a G-equivariant module over the ring of regular functions

$$S = \mathbb{C}[\mathbb{A}^n] = \mathbb{C}[x_1, \ldots, x_n].$$

This means that the S-module $\Gamma(\mathcal{F})$ is also a module over the group ring $\mathbb{C}[G]$, and these two module structures satisfy the condition that

$$g(s(m)) = g(s) \cdot g(m) \text{ for } g \in \mathbb{C}[G], s \in S \text{ and } m \in \Gamma(\mathcal{F}).$$

In other words, $\Gamma(\mathcal{F})$ is a module over the skew group ring $S \rtimes G$. Note that since $\mathcal{F}$ is coherent, $\Gamma(\mathcal{F})$ is finitely generated as an S-module.

4.4. Derived categories

In this section we introduce the derived category of an abelian category and study its structure. Derived categories of coherent sheaves are of crucial importance in string theory, where they occur as categories of branes in the B-type topologically twisted theory, as already discussed in Chapter 3.

4.4.1. Quasi-isomorphism and the derived category. Let $\mathcal{A}$ be an abelian category. The reader is advised to hold a concrete example such as $\mathcal{A} = \mathbf{Mod}(R)$ in mind when reading this chapter. A complex in $\mathcal{A}$ is a sequence of objects and morphisms in $\mathcal{A}$

$$\cdots \longrightarrow M^{i-1} \xrightarrow{d^{i-1}} M^i \xrightarrow{d^i} M^{i+1} \longrightarrow \cdots$$

such that $d^i \circ d^{i-1} = 0$ for all i. We often denote such a complex by a single symbol M.

A morphism of complexes $f\colon M \longrightarrow N$ is a sequence of morphisms $f^i\colon M^i \to N^i$ in $\mathcal{A}$, making the following diagram commute, where d_M^i, d_N^i denote the respective differentials:

$$\begin{array}{ccccccccc} \cdots & \longrightarrow & M^{i-1} & \xrightarrow{d_M^{i-1}} & M^i & \xrightarrow{d_M^i} & M^{i+1} & \longrightarrow & \cdots \\ & & \downarrow{\scriptstyle f^{i-1}} & & \downarrow{\scriptstyle f^i} & & \downarrow{\scriptstyle f^{i+1}} & & \\ \cdots & \longrightarrow & N^{i-1} & \xrightarrow{d_N^{i-1}} & N^i & \xrightarrow{d_N^i} & N^{i+1} & \longrightarrow & \cdots \end{array}$$

We let $\mathcal{C}(\mathcal{A})$ denote the category whose objects are complexes in $\mathcal{A}$ and whose morphisms are morphisms of complexes.

Given a complex M of objects of $\mathcal{A}$, the ith cohomology object is the quotient

$$H^i(M) = \ker(d^i)/\operatorname{im}(d^{i-1}).$$

This operation of taking cohomology at the ith place defines a functor

$$H^i(-)\colon C(\mathcal{A}) \longrightarrow \mathcal{A},$$

since a morphism of complexes induces corresponding morphisms on cohomology objects.

Put another way, an object of $C(\mathcal{A})$ is a $\mathbb{Z}$-graded object

$$M = \bigoplus_i M^i$$

of $\mathcal{A}$, equipped with a differential, in other words an endomorphism $d\colon M \to M$ satisfying $d^2 = 0$. The occurrence of differential graded objects in physics is well-known. In mathematics they are also extremely common. In topology one associates to a space X a complex of free abelian groups whose cohomology objects are the cohomology groups of X. In algebra it is often convenient to replace a module over a ring by resolutions of various kinds.

We would like to consider complexes only up to an equivalence relation. A topological space X may have many triangulations and these lead to different chain complexes. But we would like to associate to X a unique equivalence class of complexes. Similarly, resolutions of a fixed module of a given type will not usually be unique and one would like to consider all these resolutions on an equal footing.

The following concept is crucial in what follows.

DEFINITION 4.45. A morphism of complexes $f\colon M \longrightarrow N$ is a quasi-isomorphism if the induced morphisms on cohomology

$$H^i(f)\colon H^i(M) \longrightarrow H^i(N)$$

are isomorphisms for all i.

Two complexes M and N are said to be quasi-isomorphic if they are related by a chain of quasi-isomorphisms. In fact, as we shall see, it is sufficient to consider chains of length one, so that two complexes M and N are quasi-isomorphic if and only if there are quasi-isomorphisms

$$M \longleftarrow P \longrightarrow N.$$

For example, the chain complex of a topological space is well-defined up to quasi-isomorphism because any two triangulations have a common resolution. Similarly, all possible resolutions of a given module are quasi-isomorphic. Indeed, if

$$0 \longrightarrow S \xrightarrow{\ f\ } M^0 \xrightarrow{\ d^0\ } M^1 \xrightarrow{\ d^1\ } M^2 \longrightarrow \cdots$$

is a resolution of a module S, then by definition the morphism of complexes

$$\begin{array}{ccccccccc} 0 & \longrightarrow & S & \longrightarrow & 0 & & & & \\ \downarrow & & {\scriptstyle f}\downarrow & & \downarrow & & & & \\ 0 & \longrightarrow & M^0 & \xrightarrow{d^0} & M^1 & \xrightarrow{d^1} & M^2 & \longrightarrow & \cdots \end{array}$$

is a quasi-isomorphism.

The objects of the derived category $\mathrm{D}(\mathcal{A})$ of our abelian category $\mathcal{A}$ will just be complexes of objects of A, but morphisms will be such that quasi-isomorphic complexes become isomorphic in $\mathrm{D}(\mathcal{A})$. In fact we can formally invert the quasi-isomorphisms in $C(\mathcal{A})$ as follows.

LEMMA 4.46. *There is a category* $\mathrm{D}(\mathcal{A})$ *and a functor*

$$Q\colon C(\mathcal{A}) \longrightarrow \mathrm{D}(\mathcal{A})$$

with the following two properties:

(a) *Q inverts quasi-isomorphisms: if $s\colon a \to b$ is a quasi-isomorphism, then $Q(s)\colon Q(a) \to Q(b)$ is an isomorphism.*
(b) *Q is universal with this property: if $Q'\colon C(\mathcal{A}) \longrightarrow \mathrm{D}'$ is another functor which inverts quasi-isomorphisms, then there is a functor $F\colon \mathrm{D}(\mathcal{A}) \longrightarrow \mathrm{D}'$ and an isomorphism of functors $Q' \cong F \circ Q$.*

PROOF. First, consider the category $C(\mathcal{A})$ as an oriented graph Γ, with the objects lying at the vertices and the morphisms being directed edges. Let Γ_* be the graph obtained from Γ by adding in one extra edge $s^{-1}\colon b \to a$ for each quasi-isomorphism $s\colon a \to b$. Thus a finite path in Γ_* is a sequence of the form $f_1 \cdot f_2 \cdot \cdots \cdot f_{r-1} \cdot f_r$ where each f_i is either a morphism of $\mathcal{C}(\mathcal{A})$, or is of the form s^{-1} for some quasi-isomorphism s of $\mathcal{C}(\mathcal{A})$. There is a unique minimal equivalence relation $\sim$ on the set of finite paths in Γ_* generated by the following relations:

(a) $s \cdot s^{-1} \sim \mathrm{id}_b$ and $s^{-1} \cdot s \sim \mathrm{id}_a$ for each quasi-isomorphism $s\colon a \to b$ in $\mathcal{C}(\mathcal{A})$.

(b) $g \cdot f \sim g \circ f$ for composable morphisms $f\colon a \to b$ and $g\colon b \to c$ of $\mathcal{C}(\mathcal{A})$.

Define $\mathrm{D}(\mathcal{A})$ to be the category whose objects are the vertices of Γ_* (these are the same as the objects of $\mathcal{C}(\mathcal{A})$) and whose morphisms are given by equivalence classes of finite paths in Γ_*. Define a functor $Q\colon \mathcal{C}(\mathcal{A}) \to \mathrm{D}(\mathcal{A})$ by using the identity morphism on objects, and by sending a morphism f of $\mathcal{C}(\mathcal{A})$ to the length one path in Γ_* defined by f. The reader can easily check that the resulting functor Q satisfies the conditions of the lemma. $\square$

The second property ensures that the category $\mathrm{D}(\mathcal{A})$ of the Lemma is unique up to equivalence of categories. We define the derived category of $\mathcal{A}$ to be any of these equivalent categories. The functor $Q\colon C(\mathcal{A}) \longrightarrow \mathrm{D}(\mathcal{A})$ is called the localisation functor. Observe that there is a fully faithful functor

$$J\colon \mathcal{A} \longrightarrow C(\mathcal{A})$$

which sends an object M to the trivial complex with M in the zeroth position, and a morphism $F\colon M \to N$ to the morphism of complexes

$$\begin{array}{ccccc} 0 & \longrightarrow & M & \longrightarrow & 0 \\ \downarrow & & \downarrow{\scriptstyle f} & & \downarrow \\ 0 & \longrightarrow & N & \longrightarrow & 0 \end{array}$$

Composing with Q we obtain a functor $\mathcal{A} \longrightarrow \mathrm{D}(\mathcal{A})$ which we also denote by J. We shall see later that this functor J is also fully faithful, and so defines an embedding $\mathcal{A} \longrightarrow \mathrm{D}(\mathcal{A})$. Note also that by definition the functor $H^i(-)\colon C(\mathcal{A}) \longrightarrow \mathcal{A}$ inverts quasi-isomorphisms and so descends to a functor

$$H^i(-)\colon \mathrm{D}(\mathcal{A}) \longrightarrow \mathcal{A}.$$

Clearly the composite functor $H^0(-)\circ J$ is isomorphic to the identity functor on $\mathcal{A}$.

4.4.2. A more sophisticated approach. The construction of the derived category given in the last section is completely straightforward, and as an abstract existence result it works well, but it turns out that it gives almost no information about the derived category. For example, if one wants to compute the space of morphisms $\mathrm{Hom}_{\mathrm{D}(\mathcal{A})}(E, F)$ for two complexes E and F, the above definition will not be of much use. Similarly, it is not at all clear from the above definition what natural structure the derived category has, or even whether it is an additive category. If the reader is willing to accept without proof certain properties of the derived category, then this will not be a problem in practical applications. In this section we outline an approach which enables one to get a better handle on $\mathrm{D}(\mathcal{A})$.

First we need to define the homotopy category. Suppose $\mathcal{A}$ is an abelian category and

$$f, g\colon M \longrightarrow N$$

are morphisms of complexes. We say that f and g are homotopic if there are morphisms

$$h^i\colon M^i \to N^{i-1}$$

such that

$$g^i - f^i = d_N^{i-1} \circ h^i + h^{i+1} \circ d_M^i.$$

This can be expressed by the commutative diagram

(4.2)

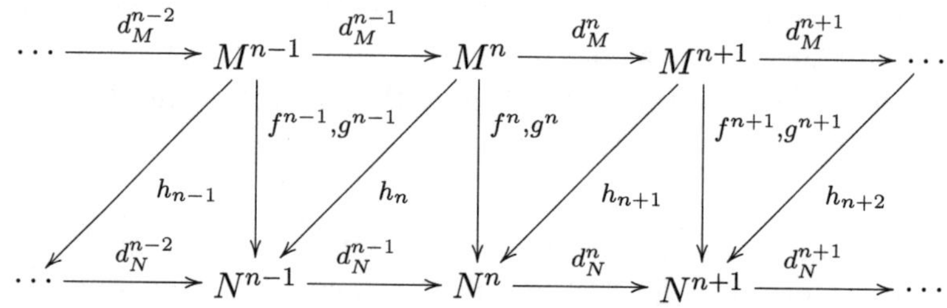

The homotopy category $K(\mathcal{A})$ is obtained from the category of complexes by identifying homotopic morphisms. Thus the objects of $K(\mathcal{A})$ are the same as those of $C(\mathcal{A})$, which is to say complexes of objects of $\mathcal{A}$, but the morphisms are homotopy equivalence classes of morphisms.

LEMMA 4.47. *Let $\mathcal{A}$ be an abelian category and let $Q\colon C(\mathcal{A}) \longrightarrow \mathrm{D}(\mathcal{A})$ be the localisation functor of Lemma 4.46. If*

$$f, g\colon M \longrightarrow N$$

are homotopic morphisms in $C(\mathcal{A})$, then $Q(f) = Q(g)$.

PROOF. For a proof see [**174**, Lemma III.4.3]. □

It follows that the localisation functor factors via the homotopy category $K(\mathcal{A})$. The key point is that the induced localisation functor

$$Q_K\colon K(\mathcal{A}) \longrightarrow \mathrm{D}(\mathcal{A})$$

has much nicer properties than the original functor Q, as we shall now try to explain.

The problem of defining the derived category is a special case of the problem of localisation of categories. Consider for a moment an analogous problem with rings. After all, as we have seen, an additive category is really just a slight generalization of a ring. Suppose then that A is a (not necessarily commutative) ring, and $S \subset A$ is a set of nonzero elements of A such that

$$1 \in S, \text{ and } s, t \in S \implies st \in S.$$

Suppose we want to construct a ring B and a homomorphism $Q\colon A \to B$ with the property that $Q(s)$ is invertible in B for all $s \in S$, and that if $Q'\colon A \to B'$ is another such homomorphism with this property then Q' factors via Q. Such a homomorphism Q is called a universal localisation.

Clearly, as in the proof of Lemma 4.46, we can define a set B with an associative multiplication, by adjoining to A symbols s^{-1} for each $s \in S$, and imposing relations

$$ss^{-1} = s^{-1}s = 1 \text{ for all } s \in S \text{ and } (st)^{-1} = t^{-1}s^{-1} \text{ for all } s, t \in S.$$

The problem is that a typical element of B then takes the form

$$f_1 s_1^{-1} f_2 s_2^{-1} f_3 \cdots f_n s_n^{-1}$$

with $f_i \in A$ and $s_i \in S$ for all i, and there is no simple way of determining when two expressions determine the same element of B. Furthermore, there is no way of adding two such expressions.

If A is a commutative ring the solution is straightforward. One has

$$fs = sf \implies s^{-1}f = fs^{-1}$$

so we can "collect denominators" and every element of B can be written (non-uniquely) in the form fs^{-1} with $f \in A$ and $s \in S$. Now we can add fractions in the usual way

$$fs^{-1} + gt^{-1} = (tf + sg)(st)^{-1}$$

and the problem is solved.

In the non-commutative case this trick still works sometimes. What we need is the following conditions on the set S, usually called the Ore conditions:

(a) For every $s \in S$ and $f \in A$ there is a $t \in S$ and a $g \in A$ such that $ft = sg$.
(b) Given $f \in A$ there is an $s \in S$ with $fs = 0$ iff there is a $t \in S$ with $tf = 0$.

We can then write $s^{-1}f = gt^{-1}$ and collect denominators as before. It is then easy to check that every element of B can be written in the form fs^{-1} with $f \in A$ and $s \in S$, and that two such expressions $f_i s_i^{-1}$ for $i = 1, 2$ define the same element of B precisely if there are elements $t_1, t_2 \in S$ such that

$$s_1 t_1 = s_2 t_2 \text{ and } f_1 t_1 = f_2 t_2.$$

Furthermore, any two elements of $b_1, b_2 \in B$ can be "put over a common denominator", which is to say that they can be written in the form $b_i = f_i s^{-1}$ for a fixed element $s \in S$. We can then add them by setting

$$b_1 + b_2 = (f_1 + f_2)s^{-1}.$$

It is easy to see that this operation makes B into a ring as required.

The remarkable fact discovered by Verdier is that inside $K(\mathcal{A})$ the set of quasi-isomorphisms satisfy the following analogue of the Ore conditions:

LEMMA 4.48. *Let $\mathcal{A}$ be an abelian category and $K(\mathcal{A})$ the homotopy category of complexes and homotopy equivalence classes of morphisms of complexes.*

(a) *If $f\colon M \to N$ and $s\colon N' \to N$ are morphisms in $K(\mathcal{A})$, with s a quasi-isomorphism, then there is a complex M' and morphisms of* *complexes, $g\colon M' \to N'$ and $t\colon M' \to M$ with t a quasi-isomorphism, so that the diagram*

$$\begin{array}{ccc} M' & \xrightarrow{g} & N' \\ {\scriptstyle t}\downarrow & & \downarrow{\scriptstyle s} \\ M & \xrightarrow{f} & N \end{array}$$

commutes.

(b) *If $f\colon M \to N$ is a morphism in $K(\mathcal{A})$, then there is a quasi-isomorphism $s\colon M' \to M$ with $f \circ s = 0$ in $K(\mathcal{A})$ precisely if there is a quasi-isomorphism $t\colon N \to N'$ with $t \circ f = 0$ in $K(\mathcal{A})$.*

PROOF. For a proof see [**174**, Theorem III.4.4]. □

Using this Lemma it follows that the localisation functor

$$Q_K\colon K(\mathcal{A}) \longrightarrow \mathrm{D}(\mathcal{A})$$

has nice properties analogous to the ones described above. Any morphism $f\colon M \longrightarrow N$ in $\mathrm{D}(\mathcal{A})$ can be represented by a "fraction" or "roof" in $K(\mathcal{A})$, which is to say by a diagram

$$\begin{array}{ccccc} & & M' & & \\ & {\scriptstyle s}\swarrow & & \searrow{\scriptstyle f} & \\ M & & & & N \end{array}$$

with $s\colon M' \to M$ a quasi-isomorphism. Two diagrams

$$\begin{array}{ccccc} & & M^i & & \\ & {\scriptstyle s_i}\swarrow & & \searrow{\scriptstyle f_i} & \\ M & & & & N \end{array}$$

with $i = 1, 2$ define the same morphism in $\mathrm{D}(\mathcal{A})$ precisely if there is a commutative diagram in $K(\mathcal{A})$ of the form

$$\begin{array}{ccccc} & & M^1 & & \\ & {\scriptstyle s_1}\swarrow & \uparrow{\scriptstyle t_1} & \searrow{\scriptstyle f_1} & \\ M & & P & & N \\ & {\scriptstyle s_2}\nwarrow & \downarrow{\scriptstyle t_2} & \nearrow{\scriptstyle f_2} & \\ & & M^2 & & \end{array}$$

with t_1, t_2 quasi-isomorphisms. Furthermore, any two morphisms

$$f, g\colon M \longrightarrow N$$

in $\mathrm{D}(\mathcal{A})$ can be put over a common denominator. In particular the category $\mathrm{D}(\mathcal{A})$ is additive.

4.4.3. The structure of the derived category. Let $\mathcal{A}$ be an abelian category. Although the derived category $\mathrm{D}(\mathcal{A})$ defined above is additive, it is not abelian. Morphisms in $\mathrm{D}(\mathcal{A})$ do not have kernels or cokernels in general. Thus there is no notion of a short exact sequence in $\mathrm{D}(\mathcal{A})$. But there is a weaker substitute, which is the notion of a distinguished triangle.

First we define operations which shift complexes up and down. Fix an integer n. If M is an object of $C(\mathcal{A})$ define a complex $M[n]$ by $M[n]^i = M^{i+n}$ and $d^i_{M[n]} = (-1)^n d^{i+n}_M$. If $f\colon M \longrightarrow N$ is a morphism in $C(\mathcal{A})$ define a morphism $f[n]\colon M[n] \to N[n]$ by setting $f[n]^i = f^{i+n}$. Clearly this defines a functor $[n]\colon C(\mathcal{A}) \longrightarrow C(\mathcal{A})$ which descends to give a functor $[n]\colon \mathrm{D}(\mathcal{A}) \to \mathrm{D}(\mathcal{A})$.

Next recall the definition of the mapping cone. Suppose $f\colon M \longrightarrow N$ is a morphism in $C(\mathcal{A})$. The mapping cone of f is the complex $C(f)$ defined by

$$C(f)^i = M^{i+1} \oplus N^i$$

with differential given by the formula

$$d^i_{C(f)}(m,n) = (-d^{i+1}_M(m), f^{i+1}(m) + d^i_N(n)).$$

There are obvious maps of complexes $\alpha(f)\colon N \to C(f)$ and $\beta(f)\colon C(f) \to M[1]$ fitting into a sequence

$$M \xrightarrow{\ f\ } N \xrightarrow{\alpha(f)} C(f) \xrightarrow{\beta(f)} M[1].$$

These are usually written in a triangle

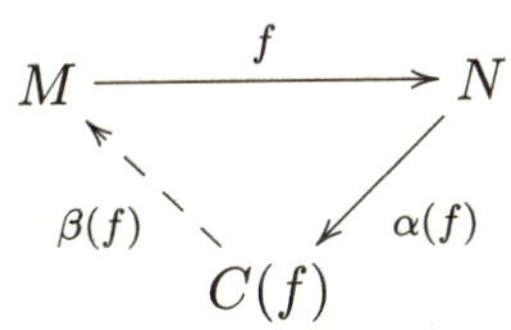

where the dashed arrow means that the given morphism is from $C(f)$ to $M[1]$ rather than to M.

EXAMPLE 4.49. Let $f\colon M \to N$ be a morphism of $\mathcal{A}$ and consider the corresponding morphism of complexes $J(f)\colon J(M) \to J(N)$ in $C(\mathcal{A})$. If f is injective with cokernel P then the mapping cone of $J(f)$ is quasi-isomorphic to $J(P)$. Similarly, if f is surjective with kernel L then the mapping cone of $J(f)$ is quasi-isomorphic to the complex $J(L)[1]$. Thus in some sense the mapping cone construction generalizes the notions of kernel and cokernel.

A distinguished triangle in $\mathrm{D}(\mathcal{A})$ is a triple of objects and morphisms

$$D \xrightarrow{\ a\ } E \xrightarrow{\ b\ } F \xrightarrow{\ c\ } D[1]$$

which is isomorphic to a triple coming from the mapping cone construction. To spell it out, a triple as above is a distinguished triangle if there is a morphism $f\colon M \to N$ and isomorphisms s, t, u in $\mathrm{D}(\mathcal{A})$ such that the diagram

$$\begin{array}{ccccccc}
D & \xrightarrow{a} & E & \xrightarrow{b} & F & \xrightarrow{c} & D[1] \\
\downarrow{\scriptstyle s} & & \downarrow{\scriptstyle t} & & \downarrow{\scriptstyle u} & & \downarrow{\scriptstyle s[1]} \\
M & \xrightarrow{f} & N & \xrightarrow{\alpha(f)} & C(f) & \xrightarrow{\beta(f)} & M[1]
\end{array}$$

commutes. Again, one usually writes a distinguished triangle as follows

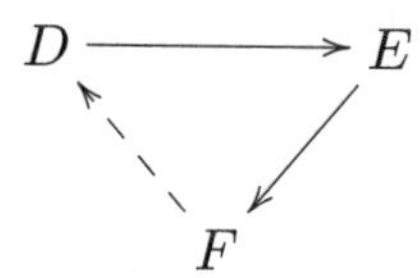

The reader should have no difficulty in verifying that given any such triangle, taking cohomology of complexes gives a long exact sequence

$$\cdots \to H^{i-1}(F) \to H^i(D) \to H^i(E) \to H^i(F) \to H^{i+1}(D) \to \cdots$$

The notion of a triangulated category is an attempt to axiomatise the properties of the shift functor $[1]\colon \mathrm{D}(\mathcal{A}) \to \mathrm{D}(\mathcal{A})$ and the distinguished triangles in $\mathrm{D}(\mathcal{A})$. It is not an entirely satisfactory definition, but as yet there is no clear idea what to replace it with.

More formally, a triangulated category $\mathcal{C}$ is an additive category together with:

(1) a *translation* functor $T : \mathcal{C} \to \mathcal{C}$ which is an isomorphism. If M is an object (or morphism) in $\mathcal{C}$ we will denote $T^n(M)$ by $M[n]$; and
(2) a set of *distinguished triangles*

$$A \xrightarrow{a} B \xrightarrow{b} C \xrightarrow{c} A[1], \tag{4.3}$$

where a morphism between two triangles is simply a commutative diagram of the form

$$\begin{array}{ccccccc}
A & \xrightarrow{a} & B & \xrightarrow{b} & C & \xrightarrow{c} & A[1] \\
\downarrow{\scriptstyle f} & & \downarrow{\scriptstyle g} & & \downarrow{\scriptstyle h} & & \downarrow{\scriptstyle f[1]} \\
A' & \xrightarrow{a} & B' & \xrightarrow{b} & C' & \xrightarrow{c} & A'[1].
\end{array} \tag{4.4}$$

This data is subject to the following axioms:

TR1: a) For any object A, the triangle

$$A \xrightarrow{1_A} A \xrightarrow{0} 0 \xrightarrow{0} A[1] \tag{4.5}$$

is distinguished;

b) If a triangle is isomorphic to a distinguished triangle then it, too, is distinguished.

c) Any morphism $a : A \to B$ can be completed to a distinguished triangle of the form (4.3).

TR2: The triangle (4.3) is distinguished if and only if

(4.6)

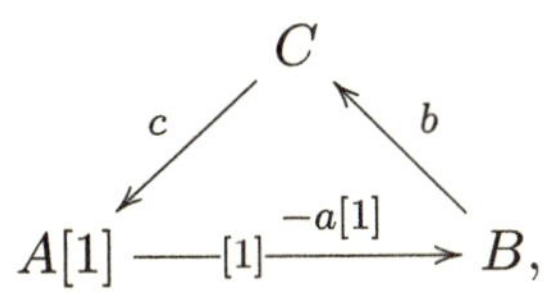

is also distinguished. That is, we may shuffle the edge containing "[1]" around the triangle, translating the objects and morphisms accordingly.

TR3: Given two triangles and the vertical maps f and g in (4.4), we may construct a morphism h to complete (4.4).

TR4: *The Octahedral Axiom:*

(4.7)

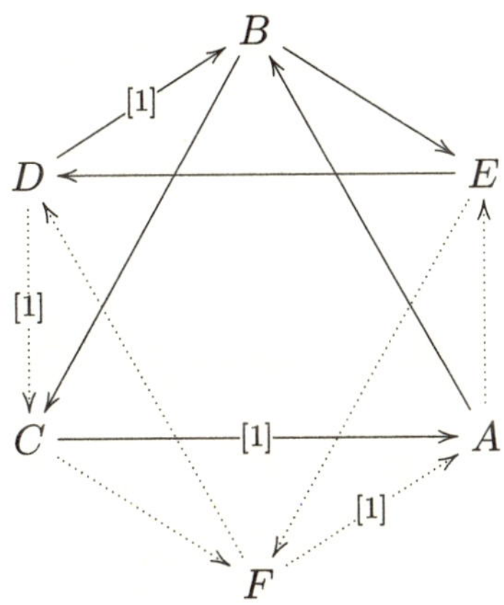

Four faces of the octahedron are distinguished triangles and the other four faces commute. The relative orientations of the arrows obviously specify which is which.

The octahedral axiom specifies that, given A, B, C, D, E and the solid arrows in the octahedron, there is an object F such that the octahedron may be completed with the dotted arrows. The pairs of maps that combine to form maps between B and F also commute.

4.4.4. More about the derived category. In this section we shall try to give a little bit more information about what objects in the derived category $D(\mathcal{A})$ look like. The basic picture to have in mind is that just as a general representation of a quiver consists of a collection of simple representations glued together by extensions, an object of $D(\mathcal{A})$ consists of its cohomology objects $H^i(E) \in \mathcal{A}$ together with some "glue" which holds them together.

Consider the operation of truncating a complex in the ith place in the following way

$$\begin{aligned} &\tau_{\leq i}\Big(\cdots \longrightarrow M^{i-1} \xrightarrow{d^{i-1}} M^i \xrightarrow{d^i} M^{i+1} \longrightarrow \cdots \Big) \\ &= \Big(\cdots \longrightarrow M^{i-1} \xrightarrow{d^{i-1}} \ker d^i \longrightarrow 0 \longrightarrow \cdots \Big) \end{aligned}$$

Note that $H^j(\tau_{\leq i}(E)) = H^j(E)$ for $j \leq i$ and $H^j(E) = 0$ for $j > i$. If $E \in C(\mathcal{A})$ is a complex, then there is an obvious morphism of complexes $\tau_{\leq i}(E) \to E$, and this map induces isomorphisms on the cohomology objects $H^j(E)$ for $j \leq i$.

Make the following definition

DEFINITION 4.50. A complex $E \in C(\mathcal{A})$ is said to be concentrated in degree i if $H^j(E) = 0$ for $j \neq i$.

The following Lemma shows that such objects can be identified with the corresponding objects of $\mathcal{A}$.

LEMMA 4.51. *The functor $J\colon \mathcal{A} \longrightarrow \mathrm{D}(\mathcal{A})$ is fully faithful and defines an equivalence of $\mathcal{A}$ with the full subcategory of $\mathrm{D}(\mathcal{A})$ consisting of objects concentrated in degree zero.*

PROOF. Take objects $A, B \in \mathcal{A}$ and consider the group homomorphism

$$J\colon \mathrm{Hom}_{\mathcal{A}}(A, B) \longrightarrow \mathrm{Hom}_{\mathrm{D}(\mathcal{A})}(J(A), J(B)).$$

This map is injective because it has a one-sided inverse obtained by applying $H^0(-)$. To prove that it is surjective, take a morphism $h\colon J(A) \longrightarrow J(B)$ such that $H^0(h) = 0$. We have to prove that $h = 0$.

The morphism f is represented by a roof of the form

$$J(A) \xleftarrow{\ s\ } P \xrightarrow{\ f\ } J(B)$$

with s a quasi-isomorphism. Since $H^i(J(A)) = 0$ for $i > 0$ the canonical morphism $\iota\colon \tau_{\leq 0}(P) \longrightarrow P$ is a quasi-isomorphism. It follows that h is represented by the roof

$$J(A) \xleftarrow{\ t\ } \tau_{\leq 0}(P) \xrightarrow{\ g\ } J(B)$$

where $g = f \circ \iota$ and $t = s \circ \iota$. But now g is a morphism of complexes

$$\begin{array}{ccccccc} \cdots & \longrightarrow & P_1 & \xrightarrow{d} & P_0 & \longrightarrow & 0 \\ & & \downarrow{\scriptstyle 0} & & \downarrow{\scriptstyle g} & & \downarrow{\scriptstyle 0} \\ \cdots & \longrightarrow & 0 & \longrightarrow & B & \longrightarrow & 0 \end{array}$$

which induces the zero map $H^0(g)\colon P_0/\operatorname{im}(d) \longrightarrow B$ on cohomology. It follows that $g = 0$ and hence $h = 0$ as required.

The final part of the statement is that if $E \in C(\mathcal{A})$ is concentrated in degree zero then E is quasi-isomorphic to $J(H^0(E))$. This is easy: the canonical map $\tau_{\leq 0}(E) \longrightarrow E$ is a quasi-isomorphism, so we can assume that E is of the form

$$\cdots \longrightarrow E_n \longrightarrow \cdots \longrightarrow E_1 \longrightarrow E_0 \longrightarrow 0.$$

But then there is clearly a quasi-isomorphism

$$\begin{array}{ccccccc} \cdots & \longrightarrow & E_1 & \xrightarrow{d} & E_0 & \longrightarrow & 0 \\ & & \downarrow & & \downarrow & & \downarrow \\ \cdots & \longrightarrow & 0 & \longrightarrow & H^0(E) & \longrightarrow & 0. \end{array}$$

since $H^0(E) = E_0/\operatorname{im}(d)$. □

If $E, F \in \mathcal{A}$ then one sets

$$\operatorname{Ext}^i_{\mathcal{A}}(E,F) := \operatorname{Hom}_{\mathrm{D}(\mathcal{A})}(E, F[i]).$$

By the above Lemma $\operatorname{Ext}^0_{\mathcal{A}}(E,F) = \operatorname{Hom}_{\mathcal{A}}(E,F)$. The same argument given in the proof shows that

$$\operatorname{Ext}^i_{\mathcal{A}}(E,F) = 0 \text{ for } i < 0.$$

The significance of the groups $\operatorname{Ext}^i_{\mathcal{A}}(E,F)$ for $i > 0$ is explained in the following examples.

From now on we shall suppress the functor J and identify an object $E \in \mathcal{A}$ with the corresponding object $J(E)$ of $\mathrm{D}(\mathcal{A})$. Conversely, if an object $E \in \mathrm{D}(\mathcal{A})$ is concentrated in degree zero we shall identify it with the corresponding object $H^0(E)$ of $\mathcal{A}$.

EXAMPLE 4.52. Suppose

$$0 \longrightarrow D \xrightarrow{f} E \xrightarrow{g} F \longrightarrow 0$$

is a short exact sequence in an abelian category $\mathcal{A}$. Then F is quasi-isomorphic to the mapping cone of f, so there is a morphism $F \to D[1]$

in $\mathrm{D}(\mathcal{A})$ such that the resulting triple

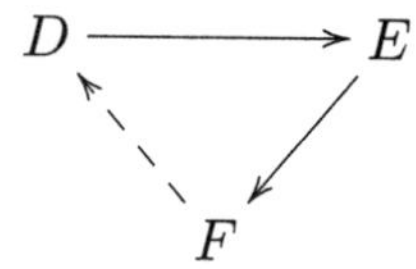

is a distinguished triangle. Thus short exact sequences provide special examples of triangles. Conversely, given a pair of objects $D, F \in \mathcal{A}$ and a morphism $F \to D[1]$, then by axiom (b) above, we can complete to a triangle as above. Applying the cohomology functor we see that E is concentrated in degree zero, and hence is quasi-isomorphic to an object of $\mathcal{A}$. Taking cohomology gives a short exact sequence

$$0 \longrightarrow D \xrightarrow{\ f\ } E \xrightarrow{\ g\ } F \longrightarrow 0$$

in $\mathcal{A}$. Thus short exact sequences

$$0 \longrightarrow D \xrightarrow{\ f\ } E \xrightarrow{\ g\ } F \longrightarrow 0$$

in $\mathcal{A}$ are classified by elements of the abelian group $\mathrm{Ext}^1_{\mathcal{A}}(F, D)$ as claimed before.

The reader will easily verify that for each i there is a triangle

$$\tau_{\le i-1}(E) \longrightarrow \tau_{\le i}(E) \longrightarrow H^i(E)[-i] \longrightarrow \tau_{\le i-1}(E)[1].$$

This suggests the idea that the objects $\tau_{\le i}(E)$ define a filtration of E whose factors are its shifted cohomology sheaves $H^i(E)[-i]$.

EXAMPLE 4.53. Consider the problem of determining objects $E \in \mathrm{D}(\mathcal{A})$ satisfying

$$H^i(E) = 0 \text{ unless } i \in \{-1, 0\}$$

up to isomorphism. By the triangle above, one sees that $\tau_{\le_{i-1}}(E) = \tau_{\le i}(E)$ unless $i = -1$ or $i = 0$. Since $\tau_{\le -i}(E)$ is quasi-isomorphic to the zero complex for large i, one has a triangle

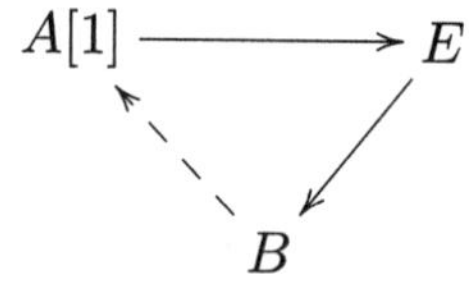

where $A = H^{-1}(E)$ and $B = H^0(E)$. Now one can see that isomorphism classes of two-step objects $E \in \mathrm{D}(\mathcal{A})$ as above are classified by triples (A, B, η) where A and B are objects of $\mathcal{A}$ and $\eta \in \mathrm{Ext}^2_{\mathcal{A}}(B, A)$.

Suppose that $E, F \in \mathrm{D}(\mathcal{A})$ are objects of the derived category. If we only know the cohomology objects $H^i(E) \in \mathcal{A}$ and $H^j(F) \in \mathcal{A}$ we cannot expect to be able to determine the group of morphisms $\mathrm{Hom}_{\mathrm{D}(\mathcal{A})}(E, F)$. Without knowing exactly how the cohomology groups are glued together we do not have enough information to specify this group. But what we do have is a spectral sequence

$$E_2^{p,q} = \bigoplus_{i\in\mathbb{Z}} \mathrm{Ext}^p_{\mathcal{A}}(H^q(E), H^{q+i}(F)) \implies \mathrm{Hom}_{\mathrm{D}(\mathcal{A})}(E, F[p+q]).$$

In certain special cases this can give useful information. See §6.4.4 for a brief explanation of spectral sequences in a different context.

Finally, consider once again the defining formula for Ext groups in our treatment:

$$\mathrm{Ext}^i_{\mathcal{A}}(E,F) = \mathrm{Hom}_{\mathrm{D}(\mathcal{A})}(E, F[i]).$$

Given a third object G of $\mathcal{A}$, we also have

$$\mathrm{Ext}^j_{\mathcal{A}}(F,G) = \mathrm{Hom}_{\mathrm{D}(\mathcal{A})}(F, G[j]) \cong \mathrm{Hom}_{\mathrm{D}(\mathcal{A})}(F[i], G[i+j])$$

and

$$\mathrm{Ext}^{i+j}_{\mathcal{A}}(E,G) = \mathrm{Hom}_{\mathrm{D}(\mathcal{A})}(E, G[i+j]).$$

Hence composition of Homs in the derived category gives rise to a product on Ext-groups, the so-called Yoneda product

$$\mathrm{Ext}^i_{\mathcal{A}}(E,F) \times \mathrm{Ext}^j_{\mathcal{A}}(F,G) \to \mathrm{Ext}^{i+j}_{\mathcal{A}}(E,G).$$

For the special case where $E = F = G$, we therefore obtain an algebra structure on the vector space $\oplus_i \mathrm{Ext}^i_{\mathcal{A}}(E,E)$.

4.4.5. Derived functors. An additive functor $F\colon \mathrm{D}_1 \longrightarrow \mathrm{D}_2$ between triangulated categories is said to be exact if it preserves the relevant structure. More precisely this means the following

(a) F commutes with the shift functors, i.e., there is an isomorphism of functors

$$\epsilon\colon F \circ [1] \longrightarrow [1] \circ F.$$

(b) F takes triangles to triangles: if

$$A \xrightarrow{\ f\ } B \xrightarrow{\ g\ } C \xrightarrow{\ h\ } A[1]$$

is a distinguished triangle in D_1 then

$$F(A) \xrightarrow{F(f)} F(B) \xrightarrow{F(g)} F(C) \xrightarrow{\epsilon(A)\circ F(h)} F(A)[1]$$

is a distinguished triangle in D_2.

Suppose for definiteness that $\mathcal{A} = \mathbf{Mod}(R)$ is the category of modules for a ring R and fix a module $P \in \mathbf{Mod}(R)$. Tensor product of modules defines a functor

$$F = - \otimes P\colon \mathbf{Mod}(R) \longrightarrow \mathbf{Mod}(R)$$

sending a module M to $M \otimes_R P$. This functor is not exact— it does not take exact sequences to exact sequences. It is however right exact, which is to say that if

$$\longrightarrow M_1 \xrightarrow{\ f\ } M_2 \xrightarrow{\ g\ } M_3 \longrightarrow 0$$

is exact, then so is

$$\longrightarrow F(M_1) \xrightarrow{F(f)} F(M_2) \xrightarrow{F(g)} F(M_3) \longrightarrow 0$$

The functor F trivially induces a functor on the category of complexes of modules

$$F\colon \mathcal{C}(\mathcal{A}) \longrightarrow \mathcal{C}(\mathcal{A})$$

but there is no reason why F should take quasi-isomorphisms to quasi-isomorphisms, and hence there is no obvious way to extend F to a functor on derived categories.

In fact, there is a way to get a tensor product on the derived category as follows. Given a module M take a resolution by free R-modules, which is to say a (possibly infinite) exact sequence

$$\cdots \longrightarrow L_n \longrightarrow \cdots \longrightarrow L_1 \longrightarrow L_0 \longrightarrow M \longrightarrow 0$$

with each $L_i \cong R^{\oplus d_i}$ a free R-module. The complex

$$L = (\cdots \longrightarrow L_n \longrightarrow \cdots \longrightarrow L_1 \longrightarrow L_0 \longrightarrow 0)$$

is trivially quasi-isomorphic to the module M considered as a complex, and the crucial fact which can be proved is that because the L_i are assumed free (in particular projective), any two such resolutions are homotopy equivalent. Now apply the functor F to the complex L to get a complex

$$\cdots \longrightarrow F(L_n) \longrightarrow \cdots \longrightarrow F(L_1) \longrightarrow F(L_0) \longrightarrow 0$$

This defines an object of $\mathrm{D}(\mathcal{A})$ which we denote $M \overset{L}{\otimes} P$. Note that if we chose a different free resolution L' then the resulting complex $F(L')$ would be homotopy equivalent to $F(L)$, and in particular quasi-isomorphic, so we would obtain an isomorphic object of the derived category.

The above construction can be made functorial without difficulty and defines a derived functor

$$\mathbf{L}F = - \overset{L}{\otimes} P\colon \mathrm{D}(\mathcal{A}) \longrightarrow \mathrm{D}(\mathcal{A}).$$

This sort of construction works much more generally with other functors, and the resulting derived functors can be shown to satisfy certain universal properties, which in particular ensure their uniqueness, so that one doesn't have to worry about the apparently arbitrary construction given above.

For most purposes the following result suffices. Suppose $F\colon \mathcal{A} \longrightarrow \mathcal{B}$ is a right exact additive functor between abelian categories. If the category $\mathcal{A}$ contains enough projective objects (meaning that every object of $\mathcal{A}$ is a quotient of a projective object) then there is a left derived functor

$$\mathbf{L}F\colon \mathrm{D}(\mathcal{A}) \longrightarrow \mathrm{D}(\mathcal{B})$$

with the following two properties. Firstly, $\mathbf{L}F$ is an exact functor, as defined above. Secondly, F is the "first approximation" to $\mathbf{L}F$, in the sense that if

E is an object of A which we consider also as a trivial complex defining an object of $\mathrm{D}(\mathcal{A})$ then

$$H^0(\mathbf{L}F(E)) = F(E).$$

The object $\mathbf{L}F(E)$ is obtained by applying the functor F to a projective resolution of E, i.e., a complex of projective objects $L = (L_i)$ with a quasi-isomorphism $L \to E$.

If F is a left exact functor there is an analogous result. One needs to assume that A has enough injective objects (meaning that every object of A is a subobject of an injective object), and the result is a right derived functor

$$\mathbf{R}F\colon \mathrm{D}(\mathcal{A}) \longrightarrow \mathrm{D}(\mathcal{B})$$

with the same properties. The object $\mathbf{R}F(E)$ is obtained by applying the functor F to an injective resolution of E, i.e., a complex of injective objects I with a quasi-isomorphism $E \to I$.

Suppose one has an exact functor $\Phi\colon \mathrm{D}(\mathcal{A}) \longrightarrow \mathrm{D}(\mathcal{B})$ and an object $E \in \mathrm{D}(\mathcal{A})$. Then there is a spectral sequence

$$E_2^{p,q} = H^p(\Phi(H^q(E))) \implies H^{p+q}(\Phi(E)).$$

In general, just knowing the cohomology objects $H^q(E)$ is not enough to determine the cohomology objects $H^i(\Phi(E))$, the point being that one has thrown away the information about how the cohomology objects $H^q(E)$ are bound together to form E, and this information is required to determine the cohomology objects of $\Phi(E)$. But nonetheless, in calculations, particularly in low-dimensional examples, the above spectral sequence can give a lot of useful information.

4.4.6. t-structures. Recall from Lemma 4.51 that an abelian category $\mathcal{A}$ sits inside its derived category $\mathrm{D}(\mathcal{A})$ as the subcategory of complexes whose cohomology is concentrated in degree zero. In the following, we shall encounter many examples of interesting algebraic and geometrical relationships which can be described by an equivalence of derived categories

$$\Phi\colon \mathrm{D}(\mathcal{A}) \longrightarrow \mathrm{D}(\mathcal{B}).$$

Such equivalences will usually not arise from an equivalence of the underlying abelian categories $\mathcal{A}$ and $\mathcal{B}$; indeed, this is why one must use derived categories. Changing perspective slightly, one could think of a derived equivalence as being described by a single triangulated category with two different abelian categories sitting inside it. The theory of t-structures is the tool which allows one to see these different abelian categories.

Given a full subcategory $\mathcal{A} \subset \mathrm{D}$, define the right-orthogonal of $\mathcal{A}$ to be the full subcategory of D with objects

$$\mathcal{A}^{\perp} = \{E \in \mathrm{D} : \mathrm{Hom}_{\mathrm{D}}(A, E) = 0 \text{ for all } A \in \mathcal{A}\}$$

DEFINITION 4.54. A t-structure on a triangulated category D is a full subcategory $\mathcal{F} \subset \mathrm{D}$ which is preserved by left-shifts, that is, $\mathcal{F}[1] \subset \mathcal{F}$, and such that for every object $E \in \mathrm{D}$ there is a triangle

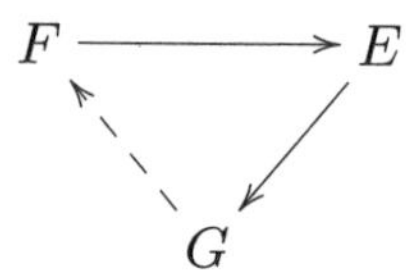

in D with $F \in \mathcal{F}$ and $G \in \mathcal{F}^{\perp}$.

The heart of a t-structure $\mathcal{F} \subset \mathrm{D}$ is the full subcategory

$$\mathcal{A} = \mathcal{F} \cap \mathcal{F}^{\perp}[1] \subset \mathrm{D}.$$

It was proved in [**37**] that $\mathcal{A}$ is an abelian category, where the short exact sequences $0 \longrightarrow a_1 \longrightarrow a_2 \longrightarrow a_3 \longrightarrow 0$ in $\mathcal{A}$ are precisely the triangles $a_1 \longrightarrow a_2 \longrightarrow a_3 \longrightarrow a_1[1]$ in D all of whose vertices a_i are objects of $\mathcal{A}$.

EXAMPLE 4.55. The basic example is the standard t-structure on the derived category $\mathrm{D}(\mathcal{A})$ of an abelian category $\mathcal{A}$, given by

$$\begin{aligned} \mathcal{F} &= \{E \in \mathrm{D}(\mathcal{A}) : H^i(E) = 0 \text{ for all } i > 0\}, \\ \mathcal{F}^{\perp} &= \{E \in \mathrm{D}(\mathcal{A}) : H^i(E) = 0 \text{ for all } i < 0\}. \end{aligned}$$

The heart is the original abelian category $\mathcal{A}$. To give another example, suppose that $\mathrm{D}(\mathcal{A}) \longrightarrow \mathrm{D}(\mathcal{B})$ is an equivalence of derived categories. Then pulling back the standard t-structure on $\mathrm{D}(\mathcal{B})$ gives a t-structure on $\mathrm{D}(\mathcal{A})$ whose heart is the abelian category $\mathcal{B}$.

A t-structure $\mathcal{F} \subset \mathrm{D}$ is said to be bounded if

$$\mathrm{D} = \bigcup_{i,j \in \mathbb{Z}} \mathcal{F}[i] \cap \mathcal{F}^{\perp}[j].$$

A bounded t-structure $\mathcal{F} \subset \mathrm{D}$ is determined by its heart $\mathcal{A} \subset \mathrm{D}$. In fact $\mathcal{F}$ is the extension-closed subcategory generated by the subcategories $\mathcal{A}[j]$ for integers $j \geq 0$. The following result gives another characterisation of bounded t-structures. The proof is a good exercise in manipulating the definitions.

LEMMA 4.56. *A bounded t-structure is determined by its heart. Moreover, if $\mathcal{A} \subset \mathrm{D}$ is a full additive subcategory of a triangulated category* D*, then $\mathcal{A}$ is the heart of a bounded t-structure on* D *if and only if the following two conditions hold:*

(a) *if A and B are objects of $\mathcal{A}$ then* $\mathrm{Hom}_{\mathrm{D}}(A, B[k]) = 0$ *for $k < 0$,*

(b) *for every nonzero object $E \in \mathrm{D}$ there are integers $m < n$ and a collection of triangles*

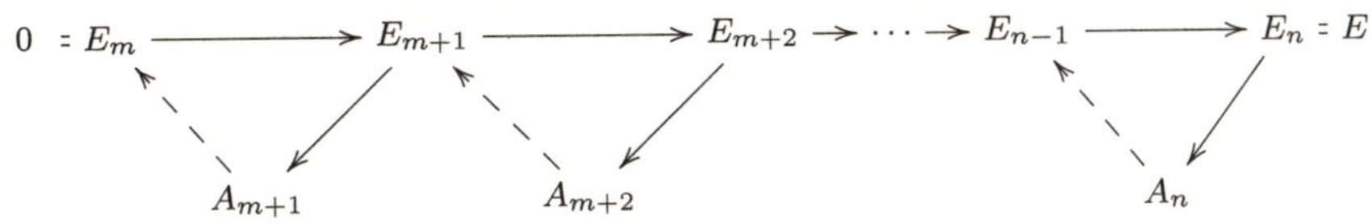

with $A_i[i] \in \mathcal{A}$ for all i. □

In analogy with the standard t-structure on the derived category of an abelian category, the objects $A_i[i] \in \mathcal{A}$ are called the cohomology objects of E in the given t-structure, and denoted $H^i(E)$.

Note that the group Auteq(D) of exact autoequivalences of D acts on the set of bounded t-structures: if $\mathcal{A} \subset \mathrm{D}$ is the heart of a bounded t-structure and $\Phi \in \mathrm{Auteq(D)}$, then $\Phi(\mathcal{A}) \subset \mathrm{D}$ is also the heart of a bounded t-structure.

4.4.7. Tilting. A very useful way to construct t-structures is provided by the method of tilting. This was first introduced in this level of generality by Happel, Reiten and Smalø [**219**], but the name and the basic idea go back to a paper of Brenner and Butler [**58**].

DEFINITION 4.57. A torsion pair in an abelian category $\mathcal{A}$ is a pair of full subcategories $(\mathcal{T}, \mathcal{F})$ of $\mathcal{A}$ which satisfy $\mathrm{Hom}_{\mathcal{A}}(T, F) = 0$ for $T \in \mathcal{T}$ and $F \in \mathcal{F}$, and such that every object $E \in \mathcal{A}$ fits into a short exact sequence

$$0 \longrightarrow T \longrightarrow E \longrightarrow F \longrightarrow 0$$

for some pair of objects $T \in \mathcal{T}$ and $F \in \mathcal{F}$.

The objects of $\mathcal{T}$ and $\mathcal{F}$ are called torsion and torsion-free, respectively. The proof of the following result [**219**, Proposition 2.1] is pretty-much immediate from Lemma 4.56.

PROPOSITION 4.58. *(Happel, Reiten, Smalø) Suppose $\mathcal{A}$ is the heart of a bounded t-structure on a triangulated category* D. *Given an object $E \in \mathrm{D}$ let $H^i(E) \in \mathcal{A}$ denote the ith cohomology object of E with respect to this t-structure. Suppose $(\mathcal{T}, \mathcal{F})$ is a torsion pair in $\mathcal{A}$. Then the full subcategory*

$$\mathcal{A}^\sharp = \{E \in \mathrm{D} : H^i(E) = 0 \text{ for } i \notin \{-1, 0\}, H^{-1}(E) \in \mathcal{F} \text{ and } H^0(E) \in \mathcal{T}\}$$

is the heart of a bounded t-structure on D. □

In the situation of this Proposition, one says that the the subcategory $\mathcal{A}^\sharp$ is obtained from the subcategory $\mathcal{A}$ by tilting with respect to the torsion pair $(\mathcal{T}, \mathcal{F})$. In fact one could equally well consider $\mathcal{A}^\sharp[-1]$ to be the tilted subcategory. Note that the pair $(\mathcal{F}[1], \mathcal{T})$ is a torsion pair in $\mathcal{A}^\sharp$ and that tilting with respect to this pair gives back the original subcategory $\mathcal{A}$ with a shift.

Now suppose $\mathcal{A} \subset \mathrm{D}$ is the heart of a bounded t-structure and is a finite length abelian category. Note that the t-structure is completely determined by the set of simple objects of $\mathcal{A}$; indeed $\mathcal{A}$ is the smallest extension-closed subcategory of D containing this set of objects. Given a simple object $S \in \mathcal{A}$ define $\langle S \rangle \subset \mathcal{A}$ to be the full subcategory consisting of objects $E \in \mathcal{A}$ all of whose simple factors are isomorphic to S. One can either view $\langle S \rangle$ as the torsion part of a torsion pair on $\mathcal{A}$, in which case the torsion-free part is

$$\mathcal{F} = \{E \in \mathcal{A} : \mathrm{Hom}_{\mathcal{A}}(S, E) = 0\},$$

or as the torsion-free part, in which case the torsion part is

$$\mathcal{T} = \{E \in \mathcal{A} : \mathrm{Hom}_{\mathcal{A}}(E, S) = 0\}.$$

The corresponding tilted subcategories are

$$\begin{aligned} \mathcal{L}_S\mathcal{A} &= \left\{ E \in \mathrm{D} \,\middle|\, \begin{array}{c} H^i(E) = 0 \text{ for } i \notin \{0,1\}, \\ H^0(E) \in \mathcal{F} \text{ and } H^1(E) \in \langle S \rangle \end{array} \right\} \\ \mathcal{R}_S\mathcal{A} &= \left\{ E \in \mathrm{D} \,\middle|\, \begin{array}{c} H^i(E) = 0 \text{ for } i \notin \{-1,0\}, \\ H^{-1}(E) \in \langle S \rangle \text{ and } H^0(E) \in \mathcal{T} \end{array} \right\}. \end{aligned}$$

We define these subcategories of D to be the left, respectively right tilts of the subcategory $\mathcal{A}$ at the simple object S. It is easy to see that $S[-1]$ is a simple object of $\mathcal{L}_S\mathcal{A}$, and that if this category is finite length, then $\mathcal{R}_{S[-1]}\mathcal{L}_S\mathcal{A} = \mathcal{A}$. Similarly, if $\mathcal{R}_S\mathcal{A}$ is finite length, then $\mathcal{L}_{S[1]}\mathcal{R}_S\mathcal{A} = \mathcal{A}$.

An extended example of tilting, based on Example 4.35, will be discussed in §5.8.3.2.

4.5. The derived category of coherent sheaves

We shall now apply the general machinery of §4.4 to the category of coherent sheaves on an algebraic variety X. This leads to the *derived category of coherent sheaves* $\mathrm{D}(X)$, the triangulated category of complexes of coherent sheaves on X. In fact, for nonsingular varieties a better behaved category is $\mathrm{D}^b(X)$, the *bounded derived category of coherent sheaves* on X, the full subcategory of $\mathrm{D}(X)$ consisting of complexes which are (quasi-isomorphic to) complexes with finitely many nonzero terms.

The category $\mathrm{D}^b(X)$ is still triangulated, and has a translation functor $[1]$, translating complexes to the right. If $\mathcal{E}, \mathcal{F}$ are coherent sheaves, we can think of them as complexes concentrated in degree zero, so that by Lemma 4.51 we have a full and faithful embedding of categories

$$\mathrm{Coh}(X) \hookrightarrow \mathrm{D}^b(X);$$

in particular,

$$\mathrm{Hom}_{\mathrm{D}^b(X)}(\mathcal{E}, \mathcal{F}) \cong \mathrm{Hom}_{\mathcal{O}_X}(\mathcal{E}, \mathcal{F}),$$

where the latter is the space of ordinary sheaf homomorphisms. As in §4.4.4, we can do more: given sheaves $\mathcal{E}, \mathcal{F}$, thought of as complexes in degree zero, we have the translation functor at our disposal, and hence we can define further

$$\mathrm{Ext}^i(\mathcal{E},\mathcal{F}) = \mathrm{Hom}_{\mathrm{D}^b(X)}(\mathcal{E},\mathcal{F}[i]),$$

the so-called coherent Ext-groups (in fact vector spaces), which we also denote by $\mathrm{Ext}^i_X(\mathcal{E},\mathcal{F})$ or even $\mathrm{Ext}^i_{\mathcal{O}_X}(\mathcal{E},\mathcal{F})$ if one wishes to be pedantic in notation. As discussed in §4.4.4, the Ext-groups are zero in negative degrees; Ext^0 is the same as Hom (sheaf homomorphisms), whereas $\mathrm{Ext}^1(\mathcal{E},\mathcal{F})$ classifies extensions

$$0 \longrightarrow \mathcal{F} \longrightarrow \mathcal{G} \longrightarrow \mathcal{E} \longrightarrow 0$$

in the category $\mathrm{Coh}(X)$ (see Example 4.52). The following basic results (for proofs, see e.g. [**222**]) are more specific to the algebraic geometric context:

PROPOSITION 4.59.

- *If X is a* smooth *variety of dimension n, then for $\mathcal{E},\mathcal{F}$ coherent sheaves,*
$$\mathrm{Ext}^i(\mathcal{E},\mathcal{F}) = 0$$
unless $0 \le i \le n$.
- *If X is a* projective *variety, then $\mathrm{Ext}^i(\mathcal{E},\mathcal{F})$ is a finite-dimensional complex vector space.*
- *If X is an* affine *variety and $\mathcal{E}$ is* locally free, *then $\mathrm{Ext}^i(\mathcal{E},\mathcal{F}) = 0$ for $i \neq 0$.*

The next foundational issue is the definition of derived operations in the geometric context. Recall that in §4.3.4 we have defined the operations tensor product with a sheaf $\otimes\mathcal{F}$, and given $f\colon X \to Y$, pullback f^* and push-forward f_* on sheaves of $\mathcal{O}_X$-modules, so that the first two are right exact, and the last one is left exact. §4.4.5 explained how this leads to derived functors on the *unbounded* derived categories

$$\overset{L}{\otimes}\mathcal{F}\colon \mathrm{D}(X) \to \mathrm{D}(X)$$

as well as, given $f\colon X \to Y$,

$$\mathbf{L}f^*\colon \mathrm{D}(Y) \to \mathrm{D}(X)$$

and, assuming f is projective (or proper),

$$\mathbf{R}f_*\colon \mathrm{D}(X) \to \mathrm{D}(Y).$$

Note though that there is a technical issue here: the category $\mathrm{Coh}(X)$ of coherent sheaves on X does not have enough injectives nor projectives. The problem with injectives is solved by going to a larger category, that of $\mathcal{O}_X$-modules without finiteness conditions, where injective resolutions exist. Projective resolutions in the definition of derived pullback and tensor product are replaced by locally free resolutions, which certainly exist and do the job just as well. Moreover, under various conditions on the varieties and sheaves involved, we actually get functors on the bounded category. Here is a sample proposition along these lines, certainly sufficient for our purposes:

PROPOSITION 4.60.

(1) *If X is smooth and $\mathcal{F} \in \mathrm{D}^b(X)$, then we have a bounded derived tensor product functor*

$$\overset{L}{\otimes} \mathcal{F} \colon \mathrm{D}^b(X) \to \mathrm{D}^b(X).$$

(2) *If $f \colon X \to Y$ is a map between smooth varieties, then we have a bounded derived pullback functor*

$$\mathbf{L}f^* \colon \mathrm{D}^b(Y) \to \mathrm{D}^b(X).$$

(3) *If f is projective (or proper), then we have a bounded derived push-forward functor*

$$\mathbf{R}f_* \colon \mathrm{D}^b(X) \to \mathrm{D}^b(Y).$$

By general theory, $\overset{L}{\otimes} \mathcal{F}$, $\mathbf{L}f^*$ and $\mathbf{R}f_*$ are all exact functors: they take distinguished triangles to distinguished triangles. The canonical reference for all their intricacies, including the proof of the following compatibility relations, which we will have occasion to use, is [**221**].

THEOREM 4.61.

(1) *Given $\mathcal{E}, \mathcal{F}, \mathcal{G} \in \mathrm{D}^b(X)$,*

$$\mathcal{E} \overset{L}{\otimes} (\mathcal{F} \overset{L}{\otimes} \mathcal{G}) \cong (\mathcal{E} \overset{L}{\otimes} \mathcal{F}) \overset{L}{\otimes} \mathcal{G} \in \mathrm{D}^b(X).$$

(2) *Given $\mathcal{E}, \mathcal{F} \in \mathrm{D}^b(Y)$,*

$$\mathbf{L}f^*(\mathcal{E}) \overset{L}{\otimes} \mathbf{L}f^*(\mathcal{F}) \cong \mathbf{L}f^*(\mathcal{E} \overset{L}{\otimes} \mathcal{F}) \in \mathrm{D}^b(X).$$

(3) Adjunction: *let $\mathcal{E} \in \mathrm{D}^b(Y)$, $\mathcal{F} \in \mathrm{D}^b(X)$; then*

$$\mathrm{Hom}_{\mathrm{D}^b(X)}(\mathbf{L}f^*\mathcal{E}, \mathcal{F}) \cong \mathrm{Hom}_{\mathrm{D}^b(Y)}(\mathcal{E}, \mathbf{R}f_*\mathcal{F}).$$

(4) Projection formula: *let $\mathcal{E} \in \mathrm{D}^b(Y)$, $\mathcal{F} \in \mathrm{D}^b(X)$, and assume f is projective (or more generally proper). Then*

$$\mathbf{R}f_* \left(\mathbf{L}f^*(\mathcal{E}) \overset{L}{\otimes} \mathcal{F} \right) \cong \mathcal{E} \overset{L}{\otimes} \mathbf{R}f_*\mathcal{F} \in \mathrm{D}^b(Y).$$

(5) Smooth base change: *Let X, Y, Z be smooth varieties, and $f : Z \to Y$ a morphism. Form the diagram*

$$\begin{array}{ccc} X \times Z & \xrightarrow{F} & X \times Y \\ {\scriptstyle q}\downarrow & & \downarrow{\scriptstyle p} \\ Z & \xrightarrow{f} & Y \end{array}$$

where F is the map induced by f, and p, q are the natural projections. Then there is a natural isomorphism of functors

$$\mathbf{L}f^* \circ \mathbf{R}p_* \cong \mathbf{R}q_* \circ \mathbf{L}F^* : \mathrm{D}^b(X \times Y) \to D^b(Z).$$

4.5.1. Sheaf cohomology. For $\mathcal{E}$ a coherent sheaf on X, the Hom-space $\mathrm{Hom}(\mathcal{O}_X, \mathcal{E})$ is nothing but the space of global sections $\Gamma(\mathcal{E})$. The higher Ext's of the pair $(\mathcal{O}_X, \mathcal{E})$ are also of importance: define

$$H^i(X, \mathcal{E}) = \mathrm{Ext}^i(\mathcal{O}_X, \mathcal{E}),$$

the *sheaf cohomology* of $\mathcal{E}$. General facts and the results of Proposition 4.59 tell us that

- if $i < 0$, $H^i(X, \mathcal{E}) = 0$;
- if X is smooth of dimension n, then $H^i(X, \mathcal{E}) = 0$ for $i > n$ (in fact this is true for any X of dimension n);
- if X is projective, then $H^i(X, \mathcal{E})$ is a finite-dimensional complex vector space for all i; finally
- if X is affine, then $H^i(X, \mathcal{E}) = 0$ for $i > 0$.

One basic property of sheaf cohomology is the existence of a long exact sequence. Suppose that we have a short exact sequence of sheaves

$$0 \longrightarrow \mathcal{E} \longrightarrow \mathcal{F} \longrightarrow \mathcal{G} \longrightarrow 0$$

on X. In the derived category $\mathrm{D}^b(X)$, this is simply a distinguished triangle

$$\mathcal{E} \longrightarrow \mathcal{F} \longrightarrow \mathcal{G} \longrightarrow \mathcal{E}[1].$$

Using the exact functor

$$\mathbf{R}\mathrm{Hom}(\mathcal{O}_X, -) \colon \mathrm{D}^b(X) \to \mathrm{D}^b(\mathbf{Mod}(\mathbb{C})),$$

from the bounded derived category of sheaves on X to the bounded derived category of vector spaces, we obtain an exact triangle

$$\mathbf{R}\mathrm{Hom}(\mathcal{O}_X, \mathcal{E}) \to \mathbf{R}\mathrm{Hom}(\mathcal{O}_X, \mathcal{F}) \to \mathbf{R}\mathrm{Hom}(\mathcal{O}_X, \mathcal{G}) \to \mathbf{R}\mathrm{Hom}(\mathcal{O}_X, \mathcal{E})[1]$$

in $\mathrm{D}^b(\mathbf{Mod}(\mathbb{C}))$. Taking cohomology leads to

THEOREM 4.62. *Given a short exact sequence of sheaves*

$$0 \longrightarrow \mathcal{E} \longrightarrow \mathcal{F} \longrightarrow \mathcal{G} \longrightarrow 0,$$

there is a corresponding long exact sequence of sheaf cohomology groups

$$\cdots \to H^i(X, \mathcal{E}) \to H^i(X, \mathcal{F}) \to H^i(X, \mathcal{G}) \to H^{i+1}(X, \mathcal{E}) \to \cdots$$

REMARK 4.63. Writing out the first few terms of the long exact sequence, we obtain

$$0 \longrightarrow H^0(X, \mathcal{E}) \longrightarrow H^0(X, \mathcal{F}) \longrightarrow H^0(X, \mathcal{G}) \longrightarrow H^1(X, \mathcal{E}).$$

As $H^0(X, -)$ is just global sections, this sequence makes precise an earlier remark that a surjective map of sheaves does not necessarily give a surjective map on local sections. Indeed, according to this long exact sequence, lack of surjectivity is measured by H^1 of the kernel $\mathcal{E}$ (as well as the rest of the long exact sequence of course). In a more traditional treatment of sheaf cohomology, this remark would be the starting point of the whole story.

The long exact sequence is very useful, but in itself usually not sufficient to compute sheaf cohomology explicitly. A tool commonly used for computations is Čech cohomology, defined using a fixed open cover.

Suppose that $\mathcal{E}$ is a coherent $\mathcal{O}_X$-module, and let $X = \bigcup_i U_i$ be a cover of the variety X by Zariski open sets $\mathcal{U} = \{U_i\}$. Consider the complex of vector spaces

$$C^0(\mathcal{U},\mathcal{E}) \xrightarrow{d^0} C^1(\mathcal{U},\mathcal{E}) \xrightarrow{d^1} C^2(\mathcal{U},\mathcal{E}) \xrightarrow{d^2} \cdots.$$

Here

$$C^p(\mathcal{U},\mathcal{E}) = \prod_{i_0<\cdots<i_p} \mathcal{E}\left(U_{i_0}\cap\cdots\cap U_{i_p}\right),$$

and the differential d^p is given by

$$(d^p a)_{i_0,\dots,i_{p+1}} = \sum_{i=0}^{p+1}(-1)^i a_{i_0,\dots,\hat{i}_k,\dots i_{p+1}}|_{U_{i_0}\cap\cdots\cap U_{i_{p+1}}}$$

for a collection of local sections $\{a_{i_0,\dots,i_p}\} \in C^p(\mathcal{U},\mathcal{E})$. An easy computation gives $d^{i+1}\circ d^i = 0$, and hence this is indeed a complex. Let $\check{H}^i_{\mathcal{U}}(X,\mathcal{E})$ denote its i-th cohomology.

PROPOSITION 4.64. *Suppose that $\mathcal{U} = \{U_i\}$ is a cover of X consisting of* affine *open sets. Then Čech cohomology computes sheaf cohomology: there is a natural isomorphism*

$$H^i(X,\mathcal{E}) \xrightarrow{\sim} \check{H}^i_{\mathcal{U}}(X,\mathcal{E}).$$

This result allows us to compute sheaf cohomology in several different contexts. One simple application is Theorem 4.66 below.

Note also that the definition of Čech cohomology makes sense for any topological space, and any sheaf of abelian groups $\mathcal{F}$ on it, not just $\mathcal{O}_X$-modules. It is also possible to define the cohomology of every such sheaf. These two constructions do not always agree, but usually do for sufficiently fine coverings. For example, if the sheaf $\mathcal{F}$ is constant, it suffices to take a covering in which all intersections $U_{i_0}\cap\cdots\cap U_{i_p}$ are contractible. The following example gives a context where this more general construction is useful.

EXAMPLE 4.65. Recall that a line bundle $\mathcal{L}$ on X is constructed by taking non-vanishing glueing functions $g_{ij} \in \mathcal{O}_X(U_i\cap U_j)$ for a suitable Zariski open cover $\{U_i\}$ of X, allowing us to glue the trivial line bundles $\mathcal{O}_{U_i}$ so long as the g_{ij} satisfy the condition $g_{ij}g_{jk}g_{ki} = 1$ on triple overlaps $U_i\cap U_j\cap U_k \neq \emptyset$. Thinking of the sheaf $\mathcal{O}_X^*$ of non-vanishing regular functions on X as a sheaf of abelian groups with the multiplication operation, the $\{g_{ij}\}$

precisely define a Čech 1-cocycle with values in this sheaf. Isomorphism of line bundles corresponds to taking the quotient by 1-coboundaries, and thus we obtain the important relation $\mathrm{Pic}(X) \cong H^1(X, \mathcal{O}_X^*)$.

Further, there is an exact sequence of sheaves of abelian groups, where we take the classical (complex) topology on X:

$$0 \longrightarrow \mathbb{Z} \longrightarrow \mathcal{O}_X \xrightarrow{\exp} \mathcal{O}_X^* \longrightarrow 0;$$

note that here the exponential map is a homomorphism of abelian groups from the additive structure on sections of $\mathcal{O}_X$ to the multiplicative structure of sections of $\mathcal{O}_X^*$. The associated long exact sequence includes a connecting homomorphism

$$\delta : H^1_{\mathbb{C}}(X, \mathcal{O}_X^*) \longrightarrow H^2(X, \mathbb{Z}).$$

The subscript on $H^1_{\mathbb{C}}(X, \mathcal{O}_X^*)$ indicates that this cohomology is to be computed in the complex topology; it classifies line bundles on X in the complex topology. However, if X is projective, then the group of line bundles in the complex and Zariski topologies coincide. Thus $H^1_{\mathbb{C}}(X, \mathcal{O}_X^*) \cong \mathrm{Pic}(X)$. Putting all of this together gives a sheaf theoretic definition for the first Chern class map $c_1 \colon \mathrm{Pic}(X) \to H^2(X, \mathbb{Z})$.

In the long exact sequence, the kernel of the map $c_1 : \mathrm{Pic}(X) \to H^2(X, \mathbb{Z})$ gets identified with $H^1_{\mathbb{C}}(X, \mathcal{O}_X)/H^1(X, \mathbb{Z})$, a complex torus of dimension $\dim_{\mathbb{C}} H^1(X, \mathcal{O}_X)$ (here again, for projective X we can take either topology). This kernel is the Picard variety $\mathrm{Pic}^0(X)$ as defined in §4.3.5; the long exact sequence thus shows the origin of the complex manifold structure on this variety.

4.5.2. Serre duality. A basic property of the derived category of coherent sheaves on a variety is Serre duality. To motivate this concept, behold the following easy but fundamental result, which computes the cohomology of line bundles on $\mathbb{P}^n$.

THEOREM 4.66. *The sheaf cohomology of line bundles on $\mathbb{P}^n$ with homogeneous coordinates $x_0, \ldots, x_n$, is computed as follows:*

(1) If $k \geq 0$, then

$$H^0(\mathbb{P}^n, \mathcal{O}_{\mathbb{P}^n}(k)) = \mathbb{C}[x_0, \ldots, x_n]^{(k)},$$

the degree-k linear subspace of the polynomial ring; for $i > 0$,

$$H^i(\mathbb{P}^n, \mathcal{O}_{\mathbb{P}^n}(k)) = 0.$$

(2) If $k < 0$, then

$$H^n(\mathbb{P}^n, \mathcal{O}_{\mathbb{P}^n}(k)) = \mathbb{C}\left\langle x_0^{i_0} \cdot \ldots \cdot x_n^{i_n} \,\middle|\, i_j < 0, \sum_{i=0}^{n} i_j = k \right\rangle;$$

for $i < n$,

$$H^i(\mathbb{P}^n, \mathcal{O}_{\mathbb{P}^n}(k)) = 0.$$

PROOF. This is a computational exercise using Čech cohomology with respect to the standard open cover of $\mathbb{P}^n$. □

To analyze this result, note first of all that if $0 > k > -n-1$, then the sheaf $\mathcal{O}_{\mathbb{P}^n}(k)$ has no cohomology, since in (2) all exponents need to be strictly negative. The first place where higher cohomology appears is $H^n(\mathbb{P}^n, \mathcal{O}_{\mathbb{P}^n}(-n-1))$, which is one-dimensional, generated by the monomial $x_0^{-1} \cdot \ldots \cdot x_n^{-1}$. Further, for any $k \geq 0$, $H^0(\mathbb{P}^n, \mathcal{O}_{\mathbb{P}^n}(k))$ has a basis consisting of monomials $x_0^{i_0} \cdot \ldots \cdot x_n^{i_n}$, $\sum i_j = k$, with non-negative exponents, whereas $H^n(\mathbb{P}^n, \mathcal{O}_{\mathbb{P}^n}(-k-n-1))$ has a dual basis consisting of monomials $x_0^{-1-i_0} \cdot \ldots \cdot x_n^{-1-i_n}$, $\sum(-1-i_j) = -k-n-1$, with negative exponents. Hence, noting that all other cohomologies are zero, we deduce

COROLLARY 4.67. *Let $\mathcal{F}$ be any line bundle on $\mathbb{P}^n$; then there is a perfect pairing*

$$H^i(\mathbb{P}^n, \mathcal{F}) \times H^{n-i}(\mathbb{P}^n, \mathcal{F}^\vee \otimes \mathcal{O}_{\mathbb{P}^n}(-n-1)) \longrightarrow H^n(\mathbb{P}^n, \mathcal{O}_{\mathbb{P}^n}(-n-1)) \cong \mathbb{C}.$$

The duality statement of this Corollary is called *Serre duality*, and it is a fundamental result in the theory of coherent cohomology. As it is formulated above, it holds in fact for all vector bundles on $\mathbb{P}^n$. To get a result that holds for all coherent sheaves and eventually extends to the derived category, and works for varieties other than $\mathbb{P}^n$, we need to make the following adjustments:

- To get a formulation for complexes of sheaves, note that if $\mathcal{F}$ is a vector bundle,
$$H^i(\mathbb{P}^n, \mathcal{F}) = \mathrm{Hom}_{\mathbb{P}^n}(\mathcal{O}_{\mathbb{P}^n}[-i], \mathcal{F}),$$
(be careful to distinguish the round brackets of twisting with a line bundle on $\mathbb{P}^n$ from the square brackets of translation in its derived category), whereas
$$H^{n-i}(\mathbb{P}^n, \mathcal{F}^\vee \otimes \mathcal{O}_{\mathbb{P}^n}(-n-1)) = \mathrm{Hom}_{\mathbb{P}^n}(\mathcal{F}, \mathcal{O}_{\mathbb{P}^n}(-n-1)[n-i]).$$
- It is possible to show that the line bundle $\mathcal{O}_{\mathbb{P}^n}(-n-1)$ is in fact the *canonical bundle* of $\mathbb{P}^n$, the highest exterior power of the sheaf of holomorphic cotangent vectors. In general, we need to use this line bundle in place of $\mathcal{O}_{\mathbb{P}^n}(-n-1)$.

Putting together these ingredients leads then to the following general result.

THEOREM 4.68. (Serre duality) *Let X be a smooth projective variety of dimension n. Then there exists a line bundle $\omega_X \in \mathrm{Pic}(X)$, such that for every pair of objects $\mathcal{E}, \mathcal{F} \in \mathrm{D}^b(X)$, there is a perfect pairing*

$$\mathrm{Hom}_{D^b(X)}(\mathcal{E}, \mathcal{F}) \otimes \mathrm{Hom}_{D^b(X)}(\mathcal{F}, \mathcal{E} \otimes \omega_X[n]) \to H^n(X, \omega_X) \cong \mathbb{C}.$$

It is easy to show that ω_X, if it exists, must be unique; thus we can take the statement of the theorem as a definition of ω_X, which from this point of view is referred to as *the dualizing sheaf* of X. Alternatively, we can extend the statement of the theorem by what was said above: the canonical bundle, the highest exterior power of the sheaf of holomorphic cotangent vectors, is a dualizing sheaf for a smooth projective variety X.

The proof of this result consists of a series of reductions, starting from the case of line bundles on $\mathbb{P}^n$ discussed above; for details, we refer to [**222**] once again. More importantly, note that the statement of the Corollary is completely categorical, and thus makes sense in any $\mathbb{C}$-linear triangulated category $\mathcal{A}$. A *Serre functor* on such a category is a functor $S_{\mathcal{A}}\colon \mathcal{A} \to \mathcal{A}$, such that for all objects $E, F \in \mathcal{A}$, there is a (bifunctorial) perfect pairing

$$\operatorname{Hom}_{\mathcal{A}}(E, F) \times \operatorname{Hom}_{\mathcal{A}}(F, S_{\mathcal{A}}(E)) \to \mathbb{C}.$$

Comparing with the above formulation, we see that $-\otimes \omega_X[n]$ is a Serre functor on the derived category $\mathrm{D}^b(X)$ of a smooth projective variety X of dimension n.

4.6. Fourier-Mukai theory

The previous section built up a large toolkit relating to the triangulated category of coherent sheaves $\mathrm{D}^b(X)$ on an algebraic variety X. In this section, we will discuss some properties of these categories; we will in particular find "generating sets" and "orthonormal bases", and discuss symmetries. A much more thorough exposition of these ideas is contained in the excellent [**250**].

4.6.1. Derived correspondences. To understand the idea of a (derived) correspondence, let us start with the example of a morphism $f\colon X \to Y$ between varieties. Then all the information about f is encoded in the graph $\Gamma_f \subset X \times Y$ of f, which (as a set) is defined as

$$\Gamma_f = \{(x, f(x)) : x \in X\} \subset X \times Y.$$

Now consider the natural projections p_X, p_Y from $X \times Y$ to the factors X, Y. Restricted to the subvariety Γ_f, p_X is an isomorphism (since f is a morphism). The fibres of p_Y restricted to Γ_f are just the fibres of f; so for example f is proper if and only if $p_Y|_{\Gamma_f}$ is.

If $H(-)$ is any reasonable covariant homology theory (say singular homology in the complex topology for X, Y compact), then we have a natural pushforward map

$$f_*\colon H(X) \to H(Y).$$

It is easy to see that this map can be expressed in terms of the graph Γ_f and the projection maps as

$$f_*(\alpha) = p_{Y*}(p_X^*(\alpha) \cup [\Gamma_f]) \tag{4.8}$$

where $[\Gamma_f] \in H(X \times Y)$ is the fundamental class of the subvariety $[\Gamma_f]$.

Generalizing this construction gives us the notion of a "multi-valued function" or *correspondence* from X to Y, simply defined to be a general subvariety $\Gamma \subset X \times Y$, replacing the assumption that p_X be an isomorphism with some weaker assumption, such as $p_X|_{\Gamma_f}, p_Y|_{\Gamma_f}$ finite or proper. Under suitable assumptions, the right hand side of formula (4.8) still makes sense, and defines a generalized pushforward map

$$\Gamma_*\colon H(X) \to H(Y).$$

In our present context of sheaves on varieties, there is a further simple generalization. A subvariety $\Gamma \subset X \times Y$ can be represented by its structure sheaf $\mathcal{O}_\Gamma$ on $X \times Y$. Associated to the projection maps p_X, p_Y, we also have pullback and pushforward operations on sheaves, and as we discussed above, they are best behaved when used on the derived category. The cup product on homology turns out to have an analogue too, namely tensor product. So, appropriately interpreted, formula (4.8) makes sense as an operation from the derived category of X to that of Y. At this point however, there is no need to restrict to structure sheaves of subvarieties. Indeed, we can make the following definition.

DEFINITION 4.69. A *derived correspondence* between a pair of smooth varieties X, Y is an object $\mathcal{F} \in \mathrm{D}^b(X \times Y)$ with support which is proper over both factors. A derived correspondence defines a functor $\Phi_\mathcal{F}$ by

$$\begin{array}{rcccl} \Phi_\mathcal{F} & : & \mathrm{D}^b(X) & \to & \mathrm{D}^b(Y) \\ & & (-) & \mapsto & \mathbf{R}p_{Y*}(\mathbf{L}p_X^*(-) \overset{L}{\otimes} \mathcal{F}) \end{array}$$

where $(-)$ could refer to both objects and morphisms in $\mathrm{D}^b(X)$. $\mathcal{F}$ is sometimes called the *kernel* of the functor $\Phi_\mathcal{F}$.

Note that the functor $\Phi_\mathcal{F}$ is exact, as it is defined as a composite of exact functors. Note also that since the projection p_X is flat, the derived pullback $\mathbf{L}p_X^*$ is the same as ordinary pullback p_X^*.

Given derived correspondences $\mathcal{E} \in \mathrm{D}^b(X \times Y)$, $\mathcal{F} \in \mathrm{D}^b(Y \times Z)$, we obtain functors

$$\Phi_\mathcal{E}\colon \mathrm{D}^b(X) \to \mathrm{D}^b(Y), \quad \Phi_\mathcal{F}\colon \mathrm{D}^b(Y) \to \mathrm{D}^b(Z),$$

which can then be composed to get a functor

$$\Phi_\mathcal{F} \circ \Phi_\mathcal{E}\colon \mathrm{D}^b(X) \to \mathrm{D}^b(Z).$$

PROPOSITION 4.70. *The composite functor $\Phi_\mathcal{F} \circ \Phi_\mathcal{E}$ is isomorphic to the functor $\Phi_\mathcal{G}$ defined by the kernel*

$$\mathcal{G} = \mathbf{R}\pi_{XZ*}(\mathbf{L}\pi_{YZ}^*(\mathcal{F}) \overset{L}{\otimes} \mathbf{L}\pi_{XY}^*(\mathcal{E})) \in \mathrm{D}^b(X \times Z)$$

where $\pi_{XY}\colon X \times Y \times Z \to X \times Y$ is the projection, and π_{YZ}, π_{XZ} are defined similarly.

PROOF. This is an easy exercise using smooth base change and the projection formula from Theorem 4.61. □

The rule $\mathcal{G} = \mathcal{E} \star \mathcal{F}$ defines a composition law directly on the set of kernels with compatible source and target. It follows easily from the definition that if $\mathcal{O}_{\Delta_X} \in \mathrm{D}^b(X \times X)$ is the structure sheaf of the diagonal of X, then $\mathcal{E} \star \mathcal{O}_{\Delta_X} \cong \mathcal{E}$ and $\mathcal{O}_{\Delta_X} \star \mathcal{F} \cong \mathcal{F}$, whenever these compositions make sense. Thus $\mathcal{O}_{\Delta_X}$ is a two-sided identity with respect to composition of kernels.

4.6.2. Beilinson's theorem. In this section, we will discuss an example which shows that even the "trivial" derived correspondence is useful in concrete situations. The result is due to Beilinson, apparently conceived during a high school exercise class.

Let $X = \mathbb{P}^n$ be projective space, and consider the natural map of sheaves

$$\mathrm{Hom}(\mathcal{O}_{\mathbb{P}^n}(-1), \mathcal{O}_{\mathbb{P}^n}) \otimes \mathcal{O}_{\mathbb{P}^n}(-1) \to \mathcal{O}_{\mathbb{P}^n}.$$

It is easy to see that this map is surjective; its kernel is the sheaf $\Omega_{\mathbb{P}^n}$ of holomorphic differential forms (the holomorphic cotangent bundle) of $\mathbb{P}^n$. Hence we get a short exact sequence

$$0 \longrightarrow \Omega_{\mathbb{P}^n} \longrightarrow \mathcal{O}_{\mathbb{P}^n}(-1)^{\oplus(n+1)} \longrightarrow \mathcal{O}_{\mathbb{P}^n} \longrightarrow 0,$$

which leads, after taking duals and tensoring by $\mathcal{O}_{\mathbb{P}^n}(-1)$, to a short exact sequence

$$0 \longrightarrow \mathcal{O}_{\mathbb{P}^n}(-1) \longrightarrow \mathcal{O}_{\mathbb{P}^n}^{\oplus(n+1)} \longrightarrow \Theta_{\mathbb{P}^n}(-1) \longrightarrow 0,$$

where $\Theta_{\mathbb{P}^n} = \Omega_{\mathbb{P}^n}^{\vee}$ is the holomorphic tangent bundle of $\mathbb{P}^n$.

Using these dual exact sequences, it is a simple matter to prove that there is a natural isomorphism between two $(n+1)$-dimensional vector spaces

$$H^0(\mathbb{P}^n, \Theta_{\mathbb{P}^n}(-1)) \cong H^0(\mathbb{P}^n, \mathcal{O}_{\mathbb{P}^n}(1))^{\vee}.$$

Now comes the trick: let $p_i \colon \mathbb{P}^n \times \mathbb{P}^n \to \mathbb{P}^n$ be the natural projections; then

$$\begin{aligned} H^0(\mathbb{P}^n \times \mathbb{P}^n, p_1^*\Theta_{\mathbb{P}^n}(-1) \otimes p_2^*\mathcal{O}_{\mathbb{P}^n}(1)) &\cong H^0(\mathbb{P}^n, \Theta_{\mathbb{P}^n}(-1)) \otimes H^0(\mathbb{P}^n, \mathcal{O}_{\mathbb{P}^n}(1)) \\ &\cong (\mathbb{C}^{n+1})^{\vee} \otimes \mathbb{C}^{n+1} \\ &\cong \mathrm{Hom}(\mathbb{C}^{n+1}, \mathbb{C}^{n+1}). \end{aligned}$$

Inside the latter space, there is a canonical element $1_{\mathbb{C}^{n+1}}$, which under the above isomorphisms corresponds to a canonical section

$$s \in H^0\left(\mathbb{P}^n \times \mathbb{P}^n, p_1^*\Theta_{\mathbb{P}^n}(-1) \otimes p_2^*\mathcal{O}_{\mathbb{P}^n}(1)\right).$$

Writing everything in explicit form (compare [**380**]), it is possible to check that s vanishes exactly along the diagonal $\Delta_{\mathbb{P}^n}$ in $\mathbb{P}^n \times \mathbb{P}^n$, and thus we get an exact sequence

$$p_1^*\Omega_{\mathbb{P}^n}(1) \otimes p_2^*\mathcal{O}_{\mathbb{P}^n}(-1) \xrightarrow{s^{\vee}} \mathcal{O}_{\mathbb{P}^n \times \mathbb{P}^n} \longrightarrow \mathcal{O}_{\Delta_{\mathbb{P}^n}} \longrightarrow 0.$$

This is very nice, since this is the beginning of a resolution of the sheaf $\mathcal{O}_{\Delta_{\mathbb{P}^n}}$ on $\mathbb{P}^n \times \mathbb{P}^n$ by locally free sheaves. A standard piece of homological algebra, use of the Koszul resolution, gives the following result:

PROPOSITION 4.71. *The following complex of sheaves on $\mathbb{P}^n \times \mathbb{P}^n$ is exact, and thus gives a resolution of $\mathcal{O}_{\Delta_{\mathbb{P}^n}}$ by locally free sheaves:*

$$0 \to p_1^*\Omega^n_{\mathbb{P}^n}(n) \otimes p_2^*\mathcal{O}_{\mathbb{P}^n}(-n) \xrightarrow{\wedge^n s^\vee} p_1^*\Omega^{n-1}_{\mathbb{P}^n}(n-1) \otimes p_2^*\mathcal{O}_{\mathbb{P}^n}(-n+1) \longrightarrow \dots$$
$$\dots \longrightarrow p_1^*\Omega_{\mathbb{P}^n}(1) \otimes p_2^*\mathcal{O}_{\mathbb{P}^n}(-1) \xrightarrow{s^\vee} \mathcal{O}_{\mathbb{P}^n \times \mathbb{P}^n} \to \mathcal{O}_{\Delta_{\mathbb{P}^n}} \to 0,$$

where $\Omega^k_{\mathbb{P}^n} \cong \bigwedge^k \Omega_{\mathbb{P}^n}$ is the sheaf of holomorphic k-differentials on $\mathbb{P}^n$.

To use this result, recall that the structure sheaf $\mathcal{O}_{\Delta_{\mathbb{P}^n}}$ is an identity for composition of correspondences, and thus the associated Fourier-Mukai functor $\Phi_{\mathcal{O}_{\Delta_{\mathbb{P}^n}}}$ is the identity on $\mathrm{D}^b(\mathbb{P}^n)$. This observation immediately leads to

THEOREM 4.72. (Beilinson's theorem) *For every sheaf $\mathcal{F}$ on $\mathbb{P}^n$, there is a spectral sequence with E_1 terms*

$$E_1^{pq} = H^q\left(\mathbb{P}^n, \mathcal{F} \otimes \Omega^{-p}_{\mathbb{P}^n}(-p)\right) \otimes \mathcal{O}_{\mathbb{P}^n}(p)$$

converging to $\mathcal{F}$ in degree zero.

PROOF. Use

$$\mathcal{F} \cong p_{2*}\left(p_1^*(\mathcal{F}) \otimes \mathcal{O}_{\Delta_{\mathbb{P}^n}}\right)$$

and replace $\mathcal{O}_{\Delta_{\mathbb{P}^n}}$ by its locally free resolution. The full details are in [**380**]. □

COROLLARY 4.73. *The set of sheaves $\{\mathcal{O}_{\mathbb{P}^n}(-n), \dots, \mathcal{O}_{\mathbb{P}^n}\}$* generates *the derived category of $\mathbb{P}^n$, i.e., the smallest full subcategory of $\mathrm{D}^b(\mathbb{P}^n)$ containing all these sheaves, as well as all translates of objects and all cones of morphisms, is $\mathrm{D}^b(\mathbb{P}^n)$ itself.*

PROOF. Let $\mathcal{A}$ be the smallest subcategory of $\mathrm{D}^b(\mathbb{P}^n)$ satisfying the conditions. If $\mathcal{F}$ is a sheaf on $\mathbb{P}^n$, then all spaces in the E_1 term of the spectral sequence are sums of copies of sheaves from the set $\{\mathcal{O}_{\mathbb{P}^n}(-n), \dots, \mathcal{O}_{\mathbb{P}^n}\}$ (since $\Omega^{-p}_{\mathbb{P}^n}$ is zero otherwise!). The computation of the various later terms in the spectral sequence involves taking kernels of morphisms between earlier spaces; as $\mathcal{A}$ is closed under taking cones, all later terms also consist of sheaves in $\mathcal{A}$ and thus $\mathcal{F}$ is in $\mathcal{A}$. If $\mathcal{F} \in \mathrm{D}^b(X)$ is an arbitrary complex, using truncations inductively shows that $\mathcal{F} \in \mathcal{A}$. Thus $\mathcal{A}$ is the whole of $\mathrm{D}^b(\mathbb{P}^n)$. □

The set $\{\mathcal{O}_{\mathbb{P}^n}(-n), \dots, \mathcal{O}_{\mathbb{P}^n}\}$ is called the "Beilinson basis" of $\mathrm{D}^b(\mathbb{P}^n)$. In linear vector spaces associated to $\mathrm{D}^b(\mathbb{P}^n)$, such as in K-theory or cohomology, the images of these sheaves indeed form a basis.

REMARK 4.74. Suppose that a variety X has "resolution of the diagonal", in other words a resolution of $\mathcal{O}_{\Delta_X}$ on $X \times X$ by a complex consisting of terms which are tensor products of a set of sheaves pulled back from the factors as in Proposition 4.71. Then the same idea can be used to study sheaves on X in terms of the given set appearing in this resolution; in particular, an analogue of Corollary 4.73 holds. Beyond $\mathbb{P}^n$, there are some other interesting varieties which satisfy this property, such as Grassmannians [**281**], and some resolutions of finite quotient singularities, to be discussed in §4.7; compare Remark 4.94.

4.6.3. Fully faithful functors on categories of sheaves.

Recall Definition 4.12, repeated here for convenience: a functor $F\colon \mathcal{A} \to \mathcal{B}$ between two categories is

(1) *fully faithful*, if for every pair of objects $C_1, C_2 \in \mathcal{A}$, the functor defines an isomorphism on the Hom-sets:

$$\mathrm{Hom}_{\mathcal{A}}(C_1, C_2) \xrightarrow{\sim} \mathrm{Hom}_{\mathcal{B}}(F(C_1), F(C_2)).$$

(2) *an equivalence of categories*, if it is fully faithful and also "surjective up to isomorphism": for every object $D \in \mathcal{B}$, there is a $C \in \mathcal{A}$ with $F(C) \cong D$ in $\mathcal{B}$.

Remember that full faithfulness actually implies "injectivity up to isomorphism": if $F(C_1) \cong F(C_2)$ in $\mathcal{B}$, then $C_1 \cong C_2$ in $\mathcal{A}$.

The following is the crucial observation:

PROPOSITION 4.75. *Let X be a smooth variety. The set $\{\mathcal{O}_P | P \in X\}$ of objects in $\mathrm{D}(X)$ consisting of the structure sheaves of points satisfies the following properties:*

(1) For all $P \in X$,

$$\mathrm{Hom}_{\mathrm{D}(X)}(\mathcal{O}_P, \mathcal{O}_P) \cong \mathbb{C}.$$

(2) For all $P \neq Q$ and $i \in \mathbb{Z}$,

$$\mathrm{Hom}_{\mathrm{D}(X)}(\mathcal{O}_P, \mathcal{O}_Q[i]) \cong 0.$$

(3) If $C \in \mathrm{D}(X)$ is an object such that for all $P \in X$ and $i \in \mathbb{Z}$,

$$\mathrm{Hom}_{\mathrm{D}(X)}(\mathcal{O}_P, C[i]) = 0,$$

then $C \cong 0$ in $\mathrm{D}(X)$.

PROOF. Denote by $i_P\colon \{P\} \to X$ the embedding of a point $P \in X$, and recall that $\mathcal{O}_P$ is just a shorthand for the pushforward $i_{P*}\mathcal{O}_P$. Thus, by adjunction, Theorem 4.61 (3),

$$\mathrm{Hom}_{\mathrm{D}(X)}\left(i_{P*}\mathcal{O}_P, i_{Q*}\mathcal{O}_Q[i]\right) \cong \mathrm{Hom}_{\mathrm{D}(Q)}\left(\mathbf{L}i_Q^*(i_{P*}\mathcal{O}_P), \mathcal{O}_Q[i]\right)$$

Now (1) follows from the standard (Koszul) resolution of $i_{Q*}\mathcal{O}_Q$ on X, whereas (2) follows simply from the fact that the support of $i_{P*}\mathcal{O}_P$ is disjoint from Q in this case. (3) is a little trickier, and we refer to [**60**, Example 2.2] for the proof. □

The point of this result is that the set $\{\mathcal{O}_P \mid P \in X\}$ can for many purposes be thought of as an "orthonormal basis" of the derived category. (1) and (2) of the Proposition express the "normalization" and "orthogonality" properties, whereas (3) states that the set $\{\mathcal{O}_P : P \in X\}$ is a so-called *spanning class*: it "spans" the derived category in a certain sense (though note that it does not generate it in the sense of Corollary 4.73!). The following theorem of Bondal-Orlov and Bridgeland is a precise translation of the statement from linear algebra that the behaviour of a linear map between inner product spaces is completely characterized by its effect on an orthonormal basis.

THEOREM 4.76. *Let $\mathcal{F} \in \mathrm{D}(X \times Y)$ be a derived correspondence between smooth projective varieties X, Y. For a point $P \in X$, let*

$$i_P : Y = \{P\} \times Y \hookrightarrow X \times Y$$

denote the inclusion of a fibre of the first projection π_X, and let $\mathcal{F}_P = \mathbf{L}i_P^*\mathcal{F}$ be the restriction (derived pullback) of $\mathcal{F}$ to the fibre. Then the functor $\Phi_{\mathcal{F}}$ is fully faithful if and only if

(1) for all $P \in X$,

$$\mathrm{Hom}_{\mathrm{D}(Y)}(\mathcal{F}_P, \mathcal{F}_P) \cong \mathbb{C};$$

(2) for all $P \neq Q$ and $i \in \mathbb{Z}$,

$$\mathrm{Hom}_{\mathrm{D}(Y)}(\mathcal{F}_P, \mathcal{F}_Q[i]) \cong 0.$$

Moreover, $\Phi_{\mathcal{F}}$ is an equivalence of categories if and only if

(3)

$$\dim X = \dim Y$$

and for all $P \in X$,

$$\mathcal{F}_P \otimes \omega_Y \cong \mathcal{F}_P$$

where ω_Y is the dualizing sheaf of Y.

PROOF. The necessity of the conditions is an easy exercise exploiting the fact that $\Phi_{\mathcal{F}}$ is fully faithful, respectively an equivalence; the latter implies in particular that it commutes with the Serre functor, hence (3). Sufficiency is nontrivial; a self-contained proof can be found in [**60**]. □

4.6.4. The original Fourier-Mukai functor and generalizations.

As a first non-trivial illustration to Theorem 4.76, let A be an abelian variety of dimension d. As discussed before, associated to A is another abelian variety $\mathrm{Pic}^0(A)$; it is called the *dual* of A and denoted $A^\vee$. Recall moreover the Poincaré line bundle $\mathcal{P}$ on $A^\vee \times A$ defined at the end of §4.3.5.

THEOREM 4.77. *For an abelian variety A, the functor $\Phi_{\mathcal{P}}$ defined by the Poincaré bundle as the (derived) correspondence*

$$\Phi_{\mathcal{P}}\colon \mathrm{D}^b(A^\vee) \xrightarrow{\sim} \mathrm{D}^b(A)$$

is an equivalence of categories.

PROOF. Let us check the conditions of Theorem 4.76. Since A is an abelian variety, its holomorphic cotangent bundle is trivial, thus so is its dualizing sheaf ω_A. As A is a complex torus, $\dim H^1(A,\mathcal{O}_A) = d$, and hence by Example 4.65 its dual $A^\vee$ is also of dimension d; hence conditions (3) are satisfied. Also, by the defining property of the Poincaré line bundle, for $Q \in A^\vee$ the restriction $\mathcal{P}_Q$ is a line bundle on A; thus (1) is also satisfied:

$$\mathrm{Hom}_{\mathrm{D}(A)}(\mathcal{P}_Q,\mathcal{P}_Q) \cong \mathrm{Hom}_A(\mathcal{P}_Q,\mathcal{P}_Q) \cong H^0(A,\mathcal{O}_A) \cong \mathbb{C}.$$

Finally, the same computation shows that (2) holds if and only if whenever $\mathcal{P}_1, \mathcal{P}_2$ are non-isomorphic degree zero line bundles on an abelian variety, then

$$\mathrm{Ext}^i(\mathcal{P}_1,\mathcal{P}_2) = H^i(\mathcal{P}_1^\vee \otimes \mathcal{P}_2) = 0$$

for all i. The latter statement is a well-known result in the theory of abelian varieties (see for example [**302**, Corollary 3.12]); hence the theorem is proved. □

With the technology developed so far, the proof is thus really easy. To see why this result was surprising, consider the simplest case $\dim A = 1$, that of genus one curves, and fix a base point $P \in A$. In this case, there is an isomorphism $A \to A^\vee$, taking a point $Q \in A$ to the degree-zero line bundle $\mathcal{O}_A(Q-P)$. Under this isomorphism, $\Phi_{\mathcal{P}}$ can be thought of as an autoequivalence of the derived category $\mathrm{D}^b(A)$ of the elliptic curve (A,P). This is a symmetry that has no counterpart in "classical" geometry; in particular, it is not induced by a classical symmetry or correspondence on the elliptic curve! §4.6.6 will elaborate on this point further.

Since one of the crucial conditions of Theorem 4.76 involves the dualizing sheaf ω_Y, it is to be expected that further interesting Fourier-Mukai functors can be found when this dualizing sheaf is trivial. Thus, let Y be a K3 surface, a simply connected smooth projective surface with trivial dualizing sheaf; for example, let Y be a smooth quartic $Y_4 \subset \mathbb{P}^3$. A fundamental discovery of Mukai was that moduli spaces of sheaves on Y are often smooth, and moreover they carry a natural holomorphic two-form. If such a moduli space M is further two-dimensional, then this natural two-form trivializes the sheaf of holomorphic two-forms, which just means that the dualizing sheaf of M is trivial. If finally M happens to be projective, we are in business:

THEOREM 4.78. (Mukai [**370**]) *Let M be a projective, two-dimensional fine moduli space of stable torsion-free sheaves on a K3 surface Y. Then* *there is an equivalence of derived categories*

$$\Phi\colon \mathrm{D}^b(M) \to \mathrm{D}^b(Y)$$

PROOF. The conditions imply that there is a *universal sheaf* $\mathcal{F}$ on $M \times Y$, whose restrictions $\mathcal{F}_P$ to $\{P\} \times Y$ are the sheaves classified by the moduli problem; this is all in perfect agreement with properties of the Poincaré sheaf. The claim is that $\mathcal{F}$ defines a Fourier-Mukai equivalence; we invoke Theorem 4.76 once again. (3) follows from the conditions of the theorem; (1) holds because stable sheaves are simple (they do not have non-trivial self-Homs). As for (2), $\mathrm{Hom}(\mathcal{F}_P, \mathcal{F}_Q) = 0$ for non-isomorphic $\mathcal{F}_P, \mathcal{F}_Q$ follows from stability, $\mathrm{Ext}^2(\mathcal{F}_P, \mathcal{F}_Q) = 0$ by Serre duality, and finally $\mathrm{Ext}^1(\mathcal{F}_P, \mathcal{F}_Q) = 0$ holds because the moduli space is 2-dimensional. □

4.6.5. Fully faithful functors from birational geometry. Let X be a smooth projective variety, containing the subvariety $Y \cong \mathbb{P}^k$ with normal bundle $N_{Y/X} \cong \mathcal{O}_{\mathbb{P}^k}^{\oplus(l+1)}(-1)$. The blowup $p = \mathrm{Bl}_Y\colon \widetilde{X} \to X$ of Y in X has an exceptional divisor $E \cong \mathbb{P}^k \times \mathbb{P}^l$, and there is a different contraction $p^+\colon \widetilde{X} \to X^+$ contracting E to $\mathbb{P}^l \subset X^+$. The birational transformation $X \dashrightarrow X^+$ is referred to as a *simple flop* if $l = k$, and a flip or antiflip if $k > l$ or $k < l$.

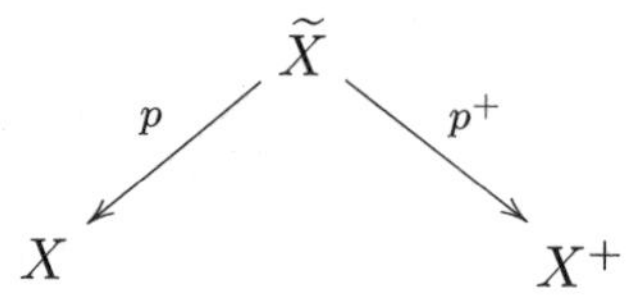

THEOREM 4.79. (Bondal-Orlov [**51**]) *If $k < l$, the functor*

$$\mathbf{R}p^+_* \circ \mathbf{L}p^*\colon \mathrm{D}^b(X) \to \mathrm{D}^b(X^+)$$

is a full and faithful embedding. If $k = l$, it is an equivalence of categories.

PROOF. Instead of attempting to apply Theorem 4.76 directly, it is better to break the proof into two steps:

(1) $\mathbf{L}p^*\colon \mathrm{D}^b(X) \to \mathrm{D}^b(\widetilde{X})$ is full and faithful;
(2) $\mathbf{R}p^+_*$ is full and faithful (respectively an equivalence) on the image of $\mathbf{L}p^*$.

For (1), Theorem 4.76 can be applied directly (exercise!). The proof of (2) is a little trickier, and uses a version of Beilinson's basis (Corollary 4.73) in the derived category of the blowup $\widetilde{X}$; the full details are in [**51**]. □

Let us look more closely at the case $k = l$. In this case, the situation is totally symmetric, in particular the pullbacks of the dualizing sheaves of X, X^+ on $\widetilde{X}$ agree: $p^*(\omega_X) \cong (p^+)^*(\omega_{X^+})$. Conversely, suppose that X_1, X_2 are two smooth projective varieties, together with a birational map $\phi\colon X_1 \dashrightarrow X_2$. By Hironaka's resolution theorem, we can assume that ϕ

factorizes as $\phi = p_2 \circ p_1^{-1}$, where $p_i\colon \widetilde{X} \to X_i$ are birational morphisms from a common smooth projective variety. We can make the following

DEFINITION 4.80. In the above context, we say X_1, X_2 are *K-equivalent* if there is an isomorphism $p_1^*\omega_{X_1} \cong p_2^*\omega_{X_2}$ between the pullbacks of the dualizing sheaves.

Note in particular that if $\omega_{X_i} \cong \mathcal{O}_{X_i}$, then this condition automatically holds.

CONJECTURE 4.81. *Suppose that X_1, X_2 are K-equivalent smooth projective varieties, in particular birationally equivalent varieties with trivial dualizing sheaf. Then there exists a Fourier-Mukai equivalence*

$$\Phi_{\mathcal{F}}\colon \mathrm{D}^b(X) \xrightarrow{\sim} \mathrm{D}^b(X^+).$$

This conjecture was made by Kawamata [**300**], after Bridgeland [**61**] had already settled the three-dimensional case, where Conjecture 4.81 is known to be a theorem. Part of the problem with the general conjecture is that there is no clear candidate for the kernel $\mathcal{F}$; simply pulling back to $\widetilde{X}$ and then pushing down is known not to work in general. For recent progress on special cases and versions for singular varieties, look in [**93, 301**].

4.6.6. Quantum symmetries: autoequivalences of the derived category.

Consider the following natural question: if X is a smooth projective variety, what are all the symmetries of its derived category? With a view to string theory, these are sometimes referred to as *quantum symmetries* of X. The first fundamental result in this direction is the following difficult theorem of Orlov [**384**]:

THEOREM 4.82. *Let X, Y be smooth projective varieties, and suppose that*

$$\Phi\colon \mathrm{D}^b(X) \to \mathrm{D}^b(Y)$$

is an exact equivalence of triangulated categories, which commutes with the Serre functor. Then there is an object $\mathcal{F} \in \mathrm{D}^b(X \times Y)$, unique up to isomorphism, such that Φ is isomorphic to the Fourier-Mukai functor $\Phi_{\mathcal{F}}$ defined by $\mathcal{F}$.

For $X = Y$, this says that every autoequivalence comes from a derived correspondence $\mathcal{P} \in \mathrm{D}^b(X \times X)$ which is invertible, in the sense that there exists another derived correspondence, with the compositions both ways being isomorphic to the structure sheaf of the diagonal $\mathcal{O}_{\Delta_X} \in \mathrm{D}^b(X \times X)$. It follows that exact self-equivalences of $\mathrm{D}^b(X)$ indeed form a group $\mathrm{Auteq}(\mathrm{D}^b(X))$.

There are three sources of obvious elements in this group. First of all, [1] generates a trivial part, the group of translations. From geometry, we get $\mathrm{Aut}(X)$, automorphisms of X acting by pullback, as well as $\mathrm{Pic}(X)$, line bundles acting by tensor product. These fit together to form a subgroup

$$(\mathrm{Pic}(X) \rtimes \mathrm{Aut}(X)) \times \mathbb{Z} < \mathrm{Auteq}(\mathrm{D}^b(X)).$$

THEOREM 4.83. (Bondal-Orlov [**51**]) *Suppose that X is smooth and projective, and moreover assume that either ω_X or ω_X^{-1} is ample (sections of some power give an embedding into projective space). Then the above inclusion is an isomorphism: X has no quantum symmetries beyond the obvious (geometric) ones.*

Thus, once again, the dualizing sheaf ω_X plays a crucial role. A Fourier-Mukai functor commutes with the Serre functor; thus, if the dualizing sheaf ω_X is ample or antiample, there is no room for quantum symmetries. On the other hand, if ω_X carries less information, for example if it is trivial, then there is room for extra autoequivalences. Indeed, we saw that this is the case for elliptic curves; in this case, the following result holds:

THEOREM 4.84. (Orlov [**384**]) *If E is a general elliptic curve, then $\mathrm{Auteq}(\mathrm{D}^b(E))$ is generated by the geometric symmetries, together with the autoequivalence $\Phi_{\mathcal{P}}$ associated to the Poincaré bundle $\mathcal{P}$ on $E \times E$.*

By mirror symmetry, it is expected that higher-dimensional varieties with trivial dualizing sheaf carry a large group of non-trivial quantum symmetries. For example, Conjecture 4.81 would imply that not only the automorphism group, but also the birational automorphism group acts on $\mathrm{D}^b(X)$ (this is known to hold if $\dim(X) \leq 3$). However, even this is not the full story. In some cases, such as for $K3$ surfaces, the group $\mathrm{Auteq}(\mathrm{D}^b(X))$ is conjectured to have a description as the fundamental group of the complement of a hyperplane arrangement; compare §5.8.2. In particular, braid groups frequently appear as subgroups of $\mathrm{Auteq}(\mathrm{D}^b(X))$ [**371, 422, 436, 447**]. The proper context for these results is that of the space of stability conditions [**65, 63**] on $\mathrm{D}^b(X)$, a subject to which we will return in Chapter 5.

4.7. The McKay correspondence

In this section we apply the derived category methods described earlier to provide an elegant explanation for the McKay correspondence in dimension $n \leq 3$. This correspondence arises naturally in mathematics via the geometry and representation theory of Gorenstein quotient singularities, and in physics in the context of D-branes on certain Calabi–Yau orbifolds.

4.7.1. The classical statement. Finite subgroups of $\mathrm{SL}(2, \mathbb{C})$ can be classified (up to conjugacy) into two infinite families and three exceptional cases:

- the cyclic group of order $n \geq 2$ generated by the transformations
$$(x, y) \to (\omega x, \omega^{n-1} y)$$
for ω a primitive nth root of unity;
- the binary dihedral group of order $4n$ ($n \geq 2$) generated by the pair
$$(x, y) \to (-y, x) \quad \text{and} \quad (x, y) \to (\omega x, \omega^{2n-1} y)$$
for ω a primitive $2n$th root of unity;
- one of three exceptional cases: the binary tetrahedral, binary octahedral and binary icosahedral groups of order 24, 48 and 120 respectively (obtained as the lift under the double cover $\mathrm{SU}(2) \to \mathrm{SO}(3)$ of the symmetry group of the corresponding Platonic solid).

In each case, the ring of G-invariant functions $\mathbb{C}[x, y]^G$ can be written in the form $\mathbb{C}[u, v, w]/\langle f \rangle$ for some polynomial $f \in \mathbb{C}[u, v, w]$. The quotient singularity $X = \mathbb{A}^2/G$ is defined by the ring of functions $\mathbb{C}[x, y]^G$, and hence is isomorphic to the hypersurface $X : (f = 0) \subset \mathbb{C}^3$ cut out by the given polynomial. The defining equation $(f = 0)$ is determined by the conjugacy class of the group G as shown in Table 1. In each case, X has an isolated singular point at the origin in $\mathbb{A}^3$.

Conjugacy class of G	Defining equation of X	Dynkin graph
cyclic $\mathbb{Z}/n\mathbb{Z}$	$u^2 + v^2 + w^n = 0$	A_{n-1}
binary dihedral $\mathbb{D}_{4n}$	$u^2 + v^2 w + w^{n+1} = 0$	D_{n+2}
binary tetrahedral $\mathbb{T}_{24}$	$u^2 + v^3 + w^4 = 0$	E_6
binary octahedral $\mathbb{O}_{48}$	$u^2 + v^3 + vw^3 = 0$	E_7
binary icosahedral $\mathbb{I}_{120}$	$u^2 + v^3 + w^5 = 0$	E_8

TABLE 1. Classification of Kleinian singularities.

The singular affine variety X has a unique resolution $\tau \colon Y \to X$ with the properties that Y has trivial canonical bundle, and the exceptional locus of τ is a tree of rational curves $C \cong \mathbb{P}^1$ intersecting transversally. We construct a graph from this tree as follows: introduce one vertex for each irreducible exceptional curve C, and join a pair of vertices by an edge if the corresponding curves intersect in Y. The resulting graph is a Dynkin graph of ADE-type. The data of the group, the defining equation and the ADE graph is recorded in Table 1.

McKay [**355**] observed that the Dynkin graph of $\mathbb{A}^2/G$ can be obtained from the quiver described in §4.2.6. Since $G \subset \mathrm{SL}(2, \mathbb{C})$, the representation W is self-dual, so every arrow $\rho\rho'$ pairs up with a unique arrow $\rho'\rho$. Replacing every such pair of arrows by a single edge produces a graph that we denote $\widetilde{\Gamma}_Q$. Let Γ_Q denote the subgraph obtained from the McKay graph by removing the vertex corresponding to the trivial representation and the edges emanating from that vertex.

THEOREM 4.85. *The McKay graph $\widetilde{\Gamma}_Q$ is an extended Dynkin graph of ADE type, and the subgraph Γ_Q is the ADE graph (tree of exceptional components) of $X = \mathbb{A}^2/G$ from Table 1, giving a one-to-one correspondence*

$$\text{basis of } H_*(Y,\mathbb{Z}) \longleftrightarrow \{\text{irreducible representations of } G\}.$$

PROOF. McKay [**355**] gives the original observation that forms the first statement. Inspecting the vertices of the graph Γ_Q establishes a one-to-one correspondence between the exceptional curves C of the resolution $\tau\colon Y \to X$ and the nontrivial irreducible representations ρ of G. The exceptional curve classes $[C]$ form a basis for the homology $H_2(Y,\mathbb{Z})$ so that, by adding the homology class of a point on one side and the trivial representation on the other, we obtain the stated one-to-one correspondence. □

The McKay correspondence admits a beautiful explanation in terms of an equivalence of derived categories, as we now describe.

4.7.2. The McKay correspondence conjecture. It is convenient to first generalise the geometric set-up to higher dimensions. For a finite subgroup $G \subset \mathrm{SL}(n,\mathbb{C})$, the singularity X is Gorenstein, i.e., the canonical sheaf ω_X is a line bundle. In fact, the form $dx_1\wedge\cdots\wedge dx_n$ on $\mathbb{A}^n$ is G-invariant and hence it descends to give a globally defined nonvanishing holomorphic n-form on X, forcing ω_X to be trivial. A resolution $\tau\colon Y \to X$ is said to be *crepant* if $\tau^*(\omega_X) = \omega_Y$; this holds here if and only if ω_Y is also trivial, in which case we call Y a (noncompact) Calabi–Yau manifold. Note that crepant resolutions need not exist, and when they do they are typically nonunique. The simplest (and in fact, the motivating) example of a crepant resolution is the minimal resolution of the singularity $X = \mathbb{A}^2/G$ arising from a finite subgroup $G \subset \mathrm{SL}(2,\mathbb{C})$, as discussed earlier.

The guiding principle behind the McKay correspondence was stated by Reid [**404**] along the following lines:

PRINCIPLE 4.86. *Let $G \subset \mathrm{SL}(n,\mathbb{C})$ be a finite subgroup. Given a crepant resolution $\tau\colon Y \to X = \mathbb{A}^n/G$, the geometry of Y should be equivalent to the G-equivariant geometry of $\mathbb{A}^n$. In particular, any two crepant resolutions of X should have equivalent geometries.*

Here, the word 'geometry' was left deliberately vague but the statement was known to hold for suitably defined notions of Euler number and Hodge numbers. More significantly, this principle, and indeed any geometric approach to the McKay correspondence owes a great debt to the pioneering work of Gonzalez-Sprinberg–Verdier [**186**]. For a finite subgroup $G \subset \mathrm{SL}(2,\mathbb{C})$ with minimal resolution $Y \to \mathbb{A}^2/G$, they constructed a collection of vector bundles $\mathcal{R}_\rho$ on Y indexed by the irreducible representations of

G, where the rank of the bundle $\mathcal{R}_\rho$ is equal to the dimension of $\rho \in \operatorname{Irr}(G)$. In a lengthy case-by-case analysis of the subgroups listed in Table 1, it was shown that the first Chern classes $c_1(\mathcal{R}_\rho)$ of the vector bundles indexed by the nontrivial irreducible representations form a basis of $H^2(Y, \mathbb{Z})$ dual to the exceptional curve classes $[C] \in H_2(Y, \mathbb{Z})$. Theorem 4.85 follows immediately in the special case when $n = 2$.

Reid [**403**] suggested that one manifestation of Principle 4.86 should be an equivalence of derived categories

$$\Phi\colon \mathrm{D}^b(Y) \longrightarrow \mathrm{D}^b_G(\mathbb{A}^n), \tag{4.9}$$

between the bounded derived category of coherent sheaves on Y and the bounded derived category of G-equivariant coherent sheaves on $\mathbb{A}^n$. The key observation, made independently by Kapranov-Vasserot [**282**] and Bridgeland-King-Reid [**67**], was to construct the derived equivalence as a Fourier-Mukai transform. We now construct the relevant integral functor.

Let $\pi\colon \mathbb{A}^n \to X = \mathbb{A}^n/G$ be the quotient morphism and $\tau\colon Y \to X$ a resolution. Consider the commutative diagram

$$\begin{array}{ccccc} & & Y \times \mathbb{A}^n & & \\ & \swarrow^{\pi_Y} & & \searrow^{\pi_{\mathbb{A}}} & \\ Y & & & & \mathbb{A}^n \\ & \searrow_{\tau} & & \swarrow_{\pi} & \\ & & X & & \end{array} \tag{4.10}$$

where π_Y and $\pi_{\mathbb{A}}$ are the projections to the first and second factors. Let G act trivially on both Y and X, so that each morphism in the diagram is G-equivariant.

By analogy with Mukai's functor on the derived category of the elliptic curve, the key step is to realize the resolution Y as a fine moduli space of certain G-equivariant coherent sheaves on $\mathbb{A}^n$. Just as with the Poincaré sheaf for the elliptic curve, this would imply that the product $Y \times \mathbb{A}^n$ comes equipped with a universal sheaf $\mathcal{F}$ such that, for each point $y \in Y$, the restriction of $\mathcal{F}$ to the fibre $\pi_Y^{-1}(y) \cong \mathbb{A}^n$ is the G-equivariant coherent sheaf $\mathcal{F}_y$ parameterised by the point $y \in Y$. Armed with this universal sheaf, one can define a functor $\Phi_{\mathcal{F}}\colon \mathrm{D}^b(Y) \to \mathrm{D}^b_G(\mathbb{A}^n)$ via

$$\Phi_{\mathcal{F}}(-) = \mathbf{R}\pi_{\mathbb{A}*}\Big(\mathcal{F} \overset{L}{\otimes} {\pi_Y}^*(- \otimes \rho_0)\Big). \tag{4.11}$$

In this formula: the tensor product with the trivial representation acknowledges that G acts trivially on Y, enabling us to take the G-equivariant pullback via π_Y; and the pullback via π_Y need not be derived since π_Y is flat by virtue of Y being a fine moduli space. Principle 4.86 suggests that $\Phi_{\mathcal{F}}$ is an equivalence of triangulated categories whenever τ is crepant.

4.7.3. Moduli interpretation. To carry out the above program, a resolution Y must be constructed as a fine moduli space of certain G-equivariant coherent sheaves on $\mathbb{A}^n$. In light of the correspondence from Proposition 4.37, G-equivariant coherent sheaves on $\mathbb{A}^n$ correspond one-to-one with representations of the McKay quiver Q satisfying the natural commutativity relations R; we call these 'representations of (Q, R)' for short. This quiver-theoretic point of view provides a nice geometric construction of the relevant moduli spaces, as we now describe.

Recall that, by definition, representations $V = \bigoplus_{i \in Q_0} V_i$ of a quiver with fixed dimension vector $\alpha = (\dim V_i)_{i \in Q_0}$ give rise to elements of the vector space

$$\bigoplus_{a \in Q_1} \mathrm{Hom}(V_{t(a)}, V_{h(a)}).$$

Since the vertex set Q_0 for the McKay quiver is the set $\mathrm{Irr}(G)$ of irreducible representations of G, we may restrict to the case where the dimension vector is $\alpha = (\dim \rho)_{\rho \in \mathrm{Irr}(G)}$.

By choosing bases for the vector spaces V_i and counting entries in the matrices corresponding to these linear maps, the dimension of this vector space is given by $d := \sum_{a \in Q_1} (\dim V_{h(a)}) \cdot (\dim V_{t(a)})$. Since quiver representations are defined independently of this choice of basis, isomorphism classes of representations are actually orbits in the vector space $\mathbb{A}^d$ under the action of the group $H := \prod_{i \in Q_0} \mathrm{GL}(V_i)$ by change of basis. Proposition 4.37 shows that we should study only those representations of the McKay quiver Q that satisfy the relations R. This forces one to work not with the entire space $\mathbb{A}^d$ of representations of Q but, rather, with a subset (in fact, subvariety or even subscheme) $\mathbb{V}(I_R) \subset \mathbb{A}^d$ cut out by an ideal of equations I_R arising from the relations R.

EXAMPLE 4.87. Consider once again the cyclic subgroup of order three embedded in $\mathrm{SL}(3, \mathbb{C})$ from Example 4.35. The corresponding quiver has nine edges which we called $x_{(j+1)j}, y_{(j+1)j}, z_{(j+1)j}$, for $j \in Q_0 = \{0, 1, 2\}$; addition is interpreted mod 3. As discussed already in Example 4.39, the ideal of relations is generated by the expressions

$$x_{(j+1)j} y_{j(j-1)} - y_{(j+1)j} x_{j(j-1)},$$

as well as the analogous expressions for the corresponding (y, z) and (z, x) pairs. Thus the entire space of representations is $\mathbb{A}^9$ with coordinates

$$\{x_{(j+1)j}, y_{(j+1)j}, z_{(j+1)j}\}_{j=0,1,2},$$

and the subscheme $\mathbb{V}(I_R) \subset \mathbb{A}^9$ is cut out by nine equations obtained by setting the relations equal to zero.

To study isomorphism classes of representations of (Q, R), we construct moduli spaces of quiver representations using Geometric Invariant Theory (GIT). Since H acts on the space $\mathbb{V}(I_R)$ of representations of (Q, R), it also acts on the coordinate ring $\mathbb{C}[z_1, \dots, z_d]/I_R$ of $\mathbb{V}(I_R)$. The simplest quotient $\mathbb{V}(I_R)/H$ is the affine variety whose coordinate ring is $(\mathbb{C}[z_1, \dots, z_d]/I_R)^H$, the subring of H-invariants. However, this variety is singular and carries too little information for our purposes. Instead we study quotients arising from 'stability parameters' in the rational vector space

$$\Theta := \Big\{\theta \in \operatorname{Hom}_{\mathbb{Z}}\big(\textstyle\bigoplus_{\rho\in\operatorname{Irr}(G)} \mathbb{Z}\cdot\rho, \mathbb{Q}\big) : \theta(\alpha) := \sum_{\rho\in\operatorname{Irr}(G)} \dim(\rho)\theta(\rho) = 0\Big\}$$

as follows. For $\theta \in \Theta$, a representation V of the quiver Q with dimension vector α is said to be *θ-stable* if every proper, nonzero subrepresentation $0 \subset V' \subset V$ of dimension vector β satisfies $\theta(\beta) > 0 = \theta(\alpha)$. Also, *$\theta$-semistable* is the same with $\geq$ replacing $>$. The subset $\mathbb{V}(I_R)^{\mathrm{ss}}_\theta \subseteq \mathbb{V}(I_R)$ parameterizing θ-semistable representations of (Q, R) forms a dense open subset, and the GIT quotient

$$\mathbb{V}(I_R)/\!\!/_\theta\, H := \mathbb{V}(I_R)^{\mathrm{ss}}_\theta/H$$

parameterizes H-orbit closures of θ-semistable representations of (Q, R). For the special case $\theta = 0$, every representation of Q is 0-semistable and we recover the affine quotient $\mathbb{V}(I_R)/\!\!/_0\, H = \mathbb{V}(I_R)/H$ as above.

In general the study of H-orbit closures is problematic, but one can do much better than this with an additional assumption on the choice of θ. More precisely, a parameter $\theta \in \Theta$ is said to be *generic* if every θ-semistable representation is θ-stable. For every such parameter it can be shown that $\mathbb{V}(I_R)/\!\!/_\theta\, H$ parameterizes genuine H-orbits in $\mathbb{V}(I_R)^{\mathrm{ss}}_\theta$ rather than orbit closures, i.e., $\mathbb{V}(I_R)/\!\!/_\theta\, H$ parameterizes isomorphism classes of θ-stable representations of (Q, R). Thus, we achieve our goal in working with isomorphism classes of representations, at the expense of having to first throw away those that are not θ-stable. In addition, work of Thaddeus [**441**] and Dolgachev-Hu [**127**] implies that the set of generic parameters $\theta \in \Theta$ decomposes into finitely many open GIT *chambers*, where the locus $\mathbb{V}(I_R)^{\mathrm{ss}}_\theta$ and hence the GIT quotient $\mathbb{V}(I_R)/\!\!/_\theta\, H$ remains unchanged as θ varies in a given chamber.

In fact a stronger statement can be made with an additional assumption on the dimension vector α:

THEOREM 4.88 (King [**304**]). *Assume that α is not a nontrivial multiple of an integer vector. Then for generic $\theta \in \Theta$, the GIT quotient $\mathcal{M}_\theta(Q, R) := \mathbb{V}(I_R)/\!\!/_\theta\, H$ is the* fine *moduli space of θ-stable representations of (Q, R) with dimension vector α.*

The fact that the moduli space $\mathcal{M}_\theta(Q, R)$ is fine means that, in addition to being a scheme, $\mathcal{M}_\theta(Q, R)$ carries a universal object. Indeed, the moduli construction determines a universal representation of Q and hence a G-equivariant coherent sheaf $\mathcal{U}_\theta$ on the product $\mathcal{M}_\theta(Q, R) \times \mathbb{A}^n$. The restriction of this sheaf to the fibre over a point $y \in \mathcal{M}_\theta(Q, R)$ is precisely the G-equivariant coherent sheaf encoded by the representation of Q corresponding to the point y. The push-forward via the projection from $\mathcal{M}_\theta(Q,R) \times \mathbb{A}^n$ to $\mathcal{M}_\theta(Q,R)$ gives the tautological bundle $\mathcal{R}_\theta$ on $\mathcal{M}_\theta(Q,R)$. Just as the regular representation $R = \oplus_{\rho \in G^*} R_\rho \otimes \rho$ of G splits into irreducibles, the bundle $\mathcal{R}_\theta$ decomposes as

$$\mathcal{R}_\theta = \bigoplus_{\rho \in \mathrm{Irr}(G)} (\mathcal{R}_\theta)_\rho, \tag{4.12}$$

where the summands $\mathcal{R}_\rho := (\mathcal{R}_\theta)_\rho$ satisfy $\mathrm{rank}(\mathcal{R}_\rho) = \dim \rho$. Without loss of generality, we normalize so that $\mathcal{R}_{\rho_0}$ for the trivial representation ρ_0 is the trivial bundle on $\mathcal{M}_\theta(Q,R)$.

REMARK 4.89. The moduli spaces $\mathcal{M}_\theta(Q,R)$ appear in the physics literature as moduli of $D0$-branes on the orbifold $\mathbb{A}^3/G$, for G a finite subgroup of $\mathrm{SL}(3,\mathbb{C})$. The parameter θ is a Fayet-Iliopoulos term for $\mathrm{U}(m)$ gauge multiplets present in the world-volume theory for $m = \dim(\rho)$, c.f. [**139**]. In this case, the ideal of relations I_R arises from the F-terms in the action functional obtained from the partial derivatives of the superpotential of the quiver gauge theory (compare Example 4.39), while the action of H on $\mathbb{V}(I_R)$ arises from the D-term (which is often described in the physics literature via a moment map). The link between the physics and mathematics literature is made transparent in the construction of the coherent component by Craw-Maclagan-Thomas [**103**].

EXAMPLE 4.90. The best-known example of a fine moduli space of θ-stable representations of (Q,R) is the G-Hilbert scheme, first studied by Ito-Nakamura [**257**]. This scheme, denoted G-Hilb, parameterizes G-invariant subschemes $Z \subset \mathbb{A}^n$ for which the space of global sections $\Gamma(\mathcal{O}_Z)$ is isomorphic as a $\mathbb{C}[G]$-module to the regular representation R of G. Ito-Nakajima [**258**] observed that there is a chamber C_0 in the space of weights Θ containing the parameters of the form

$$\{\theta \in \Theta \mid \theta(\rho) > 0 \text{ if } \rho \neq \rho_0\}$$

such that $\mathcal{M}_\theta(Q,R) = G$-Hilb for all $\theta \in C_0$.

EXAMPLE 4.91. Consider once again the cyclic group

$$G \cong \mathbb{Z}/3 \subset \mathrm{SL}(3,\mathbb{C})$$

from Example 4.35. The space

$$\Theta = \{(\theta_0,\theta_1,\theta_2) \in \mathbb{Q}^3 : \theta_0 + \theta_1 + \theta_2 = 0\} \cong \mathbb{Q}^2$$

decomposes into three GIT chambers given by

$$\begin{aligned} C_0 &= \{\theta \in \Theta : \theta_1 > 0,\ \theta_1 + \theta_2 > 0\}, \\ C_1 &= \{\theta \in \Theta : \theta_2 < 0,\ \theta_1 + \theta_2 < 0\}, \\ C_2 &= \{\theta \in \Theta : \theta_1 < 0,\ \theta_2 > 0\}. \end{aligned}$$

Since C_0 contains parameters of the form $\{\theta_+ = (\theta_0, \theta_1, \theta_2) \in \mathbb{Q}^3 : \theta_1 > 0, \theta_2 > 0\}$, we deduce from above that $\mathcal{M}_\theta(Q,R) = G\text{-Hilb}$ for all $\theta \in C_0$. It is easy to show that G-Hilb is a smooth toric variety that can be obtained as the unique crepant resolution $\tau\colon Y \to \mathbb{C}^3/G$ contracting a divisor $E \cong \mathbb{P}^2$ to the singular point. This resolution is isomorphic to the total space of the line bundle $\mathcal{O}_{\mathbb{P}^2}(-3)$.

In fact, the moduli space $\mathcal{M}_\theta(Q,R)$ is isomorphic to Y for any generic parameter $\theta \in \Theta$. Nevertheless, the moduli spaces are different for parameters lying in different chambers since the rank 3 tautological bundle $\mathcal{R}_\theta$ on $\mathcal{M}_\theta(Q,R)$ changes as θ varies between the chambers. To emphasise this point we list in Table 2 the restriction of the tautological bundles to the ex-

	$\theta \in C_0$	$\theta \in C_1$	$\theta \in C_2$
$\mathcal{R}_{\rho_0}\vert_E$	$\mathcal{O}_E$	$\mathcal{O}_E$	$\mathcal{O}_E$
$\mathcal{R}_{\rho_1}\vert_E$	$\mathcal{O}_E(2)$	$\mathcal{O}_E(-1)$	$\mathcal{O}_E(-1)$
$\mathcal{R}_{\rho_2}\vert_E$	$\mathcal{O}_E(1)$	$\mathcal{O}_E(-2)$	$\mathcal{O}_E(1)$

TABLE 2. Tautological bundles on $\mathcal{M}_\theta(Q,R)$ for $\mathbb{Z}/3 \subset \mathrm{SL}(3,\mathbb{C})$

ceptional divisor $E \subset \mathcal{M}_\theta(Q,R)$ for parameters in all three chambers. For example, parameters $\theta \in C_0$ give $\mathcal{R}_{\rho_2}|_E \cong \mathcal{O}_E(1)$ since $\mathcal{R}_{\rho_2}$ has degree one on the class of a line in E, and $\mathcal{R}_{\rho_1}|_E \cong \mathcal{O}_E(2)$.

4.7.4. The McKay correspondence via Fourier-Mukai transform. We continue to assume that $\theta \in \Theta$ is generic, so that $\mathcal{M}_\theta(Q,R)$ is the fine moduli space of θ-stable representations of (Q,R). There is a projective morphism

$$\tau\colon \mathcal{M}_\theta(Q,R) \to X = \mathbb{A}^n/G$$

sending any point of $\mathcal{M}_\theta(Q,R)$ to the G-orbit that supports the corresponding G-equivariant coherent sheaf. In general, the moduli space $\mathcal{M}_\theta(Q,R)$ may have more than one irreducible component, so to simplify matters we let $Y \subseteq \mathcal{M}_\theta(Q,R)$ denote the component containing the quiver representations arising from the structure sheaves of the free G-orbits in $\mathbb{A}^n$; this is the *coherent component* of $\mathcal{M}_\theta(Q,R)$. The restriction of the map τ to the component Y fits into a commutative diagram (4.10), and we define a functor

$$\Phi_\theta\colon \mathrm{D}^b(Y) \to \mathrm{D}^b_G(\mathbb{A}^n)$$

via the formula

$$\Phi_\theta(-) := \Phi_{\mathcal{U}_\theta}(-) = \mathbf{R}\pi_{\mathbb{A}*}\Big(\mathcal{U}_\theta \overset{L}{\otimes} (\pi_Y)^*(- \otimes \rho_0)\Big) \tag{4.13}$$

where $\mathcal{U}_\theta$ is the universal sheaf on $Y \times \mathbb{A}^n$ obtained from that on $\mathcal{M}_\theta(Q,R) \times \mathbb{A}^n$ by restriction.

The method of Bridgeland, King and Reid [**67**] generalizes from the fine moduli space G-Hilb to the fine moduli space of θ-stable representations of (Q, R) for any generic parameter $\theta \in \Theta$ as follows, see Craw-Ishii [**102**].

THEOREM 4.92. *Let $G \subset \mathrm{SL}(n, \mathbb{C})$ be a finite subgroup and let $\theta \in \Theta$ be generic. If, for the coherent component $Y \subseteq \mathcal{M}_\theta(Q, R)$, the fibre product*

$$Y \times_X Y = \{(y, y') \in Y \times Y \mid \tau(y) = \tau(y')\}$$

has dimension at most $n+1$, then:

(1) the morphism $\tau \colon Y \to X$ is a crepant resolution; and
(2) the functor Φ_θ with kernel the universal sheaf for $Y \subseteq \mathcal{M}_\theta(Q, R)$ is an equivalence of derived categories

$$\Phi_\theta \colon \mathrm{D}^b(Y) \to \mathrm{D}^b_G(\mathbb{A}^n).$$

REMARK 4.93. The condition on the dimension of the fibre product always holds for finite $G \subset \mathrm{SL}(n, \mathbb{C})$ with $n \leq 3$, because $\dim(Y \times_X Y)$ is at most twice the dimension of the exceptional locus; this equals one for $n = 2$ and two for $n = 3$. In either case, [**67**] also establishes that $Y = \mathcal{M}_\theta(Q, R)$. However, for $n \geq 4$ the dimension bound on the fibre product rarely holds, for example $\dim(Y \times_X Y) = 6$ for isolated singularities $\mathbb{A}^4/G$.

PROOF. Let $\mathrm{D}^b_0(Y)$ denote the full subcategory of $\mathrm{D}^b(Y)$ consisting of objects supported on the subscheme $\tau^{-1}(\pi(0))$ of Y, and let $\mathrm{D}^b_{G,0}(\mathbb{A}^n)$ denote the full subcategory of $\mathrm{D}^b_G(\mathbb{A}^n)$ consisting of objects supported at the origin of $\mathbb{A}^n$. Then Φ_θ restricts to a functor

$$\Phi_\theta \colon \mathrm{D}^b_0(Y) \to \mathrm{D}^b_{G,0}(\mathbb{A}^n). \tag{4.14}$$

The strategy is to prove that the set $\{\mathcal{O}_y \mid y \in \tau^{-1}(\pi(0))\}$ is a spanning class for $\mathrm{D}^b_0(Y)$, so that Theorem 4.76 can be applied. The full argument requires the intersection theorem from commutative algebra, and is somewhat technical; we refer the reader to the self-contained proof in [**67**]. □

REMARK 4.94. The equivalence Φ_θ implicitly constructs a resolution of the structure sheaf $\mathcal{O}_\Delta$ of the diagonal on $\mathcal{M}_\theta \times \mathcal{M}_\theta$.

EXAMPLE 4.95. To illustrate the functor Φ_θ, or more precisely, its restriction (4.14) to the compactly supported locus, consider once again our running example, the subgroup $G \cong \mathbb{Z}/3 \subset \mathrm{SL}(3, \mathbb{C})$ from Example 4.35. The group G acts on $\mathbb{A}^3$ and hence on $\mathcal{O}_{\mathbb{A}^3}$. For each $\rho_i \in \mathrm{Irr}(G)$, we write $\mathcal{O}_{\mathbb{A}^3} \otimes \rho_i$ for the corresponding G-eigensheaf and $\mathcal{O}_0 \otimes \rho_i$ for the corresponding simple sheaf supported at the origin. To describe the functor Φ_θ, it is enough to calculate the images under $\Psi_\theta := \Phi_\theta^{-1}$ of the objects $\mathcal{O}_0 \otimes \rho_i$ that generate $\mathrm{D}^b_{G,0}(\mathbb{A}^n)$. The results are presented in Table 3, where we write

$$E := \tau^{-1}(\pi(0)) \cong \mathbb{P}^2$$

for the exceptional divisor of the crepant resolution $\tau\colon \mathcal{M}_\theta(Q,R) \to \mathbb{A}^3/G$.

	$\theta \in C_0$	$\theta \in C_1$	$\theta \in C_2$
$\Psi_\theta(\mathcal{O}_0 \otimes \rho_0)$	$\mathcal{O}_E(-3)[2]$	$\Omega_E^2(3)$	$\Omega_E^1[1]$
$\Psi_\theta(\mathcal{O}_0 \otimes \rho_1)$	$\Omega_E^2(1)$	$\Omega_E^1(1)[1]$	$\Omega_E^2(1)[2]$
$\Psi_\theta(\mathcal{O}_0 \otimes \rho_2)$	$\Omega_E^1(-1)[1]$	$\mathcal{O}_E(-1)[2]$	$\mathcal{O}_E(-1)$

TABLE 3. Fourier-Mukai transforms on $\mathcal{M}_\theta(Q,R)$ for $\mathbb{Z}/3 \subset \mathrm{SL}(3,\mathbb{C})$

These results may be simplified via the isomorphism $\Omega_E^2 \cong \mathcal{O}_E(-3)$, but the pattern in each column is clearer in the present form. The three entries in any one of these columns generate the derived category $\mathrm{D}_0^b(\mathcal{M}_\theta(Q,R))$ for the appropriate $\theta \in \Theta$. Autoequivalences of $\mathrm{D}_0^b(\mathcal{M}_\theta(Q,R))$ are induced by moving from one chamber to another.

To illustrate the method we present two calculations in full. To perform the calculations below, we repeatedly use the formula

$$\pi_*\Phi^i(-) \cong R^i\tau_*(- \otimes \mathcal{R}_\rho) = \bigoplus_{\rho \in \mathrm{Irr}(G)} H^i(- \otimes \mathcal{R}_\rho) \otimes \rho, \tag{4.15}$$

where $\Phi^i(-)$ denotes the ith cohomology sheaf of $\Phi(-)$ and where $\{\mathcal{R}_\rho\}$ denote the tautological bundles on $\mathcal{M}_\theta(Q,R)$ (we often omit π_* from the left hand side). To begin, fix $\theta \in C_0$, hence $\mathcal{M}_\theta = G$-Hilb and write Φ_θ for the Fourier-Mukai transform. Using (4.15) and the first column of Table 2 we calculate

$$\begin{aligned}&\Phi_\theta^i(\mathcal{O}_E(-3))\\ &\quad = \big(H^i(\mathcal{O}_E(-3)) \otimes \rho_0\big) \oplus \big(H^i(\mathcal{O}_E(-1)) \otimes \rho_1\big) \oplus \big(H^i(\mathcal{O}_E(-2)) \otimes \rho_2\big).\end{aligned}$$

Since $E \cong \mathbb{P}^2$, the only nonzero vector space in this expansion is

$$H^2(\mathcal{O}_E(-3)) \cong \mathbb{C}.$$

Therefore $\Phi_\theta(\mathcal{O}_E(-3)[2]) = \Phi_\theta^2(\mathcal{O}_E(-3)) = H^2(\mathcal{O}_E(-3)) \otimes \rho_0 \cong \mathbb{C} \otimes \rho_0$. This can be written as $\Phi_\theta(\mathcal{O}_E(-3)[2]) = \mathcal{O}_0 \otimes \rho_0$ or, equivalently, as

$$\Psi_\theta(\mathcal{O}_0 \otimes \rho_0) = \mathcal{O}_E(-3)[2].$$

Similarly, fix $\theta' \in C_1$ and use (4.15) with column two of Table 2 and let $\Phi_{\theta'}$ denote the corresponding Fourier-Mukai transform. We obtain

$$\Phi_{\theta'}^i(\Omega_E^1(1)) = \big(H^i(\Omega_E^1(1)) \otimes \rho_0\big) \oplus \big(H^i(\Omega_E^1) \otimes \rho_1\big) \oplus \big(H^i(\Omega_E^1(-1)) \otimes \rho_2\big).$$

Here, only $H^1(\Omega_E^1) \cong \mathbb{C}$ is nonzero, hence

$$\Phi_{\theta'}^0(\Omega_E^1(1)[1]) = \Phi_{\theta'}^1(\Omega_E^1(1)) \cong \mathbb{C} \otimes \rho_1.$$

Write this as $\Phi_{\theta'}(\Omega^1_E(1)[1]) = \mathcal{O}_0 \otimes \rho_1$ or, equivalently, as

$$\Psi_{\theta'}(\mathcal{O}_0 \otimes \rho_1) = \Omega^1_E(1)[1].$$

The other calculations are similar.

COROLLARY 4.96. *The McKay correspondence as stated in Principle 4.86 holds on the level of derived categories in dimension two and three.*

PROOF. There's nothing to prove in dimension $n = 2$ because $\mathbb{A}^2/G$ admits a unique crepant resolution for $G \subset \mathrm{SL}(2,\mathbb{C})$. Every crepant resolution of $\mathbb{A}^3/G$ is obtained from $\mathcal{M}_\theta = G$-Hilb by a finite sequence of flops. The result follows from Bridgeland's proof of Conjecture 4.81 in dimension $n = 3$, see [**61**]. □

REMARK 4.97. Principle 4.86 has also been established as an equivalence of derived categories for finite subgroups $G \subset \mathrm{Sp}(n,\mathbb{C})$ by Kaledin-Bezrukavnikov [**45**] and for finite abelian subgroups $G \subset \mathrm{SL}(n,\mathbb{C})$ by Kawamata [**301**].

CHAPTER 5

Dirichlet branes and stability conditions

We resume the discussion of Dirichlet branes from Chapter 3, now armed with better tools. So far, we have been discussing topological Dirichlet branes, meaning boundary conditions in the topologically twisted A- and B-models, and open string correlation functions in these models. While coming out of physics, these definitions lead to structures which look very much like the Fukaya category for the A-model, and the category of coherent sheaves for the B-model.

While it is tempting to conjecture that this is indeed the correct identification, if this were true, then by mirror symmetry physics would then go on to predict that these two categories should be isomorphic. However, this is not true. The simplest sign of this is that the category of coherent sheaves is too small; it doesn't contain enough B-branes.

As Kontsevich proposed in 1994 [**309**], the correct category of B-branes is obtained by passing to the derived category of coherent sheaves. In §5.3, we will give detailed arguments for this claim within string theory.

Comparing with Chapter 4, the basic formal ingredients we need are already present, namely the facts that D-branes can naturally be thought of as objects in an abelian category, and the morphisms are cohomology classes of the Q operator with a $\mathbb{Z}$-grading. On the other hand, it is not obvious from the discussion so far why anything like the identification under quasi-isomorphisms should be relevant to physics. And, even after making this rather broad identification, consideration of simple examples shows that there are far too many objects in the derived category for all of them to correspond to physical D-branes.

A clue to the solution of this problem is to note that, even in the far smaller category of coherent sheaves, many of the objects do not correspond to D-branes. Only sheaves which admit solutions of the Hermitian Yang-Mills equations are D-branes. By the Donaldson-Uhlenbeck-Yau theorem, these are in correspondence with the subset of μ-stable sheaves.

A similar phenomenon is seen in the A-brane picture. Again, the category of Lagrangian submanifolds is far too general to have direct physical analogs. It is better to consider equivalence classes under Hamiltonian isotopy, and these again have a conjectured relation to special Lagrangian submanifolds which involves a conjectural notion of stability [**446, 262**].

	A-model	B-model
Geometry	Symplectic (no complex structure)	Algebraic (no metric)
Category	Fukaya category	Derived category
D-branes	Lagrangians	Complexes of coherent sheaves
Open strings	Floer cohomology	Ext's
Dependence	$B + iJ$	complex structure
Charges	$l_i \in H^3$	$\mathrm{ch}(\mathscr{E})\sqrt{\mathrm{td}(X)} \in H^{\mathrm{even}}(X)$ or $K(X)$
BPS A/B-branes	Special Lagrangians	Π-stable complexes
Dependence of stability	complex structure	$B + iJ$
Bound state	$A \# B$	$\mathrm{Cone}(A \to B)$

TABLE 1. Mirror symmetry. The consequences for topological branes are in the first section, those for physical (BPS) branes in the second.

This suggests that we make a distinction between "topological" and "physical" D-branes. The topological branes, defined purely in A- or B-model terms, are general objects in the Fukaya category and $\mathrm{D}(\mathrm{Coh}\, X)$ respectively. Kontsevich's mirror symmetry conjecture applies to them.

On the other hand, the physical SCFT depends on more data, and the problem of characterizing its supersymmetric boundary conditions is different. We refer to these as physical A- and B-branes, and can try to identify them with objects in the Fukaya or derived category satisfying some stability condition.

While we will be able to identify the ingredients for a stability condition within SCFT, the present state of the art does not allow deriving it within this context. However, we will have enough of its ingredients to test the general idea, by comparing predictions with those of the known stability conditions in the large volume limit, and by testing the predictions of mirror symmetry: first,

PROPERTY 5.1. The moduli spaces of mirror stability conditions are isomorphic.

Properly understanding this requires discussing "quantum equivalences" induced by monodromies in the moduli space of stability conditions. Then,

PROPERTY 5.2. Given two corresponding stability conditions on a mirror pair X and Y, the category of stable objects in the Fukaya category on X is equivalent to the category of stable objects in $\mathrm{D}(\mathrm{Coh}\,Y)$.

Certainly, the simplest way for these properties to hold would be for the stability condition to be defined purely in terms of the structures equated in Kontsevich's mirror symmetry conjecture, namely the triangulated structure and perhaps the A_∞ structures. However, we immediately run into a problem: definitions of stability based on geometric invariant theory, or simple generalizations, only make sense for abelian categories. In particular, one needs to talk about subobjects, but this does not make sense in the derived category. Solving this problem will be the main focus of the rest of the chapter.

Following these ideas, one can make a precise definition of a stability structure on triangulated categories [**65, 63**]. This definition makes no intrinsic reference to mirror symmetry or string theory, so it can be developed purely mathematically.

We begin with a brief review of the topological classification of D-branes in §5.1. In §5.2, we review the geometric definitions of stability for A- and B-branes, based on stability of special Lagrangian submanifolds and on the Donaldson-Uhlenbeck-Yau theorems respectively. While rather different, their common features suggest a way to proceed.

In §5.3, we justify the claim that the general topological B-brane is an object in the derived category of coherent sheaves. This will enable us to make contact with the underlying definitions in boundary CFT, and make general arguments. In particular, as explained in §5.3.4, the $\mathbb{Z}$-grading of the derived category is generalized to an $\mathbb{R}$-grading, which depends on the "other moduli" (complex moduli for A-branes and Kähler moduli for B-branes) ignored by the topological theory.

In §5.4, we develop the physics from the complementary "world-volume" point of view. We use this to define theories of branes on various interesting spaces, including orbifolds of $\mathbb{C}^3$ and their resolutions. We also introduce physical pictures such as tachyon condensation and Seiberg duality.

These physical arguments are brought together in §5.5, to find a set of requirements which a correct stability condition must satisfy. In §5.6 we illustrate these ideas in examples, most notably the $\mathbb{C}^3/\mathbb{Z}_3$ orbifold and its resolution.

In §5.7 we give a precise mathematical definition for a stability structure axiomatizing these ideas. It turns out that the axioms are strong enough to determine the local structure of the space of stability conditions; it is a finite dimensional manifold. We illustrate this with an explicit description of a space of stability structures for the $\mathbb{C}^3/\mathbb{Z}_3$ orbifold example.

We conclude with a discussion of some open questions in §5.9.

We can summarize a fair part of the discussion in advance in Table 1, which outlines the two mirror symmetry correspondences.

5.1. K-theory, intersection product and antibranes

We begin with physical arguments which lead to a "topological" classification of D-branes. A good general review is [**363**].

There are various ways one could try to do this. One would be to identify any pair of configurations which are related by a "continuous physical process." In particular, all relativistic quantum theories contain "antiparticles," and allow the process of particle-antiparticle annihilation. This inspires a physical version of the Grothendieck construction of K-theory [**471**].

Another approach would be to use the conserved charges of a configuration which can be measured at asymptotic infinity. By definition, these are left invariant by any physical process. Furthermore, they satisfy quantization conditions, of which the prototype is the Dirac condition on allowed electric and magnetic charges in Maxwell theory. In the case of Dirichlet branes, the conserved charges are the Ramond-Ramond charges, and the analogy between this formalism and K-theory was first noted in [**358**].

In principle, the two schemes could lead to different results – it might be that not all invariants can be measured at asymptotic infinity, or it might be that not all values of charge allowed by the quantization conditions are actually realized by physical configurations. However the simplest conjecture is that they agree.

Some of the arguments require a certain amount of physical background, which can be found in textbooks such as [**394**]. We will not assume this but only summarize the flavor and essential results of the discussion. In addition, while we will develop it mostly for the case of Calabi-Yau compactifications of type II strings, these constructions are far more general. Some interesting extensions are to the case with non-zero H-flux [**55, 344**], to the type I string (in which the bundles have real structure) [**471**], and to its generalizations to "orientifolds" [**114, 244**].

5.1.1. Physical realization of K-theory. There is an elementary construction which, given a physical theory T, produces an abelian group of conserved charges $K(T)$. Rather than consider the microscopic dynamics of the theory, we need to know a set S of "particles" described by T, and a set of "bound state formation/decay processes" by which the particles combine or split to form other particles. For brevity we will usually call these "binding processes."

Let us say that two sets of particles are "physically equivalent" if some sequence of binding processes convert the one to the other. We then define the group $K(T)$ as the abelian group $\mathbb{Z}^S$ of formal linear combinations of particles, quotiented by this equivalence relation.

As an elementary example, suppose T contains the particles

$$S = \{A, B, C\}.$$

If these are completely stable, we could clearly define three integral conserved charges, their individual numbers, so $K(T) \cong \mathbb{Z}^3$.

Suppose we now introduce a binding process

$$A + B \leftrightarrow C, \tag{5.1}$$

with the bidirectional arrow to remind us that the process can go in either direction. Clearly $K(T) \cong \mathbb{Z}^2$ in this case.

One might criticize this proposal on the grounds that we have assumed that configurations with a negative number of particles can exist. However, in all physical theories which satisfy the constraints of special relativity, charged particles in physical theories come with "antiparticles," with the same mass but opposite charge. A particle and antiparticle can annihilate (combine) into a set of zero charge particles. While first discovered as a prediction of the Dirac equation, this follows from general axioms of quantum field theory, which also hold in string theory.

Thus, there are binding processes

$$B + \bar{B} \leftrightarrow Z_1 + Z_2 + \cdots .$$

where $\bar{B}$ is the antiparticle to a particle B, and Z_i are zero charge particles, which must appear by energy conservation. To define the K-theory, we identify any such set of zero charge particles with the identity, so that

$$B + \bar{B} \leftrightarrow 0.$$

Thus the antiparticles provide the negative elements of $K(T)$.

Granting the existence of antiparticles, this construction of K-theory can be more simply rephrased as the Grothendieck construction. We can define $K(T)$ as the group of pairs $(E, F) \in (\mathbb{Z}^S, \mathbb{Z}^S)$, subject to the relations

$$(E, F) \cong (E + B, F + B) \cong (E + L, F + R) \cong (E + R, F + L)$$

where (L, R) are the left and right hand side of a binding process (5.1).

5.1.1.1. *K-theory of Dirichlet branes.* Thinking of these as particles, each brane B must have an antibrane, which we denote $\bar{B}$. If B wraps a submanifold L, one expects that $\bar{B}$ is a brane which wraps a submanifold L of opposite orientation.

A potential problem is that it is not *a priori* obvious that the orientation of L actually matters physically, especially in degenerate cases such as L a point. And in general, for example in the bosonic string, this need not be true. But fortunately, as we discussed in §3.5.3, the definition of branes in superstring theory requires a choice of orientation, to make the GSO projection.

To go further, we need some definition of binding process. The most straightforward case is boundary conditions in the sigma model, which have a geometric definition, so we can use continuous deformation.

5.1.2. A-branes. We take X a Calabi-Yau threefold for definiteness; however, most of the discussion applies for general $n \equiv \dim_{\mathbb{C}} X$.

More precisely, a physical A-brane is specified by a pair (L, E) of a special Lagrangian submanifold L with a flat bundle E. When are (L_1, E_1) and (L_2, E_2) related by a binding process?

A simple heuristic answer to this question is given by the Feynman path integral. Two configurations are connected, if they are connected by a continuous path through the configuration space; any such path (or a small deformation of it) will appear in the functional integral with some non-zero weight. Thus, the question is essentially topological.

Ignoring the flat bundles for a moment, this tells us that the K-theory group for A-branes is $H_3(Y, \mathbb{Z})$, and the class of a brane is simply $(\operatorname{rank} E) \cdot [L] \in H_3(Y, \mathbb{Z})$. This is also clear if the moduli space of flat connections on L is connected.

But suppose it is not, say $\pi_1(L)$ is torsion.

EXAMPLE 5.3. Let Y be a hypersurface in $\mathbb{CP}^4$, defined by the vanishing of the quintic polynomial $x_0^5 + x_1^5 + x_2^5 + x_3^5 + x_4^5$ (the Fermat quintic). Such a Y has a real involution σ, and let L be its fixed point set. It is not hard to check that L is a special Lagrangian with topology $\mathbb{RP}^3$, so $\pi_1(L) \cong \mathbb{Z}_2$.

In this case, we need deeper physical arguments to decide whether the K-theory of these Dirichlet branes is $H_3(Y, \mathbb{Z})$, or some larger group. But a natural conjecture, based on the ideology we just outlined, is that it will be $K_1(Y)$, which classifies bundles on odd-dimensional submanifolds.[1]

In Example 5.3, it was argued in [**69**] that two branes which differ only in the choice of flat connection are in fact connected in string theory, consistent with the K-group being $H_3(Y, \mathbb{Z})$. One can also argue using the Atiyah-Hirzebruch spectral sequence that for Y a simply connected Calabi-Yau threefold, $K_1(Y) \cong H_3(Y, \mathbb{Z})$, so the general conjecture is borne out in this case, as is true in the other examples which have been studied. However there is no general physics proof at this point.

5.1.2.1. *Intersection number.* Note that there is a natural bilinear form on $H_3(Y, \mathbb{Z})$ given by the oriented intersection number

$$I(L_1, L_2) = \#([L_1] \cap [L_2]). \tag{5.2}$$

It has symmetry $(-1)^n$; in particular, it is symplectic for $n = 3$. Furthermore, by Poincaré duality, it is unimodular, at least in our topological

[1] The physics relation between K_0 and K_1 involves T-duality on an auxiliary circle. For more detail about how this works, see [**471**]

definition of K-theory. It is a nontrivial question whether all of $H_3(X,\mathbb{Z})$ can be realized by special Lagrangians.

We can compare this with the physics expression for the "RR-charge" of the brane. Saying that a Dirichlet p-brane has RR-charge means that it is a source for an "RR potential," a generalized $(p+1)$-form gauge potential in ten-dimensional space-time [**393**]. This fact can be seen from its world-volume action, which contains a minimal coupling term,

$$\int C^{(p+1)}, \tag{5.3}$$

where $C^{(p+1)}$ denotes the gauge potential, and the integral is taken over the $(p+1)$-dimensional world-volume of the brane. For $p = 0$, $C^{(1)}$ is a one-form or "vector" potential (as in Maxwell theory), and thus the D0-brane is an electrically charged particle with respect to this 10d Maxwell theory.

Now, consider a compactification, i.e., the ten dimensions are $\mathbb{R}^4 \times X$, and a Dp-brane which wraps a p-dimensional cycle L; in other words its world-volume is $\mathbb{R} \times L$ where $\mathbb{R}$ is a time-like world-line in $\mathbb{R}^4$. We now use the Poincaré dual class $\omega_L \in H^{2n-p}(X,\mathbb{R})$ to L in X, to rewrite (5.3) as an integral

$$\int_{\mathbb{R}\times X} C^{(p+1)} \wedge \omega_L. \tag{5.4}$$

We can then do the integral over X to turn this into the integral of a one-form over a world-line in $\mathbb{R}^4$, which is the right form for the minimal electric coupling of a particle in four dimensions. Thus, such a wrapped brane carries a particular electric charge which can be detected at asymptotic infinity.

For our purposes, the RR-charge can be summarized in the formal expression

$$\int_L C = \int_X C \wedge \omega_L; \tag{5.5}$$

where $C \in H^*(X,\mathbb{R})$. In other words, it is a class in $H_p(X,\mathbb{R})$.

In particular, an A-brane (for $n = 3$) carries a conserved charge in $H_3(X,\mathbb{R})$. Of course, this is weaker than our previous statement, that $[L] \in H_3(X,\mathbb{Z})$. To see this physically, we would need to see that some of these "electric" charges are actually "magnetic" charges, and study the Dirac-Schwinger-Zwanziger quantization condition between these charges. This amounts to showing that the angular momentum J of the electromagnetic field satisfies the quantization condition $J = \hbar n/2$ for $n \in \mathbb{Z}$. Using an expression which may be familiar from undergraduate electromagnetism, $\vec{J} = \vec{E} \times \vec{B}$, this is precisely the condition that (5.2) must take an integer value. Thus the physical and mathematical consistency conditions agree.

Similar considerations apply for coisotropic A-branes. If X is a genuine Calabi-Yau 3-fold (i.e., with strict $SU(3)$ holonomy), then a coisotropic A-brane which is not a special Lagrangian must be five-dimensional, and the corresponding submanifold L is rationally homologically trivial, since $H^5(X,\mathbb{Q}) = 0$. Thus, if the bundle E is topologically trivial, the homology class of L and thus its K-theory class is torsion. We discuss the case of E nontrivial shortly, obtaining (5.8).

If X is a torus, or a K3 surface, the situation is more complicated. In that case, even rationally the charge of a coisotropic A-brane need not lie in the middle-dimensional cohomology of X. Instead, it takes its value in a certain subspace of $\bigoplus_p H^p(X,\mathbb{Q})$, where the summation is over even or odd p depending on whether the complex dimension of X is even or odd. At the semiclassical level, the subspace is determined by the condition

$$(L - \Lambda)\,\alpha = 0, \quad \alpha \in \bigoplus_p H^p(X,\mathbb{Q}),$$

where L and Λ are generators of the Lefschetz $SL(2,\mathbb{C})$ action, i.e., L is the cup product with the cohomology class of the Kähler form, and Λ is its dual [**290, 284**].

5.1.3. Intersection form as index for open strings. Although this equivalence for geometric D-branes is easy to trace through and not very deep, a deeper consequence of the physical arguments is that the same structure, an integral bilinear form on the K-theory, must be present for the boundary conditions in *any* (2, 2) SCFT. We will discuss the world-sheet computations later, but the simplest way to see this is to note that (5.2) also computes the index of the quantum theory of open strings stretched from a brane L_1 to a brane L_2,

$$I(L_1, L_2) = \mathrm{Tr}_{L_1,L_2}(-1)^F. \tag{5.6}$$

For the case of A-branes, as we discussed in §3.6.2.5, one can associate open strings to the intersection points. It can be shown [**150**] that the orientations of two points of intersection are opposite if the difference in their ghost numbers is odd. Thus the intersection number is given by the Euler characteristic of the complex (3.186). That is,

$$\begin{aligned}\#([L_1] \cap [L_2]) &= \int_Y l_1 \cdot l_2 \\ &= \sum_i (-1)^i \dim \mathrm{Hom}^i(L_1, L_2).\end{aligned} \tag{5.7}$$

5.1.4. Branes carrying non-trivial bundles. The analogous considerations for B-branes are complicated by the fact that they can carry bundles which are nontrivial in K-theory. The first part of this story to be understood was the generalization of (5.3). This can be read off from the coupling of the RR fields to the D-brane. After various partial world-sheet computations, a general expression was deduced from anomaly inflow arguments in [**190, 95, 358**]. For a Dirichlet brane (L, E), the coupling to the

RR fields C is

$$\int_L f^*C \wedge \mathrm{ch}(E)\frac{\hat{A}(TL)\cdot e^{\frac{1}{2}d}}{f^*\sqrt{\hat{A}(TM)}}. \tag{5.8}$$

Here M is ten-dimensional space-time, C is a formal sum of the even or odd RR potentials, and $\hat{A}$ is the A-roof genus (the square root of the Todd genus). As before, d is a degree two class defining a Spin^c structure on L, needed to define fermions on L.

This formula can be significantly simplified for B-branes. However the full expression is quite suggestive, as it can be rewritten as an integral over all of space-time [**95**],

$$\int_M C \wedge \mathrm{ch}(f_!E)\sqrt{\hat{A}(TM)}, \tag{5.9}$$

using the push-forward map $f_!$ of relative K-theory.

The real significance of this is that the generalization of (5.2) involves the wedge product between the charges (5.3) or (5.9) in space-time. Here it is:

$$\begin{aligned} I(L_1, L_2) &= \int_M \mathrm{ch}(f_!(E_1^\vee))\cdot \mathrm{ch}(f_!(E_2))\cdot \hat{A}(TM) \\ &= \mathrm{ind}\, \not{D}_{E_1^\vee\otimes E_2}, \end{aligned} \tag{5.10}$$

the index of the Dirac operator coupled to the product bundle. This is the signed number of fermionic (Ramond) open strings from L_1 to L_2, and thus string theory realizes (5.6) in great generality.

5.1.4.1. *B-branes.* We will use the specialization of this to B-type branes. First, for a brane (L, E) corresponding to a holomorphic vector bundle E on a complex submanifold L of a Calabi-Yau X, we use the relation $\mathrm{td} = \exp(\frac{1}{2}c_1)\hat{A}$ and the fact that $c_1(X) = 0$ to rewrite (5.9) as

$$\int_L C \wedge \mathrm{ch}(f_!E)\sqrt{\mathrm{td}(TX)}. \tag{5.11}$$

For a coherent sheaf $\mathscr{E}$ with support on a holomorphic subvariety S, this can also be rewritten as

$$\int_X C\cdot Q(i_*\mathscr{E}) = \int_S \mathrm{ch}(E')\sqrt{\frac{\hat{A}(S)}{\hat{A}(N)}}\cdot i^*C, \tag{5.12}$$

where N is the normal bundle to S in X and $E' = E\otimes K_S^{-\frac{1}{2}}$ over S. We will use this in §5.3.3.2.

5.1.5. Boundary conditions in conformal field theory. Finally, let us make contact with the discussion of Chapter 2. The formula (5.6) is in fact a special case of the Cardy conditions (2.12),

$$\theta_b(\iota_b \circ \iota^a(1)) = \sum_{\psi_\mu \in \mathcal{H}_{ab}} \psi_\mu \psi^\mu.$$

It expresses the fact that the annulus diagram with boundary conditions a and b has both closed and open string interpretations. The left-hand side is the closed string expression; in the full string theory this is the exchange of massless RR fields, which by the discussion in §3.3.4 correspond to the topological closed string states. The right-hand side is the index counting $\mathbb{Z}_2$-graded topological open string states.

The expressions above are obtained by evaluating the operations $\iota^a(1)$ and $\theta_b(\iota_b(\phi))$ on the left-hand side using the known geometric couplings between the branes and the RR fields.

In principle, given any concrete definition of an SCFT, similar expressions for the topological classes of branes and the intersection form could be derived. We will illustrate this for orbifolds in §5.4.5.6.

5.2. Preliminaries on stability

5.2.1. A-Branes. We begin by reviewing the definition of special Lagrangian submanifold. We then argue that the particular subset of Lagrangian submanifolds which is "special" depends in a nontrivial way on the complex structure of Y, by exhibiting binding processes (as in (5.1)) which can only take place on one side of a "wall of marginal stability" in complex structure moduli space.

5.2.1.1. *Special Lagrangians.* We take Y Calabi-Yau of dimension m. Recall (§1.1.4, §3.6.1) that a special Lagrangian submanifold is an oriented Lagrangian submanifold $L \subset Y$ satisfying

$$\text{Im } e^{-i\pi\xi(L)}\Omega|_L = 0 \tag{5.13}$$

for some real constant $\xi(L)$.

By "oriented" we mean that we have chosen a non-vanishing top form on L. This can be taken equal to the real volume form dV_L. Then from (1.2), we have

$$dV_L = |Z|\; e^{-i\pi\xi(L)}\Omega|_L, \tag{5.14}$$

for some positive real constant $|Z|$. This determines $\xi(L)$ up to a $2\mathbb{Z}$ constant, which we fix by taking $0 \leq \xi(L) < 2$. Note that this parameter coincides with that of (3.169). That is,

$$\xi(L) = \frac{1}{\pi} \arg \frac{\Omega|_L}{dV_L}. \tag{5.15}$$

Since $\xi(L)$ is a constant, we can also write this as

$$\xi(L) = \frac{1}{\pi} \arg \int_L \Omega, \tag{5.16}$$

which (up to the $1/\pi$) is the argument of the period of the holomorphic m-form associated to the cycle L. Thus $\xi(L)$ at this point only depends on the cohomology class of L. In §5.3.4 we will lift this to a $\mathbb{R}/2m\mathbb{Z}$-valued quantity.

Of course, Ω is only defined up to multiplication by a complex constant, and we must fix this ambiguity to define $\xi(L)$. Indeed, one standard definition of a special Lagrangian uses this freedom to fix $\xi = 0$, and then asserts that the real part of $\Omega|_L$ is zero. However, we will need the idea of comparing values of $\xi(L)$ between different special Lagrangians. One should always bear in mind that only relative values of $\xi(L)$ have any meaning (see also §5.3.4.2).

The period itself,

$$Z(L) = \int_L \Omega. \tag{5.17}$$

will also play an important role. For physics reasons Z is called the *BPS central charge*, or simply central charge, of the Dirichlet brane.[2]

EXAMPLE 5.4. *The complex torus.* Let $X \cong \mathbb{C}^n/\{z^i \sim z^i + 1, z^i \sim z^i + \tau^{ij}\}$, and take $\Omega = dz^1 \wedge \cdots \wedge dz^n$. Choose n pairs of relatively prime integers (a^i, b^i). Then the hyperplanes $z^i = a^i\sigma^i + \sum_j \tau^{ij} b^j \sigma^j$ with $\sigma^i \in [0,1)$ are special Lagrangian with $Z = \det(a^i\delta^{ij} + b^j\tau^{ij})$.

From the discussion in Chapter 1, the special Lagrangian condition implies the minimal volume condition which we discussed in §3.5.2.3. It arises physically as the condition for an A-brane to be satisfy the BPS condition [**34**]. This condition is defined using a space-time supersymmetry condition, as in (1.2),

$$\mathrm{Re}\, e^{i\pi\xi} \epsilon^t \Gamma \epsilon|_M = \mathrm{Vol}|_M.$$

The key point is that this relation must hold everywhere on L with the same covariantly constant spinor ϵ, in order for this spinor to define a supersymmetry.[3] Physics arguments then imply that the "mass" of the brane, which

---

[2]This terminology comes from the role of this term in the space-time supersymmetry algebra, which we will not use in this book. But from now on, when we say "central charge," we mean the BPS central charge, not the Virasoro central charges c and $\hat{c}$ of §3.2.2.

[3]Physically, the parameter ξ determines the $\mathcal{N} = 1$ subalgebra which is preserved out of the full $\mathcal{N} = 2$ space-time supersymmetry. Actually, since N space-time supersymmetries in $d = 4$ lead to $U(N)$ R-symmetry, one might have expected a $U(2)/U(1)$ choice here. However, the spectral flow picture of $\mathcal{N} = 2$ space-time supersymmetry only sees a $U(1) \times U(1)$ subgroup of the the R-symmetry so the parameter only lives in $U(1)$.

in the geometric limit we are discussing is the volume of L, must be greater than or equal to the absolute value of the central charge,

$$M \equiv \int_L \text{Vol} \geq |Z(L)|, \tag{5.18}$$

with equality if and only if the state is BPS. This physical argument remains valid after taking any and all corrections (in α' and g_s) into account.

Let us see the analogous condition in the sigma model. We first rewrite (5.13) as

$$\Omega|_L = \exp(2i\pi\xi)\bar{\Omega}|_L. \tag{5.19}$$

Both Ω and $\bar{\Omega}$ have operator analogs in the A-model:

$$\Omega_{IJK}\psi_+^I\psi_+^J\psi_+^K; \qquad \bar{\Omega}_{\bar{I}\bar{J}\bar{K}}\psi_-^{\bar{I}}\psi_-^{\bar{J}}\psi_-^{\bar{K}}.$$

Substituting these into (5.19) gives

$$\Omega_{IJK}\psi_+^I\psi_+^J\psi_+^K = \exp(-2i\pi\xi)\bar{\Omega}_{\bar{I}\bar{J}\bar{K}}\psi_-^{\bar{I}}\psi_-^{\bar{J}}\psi_-^{\bar{K}} \tag{5.20}$$

as a boundary condition on the fields, generalizing (3.196) in §3.6.6. In §5.3.4, we will use this identification to generalize these definitions and arguments to any $(2,2)$ SCFT, without any need to assume the geometric picture of A-branes as special Lagrangians.

Recall from Definition 3.3 (§3.6.2) that an A-brane must satisfy two additional conditions. It should be obvious that the map ξ_* in (3.170) is trivial for a special Lagrangian, and thus the Maslov class condition is automatically satisfied. On the other hand, the quantum obstruction condition is not automatically satisfied. We discuss it further in §5.4.3.5 and §8.3.3.

Note that there is no reason to suppose that all minimal 3-manifolds in a Calabi-Yau are special Lagrangians, reflecting the fact that not all stable D-branes are necessarily BPS. We will give a simple (non-smooth) example shortly.

5.2.1.2. *A geometrical decay.* In §3.6.2 we saw that, neglecting the quantum obstruction condition for the moment, an A-brane (carrying a line bundle) has a deformation space given by $H^1(L,\mathbb{C})$. As we show in §6.1.1, the infinitesimal deformation space of special Lagrangians with flat connections is also given by $H^1(L,\mathbb{C})$. Thus, locally, the moduli space of special Lagrangians agrees with the moduli space of Lagrangians modulo Hamiltonian deformation. At first sight, this might suggest that in each equivalence class of Lagrangians modulo Hamiltonian isotopy there is a unique special Lagrangian.

This is not actually true. It turns out the vast majority of Lagrangians have no special Lagrangian equivalent to them by Hamiltonian isotopy. From our perspective, the best way to see this is to consider how special Lagrangians can "disappear", or "decay", as the complex structure of the target space Y is deformed. Note that a Lagrangian submanifold is defined purely in terms of the symplectic structure of Y induced by the Kähler form and so has no dependence on the complex structure. Adding the "special" in special Lagrangian introduces the dependence on the complex structure.

As in §5.1, the inverse of a decay process is a "binding" process, in which two (or more) branes combine into one. Following Joyce [**262**], we now give an explicit picture of this process, in terms of the local behavior near a point of intersection.

We consider a family of Calabi-Yau m-folds Y_z with nearby complex structures parameterized by $z \in \mathbb{C}$ with $|z|$ small. Suppose that Y_0 contains two special Lagrangians L_1 and L_2 which intersect transversely at a point p, and such that $\xi(L_1) = \xi(L_2)$ (at $z = 0$).

One can show that, given that a smooth special Lagrangian L exists at $z = 0$, it will exist in a neighbourhood of 0 [**349, 262**]. Thus we restrict attention to a region in which both L_1 and L_2 exist, and define $\xi(L_1)$ and $\xi(L_2)$ according to (5.15). We furthermore restrict to $z \in \mathbb{R}$ such that $|\xi(L_1) - \xi(L_2)|$ is small. Then

THEOREM 5.5. *(Joyce,* [**262**, Thm 9.10]*) There exists a special Lagrangian $L_1 \# L_2 \subset Y_z$ which is close to the connected sum $L_1 \cup L_2$ if and only if $\xi(L_2) \leq \xi(L_1)$.*

Let us divide the complex structure moduli space into two regions $\mathcal{M}^+$ and $\mathcal{M}^-$ depending on the sign of $\xi(L_2) - \xi(L_1)$, separated by a "wall of marginal stability." We see that there is a special Lagrangian $L_1 \# L_2$, which only exists in $\mathcal{M}^+$. The notation is intentionally asymmetric since $L_1 \# L_2$ is quite different from $L_2 \# L_1$.

Conversely, in $\mathcal{M}^-$ there is no smooth special Lagrangian minimizing the volume of $L_1 \cup L_2$, rather $L_1 \cup L_2$ itself is volume minimizing (for small $|z|$). Physically, this is a "non-BPS" or "supersymmetry breaking" configuration.

5.2.1.3. *Special Lagrangian planes in $\mathbb{C}^m$.* Theorem 5.5 is proven by considering a neighbourhood of the intersection point p, modeling it on $\mathbb{C}^m$, and modeling the branes L_1 and L_2 there as hyperplanes.

Thus, consider a linear subspace $\mathbb{R}^m \subset \mathbb{C}^m$. We can specify its embedding in terms of angles ϕ_j, as

$$\Pi^{\vec{\phi}} = \{(e^{i\phi_1}x_1, e^{i\phi_2}x_2, \ldots, e^{i\phi_m}x_m) : x_j \in \mathbb{R}\}. \tag{5.21}$$

Using the standard holomorphic m-form $\Omega = dz_1 \wedge dz_2 \wedge \cdots \wedge dz_m$ we obtain

$$\begin{aligned} \xi(\Pi^{\vec{\phi}}) &= \frac{1}{\pi} \arg \int_{\Pi^{\vec{\phi}}} \Omega \\ &= \frac{1}{\pi} \sum_{j=1}^{m} \phi_j \pmod 2. \end{aligned} \tag{5.22}$$

We denote by $\Pi^{\vec{0}}$ the plane with orientation $dx^1 \wedge \cdots \wedge dx^m$, embedding $\phi_1 = \phi_2 = \cdots = \phi_m = 0$ and thus $\xi = 0$.

In general, the angles ϕ_j are only determined up to shifts $\phi_i \to \phi_i + 2\pi$ and the action of orientation preserving transformations

$$R_{ij} : (x^1, \dots, x^i, \dots, x^j, \dots) \to (x^1, \dots, -x^i, \dots, -x^j, \dots);$$
$$\phi_i \to \phi_i \pm \pi, \quad \phi_j \to \phi_j \pm \pi.$$

We can fix this ambiguity by requiring

$$(5.23) \qquad \begin{aligned} & 0 \le \phi_1 < 2\pi, \\ 0 \le \phi_i < \pi; \qquad & 2 \le i \le m. \end{aligned}$$

For example, we denote the brane with the same embedding as $\Pi^{\vec{0}}$ but the opposite orientation as $\Pi^{(\pi,0,0)}$. It has $\xi \equiv 1 \pmod 2$. As it is the antibrane to $\Pi^{\vec{0}}$, we could also denote it as $\bar{\Pi}^{\vec{0}}$.

5.2.1.4. *The Lawlor neck.* Consider two D-branes $\Pi^{(\pi,0,0)} \cup \Pi^{\vec{\phi}}$ intersecting transversely at the origin; this requires that all $\phi_i \neq 0$. The difference in their phases (5.22) is

$$(5.24) \qquad \Delta\xi \equiv -1 + \frac{1}{\pi} \sum_{j=1}^{m} \phi_j \in 2\mathbb{Z}.$$

Bringing this to a fundamental region using (5.23), we find $\Delta\xi \in (-1, m)$. Note in passing that this is larger than the naive range for an angular variable such as $\pi\Delta\xi$, a point we will expand on in §5.3.4.

Now assume that $\Delta\xi = 0$; clearly this is possible only for $m \geq 2$. Then, we can regard this combination of branes as *one* singular special Lagrangian. It is a scaling limit of a smooth special Lagrangian found by Lawlor [**321**], described as follows [**223, 262**]. Let

$$(5.25) \qquad P(x) = \frac{\prod_{j=1}^{m}(1 + a_j x^2) - 1}{x^2}.$$

Also fix a positive real number A, which will determine the overall scale of the solution.

The positive real numbers $a_1, \dots, a_m$ are then implicitly and uniquely determined by A and $\phi_1, \phi_2, \dots, \phi_m$ by the equations

$$(5.26) \qquad \begin{aligned} \phi_j &= a_j \int_{-\infty}^{\infty} \frac{dx}{(1 + a_j x^2)\sqrt{P(x)}} \\ A &= \frac{\omega_m}{\sqrt{a_1 \cdots a_m}}, \end{aligned}$$

where ω_m is the volume of a unit sphere in $\mathbb{R}^m$. These equations admit solutions only if $\sum \phi_j = \pi$, which is precisely the condition $\xi = 1$ we imposed above.

Now define functions $\eta_j : \mathbb{R} \to \mathbb{C}$ by

$$(5.27) \qquad \eta_j(y) = \exp\left(i a_j \int_{-\infty}^{y} \frac{dx}{(1 + a_j x^2)\sqrt{P(x)}} \right) \sqrt{\frac{1}{a_j} + y^2}.$$

This allows the Lawlor neck to be defined as

(5.28)

$$L^{\vec{\phi},A} = \{(\eta_1(y)x_1, \eta_2(y)x_2, \ldots, \eta_m(y)x_m) : y \in \mathbb{R}, x_j \in \mathbb{R}, x_1^2 + \cdots + x_m^2 = 1\}.$$

One can show that this is a smooth special Lagrangian submanifold of $\mathbb{C}^m$, with topology $S^{m-1} \times \mathbb{R}$. The region far from the origin is given by the two limits $y \to -\infty$, which asymptotes to $\bar{\Pi}^0$, and $y \to +\infty$, which asymptotes to $\Pi^{\vec{\phi}}$. As we take the limit $A \to 0$, the intersection region shrinks to zero size, producing in the limit $\Pi^{(\pi,0,0)} \cup \Pi^{\vec{\phi}}$. We sketch this in Figure 1.

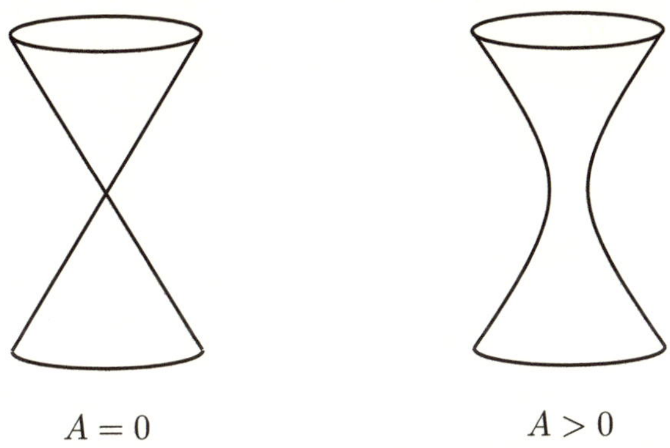

FIGURE 1. The Lawlor neck.

5.2.1.5. *The angle theorem.* Suppose now that we consider a nearby plane $\Pi^{\vec{\phi}'}$, such that $\Delta\xi' \neq 0$. Clearly there can be no smooth special Lagrangian asymptoting to the two planes in this case. This suggests that such a configuration might no longer be volume minimizing.

A construction of a submanifold with the same asymptotics but lower volume is to consider a combination of $\Pi^{(\pi,0,0)} \cup \Pi^{\vec{\phi}'}$ with a Lawlor neck $L^{\vec{\phi},A}$. We depict roughly what happens in the case $m = 1$ in Figure 2 (even though the Lawlor neck does not actually exist in this case). Intuitively, this works if the pair of planes subtends a larger angle; this is correct and the condition is

$$0 < \Delta\xi = \xi(\Pi^{\vec{\phi}'}) - \xi(\Pi^{(\pi,0,0)}). \tag{5.29}$$

Then, by replacing the interior of $\Pi^{(\pi,0,0)} \cup \Pi^{\vec{\phi}'}$ with the interior of the Lawlor neck, we would obtain a continuous submanifold with the same asymptotics. But since $L^{\vec{\phi},A}$ is calibrated while the union of planes is not, it will have lower volume in the interior. The junction could then be smoothed to obtain a smooth submanifold.

By developing this argument, one obtains

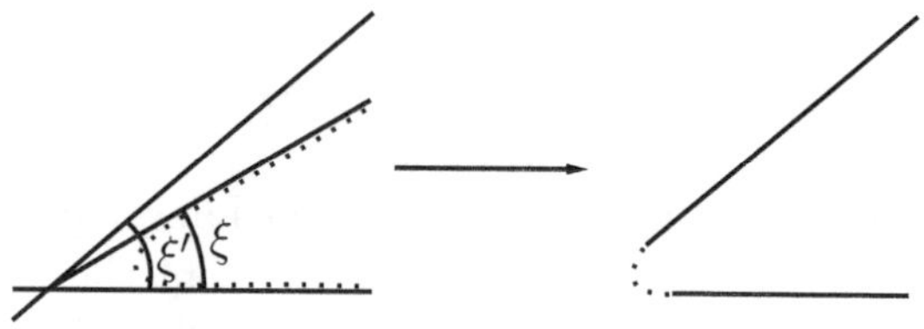

FIGURE 2. Combination of planes and Lawlor neck.

THEOREM 5.6. *(Lawlor-Nance, as in* [**223**, Theorem 7.134]*) The union of a pair of oriented p-dimensional subspaces of* $\mathbb{R}^{2p}$ *is volume minimizing if and only if the relative angles of orientation do not satisfy* (5.29) *for any choice of complex structure.*

If one's primary interest is in $\mathbb{R}^{2p}$, one can rephrase the condition not to mention complex structure, by suitably ordering the relative angles between the planes.

5.2.1.6. *Embedding in a Calabi-Yau geometry.* Now, we come back to our two A-branes L_1 and L_2 which intersect transversely at a point $p \in Y$. We may use a $U(m)$ transformation to rotate the tangent plane of L_1 into the standard plane $\Pi^{(\pi,0,0)}$ as above. This same $U(m)$ matrix rotates the tangent plane of L_2 into $\Pi^{\vec{\phi}}$.

Now, it is intuitively plausible that, by restricting attention to a small neighbourhood of p, we find precisely the situation we just described. This was proven in [**262**] leading to Theorem 5.5.

To recap, in $\mathscr{M}^+$ we have a BPS state $L_1 \natural L_2$. We also have BPS states L_1 and L_2 but the mass of $L_1 \natural L_2$ is less than the sum of the masses of L_1 and L_2. As we hit the wall, $L_1 \natural L_2$ becomes $L_1 \cup L_2$. Beyond the wall in $\mathscr{M}^-$ we only have BPS states L_1 and L_2 which together break supersymmetry. What we have just described is a *decay* of a BPS state $L_1 \natural L_2$ into its products L_1 and L_2 as we pass from $\mathscr{M}^+$ into $\mathscr{M}^-$. In $\mathscr{M}^+$ we view $L_1 \natural L_2$ as a *bound state* of L_1 and L_2. We depict this story in Figure 3.

Away from the wall, $L_1 \cup L_2$ is no longer a special Lagrangian since the two components have a different value of ξ. The smooth space $L_1 \natural L_2$ is a smooth space homological to $L_1 \cup L_2$ which minimizes the volume, i.e., energy of a D-brane wrapped around these cycles. It follows that

$$\left| \int_{L_1 \natural L_2} \Omega \right| < \left| \int_{L_1} \Omega \right| + \left| \int_{L_2} \Omega \right|, \tag{5.30}$$

and we may choose

$$\xi(L_2) < \xi(L_1 \natural L_2) < \xi(L_1). \tag{5.31}$$

Physically, this also agrees with the standard properties of BPS states in $\mathscr{N} = 2$ theories in four dimensions as studied, for example, by Seiberg

$\mathscr{M}_+$	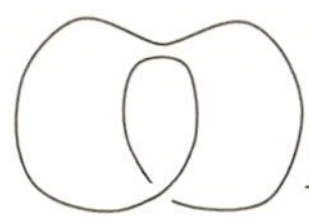	$\xi(L_2) <$ $\xi(L_1 \# L_2) <$ $\xi(L_1)+1$	$L_1 \# L_2$ is stable.
Wall	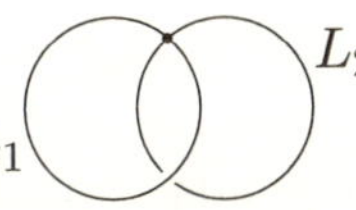	$\xi(L_2) =$ $\xi(L_1 \# L_2) =$ $\xi(L_1)+1$	$L_1 \# L_2$ is marginally stable.
$\mathscr{M}_-$	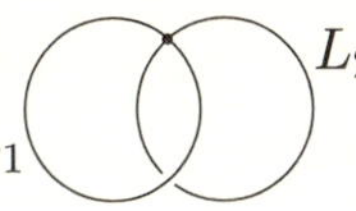	$\xi(L_2) > \xi(L_1)+1$	$L_1 \# L_2$ is unstable.

FIGURE 3. A-Brane Decay.

and Witten [**415**]. We refer to [**329**] for an introduction; see also §19.3 of MS1.

5.2.2. B-branes. The mirror of the phenomenon we just discussed would be a variation in the set of B-type branes, as the complexified Kähler structure is varied. Now, just as in the discussion of correlation functions in §3.4.1.1, §3.6.2.2, we expect this aspect of the problem to get world-sheet instanton corrections, and possibly other α' corrections. However we can begin by considering the large radius limit of the Calabi-Yau threefold X, in which the world-volume connection is governed by the Yang-Mills equations.

Let us consider a 6-brane wrapping X associated to a holomorphic vector bundle $E \to X$ with curvature (1,1)-form F. At large radius, the BPS condition reduces to the Hermitian-Yang-Mills condition [**347**]. That is, the curvature tensor obeys the relation

$$g^{j\bar{k}} F^{\beta}_{\alpha j \bar{k}} = \mu(E) \cdot \delta^{\beta}_{\alpha}, \tag{5.32}$$

where α, β are indices in the fiber of E and $\mu(E)$ is a real number called the "slope" of E. Following the analysis of [**453**], for example, one can integrate (5.32) and obtain[4]

$$\mu(E) = \frac{\deg E}{k \cdot \mathrm{Vol}(X)}, \tag{5.33}$$

[4]Here we are following the conventions of [**453**], but μ is also often defined to remove the factor of $\mathrm{Vol}(X)$ in (5.33).

where k is the rank of E and

$$\deg E = \int_X J \wedge J \wedge c_1(E), \tag{5.34}$$

is the *degree* of the bundle E.

As is usual for the condition for a supersymmetric vacuum, this is a first order differential equation for the fields (the connection) which implies that the fields solve the equation of motion (the Yang-Mills equation), a second order differential equation.

The Hermitian-Yang-Mills condition (5.32) depends explicitly on the metric and thus the Kähler form J. As such, the existence of a solution can depend upon J. In fact, it only depends on the following, algebro-geometric condition:

DEFINITION 5.7. The locally free sheaf E is said to be *μ-stable* if every subsheaf F of E satisfies

$$\mu(F) < \mu(E). \tag{5.35}$$

Then

THEOREM 5.8. *(Donaldson* [**128**]*; Uhlenbeck and Yau* [**453**]*) A bundle admits an irreducible Hermitian-Yang-Mills connection if and only if it is μ-stable.*

The stability of a BPS B-brane is thus equivalent to μ-stability in the large volume limit where α' corrections are ignored.

To begin developing the analogy with §5.2.1, a subsheaf $F \to E$ fits into a short exact sequence

$$0 \longrightarrow F \xrightarrow{f} E \xrightarrow{g} G \longrightarrow 0, \tag{5.36}$$

which allows us to regard E as a bound state of F and G. Whether this bound state exists depends on the condition (5.35). Now, looking at (5.33) suggests that for $b^{1,1} > 1$, this can depend on the ray in the Kähler cone. This is true, as was recognized in the study of the Donaldson invariants [**399**]; its physics interpretation was pointed out in [**424**]. It will turn out that this process is the mirror of the decay of §5.2.1.2.

The lower-dimensional branes can be analyzed similarly. We can focus on a subspace $S \subset X$ and look for stable vector bundles (or twisted bundles if S is not spin) within the class of bundles on S. The subspaces do not interfere with each other in the following sense. A bundle on X associated to E cannot decay into a subsheaf F supported only on S since there is no homomorphism $F \to E$. Equally, a bundle on S cannot decay into a subsheaf on X since the quotient sheaf G in (5.36) would have negative rank.

It follows that μ-stability establishes a set of stable B-branes at the large radius limit, which are a subset of the coherent sheaves, depending on the ray in the Kähler cone.

5.2.2.1. *HYM as a symplectic quotient.* A good way to think about this result, as discussed in [**129**], is to regard the left side of the HYM equation (5.32) as a moment map. We first note that the space of connections on E has a natural symplectic form. Given two tangent vectors $\beta, \gamma \in \Omega^1(X, \mathrm{ad}(E))$, its value is

$$\int_X \omega^{n-1} \wedge \mathrm{Tr}\left(\bar{\beta} \wedge \gamma\right). \tag{5.37}$$

Then, the evolution (5.75) with the "Hamiltonian"

$$H = \int_X g^{j\bar{k}}\, \mathrm{tr}\left(F_{j\bar{k}}\epsilon\right) \mathrm{vol}_X$$

generates a gauge transformation with parameter ϵ.

The problem of finding a solution modulo gauge equivalence is then an infinite-dimensional symplectic quotient. In the finite dimensional case, by geometric invariant theory, such a quotient space is equivalent to the space of orbits of a complexified symmetry group satisfying a stability condition (as in §4.7.3). By analogy, we can expect existence of HYM solutions to be governed by a stability condition as well.

One side of the Donaldson-Uhlenbeck-Yau theorem, the necessity for μ-stability, is easy to prove, using a vanishing theorem:

LEMMA 5.9. *If V admits an irreducible Hermitian-Yang-Mills connection, and has a non-zero holomorphic section s, then $c_1(V) \geq 0$.*

This follows from

$$0 \leq \int_X \omega^{i\bar{j}} (\nabla_{\bar{j}} s^*)(\nabla_i s)$$

and an integration by parts.

Then, consider a candidate destabilizing subsheaf F of E, The map f in (5.36) is a section of $V \equiv \mathrm{End}(F, E)$. Now, if E and F admit irreducible Hermitian-Yang-Mills connections, so will V, and by the lemma $c_1(V) = c_1(E) - c_1(F) \geq 0$.

A somewhat more concrete way to see this is to regard (5.36) as providing a "Jordan block" representation of a one-parameter family of connections on E, explicitly

$$A_E = \begin{pmatrix} A_F & t\delta A_{G,F} \\ 0 & A_G \end{pmatrix}$$

where $\delta A_{G,F} \in H^{0,1}(X, \mathrm{End}(G, F))$, and t is the parameter. One also notes that different choices of $t \in \mathbb{C}^*$ correspond to complex gauge transformations. GIT stability then requires stability in this one-parameter reduction of the problem,

Physically, one can compute the dependence on t of the Yang-Mills action near $t = 0$. One finds

$$S_{YM}(E) = S_{YM}(A_F) + S_{YM}(A_G) + C(\mu(F) - \mu(E))|t|^2 + \mathcal{O}(t^3)$$

for some constant C. Thus, if $\mu(F) < \mu(E)$, the deformation $t \neq 0$ lowers the energy. Conversely, if $\mu(F) > \mu(E)$, the energy is lowered by taking $t \to 0$, which in the limit $t = 0$ leads to a reducible connection.

Note the similarity with the arguments involving (5.30). Indeed, in the case in which the branes are particles in a type II compactification on Calabi-Yau threefold, the observable consequences in four dimensions are precisely the same: thus, (5.36) describes a possible process by which branes F and G can form the bound state E, or the opposite decay of E. Indeed, we will see in §5.4.4 that from the point of view of four-dimensional physics (effective field theory), the two processes are indistinguishable. This is as it must be for mirror symmetry to be a physical equivalence between theories.

5.2.2.2. *Corrections to μ-stability and the MMMSL equation.* In [**347**], the leading α' corrections to the Hermitian-Yang-Mills condition were computed, leading to the **MMMSL equation**,

$$\frac{1}{n!}(\omega \cdot 1_E + 2\pi\alpha' F)^{\wedge n} = k\frac{\omega^n}{n!} \cdot 1_E, \tag{5.38}$$

where $n = \dim_{\mathbb{C}} X$, and k is an E-dependent and α'-dependent constant which can be obtained by integrating both sides of (5.38) over X. In [**347**] this equation was derived from the supersymmetric Born-Infeld action for the gauge field on the six-brane. Recall also the world-sheet argument leading to the equivalent (3.198).

Strictly speaking, both derivations apply only for E a line bundle. In fact (5.38) is correct for an arbitrary E, as we will show in §5.5.5.

The same (5.38) was studied for algebraic geometric reasons by Leung [**330**]. He showed that for a fixed holomorphic vector bundle E, (5.38) admits a solution for all sufficiently small α' iff E is stable in the sense of Gieseker. This stability condition also takes the general form (5.35), but with a different slope function, the Gieseker slope defined by

$$\begin{aligned} p(E,r) &= \frac{1}{\operatorname{rk} E}\int_X \operatorname{ch}(E)e^{r\omega}\operatorname{Td}(X) \\ &\sim \frac{1}{\operatorname{rk} E}\dim H^0(X, E\otimes\mathcal{L}^r) \qquad r \text{ large} \end{aligned} \tag{5.39}$$

where $c_1(\mathcal{L}) = \omega$. More explicitly, E is Gieseker stable if for any subsheaf V of E one has $p(V,r) < p(E,r)$ for sufficiently large r.

Just like the HYM equation, the MMMSL equation can be interpreted as the moment map equation for a symplectic quotient of the space of connections on E by the group of gauge transformations. The only difference from the previous discussion is that (5.37) is replaced by

$$\int_X \operatorname{Tr}\left(\bar{\beta}\wedge\gamma\wedge\exp\left(\frac{F}{2\pi}+r\omega\right)\right)\operatorname{Td}(X).$$

In the limit $r\to\infty$, this approaches (5.37) up to an overall constant.

5.3. The category of B-branes

We start by reviewing the basic construction of the derived category from §4.4, rephrasing its ingredients in a more physical language. At the appropriate points we will explain how new elements enter.

We seek to define a category of B-branes in the terms of topological string theory. Objects are boundary conditions, and morphisms are elements of Q-cohomology for a twisted supercharge Q.

Recall the discussion of §3.6.3 and §3.6.4. There, we argued that the category of B-branes includes as objects the holomorphic vector bundles on X, with the morphisms being elements of the Dolbeault cohomology groups $H^{0,q}_{\bar{\partial}}(X,\operatorname{Hom}(E,F))$. The $\mathbb{Z}$-grading of this cohomology theory corresponds to R-charge (also called ghost number). As we discussed, these cohomology groups are also isomorphic to the groups $\operatorname{Ext}^q(E,F)$.

We now want to show how complexes constructed from these objects and morphisms arise from the definitions of Chapter 3. The first step is to take a direct sum of objects; clearly this is possible by the Chan-Paton construction. However, we cannot simply assert that direct sums

$$E\equiv\bigoplus_{n\in\mathbb{Z}}E^n, \tag{5.40}$$

are graded; this must somehow be implicit in the physical definitions.

Ultimately this follows from R-charge conservation, as was explained in §3.6. But let us dwell on this step a bit, as it will eventually lead to many of the new features which come out of physics.

5.3.1. The physical origin of gradings. We want to show that a boundary condition which we thought was completely specified by a choice of holomorphic bundle E, actually carries another integer quantum number n, so that different terms E^n in (5.40) are different boundary conditions. Furthermore, this quantum number must contribute to the R-charge, so that an open string in the group $\operatorname{Ext}^i(E^m,F^n)$ has R-charge

$$J=i+n-m. \tag{5.41}$$

If so, the R-charge will be a $\mathbb{Z}$-grading, in the sense used in §4.4.

Where does this quantum number come from? We are going to explain this in detail in §5.3.4, but let us give the basic idea here. As discussed in §3.6 and §3.3.4.6, the R-charge is a $U(1)$ charge, and is thus described by the factorized $U(1)$ theory of §3.2.4. This $U(1)$ theory is essentially the same as that of an S^1 target space in bosonic string theory, up to the subtleties expressed by (3.70), which we will discuss in §5.3.4. Thus, we can borrow the discussion of Dirichlet branes in the target space S^1, to understand the possible boundary conditions in this sector.

This was discussed in §3.5.4, with the conclusion that there are two possibilities: a D0-brane which is parameterized by a coordinate on S^1, and a D1-brane which is parameterized by a flat connection or equivalently a coordinate on a dual S^1. Comparing with (3.167), the B-branes are analogous to D1-branes, while the A-branes are analogous to D0-branes.

Either way, we have found a single additional quantum number, a coordinate on an S^1. While we are going to argue that this is the quantum number n we are looking for, until we have the relation straight, let us denote it by a different symbol μ. But the first check that we can identify the two is that, taking $\mu(E) = m$ and $\mu(F) = n$, (5.41) becomes identical to (3.189), so that the relations between gradings within complexes and R-charges will come out right.

Of course, this interpretation faces the problem that μ takes values in S^1, not $\mathbb{Z}$. However, depending on the periodicity of μ (circumference of the S^1), this difference might not be essential. As we discussed in §4.5, on a d-complex dimensional manifold, one only needs resolutions of length d to represent any object. Thus, we could distinguish all the required maps between complexes if the new quantum number were defined modulo N (leading to a $\mathbb{Z}/N\mathbb{Z}$-grading) for any $N \geq 2d$. In this case, by restricting to quantized values we could get a $\mathbb{Z}/N\mathbb{Z}$ quantum number.

For this interpretation to be correct, after verifying (5.41), we must check that $N \geq 2\hat{c}$. In physics terms, the shift (5.41) follows directly from the definitions (3.144) and (3.167), which show that i and n enter in the conserved R-charge $\partial_\tau \varphi$ in precisely the same way. One can make a more intuitive argument for this in the A-model picture, where R-charge is string winding number; by T-duality (or mirror symmetry) both arguments are equally valid.

Having seen this, the periodicity N of the grading should follow from the circumference $2\pi R$ of the S^1. We can compute this by carefully following the definitions in §3.3.4.6. Doing this, we find the periodicity $2\pi R = 2\hat{c}$, consistent with $N = 2\hat{c}$. Actually, this is a little imprecise as the R-charge itself is a $\mathbb{Z}$-conserved charge, rather it is the nature of the sum (3.70) which makes the grading n a periodic variable, but the conclusion will turn out to be correct.

The remaining point we need to understand is why the gradings in (5.40) were taken to be integral, when the quantum number μ need not be integral. Actually, we should take a clue from the physics at this point and accept $\mathbb{R}$-valued or $\mathbb{R}/N\mathbb{Z}$-valued gradings. Of course, the definition of a complex requires the gradings of its terms to differ by integers. With this proviso, one can check that all of the subsequent definitions make sense after this generalization. We refer to the resulting structure as an $\mathbb{R}$-graded category.

We will explain the relation between (5.41) and (3.189) in §5.3.4, by relating both n and μ to the gradings ξ of §5.2. This will lead to flow of gradings and other important ingredients of the stability proposal. But, having explained the origin and meaning of (5.40), let us postpone these developments.

5.3.2. Deformations and complexes. We have now seen that, once we take into account R-charge, boundary conditions in a topological open string theory are graded direct sums of the form (5.40),

$$E = \bigoplus_{n \in \mathbb{Z}} E^n, \tag{5.42}$$

where the E^n are A- or B-type boundary conditions as discussed in Chapter 3. We now want to allow deformations of these direct sums, and determine the full set of objects of this type. The Q-cohomology classes of maps between pairs of these objects are then the full set of topological open strings.

Let us grant that the original A- or B-type boundary conditions are objects in an additive category $\mathcal{A}$ (for concreteness, consider the locally free sheaves on X). Then

PROPOSITION 5.10. *In the topologically twisted theory with R-charge, the full set of branes corresponds to the homotopy category* $\mathbf{K}(\mathcal{A})$.

To see this, we start by considering the open strings from E to itself, i.e., linear combinations of elements of $\mathrm{Ext}^*(E,E)$. Following (5.41), the open strings of R-charge q are elements of $\mathrm{Ext}^k(E^n, E^{n-k+q})$ for any n and k.

As in §3.6.4, chiral operators of R-charge one can correspond to deformations (in general they can be obstructed). The case of $\mathrm{Ext}^1(E^n,E^n)$ was discussed there; for bundles this corresponds to varying the connection.

The case $\mathrm{Hom} \cong \mathrm{Ext}^0$ corresponds to varying the differential in the complex;

$$\begin{aligned} d &= \sum_n d_n \\ d_n &\in \mathrm{Ext}^0(E^n, E^{n+1}) = \mathrm{Hom}(E^n, E^{n+1}). \end{aligned} \tag{5.43}$$

Let $W_d^{(1)}$ be the operator obeying

$$\{Q, W_d^{(1)}\} = d_\Sigma d, \tag{5.44}$$

where d_Σ is the world-sheet de Rham operator. We deform the action by

$$S = S_0 + \oint_{\partial\Sigma} W_d^{(1)}. \tag{5.45}$$

Following the usual Noether method, we may show that this results in a change in the BRST charge

$$Q = Q_0 + d. \tag{5.46}$$

So, to maintain the relation $Q^2 = 0$, we are required to impose

$$\{Q_0, d\} + d^2 = 0. \tag{5.47}$$

Given (5.43), we know that $\{Q_0, d\} = 0$, so this condition reduces to $d^2 = 0$. Expanding this in terms of the successive maps in (5.43), this becomes

$$d_{n+1}d_n = 0 \quad \text{for all } n, \tag{5.48}$$

and thus the physical consistency condition (nilpotence of the BRST charge) is equivalent to the mathematical condition that E is a *complex*

$$\cdots \xrightarrow{d_{n-1}} E^n \xrightarrow{d_n} E^{n+1} \xrightarrow{d_{n+1}} E^{n+2} \xrightarrow{d_{n+2}} \cdots. \tag{5.49}$$

A brane is therefore more generally represented by a complex of objects from $\mathcal{A}$. The maps in the complex represent a deformation from the initial simple collection of objects. Note that an object E itself is a complex in a rather trivial way:

$$\cdots \xrightarrow{0} 0 \xrightarrow{0} E \xrightarrow{0} 0 \xrightarrow{0} \cdots \tag{5.50}$$

More generally, the position of the object in the complex is the position of the brane in the S^1 associated to R-charge, as described in §5.3.1.

5.3.2.1. *Open strings.* The deformation above will also affect the spectrum of open strings between the B-branes. In this section we compute the corresponding Hilbert spaces of open string states.

We consider open strings from a B-brane represented by a complex E, and another B-brane represented by a complex F. To be more precise, we start with a collection of objects $E^0, E^1, \ldots \in \mathcal{A}$ and another collection of locally free sheaves $F^0, F^1, \ldots \in \mathcal{A}$, where E^n and F^n have R-charge n. Now deform the theory by turning the collection of E's into a complex (5.49) with boundary maps d_n^E, and do the same for F.

Having done this, the total BRST operator becomes

$$Q = Q_0 + d^E - d^F, \tag{5.51}$$

acting on a direct sum of maps

$$f^{m,n} : E^m \to F^n$$

from the complex E to the complex F. By definition, the (topological) open strings live in the cohomology spaces of this operator.

Suppose the individual maps all satisfy $Q_0 f^{m,n} = 0$, i.e., they are elements of $\mathrm{Hom}(E^m, F^n)$. Then, taking into account signs, the condition

$Qf = 0$ is exactly the statement that f is a morphism of complexes. Furthermore, two sets of maps which differ by a BRST exact quantity as $f' = f + Qh$ are homotopic morphisms as defined in §4.4.2.

Thus, the topological open strings between complexes are precisely the morphisms of the homotopy category $\mathbf{K}(\operatorname{Coh} X)$.

Since the definitions agree, we can now adopt the notation $\operatorname{Ext}^n(E, F)$ of §4.4.4 for the cohomology space of open strings from E to F with R-charge n. We can also adopt the notation $[n]$ for an operator which changes the R-charge of a set of branes by n. More precisely, if the qth position of F is F^q, the qth position of $F[n]$ is F^{q+n}. We then have

$$\operatorname{Ext}^q(E[m], F[n]) = \operatorname{Ext}^{q-m+n}(E, F), \tag{5.52}$$

and all of the morphisms are elements of $\operatorname{Hom}(E, F[q])$.

As a simple example, we could take E and F to be injective resolutions of locally free sheaves E and F. In this case, the space $\operatorname{Ext}^n(E, F)$ is a *hyperext* group and agrees with $\operatorname{Ext}^n(E, F)$ as discussed in §4.5. Thus, we get the same open strings as we would have found between E and F by following §3.6.4.

5.3.2.2. *Other deformations.* A very broad class of deformations are as follows. Suppose we have two D-branes given by complexes E and F. Assuming the ghost numbers of the components were not affected by turning on the differentials, an open string corresponding to $f \in \operatorname{Hom}(E[-1], F)$ will have ghost number one. Thus we may consider the deformation given by f. It is easy to see that this produces the mapping cone $\operatorname{Cone}(f : E[-1] \to F)$ or just $\operatorname{Cone}(f)$ as defined in §4.4.3. Thus, it is already present.

The cone construction encompasses almost all of the other deformations we might consider. For example, a complex itself can be considered an iterated cone:

$$E = \cdots \operatorname{Cone}(d_2 : \operatorname{Cone}(d_1 : \operatorname{Cone}(d_0 : E^0 \to E^1) \to E^2) \to E^3) \cdots, \tag{5.53}$$

where we think of an object as a complex with a single entry.

5.3.2.3. *Correlation functions and A_∞ algebra.* To complete the construction of the topological open string theory including complexes, we must give the definition of correlation functions of open string operators. To define 3-point functions as in (3.176) and (3.193), we need to be able to multiply morphisms and take a trace.

If we take the B-branes to be complexes of coherent sheaves, we can use the definitions of these operations for $\mathbf{K}(\operatorname{Coh} X)$ given in Chapter 4. In particular, the definition of the trace follows from Serre duality (Theorem 4.68) and the fact that the dualizing sheaf ω_X is trivial for X a Calabi-Yau manifold. Taking the first morphism in the perfect pairing to be the identity, we get a trace

$$\operatorname{Tr} : \operatorname{Ext}^n(E, E) \to \mathbb{C}$$

unique up to overall normalization.

A simple corollary of Serre duality which we will use explicitly below is

COROLLARY 5.11. *Any morphism* $\alpha \in \mathrm{Ext}^*(E, F)$ *which is zero in all correlation functions is identically zero.*

To define higher-point correlation functions, one is best off developing further structure. In particular, one does not want to go directly to BRST homology, but rather think of the open strings as elements of a differential graded algebra. This allows showing directly that the higher-point correlation functions are related to the higher products in the associated A_∞ algebra. It also facilitates the generalization beyond coherent sheaves, by basing the discussion more directly on SCFT and its topological twisting. We refer to [**22**] for these points.

5.3.3. The derived category. We have now seen that topological open string theory contains all of the objects and morphisms of the homotopy category $\mathbf{K}(\mathcal{A})$. Following the logic of §4.4.2, if we can next give a physical argument that quasi-isomorphic objects in $\mathbf{K}(\mathcal{A})$ are indistinguishable as boundary conditions, we will have shown that the category of B-branes is the derived category $\mathrm{D}(\mathcal{A})$.

The derived category will have drastically fewer objects than $\mathbf{K}(\mathscr{A})$. On the other hand, it contains far more than the naive graded direct sums of objects from $\mathscr{A}$. To see this, suppose we try to replace a complex E with the sequence given by the cohomology itself, i.e., the following complex with zero morphisms:

$$\cdots \xrightarrow{0} H^0(E) \xrightarrow{0} H^1(E) \xrightarrow{0} H^2(E) \xrightarrow{0} \cdots, \tag{5.54}$$

In general, there will be no morphism in either direction between E and (5.54), and thus a complex need not be quasi-isomorphic to its cohomology. This very important fact leads to the intricate structure of the derived category.

5.3.3.1. *The derived category and K-theory.* Let us compare this discussion with that of §5.1. We saw in §5.3.2.2 that we could deform two B-branes $E[-1]$ and F into a single B-brane represented by the cone of a morphism $f : E[-1] \to F$. Thus the B-brane $\mathrm{Cone}(f)$ is composed of $E[-1]$ and F, where $E[-1]$ is an anti-E.

Thus, we can apply the definition of §5.1, regarding elements of $\mathrm{D}^b(X)$ as particles, and all of the deformations

$$\begin{aligned} [\mathrm{Cone}(f)] &\leftrightarrow [F] + [E[-1]] \\ &\leftrightarrow [F] - [E], \end{aligned} \tag{5.55}$$

as bound state/decay processes. The resulting group $K(X)$ is the Grothendieck group of X (see page 77 of [**295**] for more details). Its application to D-branes can be traced back to [**225, 425**].

We may naturally map the derived category to K-theory as follows. Using locally free resolutions we may replace any complex by a quasi-isomorphic complex E of locally free sheaves. We may then construct the K-theory object

$$\cdots \ominus E^{-1} \oplus E^0 \ominus E^1 \oplus E^2 \ominus \cdots, \tag{5.56}$$

where E^i is the holomorphic vector bundle associated to E^i. One can show that this leads to a well-defined map $\mathrm{D}(X) \to K(X)$. Note that this map need not be surjective. $K(X)$ is generated by all vector bundles, whereas starting only with B-branes we will get the K-theory of holomorphic vector bundles. The full K-theory might require non-holomorphic bundles, corresponding to non-BPS branes, in order to generate all possible classes [**471**].

Now we can compare the present discussion with the semisimple case discussed in Chapter 2. The categories which appeared there were abelian, with objects corresponding to complex linear vector spaces and morphisms corresponding to linear maps. In this case, there is *always* a quasi-isomorphism between a complex and its cohomology, and thus the derived category takes a simple form. Since every isomorphism class of objects is determined by its cohomology, all of its content is contained in the K-theory of §5.1.

However, in any category of nonzero homological dimension, we will not have this simplification. For example, whereas all 0-branes on X are represented by the same K-theory class, two 0-branes corresponding to distinct points in X are associated to non-isomorphic objects in $\mathrm{D}^b(X)$.

5.3.3.2. *Physical identity and quasi-isomorphisms.* From a physics point of view, why should we quotient by quasi-isomorphisms? While one can argue from the consistency of the results derived from this assumption, it would be better to find a deductive argument based on physical axioms and observables.

Definition 4.45 (a quasi-isomorphism is a morphism which preserves cohomology) is not an entirely satisfactory starting point, as the cohomology of a complex is not a physical observable.

In TFT, the physical observables are the spectrum of open strings and their correlation functions. Thus, we need to define a notion of a *physically equivalent* pair of branes, meaning a pair of branes which behave the same way in all correlation functions. The physically motivated category of branes is then

DEFINITION 5.12. The category $\mathcal{T}(\mathscr{C})$ is the quotient of $\mathbf{K}(\mathscr{C})$ by physical equivalence, where

DEFINITION 5.13. Two objects $E, E' \in \mathbf{K}(\mathscr{C})$ are *physically equivalent* if and only if

$$\begin{array}{rcl} \mathrm{Ext}^p(E,F) & \cong & \mathrm{Ext}^p(E',F) \ \text{ and} \\ \mathrm{Ext}^p(F,E) & \cong & \mathrm{Ext}^p(F,E') \end{array} \qquad \forall F \in \mathbf{K}(\mathscr{C}), \tag{5.57}$$

where these equivalences preserve the 3-point correlation functions (3.193). For example, adopting the notation of (3.192) and taking $E = E_2$, $E' = E'_2$ and

$$
\begin{aligned}
a' &\in H^{0,p}_{\bar\partial}(X, \mathrm{Hom}(E_1, E'_2)),\\
b' &\in H^{0,q}_{\bar\partial}(X, \mathrm{Hom}(E'_2, E_3)),\\
c &\in H^{0,r}_{\bar\partial}(X, \mathrm{Hom}(E_3, E_1)),
\end{aligned} \tag{5.58}
$$

we have

$$\langle W_a W_b W_c \rangle = \langle W_{a'} W_{b'} W_c \rangle \qquad \forall E_1, E_3.$$

By the OPE (or equivalently associativity) the condition on 3-point functions forces equality of all n-point functions, and thus all physical observables. This is clearly an equivalence relation.

The main result is then

THEOREM 5.14. *(Aspinwall-Lawrence* [**23**]*)* $\mathcal{T}(\mathscr{C}) \cong \mathrm{D}(\mathscr{C})$.

To prove this, recall that, by the discussion of §4.4, the derived category $\mathrm{D}(\mathscr{C})$ is the universal category in which quasi-isomorphisms of $\mathcal{C}(\mathscr{C})$ map into isomorphisms. Thus, a functor $F : \mathcal{C}(\mathscr{C}) \to \mathcal{T}(\mathscr{C})$ with this property will uniquely factor through $\mathrm{D}(\mathscr{C})$ as

$$F = G \circ Q,$$

where Q is the localization functor of Lemma 4.46. We thus want to show that the natural inclusion of $\mathbf{K}$ in $\mathcal{T}$ maps quasi-isomorphisms into isomorphisms, and that G is an equivalence of categories.

PROPOSITION 5.15. *The natural inclusion of* $\mathbf{K}$ *in* $\mathcal{T}$ *maps quasi-isomorphisms into isomorphisms.*

PROOF. Given a quasi-isomorphism $f : E \to E'$, we form its mapping cone $C(f)$ as in §4.4.3. This is acyclic (has zero cohomology), so $\mathrm{Hom}^*(C(f), F) \cong \mathrm{Hom}^*(F, C(f)) \cong 0$ for any F, and (5.57) follows. □

PROPOSITION 5.16. *G is an equivalence of categories.*

PROOF. This could only fail if some pair E, E' of inequivalent objects in $\mathrm{D}(\mathscr{C})$ were physically equivalent, in other words shared the same morphisms and correlation functions with all other objects. However, the condition (5.57) implies the existence of morphisms $\alpha \in \mathrm{Hom}(E, E')$ and $\beta \in \mathrm{Hom}(E', E)$ whose product $\beta\alpha$ behaves in correlation functions exactly like the identity $\mathrm{id}_E \in \mathrm{Hom}(E, E)$. Thus there would exist a morphism $\beta\alpha - \mathrm{id}_E$ which is zero in all correlation functions, contradicting Corollary 5.11. □

Thus, if we can construct B-branes which form an abelian category $\mathcal{A}$, these arguments define a topological open string theory corresponding to

$D(\mathcal{A})$. In some cases the B-branes clearly do form an abelian category, for example in the Landau-Ginzburg models discussed in §3.6.8.

We will also see in §5.4 that the world-volume construction of Dirichlet branes leads directly to a description of B-branes as objects in a category of quiver representations, which is an abelian category. For example, the orbifold construction of §5.4.5 is in this class.

5.3.3.3. *Resolutions and coherent sheaves.* Now let us focus on the B-model defined by twisting the sigma model on X as in §3.6.3. While we have a large supply of B-type boundary conditions, all of the holomorphic bundles on X, these do not form an abelian category, since cokernels of maps need not be bundles. Is this a problem for the argument we just gave?

Of course, there are other B-type branes, for example the D0, and those supported on holomorphic submanifolds. Including these suggests that the correct starting point for the sigma model is to take $\mathcal{A} = \operatorname{Coh} X$, the coherent sheaves. But while this is an abelian category, solving the problem, one might worry at this point that this step does not have a clear physics motivation, opening the possibility that physics might make a different choice. How do we know that coherent sheaves are the right generalization to make?

A direct approach to this question would be to define all of the B-branes as explicit boundary conditions in the sigma model, and check that they have the same properties as the coherent sheaves. We will discuss some results of this type below, but at present a completely general analysis of this type has not been made.

Nevertheless, there is a good argument that this must work, and all of the B boundary conditions are objects in $D^b(X)$. It is that within the boundary conditions we have already described how to construct (complexes of locally free sheaves), there are perfectly good candidates for the D0 and all of the other B-type branes supported on submanifolds. These are complexes arranged so that all of the structure outside of a submanifold $Z \subset X$ cancels out of the Q-cohomology.

For example, consider the complex E_Z given by

$$\mathcal{I}_Z \xrightarrow{f} \mathcal{O}_X,$$

where (as in Example 4.42), $Z \subset X$ is a subvariety defined by the vanishing of a single function f on Z. In this case, $\mathcal{I}_Z$ is a locally free sheaf.

Let us now look at the morphisms between E_Z and a D0-brane at a point $p \in X$. The D0 can be defined mathematically as the structure sheaf of the point $\mathcal{O}_p$, or physically as a purely Dirichlet boundary condition. Let us compare these. Mathematically, it is a simple exercise to check that, unless $p \in Z$, all of these morphisms cancel out of the Q-cohomology. On the other hand, we argued in §3.6.5 that the same was true from the physical definition.

Looking back at Example 4.42, we see that E_Z is a resolution of the structure sheaf $\mathcal{O}_Z$, a coherent sheaf on X which would be the natural mathematical object to identify with a physical D-brane wrapping the submanifold Z. And since the coherent sheaf and its resolution are quasi-isomorphic, we do not need to add the coherent sheaf $\mathcal{O}_Z$ explicitly in constructing our category of topological branes; it is already there. This argument generalizes to any coherent sheaf supported on any subvariety Z, and shows that a corresponding topological brane already exists and must correspond to some physical brane wrapping Z.

What is the physical D-brane corresponding to $\mathcal{O}_Z$? At first, one might say that the natural candidate is a D4-brane wrapping Z carrying a trivial bundle. However, this is not obviously right; we might need to know more about the geometry of Z to make this identification. As cited in §5.1.4, other physics arguments based on the relation to closed string theory give us an identification on the level of K-theory, (5.12), in which the normal and tangent bundles of Z enter. For present purposes, the precise form of the relation is not so important, what is important is that $\mathcal{O}_Z$ can be identified with a D4-brane wrapping Z carrying *some* bundle (which can be read off from (5.12)).

But let us now suppose for the sake of argument that this were *not* true. In other words, suppose that starting from a more direct physical definition of the D4-brane, in terms of mixed Dirichlet-Neumann boundary conditions, we obtained results which differed from those predicted by the identification with a coherent sheaf, say some morphisms were different. In this case, since we just argued that the sheaf *is* present in the topological open string theory, we would have *two* distinct objects corresponding to a brane wrapping Z, the coherent sheaf and the physical D4-brane. This would be quite surprising from both a physical point of view (since no evidence for this was seen in the many cases which have been analyzed) and from a mathematical point of view, since there is no known candidate for a category naturally associated to X, and properly containing $D^b(X)$.

Of course this argument falls short of a proof. However a proof is not at present within reach, as part of this would include an argument that all possible physically consistent boundary conditions have been considered, which (for $\hat{c} > 1$ theories) is not yet possible within physics.

5.3.3.4. *Comparison of Dirichlet branes and coherent sheaves.* Let us summarize the conclusion of the previous subsection. We consider the full subcategory of $\mathrm{D}^b(X)$ whose objects are complexes of locally free sheaves. This subcategory (let us call it $\mathrm{D}^b_0(X)$) is equivalent to $\mathrm{D}^b(X)$, since any coherent sheaf has a bounded locally free resolution. On the other hand, there is a functor F from $\mathrm{D}^b_0(X)$ to the category of B-branes which assigns to a complex the corresponding B-brane and identifies the corresponding spaces of morphisms. This functor, by definition, is full and faithful.

The claim of §5.3.3.3 is that it is also essentially surjective, i.e., that any B-brane is isomorphic, as an object of the category of B-branes, to an object in the image of F. If this were the case, the category $\mathrm{D}_0^b(X)$ (and therefore $\mathrm{D}^b(X)$) would be equivalent to the category of B-branes.

In some cases, this has been checked by computing spaces of morphisms between particular B-branes and comparing with spaces of morphisms between corresponding objects in $\mathrm{D}^b(X)$. We discussed the case of the D0-brane in §3.6.5, concluding that it corresponds to $\mathcal{O}_p$, the skyscraper sheaf supported at p. The endomorphism algebra of the skyscraper sheaf in $\mathrm{D}^b(X)$ is the exterior algebra of TX_p, just as was obtained in (3.194).

An even simpler check is to compute the space of morphisms from a B-brane corresponding to a holomorphic vector bundle E to the D0-brane at p. In this case, the boundary conditions force all fermions to vanish, and boundary operators can be identified with elements of the vector space E_p^*. This again agrees with morphisms between the corresponding objects in the derived category.

One can easily generalize this to the case of B-branes corresponding to complex submanifolds in X. Let Y be such a submanifold. The boundary conditions corresponding to such a B-brane force the bosonic fields to take values in Y, and force $\eta^{\bar{i}}$ and θ_i to take values in $TY^{0,1} \subset TX^{0,1}|_Y$ and $NY^* \subset T^*X^{1,0}|_Y$, respectively. Candidate BRST invariant boundary operators are therefore sections of

$$\bigoplus_{p,q} \Lambda^p NY \otimes \Omega^{0,q}(Y).$$

One can show that the BRST operator acts as the Dolbeault operator $\bar{\partial}$. Thus the space of endomorphisms of such a B-brane is the Dolbeault cohomology of the holomorphic vector bundle ΛNY. This again agrees with morphisms in $\mathrm{D}^b(X)$ if we assume that this B-brane corresponds to the sheaf $i_*\mathcal{O}_Y$, where i is the embedding $i : Y \hookrightarrow X$.

There is a slight surprise when one computes morphisms between a B-brane corresponding to a holomorphic vector bundle E and a B-brane corresponding to a complex submanifold Y [**298**]. The answer expected from the derived category is the Dolbeault cohomology of the holomorphic vector bundle $E^*|_Y$, while the physical computation gives the Dolbeualt cohomology of $E^*|_Y \otimes K_Y^{1/2}$. This suggests that the B-brane corresponding to Y should be identified not with $i_*\mathcal{O}_Y$, but with $i_*K_Y^{1/2}$, where K_Y is the canonical line bundle of Y. Note that the square root of K_Y exists only if Y is spin, and is not unique if Y is not simply-connected. This seems to run counter to our desire to identify B-branes with objects of the derived category, since arbitrary (not necessarily spin) complex submanifolds Y are valid objects of $\mathrm{D}^b(X)$. The way out of this paradox has been pointed out in [**156, 298**] and involves the possibility of having a nontrivial gauge field on the brane. To see how this works, consider first the situation where Y is spin, but carries a vector bundle E'. Then we must assign to it a coherent sheaf $i_*(E' \otimes K_Y^{1/2})$, where E' is the locally free sheaf on Y correponding to the vector bundle E'. Now, if Y is not spin, then according to [**156, 298**] both E' and $K_Y^{1/2}$ are twisted coherent sheaves such that $E' \otimes K_Y^{1/2}$ is an ordinary coherent sheaf on Y. The object in $\mathrm{D}^b(X)$ corresponding to such a B-brane is $i_*(E' \otimes K_Y^{1/2})$.

Further computations in [**298**] and [**289**] confirm that morphisms between B-branes agree with morphisms between the corresponding objects in $\mathrm{D}^b(X)$. While all these computations do not constitute a proof, they strongly suggest that the category of B-branes is equivalent to $\mathrm{D}^b(X)$.

5.3.3.5. *Mirror symmetry restored?* If the A-model on Y is "the same" as the B-model on X, then it would appear that we have motivated the proposal that the Fukaya category on Y is equivalent to $\mathrm{D}^b(X)$, the derived category on X. That was Kontsevich's original proposal. It turns out that there is still a small fly in the ointment, as we discuss in §5.5.2, but this proposal seems to be very close to the truth. Let us conclude by noting a few miscellaneous features of how well this mirror symmetry works.

The tadpole cancellation condition in §3.6.2.5 produces two interesting aspects of the moduli space of A-branes:

(1) First-order deformations of the Lagrangian, which correspond to $H^1(L)$, may be obstructed and do not lead to genuine A-brane deformations.
(2) Some A-branes may depend on very special values for $B + iJ$ and disappear completely for generic $B + iJ$.

The mirror statements in the B-model are both true:

(1) The first-order deformations of coherent sheaves, which correspond to $\mathrm{Ext}^1(E, E)$ can be obstructed. We refer to [**444**] for examples.
(2) There are some sheaves which only exist for special values of complex structure. An example of this is given by 4-branes wrapped around a surface ruled over an algebraic curve of positive genus in X [**462**].

This subject was also analyzed in [**276, 277**]. Note that this is a typical example of mirror symmetry in that instanton effects (i.e., tadpoles) in the A-model are mapped to effects in the B-model that can be understood from classical geometry.

5.3.4. Gradings and boundary conditions. Let us now explain the physical origin of the gradings μ we used to define complexes in more detail, and their relation to the gradings ξ which appeared in §5.2.1. We should say at the start that this topic touches on a great deal of physics, such as supersymmetry, BPS bounds, and bosonization. To make the presentation reasonably self-contained, we focus on what is essential for the purpose at hand.

Clearly, a correct definition should make sense in both the A-brane and the B-brane language. Indeed, since it must be mirror symmetric, we should look in the underlying SCFT structure responsible for mirror symmetry.

Recall from §3.2.4 and §3.4.3.2 that the SCFT origin of mirror symmetry is in the factorization of the $U(1)$ subalgebra of the $\mathcal{N} = 2$ SCA. In particular, recall from §3.3.4.6 that the $\mathbb{Z}$-gradings are defined entirely within the $U(1)$ subalgebra. Factorization thus tells us that the theory of these gradings is universal, and can be understood by appealing to the theory of any $U(1)$ CFT, including the simple free boson theory with target space S^1 which we discussed at length in Chapter 3.

We know the complete set of boundary conditions in this theory. They are Dirichlet and Neumann, in each case described by a single real parameter. Comparing (3.165) and (3.166) with (3.139), we see that an A-type brane corresponds to the Dirichlet case, while B-type corresponds to Neumann. Actually, since the action of mirror symmetry (3.134) is so simple in this description, the discussion in the two cases is entirely parallel.

As we suggested in §5.3.1, in either case, the grading μ is (up to a multiplicative factor) the real parameter in the boundary condition. In the A-brane or Dirichlet case, we can picture this parameter as the position of a D0-brane on the S^1 target space of the free boson. For the Neumann case, we can think of it as either the holonomy of a $U(1)$ connection, or as a position on a T-dual S^1, which may be simpler to visualize.

The most basic check of this claim is that the R-charges of open strings are given by the formulae (3.189) and (5.41), where μ is the parameter defined here. Given this fact, since R-charge is conserved, the resulting structure will be a graded category.

The argument for this fact is slightly simpler in the Dirichlet picture, so let us give it there. We use the bosonized description of the $U(1)$ current algebra from §3.2.3.5, as applied in §3.3.4.6. From (3.167) and (3.102), the R-charge of an open string from a brane E to a brane F is the expectation value of the operator

$$J_0 = \sqrt{\hat{c}} \int_0^\pi \partial_\sigma \varphi$$

integrated across the open string, with boundary conditions $\varphi(0)$ and $\varphi(\pi)$ determined by the real parameters of the E and F boundary conditions respectively.

Since this expression is the integral of a local density, it is intuitively clear that when two strings attach at their endpoints, their R-charges add. Furthermore, since the density is a total derivative, it can be integrated. To do this, we decompose φ as in (3.54), into an $\mathbb{R}$-valued function and a quantized "winding number" $p = m$; then

$$J_0 = \sqrt{\hat{c}}\,(p + \varphi(\pi) - \varphi(0))\,. \tag{5.59}$$

This has the form (5.41) with $q = \sqrt{\hat{c}}m$ and

$$\mu(B) = \sqrt{\hat{c}}\varphi(B) \tag{5.60}$$

for $B = E, F$. In other words, the particular Dirichlet boundary condition on φ for a brane B is just its grading (up to an overall conventional factor).

5.3.4.1. *Periodicity of the grading.* Let us begin with the case $E = F$ and $\varphi(0) = \varphi(\pi)$. In this case we recover $q = \sqrt{\hat{c}}p$ as in (3.103). Thus, p must be quantized in units of $1/\hat{c}$ to agree with the R-charge conventions of §3.3.2 (in particular, Q has charge 1). From §3.2.3.5, this implies that the boson φ has the action (3.59) with radius $R = 1/\sqrt{\hat{c}}$.

Combining this with (5.59), we see that the parameter $\mu(B) = \sqrt{\hat{c}}\varphi(B)$ has an effective periodicity $2\hat{c}$, meaning that such a shift is a symmetry of the SCFT. In other words,

$$\mu(B) \in \mathbb{R}/2\hat{c}\mathbb{Z}. \tag{5.61}$$

The same holds for $E \neq F$. The only change this makes is to produce a constant overall shift in (5.59), corresponding to the difference of gradings in (5.41).

A careful reader may be a bit confused at this point, as previously we argued that mirror symmetry included the operation of T-duality on this boson, which by §3.2.3.6 acts as $R \to 1/R$. How can mirror symmetry be consistent with any value other than $R = 1$?

The answer is that the $U(1)$ sector is not independent of the rest of the (2, 2) SCFT. Rather, factorization implies a correlation between the $U(1)$ charges and the operator spectrum of the rest of the theory, as expressed in (3.70). In an open string sector, there is a single conserved $U(1)$ charge, so this takes the form

$$\mathcal{H}_{E,F} = \bigoplus_p \mathcal{H}^{X'}_{E,F,p} \otimes V_p. \tag{5.62}$$

In doing the T-duality transform, we must take these correlations into account. While operator relations such as those of §3.2.3.7 are still valid, this changes (3.66) and subsequent steps of the analysis.

Indeed, because of the correlations in (5.62), it is not really correct to think of φ as taking values in an S^1 of radius $R = 1/\sqrt{\hat{c}}$. Rather, the effective radius is the parameter R_{eff} for which we have an isomorphism

$$\mathcal{H}^{X'}_{E,F,p} \cong \mathcal{H}^{X'}_{E,F,p+R_{\text{eff}}},$$

as this is the shift $\varphi \to \varphi + R_{\text{eff}}$ which is a symmetry of the theory. From the discussion at the end of §3.3.5, the minimal shift which is guaranteed to be a symmetry is the spectral flow with $\Delta q = 2\hat{c}$, and thus we conclude that $R_{\text{eff}} = 2\sqrt{\hat{c}}$.

5.3.4.2. *Gauge symmetry of grading.* Although we went into it in some detail, it is believed that the periodicity $\mu \to \mu + 2\hat{c}$ is not that important, and one can think of the gradings μ as real-valued. This is because the entire SCFT has a $U(1)_R$ symmetry, generated by the current J preserved by the boundary conditions of §3.6. This acts as an overall shift of the boson φ and thus of all of the gradings, say

$$\mu(B) \to \mu(B) + n$$

for any $n \in \mathbb{R}$. Consistent with this, our mathematical framework is left essentially unchanged by an overall shift: for example

$$\text{Hom}(E, F) = \text{Hom}(E[n], F[n]), \tag{5.63}$$

with all products unaffected.

Physically, one says that this global shift of the complexes is a gauge symmetry of the theory. Once we set the grading of a single brane, the gauge symmetry is "fixed," and there is no more ambiguity.

As we discussed in §3.3.5, non-vanishing open string correlation functions must have total $U(1)_R$-charge $\hat{c}$. Recalling from (3.97) and (3.98) that the individual charges satisfy $0 \leq q \leq \hat{c}$, no correlation function can actually see the $2\hat{c}$ periodicity, nor is any other effect of it known.[5] Thus we generally neglect the periodicity from now on; it would be desirable to have a proof that this is valid.

5.3.5. Gradings and central charge. We now have two numbers associated with a boundary condition. On the one hand, we have the parameter μ of (5.60), the $U(1)_R$ part of the boundary condition. On the other, the discussion of §3.6.6 (see also §5.2.1.1) involved matching conditions (3.196) and (3.197), depending on a parameter ξ. This appeared in the special Lagrangian condition as (5.19), while in SCFT terms, both the A and B cases can be summarized in the expression

$$\Omega = \exp(-2i\pi\xi)\bar{\Omega}. \tag{5.64}$$

In fact, both of these parameters are the same,

$$\xi = \mu. \tag{5.65}$$

This has several important consequences. The first is that, while the definition (5.16) for ξ suggests that ξ is only defined modulo 2, from (5.61) it must in fact be defined modulo $2\hat{c}$.

Formally, this is easily done by continuity. Recall from §5.2.1.2 that, considering variations of complex structure, a special Lagrangian exists in an open region of complex structure moduli space. Furthermore it is clear

[5] The claim in [**123**] is incorrect, see [**18**].

from (5.18) that $|Z| \neq 0$ throughout this region, as this is its volume. Thus a choice of branch of (5.16) at one point in the region determines it throughout.

For special Lagrangians, one can justify this formal argument by local models as in §5.2.1.2, but this is not obviously a correct argument in SCFT. The combination of (5.65) with the arguments of §5.3.4.1 is valid in SCFT.

5.3.5.1. *Physical arguments.* (5.65) says that the boundary condition, which must "reflect" left movers to right movers as in §3.6, also does a $U(1)_R$ transformation parameterized by ξ. The simplest way to prove this uses the vertex operator description of bosonization. We leave the detailed explanation of this to [**112**] and MS1 §11.3, as it requires more CFT background, but the argument is short and should be intuitively plausible. One can write the spectral flow operators in terms of a decomposition of the boson φ into left and right movers as

$$\Omega =: e^{i\sqrt{\hat{c}}\varphi_L} : , \qquad \bar{\Omega} =: e^{-i\sqrt{\hat{c}}\varphi_R} : . \tag{5.66}$$

The left and right movers $\varphi_{L,R}$ are defined by the relations

$$\partial_- \varphi_L = \partial_+ \varphi_R = 0; \qquad \varphi_L + \varphi_R = \varphi.$$

These determine them up to an overall constant, the boundary conditions (5.60) for φ in terms of μ. Using (5.66), this determines ξ in (5.64).

Recall that, in the special Lagrangian context, ξ was also the complex argument of the central charge (5.16). While this equation and the definition (5.17) of the BPS central charge involve geometry, if we combine the two as

$$\xi(L) = \frac{1}{\pi} \arg Z(L) \pmod 2, \tag{5.67}$$

this statement will generalize to SCFT without change.

This claim directly implies the previous one, that the gradings of open strings in the A-model depends on complex structure, and the mirror statement.

5.3.5.2. *Computation of central charge in SCFT.* This can be done by starting from the definition (5.17) and translating the computation into sigma model language, then generalizing.

Let us consider an A-type brane for definiteness. It should be plausible (and is easily checked) that the holomorphic $(n, 0)$-form Ω in (5.17) is represented in the sigma model by the operator Ω of (3.196).

Then, the integral of a form over a brane world-volume translates into an expectation value on a disk diagram with the brane as boundary condition. In the sigma model, this is clear because the functional integral over the disk world-sheet includes an integral over the zero modes (or center of mass), constrained to respect the boundary conditions as in §3.5. This expectation value makes sense in general SCFT, and there is no other candidate generalization.[6]

[6]One can give a better proof but at the cost of introducing far more physics.

Thus, $Z(L)$ is an expectation value computed on a disk world-sheet with boundary L, and the operator Ω inserted at the puncture. One can check that this respects the R-charge anomaly (3.110).

5.3.5.3. *The $A \leftrightarrow B$ switch.* Now, the main point leading to (5.67) is that, although we started out with an A-type brane, this is actually a calculation in the B-twisted model. Given the claim that $Z(L)$ is computable in a topologically twisted model at all, this is inevitable as we know that it depends only on complex structure moduli in the large volume limit. Nevertheless it is reassuring to show this explicitly on the world-sheet, as was done in [**382**].

Unlike our previous computations involving A-type branes, this is actually a closed string calculation. This is evident from the usual physical starting point, which is the space-time (supergravity) definition of the BPS central charge, because the supergravity fields which enter into this definition arise in the closed string sector. To see it from the world-sheet, we redraw the disk, enlarging the puncture to a second boundary as in Figure 4. The result is naturally thought of as a closed string diagram, the annulus.

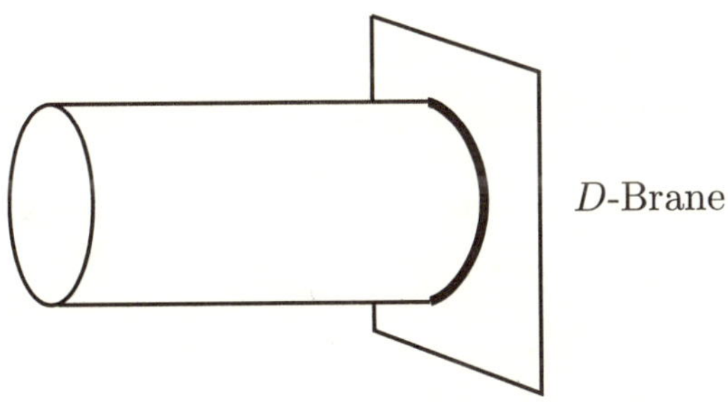

FIGURE 4. The annulus with a D-brane boundary and an external closed string.

Now, in going from an open string diagram to a closed string diagram, we need to change our interpretation of the boundary conditions from having a definite "spatial" coordinate (the end of the open string moves in time and is thus fixed in space), to a boundary in "time" (the origin of a loop of closed string emitted by the D-brane). Then, if one considers the transformation laws of the operators in the SCA under such a rotation, one finds that this rotation actually *exchanges* the A and B boundary conditions.

The simplest way to see this is to consider the A-brane boundary condition on the $U(1)$ current (3.165), which was $J_L = -J_R$. In the open string sector, this is consistent with (3.100), $Q_A = G_{+,0} + \tilde{G}_{-,0}$.

On going to the closed string sector, we reverse the relation between tangential and normal directions, and time and space directions in (3.138).

Since it is the latter which defines "left" and "right" movers, this reverses the sign of J_R compared to the closed string conventions, and we now have $J_L = J_R$.[7]

Thus, for the purpose of computing a central charge, the A boundary condition is used with the topological B-model, and vice versa.

5.3.5.4. *Central charges for A-branes.* The most important consequence of this follows from the fact that B-model closed string amplitudes are unaffected by quantum corrections. Thus, the large volume definition (5.17) of the central charge for A-branes is exact, receiving no α' corrections.

While the meaning of (5.17) is clear for an A-brane defined as a special Lagrangian manifold, one might wonder how to interpret it for the more general A-type boundary conditions of §3.6.1, or for possible "non-geometric" boundary conditions (say those defined in a Gepner model).

Physically, the answer to this question is again clearer from space-time (supergravity) considerations. If we consider a type IIB string theory compactification on the Calabi-Yau X, we obtain a $d = 4$, $\mathcal{N} = 2$ supergravity theory. In such a theory, there is a general formula for the central charge of a BPS particle, generalizing (5.17),

$$Z(L) = Q_i \Pi^i. \tag{5.68}$$

Here Q_i is a vector of electric and magnetic charges, while Π^i is a complex vector determined by the "vector multiplet" structure of the supergravity theory.

Now, each A-type D-brane will correspond to a BPS particle in this theory (see §5.4.2.1), and thus we can make contact with the general discussion of §5.1.2. There we argued that the RR-charge (or K-theory class) of an A-brane is an element $[L] \in H_3(X, \mathbb{Z})$. The formula (5.68) is obtained by expressing (5.17) using an explicit basis Σ^i for $H_3(X, \mathbb{Z})$, and writing

$$[L] = \sum Q_i \Sigma^i \tag{5.69}$$

and

$$\Pi^i = \int_{\Sigma^i} \Omega. \tag{5.70}$$

For a special Lagrangian L, this agrees with (5.17). But this argument implies that *all* A-type branes in any SCFT admit such a representation for the central charge.

The relation between (5.68) and the world-sheet computation we discussed in §5.3.5.3 is that Q_i expresses the K-theory class of the boundary condition on the disk, while the vector Π^i is a complete basis for the one-point functions of the operator Ω. Thus, we have translated this world-sheet computation into geometric terms. Furthermore, given that these one-point

[7] A more formal argument for this uses the tensorial transformation properties of $J_L dz$ and $J_R d\bar{z}$ under the rotation $z \to e^{i\pi/2} z$.

functions are independent of Kähler moduli, if there is a large volume limit, we can evaluate them there. Thus the definition of Π^i as a complete set of periods of Ω can be used in SCFT as well, at least if it has a large volume limit.

Actually, there are $(2,2)$ SCFT's without a strict large volume limit, because they have no Kähler moduli at all. These are the mirrors of rigid Calabi-Yau manifolds.[8] Such an SCFT cannot be a nonlinear sigma model, since the target space would have to have $b^{1,1}=0$. On the other hand, the central charges (5.70) still satisfy the relations they must to arise as periods of a complex variety, and it has been argued that in such cases they can be represented as periods for a higher dimensional complex variety [**83**].

Giving a non-geometric version of (5.69) requires discussing a particular SCFT. Examples in which this has been done include Gepner models [**69**], and orbifolds [**115, 137**].

5.3.5.5. *Central charges for B-branes.* These will be computed in the topological A-model, and as such can get stringy α' corrections. However, let us begin by ignoring these, which should be a good approximation in the large volume limit. In this case, we have the approximate expression

$$Z(E) = \int_X e^{-(B+iJ)}\mathrm{ch}(E)\sqrt{\mathrm{td}(X)} + \mathcal{O}(\alpha') \tag{5.71}$$

where $\mathcal{O}(\alpha')$ represents the stringy corrections.

The dependence on the K-theory class of the brane is clear from §5.1.4.1 and the formula (5.69). In addition we are asserting that

$$\Pi_i = \int_{\Sigma_i} e^{-(B+iJ)} + \mathcal{O}(\alpha'), \tag{5.72}$$

where Σ_i is a complete basis of $H_{2p}(X,\mathbb{Z})$.

Granting various facts from string theory, there is a simple argument for this. First, we saw in §3.5.2.7 and §3.6.1 that the curvature F of a bundle associated to a D-brane must always come in the combination $B-F$ with the two-form potential B, by the gauge symmetry (3.150). Thus, since $\mathrm{ch}(E) = \mathrm{Tr}\, e^F$, (5.71) must contain the combination $e^{-B}\mathrm{ch}(E)$. Furthermore, by general properties of $\mathcal{N}=2$ supergravity, the central charge $Z(E)$ must be a holomorphic function of the complexified Kähler moduli $B+iJ$. This forces (5.71) and its corollary (5.72).

Later we will derive a formula (5.142) for $\xi(E)$ in the large volume limit, by substituting (5.71) into (5.67) and expanding in $1/J$. Using continuity this will determine $\xi(E)$ modulo $2\hat{c}$ whenever α' corrections can be neglected.

---

[8]It is conceivable that there are $(2,2)$ SCFT's with both types of moduli but without a large volume limit, but none are known.

In fact the physics argument prohibits almost all power-like α' corrections to (5.71) (equivalently (5.72)). However it is clear from mirror symmetry (or the arguments in §3.4) that there still must be stringy corrections. These were first discussed in [**85**] and appear in two ways:

(1) The perturbative corrections will be due to the 4-loop correction in the non-linear sigma model as analyzed in [**198**]. These corrections will be three powers of $B + iJ$ less than the leading term in (5.71). Thus, in the case of Calabi-Yau threefolds this produces at most a constant term.

(2) The nonperturbative corrections will produce a power series, with no constant term, in $q_i = \exp 2\pi i(\int_{C_i} B + iJ)$ for some basis $\{C_i\}$ of $H_2(X)$.

Rather than go into the details of how to compute these corrections, the simplest way to deal with this point is to finesse it by simply *granting* mirror symmetry, and using it to define B-brane central charges in terms of the A-brane definition we just gave, as outlined in §3.4.3. While at first this might sound like cheating, from a physical point of view (as we discussed in §3.4.3.2) mirror symmetry is manifest in SCFT, so once we start basing our considerations on SCFT, this is entirely legitimate. Of course, we must still justify each step in which we identify some SCFT definition with one made in more conventional geometry.

Thus, we adopt as the primary definition of a B-brane central charge the formula (5.68). Besides mirror symmetry, this also follows from considering the B-branes as BPS particles in type IIA string theory compactification on the Calabi-Yau X. Many of the physical arguments are done using this language, but we will not rely on it here.

From closed string mirror symmetry, we infer that the periods Π^i can be defined as a complete basis of periods (5.70) for the holomorphic n-form Ω on the *mirror* Calabi-Yau Y. Once one sets up the SCFT computation, this is evident, as mirror symmetry simply turns one set of vertex operators into another.

5.3.5.6. *Central charges for complexes.* Having defined central charge for individual A- and B-branes, we can infer it for a complex by linearity:

$$Z(E) = \sum_n (-1)^n Z(E_n).$$

This follows directly from (5.68), the fact that the charges Q_i are additive conserved charges, and the fact that the grading only enters the computation of Q_i through its role in defining the orientation of a brane.

This determines $\xi(E)$ only modulo 2. Again, to lift it to $[0, 2\hat{c})$, we do this at a point, and then use continuity.

5.3.6. Spectral flow and gradings. All of the above was within a single SCFT, indeed a single open string sector. We now argue that there are variations of the SCFT which modify the grading of the boundary conditions, but do not affect the properties of the open strings within the topological theory.

For example, since we know that the topological B-model only depends on complex structure data, we could vary the Kähler class of the Calabi-Yau in the non-linear sigma model, without changing the spectrum or correlation functions of the open strings. The new claim is that, nevertheless, it affects their gradings. The mirror statement is that varying the complex structure in the A-model has the same effect.

Let us make such a variation of the SCFT, and consider its effect on the open strings. For example, we could take the grading a_0 of A fixed, and the grading a_1 of B to slowly increase to $a_1 + x$. In this case, we can consider the Hilbert space $\mathcal{H}_{AB}$ and any particular open string $\psi \in \mathcal{H}$ as remaining essentially unchanged during this process, except in one respect: as the endpoint of the string moves, so its total winding number increases by x.

In particular, there is no *a priori* reason that x must be an integer. And, in physical SCFT, it can be an arbitrary real number. Even from a geometric point of view, the most natural definition of grading for A-branes has this property, as we review shortly.

Let us continue to compare the notation we just introduced in SCFT with the notations we introduced on mathematical grounds in Chapter 4. Any explicit CFT definition of a boundary condition B, for example as a D-brane carrying a bundle, a fractional brane or otherwise, will come with some prescription for computing the grading of all open strings which start or end on it. We then associate an arbitrary "grading" x to B to be used **with that definition**; in other words an open string from A to B, with fermion number n computed in that definition of B, would be denoted as

$$\psi \in \mathrm{Ext}^n(A, B[x]).$$

Then, if we carry out an operation which varies the grading of B to (say) x', the $U(1)_R$-charge of the open string ψ "flows" to $n+x'-x$, and we write

$$\psi \in \mathrm{Ext}^n(A, B[x']).$$

If the grading of A shifted from (say) y to y', a similar flow would take place, now to $n + y - y'$.

Furthermore, since there is no difference in how the various ingredients a_0, a_1 and w appear in (5.41), we can if we choose follow the convention of Chapter 4 and write

$$\psi \in \mathrm{Hom}(A, B[n + x]),$$

varying all the gradings into the choice of "image" of B which the open string ends on (or begins on, for A). This is not to say that the spectrum of open strings is invariant under integer shifts of the grading. Because of the correlation (5.62), one is only guaranteed invariance under shifts by $2\hat{c}$.

5.3.6.1. *Special Lagrangians.* In this case, the flow of gradings can also be derived from the properties of Floer cohomology cited in §3.6.2.5. Recall the discussion from §5.2.1.2 of a pair of transversely intersecting special Lagrangians L_1 and L_2. We argued that there was a region in complex structure moduli space for which both branes existed, with $\xi(L_i)$ given by (5.22).

Now, if L_1 and L_2 intersect at a point p with Floer index (i.e., ghost number) $\mu(p)$, then

$$\xi(L_2) - \xi(L_1) + \mu(p) = \frac{1}{\pi}\sum_{j=1}^{m} \phi_j. \tag{5.73}$$

This follows from continuity and the following fact. Suppose L_1 and L_2 intersect at two points p_1 and p_2 and that each Lagrangian has a trivial Maslov class as in §3.6.1. Using the arguments of [**150**], one can show that the difference in $\mu(p_1)$ and $\mu(p_2)$ is equal to the difference in $\sum \phi_j$ for each point.

The equation (5.73) ties the ambiguity in defining the ghost number $\mu(p)$ discussed in §3.6.2.5, to the ambiguity in the definition of ξ. The ambiguity in $\mu(p)$ was fixed by labeling each A-brane L with some integral ghost number $\mu(L)$. Now let $L[n]$ be exactly the same A-brane as L except that we have increased its ghost number by n. It follows from (3.189) and (5.73) that

$$\xi(L[n]) = \xi(L) + n. \tag{5.74}$$

Of course the main conclusion of §5.2.1.2 was that the formation of a connected sum (or bound state) $L_1 \# L_2$ was controlled by the difference $\xi(L_1) - \xi(L_2)$. From §5.3.6, this will translate into a condition on the $U(1)_R$-charge of the open string responsible for the formation of this bound state, as we discuss in §5.4.4.

5.4. Effective world-volume theories

So far, we have concentrated on world-sheet approaches to string theory and Dirichlet branes, such as the sigma model and SCFT. In these approaches, one starts with the the two dimensions of the string, and derives the rest.

There is a rather different physical approach in which one starts with consistency conditions on the physics in four space-time dimensions, classifies their possible realizations, and then tries to identify those which satisfy some other constraints. These might be constraints from experimental data, or perhaps theoretical constraints such as known symmetries of the theory.

This approach generally goes under the name of "effective field theory" or EFT.

While this is a vast subject, the part which is directly relevant for us here is the case of four-dimensional $\mathcal{N} = 1$ supersymmetric quiver gauge theories.[9] One can argue on very general physical grounds that a set of A- or B-type Dirichlet branes must correspond to such a theory. Thus, the EFT approach gets very quickly to one of the primary questions, which is to get a simple general description of the possibilities.

We will mostly discuss the classical theory (before quantization). To a large extent, this can be thought of as an alternate language for the quivers with relations discussed in §4.2. This is rather striking as the original motivations for the study of these concepts in mathematics and physics were quite different.

One point particular to $\mathcal{N} = 1$ supersymmetric quiver EFT is that the obstruction theory is determined by a superpotential, the same one introduced in §4.2.5. Now, the physics arguments for this have no direct connection to string theory. Nevertheless, mathematically this implies that the path algebra of the quiver is Calabi-Yau with homological dimension 3. Since this algebra represents the physics derived from the compactified dimensions, already there is something "ten-dimensional" about this class of theories.

Additional physical content, as yet less well formulated mathematically, emerges when we quantize the theories. This is the focus of most of the physics work, and we briefly survey a few aspects of it in §5.4.6.

5.4.1. $\mathcal{N} = 1$ supersymmetric gauge theory. The standard physics reference is [**460**], while [**153**] is an introduction for mathematicians. We give a very schematic description adapted to our purposes.

A four-dimensional effective field theory (EFT) is a quantum field theory of maps from a four-dimensional space-time M (say, $M \cong \mathbb{R}^{3,1}$), to some configuration space $\mathcal{C}$. In an $\mathcal{N} = 1$ supersymmetric EFT, the fields must admit an action of the $\mathcal{N} = 1$ supersymmetry algebra, somewhat similar to (3.87). This places constraints on the fields and the action, whose form and general solution is described in [**460**].

The result is that a $d = 4$, $\mathcal{N} = 1$ EFT is essentially a generalization of the supersymmetric nonlinear sigma model defined in §3.3.2 to incorporate gauge invariance.[10] For our purposes we simplify this to

---

[9]We reserve the notation $\mathcal{N}$ for the number of supersymmetries, leaving N free for other purposes (such as a rank of a gauge group).

[10]This is oversimplified, but the essential truth in it is that both classes of theory have 4 supersymmetries, the $(2, 2)$ left and right moving supersymmetries of the two-dimensional case discussed in §3.3.2, and the 4 real components of an irreducible spinor representation of the four-dimensional Lorentz (or Euclidean) groups. Furthermore, one can always take a $d = 4$, $\mathcal{N} = 1$ theory and dimensionally reduce it to get a $\mathcal{N} = (2, 2)$

DEFINITION 5.17. An $\mathcal{N} = 1$ EFT is a tuple $(\mathcal{C}, H, \mu, W)$, where

- $\mathcal{C}$ (the configuration space) is a complex manifold with Kähler metric. We denote the metric as $g_{i\bar{j}}$ and the Kähler form as ω.
- H (the gauge group) is a compact Lie group which acts by isometries on $\mathcal{C}$, and its complexification $H_{\mathbb{C}}$ acts holomorphically.
- μ is a moment map for the H action.
- W is a holomorphic H-invariant function on $\mathcal{C}$, the superpotential.

5.4.1.1. *Moment map.* A good math-oriented reference for group actions and moment maps in supersymmetric theories is [**239**]. We recall that a moment map for a group action on a symplectic manifold is a function $\mu : \mathcal{C} \to (\mathrm{Lie}\, H)^*$, which determines the infinitesimal group action $h : \mathrm{Lie}\, H \to T\mathcal{C}$ as follows. Any $t \in \mathrm{Lie}\, H$ determines a function F_t by

$$F_t(x) = \langle \mu(x), t \rangle,$$

and this function satisfies

$$\iota(h(t))\omega = dF_t, \tag{5.75}$$

where $\iota(h(t))\omega$ denotes contraction of the vector field $h(t)$ with the 2-form ω. If H is semisimple, the moment map μ is determined by ω and h, along with the additional condition that $F_{[t_1,t_2]} = \{F_{t_1}, F_{t_2}\}$, where $\{\cdot, \cdot\}$ denotes Poisson bracket. If H contains an abelian factor H_{abelian}, it is determined up to the freedom of adding an element of $(\mathrm{Lie}\, H_{\text{abelian}})^*$. This is referred to in physics as the possibility of adding "Fayet-Iliopoulos terms."

Very often, $\mathcal{C}$ is Euclidean and the H action on it is linear. In this case, one can equivalently specify $\mathcal{C} \cong R$, where R is a linear representation of $H_{\mathbb{C}}$. The moment map μ is then quadratic in the natural coordinates on $\mathcal{C}$. This is also a good approximate description of a small neighbourhood in $\mathcal{C}$.

The sigma model of §3.3.2 is the special case in which H is trivial and $W = 0$, reduced to $d = 2$. As we are working on the classical level, we are not imposing the constraint that $\mathcal{C}$ be Ricci-flat.

While we will not give the general Lagrangian analogous to (3.86), the additional data (H, μ, W) is used as follows. Let M be space-time (say $M \cong \mathbb{R}^{3,1}$), then as in Chapter 3 we introduce "fields" which are maps $M \to \mathcal{C}$, along with fermionic "superpartners." One then introduces an H-connection on M, and uses it to covariantize all space-time derivatives in the Lagrangian. One then looks for Lagrangians with supersymmetry transformations analogous to (3.87). This is made fairly simple by using a superspace formalism, in which the Lagrangian is determined by doing superspace integrals over the functions W and μ as well as the Kähler potential on $\mathcal{C}$; see [**460**] for details.

$d = 2$ theory. However, the $d = 2$ theories have additional possibilities which do not come from $d = 4$.

5.4.1.2. *Scalar potential.* A key constraint of supersymmetry is that the scalar potential, which for the most general EFT would be a real-valued function on $\mathcal{C}$, here must take the form

$$V = ||\partial W||^2 + \frac{1}{2}||\mu||^2. \tag{5.76}$$

In other words, it is a sum of the norm squared of the gradient of the holomorphic superpotential W and the norm squared of the moment map.

DEFINITION 5.18. A *vacuum* is an H-orbit of local minima $p \in \mathcal{C}$ of V,

$$\partial V|_p = 0. \tag{5.77}$$

Note that the minimizing H-orbit need not be isolated. If it is not, physicists refer to a path in $\mathcal{C}/H$ on which V is constant as a "flat direction."

A particular case of this is

DEFINITION 5.19. A *supersymmetric vacuum* is an H-orbit of p satisfying

$$\partial W|_p = 0 \qquad \text{F-term conditions} \tag{5.78}$$

and

$$\mu_p = 0 \qquad \text{D-term conditions.} \tag{5.79}$$

The F and D nomenclature is universal in the physics literature. In general, there can be other local minima, called non-supersymmetric vacua.

We will restrict our interest to the supersymmetric vacua. The moduli space of these is the symplectic H-quotient of the variety of critical points of W,

$$\mathcal{M} = \{p \in \mathcal{C} : \partial W|_p = 0\}//(H, \mu). \tag{5.80}$$

This is of course a classical definition. We discuss some aspects of quantization and the related consistency conditions in §5.4.6.1.

5.4.1.3. *Supersymmetric quiver gauge theories.* These are the particular case in which $H_{\mathbb{C}}$ is the automorphism group of a set of vector spaces V_i associated to a quiver as in §4.2,

$$H = \prod_{i=1}^{r} U(\alpha_i), \tag{5.81}$$

and $\mathcal{C}$ is the space of linear maps introduced there,

$$\mathcal{C} = \oplus_a \mathrm{Hom}(V_{t(a)}, V_{h(a)})$$

with the Euclidean metric. Given this H, the moment map is determined by a vector of weights $\vec{\theta}$, one for each $U(n)$ factor.

Not all algebras correspond to quiver gauge theories; the relations on the quiver must be expressible as the gradients (5.78) of a single function. Such algebras are referred to as "algebras with superpotential."

Conversely, not all superpotentials W lead to quiver relations. This will be true if W is of "single-trace" form, i.e., it is a trace of a weighted sum of monomials corresponding to closed loops in the quiver.

We will see in §5.4.5.5 that the D-term conditions are precisely those discussed in Theorem 4.88. Thus

PROPOSITION 5.20. *A supersymmetric quiver gauge theory corresponds to a quiver with relations which follow from a superpotential. Its supersymmetric vacua are the isomorphism classes of θ-semistable representations of the quiver.*

EXAMPLE 5.21. A very basic example is $\mathcal{N} = 4$ super Yang-Mills (SYM) theory. Here we have $H = U(N)$, $\mathcal{C} \cong \mathbb{C}^3 \otimes \mathrm{Mat}_N$ with the Euclidean metric, and

$$W = \mathrm{Tr}[Z^1, Z^2]Z^3, \tag{5.82}$$

where the Z^i are an explicit $N \times N$ complex matrix parametrization of $\mathcal{C}$.[11] The F-term equations are

$$[Z^i, Z^j] = 0$$

and thus it is easy to check that the moduli space of supersymmetric vacua is

$$\mathcal{M} \cong \mathbb{C}^3 \otimes \mathbb{C}^N / S_N,$$

the nth symmetric product of $\mathbb{C}^3$, for $\theta = 0$, and empty otherwise. Compare Example 4.33.

EXAMPLE 5.22. Consider the theory corresponding to the quiver of Example 4.31, with two nodes connected by n arrows, and dimension vector $(1,1)$. Thus $H = U(1) \times U(1)$, and $\mathcal{C} \cong \mathbb{C}^n$. We take the Euclidean metric on $\mathcal{C}$, and introduce explicit coordinates ψ^i.

Looking back at Definition 5.17, we have now specified the EFT almost uniquely. The remaining datum is the moment map for H, which is a pair of real numbers θ_1 and θ_2. Supposing we know these numbers, then the supersymmetric vacua are the solutions of the moment map conditions (5.79) modulo gauge equivalence. Explicitly, these are

$$0 = \mu = \theta_1 - \theta_2 + \sum_i |\psi^i|^2. \tag{5.83}$$

Thus, we see that vacua exist if and only if $\theta_2 \geq \theta_1$. For $\theta_2 = \theta_1$, there is a unique vacuum, while for $\theta_2 > \theta_1$ the moduli space is $\mathbb{CP}^{n-1}$.

We see that the supersymmetric vacua are the isomorphism classes of quiver representations in the language of Chapter 4. The indecomposibility of a representation corresponds in physics language to the breaking of gauge

[11]While this is the case related to quivers, actually H can be any compact Lie group. In this case, the Z^i are taken in the complexified adjoint representation, and the superpotential (5.82) is the cubic antisymmetric invariant.

symmetry. More precisely, given a representation R, the maximal compact subgroup $H_R \subset \operatorname{End} R$ is the *unbroken gauge group*. Thus the unique vacuum for $\theta_2 = \theta_1$ has unbroken $U(1)^2$, while for $\theta_2 > \theta_1$ one says that "the expectation value of ψ breaks the gauge symmetry to $U(1)$."

EXAMPLE 5.23. The $\mathcal{N} = 2$ supersymmetric Yang-Mills theories. These are the particular cases of $\mathcal{N} = 1$ in which

$$\mathcal{C} \cong (\operatorname{ad} H \otimes \mathbb{C}) \oplus R \oplus R^*,$$

where $R \cong \mathbb{C}^N$ is a linear H representation and R^* is its dual (since H is unitary, this is equivalent to the complex conjugate). Taking $\mathcal{C}$ Euclidean and denoting the natural coordinates on the three factors as A, Z and $\tilde{Z}$, we have

$$W = \operatorname{tr}(\iota(A)Z)\tilde{Z}.$$

For example, $\mathcal{N} = 4$ SYM is the special case $R \cong \operatorname{ad} H \otimes \mathbb{C}$.

EXAMPLE 5.24. Supersymmetric QCD is the quiver of Figure 5. We discuss it in more detail in §5.4.4.5.

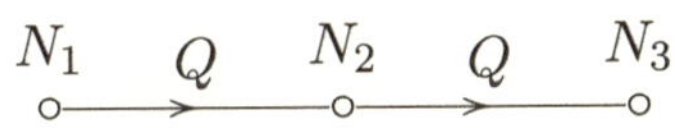

FIGURE 5. $N = 1$ supersymmetric QCD.

5.4.1.4. *Masses.* Let us now state some elementary physics definitions and arguments for the benefit of more mathematical readers.

DEFINITION 5.25. The *mass matrix* M^2 for a vacuum p is the Hessian $\partial_i \partial_j V|_p$ expressed in an orthonormal frame.

Since $\partial V|_p = 0$, it is a tensor. Also, since it is a real symmetric matrix, it can be diagonalized. Its eigenvalues are known as *squared masses*, and its kernel is the space of "massless fields."

Thus the tangent space to the moduli space of vacua is contained in the subspace of massless fields. Of course it can be a subspace if V contains cubic or higher order nonlinearities.

5.4.1.5. *Integrating out.* Suppose we are in a region of $\mathcal{C}$ in which some field ϕ is always massive; then we can solve

$$0 = \partial V/\partial\phi = M\phi + \mathcal{O}(\phi^2)$$

for ϕ and eliminate it. This is called *integrating out* the field ϕ, as it is a classical approximation to the quantum procedure of doing the functional integral over a subset of the fields.

In an $\mathcal{N} = 1$ theory, the same procedure can be applied to the superpotential, if $\partial^2 W/\partial\phi^2 \neq 0$. Of course this is most useful if ϕ appears at most quadratically in W, so that the solution is unique.

In the special case in which ϕ appears linearly in W (or V), it is referred to as a Lagrange multiplier for the constraint $0 = \partial W/\partial\phi$. If the subvariety S defined by this constraint is a submanifold, the supersymmetric vacua are the critical points of $W|_S$.

5.4.2. General remarks on EFT and Dirichlet branes. We consider type II string theory compactified on the product of Minkowski space-time $\mathbb{R}^{3,1}$ with a Calabi-Yau X, or more generally a (2, 2) SCFT with $\hat{c} = 3$. We now include a set of boundary conditions, embedded in $\mathbb{R}^{3,1} \times L$ for some n-dimensional submanifold $L \subset X$. Since they fill the Minkowski dimensions, such branes are often referred to as "space-filling branes."

In this book, we will almost always refer to Dirichlet branes by the dimension of the submanifold $L \subset X$. But since the branes we discuss now have world-volumes of $3 + n$ spatial dimensions, in the standard physics nomenclature these are Dirichlet $(3 + n)$-branes.

If all volumes are large compared to the string scale, we can describe this system using ten-dimensional supergravity (for the closed strings) coupled to a $(4 + n)$-dimensional supersymmetrized Born-Infeld action of the general form (3.142). If L is a minimal submanifold and the Yang-Mills field strengths are small as well, we can further approximate this as a super Yang-Mills theory on the brane world-volume,

$$S = \int_{\mathbb{R}^{3,1}\times L} d^{4+n}x \ \operatorname{tr} F^2 + (DX_N)^2 + \text{fermions}. \tag{5.84}$$

Here X_N are scalar fields taking values in the normal bundle to L (again, in a small field approximation).

Of course, the underlying ten-dimensional nature of space-time predicted by string theory remains a hypothesis. To see this directly, we would need to do experiments at a very high energy $E \sim \hbar c/R$, where R is the diameter of X or L. Unless and until advances in technology make it possible to do this, the primary interest of string theorists will not be in the full $(4 + n)$-dimensional field theory, but rather in its "Kaluza-Klein reduction" (or simply KK reduction) to $\mathbb{R}^{3,1}$.

For present purposes, the KK reduction of a $(d + 4)$-dimensional field theory is a 4-dimensional field theory in which $\mathcal{C}$ is the *moduli space* of brane configurations. As a basic example, consider the case of L a point on X, a D0-brane. In this case, the moduli space of a single brane is simply X itself. As another general class of examples, given the $(4 + n)$-dimensional theory with action (5.84), the KK reduction is a theory for which $\mathcal{C}$ is the moduli space of minimal volume manifolds carrying solutions of the Yang-Mills equations.

Suppose we take N identical copies of one of these branes. Since the branes are identical objects (bosons in physics terms), we might intuitively expect their moduli space to be the symmetrized orbifold product $\mathrm{Sym}^N \mathcal{C}$. For example, the moduli space of N identical D0-branes is $\mathrm{Sym}^N X$.

However, as we know that such a product is not a manifold, we should ask how the coincident singularities are resolved. The physical answer is that when a subset $n \leq N$ of the branes coincide, the four-dimensional field theory has a larger gauge symmetry, enhanced from $U(1)^N$ to $U(n) \times U(1)^{N-n}$. If multiple subsets coincide, the gauge group will be the unitary subgroup of the endomorphism group of the corresponding direct sum of sheaves on points. This follows from a simple physical argument involving the Higgs mechanism, which we leave as an exercise.

Thus, even though the moduli space can be singular, the full gauge theory will be non-singular. This "mechanism" is the general way in which singularities in D-brane moduli spaces are resolved, at least at string tree level. In physical terms, one of the essential aspects of this is that, at the singular point in moduli space, additional fields become massless, here the "off-diagonal gauge bosons" of enhanced gauge symmetry. This observation can be generalized to other situations in which fields become massless. For example, there are other mechanisms for singularity "resolution" in which wrapped D-branes become massless [**432**], but seeing how this works requires going beyond string perturbation theory.

Since this mechanism is mathematically natural, we should not be surprised to see that it agrees in examples with previous mathematical discussions of resolution of singularities.

Now, as we move away from the large volume limit, the equations (5.84) will get corrections in α', and perhaps the string coupling g_s as well if we consider higher genus surfaces. But as long as we only consider maps which vary slowly in $\mathbb{R}^{3,1}$ (meaning that $\alpha'|\partial\phi|^2 \ll 1$ where ∂ is the derivative in $\mathbb{R}^{3,1}$), a similar four-dimensional field theory can be derived, the effective field theory. Thus whatever the physics might be of branes in X, there is a subset of it which can be described within the EFT framework set out in this section. This subset includes the scalar potential which determines the possible vacua, the spectrum of light particles and their couplings, so it is of great physical interest.

5.4.2.1. *Dirichlet branes and quivers.* If we restrict ourselves to string tree level (the sphere and disc diagrams), then a set of Dirichlet branes will be described by a quiver gauge theory. To see this starting from the Chan-Paton prescription of §3.5.2.8, we introduce a vector space $V = \bigoplus_{i \in I} \mathbb{C}^I$, where I is the set of Chan-Paton labels. We then identify the vector spaces V_i of §4.2 with the subspaces of V corresponding to all branes of a given type B_i. Following the Chan-Paton prescription, we can obtain a small (linear) patch of $\mathcal{C}$ as the space of near-marginal boundary operators.

Terms in the gauge theory Lagrangian correspond to correlation functions of these boundary operators, whose Chan-Paton factors are contracted in sequence along the boundary. For a Riemann surface with a single boundary, such as the disc, the result will be an interaction which satisfies the single trace property.

Comparing with the definitions of Chapter 4, we obtain the the dictionary we have been using throughout our presentation: boundary conditions become objects in the category of quiver representations, and open strings become morphisms. This argument does not require supersymmetry or a specific number of dimensions.

We still need to see that gauge fields only arise from operators which preserve a boundary condition, i.e., transform in $\mathrm{End}(V_i)$, and not boundary changing operators. This uses the decomposition of the CFT into a product of an $\mathbb{R}^{3,1}$ factor with the "internal" CFT, and the fact that the gauge field boundary operator (3.144) is already marginal. Thus, to get a marginal operator, the internal part of the boundary operator must have dimension zero. In a unitary theory, this implies that it is the identity and preserves the boundary condition.

A marginal operator in the internal CFT corresponds to a tangent vector to the moduli space of vacua, which as we discussed above is a massless scalar field (in the linearized theory). More generally [**394**], a mass squared in EFT is related to the dimension h of the corresponding boundary operator in CFT as[12]

$$m^2 = \frac{h-1}{\alpha'}. \tag{5.85}$$

The three cases $m^2 = 0$, $m^2 > 0$ and $m^2 < 0$ correspond directly to the three cases of the discussion in §3.2.5. Varying a massive field, with $m^2 > 0$, takes us out of the set of vacua. We discuss the case $m^2 < 0$ in §5.4.4.

None of these arguments require supersymmetry, so they also do not require X to be Calabi-Yau. On the other hand, the other known Ricci-flat manifolds are also related to supersymmetry, as we discuss below. A more general class of conformal field theories can be defined by turning on H flux and non-constant dilaton fields. Some results for Dirichlet branes in these backgrounds can be found in [**363, 440**].

5.4.2.2. *D0 quantum mechanics and bound states.* Certain arguments are simpler if we take not space-filling branes but other embeddings in Minkowski space-time. For example, in §5.3.4, we used the language of "BPS particle," obtained by embedding the brane in $\mathbb{R}^{0,1} \times L$. While these are two different branes with very different physical behavior, for the purpose of deriving the EFT at string tree level they are essentially the same. While the BPS particle language was advantageous for reasoning about central

[12]For closed strings, h is the left (or right) dimension; also there is a different constant of proportionality. Similar formulas apply for the superstring, with the -1 replaced by $-1/2$ (NS sector) or $-5/8$ (R sector).

charges, since the constraints of supersymmetry are stronger in four dimensions, the language of space-filling branes is advantageous for the purposes of deriving the EFT.

By "essentially the same," we mean the following. Suppose we have a world-volume theory for D3-branes, in the explicit form used above. In particular, it contains gauge connections A, which are one-forms on $\mathbb{R}^{3,1}$ taking values in the Lie algebra of H. Write these in terms of an explicit basis t_a for the Lie algebra and dx^i for one-forms,

$$A = \sum_{1 \leq a \leq \dim H; 0 \leq i \leq 3} A_i^a t_a dx^i.$$

Now, if one simply drops all spatial derivatives in $\mathbb{R}^3$, and relabels the space-like components A_i^a for $i = 1, 2, 3$ as X_i^a, one has the quantum mechanics of D0-branes.

Applying this prescription to the gauge theory action to compute the scalar potential, one finds three sets of terms:

$$V = V_1(\phi) - \sum_i \operatorname{tr}[X_i, \phi]^2 - \sum_{i<j} [X_i, X_j]^2. \tag{5.86}$$

Here $V_1(\phi)$ is (5.76). The second term comes from the covariant derivatives $D\phi$, while the third from the Yang-Mills action.[13]

Let us work out the possible time-independent solutions of this D0 quantum mechanics. First, we take ϕ corresponding to a quiver representation R, possibly reducible, to minimize $V_1(\phi)$. Then, the other terms are minimized if the X_i take values in $\mathbb{R}^3 \otimes C_R$, where C_R is the Cartan subalgebra of the unbroken gauge group H_R. Finally, we quotient by the Weyl group of H_R.

Thus, for a representation R with

$$H_R = \prod_{i=1}^{r} U(N_i), \tag{5.87}$$

we find a moduli space

$$\mathscr{M}_R = \bigoplus_i \left(\mathbb{R}^3 \otimes \mathbb{R}^{N_i}\right) / S_{N_i}.$$

This is naturally interpreted as the configurations of a set of N_1 identical particles of one type, N_2 of a second type, and so on.

In particular, one can consider the region in which all of these positions $\vec{X}^a = X_i^a$ are widely separated in $\mathbb{R}^3$. One might expect, and it can be shown, that in this region the interactions between the particles can be neglected.[14] Thus the physics of the EFT becomes that of a collection of

[13]The signs are the ones for which $V \geq 0$.

[14]While the open strings in $\operatorname{Hom}(B_i, B_j)$ are still present, the second term in (5.86) gives them large masses $m^2 > 0$, so they can be neglected in finding classical vacua. This argument is useful in light of §5.4.4 as it provides a precise way to define "tachyonic fields,"

independent EFT's, one for each $U(1)$ factor in H_R. This is the precise sense in which a representation R satisfying (5.87) describes $N = \sum N_i$ particles.

This structure provides the physical definition of "bound state" and "binding process," at least in the classical limit of the quantum mechanics. An indecomposable representation R corresponds to a single particle, because it sits at a point in $\mathbb{R}^3$. Suppose it is possible to continuously vary this to another representation R' with $H_{R'} \cong U(1)^2$; such a representation corresponds to two particles, so this is an explicit model of particle decay.

5.4.3. Supersymmetric Dirichlet branes. We now assume that our boundary conditions are of A- or B-type as defined as in §3.6.6. These definitions, particularly (3.196) and (3.197), originated physically in the requirement that the boundary conditions preserve $\mathscr{N} = 1$ supersymmetry in four dimensions. Thus, the general arguments we just gave imply that these branes will be described by an $\mathscr{N} = 1$ supersymmetric EFT, with H as in (5.81).

The general constraints on supersymmetric EFT provide short arguments for a variety of general aspects of these problems. For example, they imply that the moduli space of A-branes is a Kähler manifold, as we will verify directly in Chapter 6.

5.4.3.1. *BPS states.* In general terms, a BPS state is a localized configuration in a supersymmetric theory which preserves some supersymmetry.

For the purposes of this book, a BPS state is a supersymmetric vacuum of the supersymmetric quiver gauge theory of a collection of D-branes, as summarized in Proposition 5.20. A BPS particle is a vacuum which breaks the gauge group to $U(1)$, in other words an indecomposable representation.

We would be remiss not to mention that this is a "classical" definition, whereas in much of the string theory literature, the term "BPS state" refers to its *quantum* analog.[15] In particular, by a "BPS Dirichlet brane state" one usually means a vector in the Hilbert space $\mathscr{H}$ of D0-brane quantum mechanics. Following the dimensional reduction we just discussed, this is an $\mathscr{N} = (2,2)$ supersymmetric quantum mechanics. By the arguments of §3.1.1, if the moduli space of supersymmetric vacua $\mathscr{M}$ is a nonsingular manifold, then $\mathscr{H}$ is its de Rham cohomology.

One can define an index counting BPS states as in §3.1.3; it is implicitly a function of the dimension vector of the quiver. If $\mathscr{M}$ is a nonsingular manifold, then the index is given by its Euler characteristic.

Typically $\mathscr{M}$ will be singular. However, as it is obtained as an explicit quotient (5.80), the quantum mechanics provides an explicit definition of $\mathscr{H}$. Presumably this corresponds to some known cohomology theory [**225, 110**], but to our knowledge this has not been well understood. Rather, at present

---

which otherwise would lead to instabilities. The massive fields can also affect the quantum theory.

[15] This is "quantum" in the sense of the genus (string coupling) expansion.

one defines this index by generalizing the Euler characteristic to moduli spaces with singularities and obstructions, using the idea of the "virtual fundamental class" (MS1, §27).

However, apart from some brief comments in §5.9.2, for the rest of the book we will stick to the "classical" definition of BPS state.

5.4.3.2. *Derivation of EFT from the world-sheet.* Let us now suppose that we have an $\mathcal{N} = 2$ SCFT X such that the closed string (c, c) and (a, c) algebras have integral $U(1)_R$ charges, and a set of distinct boundary states B_i, all of A or B type. Then $\mathcal{C}$ is the GSO projected space of chiral operators (as in §3.5.3.2), W is the generating function of the disk correlation functions defined in §3.6.4, and μ is determined by the gradings (5.67) of the branes. Let us proceed to explain and justify these statements.

We begin with the special case in which the B_i preserve the same supersymmetry, i.e., the phases (5.67) are all equal (modulo 2). In this case, the $U(1)_R$ charges of all open string chiral operators will be integral, and the combined world-volume theory will have $\phi = 0$ as a supersymmetric vacuum.

We then have

$$\begin{aligned} \text{Lie}\, H &= \bigoplus_{i,j} \text{Hom}(B_i, B_j), \\ \mathcal{C} &= \bigoplus_{i,j} \text{Ext}^1(B_i, B_j), \end{aligned} \tag{5.88}$$

the superpotential

$$W[\phi] = \langle e^{\int \phi^i \int O_i} \rangle \tag{5.89}$$

interpreted as in (3.180), and zero Fayet-Iliopoulos terms, *i.e.*, μ homogeneous of degree two (in the approximation in which the metric $\mathcal{C}$ is taken to be Euclidean).

The GSO projection has correlated the $U(1)$ charge to the type of space-time field: even is a vector field and thus related to gauge symmetry, while odd is a scalar field. The chiral operators with $q \gg 1$ do not give massless fields, by the supersymmetric analog of (5.85).

EXAMPLE 5.26. The D0-brane in $\mathbb{C}^3$. The open string chiral ring is generated by the free fermions ψ^i with $1 \leq i \leq 3$ of §3.3.2, satisfying the boundary conditions (3.156). Thus it is the Grassmann algebra $\Lambda\mathbb{C}^3$ with the obvious trace. Taking $O_i = \psi^i$, $\phi^i = Z^i$ and evaluating W, reproduces (5.82).

5.4.3.3. *Superpotential and obstruction theory.* Not every element of

$$\text{Ext}^1(B, B)$$

corresponds to a deformation of an object B; in general deformations can be obstructed.

In $\mathcal{N} = 1$ EFT terms, this is to say that not every non-zero point in $\mathcal{C}$ is a supersymmetric vacuum, rather the equations $\partial W = 0$ must be satisfied. Thus, the obstruction theory of A- and B-branes on a threefold is always described by a superpotential.

This is rather non-trivial and is only generally true for a Calabi-Yau threefold. To see why this is so, recall that obstructions correspond to elements of a second cohomology (or Ext) group. For example, the obstruction to combining two deformations of a holomorphic bundle V on X, call these $\alpha, \beta \in H^1(X, \mathrm{End}\, V)$, is a class

$$\alpha \wedge \beta \in H^2(X, \mathrm{End}\, V).$$

Now, if all the obstructions come from gradients $\partial W/\partial \phi$ of a single holomorphic function, the dimension of this group can be at most $\dim H^1(X, \mathrm{End}\, V)$. Furthermore, there will be a correspondence between obstructions and deformations. While in general neither of these is true, if we are on a Calabi-Yau threefold, they both follow from Serre duality.

A general proof of this fact for holomorphic curves was given in [**98, 296**]. More general arguments appear in [**22**].

5.4.3.4. *Superpotential for B-branes.* For holomorphic bundles on X a threefold, there is a "universal expression" for the superpotential in terms of the holomorphic Chern-Simons action. This is a functional of the connection, whose critical point condition is precisely $F^{0,2} = 0$. Writing the $(0,1)$-component of the connection in terms of a reference connection and a $(0,1)$-form as

$$\bar{\nabla} = \bar{\nabla}_0 + A^{0,1},$$

we have

$$S_{\text{holo CS}} = \int_X \Omega \wedge \mathrm{Tr}\left(A^{0,1}\bar{\partial}A^{0,1} + \frac{2}{3}A^{0,1} \wedge A^{0,1} \wedge A^{0,1} \right). \tag{5.90}$$

One can verify that $\delta S_{\text{holo CS}}/\delta A^{0,1} = F^{0,2}$.

In itself, this is not the superpotential, which should be a function on a linearized deformation space dual to $H^1(X, \mathrm{End}\, V)$. But by evaluating it on a parameterized family of connections, one can obtain the cubic term in the superpotential, as

$$W(t) = S_{\text{holo CS}}[\sum_\alpha t^\alpha A^{0,1}_\alpha] + \mathcal{O}(t^4).$$

The higher order terms can be obtained using the relation to A_∞ algebras discussed in §8.2.1.

5.4.3.5. *Superpotential for A-branes.* Here the superpotential is the generating function for the disk correlation functions of §3.6.2.5. These are expected to reproduce the A_∞ structure of the Fukaya category discussed in §8.3.3. As the superpotential is a single trace, this requires the A_∞ structure to be cyclic. The tadpole condition is then (5.78).

5.4.3.6. *Branes and antibranes.* Let us next consider the case in which there are two sets of branes with opposite signs of the central charge, say

$$\xi(B_i) \equiv 0; \xi(\bar{B}_a) \equiv 1 \pmod 2.$$

In this sense, the latter set are "antibranes."

As explained in §3.5.3.2, the only difference between a brane and antibrane in the world-sheet definitions is the GSO projection. Thus we can obtain the EFT in this case by taking the direct sum of the separate EFT's for the branes B_i and antibranes $\bar{B}_a$, and adding the new fields ("brane-antibrane open strings")

$$\mathcal{C}' = \bigoplus_{i,a} \left(\operatorname{Hom}(B_i, \bar{B}_a) \oplus \operatorname{Hom}(\bar{B}_a, B_i) \right).$$

In other words, $\mathcal{C}$ for the combined theory is the direct sum (as a vector space) of those for the separate theories, with the new factor $\mathcal{C}'$. Note that there is no new factor in H from $\operatorname{Ext}^1(B_i, \bar{B}_a)$, by the argument of §5.4.2.1 (these are massive gauge bosons).

What about W and μ for the new degrees of freedom? Recalling that the discussion of topological correlation functions in §3.6.2.2 and §3.6.4 did not depend on the GSO projection at all, we see that the same expression for the superpotential (5.89) applies in this case, under the interpretation that we keep all terms with total $U(1)_R$-charge 3.

In particular, W has no quadratic terms. This also follows from the following physics argument. According to the discussion in §3.3.4.1, all Ramond ground states have the same dimension $h = \hat{c}/8$. Using the appropriate version of (5.85), this implies that Ramond ground states always correspond to *massless* fermions in $d = 4$. On the other hand, fermion masses in $\mathcal{N} = 1$ EFT are precisely the quadratic terms in W.

The bosonic components of the fields in $\mathcal{C}'$ are not massless. In fact they have mass squared $m^2 < 0$ and are thus "tachyons" as explained in §5.4.4. Once we have made this discussion, it will become apparent that all of this holds for a general set of branes with general gradings.

5.4.3.7. *Supersymmetric branes in other backgrounds.* These include hyperkähler manifolds, and manifolds of G_2 and Spin(7) holonomy. A good introduction is [**172**].

Dirichlet branes in type II string compactification on a G_2 manifold are described by three-dimensional EFT with two supercharges, (usually called $\mathcal{N} = 1$ in $d = 3$). This is comparable to $(1,1)$ supersymmetry in $d = 2$. The target space is a real manifold, and the potential arises from a real superpotential analogous to (5.76). A few studies have been done, for example [**148, 228, 214**] which study the orbifold construction of §5.4.5 for $G \subset G_2$.

Dirichlet branes in type II string compactification on a Spin(7) manifold are described by two-dimensional EFT with $(0,1)$ supersymmetry. So far this subject is *terra incognita*, although orbifold examples could be worked out.

5.4.4. Tachyon condensates. In §5.2.1.2 we saw a fairly concrete geometrical picture for A-brane bound state formation and decay. The same process can be motivated from the $\mathcal{N} = 1$ effective field theory point of view, by using the physical idea of tachyon condensation ([**423**]; see also [**363**] and references therein).

5.4.4.1. *Combinations of branes.* We begin by recalling Example 5.22, a theory based on the quiver with two nodes connected by n arrows. In EFT terms, this theory has $U(1)^2$ gauge group, n charged matter multiplets, and Fayet-Iliopoulos terms (moment map parameters) θ_1 and θ_2 entering as in (5.83).

Following the logic of §5.4.2 and §5.4.3, this EFT can arise as the world-volume theory of a pair of Dirichlet branes B_1 and B_2 connected by n open strings, i.e., $\dim \mathrm{Ext}^1(B_1, B_2) = n$. Thus we can use it as a general model for the bound state formation of supersymmetric Dirichlet branes.

A bound state is by definition a single Dirichlet brane and thus has unbroken gauge group $U(1)$. This will be the case if any matter field $\psi^i \neq 0$. From (5.83) we see that such vacuum configurations will exist if and only if $\theta_2 > \theta_1$.

For $\theta_1 = \theta_2$, the moment map condition (5.83) is satisfied by $\psi = 0$. This configuration preserves $U(1)^2$ gauge symmetry and thus we have a supersymmetric combination of the two branes which is not a bound state.

What happens if $\theta_2 < \theta_1$? To understand this, we consider the scalar potential (5.76). This is

$$V = \frac{1}{2}\left(\theta_1 - \theta_2 + |\psi|^2\right)^2 \tag{5.91}$$

$$= V_0 + \frac{M^2}{2}|\psi|^2 + \frac{1}{2}|\psi|^4 \tag{5.92}$$

with

$$M^2 = 2(\theta_1 - \theta_2) \tag{5.93}$$

and $V_0 = \frac{1}{2}(\theta_1 - \theta_2)^2$.

From this we see that $\psi = 0$ is a vacuum for any $\theta_2 \leq \theta_1$. The difference is that, for $\theta_2 < \theta_1$, the vacuum is not supersymmetric; for example the energy of the vacuum is nonzero. Thus, while B_1 and B_2 separately preserve supersymmetry, in combination supersymmetry is broken.

Usually, vacua which break supersymmetry are local but not global minima of the potential. And, although it is easy to check that (for $\theta_2 < \theta_1$) $\psi = 0$ is a global minimum of (5.91), in more realistic problems we will not have an exact expression for the potential. Thus we would like to better understand this case.

Note also that, even if $\theta_2 > \theta_1$, the configuration $\psi = 0$ is still a critical point of (5.91), again with nonzero energy. This configuration again represents a combination of B_1 and B_2 and thus for any $\theta_1 \neq \theta_2$ the combination breaks supersymmetry. One wants additional physical arguments to better understand what distinguishes the two cases.

5.4.4.2. *Tachyon condensation.* Let us now follow the physical arguments of §5.4.2 by starting with D3-branes B_1 and B_2 but separated by a large distance ΔX in the Minkowski space dimensions. By locality, any interaction between the branes would be small (here it first appears at genus one, and we will ignore it) and we would expect the combination to be a stable configuration. This can be seen in the potential (5.91) as a large additional term $(\Delta X)^2|\psi|^2/2$ which forces $\psi = 0$.

We then bring the branes together. What happens next depends on the sign of M^2 in (5.93). If it is positive, the configuration $\psi = 0$ remains the only critical point, and the combination of branes is physically stable, despite breaking supersymmetry.

If $M^2 < 0$, then as we pass through $|\Delta X| < M$, the configuration $\psi = 0$ changes from a minimum to a maximum or saddle point. The linearized equation of motion now predicts that any small variation of the field ψ will grow exponentially with time. In real life, such fluctuations are ubiquitous (both because of quantum mechanics and small interactions with the environment), and thus one speaks of such a configuration as "becoming unstable."

Very generally, a linearized stability analysis (in the sense of physics and engineering) amounts to finding the masses squared for all bosonic fields. If any bosonic field has $M^2 < 0$, a critical point will be unstable, in the physics sense that an arbitrarily small perturbation will drive the system away from the critical point.

In particle physics, such fields with $M^2 < 0$ are often called *tachyons*,[16] and the dynamics which leads to a configuration with $\psi \neq 0$ is called *tachyon condensation.*

Almost always, the linearized analysis is only approximate, and one needs to perform an exact (nonlinear) analysis to determine the fate of an unstable system. However, in most real physical systems, this is not done by analyzing the equations of motion. Rather, the following simple argument is applied. Suppose, as is usually the case, that the system has small interactions with other degrees of freedom (" the environment"), that allow exchanging energy. Then, maximization of entropy will favor configurations

[16] As in §3.3.5.4, the science fiction terminology has a historical explanation. In this case, it comes by applying the standard relation between the energy, momentum and mass of a particle, $E^2 = \vec{P}^2 + M^2$. This leads to a naive interpretation of $M^2 < 0$ as describing particles which travel faster than light. Actually, sensible physical theories never contain such particles – rather, as we just explained, $M^2 < 0$ leads to instability. Thus, one appropriates the name tachyon for a field which drives an instability.

in which all available energy is "radiated" to the environment, while the original system is in its ground state. Thus the eventual fate of the system will be to reach a vacuum near the original configuration.

This type of dynamics can be usefully approximated as a gradient flow controlled by the potential energy function,

$$\frac{\partial}{\partial t}\psi^i = G^{ij}\frac{\partial V}{\partial \psi^j}.$$

where G^{ij} is the (inverse) metric on configuration space. Various other physics arguments, such as the renormalization group of §3.2.6.2, lead to essentially the same result.

The conclusion for the case at hand is that, when $M^2 < 0$, tachyon condensation leads to a stable configuration which is the minimum of $V(\psi)$ given by (5.83), a supersymmetric bound state of the branes B_1 and B_2.

5.4.4.3. *Special Lagrangians.* Now, recall the discussion from §5.2.1.2 of a pair of intersecting A-branes L_1 and L_2. In effective field theory, this will be described by an $H \cong U(1)^2$ supersymmetric gauge theory, the direct sum of the world-volume theories of the individual branes. Let us generalize slightly by taking the two branes to have intersection number $+n$, so that the open strings from L_1 to L_2 predicted by (5.7) are elements

$$\psi \in \text{Hom}(E_1, E_2) \cong \mathbb{C}^n.$$

Then, the EFT is precisely the one we just discussed.

Thus, §5.4.4.1 is the EFT description of something we saw for A-branes in §5.2.1.1. Recall that our statement of the special Lagrangian condition, (5.13), depends on a parameter ξ. Suppose we have two special Lagrangians L_1 and L_2; if $\xi(L_1) \neq \xi(L_2)$ their union will not be special Lagrangian. This corresponds to the discussion in §5.4.4.1.

Furthermore, by approximating the branes near the intersection as special Lagrangian planes $\Pi^{(\pi,0,0)}$ and $\Pi^{\vec{\phi}}$ in $\mathbb{C}^m$, we showed that a special Lagrangian $L_1 \# L_2$ which is close to the union $L_1 \cup L_2$, exists if and only if

$$\xi(L_2) < \xi(L_1).$$

This corresponds to the discussion in §5.4.4.2, if we make the identification

$$\xi(L_2) - \xi(L_1) = c\,(\theta_1 - \theta_2) \tag{5.94}$$

for some real constant $c > 0$. Thus, we can tentatively identify the components of the moment map θ_i, as linear functions of the gradings $\xi(L_i)$.

Of course, this argument does not prove that the relation between ξ and θ takes the simple linear form in (5.94). While computing the entire potential is difficult, the mass formula (5.93) can be verified by direct computation in world-sheet string theory.[17] One may analyze the masses of the open

[17]This is why one refers to the process as tachyon condensation as opposed to simply finding the minimum of a potential.

strings which begin on Π^0 and end on $\Pi^{\vec{\phi}}$ following [**42**] or [**394**, 13.4] (see also MS1). The result is that there are R-sector strings which are always massless, and NS-sector scalar fields with mass squared (in string units $1/\alpha'$)

$$M^2 = \frac{1}{2\pi}\left(\sum_{j=1}^{m}\phi_j - \pi\right). \tag{5.95}$$

These scalar fields are not projected out by the GSO process if one of the D-branes is viewed as an anti-D-brane.

If we consider open strings with ghost number n, i.e., $\mathrm{Ext}^n(L_1, L_2)$ in the Fukaya category, we find that

$$M^2 = \frac{1}{2}\left(\xi(L_2) - \xi(L_1) - n\right) \tag{5.96}$$

Thus, comparing to the last section, in $\mathscr{M}^+$ we have $M^2 < 0$ and so the open string in $\mathrm{Hom}(L_1, L_2)$ is tachyonic. This is entirely consistent with the fact that there is a ground state $L_1 \# L_2$ lower in energy than $L_1 \cup L_2$. This tachyon condenses to form $L_1 \# L_2$. In $\mathscr{M}^-$ the open string is not tachyonic and no condensation occurs.

The tachyonic condensation picture therefore gives a very simple description of the hard analysis performed by Joyce and reviewed in §5.2.1.2. It is natural to conjecture that this tachyon picture gives a complete criterion for how A-branes decay as one moves in the moduli space of complex structures. This is certainly well-motivated from a physics point of view, but the differential geometry required to make such a statement rigorous is difficult. Progress has been made in this direction in [**446, 442**].

Note that since $0 \le \phi_j < \pi$, it follows from (5.73) that, if Y is a Calabi-Yau m-fold,

$$0 \le \xi(L_2) - \xi(L_1) + \mu(p) < m. \tag{5.97}$$

This relation is nicely consistent with the bound (3.97) for the dimension of a chiral primary field in a unitary SCFT, discussed in §3.3.4. A vertex operator for a primary chiral field of conformal weight h is associated to a space-time bosonic field with mass $M^2 = h - \frac{1}{2}$. Comparing to (5.95) and (5.73), we see agreement.

Finally in this section we recall from standard string theory analysis that there are open string states in the NS-sector corresponding to vector particles in the uncompactified dimensions. These have mass

$$M^2 = \frac{1}{2}\left(\xi(L_2) - \xi(L_1)\right). \tag{5.98}$$

These are therefore always massless when $L_1 = L_2$. In other words we have vectors associated to L_1 given by $\mathrm{Hom}(L_1, L_1)$. These are the vectors associated to the gauge group present in the D-brane – to be precise, $\mathrm{Hom}(L_1, L_1)$

is the complexification of the gauge algebra. In the case of a single irreducible D-brane we expect a $U(1)$ gauge group and thus $\mathrm{Hom}(L_1, L_1) = \mathbb{C}$. If the gauge group is enhanced, either because we have two distinct D-branes, or because we have coincident D-branes, the gauge group, and thus $\mathrm{Hom}(L_1, L_1)$, will be bigger. The fact that the irreducibility of a D-brane is equivalent to $\mathrm{Hom}(L_1, L_1) = \mathbb{C}$ may be viewed as a version of Schur's Lemma in representation theory.

5.4.4.4. *Brane-antibrane annihilation.* One may ask why we began with special Lagrangians – is not the simplest decay process the annihilation of a brane with its own antibrane? Actually, it is not possible to fully understand this process in EFT terms; one must bring in ingredients from string theory. Still, it is useful to develop such a description, keeping in mind its shortcomings.

The basic picture uses the same EFT we just described. By string world-sheet computation, one finds that the bosonic open string between a brane and its antibrane always has

$$M^2 = -\frac{1}{2},$$

and thus bound state formation always takes place.

Another way to check this result is to note that, for $X \cong \mathbb{C}^3$, the difference between the flat A-brane Π_0 we considered and its antibrane is simply an overall rotation in space. Consider the family of $SU(n)$ transformations

$$z_i \to e^{i\pi\alpha} z_i.$$

These act on the grading as

$$\phi \to \phi + n\alpha.$$

For $\alpha = 1$ and n odd, the result is simply an orientation reversal.

Then, tracing the effect of this on the open string computation, an NS scalar open string field with $M^2 = 1$ decreases continuously to $M^2 = -\frac{1}{2}$.

While this implies that brane-antibrane annihilation is indeed tachyon condensation, trying to describe this with the EFT of Example 5.22 with $n = 1$ does not really work. The problem is that, after the tachyon condensation, one is left not with the vacuum, but with a $U(1)$ EFT describing a non-trivial D-brane. This is inevitable in quiver gauge theory as all bifundamental fields are uncharged under the diagonal $U(1)$, so there is no way to fix this within the EFT physics we discuss.

A full description of this process must involve either string world-sheet considerations, or some sort of singular EFT, or else string field theory. We briefly discuss some of these ideas in §5.9. However, there is a way to get some of this physics by incorporating one more ingredient in the EFT.

5.4.4.5. *Seiberg duality and tilting.* Let us consider supersymmetric QCD (Example 5.24), with $H \cong U(N_1) \times U(N_2) \times U(N_3)$, and

$$R = (V_1^\vee \otimes V_2) \oplus (V_2^\vee \otimes V_3), \tag{5.99}$$

where $V_i \cong \mathbb{C}^{N_i}$ is the defining representation of $U(N_i)$. We denote the fields (coordinates on $\mathcal{C}$) by $\tilde{Q}^i$ and Q^j respectively.

By our previous arguments, this would be the universal theory describing a set of three Dirichlet branes B_1, B_2 and B_3, with intersection numbers $B_1 \cap B_2 = B_2 \cap B_3 = +1$ and $B_1 \cap B_3 = 0$.

The moduli space of supersymmetric vacua depends on the moment map, in a way we will systematize below. However, it is simple to see (as in Example 5.22) that if $\theta_1 < \theta_2 < \theta_3$, a representation is a generic orbit of H. For representations without automorphisms (breaking to $U(1)$), the expected complex dimension of the moduli space is[18]

$$\begin{aligned} \dim \mathcal{M} &= 1 + \dim \mathcal{C} - \dim H_{\mathbb{C}} \\ &= 1 + N_1 N_2 + N_2 N_3 - N_1^2 - N_2^2 - N_3^2, \end{aligned} \tag{5.100}$$

where the extra 1 arises because an overall $\mathbb{C}^* \subset H$ (the "diagonal $U(1)$") acts trivially. If it is non-negative, it can be shown to be the actual dimension of $\mathcal{M}$.

Now, the physics arguments of [**414**] show that, when this theory makes sense as a quantum theory[19], it is equivalent (in the IR) to a different *dual* quantum theory with the quiver of Figure 6, the gauge group $H \cong U(N_1) \times U(N_1 - N_2) \times U(N_3)$, and the superpotential

$$W_{\text{dual}} = \text{Tr}\, M \tilde{q} q. \tag{5.101}$$

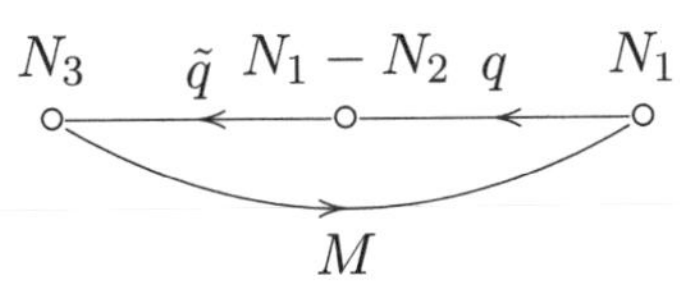

FIGURE 6. The Seiberg dual of supersymmetric QCD.

More generally, Seiberg duality can be applied to any subgroup within a general quiver theory with a fairly general superpotential. Thus, we choose a node V_i and identify this with the central node V_2 of of Figure 5; all incoming and outgoing arrows for this node are identified with the $\tilde{Q}$ and Q

[18]Note that a similar formula applies to any quiver without relations, but stability can be more complicated.

[19]As we discuss in §5.4.6.1, this requires $N_1 = N_3 \leq 3N_2$, and that we only quantize the $U(N_2)$ factor. The considerations here are classical and do not require this condition.

above (respectively).[20] We then replace this subquiver with the subquiver of Figure 6, and take as superpotential

$$W_{\text{new}} = W_{\text{original}}/\{Q\tilde{Q} \to M\} + W_{\text{dual}}(M, q, \tilde{q}), \tag{5.102}$$

where the operation / denotes substitution, and W_{dual} is as in (5.101), to obtain a dual theory.

By successively applying these dualities, one obtains a family of dual theories. Naively, applying the duality twice on the same node leads to a theory different from the starting point. For example, starting with supersymmetric QCD, one obtains a theory with additional fields M_1 and M_2, and a superpotential $W = \operatorname{tr} M_1 M_2$. However, this is equivalent to the original theory after integrating out the fields M_1 and M_2, as defined in §5.4.1.5.

Following [**40**], Seiberg duality can be understood as following from the fact that the dual theory describes the *same* configurations of Dirichlet branes, but as the bound states of a different "basis" of elementary branes. In particular, it does not require quantum field theory, and is also true on the classical level of our current discussion.

Suppose we apply the brane combination process of §5.4.4.3 to the branes B_1 and B_2, to obtain a bound state B_4. The K-theory relation

$$[B_1] = [B_4] - [B_2]$$

then suggests that we could replace our original basis (B_1, B_2, B_3) with $(B_4, \bar{B}_2, B_3)$, at least for a subset of the quiver representations, those for which the ranks N_i are non-negative in both bases.

It is easy to see that this transformation corresponds to the K-theory equivalence

$$N_1[B_1] + N_2[B_2] + N_3[B_3] = N_1[B_4] + (N_1 - N_2)[\bar{B}_2] + N_3[B_3],$$

thus reproducing the gauge group of the dual theory. With a bit more work, one can verify the equivalence of the moduli spaces, granting the superpotential (5.101).

In this sense, we can treat both the brane B_2 and the antibrane $\bar{B}_2$ in EFT, but not by using the same action, rather two different actions describing the same EFT.

The mathematics underlying this is the formalism of tilting in §4.4.7. The general conjecture which comes out of this physics is that tilting an algebra with superpotential produces a new algebra with superpotential, given by the explicit formula (5.102). This has recently been proven in [**457**].

---

[20]We discuss the case without adjoint matter, i.e., trivial $\operatorname{Ext}^1(V_i, V_i)$. This can be generalized [**317**].

5.4.5. Branes at orbifold singularities. The original and most direct physics way to see how quivers arise in the theory of D-branes is to consider an orbifold background [**141**].

Let us consider X, an orbifold of $\mathbb{C}^3$ by a finite subgroup $G \subset SU(3)$. We also write $W \cong \mathbb{C}^3$ for the covering space $\mathbb{C}^3$, and the G-action as ρ_W.

We now want to derive the EFT for a collection of N D0-branes; in other words $L \subset X$ is a point. A natural way to do this in physics is to analyze a collection of D-branes in $\mathbb{C}^3$ corresponding to the preimage of the quotient map, and then impose G-invariance.

As we know, D-branes in $\mathbb{C}^3$ are described by the $\mathcal{N} = 4$ supersymmetric gauge theory in four dimensions of Example 5.21. This has various fields: the scalars Z^i, fermions, and the four-dimensional gauge connection A. We can infer the action on the fermions at the end using $\mathcal{N} = 1$ supersymmetry. The gauge fields are scalars in $\mathbb{C}^3$ and thus are the simplest to start with.

Naively, a scalar on $\mathbb{C}^3$ is left invariant by G. However, since the D-brane theory has a gauge invariance, we should allow G to act by gauge transformations. Given that $\mathbb{C}^3$ is contractible, we can restrict attention to global gauge transformations, which act on the connection A as

$$g(A) = \rho_V(g) \cdot A \cdot \rho_V(g)^{-1} \tag{5.103}$$

for some representation ρ_V of G.

In other words, A transforms in the representation $\mathrm{End}(V) = V \otimes V^*$ and the G-invariant part of this can be written $\mathrm{Hom}_G(V, V)$. The resulting gauge group is the commutant of ρ_V. Decomposing this into irreducibles ρ_k as

$$\rho_V = \bigoplus_k m_k \rho_k,$$

by Schur's lemma the commutant is $U(m_1) \times U(m_2) \times \cdots$.

The three scalar fields Z^i parameterize the normal bundle, which in this case is canonically isomorphic to $\mathbb{C}^3$. Thus G acts on these as the representation ρ_W, tensored with the gauge action as $\rho_W \otimes \rho_V$. In other words, the G-invariant subspace of invariant scalar fields is given by $\mathrm{Hom}_G(V, W \otimes V)$.

Looking back at §4.2.6, we have simply reproduced the definitions entering into the McKay quiver of G. The factors of the gauge group are nodes, labelled by representations, while the Z's are the matrices associated to the arrows in a quiver representation.

Finally, the relations on the quiver are obtained from a superpotential, which is the restriction of the $\mathcal{N} = 4$ superpotential (5.82) to the G-invariant subspace. Its critical points are precisely those satisfying the relations given in §4.2.6.

Thus, we have a family of quiver gauge theories parameterized by the multiplicities m_k of the irreducibles ρ_k, or equivalently the gauge group ranks associated to the nodes. The case which corresponds directly to a D0-brane on X is to take ρ_V to be the regular representation. But one can take any representation; all of these theories are physically sensible.

A generic quiver representation R in this case will have $H_R \cong U(1)^{\sum_k m_k}$. Following the discussion of §5.4.2.1, we can interpret the individual $U(1)$ factors as describing individual particles, out of which all others can be formed as bound states. It follows that we have a distinguished set of D-branes F_k on the orbifold associated to the *irreducible* representations ρ_k of G. These branes were dubbed *fractional branes* in [**116**].

By finite group theory, for a D0-brane (i.e., the regular representation), the integer m_i attached to each node in the quiver representation is equal to the dimension of the corresponding irreducible representation. The location of the 0-brane will be dictated by the matrices associated to the arrows of the quiver. Once we study the stability of these B-branes we will see that the moduli space of such stable quiver representations is equal to X, as expected.

It also follows that the 0-brane is always composed of a nontrivial sum of fractional branes (hence the name). We will see that, at the orbifold point, the 0-brane is always marginally stable against decay into this set of underlying fractional branes.

5.4.5.1. *Other interpretations.* Mathematically, one can obtain the same quivers as the moduli space of translation-invariant G-equivariant holomorphic bundles on $Q = \mathbb{C}^d$ [**409**]. Let the fiber of a vector bundle be given by a representation V of G. Then, for G-equivariance, the connection on this bundle transforms yet again in $\mathrm{Hom}_G(V, Q \otimes V)$. Let us write this connection in the form

$$\sum_{\mu=1}^{d} Z_\mu dz_\mu - Z_\mu^\dagger d\bar{z}_\mu. \tag{5.104}$$

The $(2,0)$-part of the curvature is then

$$\sum_{\mu,\nu} \left(-\frac{\partial Z_\mu}{\partial z_\nu} + \tfrac{1}{2}[Z_\mu, Z_\nu] \right) dz_\mu \wedge dz_\nu. \tag{5.105}$$

If we impose translation invariance, we demand that the derivatives of Z_μ vanish. The condition that the bundle be holomorphic then amounts to $[Z_\mu, Z_\nu] = 0$, which again imposes the relations on the quiver representation.

Thus we have three interpretations for the quiver representation:

(1) A G-equivariant sheaf.
(2) The scalar fields on the world-volume of a D-brane on an orbifold.
(3) A connection on a translation-invariant holomorphic G-equivariant vector bundle.

5.4.5.2. *Properties of fractional branes.* It is of course tempting to immediately identify the fractional branes with coherent sheaves as in §4.7.

However we should check this with physics arguments. We will do this in §5.4.5.6 once we have the basic picture.

The scalar fields Z_μ in our world-volume theory must arise as open string states in the world-sheet description. That is, they occur as certain Ext's between the D-branes. The Ext-quiver language immediately tells us, of course, that these scalars are actually associated to Ext^1's between the fractional branes. Thus we can compute the intersection product (5.6) as the total signed number of arrows between nodes F_j and F_k,

$$I(F_j, F_k) = \dim_{\mathbb{C}} \mathrm{Hom}_G(\rho_j, W \otimes \rho_k) - \dim_{\mathbb{C}} \mathrm{Hom}_G(\rho_k, W \otimes \rho_j). \tag{5.106}$$

Furthermore, the discussion around equation (5.96) and §5.5.4 tells us that a scalar state associated with Ext^1 is massless if and only if the ξ-gradings at the end of the string are equal. Thus we have proven:

CLAIM 5.27. The gradings are equal for all the fractional branes at the orbifold point.

A consequence of this theorem is that the central charges of fractional D-branes must align to have the same argument at the orbifold point. We will see this explicitly in an example in §5.6.3.

More generally, alignment of the gradings guarantees that a system of D-branes will have a description purely in terms of EFT, and avoid the subtleties mentioned in §5.4.4.4. This generalizes to an approximate alignment $|\xi - \xi'| \ll 1$ as we explained in §5.4.4.

5.4.5.3. *Resolution of singularities.* We have established that the category of topological B-branes for an orbifold $\mathbb{C}^d/G$ corresponds to the derived category of quiver representations with relations. Furthermore, in this case we also know the stability condition for physical branes. From the basic definitions in §5.4.1, these are obtained by doing the symplectic reduction by the gauge group H, with a given moment map. As we mentioned in §5.4.1.3, the resulting moduli space is the set of θ-stable configurations as defined in §4.7.3. Thus, we can get a quite explicit picture in this case, studied in [**138, 137, 149**]. (Compare §4.7.3).

It is natural to start with the case $\theta = 0$, for which the moment map conditions (5.79) are

$$\sum_i [Z^i, (Z^i)^\dagger] = 0. \tag{5.107}$$

In §4.7 we saw that the regular representation of G should correspond to the skyscraper sheaf $\mathcal{O}_x$. One can easily show that the moduli space of Z's associated to the regular representation that satisfy (5.107) includes the variety $\mathbb{C}^d/G$.[21] Thus, this is the "orbifold point" in the moduli space, as can be checked by explicit world-sheet computations [**141**].

[21] In general it can have other branches as well, which correspond physically to configurations of fractional branes. See [**103**] for a more complete analysis.

Now we want to resolve the orbifold a little. This means that we want to modify our theory in such a way that the moduli space of Z's associated to the regular representation becomes X, a resolution of the orbifold. Since this preserves the complex structure, it is natural to conjecture that this is done by varying the components of the moment map (adding Fayet–Iliopoulos terms to the action) [**141, 139, 194**]. One such term may be added for each unbroken $U(1)$ of the gauge group, i.e., one for each irreducible representation of G. Let us call the coefficients of these terms ζ_i, where i is an index running over the irreducible representations of G.

The result is that (5.107) becomes

$$\sum_{\mu}[Z_\mu, Z_\mu^\dagger] = \mathrm{diag}(\underbrace{\zeta_1, \zeta_1, \ldots, \zeta_1}_{\dim(V_1)}, \underbrace{\zeta_2, \zeta_2, \ldots, \zeta_2}_{\dim(V_2)}, \ldots). \tag{5.108}$$

Physically, one expects the resulting moduli space of the 0-brane to be a resolution of $\mathbb{C}^d/G$ [**141, 139**]; this follows (for $d \le 3$) from [**67**]. It was first shown for $d = 2$ and arbitrary G in[**316**]. For $d = 3$ and abelian G, some examples were treated in [**409**], and a general proof given in [**256**].

Since the blow-up of an orbifold singularity is obtained by a deformation of $B + iJ$, which should be localized near the singularity, the resolution should be produced by adding twisted closed string marginal operators. These are labelled by nontrivial conjugacy classes in G [**125**] which we denote C. Let ϕ_C be a twisted operator present in the topological A-model for closed strings for the conjugacy class C, and consider a deformation of $B + iJ$ produced by adding a term

$$a_C \int_\Sigma d^2z\, \phi_C, \tag{5.109}$$

to the action. Turning on the ζ_i's should be equivalent to turning on some a_C's, which implies that ϕ_C should acquire a nonzero one-point function in the D-brane background. This one-point function was computed in [**141, 139**]. The result is, at least in a linear approximation for very small blow-ups,

$$a_C = \sum_i \chi_i(C)\, \zeta_i, \tag{5.110}$$

where χ_i's are the nontrivial characters of the group G. This can be inverted to determine the ζ_i in terms of the original $B + iJ$.

Note that, since the trivial representation is not present (it is not a twisted sector),

$$0 = \sum_i \zeta_i. \tag{5.111}$$

5.4.5.4. *Partial resolution.* We note in passing that, having a concrete definition of the world-volume theory of D-branes near a threefold quotient singularity, it is straightforward (in principle) to derive the world-volume theories for the large class of threefold singularities which can be obtained by partial resolution of an orbifold singularity, as first pointed out in [**368**]. Many examples are worked out there and in the subsequent literature.

EXAMPLE 5.28. The ODP or "conifold" singularity.
We consider a hypersurface X defined by

$$z_1 z_2 = z_3 z_4 \tag{5.112}$$

in $\mathbb{C}^4$. Such a singularity can be modified to produce a smooth threefold in two ways. The evident one is to deform the equation (5.112). This leads to a space which we will describe in Example 7.3; topologically it is $\mathbb{R}^3 \times S^3$, so has $b_2 = 0$ and no Kähler parameters.

There is also a less evident possibility, the so-called small resolution. A simple way to get this is to solve (5.112) in terms of new variables $a_1, a_2, b_1, b_2 \in \mathbb{C}$, by writing

$$z_1 = a_1 b_1; \qquad z_2 = a_2 b_2; \qquad z_3 = a_2 b_1; \qquad z_4 = a_1 b_2.$$

This provides a map from $\mathbb{C}^4$ to X for which the pre-image of a point is the orbit of the $\mathbb{C}^*$ action

$$a_i \to \lambda a_i; \qquad b_i \to \lambda^{-1} a_i.$$

Thus we have

$$X \cong \mathbb{C}^4 // H \tag{5.113}$$

with $H = U(1)$ and the homogeneous moment map. The small resolution is now obtained by turning on the "Fayet-Iliopoulos term" $\zeta > 0$ in this theory (varying the moment map). Topologically this space is $\mathbb{R}^4 \times S^2$; it is easy to show that $\zeta = C \int \omega$ for some real constant C.

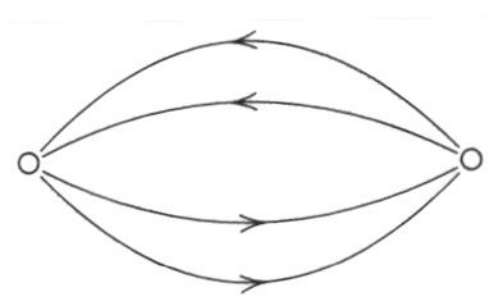

FIGURE 7. The quiver gauge theory of the conifold.

Note in passing that the same topology is obtained by taking $\zeta < 0$, but now with $\zeta = -C \int \omega$. However, if one identifies this noncompact geometry as a subregion of a compact threefold, one finds that the topology on the two sides is different, and related by a flop. We will return to this in §5.5.7.

To interpret the $U(1)$ quotient (5.113) as a quiver gauge theory, we introduce a second $U(1)$ acting trivially, to obtain the quiver in Figure 7.

This is a good description of a single D0-brane in the resolved geometry, but we are not finished, as we would also like to describe a set of N D0-branes using an analogous theory with $U(N)$ gauge groups.

To do this, we must also postulate relations, which follow from the superpotential [**305**]:

$$W = \sum_{1 \leq i,j,k,l \leq 2} \epsilon^{ij} \epsilon^{kl} \mathrm{Tr}\, A_i B_k A_j B_l, \tag{5.114}$$

where ϵ^{ij} is the antisymmetric invariant tensor.

To justify this in a systematic way, one can obtain the singularity by partial resolution, for example from the orbifold $\mathbb{C}^3/\mathbb{Z}_2 \times \mathbb{Z}_2$. Working out the resolved moduli space, one finds that certain fields are non-zero everywhere, which after substitution into the orbifold superpotential (the G-invariant projection of (5.82)) make other fields massive (analogous to the discussion in §5.4.4.5). Solving for the massive fields reproduces (5.114); we refer to [**368**] for more details.

5.4.5.5. *θ-stability.* Let us come back to the discussion of $\mathbb{C}^3/G$ and now consider a general representation $V = \bigoplus_i m_i V_i$. The right-hand side of (5.108) becomes a diagonal matrix with each ζ_i appearing m_i times. Taking the trace of this equation, we find that supersymmetric vacua exist only if

$$\sum_i \dim(V_i)\, \zeta_i = 0. \tag{5.115}$$

In general, this is incompatible with (5.111), in which case we cannot solve the D-term conditions (5.79).

At first sight, this would appear to break supersymmetry. However, looking more closely at the world-volume theory, there is a second supersymmetry, associated to the diagonal $U(1)$ (it is the constant shift of its gaugino). This allows solutions in which μ is non-zero but proportional to an overall constant vector $e = (1, 1, \ldots, 1)$ to preserve a linear combination of the two supersymmetries, and thus these are BPS states [**116, 138**].

Consider a configuration for which

$$\sum_\mu [Z_\mu, Z_\mu^\dagger] = \mathrm{diag}(\underbrace{\theta_1, \theta_1, \ldots, \theta_1}_{m_1}, \underbrace{\theta_2, \theta_2, \ldots, \theta_2}_{m_2}, \ldots), \tag{5.116}$$

for some real numbers θ_i. The scalar potential $|\mu|^2$ is then

$$\sum_i (\zeta_i - \theta_i)^2, \tag{5.117}$$

which is minimized subject to the condition (5.115) by

$$\theta_i = \zeta_i - \frac{\sum_j m_j \zeta_j}{\sum_j m_j}. \tag{5.118}$$

This solves (5.108) up to an overall constant shift and is thus a BPS state.

The equation (5.116) may be written in a more quiver-friendly way as follows. Let a be an arrow in the quiver with head $h(a)$ and tail $t(a)$. Let Z_a be the $m_{h(a)} \times m_{t(a)}$ matrix associated with this arrow in a given quiver representation. Then (5.116) becomes

$$\sum_{h(a)=i} Z_a Z_a^\dagger - \sum_{t(a)=i} Z_a^\dagger Z_a = \theta_i \,\mathrm{id}\,. \tag{5.119}$$

This is exactly the equation studied by King [**304**], and discussed in Theorem 4.88. We recall the definition of θ-stability made there. Fix a representation of the quiver associated to a representation $V = \bigoplus_i m_i V_i$ of G. For any representation $W = \bigoplus_i n_i V_i$ of G we define

$$\theta(W) = \sum_i \theta_i n_i. \tag{5.120}$$

Thus, by the tracelessness of (5.116), we see $\theta(V) = 0$. We say that the quiver representation is *θ-stable* if, for any nontrivial quiver subrepresentation associated to a representation W of G, $\theta(W) > 0$.

Now, under the conditions stated in Theorem 4.88, a quiver representation satisfies (5.119) (with an inner product unique up to obvious automorphisms) if and only if it is a direct sum of θ-stable representations.

Thus, very close to the orbifold point we have a stability condition expressed purely in terms of quivers. As we vary the Kähler moduli away from this point, eventually we expect to see the more nontrivial brane-antibrane phenomena discussed in §5.4.4.4 and §5.4.4.5.

5.4.5.6. *Relation to McKay correspondence.* Let us finally justify the identification of the fractional branes with the elementary objects of the McKay correspondence. To do this we add a D6-brane to the theory. We need to choose a G-action; let V_k be the "fractional" D6 for which this is the irrep ρ_k. The combined theory has D0-D6 open strings; one can check by world-sheet arguments that these are massless chiral fermions which transform as scalars in $\mathbb{C}^3$, along with massive scalars (this also follows from (5.96) and (5.142)). The analog of the projection (5.103) on the fermions is then

$$\psi = \rho_k(g) \cdot \psi \cdot \rho_V(g)^{-1}, \tag{5.121}$$

which has one solution for each irreducible $\rho_k \in V$.

By (5.6), the index of these massless fermions is the K-theoretic intersection number between the D0- and D6-branes. Thus the fractional D0's and D6's form dual bases of the K-theory (to be precise, the fractional D0's provide a basis for K-theory with compact support). Indeed, we have the stronger statement

$$\dim \mathrm{Ext}^*(V_k, F_i) = \delta_{i,k} \tag{5.122}$$

(5.121) also tells us the geometric interpretation of V_k – it is a G-equivariant line bundle. In physics terminology, it is a line bundle with a discrete Wilson line. This comes about because the asymptotic region (say $|z| > R$) has $\pi_1 \cong G$, and the representation $\rho_k(g)$ can be interpreted as the holonomy $\pi_1 \to H$.

Now, if we deform the Kähler moduli and resolve the orbifold, the G-equivariant line bundles V_k must each correspond to some sheaf $\tilde{V}_k$ on the resolution. The specific correspondence will depend on which resolution we take, but once we have it, the dual relation (5.122) and the mathematics of §4.7 uniquely determine the identification of the fractional branes.

To find this correspondence, we can consider a regular D0, as its moduli space for nonzero FI terms now provides a specific resolution. Now the D0-D6 strings are evaluation maps from sections of V_k to a point, the location of the D0. Tracing through the definitions, we have precisely reproduced the definition of the tautological bundles $\mathcal{R}_\rho$ from (4.12), and thus we make complete contact with the mathematics.

The results of §4.7.3 now give us the identification of the fractional branes for each chamber in the space of FI terms, as shown for the $\mathbb{C}^3/\mathbb{Z}_3$ example in Table 3. This is not a complete answer yet as we also need to continue this identification beyond the regime of validity of the orbifold description, for example to the large volume limit. This can be done by considering monodromy, as we explain in §5.6.

5.4.6. Brief overview of other physics results. A great deal of work has been done on $\mathcal{N} = 1$ EFT's of Dirichlet branes, because they provide a fairly direct route from superstring compactification to theories of direct physical interest, such as quantum chromodynamics and the Minimal Supersymmetric Standard Model. While we cannot discuss this in depth, we will now try to give a brief (incomplete) guide with some pointers to the literature.

5.4.6.1. *Comments on quantization.* A standard (although by no means the only) way to obtain a quantum field theory in $n+1$ space-time dimensions from a superstring compactification is to consider a collection of D$(n+p)$-branes such that $n+1$ of the world-volume dimensions are embedded in a common hyperplane,

$$\mathbb{R}^{n,1} \times L_i \subset \mathbb{R}^{3,1} \times X; \qquad n \leq 3.$$

If we then assume that the cycles L_i are (metrically) small compared to the length scales of interest in $\mathbb{R}^{n,1}$, a good description can be obtained by first reducing to $(n+1)$-dimensional quantum field theory (along the lines of §5.4.2.1) and then quantizing.

Normally the resulting quantum field theory depends only on the local geometry near the cycles; if this only covers a small region $U \subset X$, for low energy questions we can think of the string theory as effectively "compactified" on a noncompact threefold as $\mathbb{R}^{3,1} \times U$. This is the primary reason why so much attention is paid in the physics literature to the noncompact case, and to resolution and deformation of singularities.

The discussion of quantization depends very much on n. We have already discussed $n = 0$ in §3.1.1 and §5.4.2.1, and $n = 1$ in §3.2. We will focus here on $n = 3$, but much work has also been done on $n = 2$, for which [**254, 5**] might be a good starting point.

One might ask about $n > 3$. By considering branes in string compactifications with $\dim_{\mathbb{R}} X < 6$ (take say $X \cong T^4$ or K3), one can obtain quantum field theories with $n > 3$. For a long time, it was believed that the only such quantum theories were non-interacting theories, since they are non-renormalizable in perturbation theory. However, in recent years fairly convincing arguments based on string theory have been made that a few interacting supersymmetric theories exist in $4+1$ and $5+1$ dimensions. These have no classical limit and their study is in its infancy. Perhaps the most interesting example in the present context is obtained, not from Dirichlet branes, but rather by compactifying M-theory on $\mathbb{R}^{4,1} \times X$ with X a Calabi-Yau threefold. Their connection with resolution of singularities is discussed in [**252**].

5.4.6.2. $3+1$ *dimensions.* We now fix on $n = 3$ and outline some of the general results. A good introduction is [**253**].

First, a gauge theory with fermions in even space-time dimensions generally has anomalies which prohibit quantization [**224**]. In $3+1$ dimensions, these are controlled by the symmetric invariant $d : (\mathrm{ad}^* H)^3 \to \mathbb{R}$ of H, present for $SU(n)$ with $n \geq 3$. In terms of an explicit matrix representation $t_R : \mathrm{ad}\, H \to \mathrm{Mat}(\mathbb{C})$, anomaly cancellation requires

$$\mathrm{Tr}\; t_R = 0; \qquad \mathrm{Tr}\; t_R^3 = 0.$$

This is automatic for R real, but otherwise is very restrictive.

Considering the examples of §5.4.1.3, Example 5.22 is always anomalous, while supersymmetric QCD (Example 5.24) is a consistent quantum field theory if $N_1 = N_3$ and we only quantize the gauge group $U(N_2)$, leaving the others out of the path integral (treating them as global symmetries). For the $\mathbb{C}^3$ McKay quivers, typically the only anomaly-free theories are the multiples of the regular representation.

Next, we might ask that our theory be renormalizable. While this is not strictly necessary, if it is not true the physics of the theory will depend on more than the data (H, R, μ, W) that we specified in Definition 5.17. Renormalizability requires that $\mathcal{C}$ be complex Euclidean space, that the action of H be linear, and that W be an inhomogeneous cubic in the Euclidean coordinates on $\mathcal{C}$.

Next, one of the most important considerations determining the physics of the theory is the RG and the beta function, along the lines we discussed in §3.2.5. Recall that if the beta function for a coupling is positive (the corresponding operator is irrelevant), it will be driven to zero at long distances.

5.4.6.3. *Non-renormalization of the superpotential.* The most important result on the beta function for supersymmetric gauge theory is that it vanishes for the superpotential W,[22] to all orders in perturbation theory. This is the famed "non-renormalization theorem" of [**170**].

The non-renormalization theorem still allows a non-trivial beta function for W and thus quantum corrections, but it implies that these are exponentially small in the gauge coupling. This is extremely interesting physically as it provides a natural way to produce very small ratios, such as the observed ratio $\sim 10^{-40}$ between the strength of the gravitational force and that of the other forces between particles.

For our formal discussion, the primary importance of the non-renormalization theorem is that it allows non-trivial moduli spaces of vacua to exist in many of these quantum theories, those in which the exponentially small corrections also vanish (or are appropriately constrained). This can be shown to be the case using "non-perturbative non-renormalization theorems" as discussed in [**253**].

5.4.6.4. *Beta functions for the gauge couplings.* These are (approximately, and up to an overall multiplicative factor)

$$\beta = \Lambda \frac{\partial}{\partial \Lambda} \frac{1}{g_{YM}^2} = C_2(R) - 3C_2(\text{adjoint}), \tag{5.123}$$

where $C_2(R)$ is the "second Casimir" of the representation (the quadratic invariant in the enveloping algebra of Lie H). This result is exact for $\mathcal{N} = 2$ and $\mathcal{N} = 4$; exact results for $\mathcal{N} = 1$ appear in [**253**] and references there.

As an example, using $R \cong (\text{ad}\, H)^{\oplus 3}$ for the $\mathcal{N} = 4$ theory, we find that $\beta = 0$ in this case.

5.4.6.5. *Superconformal field theory.* As in Chapter 3, the case $\beta = 0$ is particularly interesting, as in this case the quantum field theory has scale and conformal invariance. We just saw that this was true for the $\mathcal{N} = 4$ theory, and it is true for certain choices of H and R in the cases $\mathcal{N} = 1, 2$. This includes the $\mathbb{C}^3/\Gamma$ McKay quivers with multiples of the regular representation.

In general, the superpotential enters in showing superconformal invariance. One necessary condition follows from the fact that the $d = 4$ SCA contains a $U(1)_R$ subalgebra, analogous to the $U(1)_R$ of §3.3.3 which played such a central role in Chapter 3. This acts on $\mathcal{C}$ in the neighbourhood of the superconformal point, and thus on the superpotential; one can show that

[22] By this we mean that the beta function for the coefficient of each monomial in W vanishes.

W must be homogeneous of weight 2 under this action. Coordinates on $\mathcal{C}$ can be regarded as chiral fields, so this leads to direct analogs of the chiral algebra and bounds on $U(1)_R$ charge discussed in §3.3.4. On the other hand, while it is possible to use the $U(1)_R$ to twist these theories, one gets only a subset of the topological theories we already saw in $d = 2$, since $U(1)$ holonomy in $d = 4$ forces $M^4 \cong \Sigma \times M^2$ with M^2 flat.

Various of the other $d = 2$, $\mathcal{N} = 2$ concepts we discussed are known to have analogs in this case. For example, one can define central charges somewhat analogous to the Virasoro central charge, in terms of the two-point functions of the $U(1)$ current and stress tensor. These have been conjectured to satisfy minimization principles [**251**], which go some ways towards systematizing the construction and RG flows of these theories.

5.4.6.6. *Seiberg duality and cascades.* A noteworthy feature of solutions of (5.123) is that $1/g^2_{YM}$ can pass through zero. This corresponds to "infinite coupling" and would at first seem problematic.

Nevertheless, it is possible to define an "RG flow" or "transition" which continues through this apparent singularity. The result is a *different* EFT, but with (G, R, μ, W) simply related to that of the original theory. Continuing the flow in the new theory, one can find a sequence or "cascade" of transitions.

Because the new theories are associated to the same collection of Dirichlet branes, the moduli spaces of quiver representations must be the same, at least for a large subset of dimension vectors. Thus one might expect, and it indeed turns out, that the relation is Seiberg duality, as discussed in §5.4.4.5.

See [**217**] for an overview of duality cascades, with explicit examples from the $\mathbb{C}^3/\mathbb{Z}_3$ models.

5.4.6.7. *Gauge-gravity dualities.* The most far-reaching connections provided by physics are between the field theories we discussed, and their "gravitational" or string theoretic duals. These arise by comparing the field theoretic description of a collection of D-branes with the original description as a solution of string theory. If we can get an independent description of the latter, we can propose a "duality," a precise translation between certain field theory problems and dual string theory problems.

Generally speaking, a duality is a statement about the full quantum field theory. Thus, the consistency conditions above (anomalies and renormalizability) are required. It will relate strong coupling problems in this theory to weakly coupled (classical or semiclassical) problems in the dual description.

There are numerous variations on this theme. One of them is to reinterpret Dirichlet branes as branes in a dual string theory. For example, one can "derive" the Seiberg-Witten solution of $\mathcal{N} = 2$ SYM theory in this way [**470**]. Another is to set up an intersecting configuration of branes of different dimensions, with multiple world-volume interpretations: this leads to a rederivation of the Nahm construction of monopole moduli spaces [**113**], more solutions of $\mathcal{N} = 2$ theories [**218**], and the like.

Perhaps the best studied duality is the AdS/CFT correspondence [**343, 4**]. Consider $\mathcal{N} = 4$ SYM with gauge group $U(N)$ (Example 5.21). As we argued in §5.4.2, this is the world-volume theory of N D3-branes at points in $\mathbb{C}^3$. The simple observation that the moduli space of vacua of the gauge theory must be that of configurations of points or $\mathbb{C}^{3N}/S_N$ is an elementary example.

Let us look more closely at the supergravity solution corresponding to these Dirichlet branes. This consists of a ten-dimensional metric $g_{\mu\nu}$, along with the other fields of supergravity. It can be found explicitly: the metric is

$$ds^2 = (1 + \frac{g_s N}{r^4})^{1/2}(-dt^2 + d\vec{x}^2) + (1 + \frac{g_s N}{r^4})^{-1/2}(dr^2 + r^2 d\Omega^2_{S^5}),$$

where $(t, \vec{x})$ are coordinates on $\mathbb{R}^{3,1}$, and we write $\mathbb{C}^3 \cong \{0\} \cup \mathbb{R}^+ \times S^5$ with $r \in \mathbb{R}^+$ and $d\Omega^2_{S^5}$ the round metric on S^5.

Now, consider the limit $r \to 0$, in which the "1" terms in the coefficients can be dropped. The limiting metric factorizes into a direct product of two homogeneous metrics, one on S^5 with one on the so-called anti-de Sitter space (an analytic continuation of hyperbolic space).

While this particular example does not make contact with mirror symmetry, one can do the same thing for D3-branes at an orbifold singularity, and obtain a dual description of the theories we discussed in §5.4.5, in terms of supergravity on $\text{AdS}_5 \times S^5/G$. This was first studied in [**278**] and [**368**], and has inspired a good deal of physics work. In mathematical terms, this approach makes contact with the metric geometry on the space S^5/G. For example, a large family of new explicit Sasaki-Einstein metrics has been obtained and studied in this way [**350, 351**].

Let us also mention [**151**] as a representative work tying together many of the developments we mentioned, along with others such as the relation to dimer models.

5.5. Stability

We now begin the discussion of how to identify "physical" D-branes, i.e., boundary conditions in $(2, 2)$ superconformal field theory, as a subset of the boundary conditions in the A- and B-models obtained by twisting. We saw in §5.2 that in the large volume limits, this subset varies with the "other" moduli (complex for the A-model and Kähler for the B-model), so clearly we must expect this for the full SCFT. In the large volume limit of the B-model, the subset is the subset of μ-stable coherent sheaves, and it is plausible that special Lagrangians are (in some sense yet to be made precise) a stable subset of the A-model isotopy classes of Lagrangian submanifolds.

Furthermore, granting mirror symmetry (which we know is true in $(2,2)$ SCFT), one can relate a highly "stringy" regime in one sigma model to the large volume limit in the mirror sigma model. Thus, the relation between physical branes and stability must make sense beyond the large volume limit. This makes it very plausible that physical D-branes in a general $(2,2)$ SCFT, making no assumption about the large volume limit, are also determined by some stability condition.

In principle this stability condition might depend on all sorts of SCFT data. But the simplest hypothesis is that it only depends on the data of the topologically twisted models; but now on *both* the A- and B-models. (Recall Table 1 from the introduction to the chapter.) This data includes the categorical structure, which summarizes the topological branes and open string correlation functions; we will use the B-model language and refer to this as a choice of derived category $\mathrm{D}(\mathrm{Coh}\, X)$. It also includes the A-model on X, but only that part of it which entered the discussion in §5.3.4, namely the topological closed string theory which determined one-point functions on the disk.

The first test of this idea is in the large volume limit: we must see how the ingredients of μ-stability (Definition 5.7) come out of SCFT data. This is relatively easy for the slope μ, as we will see in §5.5.5. What is harder to understand is how to decide when one object is a subobject of another, as this concept simply does not generalize to the derived category. Thus let us begin with this point.

5.5.1. Triangles. Just like the A-model, the B-model itself does not know about stability. What we do demand from the B-model though is some criterion of whether a given object in the derived category can *potentially* decay into two other objects.

The discussion of A-brane decay via tachyon condensates in §5.4.4 had shown that, when we were on the wall of marginal stability, the open string was massless. Thus it acts like a marginal (but probably not truly marginal) operator in the conformal field theory. In this sense a binding process looks like a deformation. Thus it is the mapping cone of §4.4.3 which defines a potential bound state of two D-branes.

The mapping cone construction in the derived category gives rise to a *triangulated* structure on the category. This mathematical structure turns out to be central to the notion of D-brane stability.

We recall the definition of a triangulated category from §4.4.3. For any abelian category $\mathcal{C}$, the derived category $\mathrm{D}(\mathcal{C})$ is a triangulated category. The translation functor is the same as the shift functor, while the distinguished triangles are provided by the mapping cones. At first the definition of mapping cone may seem less symmetric than that of the triangle, but it is not — any vertex of a distinguished triangle is isomorphic to the mapping cone of the opposite edge when the "[1]" is shuffled around to the appropriate edge.

Let us begin with Example 4.52. Recall that if B is an extension of C by A in $\mathcal{C}$, the short exact sequence

$$0 \longrightarrow A \xrightarrow{a} B \xrightarrow{b} C \longrightarrow 0, \tag{5.124}$$

induces a distinguished triangle in $\mathrm{D}(\mathcal{C})$,

$$\begin{array}{ccc} & C & \\ {}^{c}_{[1]}\swarrow & & \nwarrow^{b} \\ A & \xrightarrow{\quad a \quad} & B, \end{array} \tag{5.125}$$

whose new edge is an element $c \in \mathrm{Ext}^1(C, A) \cong \mathrm{Hom}(C, A[1])$.

Now since $\mathrm{D}(\mathcal{C})$ is not an abelian category, there are no short exact sequences or subobjects, as were called upon in the definitions of μ-stability and θ-stability. Everywhere we used subobjects we will need to find some replacement which uses only the triangulated structure.

Physically, we want to read the triangle (5.125) as *the D-branes A and C may bind via the potentially tachyonic open string c to form B.* We may then go through each axiom in turn and say what it means:

TR1: a) A binds with 0 (the empty brane) to produce A.
  b) We consider two objects in $\mathrm{D}(X)$ which are isomorphic to be the same D-brane. Thus this rule is required for consistency.
  c) The existence of an open string from A to B means that B can potentially decay into A and some other decay product C. This is not obvious but this axiom may be rephrased after the following.

TR2: If B can potentially decay into A and C then C can potentially decay into $A[1]$ and B. This is consistent with the observation in §5.3.3.1 that $A[1]$ could be interpreted as an anti-A.

Note that using this axiom we may now rephrase **TR1:** c) as follows. Given an open string from A to B we may potentially form a bound state of these two D-branes.

TR3: Given open strings between D-branes A and A' and between B and B', we may construct open strings between the corresponding bound states.

TR4: This formidable looking axiom is little more than a statement of associativity in the rules for combining D-branes. If we crudely write addition to represent rules for combining, the distinguished triangles in (4.7) can be read (using **TR2**) as

$$\begin{aligned} C &= A[1] + B \\ &= A[1] + (E + D[-1]) \\ &= (A[1] + E) + D[-1] \\ &= F + D[-1]. \end{aligned} \tag{5.126}$$

One may choose to regard these rules for D-brane decay as self-evident, or as proven since we have proven that the category of B-branes is the derived category and therefore triangulated.

5.5.2. Categorical mirror symmetry at last. Of course, there is some vagueness in the word "potentially" whenever we refer to binding or decay in §5.5.1. We have stated explicitly above that if there is an open string from A to B then we regard $A+B$ as a potential bound state. In order for this to actually happen there must be some region of moduli space where A and B are both themselves stable and the open string from A to B is tachyonic. This is not guaranteed. Thus, the triangulated structure appears when one has an optimistic view (which is as much as the topological field theory can know) about what can bind to what.

Our discussion of A-brane stability in §5.2.1 was approached directly rather than using the topological field theory language. Because of this the Fukaya category need not have a triangulated structure – it certainly knows about the A-branes which really are stable but it need not include the potentially stable branes in the topological field theory which never actually make it to stability. In particular there is no reason to suppose that the Fukaya category is actually triangulated. That is, it may well violate axiom **TR1:** c).

If the Fukaya category is not triangulated then the mirror symmetry proposal in §5.3.3.5 cannot possibly be correct. The derived category $\mathrm{D}(X)$ is triangulated and thus cannot be equivalent to a category which is not triangulated. The solution, of course, is to add the extra "potentially stable" A-branes to the Fukaya category so that the result is triangulated. This can be done by following the procedure of Bondal and Kapranov [**50**], as we discuss in §8.3.4.

The current state-of-the-art conjecture for mirror symmetry which follows from our topological field theory constructions is then:

CONJECTURE 5.29. *If X and Y are mirror Calabi-Yau threefolds then the category $\mathrm{D}(X)$ is equivalent to the category $\mathrm{Tw}\,\mathcal{F}(Y)$.*

We will discuss the present status of this conjecture, as well as the definition of $\mathrm{Tw}\,\mathcal{F}(Y)$, at length in Chapter 8.

5.5.3. Monodromy. Before we go more deeply into stability, we should realize that there is a fundamental obstacle to identifying a unique set of stable B-branes at a given point in the moduli space of $B + iJ$. The true string theory version of this moduli space has non-contractible closed loops, which turn out to be associated to nontrivial monodromy in $\mathrm{D}(X)$. This monodromy changes the set of stable objects as we go around loops in the moduli space.

This is easier to see in the mirror A-model moduli space, which is the moduli space of complex structures of the mirror manifold Y. We now consider a limit in which the complex structure degenerates. Consider for example a hypersurface of a projective variety defined by the vanishing of a section f; the non-degeneracy condition $\partial f \neq 0$ will typically be violated in families at complex codimension 1. Excising these points produces a non-simply connected moduli space.

These degenerate limits correspond to "singular CFT's" which (in an only partially understood way) violate the axioms of Chapter 3. One very common case is when some physical A-brane becomes massless, i.e., $Z(B) \to 0$. Following the arguments of [**432, 195**], this induces a singularity in the CFT.

From (5.128), near such a singular CFT $\xi(B)$ is multivalued. Typically $Z(B)$ is a good coordinate, so a closed loop around the singular point induces a monodromy $\xi(B) \to \xi(B) + 2$. We will discuss this in much more depth shortly.

For some purposes, it is simplest to remove this ambiguity by making

DEFINITION 5.30. The Teichmüller space $\mathscr{T}(Y)$ is the universal cover of the moduli space of (non-singular) complex manifolds Y.

5.5.3.1. *Monodromy action on the derived category.* Another way to understand the monodromy is by comparison to the picture for A-branes, where it is purely classical. The periods of the holomorphic 3-form over integral 3-cycles undergo non-trivial monodromy as we go around a non-contractible loop in the moduli space of complex structures. Such monodromy may be interpreted as an automorphism T of $H_3(Y, \mathbb{Z})$ which preserves the intersection form between 3-cycles.

More precisely, if one considers the universal family of Calabi-Yau threefolds over the complex moduli space, along with a symplectic structure on the Calabi-Yau manifolds in this family, following a path in the moduli space should yield a symplectomorphism between two Calabi-Yau threefolds. This symplectomorphism takes A-branes to A-branes, but individual A-branes will undergo a complicated variation of stability. But as closing the loop restores the original metric geometry, the original set of stable A-branes must be restored. This can be described by expressing the set of homology classes of stable branes in an explicit basis. Furthermore, we obtain a symplectic automorphism on the initial Calabi-Yau manifold which induces the monodromy operator T. But it also induces an automorphism of the Fukaya category, since it takes A-branes to A-branes.

This monodromy action on the homology can be copied from the A-model to the mirror B-model by using the mirror map of §3.4.3. One can then get some picture of the B-model monodromy action by considering examples. In fact it is not hard to find examples in which one starts with a coherent sheaf, and produces an element of $H^{\mathrm{even}}(X,\mathbb{Z})$ which cannot correspond to the Chern character of any sheaf (for example, the rank of the bundle might be negative). This is another indication that the derived category must enter the discussion.

Since the B-model is invariant under these variations of moduli, we conclude that each such monodromy must correspond to an autoequivalence on $\mathrm{D}(X)$ as discussed in §4.6.6. Recall that by Theorem 4.82, any such autoequivalence on $\mathrm{D}(X)$ is induced by a Fourier-Mukai transform. Thus, there is a homomorphism

$$\pi_1(\mathscr{M}(Y)) \to \mathrm{Auteq}(\mathrm{D}(X)) \tag{5.127}$$

of the fundamental group of the complex structure moduli space of the mirror Y into the autoequivalence group of self-equivalences of the derived category of coherent sheaves on X. The subcategories of stable objects for two points in $\mathscr{T}(Y)$ related by monodromy must then be related by the corresponding autoequivalence. We will see examples of this in §5.6.2 and §5.8.3.1.

5.5.4. Π-Stability. Assuming mirror symmetry to be true we may now copy the description of the stability of A-branes in §5.2.1 over to the case of B-branes. We recall the definition of the central charge $Z(E)$ from §5.3.5.5. Given E, we may choose $\xi(E)$ such that

$$\xi(E) = \frac{1}{\pi} \arg Z(E) \pmod 2 \tag{5.128}$$

and demand that $\xi(E)$ vary continuously with $B+iJ$ so long as E is stable. Following (5.74) we have

$$\xi(E[n]) = \xi(E) + n. \tag{5.129}$$

Finally we copy the picture in Figure 3 by asserting that, if we have a distinguished triangle in $\mathrm{D}(X)$ of the form

$$\begin{array}{ccc} & C & \\ {}^{c}\swarrow_{[1]} & & \nwarrow^{b} \\ A & \xrightarrow{\quad a \quad} & B, \end{array} \tag{5.130}$$

with A and B stable, then C is stable with respect to the decay represented by this triangle if and only if $\xi(B) < \xi(A) + 1$. Also, if $\xi(B) = \xi(A) + 1$

then C is marginally stable and we may state that

$$\xi(C) = \xi(B) = \xi(A) + 1. \tag{5.131}$$

We may use axiom **TR2** in §5.5.1 to rephrase this as follows.

DEFINITION 5.31. [**133**] A brane B is Π-stable if, for all triangles of the form (5.130) with A and C Π-stable, we have

$$\xi(A) < \xi(B) < \xi(C). \tag{5.132}$$

Conversely, if (5.132) is violated for any such triangle, then B is Π-unstable.

The necessity of this condition can also be seen directly within SCFT. For any two physical boundary conditions A and B, we can define a Hilbert space $\mathcal{H}_{AB}$ of open strings, with an action of the $\mathcal{N} = 2$ SCA. These must satisfy the constraints derived in §3.3.3, in particular $U(1)$ charges must satisfy (3.97). Using the relation (5.41) between $U(1)$ charges and gradings, this implies that

$$\xi(A) > \xi(B) \Rightarrow \mathrm{Hom}(A, B) = 0. \tag{5.133}$$

This is directly analogous to Lemma 5.9 in the discussion of μ-stability, and reduces to it in the large volume limit.

5.5.4.1. *Analysis of* Π*-stability.* Upon varying the gradings as in §5.3.6, in general we will find violations of (5.133), unless the set of Π-stable branes changes in some way. One might hope to understand this by analogy to the discussion of μ-stability, by regarding the set of inequalities (5.132) for all triangles involving stable branes A and C as necessary conditions for the existence of B, and then proving that these conditions are sufficient. However, the inherent circularity of this definition makes it unclear how to proceed.

Following [**138, 137, 149, 133, 18**], one can explore to what extent Definition 5.31 uniquely determines a consistent set of stable B-branes. While it is not obvious how to do this directly, if we know the set of stable B-branes (including their ξ's) at some basepoint in the moduli space, then we can apply the following rules to determine how the stable spectrum changes as we move along a path in the moduli space.

- We begin with a stable set of B-branes together with a value of ξ for each B-brane. This set must be consistent with the rules of Π-stability. That is, no distinguished triangle may allow a stable B-brane to decay into two other stable B-branes.
- As we move along a path in moduli space the ξ's will change continuously.
- Two stable B-branes may bind to form a new stable state.
- A stable B-brane may decay into other (marginally) stable states.

Note in the last case that a brane may decay into another state which becomes unstable at exactly the same point in moduli space. Note also that these rules make no reference to a value of ξ for an unstable object. This is probably not defined.

The conjecture is then that these rules determine a unique Π-stable subset at the end of the path. Such a formulation in terms of paths avoids the difficulties raised by monodromy. On the other hand, one needs to prove that the result is invariant under continuous deformations of the path. Arguments to this effect were given in [**18**], but are subsumed by the formulation we discuss next. Granting this point, we can expect the set of stable B-branes on X to be uniquely defined for any point in $\mathscr{T}(Y)$.

5.5.4.2. *Multiple decays.* Every object in $\mathrm{D}(X)$ is either stable or unstable for a given point in $\mathscr{T}(Y)$. Now, presumably, a particle which is unstable must decay into a set of stable particles. Thus the set of stable objects must be big enough to describe this. This puts a stronger constraint on stability than the previous subsection. For example, having no stable objects at all would have been consistent with our considerations so far.

If an unstable object decays into 2 stable objects we know how to describe the decay by a distinguished triangle. We now want to describe a decay of an object into 3 stable objects. We use the following octahedron to describe the process [**18**]:

(5.134)

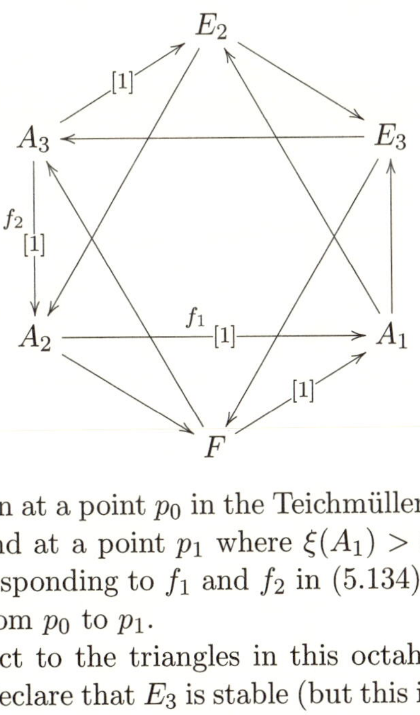

Suppose that we begin at a point p_0 in the Teichmüller space where $\xi(A_1) < \xi(A_2) < \xi(A_3)$ and end at a point p_1 where $\xi(A_1) > \xi(A_2) > \xi(A_3)$. Thus the open strings corresponding to f_1 and f_2 in (5.134) go from tachyonic to massive as we pass from p_0 to p_1.

At p_0, with respect to the triangles in this octahedron, E_2 and F are stable. We may also declare that E_3 is stable (but this isn't really necessary). Suppose there are two walls W_1 and W_2 between p_0 and p_1 such that $\xi(A_i) - \xi(A_{i+1})$ is negative on the p_0 side of W_i and positive on the p_1 side of W_i. Then there are two possibilities to consider as we move from p_0 to p_1:

(1) We cross W_1 and then W_2. As we cross W_2 the object F will decay into A_2 and A_3. At this instant $\xi(F) = \xi(A_2) < \xi(A_1)$ so we know

that E_3 must have already decayed into F and A_1. Thus E_3 decays into A_1, A_2 and A_3 by the time we reach p_1.

(2) We cross W_2 and then W_1. As we cross W_1 the object E_2 will decay into A_1 and A_2. At this instant $\xi(E_2) = \xi(A_2) > \xi(A_3)$ so we know that E_3 must have already decayed into E_2 and A_3. Thus E_3 decays into A_1, A_2 and A_3 by the time we reach p_1.

Either way, the condition for E_3 to decay into A_1, A_2 and A_3 is that

$$\xi(A_1) > \xi(A_2) > \xi(A_3). \tag{5.135}$$

We may generalize this to the case of decays into any number of objects. For any object E we define the following set of distinguished triangles

(5.136)

$$\begin{array}{ccccccccc} 0 = E_0 & \longrightarrow & E_1 & \longrightarrow & \cdots & \longrightarrow & E_{n-1} & \longrightarrow & E_n = E \\ & {\scriptstyle [1]}\nwarrow \swarrow & & {\scriptstyle [1]}\nwarrow \swarrow & & {\scriptstyle [1]}\nwarrow \swarrow & & {\scriptstyle [1]}\nwarrow \swarrow & \\ & A_1 & & A_2 & & A_{n-1} & & A_n & \end{array}$$

Then E decays into $A_1, A_2, \dots, A_n$ so long as

$$\xi(A_1) > \xi(A_2) > \cdots > \xi(A_n). \tag{5.137}$$

This suggests the following

CONJECTURE 5.32. **[65]** *For any point in* $\mathscr{T}(Y)$, *the subset of stable objects in* $\mathrm{D}(X)$ *is such that any object E can be decomposed (as in* (5.136)*) into a unique set of n stable objects A_k satisfying (5.137).*

While this is physically reasonable, it is not completely manifest, as it is not obvious what one means in SCFT by the unstable particles under discussion. By analogy to the Donaldson-Uhlenbeck-Yau theorem, a natural definition would be that these are boundary conditions defined using the correct holomorphic structure of the objects involved, but without knowing the actual solutions of the beta function conditions. One could then try to do RG flow to construct the conformal boundary condition; an unstable starting point would presumably flow to the direct sum of the A_k. We will discuss this idea further in §5.9.

In any case, granting Conjecture 5.32 leads to a set of axioms which determine unique sets of stable objects, as we will show in §5.7.

5.5.5. Comparison with μ-stability. In order to determine the set of Π-stable objects, it is best if we choose a basepoint in the moduli space of $B + iJ$ at which we know the stable objects *a priori*, from which we then follow paths as in §5.5.4. The obvious place to put the basepoint is near the large radius limit of the Calabi-Yau threefold X, since we expect B-branes there to correspond to vector bundles supported on (or "wrapping") submanifolds $L \subset X$.

Let us consider a brane wrapping L associated to a holomorphic vector bundle $E \to L$ with curvature (1,1)-form F. At large radius, the BPS condition reduces to the Hermitian-Yang-Mills condition, and thus such a bundle will exist if E is μ-stable (§5.2.2), meaning that for any subsheaf F of E the slope (5.33) satisfies

$$\mu(F) < \mu(E). \tag{5.138}$$

Of course, if F is a subsheaf of E then we have the short exact sequence ((5.36))

$$0 \longrightarrow F \longrightarrow E \longrightarrow G \longrightarrow 0, \tag{5.139}$$

for some sheaf G, and thus the description of this process using Π-stability should involve the corresponding triangle.

One might wonder whether triangles involving branes of different dimension might also be relevant. In the large volume limit they are not. A bundle on X associated to E cannot decay into a subsheaf F supported only on S since there is no homomorphism $F \to E$. Equally, a bundle on S cannot decay into a subsheaf on X since the quotient sheaf G in (5.139) would have negative rank.

Thus, we will recover μ-stability as a limit of Π-stability if the condition (5.132) reduces to (5.138) in the large volume limit.

From (5.71) we see that, for large J, the leading contribution to the central charge $Z(A)$ will be given by the lowest degree differential form in $\mathrm{ch}(A)$. As shown in §5.1.4.1, the lowest component of $\mathrm{ch}(i_*F)$ is given by the $(6-2\dim(S))$-form s which is Poincaré dual to S. Note that $\dim(S)$ is the complex dimension of S. Thus, for large J,

$$\begin{aligned} Z &\sim \int_X (-iJ)^{\dim(S)} \wedge s \\ &\sim \int_S (-iJ|_S)^{\dim(S)} \\ &\sim (-i)^{\dim(S)} \mathrm{Vol}(S), \end{aligned} \tag{5.140}$$

yielding

$$\xi(i_*E) \equiv -\tfrac{1}{2}\dim(S) \pmod 2. \tag{5.141}$$

If we choose the values of ξ to fix the mod 2 ambiguity arbitrarily we will violate the unitarity condition (5.97). For example, let $\mathcal{O}_X$ be the 6-brane wrapping X and let $\mathcal{O}_p$ be the 0-brane (skyscraper sheaf) at a point $p \in X$. Thus $\xi(\mathcal{O}_X) = -\frac{3}{2} \pmod 2$ and $\xi(\mathcal{O}_p) = 0 \pmod 2$. By restricting the value of a function on X to its value at p we see that $\mathrm{Hom}(\mathcal{O}_X, \mathcal{O}_p) = \mathbb{C}$ and so we must insist that $\xi(\mathcal{O}_X) < \xi(\mathcal{O}_p)$ if these B-branes are stable. Furthermore, by Serre duality, $\mathrm{Hom}(\mathcal{O}_p, \mathcal{O}_X[3]) = \mathbb{C}$ and so $\xi(\mathcal{O}_p) < \xi(\mathcal{O}_X) + 3$.[23] So the only possibility at large radius is that $\xi(\mathcal{O}_X) = \xi(\mathcal{O}_p) - \frac{3}{2}$.

A consistent choice is to set

$$\xi(i_* E) = -\tfrac{1}{2}\dim(S). \tag{5.142}$$

The corresponding shift of $U(1)_R$-charges of open strings can be seen directly in the nonlinear sigma model as well, as a consequence of the quantization of Dirichlet-Neumann open strings.

Let us see what happens at the subleading order in J. We restrict attention to the case of 6-branes, i.e., locally free sheaves. Now $\mathrm{ch}(E) = k + c_1 + \cdots$, where k is the rank of the associated vector bundle. Applying (5.71) we now obtain

$$\xi(E) = -\frac{3}{2} + \frac{1}{\pi}\tan^{-1}\frac{\mu(E)}{2} + \cdots, \tag{5.143}$$

where $\mu(E)$ is the slope defined in (5.33). Note that $|\mu(E)| \ll 1$ in the large volume limit.

The short exact sequence (5.139) induces the triangle

$$\begin{array}{ccc} & G & \\ {}^{[1]}\swarrow & & \nwarrow \\ E & \longrightarrow & F, \end{array} \tag{5.144}$$

in $\mathrm{D}(X)$, as explained in §5.5.1. If we are near the large radius limit $\xi(E)$ and $\xi(F)$ are both very close to $-\frac{3}{2}$ since they are locally free sheaves. G is either locally free or supported on a complex codimension one subspace. Thus $\xi(G)$ is close to either $-\frac{3}{2}$ or -1. Since the D-brane charges add according to §5.1, we have $Z(F) = Z(E) + Z(G)$, which implies $\xi(F)$ lies between $\xi(E)$ and $\xi(G)$. The Π-stability condition for F is $\xi(E) < \xi(G)$, which is therefore equivalent to $\xi(E) < \xi(F)$. By (5.143) this, in turn, is equivalent to $\mu(E) < \mu(F)$. Thus Π-stability reduces to μ-stability in the large volume limit, as first observed in [**138**].

5.5.6. Examples on the quintic. Let us now give some concrete examples of variation of stability. Perhaps the most important place to start is to give an example of a B-brane which is not a coherent sheaf, thus justifying the rather complicated formalism we have introduced (at least from the physics point of view). The original example of such an "exotic" brane was found in a study of the quintic threefold in [**134**].

Thus, let X be a quintic hypersurface in $\mathbb{P}^4$. This has $b^{1,1} = 1$ and thus $K(X) \cong \mathbb{Z}^4$. We denote the generator of $H^2(X, \mathbb{Z})$ by e.

---

[23] $\mathcal{O}_X[3]$ is the complex with $\mathcal{O}_X$ in position -3 and zero elsewhere.

Let us consider coherent sheaves on X obtained by restriction from $\mathbb{P}^4$; while this is not all, it spans $K(X) \otimes \mathbb{R}$ and is a large enough supply for present purposes. Let $\mathcal{O}_X(m) \cong \mathcal{O}_{\mathbb{P}^4}(m)|_X$; these are locally free rank one sheaves.

The essential point can be seen without introducing all of the formalism, but rather by using the large volume approximation to the B-brane central charges, with the perturbative α' corrections. A more complete discussion using the exact periods can be found in [**17**]. Recall from (5.71) that

$$Z(E) = \int_X e^{-(B+iJ)} \mathrm{ch}(E) \sqrt{\mathrm{td}(X)} \tag{5.145}$$

up to exponentially small corrections, which we drop. We can also drop $\sqrt{\mathrm{td}(X)} = \sqrt{1 + \frac{5}{6}e^6}$ for the present discussion.

Thus

$$\begin{aligned} \xi(\mathcal{O}_X(m)) &= \frac{1}{\pi} \arg \int_X \exp((m - B - iJ)e) \\ &= \frac{1}{\pi} \arg\big(5(m - B - iJ)^3\big) \\ &= \frac{3}{\pi}\theta_m - 3, \end{aligned} \tag{5.146}$$

where θ_m is the angle in the complex $(B + iJ)$-plane between the positive real axis, and a straight line connecting $m + i0$ to $B + iJ$.

5.5.6.1. *Stability of 4-branes on the quintic.* Consider the following short exact sequence of sheaves:

$$0 \longrightarrow \mathcal{O}_X(a) \xrightarrow{f} \mathcal{O}_X(b) \longrightarrow \mathcal{O}_S(b) \longrightarrow 0, \tag{5.147}$$

where $b > a$. All 4-branes corresponding to B-branes on the quintic correspond to $\mathcal{O}_S(b)$ for some f and some b.

Applying the Π-stability criterion to the distinguished triangle associated to (5.147) gives $\xi(\mathcal{O}_X(b)) - \xi(\mathcal{O}_X(a)) < 1$, which yields

$$\theta_b - \theta_a < \frac{\pi}{3}. \tag{5.148}$$

When this is satisfied, the open string corresponding to f in (5.147) is tachyonic. Simple geometry yields that this corresponds to the points above a circular arc in the upper $(B + iJ)$-plane with center $\frac{1}{2}(a + b) + \frac{1}{2\sqrt{3}}(b - a)i$ and radius $\frac{1}{\sqrt{3}}(b - a)$.

Thus, these 4-branes are stable in the large radius limit, as expected. But below this arc of marginal stability the 4-brane decays into $\mathcal{O}_X(b)$ and $\mathcal{O}_X(a)[-1]$. Physically, this is a 6-brane and anti-6-brane with some 4-brane charges. On the other hand, a 4-brane is always μ-stable since it has no subobjects. Thus this is a simple example for which the predictions of the two conditions differ.

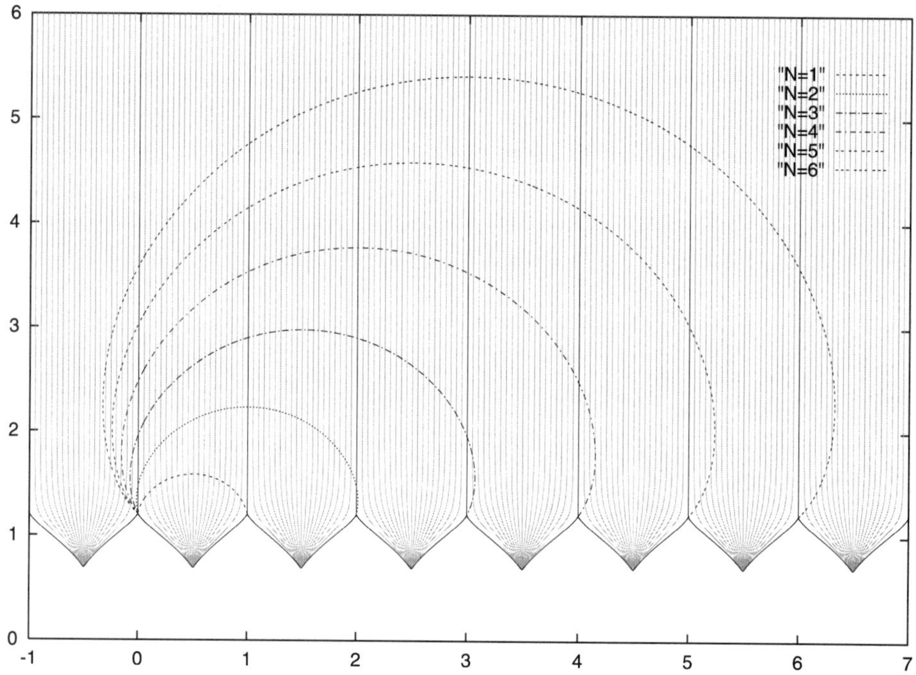

FIGURE 8. Stability of various D4-branes in the $(B+iJ)$-plane. As explained in §5.5.6.1, below the Nth arc the D4-brane $\mathcal{O}_S(N)$ decays according to the exact sequence (5.147). From [**18**].

One can check that using the exact (instanton corrected) expression for (5.146) predicts the same decay (at a slightly different $B+iJ$). Thus the problem with μ-stability is *not* that we have not taken enough α' corrections into account. Rather μ-stability fails because it is only appropriate when the subobject relation is that of coherent sheaves. In particular, it can never see decays caused by anti-branes.

Note also that Π-stability corrections can sometimes be relevant even at large radius. For very large values of $b-a$, this line of marginal stability can extend to large values of J. The condition for μ-stability to hold is rather that $J \gg |b-a|$, or more generally all the Chern classes of the sheaves involved.

Of course, we have not found the precise form of the line of marginal stability since we used the large radius approximation in (5.146). The precise curves for $a=0$ and $b=N$, found using the exact periods from [**85**], are shown in Figure 8. This figure depicts part of the Teichmüller space, with each enclosed region representing a fundamental region. The coordinate $B+iJ$ is defined using the mirror map of [**85**]. We can see that the exact lines of marginal stability are not far from being circular arcs.

We should emphasize that the considerations so far prove neither that the 4-branes are stable above the lines in Figure 8 nor that the 4-branes are unstable below the lines. The 4-branes might decay by other channels before these lines are reached, and the decay products might also be unstable by the time we reach a line of marginal stability. We will come back to this point shortly.

5.5.6.2. *Conifold points.* Note that the lines of marginal stability in Figure 8 have endpoints. These are "conifold points," at which one of the decay products $\mathcal{O}_X(a)$ or $\mathcal{O}_X(b)$ becomes massless (has $Z = 0$), and thus the grading becomes ill-defined.

These points are beyond the regime of validity of (5.71), and thus we would need an exact analysis of the periods of the mirror quintic to properly discuss them. Since we will make such an analysis for the $\mathbb{C}^3/\mathbb{Z}_3$ orbifold shortly, leading to very similar results, we omit this here, referring to [**17**].

The underlying reason for the existence of such massless branes is clear in the mirror A-model. They are special Lagrangians on vanishing cycles associated to a degeneration of complex structure. One might try to prove mathematically that these exist by using an approximation to the Ricci-flat metric valid near the degeneration,[24] and physically by appealing to the principle that singularities in the SCFT (here degeneration of complex structure) must be associated with massless particles.

Thus, we know that the B-branes $\mathcal{O}_X(m)$ are stable near the endpoints of the claimed line of marginal stability, as well as near large volume. This goes some distance towards justifying the claimed lines of marginal stability. One can also hypothesize other, more contrived modes of decay of 4-branes, and check that these decay at smaller radii. Thus it seems hard to imagine that Figure 8 is incorrect.

More generally, branes which can become massless play a very central role in the discussion, determining monodromies and a decomposition of the stringy Kähler moduli space into regions governed by stability conditions on abelian subcategories. We will develop this idea shortly.

5.5.6.3. *An exotic B-brane.* Let us now apply Serre duality to the 4-brane decay we just discussed [**134**]. The potential tachyons of that discussion lie in $\mathrm{Hom}(\mathcal{O}_X(a), \mathcal{O}_X(b))$ which is nonzero for $b > a$. By Serre duality, this Hilbert space of open strings is isomorphic to $\mathrm{Hom}(\mathcal{O}_X(b), \mathcal{O}_X(a)[3])$. Thus, we may consider the distinguished triangle

$$\begin{array}{ccc} & \mathscr{X}_{a,b} & \\ {}^{[1]}\swarrow & & \nwarrow \\ \mathcal{O}_X(b) & \xrightarrow{\quad g \quad} & \mathcal{O}_X(a)[3], \end{array} \tag{5.149}$$

where $\mathscr{X}_{a,b}$ is defined as the cone of the map g.

[24]As of 2008, this has not yet been done.

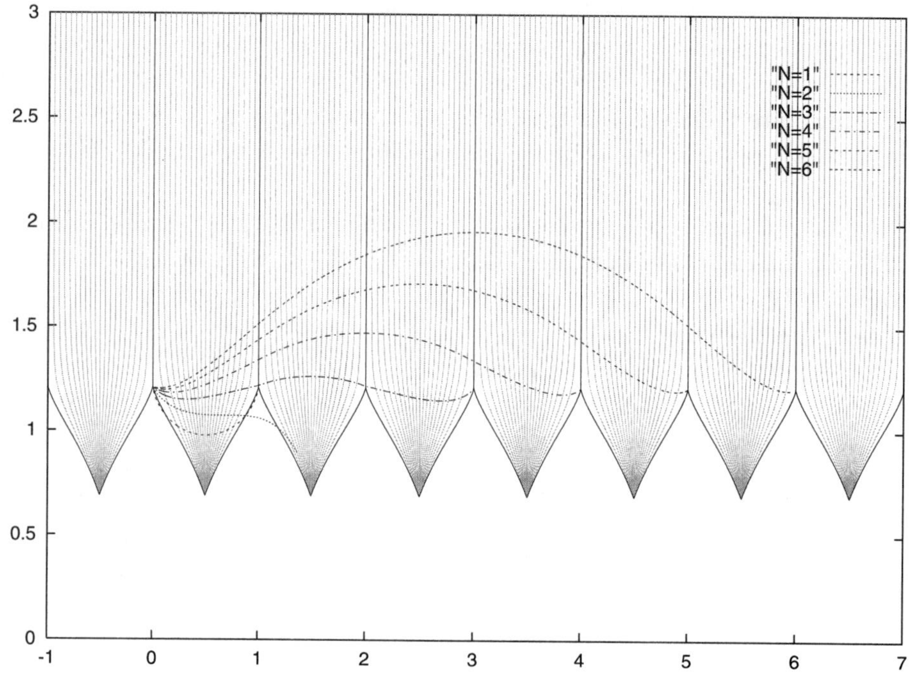

FIGURE 9. Stability of the exotic objects $\mathscr{X}_{0,N}$ in the $(B+iJ)$-plane. The Nth arc is associated to a decay described by (5.149).

The stability of $\mathscr{X}_{a,b}$ can now be determined (at least relative to the triangle (5.149)) from the analysis above together with the relation (5.129). For stability we require $\xi(\mathcal{O}_X(a)[3])-\xi(\mathcal{O}_X(b)) = 3+\xi(\mathcal{O}_X(a))-\xi(\mathcal{O}_X(b)) < 1$. Using the approximation (5.146) this yields that $\mathscr{X}_{a,b}$ is stable *below* a circular arc in the upper half-plane with center $\frac{1}{2}(a+b) - \frac{1}{2\sqrt{3}}(b-a)i$ and radius $\frac{1}{\sqrt{3}}(b-a)$. In particular, $\mathscr{X}_{a,b}$ is always unstable in the large radius limit, although we may make it stable at any given arbitrarily large radius by choosing a large enough value of $b-a$.

Again we may use numerical techniques to plot a more precise version of the lines of marginal stability. In Figure 9 (again taken from [**18**]) we plot some examples. We should note that something very interesting happens for $\mathscr{X}_{0,2}$. It is an example of a case where a decay product can itself decay, forcing the line of marginal stability to end on another line of marginal stability rather than at a conifold point. We refer to [**18**] for a full discussion.

So, what exactly is $\mathscr{X}_{a,b}$? It is a B-brane that does not exist in the large radius limit and does not have a direct sigma model interpretation, but can be constructed by starting from a combination of sigma model boundary states and varying the Kähler moduli. Let's examine this a little more closely. Suppose we have an injective resolution for $\mathcal{O}_X$:

$$0 \longrightarrow \mathcal{O}_X \longrightarrow \mathcal{I}^0 \xrightarrow{i_0} \mathcal{I}^1 \xrightarrow{i_1} \mathcal{I}^2 \xrightarrow{i_2} \mathcal{I}^3 \xrightarrow{i_3} \cdots \tag{5.150}$$

Referring to Chapter 4, an element of $\mathrm{Ext}^3(\mathcal{O}_X(1), \mathcal{O}_X)$ corresponds to a map $g : \mathcal{O}_X(1) \to \mathcal{I}_3$ such that $i_3 g = 0$. Thus $\mathcal{X}_{0,1} = \mathrm{Cone}(g)$ corresponds to the complex

(5.151)

$$\cdots \longrightarrow 0 \longrightarrow \mathcal{I}^0 \xrightarrow{i_0} \mathcal{I}^1 \xrightarrow{\binom{0}{i_1}} \begin{matrix}\mathcal{O}_X(1)\\ \oplus\\ \mathcal{I}^2\end{matrix} \xrightarrow{(g\ i_2)} \underline{\mathcal{I}^3} \xrightarrow{i_3} \cdots,$$

where we denote the zero position with a dotted underline.

The cohomology sheaves of this complex look like $\mathcal{H}^{-3} = \mathcal{O}_X$ and $\mathcal{H}^{-1} = \mathcal{O}_X(1)$, so naively this B-brane looks like an anti-6-brane added to another anti-6-brane with a 4-brane charge. This is certainly true if g is zero. More precisely, (5.151) is quasi-isomorphic to

$$\cdots \longrightarrow \mathcal{O}_X \longrightarrow 0 \longrightarrow \mathcal{O}_X(1) \longrightarrow \underline{0} \longrightarrow \cdots, \tag{5.152}$$

if and only if $g = 0$. One can confirm this by explicitly computing

$$\mathrm{Hom}(\mathcal{X}_{0,1}, \mathcal{X}_{0,1})$$

using long exact sequences, to obtain

$$\mathrm{Hom}(\mathcal{X}_{0,1}, \mathcal{X}_{0,1}) = \begin{cases} \mathbb{C}^2 & \text{if } g = 0, \\ \mathbb{C} & \text{if } g \neq 0. \end{cases} \tag{5.153}$$

This ties in nicely with the comments at the end of §5.4.4. If $g \neq 0$ we have a single irreducible D-brane. When $g = 0$ we have a direct sum of two D-branes and thus a gauge group $U(1) \times U(1)$. $\mathcal{X}_{0,1}$ thus becomes a distinct object when the tachyon is turned on.

All this shows that $\mathcal{X}_{0,1}$ is a truly exotic object from a classical geometric point of view. It is not quasi-isomorphic to a complex with a single coherent sheaf and so it cannot be viewed as a vector bundle supported on some subspace of X.

Note that we were able to build these exotic D-branes because we were able to use Ext^n's for $n > 1$. This separated terms in the complexes far enough to avoid everything collapsing back to a single term complex. Thus we have an explicit example which depends on the considerations made in §5.3.4.

In fact, certain D-brane states at the Gepner point have been explicitly constructed as boundary conformal field theories [**402, 69, 163**]. These include a boundary state with all of the properties (the charge and no chiral matter) of this exotic state. This identification can be confirmed using an identification between the B-twist of the Gepner model and a derived category constructed from the matrix factorizations discussed in §3.6.8.

5.5.7. Flops. Simpler examples of stability are provided by noncompact Calabi-Yau threefolds. A very instructive case is the flop. Its role in mirror symmetry for threefolds has been much studied in physics and we refer to [**192**] for an overview.

The basic example is obtained by considering a singular algebraic variety X_0 containing the conifold point of Example 5.28. As we saw there, such a singular space may be resolved by replacing the conifold point by a $\mathbb{P}^1$ in two different ways to form smooth Calabi-Yau manifolds X or X'. Generally X and X' are topologically distinct, and related by a flop.

Geometrically the process of blowing down X or X' back to X_0 may be viewed as a deformation of the Kähler form J. Indeed, in the space of Kähler forms, X and X' may be considered to live on the two sides of a wall corresponding to X_0. Let C be the $\mathbb{P}^1$ inside X. As we approach the wall from the X side, the area of C shrinks down to zero. While continuing past this wall would give C negative area, by reinterpreting the geometry in terms of X' we give positive area to the new $C' \subset X'$.

In the nonlinear sigma model, the Kähler class is promoted to a complex modulus $B + iJ$. This has a profound effect as described in [**20**]. The conformal field theory associated to the singular target space X_0 is nonsingular so long as the component of B associated to C (or C') is nonzero. Thus, rather than having a real codimension one wall of singular spaces in the moduli space of J, we have a *complex* codimension one subspace of singular conformal field theories, as required by mirror symmetry and the discussion of §5.5.3. It follows that one can pass from X to X' smoothly by going around this singular subset, obtaining a connected moduli space of SCFT's.

However, once we bring D-branes into the picture, we see that we must have a jump in some sense. At least to some degree of approximation, the Calabi-Yau target space is the moduli space of 0-branes. Thus the moduli space of 0-branes must undergo some transition during the flop even if we avoid the singular conformal field theory at $B = 0$. As shown in [**61**] (building on a construction in [**16**]), this discontinuity is provided by 0-brane stability considerations.

We imagine that we are in a Calabi-Yau threefold X with all $J \gg 1$ except for that of the flopping $\mathbb{P}^1$. The periods on its mirror may be analyzed simply in this limit as explained in [**19**]. Essentially the only component of $B + iJ$ of interest is given by

$$t = \int_C B + iJ. \tag{5.154}$$

This has a moduli space given by $\mathbb{P}^1$ as shown in Figure 10. The flop takes place on the equator and the singular conformal field theory is at $t = O$. Let z be an affine coordinate on this $\mathbb{P}^1$ so that $z = 0$ is the large C limit,

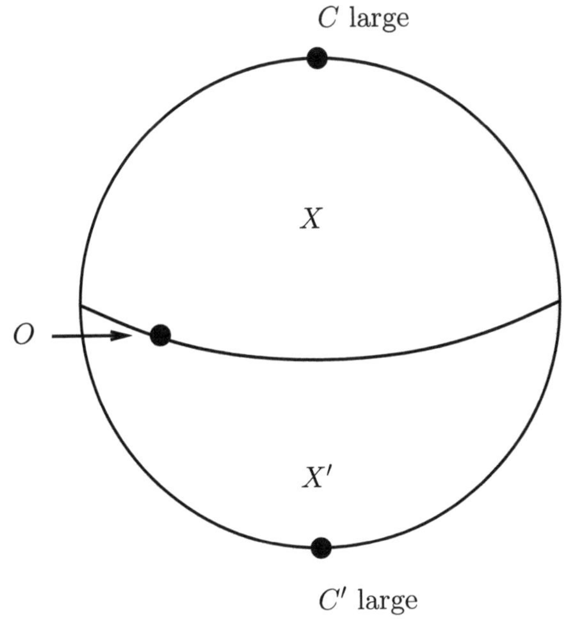

FIGURE 10. The moduli space of $B + iJ$ for the flop.

$z = 1$ corresponds to $t = O$ and $z = \infty$ gives the large C' limit after the flop transition.

The periods on the mirror take the general form [**19**]

$$\Phi = A_1 + A_2 \log(z). \tag{5.155}$$

Thus we have the *exact* relation given by the mirror map

$$t = \frac{1}{2\pi i} \log(z). \tag{5.156}$$

Let $i : C \to X$ be the inclusion map and let $\mathcal{O}_C(m)$ denote $i_*\mathcal{O}(m)$. We will use $\mathcal{O}_x$ to denote the skyscraper sheaf, i.e., 0-brane, associated with the point $x \in X$. Using the Grothendieck-Riemann-Roch theorem (§5.1),

$$\int_X e^{-(B+iJ)} \mathrm{ch}(\mathcal{O}_C(m)) \sqrt{\mathrm{td}(T_X)} = -t + m + 1. \tag{5.157}$$

Therefore $Z(\mathcal{O}_C(m)) = -t + m + 1$ exactly since t and 1 are periods. The most natural statement would seem to be therefore that the brane corresponding to $\mathcal{O}_C(-1)$ becomes massless at O. Actually, we are free to choose the branch of the logarithm; however, we feel that we could, instead, say that $\mathcal{O}_C$ becomes massless for simplicity. This is equivalent to choosing a basepoint near the large radius limit but then going once around this large radius limit before heading towards O. With this choice, we focus on this neighbourhood of O by putting $t = 1 + \epsilon e^{i\theta}$ for a small real and positive ϵ. We sketch this neighbourhood in Figure 11.

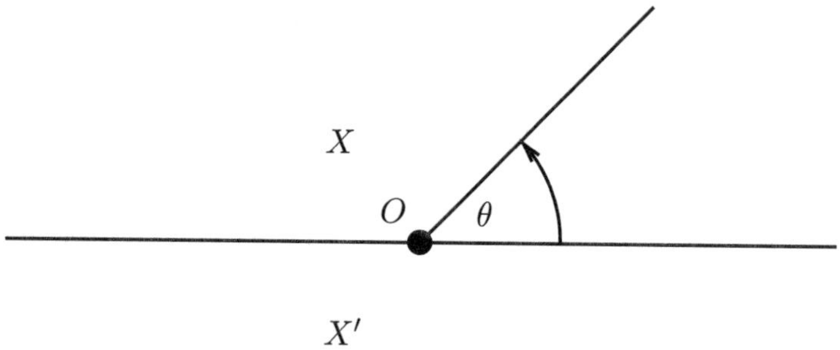

FIGURE 11. The neighbourhood of O.

Suppose $x \in C$. Then we have a short exact sequence

$$0 \longrightarrow \mathcal{O}_C(-1) \longrightarrow \mathcal{O}_C \longrightarrow \mathcal{O}_x \longrightarrow 0. \tag{5.158}$$

This leads to a distinguished triangle which we write in the form

$$\begin{array}{ccc} & \mathcal{O}_C(-1)[1] & \\ {\scriptstyle \frac{\theta}{\pi}\,[1]} \swarrow & & \nwarrow {\scriptstyle 0} \\ \mathcal{O}_C & \xrightarrow[1-\frac{\theta}{\pi}]{} & \mathcal{O}_x. \end{array} \tag{5.159}$$

Near O, $Z(\mathcal{O}_C)$ is very small and $Z(\mathcal{O}_x) = 1$. It follows that $\xi(\mathcal{O}_x) = \xi(\mathcal{O}_C(-1)[1]) = 0$ and $\xi(\mathcal{O}_C) = \theta/\pi - 1$. The morphisms in (5.159) are labelled by the differences in the ξ's between the head and tail of the arrow. Thus a vertex is Π-stable (with respect to that triangle) when and only when the label on the opposite edge is < 1. Therefore, the 0-brane $\mathcal{O}_x$ decays into $\mathcal{O}_C$ and $\mathcal{O}_C(-1)[1]$ as θ increases beyond π.

We also have a distinguished triangle

$$\begin{array}{ccc} & \mathcal{O}_C[1] & \\ {\scriptstyle 1-\frac{\theta}{\pi}\,[1]} \swarrow & & \nwarrow {\scriptstyle 0} \\ \mathcal{O}_C(1) & \xrightarrow[\frac{\theta}{\pi}]{} & \mathcal{O}_x, \end{array} \tag{5.160}$$

which shows that $\mathcal{O}_x$ decays into $\mathcal{O}_C(1)$ and $\mathcal{O}_C[1]$ for $\theta < 0$. Either way, we see that $\mathcal{O}_x$ decays as we move from the X phase into the X' phase in Figure 11.

Suppose $y \notin C$. Then, even though $\mathcal{O}_y$ has exactly the same D-brane charge as $\mathcal{O}_x$ it does not decay by (5.159) or (5.160) since there are no morphisms from $\mathcal{O}_C$ or $\mathcal{O}_C(1)$ to $\mathcal{O}_y$. Indeed, we would not expect 0-branes away from C to be affected by the flop transition.

We should also be able to see the objects in the derived category of X which play the role of points on C' *after* we do the flop. Before we do this we need to introduce a method of computing some of the relevant Ext's. Suppose we are given sheaves E and F on C. Given the embedding $i : C \to X$ with a normal bundle N on C, there is a spectral sequence with

$$E_2^{p,q} = \operatorname{Ext}_C^p(E, F \otimes \bigwedge^q N), \tag{5.161}$$

converging to $E_\infty^{p,q}$ with $\bigoplus_{p+q=n} E_\infty^{p,q} = \operatorname{Ext}_X^n(i_*E, i_*F)$.[25]

For example, consider $\operatorname{Ext}_X^n(\mathcal{O}_C, \mathcal{O}_C(-1))$. Since $N = \mathcal{O}(-1) \oplus \mathcal{O}(-1)$, we have an E_2 stage of the spectral sequence given by

$$E_2^{p,q} = H^p(C, \mathcal{O}_C(-1) \otimes \bigwedge^q N) :$$

$$\begin{array}{c|cc} & 0 & \mathbb{C}^2 \\ q & 0 & \mathbb{C}^2 \\ & 0 & 0 \\ \hline & & p \end{array} \tag{5.162}$$

and thus

$$\operatorname{Ext}_X^n(\mathcal{O}_C, \mathcal{O}_C(-1)) = \begin{cases} \mathbb{C}^2 & \text{if } n = 2 \text{ or } 3, \\ 0 & \text{otherwise.} \end{cases} \tag{5.163}$$

Thus we may use morphisms $f \in \operatorname{Hom}(\mathcal{O}_C[-1], \mathcal{O}_C(-1)[1]) \cong \mathbb{C}^2$ to form new objects D_f:

(5.164)

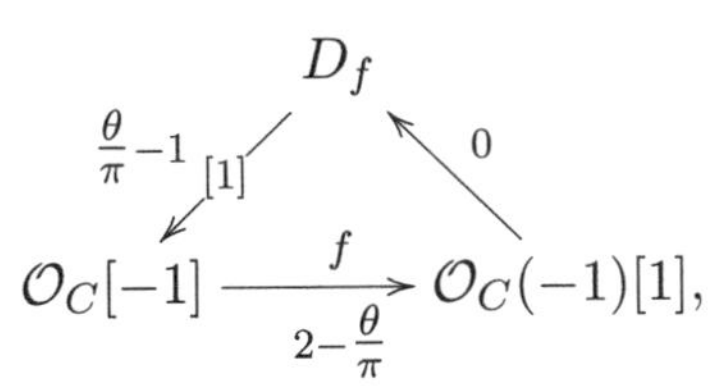

which become stable for $\theta > \pi$. As noted in §5.3.2.2, rescaling f by a complex number has no effect on D_f, so we have a $\mathbb{P}^1$'s worth of D_f's. These indeed represent the points on C' which become stable as we flop into X' by increasing θ through π, as argued in [**16**].

The objects D_f are exotic in the same sense as those in §5.5.6.3. They were described as "perverse sheaves" in [**16**].

We can compare with Conjecture 4.81 (proven for threefolds) which implies that $\mathrm{D}(X)$ is equivalent to $\mathrm{D}(X')$. The physics arguments imply that this will be true whenever the nonlinear sigma models on X and X' are continuously connected by varying $B+iJ$, since they lead to the same B-model.

[25] A quick way of proving this is to use the "right adjoint" functor $i^!$ of i_*.

5.6. The orbifold $\mathbb{C}^3/\mathbb{Z}_3$ and its resolution

We now give an extended analysis for this case. Most of the original analysis of D-branes on threefold orbifolds was done for this simplest example (e.g., [**139, 116, 115, 137**]).

We recall the discussion of Example 4.35. Take g to be a generator of $\mathbb{Z}_3$ acting as $g : (z_1, z_2, z_3) \mapsto (\omega z_1, \omega z_2, \omega z_3)$ for $\omega = \exp(2\pi i/3)$. Let V_i, $i = 0, \ldots, 2$ be the one-dimensional irreducible representations of $\mathbb{Z}_3$ given by $\rho(g) = \omega^i$. Then a representation $V = \bigoplus_i m_i V_i$ is associated to the quiver

(5.165)

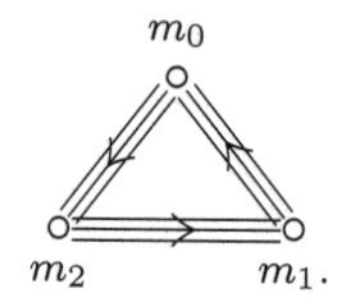

Let $\Delta_{m_0 m_1 m_2}$ denote a quiver representation of the form (5.165); for example the fractional branes are $F_0 = \Delta_{100}$, $F_1 = \Delta_{010}$ and $F_2 = \Delta_{001}$.

As is well-known, this orbifold is resolved with an exceptional divisor $E \cong \mathbb{P}^2$. In this case, X may be viewed as the total space of the line bundle corresponding to the sheaf $\mathcal{O}_E(-3)$. This space has $h^{1,1} = 1$ and thus the stringy moduli space is one-complex-dimensional. By arguments we discuss shortly, it can be viewed as a $\mathbb{P}^1$ with three punctures. One of these corresponds to the large radius limit where the exceptional divisor E is infinitely large. At the other extreme, we have the orbifold point where we have no blow-up. At a third point on the $\mathbb{P}^1$, which we denote P_0, the SCFT is believed to become singular, for reasons we will discuss.

5.6.1. Periods.

Again we may use mirror symmetry to analyze this one-dimensional moduli space. The mirror to $\mathbb{C}^3/\mathbb{Z}_3$ and its resolution can be described as the non-compact threefold $xy = 1 + s + t + z/st$ inside $\mathbb{C}^2 \times (\mathbb{C}^*)^2$, with x, y coordinates on $\mathbb{C}^2$ and s, t coordinates on $(\mathbb{C}^*)^2$. In addition, z provides the one-parameter complex structure moduli mirror to the volume of the exceptional divisor E in the resolution of $\mathbb{C}^3/\mathbb{Z}_3$. (See for example [**243**]). Now analyze this moduli space of complex structures and, in particular the Gauss-Manin connection on this moduli space to find the periods.

The Picard-Fuchs equations in question are

$$\left(z\frac{d}{dz}\right)^3 \Phi + 27z\left(z\frac{d}{dz}\right)\left(z\frac{d}{dz} + \tfrac{1}{3}\right)\left(z\frac{d}{dz} + \tfrac{2}{3}\right)\Phi = 0. \tag{5.166}$$

Clearly any constant solves this differential equation. Putting

$$z = (3e^{-\pi i}\psi)^{-3}$$

we may write a basis for the remaining solutions near $\psi = 0$ as

$$\varpi_j = \frac{1}{2\pi i}\sum_{n=1}^{\infty}\frac{\Gamma(\frac{n}{3})\omega^{nj}}{\Gamma(n+1)\Gamma(1-\frac{n}{3})^2}(3\psi)^n, \tag{5.167}$$

where $\omega = \exp(2\pi i/3)$. Thus 1, ϖ_0 and ϖ_1 form a basis for the solutions of the Picard-Fuchs equation. We then make an analytic continuation to a basis for small z (valid for $|\arg(z)| < \pi$),

$$\begin{aligned}
\Phi_0 &= 1,\\
\Phi_1 &= \frac{1}{2\pi i}\cdot\frac{3}{2\pi i}\int\frac{\Gamma(3s)\Gamma(-s)}{\Gamma(1+s)^2}z^s\,ds\\
&= \frac{1}{2\pi i}\log z + O(z)\\
&= t\\
&= \varpi_0,\\
\Phi_2 &= -\frac{1}{4\pi^2}\cdot\frac{-6}{2\pi i}\int\frac{\Gamma(3s)\Gamma(-s)^2}{\Gamma(s+1)}(e^{-\pi i}z)^s\,ds\\
&= -\frac{1}{4\pi^2}(\log z - i\pi)^2 - \tfrac{5}{12} + O(z)\\
&= t^2 - t - \tfrac{1}{6} + O(e^{2\pi i t})\\
&= -\tfrac{2}{3}(\varpi_0 - \varpi_1),
\end{aligned} \tag{5.168}$$

where the mirror map is given by $t = \int_C B + iJ = \frac{1}{2\pi i}\log(z) + O(z)$, and C is a $\mathbb{P}^1$ hyperplane in E.

A picture of the stringy Kähler moduli space near the $\mathbb{C}^3/\mathbb{Z}_3$ orbifold point is given in Figure 12. Note that the orbifold point lies at exactly $B = J = 0$.[26]

Now, using (5.12) and (5.71), we can compute exact central charges. For example, consider the 4-branes $\mathcal{O}_E(m)$ wrapping the exceptional divisor:

$$Z(\mathcal{O}_E(m)) = -(m+\tfrac{4}{3})\varpi_0 + \tfrac{1}{3}\varpi_1 + \tfrac{1}{2}m^2 + \tfrac{3}{2}m + \tfrac{4}{3}. \tag{5.169}$$

5.6.1.1. *Massless branes.* Consider the point P_0 in the moduli space where $\psi = 2\pi i/3$ and we have a singular conformal field theory. At P_0 we have $\varpi_0 = t = \frac{1}{2} + iJ_0$ where $J_0 \approx 0.4628$. From (5.167) the value of ϖ_1 at P_0 will clearly be equal to the value of ϖ_0 at $\psi = 4\pi i/3$, namely $-\frac{1}{2} + iJ_0$. Thus $\varpi_0 - \varpi_1 = 1$ at P_0. It follows from (5.169) that $\mathcal{O}_E(-1)$ becomes massless at P_0. Similarly $\mathcal{O}_E(-2)$ becomes massless at the point $\psi = 4\pi i/3$.

---

[26]There is a false assumption in [**19**] which shifts B by $\frac{1}{2}$. The correct argument appears in [**367**].

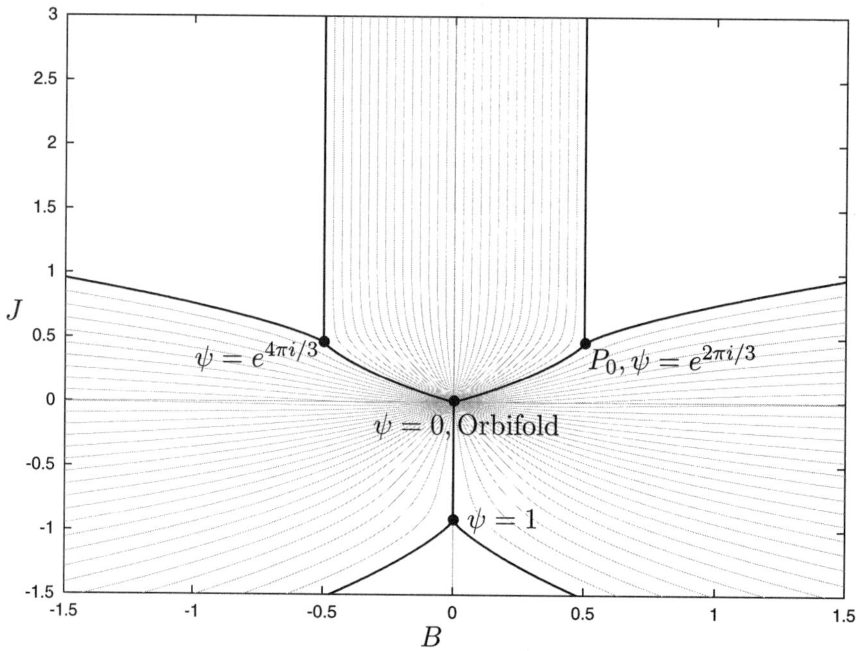

FIGURE 12. Fundamental regions for the moduli space of the $\mathbb{Z}_3$-orbifold. Each shaded region (bordered by solid lines) is an image under the map $\psi \to t$ of (5.168) of the plane $\psi^3 \in \mathbb{C}$. The three shaded regions are permuted by the orbifold monodromy, while the unshaded areas are divided into (infinitely many) additional fundamental regions related by the other monodromies.

Let us tentatively assume that the singularity in the conformal field theory at P_0 is caused by the stable B-brane $\mathcal{O}_E(-1)$ becoming massless. We will check this momentarily.

5.6.2. Monodromy. Recall from §5.5.3.1 that following a nontrivial closed path in the moduli space produces an autoequivalence on the derived category and the set of stable objects. Now that we have an explicit example of a multiply connected moduli space, we can explore this in some depth.

We first discuss monodromy around the large radius limit. This corresponds to $B \to B + \omega$ for $\omega \in H^2(X, \mathbb{Z})$ dual to some divisor D. By §3.5.2.7 this is equivalent to $F \to F + \omega$ of the curvature of the bundles, and can be achieved by the tensor product

$$E \to E \otimes \mathcal{O}_X(D).$$

In terms of the Fourier-Mukai transform, this may be achieved by a kernel $\mathcal{L} = \Delta_* \mathcal{O}_X(D) = \mathcal{O}_{\Delta X}(D)$. We will denote this transform by $\mathcal{L}$.

For our noncompact example, we take D Poincaré dual to the component of the Kähler form which is being taken to be very large. Thus we require D to intersect C (the $\mathbb{P}^1$ hyperplane of E) in one point. We will also denote $\mathcal{O}_X(D)$ by $\mathcal{O}_X(1)$. This notation is also consistent with the fact that $\mathcal{O}_X(1)\otimes \mathcal{O}_E = \mathcal{O}_E(1)$. Note also that since the normal bundle of E corresponds to $\mathcal{O}_E(-3)$, we have a short exact sequence

$$0 \longrightarrow \mathcal{O}_X(3) \longrightarrow \mathcal{O}_X \longrightarrow \mathcal{O}_E \longrightarrow 0. \tag{5.170}$$

5.6.2.1. *Monodromy around the orbifold point $\psi = 0$.* In general, orbifolds in string theory may exhibit a "quantum symmetry" which acts on a state twisted by $g \in G$ by multiplication by $q(g)$, where

$$q \in \mathrm{Hom}(G, \mathbb{C}^*). \tag{5.171}$$

The group $\mathrm{Hom}(G, \mathbb{C}^*)$ is thus the group of quantum symmetries. It is not hard to show that this is isomorphic to the *abelianization* of G, i.e., $G/[G, G]$. Yet another interpretation of $\mathrm{Hom}(G, \mathbb{C}^*)$ is the group of one-dimensional representations of G, where the group operation is the tensor product. Given a one-dimensional representation U of G we may act on the set of representations by $U \otimes -$. This gives us a symmetry of the McKay quiver and thus shows exactly how the quantum symmetries of an orbifold act on the category of D-branes.

For G abelian, the quantum symmetry group is isomorphic to G and acts transitively on the nodes of the McKay quiver. In our case, the quantum symmetry group $\mathbb{Z}_3$ acts by tensoring V by the representation F_1, in other words rotating the McKay quiver by $2\pi/3$. Denoting this transform as $\mathcal{G}$, we have

$$\mathcal{G} : (F_0, F_1, F_2) \to (F_1, F_2, F_0).$$

Now recall the discussion of the identification of the fractional branes from §5.4.5.6. This depended on a choice of chamber, which is also permuted by the $\mathbb{Z}_3$ action. This choice corresponds to a specific analytic continuation of the periods (5.168) from large volume to the orbifold point. To test whether we have correctly matched up these choices, we can take a proposed identification of F_i as large volume sheaves, compute their large volume central charges from their K-theory classes, and continue this expression back to the orbifold point. The correct identification will then respect the quantum symmetry.

For example, let us adopt the correspondence of chamber $\theta \in C_2$ of Table 3 in Chapter 4,

(5.172)

$F_0 =$ 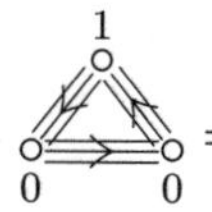 $= \mathcal{O}_E(-1)$

$F_1 =$ 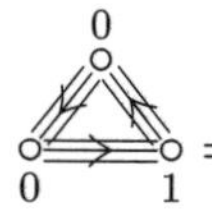 $= \Omega_E[1]$

$F_2 =$ 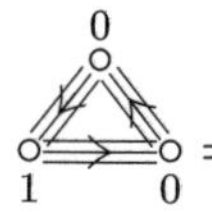 $= \mathcal{O}_E(-2)[2].$

Using this in (5.168) and continuing in ψ along the negative real axis, near the orbifold point we obtain

$$
\begin{aligned}
Z(F_0) &= \tfrac{1}{3}(1 - \varpi_0 + \varpi_1) \\
Z(F_1) &= \tfrac{1}{3}(1 - \varpi_0 - 2\varpi_1) \qquad (5.173) \\
Z(F_2) &= \tfrac{1}{3}(1 + 2\varpi_0 + \varpi_1).
\end{aligned}
$$

Thus, at the orbifold point the F_i's all have central charge $\frac{1}{3}$, consistent with the quantum symmetry. This also confirms the identification of the massless branes in §5.6.1.1.

Note that specifying $\mathcal{G}$ as the transform that permutes the F_j's comes quite close to specifying an automorphism of the derived category uniquely. The full triangulated subcategory that contains $\{F_0, F_1, F_2\}$ is equivalent to the derived category of finite dimensional matrix factorizations and this includes all sheaves with compact support.[27] So $\mathcal{G}$ is at least uniquely specified on this subcategory.

5.6.2.2. *Monodromy around the conifold point.* We now want to find the Fourier-Mukai transform for this case, call it $\mathcal{K}$. Its action on the D-brane charges can be found from the monodromy of the periods (5.168), and is

$$\mathrm{ch}(\mathcal{K}(F)) = \mathrm{ch}(F) - \langle \mathcal{O}_E, F \rangle \mathrm{ch}(\mathcal{O}_E), \qquad (5.174)$$

where $\langle\, , \rangle$ is the intersection form (5.106).

Unlike the other cases, we have no direct physics definition of the SCFT near this point, and thus no physics argument that yields $\mathcal{K}$ in a concrete way. Perhaps the most direct way to proceed is to use our previous results and the relations on the fundamental group of the thrice-punctured sphere, and compute

$$\mathcal{K} = \mathcal{L} \circ \mathcal{G}. \qquad (5.175)$$

Instead we will follow the historical development by making a conjecture for this monodromy [**310**], based on the theory of mutations (see, for

[27] While we will not do it, one can obtain the full derived category by considering quiver representations which are finitely generated but not finite dimensional.

example, [**406**]). This was subsequently studied extensively in [**422**], and significantly generalized in [**245, 21**]. We will then verify it with (5.175).

The idea applies to any monodromy induced by a massless brane E, so we define it in this generality, replacing $\mathcal{O}_E$ in (5.174) with an object E. Let us define the notation $A \boxtimes B$ to mean $p_1^* A \otimes p_2^* B$ for $A, B \in \mathrm{D}(X)$, where $p_1, p_2 : X \times X \to X$ are the projections on the first and second factor of the direct product.

DEFINITION 5.33. The twist functor Φ_E is defined as

$$\Phi_E(F) = \mathrm{Cone}\big(\mathrm{Hom}(E, F) \otimes E \xrightarrow{r} F\big), \tag{5.176}$$

where r is the evaluation map (or "tautological" map) which acts on E by an element of $\mathrm{Hom}(E, F)$.

This corresponds to a Fourier-Mukai transform with kernel

$$\mathcal{K} = \mathrm{Cone}\big(E^\vee \boxtimes E \xrightarrow{r} \mathcal{O}_{\Delta X}\big), \tag{5.177}$$

where $E^\vee$ is the dual of E defined in the derived category as $\mathbf{R}\mathscr{H}om(E, \mathcal{O}_X)$, and r is the obvious restriction map. Let us denote also by $\mathcal{K}$ the Fourier-Mukai transform induced by $\mathcal{K}$, and see that this gives the same thing as the twist functor Φ_E. We see that

$$\begin{aligned} \mathcal{K}(F) &= p_{2*}\,\mathrm{Cone}((F \otimes E^\vee) \boxtimes E \to \mathcal{O}_{\Delta X} \otimes p_1^* F) \\ &\cong p_{2*}\,\mathrm{Cone}(\mathbf{R}\mathscr{H}om(E, F) \boxtimes E \to \mathcal{O}_{\Delta X} \otimes p_1^* F) \\ &\cong \mathrm{Cone}(\mathrm{Hom}(E, F) \otimes E \to F). \end{aligned}$$

Thus

$$\mathcal{K} = \Phi_E. \tag{5.178}$$

Note that the Fourier-Mukai transform $\mathcal{K}$ gives by construction

$$\mathrm{ch}(\mathcal{K}(F)) = \mathrm{ch}(F) - \langle E, F \rangle \mathrm{ch}(E),$$

generalizing (5.174), which we obtain if we replace the complex E with $\mathcal{O}_E$, the structure sheaf of the exceptional divisor E.

Another simple test of this identification is that under the action of $\mathcal{K}$, $\xi(E)$ must change by 2; indeed

$$\mathcal{K}[E] = E[-2]$$

(or perhaps $E[2]$ depending on conventions). This requires a further condition on E, namely

$$H^k(X, E) = \begin{cases} \mathbb{C} & \text{if } k = 0 \text{ or } 3 \\ 0 & \text{otherwise,} \end{cases} \tag{5.179}$$

Such objects are sometimes referred to as "spherical" as this is the cohomology of S^3. This implies that $\mathrm{Hom}(\mathcal{O}_X, E)$ can be represented by a complex of vector spaces

$$\cdots \longrightarrow 0 \longrightarrow \underset{\cdots\cdots}{\mathbb{C}} \longrightarrow 0 \longrightarrow 0 \longrightarrow \mathbb{C} \longrightarrow 0 \longrightarrow \cdots. \tag{5.180}$$

Tensoring by E simply replaces the $\mathbb{C}$'s by E's. The cone construction then shifts this one place left and the first E cancels with the E on the right of the cone (since the evaluation map is the identity) leaving $E[-2]$.

With this condition, one can prove that Φ_E is an autoequivalence [**422**], and thus (5.178) is a candidate for the final monodromy. To confirm this one can compute the action of $\mathcal{L}^{-1} \circ \mathcal{K}$ on $\{F_0, F_1, F_2\}$. It is not hard to show that it permutes these objects in the exact same way as $\mathcal{G}$. Thus we confirm (5.175) (for finite dimensional quiver representations).

It is perhaps worth mentioning that a similar analysis in the case of the quintic threefold of §5.5.6 has a subtle difference. One can again define the monodromies $\mathcal{L}$ and $\mathcal{K}$ around the large radius limit and the conifold point respectively and put $\mathcal{G} = \mathcal{L}^{-1} \circ \mathcal{K}$. One might expect the relation $\mathcal{G}^5 = 1$ due to the quantum $\mathbb{Z}_5$ symmetry of the Gepner point. Instead one finds that $\mathcal{G}^5$ corresponds to a shift "[2]" in the derived category. This can be understood in terms of matrix factorizations [**459**] but we will not pursue this here.

5.6.3. Examples of stability. In the case of the quintic we used the large radius limit as our base point for determining stability. In the case of the orbifold, we can utilize the quantum symmetry to use the orbifold point as the base point.

First, from (5.173) we can compute ξ near the orbifold point. As argued at the end of §5.4.5, all fractional branes have the same value of ξ at the orbifold point, which we declare to be zero. Nearby, to linear order in ψ we have

$$\begin{aligned} \xi(\Delta_{m_0 m_1 m_2}) &= -c\frac{(-m_0 - m_1 + 2m_2)\,\mathrm{Re}(\psi) + (m_0 - 2m_1 + m_2)\,\mathrm{Re}(\omega\psi)}{m_0 + m_1 + m_2} \\ &= c\frac{\sum_i m_i \zeta_i}{\sum_i m_i}, \end{aligned} \tag{5.181}$$

for a positive constant c and we *define* the ζ_k by

$$\zeta_k = \sqrt{3}\,\mathrm{Re}(e^{\frac{\pi i}{6}(4k-1)}\psi), \tag{5.182}$$

so that

$$\begin{aligned} \zeta_0 + \zeta_1 + \zeta_2 &= 0 \\ \zeta_0 + \omega\zeta_1 + \omega^2\zeta_2 &= \tfrac{3\sqrt{3}}{2} e^{\frac{\pi i}{6}} \bar{\psi} \\ \zeta_0 + \omega^2\zeta_1 + \omega\zeta_2 &= \tfrac{3\sqrt{3}}{2} e^{-\frac{\pi i}{6}} \psi. \end{aligned} \tag{5.183}$$

Clearly this is the analogue of (5.110) and the ζ's we have just introduced here correspond to those of §5.4.5.5. Indeed, we may now explicitly show that θ-stability is a limiting form of Π-stability near the orbifold point. Suppose we have a short exact sequence of quiver representations

(5.184)

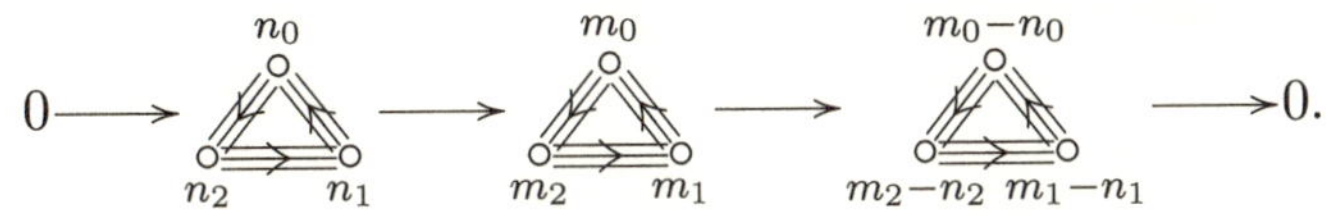

Near the orbifold point, the ξ's of these 3 D-branes will be very close to zero. Thus, by the way central charges add, the ξ of the middle entry in (5.184) must lie between the ξ's of the other two. For Π-stability of the middle entry we draw the distinguished triangle

(5.185)

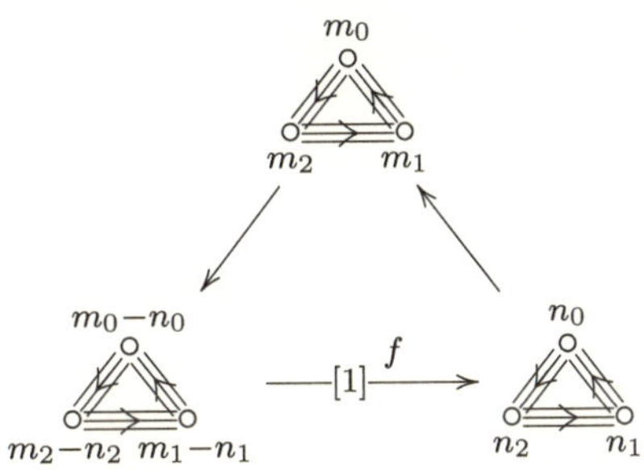

and look for the condition that f be tachyonic. From (5.181) this is precisely

$$\frac{\sum_i n_i\zeta_i}{\sum_i n_i} < \frac{\sum_i m_i\zeta_i}{\sum_i m_i}, \tag{5.186}$$

which, from (5.118) is equivalent to King's θ-stability statement of §5.4.5.5.

This θ-stability formulation allows us to completely classify the stable irreducible B-branes near the orbifold point. As mentioned in §5.4.4, the irreducibility for an object A amounts to $\mathrm{Hom}(A, A) = \mathbb{C}$. A quiver representation satisfying this condition is known as a "Schur representation". The problem of finding such representations was discussed in [**137**].

Determining whether a quiver representation (with relations) is Schur is purely a question of algebra, but can be awkward. In many cases it is actually more convenient to use the equivalence of §4.7 to rephrase the question in terms of coherent sheaves.

As an example of a non-Schur quiver representation, consider Δ_{211} with *generic* maps on the arrows in the quiver. With some effort one can show that the short exact sequence

(5.187)

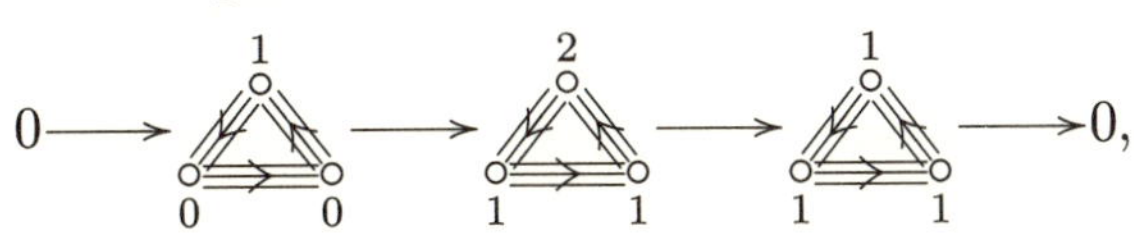

is *split*. This immediately implies that $\mathrm{Hom}(\Delta_{211}, \Delta_{211}) \supset \mathbb{C}^2$. This fact becomes more obvious when written in terms of sheaves. Δ_{111} is a 0-brane which is generically nowhere near the exceptional divisor E, whereas Δ_{100} is the 4-brane $\mathcal{O}_E(-1)$ wrapping E. Thus Δ_{211} is a sum of two quite disjoint D-branes and is obviously reducible. Note that when the maps on the arrows are not generic, this quiver representation might actually be Schur. This would correspond to the 0-brane being on E, leading to a possible irreducible bound state with the 4-brane.

We will not attempt to explicitly provide a complete solution to the classification problem here, but some Schur representations of interest are Δ_{111}, and Δ_{abc}, where $\{a, b, c\}$ is any permutation of $\{0, 1, n\}$ for $n \leq 3$.

The fractional branes F_k are obviously always stable near the orbifold point since they have no nontrivial subobject in the category of quiver representations. Let us next focus on Δ_{111}, some of which correspond to 0-branes. The quiver

(5.188)

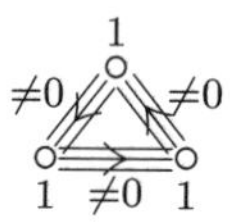

with at least one nonzero map between each pair of nodes is stable since there is no injective map from any possible subobject to it. Such a quiver represents a 0-brane away from the exceptional divisor E (or orbifold point if we haven't blown up). Indeed, one would expect that the stability of such a 0-brane should not be affected by orbifold-related matters.

Now consider the following short exact sequence:

(5.189)

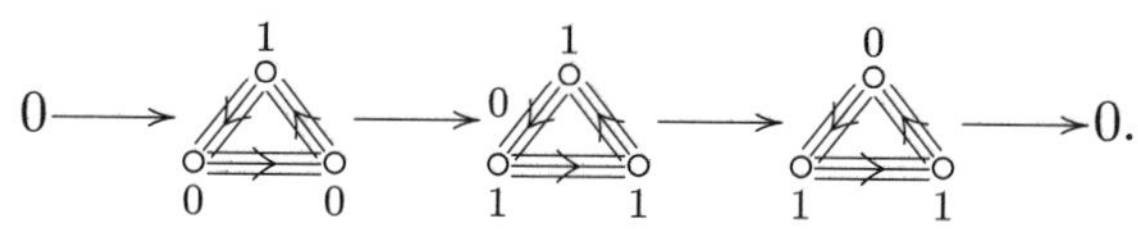

This Δ_{111} is stable against decay to Δ_{100} by θ-stability if $\zeta_0 < 0$. We also have the sequence

(5.190)

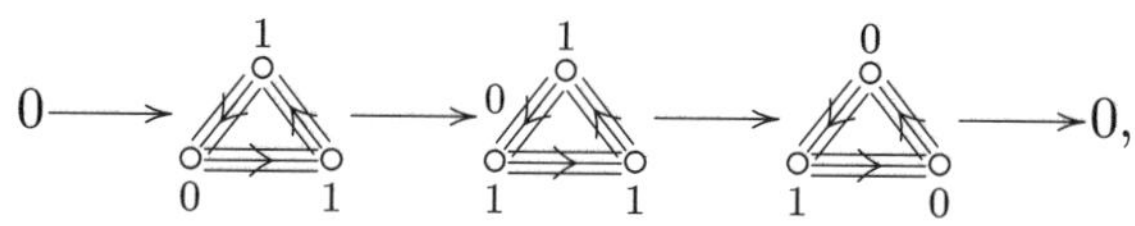

giving a further constraint $\zeta_2 > 0$ on the stability of this 0-brane. Thus this 0-brane is stable in a $2\pi/3$ wedge coming out of the orbifold point. After blowing-up a little into this wedge, this 0-brane corresponds to a point on the exceptional divisor. Obviously a cyclic permutation of the zero to another edge of the quiver results in similar statements with the ζ's permuted accordingly. Thus, all other quivers Δ_{111} are unstable in the wedge $\zeta_2 > 0, \zeta_0 < 0$ and do not correspond to 0-branes at all.

The quiver cho on the left edge have a close connection to sheaves on E as can be seen as follows. The general representation Δ_{abc} falls into the sequence

(5.191)

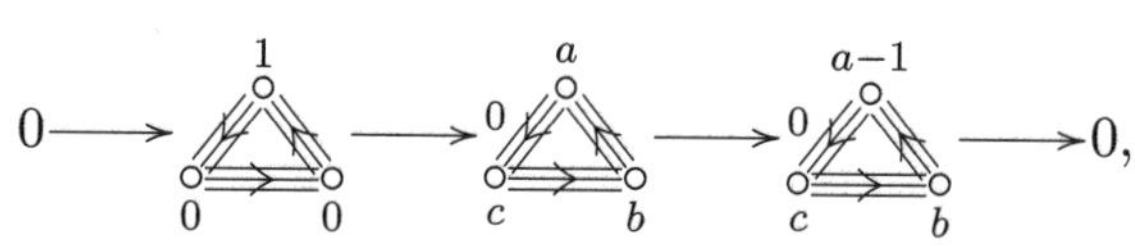

and thus we iterate

(5.192)

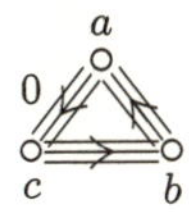 $= \mathrm{Cone}\Big($ 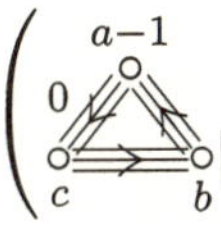 $[-1] \longrightarrow \mathcal{O}_E(-1)\Big)$

$= \mathrm{Cone}\Big($ 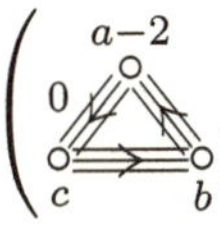 $[-1] \longrightarrow \mathcal{O}_E(-1)^{\oplus 2}\Big)$

$= \mathrm{Cone}\Big($ 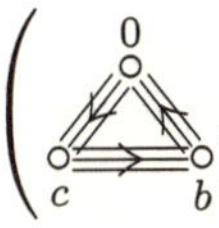 $[-1] \longrightarrow \mathcal{O}_E(-1)^{\oplus a}\Big).$

Continuing this process yields

(5.193)

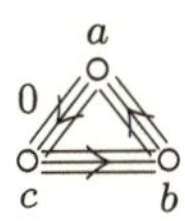 $= \left(\mathcal{O}_E(-2)^{\oplus c} \longrightarrow \Omega_E^{\oplus b} \longrightarrow \underline{\mathcal{O}_E(-1)^{\oplus a}}\right),$

explicitly mapping this quiver representation into the derived category of coherent sheaves on X (or E). This is precisely Beilinson's construction of sheaves on $\mathbb{P}^2$ as in §4.6.2, a correspondence identified in [**137**]. Thus, quiver representations with zero maps on the left edge are seen to be associated to D-branes on E. We will get a pure sheaf of course only if the cohomology of the complex in (5.193) is concentrated at one term.

We denote some lines of marginal stability for Π-stability in Figure 13. In each case, the arrow denotes the direction you cross the line to cause the relevant object to decay. Naturally this agrees with θ-stability near the origin. The figure shows the moduli space in the form of the complex $(-i\psi)$-plane. We make this choice so that the picture is aligned with Figure 12, i.e., the large radius limit is upwards. Note that the lines of marginal stability corresponding to $\mathcal{O}_p$, $p \in E$, (i.e., the corresponding Δ_{111} quivers above) follow the lines of constant $\arg(\psi)$ even when the non-perturbative effects of Π-stability are taken into account.

Some decays of note are the following:

(1) $\mathcal{O}_C(-1)$: This sheaf fits into the exact sequence

$$0 \longrightarrow \mathcal{O}_E(-2) \longrightarrow \mathcal{O}_E(-1) \longrightarrow \mathcal{O}_C(-1) \longrightarrow 0, \tag{5.194}$$

and thus decays by Π-stability in a way essentially identical to the 4-branes in the quintic as in §5.5.6.1. Thus, these 2-branes are stable at large radius but decay before the orbifold point is reached. Note that (5.194) implies that, in the derived category of quiver representations, we have

$$\mathcal{O}_C(-1) = \mathrm{Cone}(F_2[-2] \to F_0). \tag{5.195}$$

That is, this D-brane is essentially a complex of quivers and cannot be written in terms of a single quiver. In other words, it is not in the

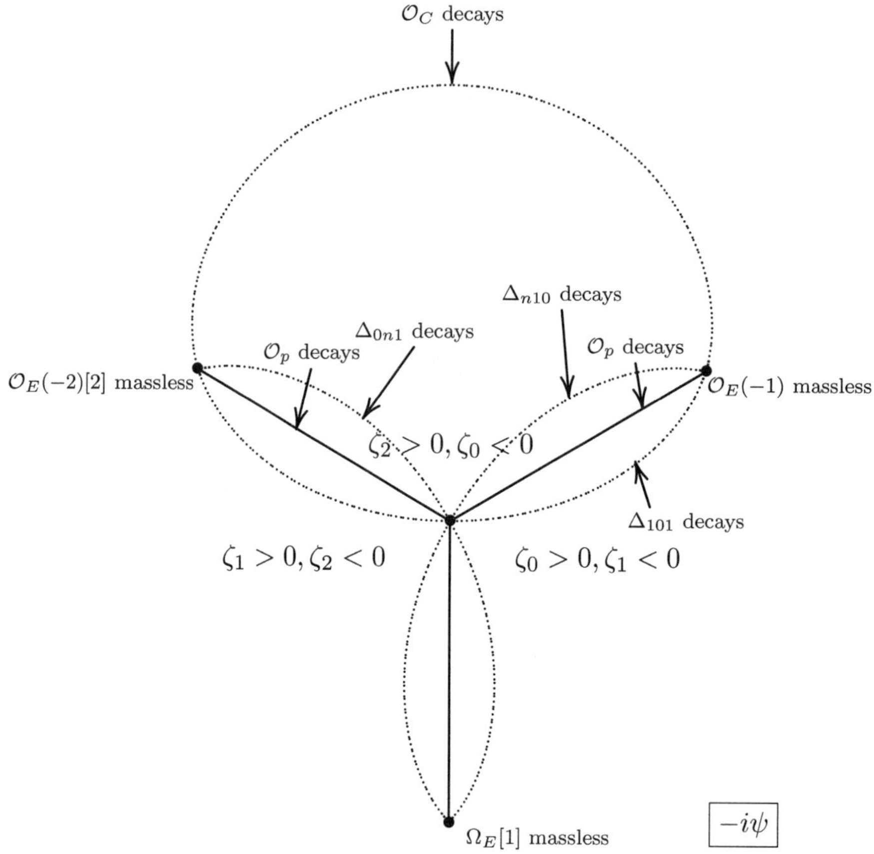

FIGURE 13. Some lines of marginal stability for the $\mathbb{Z}_3$-orbifold.

abelian category of quiver representations. It is therefore consistent with our picture that it decays before we get close to the orbifold point.

(2) Δ_{101}: Following the logic of §5.5.6.3 we can now look for an "exotic" D-brane by taking the "Serre dual" of (5.195). This gives $\operatorname{Cone}(F_0[-1] \to F_2)$, i.e., an extension of F_0 by F_2. This is precisely Δ_{101}. As expected from §5.5.6.3, these objects should not be stable at large radius but can become stable as we shrink the exceptional divisor down. The line of marginal stability is shown in Figure 13. Note that they do not actually become stable until we shrink down to, or beyond, the orbifold point. These objects generically have nonzero maps along the left edge of the triangle and so are not classified by Beilinson's construction (5.193).

We see a nice complementarity between the D-branes Δ_{101} and $\mathcal{O}_C(-1)$. $\mathcal{O}_C(-1)$ is an object in the category of coherent sheaves but is a complex in terms of quivers. Δ_{101} is an object in the category of quiver representations but becomes an exotic complex $\mathrm{Cone}(\mathcal{O}_E(-1)[-1], \mathcal{O}_E(-2)[2])$ in the derived category of sheaves.

(3) Δ_{n10}: This fits into the sequence

(5.196)

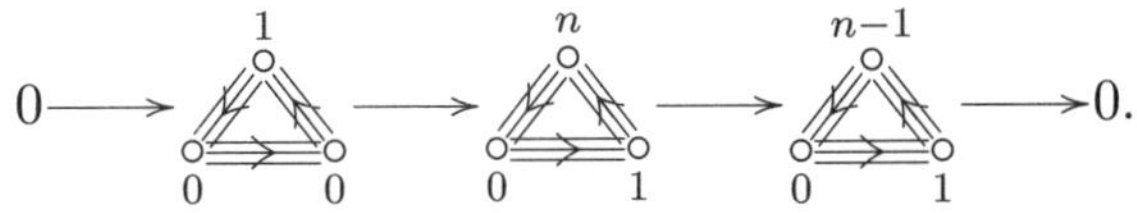

It produces a decay as shown in Figure 13. The identification (5.193) together with the short exact sequence

$$0 \longrightarrow \Omega_E \longrightarrow \mathcal{O}_E(-1)^{\oplus 3} \longrightarrow \mathcal{O}_E \longrightarrow 0, \tag{5.197}$$

may be used to show that

$$\begin{aligned} \Delta_{210} &\cong \mathcal{O}_C \\ \Delta_{310} &\cong \mathcal{O}_E. \end{aligned} \tag{5.198}$$

These D-branes are therefore simply objects both in terms of quivers and sheaves. It is not surprising therefore that they are both stable at large radius limit and near the orbifold point.

On the other hand Δ_{110} does not correspond to a sheaf since the complex in (5.193) has cohomology in more than one place. In this case we have a sequence

(5.199)

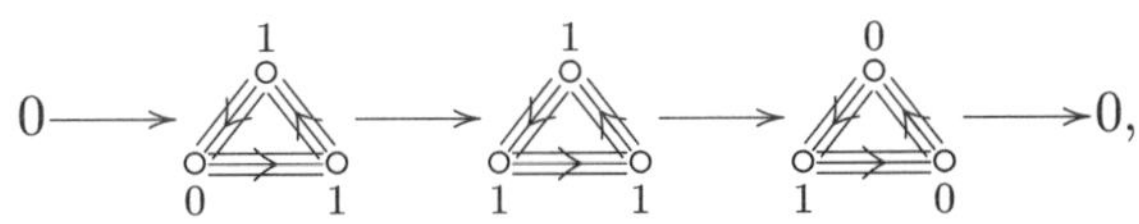

which makes Δ_{110} only marginally stable at the large radius limit.

(4) Δ_{0n1}: This is similar to the Δ_{n10} case and again we plot the line of marginal stability in Figure 13. This time Δ_{011} corresponds to the ideal sheaf of a point $\mathscr{I}_{E,p}[1]$ and is again only marginally stable at the large radius limit.

5.7. Stability structures

We now have a variety of requirements that the subset of the topological boundary conditions which are physical D-branes must satisfy. In [**65**], Bridgeland proposed a precise mathematical definition of a stability condition. Although inspired by the physical discussion, it is self-contained and thus can be analyzed purely mathematically. In particular, one can show that (under certain reasonable assumptions) the space of stability conditions on a reasonable triangulated category is a finite dimensional complex manifold, thereby providing an interesting geometric invariant of such categories.

While so far the evidence suggests that this notion of stability does describe the "physical D-branes" or SCFT boundary states we have been discussing, this is not yet proven. Indeed, there are some intriguing differences from the string theoretic discussion – the definition does not require the Calabi-Yau condition, and the space of stability conditions generally has higher dimension than the SCFT Kähler moduli space. We will return to these questions in §5.9. On the other hand, this notion has proved immensely fruitful in mathematics, its ubiquity connecting seemingly different fields.

Let us turn to Bridgeland's definition:

DEFINITION 5.34. A stability condition $\sigma = (Z, \mathcal{P})$ on a triangulated category D consists of a group homomorphism $Z\colon K(\mathrm{D}) \to \mathbb{C}$ called the central charge, and full additive subcategories $\mathcal{P}(\phi) \subset \mathrm{D}$ for each $\phi \in \mathbb{R}$, satisfying the following axioms:

(a) if $E \in \mathcal{P}(\phi)$ then $Z(E) = m(E)\exp(i\pi\phi)$ for some $m(E) \in \mathbb{R}_{>0}$,
(b) for all $\phi \in \mathbb{R}$, $\mathcal{P}(\phi + 1) = \mathcal{P}(\phi)[1]$,
(c) if $\phi_1 > \phi_2$ and $A_j \in \mathcal{P}(\phi_j)$ then $\mathrm{Hom}_{\mathrm{D}}(A_1, A_2) = 0$,
(d) for each nonzero object $E \in \mathrm{D}$ there is a finite sequence of real numbers

$$\phi_1 > \phi_2 > \cdots > \phi_n$$

and a collection of triangles

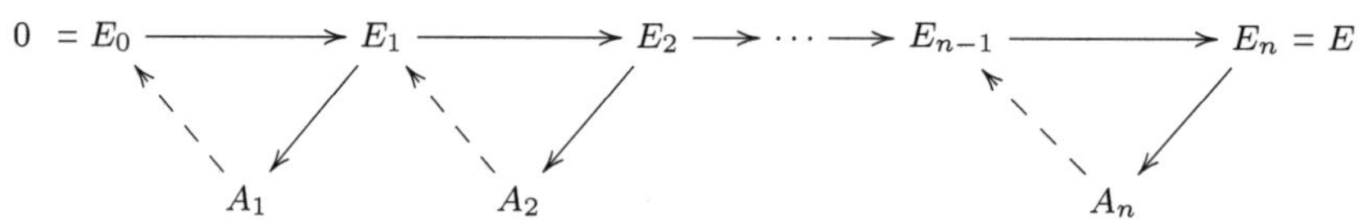

with $A_j \in \mathcal{P}(\phi_j)$ for all j.

Comparing with the discussion of §5.5, perhaps the main new ingredient is axiom (d), which asserts the existence of a Harder-Narasimhan filtration analogous to that in the discussion of stable coherent sheaves [**220, 327**]. While this claim is sensible physically (as in Conjecture 5.32), it is not completely manifest, as we discuss in §5.9.3.

Given a stability condition $\sigma = (Z, \mathcal{P})$ as in the definition, each subcategory $\mathcal{P}(\phi)$ is abelian, as we argue below. The nonzero objects of $\mathcal{P}(\phi)$ are said to be semistable of phase ϕ in σ, and the simple objects of $\mathcal{P}(\phi)$ are said to be stable. It follows from the other axioms that the decomposition of an object $0 \neq E \in \mathrm{D}$ given by axiom (d) is uniquely defined up to isomorphism. Write $\phi^+_\sigma(E) = \phi_1$ and $\phi^-_\sigma(E) = \phi_n$. The *mass* of E is defined to be the positive real number

$$m_\sigma(E) = \sum_i |Z(A_i)|.$$

By the triangle inequality, for every nonzero E, we have

$$m_\sigma(E) \geq |Z(E)|,$$

with equality if and only if E is semistable; this is just the BPS condition (5.18).

For any interval $I \subset \mathbb{R}$, define $\mathcal{P}(I)$ to be the extension-closed subcategory of D generated by the subcategories $\mathcal{P}(\phi)$ for $\phi \in I$. Thus, for example, the full subcategory $\mathcal{P}((a,b))$ consists of the zero objects of D together with those objects $0 \neq E \in \mathrm{D}$ which satisfy $a < \phi_\sigma^-(E) \le \phi_\sigma^+(E) < b$.

5.7.1. Relation to stability in abelian categories. The physics discussion was mostly based on the idea that a particular abelian subcategory of the derived category would be preferred in a particular region of stringy Kähler moduli space, for example coherent sheaves in the large volume limit, or quiver representations in a field theoretic limit. In the present setup, this is made precise in

PROPOSITION 5.35. [**65**, 5.3] *To give a stability condition on* D *is equivalent to giving a bounded t-structure on* D *and a stability function on its heart with the Harder-Narasimhan property.*

As in §4.4.6, a t-structure determines an abelian subcategory $\boldsymbol{A}$ of D, its *heart.* We define a stability function on an abelian category $\boldsymbol{A}$ to be a group homomorphism $Z\colon K(\boldsymbol{A}) \to \mathbf{C}$ such that

$$0 \neq E \in \boldsymbol{A} \implies Z(E) \in \mathbb{R}_{>0}\exp(i\pi\phi(E)) \text{ with } 0 < \phi(E) \le 1.$$

The real number $\phi(E) \in (0,1]$ is called the phase of the object E. It allows us to order the objects and thus define a notion of stability: a nonzero object $E \in \boldsymbol{A}$ is *semistable* with respect to Z if every subobject $0 \neq A \subset E$ satisfies $\phi(A) \le \phi(E)$.

Now, a *Harder-Narasimhan filtration* of an object $0 \neq E \in \boldsymbol{A}$ is a finite chain of subobjects

$$0 = E_0 \subset E_1 \subset \cdots \subset E_{n-1} \subset E_n = E$$

whose factors $F_j = E_j/E_{j-1}$ are semistable objects of $\boldsymbol{A}$ with

$$\phi(F_1) > \phi(F_2) > \cdots > \phi(F_n).$$

Note that if $f\colon E \to F$ is a nonzero map between semistable objects then by considering $\operatorname{im} f \cong \operatorname{coim} f$ in the usual way, one sees that $\phi(E) \le \phi(F)$. It follows easily from this that Harder-Narasimhan filtrations (when they exist) are unique.

DEFINITION 5.36. The stability function Z is said to have the *Harder-Narasimhan property* if every nonzero object of $\boldsymbol{A}$ has a Harder-Narasimhan filtration.

It can be shown [**65**, 2.4], following ideas of Rudakov [**407**]) that this follows from a rather weak condition on $\boldsymbol{A}$, prohibiting infinite ascending or descending chains of this type.

We can now prove Proposition 5.35:

PROOF. Given a stability condition on D, the central charge defines a stability function on its heart $\boldsymbol{A} = \mathcal{P}((0,1]) \subset \mathrm{D}$, and the decompositions of axiom (d) give Harder-Narasimhan filtrations.

Conversely, given a bounded t-structure on D together with a stability function Z on its heart $\boldsymbol{A} \subset \mathrm{D}$, we can define subcategories $\mathcal{P}(\phi) \subset \boldsymbol{A} \subset \mathrm{D}$ to be the semistable objects in $\boldsymbol{A}$ of phase ϕ for each $\phi \in (0, 1]$. Axiom (b) then fixes $\mathcal{P}(\phi)$ for all $\phi \in \mathbb{R}$. □

By the same argument, any $\boldsymbol{A} = \mathcal{P}((\phi - 1, \phi])$ is an abelian category. Furthermore, the individual subcategories $\mathcal{P}(\phi)$ are abelian [**65**, 5.2]. To show this, it is enough to show that if $f \colon E \to F$ is a morphism in $\mathcal{P}(\phi)$ then the kernel and cokernel of f, considered as a morphism of $\boldsymbol{A} = \mathcal{P}((\phi-1, \phi])$, actually lie in $\mathcal{P}(\phi)$. But if two elements of a short exact sequence lie in $\mathcal{P}(\phi)$, so must the third, by the additivity of central charges. On the other hand, the $\mathcal{P}(I)$ for an interval I of length < 1 need only be a *quasi-abelian category*, as defined in [**411, 65**].

A stability condition is called *locally finite* if there is some $\epsilon > 0$ such that each quasi-abelian category $\mathcal{P}((\phi-\epsilon, \phi+\epsilon))$ is of finite length, i.e., any infinite chain of subobjects or quotients must stabilize. This condition is assumed for some of the results below.

While one can construct stability conditions which violate local finiteness [**65**, 5.6], these are somewhat pathological. Physically, local finiteness should follow from the claim of §5.4.4 that the stable objects with $\phi \in I$, a small interval, can be described by an effective field theory. It is also related to the idea that non-singular string compactifications, in the $g_s \to 0$ limit we are considering, have a *mass gap*, in other words the set of central charges of stable objects is bounded away from zero.

Let us compare with the discussion of §5.3.3 and §5.5. We saw there that the natural physical definition is to start with an SCFT and an associated set of physical branes, and use this to build the homotopy category and ultimately the derived category. This procedure is unambiguous if we know that the original set of physical branes form an abelian category. Conversely, the results we just described imply that putting a stability structure on the derived category gives us a set of preferred abelian categories, the $\mathcal{P}((\phi - 1, \phi])$ for each ϕ. Furthermore, as we move through the space of stability structures, by axiom (c) these abelian categories will vary according to the rules of Π-stability. We can now answer the question raised in §5.5, namely to what extent does this condition determine the sets of stable branes?

5.7.2. The space of stability conditions. To define this space, we start by putting a topology on the set Stab(D) of locally finite stability conditions on D. This is induced by the metric

$$d(\sigma_1, \sigma_2) = \sup_{0 \neq E \in \mathrm{D}} \left\{ |\phi^-_{\sigma_2}(E) - \phi^-_{\sigma_1}(E)|, |\phi^+_{\sigma_2}(E) - \phi^+_{\sigma_1}(E)|, |\log \frac{m_{\sigma_2}(E)}{m_{\sigma_1}(E)}| \right\} \in [0, \infty].$$

The essential point is that any small deformation of the central charge in this metric can be lifted to a unique deformation of the stability condition. This is made precise by the following basic result of Bridgeland.

THEOREM 5.37. [**65**] *For each connected component* $\Sigma \subset \mathrm{Stab}(\mathrm{D})$ *there is a linear subspace* $V(\Sigma) \subset \mathrm{Hom}_{\mathbb{Z}}(K(\mathrm{D}), \mathbb{Z})$ *with a well-defined linear topology and a local homeomorphism* $\mathcal{Z}\colon \Sigma \to V(\Sigma)$, *which maps a stability condition* $(Z, \mathcal{P})$ *to its central charge* Z.

If X is a smooth projective variety, we write $\mathrm{Stab}(X)$ for the set of locally finite stability conditions on $\mathrm{D}(X)$ for which the central charge Z factors via the Chern character map $\mathrm{ch}\colon K(X) \to H^*(X, \mathbb{Q})$. Theorem 5.37 immediately implies that $\mathrm{Stab}(X)$ is a finite dimensional complex manifold. By Proposition 5.35, the points of $\mathrm{Stab}(X)$ parameterize bounded t-structures on $\mathrm{D}(X)$, together with the extra data of the map Z.

EXAMPLE 5.38. Let $\boldsymbol{A}$ be the category of coherent $\mathcal{O}_X$-modules on a nonsingular projective curve X over $\mathbb{C}$, and set $Z(E) = -\deg E + i \operatorname{rk} E$. Applying Proposition 5.35 gives a stability condition on the bounded derived category $\mathrm{D}(\boldsymbol{A})$.

5.7.2.1. *Natural group actions.* These exist on the space of stability conditions of any triangulated category. The first generalizes the gauge invariance of §5.3.4.2, while the second expresses the monodromy action of §5.6.2.

LEMMA 5.39. *The space* $\mathrm{Stab}(\mathrm{D})$ *carries a right action of the group* $\widetilde{\mathrm{GL}^+}(2, \mathbb{R})$, *the universal covering space of* $\mathrm{GL}^+(2, \mathbb{R})$, *and a left action of the group* $\mathrm{Aut}(\mathrm{D})$ *of exact autoequivalences of* D. *These two actions commute.*

Proof. First note that the group $\widetilde{\mathrm{GL}^+}(2, \mathbb{R})$ can be thought of as the set of pairs (T, f) where $f\colon \mathbb{R} \to \mathbb{R}$ is an increasing map with $f(\phi+1) = f(\phi)+1$, and $T\colon \mathbb{R}^2 \to \mathbb{R}^2$ is an orientation-preserving linear isomorphism, such that the induced maps on $S^1 = \mathbb{R}/2\mathbb{Z} = (\mathbb{R}^2 \setminus \{0\})/\mathbb{R}_{>0}$ are the same.

Given a stability condition $\sigma = (Z, \mathcal{P}) \in \mathrm{Stab}(\mathrm{D})$, and a pair $(T, f) \in \widetilde{\mathrm{GL}^+}(2, \mathbb{R})$, define a new stability condition $\sigma' = (Z', \mathcal{P}')$ by setting $Z' = T^{-1} \circ Z$ and $\mathcal{P}'(\phi) = \mathcal{P}(f(\phi))$. Note that the semistable objects of the stability conditions σ and σ' are the same, but the phases have been relabelled.

For the second action, note that an element $\Phi \in \mathrm{Aut}(\mathrm{D})$ induces an automorphism ϕ of $K(\mathrm{D})$. If $\sigma = (Z, \mathcal{P})$ is a stability condition on D define $\Phi(\sigma)$ to be the stability condition $(Z \circ \phi^{-1}, \mathcal{P}')$, where $\mathcal{P}'(t) = \Phi(\mathcal{P}(t))$. □

Neither of the two group actions of Lemma 5.39 will be free in general. In particular, if $\sigma = (Z, \mathcal{P})$ is a stability condition in which the image of the central charge $Z\colon K(\mathrm{D}) \to \mathbb{C}$ lies on a real line in $\mathbb{C}$ then σ will be fixed by some subgroup of $\widetilde{\mathrm{GL}^+}(2, \mathbb{R})$. However there is a subgroup $\mathbb{C} \subset \widetilde{\mathrm{GL}^+}(2, \mathbb{R})$ which does act freely. If $\lambda \in \mathbb{C}$ then λ sends a stability condition $\sigma = (Z, \mathcal{P})$

to the stability condition $\lambda(\sigma) = (Z', \mathcal{P}')$, where $Z'(E) = e^{-i\pi\lambda} Z(E)$ and $\mathcal{P}'(\phi) = \mathcal{P}(\phi + \mathrm{Re}(\lambda))$. Note that for any integer n the action of the shift functor $[n]$ on Stab(D) coincides with the action of $n \in \mathbb{C}$.

5.8. Examples of spaces of stability conditions

5.8.1. The elliptic curve. Let X be a nonsingular genus one curve, and D the bounded derived category of coherent sheaves on X. The Grothendieck group $K(\mathrm{D}) \cong \mathbb{Z} \oplus \mathbb{Z}$; for a sheaf these are the rank and degree.

It follows from Lemma 5.39 and Theorem 5.37 that there is a local homeomorphism

$$Z\colon \mathrm{Stab}(X) \to \mathrm{Hom}_{\mathbb{Z}}(K(\mathrm{D}), \mathbb{C})$$

whose image is some open subset of the two-dimensional vector space

$$\mathrm{Hom}_{\mathbb{Z}}(K(\mathrm{D}), \mathbb{C}).$$

In fact,

THEOREM 5.40. **[65]** *The action of the group* $\widetilde{\mathrm{GL}^+}(2, \mathbb{R})$ *on* $\mathrm{Stab}(X)$ *is free and transitive, so that*

$$\mathrm{Stab}(X) \cong \widetilde{\mathrm{GL}^+}(2, \mathbb{R}) \cong \mathbb{C} \times \mathcal{H}.$$

Proof. First note that if E is an indecomposable sheaf on X then E must be semistable in any stability condition $\sigma \in \mathrm{Stab}(X)$ because otherwise there is a nontrivial triangle $A \to E \to B$ with $\mathrm{Hom}_{\mathrm{D}}(A, B) = 0$, and then Serre duality gives

$$\mathrm{Hom}^1_{\mathrm{D}}(B, A) = \mathrm{Hom}_{\mathrm{D}}(A, B)^* = 0,$$

which implies that E is a direct sum $A \oplus B$.

Take an element $\sigma = (Z, \mathcal{P}) \in \mathrm{Stab}(X)$. Suppose for a contradiction that the image of the central charge Z is contained in a real line in $\mathbb{C}$. Since σ is locally finite, the heart $\boldsymbol{A}$ of σ must then be of finite length. If A and B are simple objects of $\boldsymbol{A}$ then

$$\mathrm{Hom}_{\mathrm{D}}(A, B) = \mathrm{Hom}_{\mathrm{D}}(B, A) = 0,$$

and it follows from this that $\chi(A, B) = 0$. But this implies that all simple objects of $\boldsymbol{A}$ lie on the same line in $K(\mathrm{D})$, and hence that all objects of D do too, which gives a contradiction. Thus Z, considered as a map from $K(D) \otimes \mathbb{R} = \mathbb{R}^2$ to $\mathbb{C} \cong \mathbb{R}^2$ is an isomorphism, and it follows that the action of $\widetilde{\mathrm{GL}^+}(2, \mathbb{R})$ on $\mathrm{Stab}(X)$ is free.

Suppose A and B are line bundles on X with $\deg A < \deg B$. Since A and B are indecomposable they are semistable in σ with phases ϕ and ψ, say. The existence of maps $A \to B$ and $B \to A[1]$ gives inequalities $\phi \leq \psi \leq \phi + 1$, which implies that Z is orientation preserving. Thus, acting by an element of $\widetilde{\mathrm{GL}^+}(2, \mathbb{R})$, one can assume that $Z(E) = -\deg E + i \operatorname{rk} E$, and that for some point $x \in X$ the skyscraper sheaf $\mathcal{O}_x$ has phase 1. Then all semistable vector bundles on X are semistable in σ with phase in the interval $(0,1)$, and it follows quickly from this that σ is the standard stability condition described in Example 5.38. □

One can easily show that the autoequivalences of D are generated by shifts, automorphisms of X and twists by line bundles together with the Fourier-Mukai transform. Automorphisms of X and twists by line bundles of degree zero act trivially on $\mathrm{Stab}(X)$ and one obtains

$$\mathrm{Stab}(X)\,/\mathrm{Aut}\,\mathrm{D} \cong \mathrm{GL}(2,\mathbb{R})\,/\mathrm{SL}(2,\mathbb{Z}). \tag{5.200}$$

This is easily seen to be a $\mathbb{C}^*$-bundle over the moduli space of elliptic curves, parameterizing equivalence classes of data consisting of a complex structure on X together with a non-zero holomorphic 1-form.

We can now check whether this agrees with expectations from string theory. By Property 5.1, the space of stability conditions should be the moduli space of complex structures of the mirror Y, along with a choice of holomorphic one-form Ω. Since Y is again a two-torus, we have complete agreement.

Other dimension one examples, not connected to string theory, have also been worked out. In [**78**], the case of irreducible singular curves of arithmetic genus one was treated, again obtaining (5.200). S. Okada [**379**] proved that

$$\mathrm{Stab}(\mathbb{P}^1) \cong \mathbb{C}^2$$

and E. Macri [**342**] proved that for any curve C of genus $g \geq 2$ one has

$$\mathrm{Stab}(C) \cong \widetilde{\mathrm{GL}^+}(2,\mathbb{R}) \cong \mathbb{C}\times\mathcal{H}.$$

5.8.2. K3 surfaces. Let X be a K3 surface, a simply connected smooth projective surface with trivial canonical bundle. Cup product defines a symmetric pairing on the Picard group $\mathrm{Pic}(X)$, the algebraic part of the second cohomology. Mukai extended this to a pairing on

$$\mathcal{N}(X) = \mathbb{Z}\oplus\mathrm{Pic}(X)\oplus\mathbb{Z}$$

by defining

$$\langle (r_1, D_1, s_1), (r_2, D_2, s_2)\rangle = D_1\cdot D_2 - r_1 s_2 - r_2 s_1.$$

The lattice $\mathcal{N}(X)$ is even and unimodular of signature $(2,\rho)$, where ρ is the rank of the Picard group of the K3 surface X.

An object $E \in \mathrm{D}(X)$, the bounded derived category of coherent sheaves on X, has an associated Mukai vector

$$v(E) = \mathrm{ch}(E)\sqrt{\mathrm{td}_X} \in \mathcal{N}(X),$$

where td_X is the Todd class of X. The product on $\mathcal{N}(X)$, the Mukai vector of objects and the structure of $\mathrm{D}(X)$ are related by the beautiful formula (a simple consequence of the Riemann-Roch theorem)

$$-\langle v(E), v(F)\rangle = \sum_{i\in\mathbb{Z}}(-1)^i \dim \mathrm{Hom}_{\mathrm{D}(X)}(E, F[i]) \text{ for } E, F \in \mathrm{D}(X).$$

Recall that natural functors on $\mathrm{D}(X)$ are given by twist functors, as defined in Definition 5.33. A twist functor Φ_E defined by an object $E \in \mathrm{D}(X)$ is an autoequivalence of $\mathrm{D}(X)$ if and only if it is spherical [**422**]; in two dimensions, the spherical condition reads

$$\dim \mathrm{Hom}_{\mathrm{D}(X)}(E, E[i]) = \begin{cases} 1 & \text{for } i = 0, 2, \\ 0 & \text{otherwise.} \end{cases}$$

Substituting these values in the formula above, we obtain that the Mukai vectors of spherical objects $E \in \mathrm{D}(X)$ satisfy

$$\langle v(E), v(E)\rangle = -2.$$

In other words, the Mukai vectors of spherical objects are roots of the lattice $\mathcal{N}$. Thus the root system of $\mathcal{N}$ plays an important role in the description of the autoequivalence group of $\mathrm{D}(X)$. Given this fact, it may not come as a surprise that the root system plays a fundamental role in describing $\mathrm{Stab}(X)$ as well.

Since the lattice $\mathcal{N}$ is unimodular, and it is the image of the Chern character map from the K-group of the category $\mathrm{D}(X)$, the central charge map of Theorem 5.37 can simply be considered as a map

$$Z : \mathrm{Stab}(X) \to \mathcal{N} \otimes \mathbb{C}.$$

Inside $\mathcal{N} \otimes \mathbb{C}$, consider one of the two components $\mathcal{P}$ of the open subset consisting of vectors which span positive definite two-planes in $\mathcal{N} \otimes \mathbb{R}$. For each root $\delta \in \mathcal{N}$, let $\delta^\perp$ denote its perpendicular in $\mathcal{N} \otimes \mathbb{C}$ with respect to the Mukai product. Then we have the following result of Bridgeland.

THEOREM 5.41. [**66**] *There is a connected component of* $\mathrm{Stab}(X)$ *which is mapped by the central charge map* Z *as a regular covering onto the open subset*

$$\mathcal{P}^0 = \mathcal{P} \setminus \bigcup_\delta \delta^\perp \subset \mathcal{N} \otimes \mathbb{C}.$$

Conjecturally, much more should be true: one expects this component of $\mathrm{Stab}(X)$ to be simply connected, and preserved by autoequivalences of $\mathrm{D}(X)$. Given this, one would have a complete geometric description of the autoequivalence group of $\mathrm{D}(X)$, essentially as the fundamental group of the domain $\mathcal{P}^0$. We refer to [**66**] for a proof of Theorem 5.41 and further discussion.

5.8.3. A three-dimensional example: $\mathcal{O}_{\mathbb{P}^2}(-3)$. We take D to be the subcategory of the bounded derived category of coherent sheaves on $X = \mathcal{O}_{\mathbb{P}^2}(-3)$ consisting of complexes whose cohomology sheaves are supported on the zero section $\mathbb{P}^2 \subset X$. Note that, by the derived McKay correspondence, this derived category is equivalent to the derived category of orbifold sheaves supported over the origin in $\mathbb{C}^3/\mathbb{Z}_3$, studied in §5.6 via mirror symmetry.

The Grothendieck group $K(\mathrm{D})$ is a free abelian group of rank three. Following [**64**], we explicitly describe a large connected component of Stab(D) (conjecturally, the only one). It is a union of regions $D(g)$, each labelled by an element $g \in G$, where G is the affine braid group with presentation

$$G = \langle \tau_0, \tau_1, \tau_2 \mid \tau_i \tau_j \tau_i = \tau_j \tau_i \tau_j \text{ for all } i, j \rangle.$$

In each region, the stable objects live in a single category of quiver representations $\boldsymbol{A}(g)$. These regions are patched up into a single component as follows:

THEOREM 5.42. *There is a connected open subset* $\mathrm{Stab}^0(X) \subset \mathrm{Stab}(X)$ *which can be written as a disjoint union of regions*

$$\mathrm{Stab}^0(X) = \bigsqcup_{g \in G} D(g).$$

Each region $D(g)$ is mapped isomorphically by the central charge map Z onto a locally-closed subset of the three-dimensional vector space $\mathrm{Hom}_{\mathbb{Z}}(K(\mathrm{D}), \mathbb{C})$, *and the closures of two regions $D(g_1)$ and $D(g_2)$ intersect in* $\mathrm{Stab}^0(X)$ *precisely if $g_1 g_2^{-1} = \tau_i^{\pm 1}$ for some $i \in \{0, 1, 2\}$.*

5.8.3.1. *Quiver subcategories.* Let (E_0, E_1, E_2) be an exceptional collection of vector bundles on $\mathbb{P}^2$. Any exceptional collection in $\mathrm{D}^b(\mathbb{P}^2)$ is of this form up to shifts. It was proven in [**62**] that there is an equivalence of categories

$$\mathrm{Hom}^{\bullet}\Big(\bigoplus_{i=0}^{2} \pi^* E_i, -\Big) \colon \mathrm{D}^b(X) \longrightarrow \mathrm{D}^b\,\mathbf{Mod}(B),$$

where $\mathbf{Mod}(B)$ is the category of finite dimensional right modules for the algebra

$$B = \mathrm{End}_X\Big(\bigoplus_{i=0}^{2} \pi^* E_i\Big).$$

The algebra B can be described as the path algebra of a quiver with relations taking the form

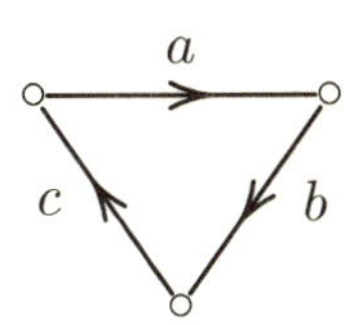

Pulling back the standard t-structure on $\mathrm{D}^b\,\mathbf{Mod}(B)$ gives a bounded t-structure on D whose heart is equivalent to the category of nilpotent modules of B. The abelian subcategories $\boldsymbol{A} \subset \mathrm{D}$ obtained in this way are called exceptional. An abelian subcategory of D is called a *quiver subcategory* if it is of the form $\Phi(\boldsymbol{A})$ for some exceptional subcategory $\boldsymbol{A} \subset \mathrm{D}$ and some autoequivalence $\Phi \in \mathrm{Aut}(\mathrm{D})$.

Any quiver subcategory $\boldsymbol{A} \subset \mathrm{D}$ is equivalent to a category of nilpotent modules of an algebra of the above form. As such it has three simple objects $\{S_0, S_1, S_2\}$ corresponding to the three one-dimensional representations of the quiver. These objects S_i are spherical in the sense of (5.179) and thus give rise to autoequivalences (5.176)

$$\Phi_{S_i} \in \mathrm{Aut}(\mathrm{D}).$$

Note that the three simple objects S_i completely determine the corresponding quiver subcategory $\boldsymbol{A} \subset \mathrm{D}$. The Ext groups between them can be read off from the quiver

$$\mathrm{Hom}^1_{\mathrm{D}}(S_0, S_1) = \mathbb{C}^a, \quad \mathrm{Hom}^1_{\mathrm{D}}(S_1, S_2) = \mathbb{C}^b, \quad \mathrm{Hom}^1_{\mathrm{D}}(S_2, S_0) = \mathbb{C}^c$$

with the other Hom^1 groups being zero. Serre duality then determines the other groups.

EXAMPLE 5.43. Take $\boldsymbol{A}$ to be the exceptional subcategory of D corresponding to the exceptional collection $(\mathcal{O}, \mathcal{O}(1), \mathcal{O}(2))$ on $\mathbb{P}^2$. Its simple objects are

$$S_0 = i_*\mathcal{O}, \quad S_1 = i_*\Omega^1(1)[1], \quad S_2 = i_*\mathcal{O}(-1)[2],$$

where $i\colon \mathbb{P}^2 \hookrightarrow X$ is the inclusion of the zero section, and Ω denotes the cotangent bundle of $\mathbb{P}^2$. We have $(a, b, c) = (3, 3, 3)$.

Let us compute the automorphisms ϕ_{S_i} of $K(\mathrm{D})$ induced by the autoequivalences Φ_{S_i}. The twist functor Φ_S is defined by the triangle

$$\mathrm{Hom}^{\bullet}_{\mathrm{D}}(S, E) \otimes S \longrightarrow E \longrightarrow \Phi_S(E)$$

so that, at the level of K-theory,

$$\phi_S([E]) = [E] - \chi(S, E)[S].$$

If we write P_i for the matrix representing the transformation ϕ_{S_i} with respect to the basis $([S_0], [S_1], [S_2])$ of $K(\mathrm{D})$ then

$$P_0 = \begin{pmatrix} 1 & 3 & -3 \\ 0 & 1 & 0 \\ 0 & 0 & 1 \end{pmatrix}, \quad P_1 = \begin{pmatrix} 1 & 0 & 0 \\ -3 & 1 & 3 \\ 0 & 0 & 1 \end{pmatrix}, \quad P_2 = \begin{pmatrix} 1 & 0 & 0 \\ 0 & 1 & 0 \\ 3 & -3 & 1 \end{pmatrix}.$$

5.8.3.2. *Braid group action.* It was shown in [**62**] that if one tilts a quiver subcategory $\boldsymbol{A} \subset \mathrm{D}$ at one of its simple objects one obtains another quiver subcategory. To describe this process in more detail we need to define a certain braid group which acts on triples of spherical objects.

The three-string annular braid group CB_3 is the fundamental group of the configuration space of three unordered points in $\mathbb{C}^*$. It is generated by three elements τ_i indexed by the cyclic group $i \in \mathbb{Z}_3$ together with a single element r, subject to the relations

$$\begin{aligned} r\tau_i r^{-1} &= \tau_{i+1} \text{ for all } i \in \mathbb{Z}_3, \\ \tau_i\tau_j\tau_i &= \tau_j\tau_i\tau_j \text{ for all } i, j \in \mathbb{Z}_3. \end{aligned}$$

For a proof of the validity of this presentation see [**303**]. If we take the base point to be defined by the three roots of unity, then the elements τ_1 and r correspond to the loops obtained by moving the points as follows:

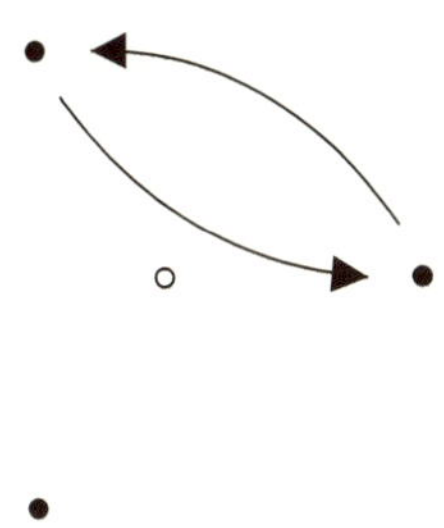

τ_1

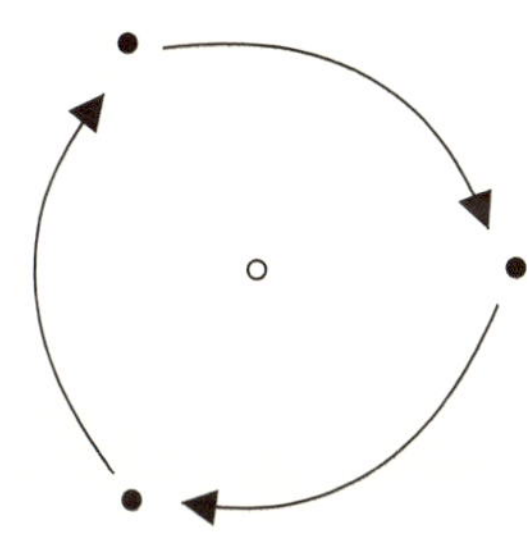

r

We write $G \subset CB_3$ for the subgroup generated by the three braids τ_0, τ_1, τ_2, the group already introduced above.

Define a spherical triple in D to be a triple of spherical objects (S_0, S_1, S_2) of D. The group CB_3 acts on the set of spherical triples in D by the formulae

$$\tau_1(S_0, S_1, S_2) = (S_1[-1], \Phi_{S_1}(S_0), S_2), \quad r(S_0, S_1, S_2) = (S_2, S_0, S_1).$$

The following result allows one to completely understand the process of tilting for quiver subcategories of D.

PROPOSITION 5.44. *Let $\boldsymbol{A} \subset \mathrm{D}$ be a quiver subcategory with simple objects (S_0, S_1, S_2). Then for each $i = 0, 1, 2$ the three simple objects of the tilted quiver subcategory $L_{S_i}(\boldsymbol{A})$ are given by the spherical triple*

$$\tau_i(S_0, S_1, S_2).$$

For each $g \in G$ we then have a quiver subcategory $\boldsymbol{A}(g) \subset \mathrm{D}$ obtained by repeatedly tilting starting at $\boldsymbol{A}$. Its three simple objects are given by the spherical triple

$$(S_0(g), S_1(g), S_2(g)) = g(S_0, S_1, S_2).$$

Note that the three simple objects of an arbitrary quiver subcategory have no well-defined ordering, but the above definition gives a chosen order for the simple objects of the quiver subcategories $\boldsymbol{A}(g)$.

Let $P_i(g) \in \mathrm{SL}(3,\mathbb{Z})$ be the matrix representing the automorphism of $K(\mathrm{D})$ induced by the twist functor $\Phi_{S_i(g)}$ with respect to the fixed basis $([S_0],[S_1],[S_2])$. The formulae defining the action of the braid group on spherical triples show that this system of matrices has the following transformation laws

$$\begin{aligned} P_0(\tau_1 g) &= P_1(g), & P_1(\tau_1 g) &= P_1(g)P_0(g)P_1(g)^{-1}, & P_2(\tau_1 g) &= P_2(g), \\ P_0(rg) &= P_2(g), & P_1(rg) &= P_0(g), & P_2(rg) &= P_0(g). \end{aligned}$$

Introduce a graph $\Gamma(\mathrm{D})$ whose vertices are the quiver subcategories of D, and in which two subcategories are joined by an edge if they differ by a tilt at a simple object. It was shown in [**62**] that distinct elements $g \in G$ define distinct subcategories $\boldsymbol{A}(g) \subset \mathrm{D}$. It follows that each connected component of Γ is just the Cayley graph of G with respect to the generators τ_0, τ_1, τ_2.

5.8.3.3. *Stability conditions on X.* Given an element $g \in G$ let $\boldsymbol{A}(g) \subset \mathrm{D}$ be the corresponding quiver subcategory. The class of any nonzero object $E \in \boldsymbol{A}(g)$ is a strictly positive linear combination:

$$[E] = \sum n_i [S_i(g)] \text{ with } n_1, n_2, n_3 \geq 0 \text{ not all zero.}$$

It follows that to define a stability condition on D we can just choose three complex numbers z_i in the strict upper half-plane

$$H = \{z \in \mathbb{C} : z = r\exp(i\pi\phi) \text{ with } r > 0 \text{ and } 0 < \phi \leq 1\}$$

and set $Z(S_i(g)) = z_i$. The Harder-Narasimhan property is automatically satisfied because $\boldsymbol{A}(g)$ has finite length. We shall denote the corresponding stability condition by $\sigma(g, z_0, z_1, z_2)$.

LEMMA 5.45. *If $\sigma = \sigma(g, z_0, z_1, z_2)$ is a stability condition on* D *of the sort defined above, and $E \in \mathrm{D}$ is stable in σ, then there is an open subset $U \subset \mathrm{Stab}(\mathrm{D})$ containing σ such that E is stable for all stability conditions in U.*

Proof. This follows from the arguments of [**65**, Section 8]. It is enough to check that the set of classes $\gamma \in K(\mathrm{D})$ such that there is an object $F \in \mathrm{D}$ with class $[F] = \gamma$ such that $m_\sigma(F) \leq m_\sigma(E)$ is finite. This is easy to see because the heart of σ has finite length. □

To each element $g \in G$ there is an associated set of stability conditions

$$\begin{aligned} D(g) &= \{\sigma(g, z_0, z_1, z_2) : (z_0, z_1, z_2) \in H^3 \text{ with at most one } z_i \in \mathbb{R}\} \\ &\subset \mathrm{Stab}(X). \end{aligned}$$

By definition these subsets of $\mathrm{Stab}(X)$ are disjoint since they correspond to stability conditions with different hearts.

PROPOSITION 5.46. *There is an open subset*

$$\operatorname{Stab}^0(X) = \bigsqcup_{g \in G} D(g) \subset \operatorname{Stab}(X).$$

If $g_1, g_2 \in G$ then the closures of the regions $D(g_i)$ intersect in $\operatorname{Stab}^0(X)$ *precisely if $g_1 = \tau_i^{\pm 1} g_2$ for some $i \in \{0,1,2\}$.*

Proof. Suppose a point $\sigma = \sigma(g, z_0, z_1, z_2)$ lies in $D(g)$. We must show that there is an open neighbourhood of σ contained in the subset $\operatorname{Stab}^0(X)$. The simple objects $S_i = S_i(g) \in \boldsymbol{A}(g)$ are stable in σ. They remain stable in a small open neighbourhood U of σ in $\operatorname{Stab}(X)$. We repeatedly use the easily proved fact that if $\boldsymbol{A}, \boldsymbol{A}' \subset \mathrm{D}$ are hearts of bounded t-structures and $\boldsymbol{A} \subset \boldsymbol{A}'$ then $\boldsymbol{A} = \boldsymbol{A}'$.

Suppose first that $\operatorname{Im}(z_i) > 0$ for each i. Shrinking U we can assume each S_i has phase in the interval $(0,1)$ for all stability conditions $(Z, \mathcal{P})$ of U. Since $\boldsymbol{A}(g)$ is the smallest extension-closed subcategory of D containing the S_i it follows that $\boldsymbol{A}(g)$ is contained in the heart $\mathcal{P}((0,1])$ of all stability conditions in U. This implies that $\mathcal{P}((0,1]) = \boldsymbol{A}(g)$ and so U is contained in $D(g)$.

Suppose now that one of the z_i, without loss of generality z_0, lies on the real axis, so that σ lies on the boundary of $D(g)$. Thus $z_0 \in \mathbb{R}_{<0}$, and $\operatorname{Im}(z_i) > 0$ for $i = 1,2$. Shrinking U we can assume that $\operatorname{Re} Z(S_0) < 0$ and $\operatorname{Im} Z(S_i) > 0$ for $i = 1,2$ for all stability conditions $(Z, \mathcal{P})$ of U.

The object $S' = \Phi_{S_0}(S_2) \in \mathrm{D}$ lies in $\boldsymbol{A}(g)$, and is in fact a universal extension

$$0 \longrightarrow S_2 \longrightarrow S' \longrightarrow S_0^{\oplus a} \longrightarrow 0$$

where $a = \dim \operatorname{Hom}^1_{\mathrm{D}}(S_0, S_2)$. Since $\operatorname{Hom}_{\mathrm{D}}(S_0, S') = 0$ the object S' lies in $\mathcal{P}((0,1))$ and shrinking U we can assume that this is the case for all stability conditions $(Z, \mathcal{P})$ of U.

We split U into the two pieces $U_+ = \operatorname{Im} Z(S_0) \geq 0$ and $U_- = \operatorname{Im} Z(S_0) < 0$. The argument above shows that $U_+ \subset D(g)$. On the other hand, for any stability condition $(Z, \mathcal{P})$ in U_- the object S_0 is stable with phase in the interval $(1, 3/2)$. Thus the heart $\mathcal{P}((0,1])$ contains the objects $S_0[-1], S'$ and S_1. Since these are the simple objects of the finite length category $\boldsymbol{A}(\tau_0 g)$ it follows that $U_- \subset D(\tau_0 g)$. $\square$

Putting this together, this proves Theorem 5.42.

5.9. Further directions and open questions

5.9.1. Relation between string theory and mathematics. Here we summarize a variety of loose ends and more substantive questions whose resolution would significantly clarify the picture.

5.9.1.1. *Categorical structure.* There are many constructions from homological algebra which played little or no role in our physics discussion, such as the derived versions of tensor product, push-forward and pullback. Furthermore, while we made essential use of correspondences and derived autoequivalences, this was as monodromies, which are rather complicated operations from a physical point of view.

It might be useful to give more direct physical definitions of these constructions, both to flesh out the picture and to guide physics work in related contexts, such as the theory of Dirichlet branes with less or no supersymmetry.

One example for which the physics construction is known is that of the correspondences discussed in §4.6.1. The conformal field theory counterpart of the kernel $\mathcal{F}$ used there is a boundary state $\mathcal{F}$ in the product SCFT $X \times Y$. One can reinterpret such a state as a "defect line" or "interface" separating a left half of the world sheet, described by the SCFT X, from a right half described by the SCFT Y. A boundary state B for Y can then be placed to the right, and a limit taken in which it approaches $\mathcal{F}$. The result is a corresponding boundary state $\Phi_{\mathcal{F}}(B)$ in SCFT X. This construction has been used to formulate dualities in rational conformal field theory [**161, 162**].

After topologically twisting the SCFT, the limiting operation can be replaced by an explicit trace over Hilbert space. Thus, the transform $\Phi_{\mathcal{F}}(B)$ of a boundary state B in X is obtained by regarding $\mathcal{F}$ as a linear superposition of boundary states in Y, and computing

$$\Phi_{\mathcal{F}}(B) = \mathrm{Tr}_{\mathcal{H}_{B,\mathcal{F}}}(-1)^F.$$

The Hilbert space is derived equivalent to $\mathrm{Hom}(B, \mathcal{F})$, while taking the index in the X theory computes cohomology. Thus the derived correspondence can be regarded as a topologically twisted version of the defect line construction. This might have interesting consequences as the physics understanding of this construction develops.

5.9.1.2. *Stability conditions and the stringy Kähler moduli space.* Let X be a simply-connected Calabi-Yau threefold and $\mathrm{D} = \mathrm{D}^b(X)$. According to Property 5.1, we might expect that

$$\mathrm{Aut}(\mathrm{D})\backslash\mathrm{Stab}(X)/\mathbb{C}$$

should be identified with the moduli space of complex structures of the mirror Y with a choice of holomorphic three-form.

In fact, it is easy to see using Theorem 5.37 that this could never be the case. Put simply, the space $\mathrm{Stab}(X)$ is too big and too flat.

For concreteness let us take X to be the quintic threefold. The stringy Kähler moduli space $\mathcal{M}_K(X)$ is, more or less by definition, the complex moduli space of the mirror threefold Y. As is well-known this is a twice-punctured two-sphere with a special point. The punctures are the large volume limit point and the conifold point, and the special point is called the Gepner point. The periods of the mirror Y define holomorphic functions on $\mathcal{M}_{\mathbb{C}}(Y)$ which satisfy a fourth order Picard-Fuchs equation which has regular singular points at the special points.

Under mirror symmetry the periods of Lagrangian submanifolds of Y correspond to central charges of objects of D. Thus we see that the possible maps $Z\colon K(\mathrm{D}) \to \mathbb{C}$ occurring as central charges of stability conditions coming from points of $\mathcal{M}_K(X)$ satisfy the Picard-Fuchs equation for Y. Since these satisfy no linear relation, comparing with Theorem 5.37 we see that the space $\mathrm{Stab}(X)$ must be four-dimensional and the double quotient above is a three-dimensional space containing $\mathcal{M}_K(X)$ as a one-dimensional submanifold. The embedding of this submanifold in $\mathrm{Stab}(X)$ is highly transcendental.

More generally, for a simply-connected Calabi-Yau threefold X we would guess that the space $\mathrm{Stab}(X)$ is *not* the stringy Kähler moduli space, whose tangent space can be identified with $H^{1,1}(X)$, but rather some extended version of it, whose tangent space is

$$\bigoplus_p H^{p,p}(X).$$

To pick out the Kähler moduli space as a submanifold of

$$\mathrm{Aut}(\mathrm{D})\backslash \mathrm{Stab}(X)/\mathbb{C},$$

we would need to define some extra structure on the space of stability conditions. Clues as to the nature of this extra structure can be obtained by studying other situations where extended moduli spaces occur in the mirror symmetry story.

5.9.1.3. *Extended moduli spaces.* There are (at least) three places where extended moduli spaces crop up in algebraic geometry: universal unfolding spaces, big quantum cohomology and the extended moduli spaces of Barannikov-Kontsevich. All these spaces carry rich geometric structures closely related to Frobenius structures and all of them are closely related to moduli spaces of SCFTs. In each case one can make links with spaces of stability conditions, although none of these are close to being made precise. We content ourselves with a brief outline of the connections, with some references.

As explained by Takahashi [**439, 280**], the unfolding space T of an isolated hypersurface singularity X_0 of dimension n should be related to the space of stability conditions on the Fukaya category of the Milnor fiber X_t of the singularity. Note that

$$\mu = \dim_{\mathbb{C}} H_n(X_t, \mathbb{C}) = \dim_{\mathbb{C}} T.$$

Given a basis $L_1, \ldots, L_\mu$ of $H_n(X_t, \mathbb{C})$, K. Saito's theory of primitive forms shows that for a suitable family of holomorphic n-forms Ω_t on the fibers X_t

the periods

$$Z(L_i) = \int_{L_i} \Omega_t,$$

form a system of flat coordinates on the unfolding space T. Since these periods are the analogs of central charges, this is exactly what one would expect from Theorem 5.37.

The big quantum cohomology of a Fano variety Z seems to be related to the space of stability conditions on the derived categories of Z and of the corresponding local Calabi-Yau variety ω_Z. In the case when the quantum cohomology of Z is generically semisimple, Dubrovin showed how to analytically continue the prepotential from an open subset of $H^*(X, \mathbb{C})$ to give a Frobenius structure on a dense open subset of the configuration space

$$M \subset \mathrm{Con}_n(\mathbb{C}) = \{(u_1, \ldots, u_n) \in \mathbb{C}^n : i \neq j \implies u_i \neq u_j\}.$$

According to a conjecture of Dubrovin [**143**] the quantum cohomology of Z is generically semisimple (so that one can define the above extended moduli space M) iff the derived category $\mathrm{D}^b(Z)$ has a full, strong exceptional collection $(E_0, \ldots, E_{n-1})$ (so that one can understand the space of stability conditions by tilting as in Theorem 5.42). Moreover, in suitable coordinates, the Stokes matrix of the quantum cohomology S_{ij} (which controls the analytic continuation of the Frobenius structure on M) is equal to the Gram matrix $\chi(E_i, E_j)$ (which controls the tilting or mutation process). For more on this see [**64, 59**].

Finally, Barannikov and Kontsevich [**30**] showed that if X is a complex projective variety then the formal germ to deformations of X, whose tangent space has dimension $H^1(X, T_X)$, is contained in a larger formal germ whose tangent space has dimension

$$\bigoplus_{p,q} H^p(X, \bigwedge^q T_X),$$

and which describes A_∞ deformations of the category $\mathrm{D}^b(X)$. Suppose X_1 and X_2 are a mirror pair of Calabi-Yau threefolds. Complex deformations of X_1 correspond to Kähler deformations of X_2. Passing to extended moduli spaces one might imagine that some global form of Barannikov and Kontsevich's space parameterizing deformations of $\mathrm{D}^b(X_1)$ should be mirror to the space of stability conditions on X_2. Very schematically we might write

$$\mathrm{Def}(\mathrm{D}^b(X_1)) \cong \mathrm{Stab}(\mathrm{D}^b(X_2)),$$

although to make the dimensions add up one should extend $\mathrm{Stab}(X_2)$ so that its tangent space is the whole cohomology of X_2. Such an isomorphism would be a mirror symmetry statement staying entirely within the realm of algebraic geometry.

5.9.2. Counting stable objects and wall crossing. Given an object $E \in \mathrm{D}$ in a three-dimensional Calabi-Yau category (for example, a vector bundle on a Calabi-Yau threefold), the Ext groups satisfy the duality

$$\mathrm{Ext}^i_{\mathrm{D}}(E, E) \cong \mathrm{Ext}^{3-i}_{\mathrm{D}}(E, E).$$

In particular, the space of deformations $\mathrm{Ext}^1_{\mathrm{D}}(E, E)$ is dual to the space of obstructions $\mathrm{Ext}^2_{\mathrm{D}}(E, E)$. This means that E, locally, sits in a moduli space of objects which is cut out in a space of some dimension by the same number of equations, and therefore the moduli space should locally be an isolated point $[E]$. Therefore, fixing some further conditions which would ensure a compact moduli space, we should just be able to count these points, i.e., objects E satisfying these conditions.

It is of course too much to expect in general that the equations defining E should always be transversal, leading to an honest zero-dimensional moduli space. However, the technology of the virtual cycle [**333, 35, 443**] allows one to nonetheless associate an integer to a compact moduli space of objects E, which is to be interpreted as the "number" of objects of a fixed type.

The topological type of E is given by its class $[E] \in K(\mathrm{D})$ in the (numerical) K-group of D. As usual in sheaf theory, compactness needs more; this is of course where stability comes in. Given a stability condition $\sigma = (Z, \mathcal{P})$ on D, one expects to be able to form moduli spaces of σ-semistable objects of class $\alpha \in K(\mathrm{D})$, and, under the assumption that there are no strictly semistable objects of the given type, use the virtual cycle theory to get integers $n_\sigma(\alpha)$ counting σ-stable objects of class α.

The technical constructions making the previous paragraph precise are difficult; one of the tricky issues is the exact nature of the moduli "space" of objects in a general category D. Vector bundles on varieties form quasi-projective moduli spaces, but the construction involves Invariant Theory and embedding vector bundles into Grassmannians, a method which does not readily generalize to arbitrary objects. Different approaches are employed in the literature, the most general being algebraic stacks [**269**] and formal moduli spaces [**313**].

A very interesting question is what happens to the invariants $n_\sigma(\alpha)$ as one changes the stability condition σ. When two non-zero objects cross phases along a wall in the space of stability conditions, the sets of stable objects change dramatically (as discussed in several places above), hence so do the invariants $n_\sigma(\alpha)$. However, the set of all invariants $\{n_\sigma(\alpha)\}_{\alpha \in K(\mathrm{D})}$ changes in a universal fashion, studied in a series of papers by Joyce [**269**], as well as in very recent work by Kontsevich and Soibelman [**313**]. Wall crossing is studied by Denef and Moore from the physical point of view in [**111**]. Joyce [**272**] used these ideas to define a flat connection on the space of stability conditions on an abelian category satisfying the Calabi-Yau condition, though extending this work to the derived category seems to be problematic. Bridgeland and Toledano Laredo [**59**] interpret Joyce's construction as one of an isomonodromic family of irregular connections on the space of stability conditions of the abelian category.

This is a rapidly developing area. Without making further technical points, let us just mention that the theory also has connections to topological strings [**381, 354, 386, 437**], modular forms [**111**], Ringel-Hall algebras [**269, 270, 271, 273, 59**] as well as the combinatorics of mutations on quivers [**313**].

5.9.3. Direct construction of boundary states. The most satisfactory way to determine the set of physical D-branes, and see whether this is determined by a stability condition, would be to formulate and prove a physical analog of the Donaldson and Uhlenbeck-Yau Theorem 5.8 which governs this problem for large volume B-branes.

Perhaps the best way to do this would be to use the boundary renormalization group. Thus, given an object $\mathcal{E}$ in the topological theory, we formulate a simple boundary condition E' in the SCFT, which is the same as $\mathcal{E}$ after topological twisting, but need not satisfy the constraints of boundary conformal invariance.

We then apply the boundary RG to flow to a fixed point. Physically, this amounts to taking the long distance limit of correlation functions on the world-sheet. Since the topologically twisted theory is independent of the world-sheet metric, this operation produces a boundary state E', still with the same topological twisting $\mathcal{E}$, but now conformally invariant.

If E is a single brane (has a unique identity operator, or in world-volume terms is a $U(1)$ gauge theory), then the result will be the physical brane corresponding to the object $\mathcal{E}$. In general, however, we end up with a set E_i of corresponding physical branes. Presumably, this will always be the set required by Conjecture 5.32). If so, this will imply that the physical branes are the stable objects of §5.7.

This general recipe includes as a particular case the Yang-Mills flow used by Donaldson to prove the Donaldson-Uhlenbeck-Yau theorem. Starting with the nonlinear sigma model definition of the SCFT and a bundle E, we choose a generic connection, and use it to define a boundary condition as in §3.5.2.2. It has then been shown that, to leading order (and all orders for rk $E = 1$), the RG flow is [**451**]:

$$\Lambda \frac{\partial}{\partial \Lambda} A_i = \frac{\delta S_{BI}}{\delta A_i}$$

where S_{BI} is the Born-Infeld action of (3.142). Taking the large volume limit, this becomes the Yang-Mills flow.

Many ingredients would need to be developed in order to make this type of argument convincing in more general cases. Although particular examples of tachyon condensation have been understood physically, *a priori* this involves large variations of the couplings, which can cause corresponding large changes in operator dimensions and thus in the flow.

One simplification would be to grant the implication of §5.7 that, for a fixed stability structure, the stable objects live in an abelian subcategory. We then take as the starting point for the flow only boundary conditions corresponding to objects in the subcategory. This would presumably avoid the need for general results on tachyon condensation. The most concrete version of this is again to work with the nonlinear sigma model and bundles, and show that the deformations of stability due to α' corrections and world-sheet instantons are as predicted. Even this appears hard at present.

One of the main questions this argument raises is whether there is an analog of the symplectic quotient interpretation of the RG fixed point condition. This is rather important in Donaldson's proof as it implies that the action is convex along an orbit and thus the end point of a flow will be independent of the starting point. This is hardly clear in the proposed generalization; if not it could be that some flows end at non-supersymmetric but conformal boundary conditions, which would significantly complicate the story.

5.9.4. Application to Geometric Langlands Duality. Topological open string theory has an important application to Geometric Langlands Duality. On the most elementary level, Geometric Langlands Duality can be formulated as follows. Let C be a Riemann surface (compact, without boundary), G be a compact reductive Lie group, $G_{\mathbb{C}}$ be its complexification, and $\mathscr{M}_{\text{flat}}(G, C)$ be the moduli space of stable flat $G_{\mathbb{C}}$-connections on C. The Langlands dual of G is another compact reductive Lie group ${}^LG$ defined by the condition that its weight and coweight lattices are exchanged relative to G. Let $\text{Bun}({}^LG, C)$ be the moduli stack of holomorphic ${}^LG$-bundles on C. One of the statements of Geometric Langlands Duality is that the derived category of coherent sheaves on $\mathscr{M}_{\text{flat}}(G, C)$ is equivalent to the derived category of D-modules over $\text{Bun}({}^LG, C)$.

As explained in **[294]**, this statement can be deduced from gauge theory in the following way. First, one shows that $\mathscr{M}_{\text{flat}}(G, C)$ is mirror to another moduli space which, roughly speaking, can be described as the cotangent bundle to $\text{Bun}({}^LG, C)$. Second, one shows that the category of A-branes on $T^*\text{Bun}({}^LG, C)$ (with the canonical symplectic form) is equivalent to the category of B-branes on a noncommutative deformation of $T^*\text{Bun}({}^LG, C)$. The latter is the same as the category of (analytic) D-modules on $\text{Bun}({}^LG, C)$.

Explaining the first part of the argument requires familiarity with supersymmetric gauge theories and would take us far beyond the scope of this book. Our goal here is to explain the second part of the argument, that is, the relationship between A-branes and noncommutative B-branes. This relationship arises whenever the target space X is the total space of the cotangent bundle to a complex manifold Y. It is understood that the symplectic form ω is proportional to the canonical symplectic form on T^*Y. (We will also assume that the B-field vanishes). Since Y is complex, we may regard ω as the real part of a holomorphic symplectic form Ω. If q^i are holomorphic coordinates on Y, and p_i are dual coordinates on the fibers of T^*Y, Ω can be written as

$$\Omega = \frac{1}{\hbar} dp_i \wedge dq^i = d\Theta.$$

Since ω (as well as Ω) is exact, the closed A-model of X is rather trivial: there are no nontrivial instantons, and the quantum cohomology ring is isomorphic to the classical one.

We would like to understand the category of A-branes on $X = T^*Y$. The key observation is that there exists a natural coisotropic A-brane on X well-defined up to tensoring with a flat line bundle on X. Its curvature 2-form is exact and given by

$$F = \text{Im }\Omega.$$

If we denote by I the natural almost complex structure on X coming from the complex structure on Y, we have $F = \omega I$, and therefore the endomorphism $\omega^{-1}F = I$ squares to -1. Therefore any unitary connection on a trivial line bundle over X whose curvature is F defines a coisotropic A-brane. For example, we can take the connection 1-form to be

$$A = \text{Im }\Theta$$

We will call this A-brane the canonical coisotropic brane.

Next we would like to understand the endomorphisms of the canonical coisotropic A-brane, i.e., the algebra of BRST-closed open string vertex operators. This is easy to do in the case when Y is an affine space. To understand the general case, one would like to cover Y with charts each of which is an open subset of $\mathbb{C}^n$, and then argue that the computation can be performed locally on each chart and the results "glued together". More precisely, one would like to argue that the algebra in question is the cohomology of a certain sheaf of algebras, whose local structure is the same as for $Y = \mathbb{C}^n$.

In general, the path integral defining the correlators of vertex operators does not have any locality properties in the target space. For example, contributions of disc instantons in the A-model clearly depend on the whole of X. However, one can show that there are no instantons which satisfy boundary conditions corresponding to the canonical coisotropic brane.[28] Further, sigma model perturbation theory is based on Taylor-expanding the integrand of the path integral around constant maps to X. It is easy to show that each term in perturbation theory depends only on the infinitesimal

[28]In fact, this is true for any coisotropic A-brane of codimension 0 on any symplectic manifold X.

neighbourhood of a point on X. These observations can be combined to show that the algebra of open-string vertex operators, regarded as a formal power series in $\hbar$, is the cohomology of a sheaf of algebras, which is locally isomorphic to a similar sheaf for $X = \mathbb{C}^n \times \mathbb{C}^n$.

Let us apply these observations to the canonical coisotropic A-brane on $X = T^*Y$. Locally, we can identify Y with a region in $\mathbb{C}^n$ by means of holomorphic coordinate functions $q^1, \dots, q^n$. Up to BRST-exact terms, the action of the A-model on a disc Σ takes the form

$$S = \frac{1}{\hbar} \int_{\partial\Sigma} \phi^*(p_i dq^i),$$

where ϕ is a map from Σ to X. This action is identical to the action of a particle on Y with zero Hamiltonian, except that q^i are holomorphic coordinates on Y rather than ordinary coordinates. Furthermore, one can show that BRST-invariant open-string vertex operators can be taken to be holomorphic functions of p, q. Therefore quantization is locally straightforward and gives a noncommutative deformation of the algebra of holomorphic functions on T^*Y corresponding to a holomorphic Poisson bivector

$$P = \hbar \frac{\partial}{\partial p_i} \wedge \frac{\partial}{\partial q^i}.$$

One can write an explicit formula for the deformed product:

$$(f \star g)(p,q) = \exp\left(\frac{\hbar}{2}\left(\frac{\partial^2}{\partial p_i \partial \tilde{q}^i} - \frac{\partial^2}{\partial q^i \partial \tilde{p}_i}\right)\right) f(p,q) g(\tilde{p},\tilde{q})|_{\tilde{p}=p,\tilde{q}=q}.$$

This product is known as the Moyal-Wigner product.

The Moyal-Wigner formula is a formal power series in $\hbar$ which may have zero radius of convergence. To rectify the situation, one can restrict to functions which are polynomial in the fiber coordinates p_i. Such locally-defined functions on T^*Y can be thought of as symbols of differential operators; the Moyal-Wigner product in this case reduces to the product of symbols and is a polynomial in $\hbar$. Thus locally the sheaf of open-string vertex operators is modelled on the sheaf of holomorphic differential operators on Y (provided we restrict to operators polynomial in p_i).

Locally, there is no difference between the sheaf of holomorphic differential operators $\mathscr{D}(Y)$ and the sheaf of holomorphic differential operators $\mathscr{D}(Y, \mathscr{L})$ on a holomorphic line bundle $\mathscr{L}$ over Y. Thus the sheaf of open-string vertex operators could be any of the sheaves $\mathscr{D}(Y, \mathscr{L})$. It is shown in [**294**] that the choice of $\mathscr{L}$ is the only ambiguity in quantization. Moreover, the classical problem is symmetric under $p_i \to -p_i$ combined with the orientation reversal of Σ; if we require that quantization preserve this symmetry, then the algebra of open-string vertex operators must be isomorphic to its opposite algebra. It is well known [**52**] that the opposite of the sheaf $\mathscr{D}(Y, \mathscr{L})$ is the sheaf $\mathscr{D}(Y, \mathscr{L}^{-1} \otimes K_Y)$, so symmetry under $p_i \to -p_i$ requires $\mathscr{L}$ to be a square root of the canonical line bundle K_Y. It does not matter which square root one takes, since they all differ by flat line bundles on Y, and tensoring $\mathscr{L}$ by a flat line bundle does not affect the sheaf $\mathscr{D}(Y,\mathscr{L})$.

The conclusion is that the sheaf of open-string vertex operators for the canonical coisotropic A-brane α on $X = T^*Y$ is isomorphic to the sheaf of noncommutative algebras $\mathscr{D}(Y, K_Y^{1/2})$. One can use this fact to associate to any A-brane β on X a twisted D-module, i.e., a sheaf of modules over $\mathscr{D}(Y, K_Y^{1/2})$. Consider the A-model with target X on a strip $\Sigma = I \times \mathbb{R}$, where I is a unit interval, and impose boundary conditions corresponding to branes α and β on the two boundaries of Σ. Upon quantization of this model, one gets a sheaf on vector spaces on Y which is a module over the sheaf of open-string vertex operators inserted at the α boundary. A simple example is to take β to be the zero section of T^*Y with a trivial line bundle. Then the corresponding sheaf is simply the sheaf of sections of $K_Y^{1/2}$, with a tautological action of $\mathscr{D}(Y, K_Y^{1/2})$.

One can argue (nonrigorously) that the map from A-branes to (complexes of) D-modules can be extended to an equivalence of categories of A-branes on X and the derived category of D-modules on Y. The argument relies on the conjectural existence of the category of generalized complex branes for any generalized Calabi-Yau [**284, 390**]. We will not go into this here.

One simple application of these ideas is a physical explanation of a theorem by G. Laumon and M. Rothstein on Fourier transform for D-modules [**320, 405**]. Let Y be an abelian variety of dimension n, which we think of as a quotient of a vector space V by a lattice Γ. Let $Y^\vee$ be the dual torus, and let $X = T^*Y$ as above. X is isomorphic to $V^\vee \times V/\Gamma$. By performing T-duality on Y, we get $X^\vee = V^\vee \times V^\vee/\Gamma^\vee$. Note that Y is a Lagrangian submanifold of X, and therefore this T-duality is a mirror symmetry. That is, $X^\vee$ is mirror to X. As a real manifold, $X^\vee$ is isomorphic to the tangent bundle of $Y^\vee = V^\vee/\Gamma^\vee$, but it is easy to show that the complex structure T-dual to the canonical symplectic form on X is not a product complex structure. One way to describe this complex structure is to choose real coordinates v^j on $V^\vee$; if y^j are the corresponding coordinates on $V^\vee/\Gamma^\vee$, then the complex coordinates on $X^\vee$ are $y^j + iv^j$. In [**320**] and [**405**] this is called the twisted tangent bundle of $Y^\vee$. We conclude that that the category of A-branes on X is equivalent to the category of B-branes on the twisted tangent bundle of $Y^\vee$. Assuming the relationship between A-branes and D-modules, we infer that T-duality on Y induces an equivalence between the derived category of D-modules on Y and the derived category of coherent sheaves on the twisted cotangent bundle of $Y^\vee$. This is the main theorem of [**320**] and [**405**].

One can regard the Geometric Langlands Duality as a nonabelian generalization of this example. Instead of T^*Y one considers the moduli space $\mathscr{M}_{\text{Higgs}}({}^LG, C)$ of Higgs ${}^LG$-bundles on a Riemann surface C. By definition, a Higgs ${}^LG$-bundle is a pair (E, φ), where E is a holomorphic ${}^LG$-bundle on C and φ is a holomorphic section of $\mathrm{ad}(E) \otimes K_C$. The analog of the twisted tangent bundle of $Y^\vee$ is $\mathscr{M}_{\text{flat}}(G, C)$, the moduli space of flat $G_{\mathbb{C}}$ connections on C. As explained in [**294**], the Montonen-Olive conjecture implies that these two moduli spaces are mirror to each other, for a certain canonical choice of symplectic structure on $\mathscr{M}_{\text{Higgs}}({}^LG, C)$. This symplectic structure is the real part of a holomorphic symplectic form Ω on $\mathscr{M}_{\text{Higgs}}({}^LG, C)$. This implies that on $\mathscr{M}_{\text{Higgs}}({}^LG, C)$ there exists a canonical coisotropic A-brane.

While $\mathscr{M}_{\text{Higgs}}({}^LG, C)$ is not exactly a cotangent bundle, it is birational to one: there exists a partially-defined map from $\mathscr{M}_{\text{Higgs}}({}^LG, C)$ to the moduli space of stable ${}^LG$-bundles which sends a pair (E, φ) to E. The preimage of this map is the cotangent bundle of the moduli space of stable ${}^LG$-bundles $\mathscr{M}({}^LG, C)$. In other words, the open piece of $\mathscr{M}_{\text{Higgs}}({}^LG, C)$ consisting of pairs (E, φ) where E is stable can be identified with $T^*\mathscr{M}({}^LG, C)$; one can show that this identification maps the holomorphic symplectic form Ω to the canonical symplectic form on $T^*\mathscr{M}({}^LG, C)$. Then the preceding arguments apply, and we conclude that mirror symmetry relates B-branes on $\mathscr{M}_{\text{flat}}(G, C)$ with (twisted) D-modules on $\mathscr{M}({}^LG, C)$. Strictly speaking, the physical argument applies only to those B-branes whose mirrors belong to the subset $T^*\mathscr{M}({}^LG, C) \subset \mathscr{M}_{\text{Higgs}}({}^LG, C)$; the mathematical formulation of the geometric Langlands duality avoids this restriction by replacing $\mathscr{M}({}^LG, C)$ with the moduli stack of ${}^LG$-bundles on C.

CHAPTER 6

The Strominger-Yau-Zaslow Picture of Mirror Symmetry

Recall that we gave an informal introduction to the physical background of the Strominger-Yau-Zaslow (SYZ) conjecture in the overview Chapter 1, where we stated it in Conjecture 1.5. We shall now explore the picture of mirror symmetry which has grown out of the SYZ conjecture in a systematic way. First, we study McLean's theory of the moduli space of special Lagrangian submanifolds. This leads to certain key structures on the base of an SYZ fibration; eventually, it will be clear that these structures on the base are in fact more important than the conjectured special Lagrangian fibrations. We will explain a toy version of mirror symmetry which can be derived from these structures, explaining various aspects of mirror symmetry from this point of view. In particular, the exchange of complex and Kähler structures becomes clear, and this can then be generalized to an exchange of Hitchin's generalized complex structures. We also explore the interchange of A- and B-branes in this semi-flat context. Finally, we consider more interesting examples of torus fibrations with singular fibers; this is necessary to incorporate the most interesting examples of mirror symmetry. In particular, we give a detailed description at a topological level of SYZ duality.

6.1. Moduli of special Lagrangian submanifolds

As we have seen in §1.3, the physics of D-branes has led to the Strominger-Yau-Zaslow conjecture, which is now a more or less precise mathematical statement. To start to think about this conjecture mathematically, we need to explore the natural structures which appear on the moduli spaces of special Lagrangian submanifolds. These were first examined in fundamental works of McLean [**356**] and Hitchin [**232**]. In what follows, X will be an n-dimensional Calabi-Yau manifold with a nowhere vanishing holomorphic n-form Ω, and Kähler form ω corresponding to a Ricci-flat metric. Recall that a submanifold $M \subseteq X$ is *special Lagrangian* if $\dim_{\mathbb{R}} M = \dim_{\mathbb{C}} X$, i.e., M is half the dimension of X, $\omega|_M = 0$, i.e., M is Lagrangian, and $\operatorname{Im} \Omega|_M = 0$.

6.1.1. McLean's Theorem. In [**356**], McLean proved that the moduli space of special Lagrangian submanifolds was unobstructed, and identified its tangent space. Explicitly, let X be a Calabi-Yau manifold and $M \subseteq X$ be a special Lagrangian submanifold. The tangent space to the space of all deformations of M inside X is given by global sections of the normal bundle of M in X, $\Gamma(M, N_{M/X})$. Because M is Lagrangian, given a normal vector field $\nu \in \Gamma(M, N_{M/X})$, one obtains a well-defined 1-form $\iota(\nu)\omega$ on M, with ι denoting contraction of the vector field ν with the form ω. Explicitly

$$(\iota(\nu)\omega)(w) := \omega(\nu, w)$$

for w a tangent vector to M. This is independent of a lift of ν to a vector field in X. Similarly, because $\operatorname{Im}\Omega|_M = 0$, $\iota(\nu)\operatorname{Im}\Omega$ is a well-defined $(n-1)$-form on M.

An important property of special Lagrangian submanifolds is derived from the following simple linear algebra fact:

EXERCISE 6.1. Consider $\mathbb{C}^n$ with the symplectic form

$$\omega = \frac{\sqrt{-1}}{2}\sum_{i=1}^{n} dz_i \wedge d\bar{z}_i$$

and holomorphic n-form $\Omega = dz_1 \wedge \cdots \wedge dz_n$. Note that $\mathrm{SU}(n)$ is the subgroup of $\mathrm{GL}(n, \mathbb{C})$ preserving ω and Ω. Show that if $V \subseteq \mathbb{C}^n$ is an n-dimensional real subspace with $\omega|_V = \operatorname{Im}\Omega|_V = 0$, then V is the image under some element of $\mathrm{SU}(n)$ of the subspace $\mathbb{R}^n \subseteq \mathbb{C}^n$ (i.e., the subspace given by $\operatorname{Im} z_i = 0$ for $i = 1, \ldots, n$).

From this comes a crucial calculation. If $M \subseteq X$ is a special Lagrangian submanifold, one can find for any point $m \in M$ an isomorphism $u : T_mX \to \mathbb{C}^n$ between the tangent space of X at m and $\mathbb{C}^n$ such that $u^*\left(\frac{\sqrt{-1}}{2}\sum_{i=1}^n dz_i \wedge d\bar{z}_i\right) = \omega$, $u^*(dz_1 \wedge \cdots \wedge dz_n) = \Omega$ at m, and $u(T_mM) = \mathbb{R}^n \subseteq \mathbb{C}^n$. Writing $z_i = x_i + \sqrt{-1}y_i$ and a normal vector field to $u(T_mM)$ as $\nu = \sum_{i=1}^n a_i\partial/\partial y_i$, we see that at m

$$(\iota(\nu)\omega)|_M = -\sum_{i=1}^{n} a_i dx_i$$

while

$$\begin{aligned}(\iota(\nu)\operatorname{Im}\Omega)|_M &= \sum_{i=1}^{n}(-1)^{i+1}a_i dx_1 \wedge \cdots \wedge \widehat{dx_i} \wedge \cdots \wedge dx_n \\ &= \sum_{i=1}^{n} a_i * dx_i,\end{aligned}$$

where $*$ is the Hodge star operator. Thus we obtain

$$*\iota(\nu)\omega = -\iota(\nu)\operatorname{Im}\Omega.$$

Now by identifying a small tubular neighbourhood of M in X with a tubular neighbourhood of the zero section of $N_{M/X}$, there is an open neighbourhood of the zero section $U \subseteq \Gamma(M, N_{M/X})$ of normal vector fields to M for which one can define a map $\exp_\nu : M \to X$. This map takes $m \in M$ to $\nu(m)$, i.e., deforms M in the direction specified by ν. We can then define

$$F : U \to \Gamma(M, \Omega^n_M) \oplus \Gamma(M, \Omega^2_M),$$

the range being the direct sum of the space of n-forms and 2-forms respectively, by

$$F(\nu) = (\exp_\nu^*(\operatorname{Im}\Omega), \exp_\nu^*(\omega)).$$

The local moduli space of special Lagrangian submanifolds near M is given by $F^{-1}(0)$.

We can compute the tangent space to $F^{-1}(0)$ by computing $F'(0)$, which is a linear map

$$F'(0) : \Gamma(M, N_{M/X}) \to \Gamma(M, \Omega^n_M) \oplus \Gamma(M, \Omega^2_M),$$

with

$$\begin{aligned} F'(0)(\nu) &= \left.\frac{d}{dt}F(t\nu)\right|_{t=0} \\ &= \big((\mathcal{L}_\nu(\operatorname{Im}\Omega))|_M, (\mathcal{L}_\nu(\omega))|_M\big) \\ &= \big((\iota(\nu)d(\operatorname{Im}\Omega) + d(\iota(\nu)\operatorname{Im}\Omega))|_M, (\iota(\nu)d(\omega) + d(\iota(\nu)\omega))|_M\big) \\ &= \big((d(\iota(\nu)\operatorname{Im}\Omega)|_M, (d(\iota(\nu)\omega)|_M\big) \end{aligned}$$

Here $\mathcal{L}_\nu$ is the Lie derivative with respect to ν, and the third line is the Cartan formula for the Lie derivative. Thus we see that $\nu \in \ker(F'(0))$ if and only if on M we have

$$\begin{aligned} d(\iota(\nu)\omega) &= 0 \\ d(\iota(\nu)\operatorname{Im}\Omega) &= 0, \end{aligned}$$

the latter being equivalent to $d^*(\iota(\nu)\omega) = 0$. Thus $\iota(\nu)\omega$ must be a *harmonic* 1-form on M. So the tangent space to $F^{-1}(0)$ is $\mathcal{H}^1(M, \mathbb{R})$, the space of harmonic 1-forms on M. Without going into the analytic details, an application of the implicit function theorem then yields McLean's theorem: [**356**]

THEOREM 6.2. *Let $M \subseteq X$ be a compact special Lagrangian submanifold. The space of special Lagrangian deformations of M in X is a manifold B, with tangent space at the point $[M] \in B$ corresponding to M isomorphic to $\mathcal{H}^1(M, \mathbb{R})$, the space of harmonic 1-forms on M.*

Now let us consider families of special Lagrangian submanifolds parameterized by a manifold B as follows. Let $\mathcal{U} \subseteq B \times X$ be a submanifold, and let $p_1 : \mathcal{U} \to B$, $p_2 : \mathcal{U} \to X$ be the projections. Assume $p_2(p_1^{-1}(y))$ is a special Lagrangian submanifold of X for all $y \in B$. For each $y \in B$, one has a map $T_yB \to \mathcal{H}^1(p_1^{-1}(y), \mathbb{R})$, essentially as described above. Explicitly, given $\nu \in T_yB$, lift ν to a normal vector field ν to $p_1^{-1}(y)$, and then $p_{2*}\nu$ is the normal vector field to $p_2(p_1^{-1}(y))$ corresponding to the deformation direction ν. Then $\iota(p_{2*}\nu)\omega = \iota(\nu)p_2^*\omega$ is the harmonic 1-form corresponding to ν. We *now make the assumption that this map is an isomorphism for each* $y \in B$. We then obtain the following structures on B.

The McLean metric. The Hodge metric on $\mathcal{H}^1(p_1^{-1}(y), \mathbb{R})$ given by

$$g(\alpha, \beta) = \int_{p_1^{-1}(y)} \alpha \wedge *\beta$$

induces a metric on B, which can be written as

$$g(\nu_1, \nu_2) = -\int_{p_1^{-1}(y)} \iota(\nu_1)p_2^*\omega \wedge \iota(\nu_2)p_2^* \operatorname{Im}\Omega.$$

The Hodge metric is positive definite, so we obtain a Riemannian metric on B.

Affine coordinates 1. Let $U \subseteq B$ be a small open set, and suppose we have submanifolds $\gamma_1, \ldots, \gamma_s \subseteq \mathcal{U}$ which are families of 1-cycles over U, i.e., $\gamma_i \cap p_1^{-1}(y)$ is a 1-cycle in $p_1^{-1}(y)$ for $y \in U$. Assume further $\{\gamma_i \cap p_1^{-1}(y)\}$ form a basis for $H_1(p_1^{-1}(y), \mathbb{Z})/\text{tors}$. Then we can consider the 1-forms

$$\omega_i = p_{1*}(\omega|_{\gamma_i})$$

on U. This is obtained by fiberwise integration, with

$$\omega_i(\nu) = \int_{p_1^{-1}(y)\cap\gamma_i} \iota(\nu)\omega$$

for $\nu \in T_yB$. Since ω is closed, so is each ω_i. Thus, locally, we can write $\omega_i = dy_i$ for functions $y_1, \ldots, y_s$. These have the following properties:

- The forms $\omega_1, \ldots, \omega_s$ are linearly independent at each $y \in U$; if not, there exist $a_1, \ldots, a_s \in \mathbb{R}$ such that $\int_{p_1^{-1}(y)\cap\sum a_i\gamma_i} \iota(\nu)\omega = 0$ for all $\nu \in T_yB$. But $\iota(\nu)\omega|_{p_1^{-1}(y)}$ runs over all elements of $\mathcal{H}^1(p_1^{-1}(y), \mathbb{R})$, and hence $\sum a_i\gamma_i = 0$ in $H_1(p_1^{-1}(y), \mathbb{R})$. But the γ_i are linearly independent by assumption. Hence the functions $y_1, \ldots, y_s$ give a local coordinate system near y.
- $y_1, \ldots, y_s$ are well-defined up to a choice of constants, and if a different choice of basis $\gamma_1, \ldots, \gamma_s$ is used, $y_1, \ldots, y_s$ are replaced with $y_i' = \sum a_{ij}y_j + b_i$, with $(a_{ij})_{1\le i,j\le s} \in \mathrm{GL}(s, \mathbb{Z})$ and $b_i \in \mathbb{R}$. Thus we obtain well-defined *affine coordinates* (see Definition 6.3).

Affine coordinates 2. We repeat the same process with families

$$\Gamma_1, \ldots, \Gamma_s \subseteq \mathcal{U}$$

of $(n-1)$-cycles over U with $\{\Gamma_i \cap p_1^{-1}(y)\}$ forming a basis for

$$H_{n-1}(p_1^{-1}(y), \mathbb{Z})/\text{tors}.$$

Then we obtain the 1-forms $\lambda_i = -p_{1*}(\operatorname{Im}\Omega|_{\Gamma_i})$, or equivalently,

$$\lambda_i(\nu) = -\int_{p_1^{-1}(y)\cap\Gamma_i} \iota(\nu)\operatorname{Im}(\Omega).$$

Again, $\lambda_1, \ldots, \lambda_s$ are closed 1-forms, linearly independent at each point. Thus we obtain functions $\check{y}_1, \ldots, \check{y}_s$ forming a local coordinate system, with $d\check{y}_i = \lambda_i$.

6.1.2. Affine manifolds and the Legendre transform. Let us focus on the structure given by the affine coordinates which emerged in the previous section more abstractly. We fix $M = \mathbb{Z}^n$, $N = \mathrm{Hom}_{\mathbb{Z}}(M, \mathbb{Z})$, $M_{\mathbb{R}} = M \otimes_{\mathbb{Z}} \mathbb{R}$, $N_{\mathbb{R}} = N \otimes_{\mathbb{Z}} \mathbb{R}$. We set

$$\mathrm{Aff}(M_{\mathbb{R}}) = M_{\mathbb{R}} \rtimes \mathrm{GL}_n(\mathbb{R})$$

to be the group of affine transformations of $M_{\mathbb{R}}$, i.e., maps $T : M_{\mathbb{R}} \to M_{\mathbb{R}}$ of the form $T(x) = Ax + b$ for $A : M_{\mathbb{R}} \to M_{\mathbb{R}}$ invertible and linear, and $b \in M_{\mathbb{R}}$. This has a subgroup

$$\mathrm{Aff}(M) = M \rtimes \mathrm{GL}_n(\mathbb{Z}).$$

DEFINITION 6.3. Let B be an n-dimensional manifold. An *affine structure* on B is given by an open cover $\{U_i\}$ along with coordinate charts $\psi_i : U_i \to M_{\mathbb{R}}$, whose transition functions $\psi_i \circ \psi_j^{-1}$ lie in $\mathrm{Aff}(M_{\mathbb{R}})$. The affine structure is *integral* if the transition functions lie in $\mathrm{Aff}(M)$. If B and B' are affine manifolds of the same dimension, and $f : B \to B'$ is an immersion, then we say f is an *(integral) affine map* if f is given by (integral) affine maps on each (integral) affine coordinate chart.

Thus we see in fact that in the situation of §6.1.1, B carries two different affine structures. We call these two affine structures the *symplectic* and *complex* affine structures respectively, as they are induced by the symplectic form ω on X and the holomorphic n-form Ω respectively.

The last point of **[232]** we wish to make here is that these two affine structures and the McLean metric are compatible in the following sense, and in particular the two affine structures are related by a *Legendre transform*.

PROPOSITION 6.4. *Let B be a moduli space parameterizing special Lagrangian deformations of some special Lagrangian submanifold $M \subseteq X$ as in §6.1.1. Let $y \in B$. Then there is an open neighbourhood U of y with symplectic affine coordinates $y_1, \ldots, y_s$ and a convex function K on U such that*

$$g(\partial/\partial y_i, \partial/\partial y_j) = \partial^2 K/\partial y_i \partial y_j.$$

Furthermore, $\check{y}_i = \partial K/\partial y_i$ form a system of complex affine coordinates, and if

$$\check{K}(\check{y}_1, \ldots, \check{y}_s) = \sum \check{y}_i y_i - K(y_1, \ldots, y_s)$$

is the Legendre transform of K, then

$$y_i = \partial \check{K}/\partial \check{y}_i$$

and

$$\partial^2 \check{K}/\partial \check{y}_i \partial \check{y}_j = g(\partial/\partial \check{y}_i, \partial/\partial \check{y}_j).$$

PROOF. Take families $\gamma_1,\ldots,\gamma_s,\Gamma_1,\ldots,\Gamma_s$ as in §6.1.1 over an open neighbourhood U with the two bases being Poincaré dual, i.e., $\gamma_i\cdot\Gamma_j=\delta_{ij}$ for $y\in U$. Let $\gamma_1^*,\ldots,\gamma_s^*$ and $\Gamma_1^*,\ldots,\Gamma_s^*$ be dual bases for $H^1(p_1^{-1}(y),\mathbb{R})$ and $H^{n-1}(p_1^{-1}(y),\mathbb{R})$ respectively. This gives local coordinates $y_1,\ldots,y_s$ with $dy_i=\omega_i$, so in particular

$$\delta_{ij}=\omega_i(\partial/\partial y_j)=\int_{\gamma_i\cap p_1^{-1}(y)}\iota(\partial/\partial y_j)\omega,$$

so $\iota(\partial/\partial y_j)\omega$ defines the cohomology class γ_j^* in $H^1(p_1^{-1}(y),\mathbb{R})$. Similarly, let

$$g_{ij}=-\int_{\Gamma_i\cap p_1^{-1}(y)}\iota(\partial/\partial y_j)\operatorname{Im}\Omega;$$

then $-\iota(\partial/\partial y_j)\operatorname{Im}\Omega$ defines the cohomology class

$$\sum g_{ij}\Gamma_i^*\in H^{n-1}(p_1^{-1}(y),\mathbb{R}),$$

and $\lambda_i=\sum g_{ij}dy_j$. Thus

$$\begin{aligned} g(\partial/\partial y_j,\partial/\partial y_k) &= -\int_{p_1^{-1}(y)}\iota(\partial/\partial y_j)\omega\wedge\iota(\partial/\partial y_k)\operatorname{Im}\Omega\\ &= g_{jk}.\end{aligned}$$

On the other hand, let $\check y_1,\ldots,\check y_s$ be coordinates with $d\check y_i=\lambda_i$. Then

$$\partial\check y_i/\partial y_j=g_{ij}=g_{ji}=\partial\check y_j/\partial y_i,$$

so $\sum\check y_i dy_i$ is a closed 1-form. Thus there exists locally a function K such that $\partial K/\partial y_i=\check y_i$ and $\partial^2K/\partial y_i\partial y_j=g(\partial/\partial y_i,\partial/\partial y_j)$. A simple calculation then confirms that $\partial\check K/\partial\check y_i=y_i$. On the other hand,

$$\begin{aligned} g(\partial/\partial\check y_i,\partial/\partial\check y_j) &= g\left(\sum_k\frac{\partial y_k}{\partial\check y_i}\frac{\partial}{\partial y_k},\sum_l\frac{\partial y_l}{\partial\check y_j}\frac{\partial}{\partial y_l}\right)\\ &= \sum_{k,l}\frac{\partial y_k}{\partial\check y_i}\frac{\partial y_l}{\partial\check y_j}g(\partial/\partial y_k,\partial/\partial y_l)\\ &= \sum_{k,l}\frac{\partial y_k}{\partial\check y_i}\frac{\partial y_l}{\partial\check y_j}\frac{\partial\check y_k}{\partial y_l}\\ &= \frac{\partial y_j}{\partial\check y_i}=\frac{\partial^2\check K}{\partial\check y_i\partial\check y_j}.\end{aligned}$$

□

It is useful to be able to do things in a more coordinate-independent manner, as follows.

PROPOSITION 6.5. *Let $\pi:\tilde B\to B$ be the universal covering of an (integral) affine manifold B, inducing an (integral) affine structure on $\tilde B$. Then* *there is an (integral) affine map $\delta : \tilde{B} \to M_{\mathbb{R}}$, called the developing map, and any two such maps differ only by an (integral) affine transformation.*

PROOF. This is standard, see [**184**], pg. 641 for a proof. One simply patches together affine coordinate charts. □

Note that there is no need for the developing map to be injective or a covering space; it is only an immersion in general.

DEFINITION 6.6. The fundamental group $\pi_1(B)$ acts on $\tilde{B}$ by deck transformations; for $\gamma \in \pi_1(B)$, let $\Psi_\gamma : \tilde{B} \to \tilde{B}$ be the corresponding deck transformation. Then by the uniqueness of the developing map, there exists a $\rho(\gamma) \in \mathrm{Aff}(M_{\mathbb{R}})$ such that $\rho(\gamma) \circ \delta \circ \Psi_\gamma = \delta$. The map $\rho : \pi_1(B) \to \mathrm{Aff}(M_{\mathbb{R}})$ is called the *holonomy representation*. If the affine structure is integral, then $\mathrm{im}\,\rho \subseteq \mathrm{Aff}(M)$.

We use the convention that $\Psi_{\gamma_1\gamma_2} = \Psi_{\gamma_2} \circ \Psi_{\gamma_1}$, from which it follows we have defined ρ to be a group homomorphism. The holonomy representation can be viewed in a slightly different way. There is a flat connection on TB such that $\partial/\partial y_1, \ldots, \partial/\partial y_n$ are flat sections whenever $y_1, \ldots, y_n$ is an affine coordinate system. Then the linear part of ρ is just the holonomy of this affine connection, i.e., the linear part of $\rho(\gamma)$ is given by parallel transport in TB around γ^{-1}. The representation ρ itself can be viewed as the holonomy of a natural *affine* connection on TB: see [**184**] for details. Note that the linear part of ρ is precisely the monodromy of Λ.

EXAMPLE 6.7. (1) A torus $T^n = M_{\mathbb{R}}/\Lambda$ for a lattice $\Lambda \subseteq M_{\mathbb{R}}$ has a natural affine structure induced by the developing map the identity $\mathrm{id} : M_{\mathbb{R}} \to M_{\mathbb{R}}$. Here for $\lambda \in \pi_1(T^n) = \Lambda$, $\rho(\lambda)$ is translation by $-\lambda$. Thus the affine structure is integral if and only if $\Lambda \subseteq M$.

(2) This example is from [**33**]. Take $M = \mathbb{Z}^2$, and consider the subgroup $G \subseteq \mathrm{Aff}(M_{\mathbb{R}})$ defined by

$$G = \left\{ A \in \mathrm{Aff}(M_{\mathbb{R}}) \middle| \begin{array}{c} A(m_1, m_2) = (m_1 + vm_2 + u + v(v-1)/2, m_2 + v) \\ \text{for } u, v \in \mathbb{R} \end{array} \right\}.$$

(This is not quite the form given in [**33**], but rather G has been conjugated by translation by $(0, -1/2)$ to obtain better integrality properties). G is isomorphic to $\mathbb{R}^2$, and if we choose any lattice $\Gamma \subseteq G$, then Γ acts properly and discontinuously, so that $M_{\mathbb{R}}/\Gamma$ is an affine manifold, topologically a two-torus. This is the only other affine structure on the two-torus obtained from $M_{\mathbb{R}}$ by dividing out by a properly discontinuous group action. (See [**33**], Theorem 4.5). Note that the affine structure is integral with respect to the integral structure $M \subseteq M_{\mathbb{R}}$ if and only if

$$\Gamma \subseteq \left\{ A \in \mathrm{Aff}(M_{\mathbb{R}}) \middle| \begin{array}{c} A(m_1, m_2) = (m_1 + vm_2 + u + v(v-1)/2, m_2 + v) \\ \text{for } u \in \mathbb{Z},\ v \in \mathbb{Z} \end{array} \right\}.$$

REMARK 6.8. Note in fact that we can specify an affine structure on B by giving the developing map. In other words, if we have a map $\delta : \tilde{B} \to M_{\mathbb{R}}$ which is an immersion, and δ satisfies

$$\rho(\gamma) \circ \delta \circ \Psi_\gamma = \delta$$

for all $\gamma \in \pi_1(B)$ for some representation

$$\rho : \pi_1(B) \to \mathrm{Aff}(M_{\mathbb{R}}),$$

then δ determines an affine structure on B. Furthermore, if the image of ρ is contained in $\mathrm{Aff}(M)$ then the affine structure is integral.

We now describe the Legendre transform in a coordinate-free manner. Suppose now B is an affine manifold carrying a metric of Hessian form, i.e., locally there exists a function K such that $g_{ij} = \partial^2 K/\partial y_i \partial y_j$, where $y_1, \ldots, y_n$ are local affine coordinates. Note that K does not depend on the particular choice of coordinates. However, it is only well-defined up to a choice of affine linear function, i.e., a function of the form $\sum a_i y_i + b$. Any two such functions K differ by an affine linear function, so we can patch to get a well-defined $K : \tilde{B} \to \mathbb{R}$. Note that $K - K \circ \Psi_\gamma := \alpha(\gamma)$ is an affine linear function $\alpha(\gamma) : \tilde{B} \to \mathbb{R}$; in other words $\alpha(\gamma)$ is the composition of the developing map with an affine linear function $M_{\mathbb{R}} \to \mathbb{R}$.

The differential of the developing map $\delta : \tilde{B} \to M_{\mathbb{R}}$ yields an isomorphism of cotangent bundles

$$\delta^* : \delta^* T^* M_{\mathbb{R}} \to T^* \tilde{B}.$$

Since $T^* M_{\mathbb{R}} = M_{\mathbb{R}} \times N_{\mathbb{R}}$, δ^* gives an isomorphism of $\tilde{B} \times N_{\mathbb{R}}$ with $T^* \tilde{B}$. Let $q : \tilde{B} \times N_{\mathbb{R}} \to N_{\mathbb{R}}$ be the projection. The differential dK of K is a section of $T^* \tilde{B}$, i.e., a map $dK : \tilde{B} \to T^* \tilde{B}$, and thus we can view $(\delta^*)^{-1}(dK)$ as a section of $\delta^* T^* M_{\mathbb{R}}$. Hence we can view

$$\check{\delta} := q \circ (\delta^*)^{-1}(dK)$$

as a function $\check{\delta} : \tilde{B} \to N_{\mathbb{R}}$. This is just the differential dK under these identifications. Because the Hessian of K is positive definite, $\check{\delta}$ is an immersion.

PROPOSITION 6.9. *Let $\tilde{\rho}$ be the composition of ρ with the natural projection* $\mathrm{Aff}(M_{\mathbb{R}}) \to \mathrm{GL}(M_{\mathbb{R}})$ *(this is just the linear part of ρ). Then*

$${}^t\tilde{\rho}(\gamma^{-1}) \circ \check{\delta} \circ \Psi_\gamma + d\alpha(\gamma) = \check{\delta},$$

and $\pi_1(B)$ acts on the new affine structure on $\tilde{B}$ by affine transformations. Dividing out by this action, we obtain a new affine structure on B, which we denote by $\check{B}$. So the holonomy representation $\check{\rho} : \pi_1(\check{B}) \to$ $\mathrm{Aff}(N_{\mathbb{R}})$ *of the affine structure given by $\check{\delta}$ has linear part $\tilde{\rho}^\vee$ dual to the representation $\tilde{\rho}$.*

PROOF. As $\delta \circ \Psi_\gamma = \rho(\gamma^{-1}) \circ \delta$, we have a commutative diagram

$$\begin{array}{ccccc} \Psi_\gamma^*\delta^*T^*M_{\mathbb{R}} & \xrightarrow{=} & \delta^*\rho(\gamma^{-1})^*T^*M_{\mathbb{R}} & \xrightarrow{\rho(\gamma^{-1})^*} & \delta^*T^*M_{\mathbb{R}} \\ \downarrow{\scriptstyle \delta^*} & & & & \downarrow{\scriptstyle \delta^*} \\ \Psi_\gamma^*T^*\tilde{B} & & \xrightarrow[\Psi_\gamma^*]{} & & T^*\tilde{B} \end{array}$$

But $\rho(\gamma^{-1})^*$ acts on $N_{\mathbb{R}}$ simply as ${}^t\tilde{\rho}(\gamma^{-1})$, so for any one-form ω on $\tilde{B}$,

$$q \circ (\delta^*)^{-1}(\Psi_\gamma^*(\omega)) = q \circ \rho(\gamma^{-1})^* \circ (\delta^*)^{-1}(\omega) = {}^t\tilde{\rho}(\gamma^{-1}) \circ q \circ (\delta^*)^{-1}(\omega).$$

Now by the chain rule, we have the equality of forms

$$\Psi_\gamma^*(dK) = d(K \circ \Psi_\gamma).$$

Thus the form dK at a point $\Psi_\gamma(b)$ coincides with the pull-back $\Psi_{\gamma^{-1}}^*$ applied to the form $d(K \circ \Psi_\gamma)$ at $b \in \tilde{B}$. This translates into the identity

$$\begin{aligned} \check{\delta} \circ \Psi_\gamma &= q \circ (\delta^*)^{-1}(\Psi_{\gamma^{-1}}^*(d(K \circ \Psi_\gamma))) \\ &= {}^t\tilde{\rho}(\gamma) \circ q \circ (\delta^*)^{-1}(dK - d\alpha(\gamma)) \\ &= {}^t\tilde{\rho}(\gamma) \circ (\check{\delta} - d\alpha(\gamma)). \end{aligned}$$

Here $d\alpha(\gamma)$ is naturally identified with an element of $N_{\mathbb{R}}$. □

We can also define the Legendre transform of K as, for $b \in \tilde{B}$,

$$\check{K}(b) = \langle \check{\delta}(b), \delta(b) \rangle - K(b).$$

Then the Hessian of $\check{K}$ on $\check{B}$ defines the same metric as K on B, and $d\check{K} = \delta$. We call $(\check{B}, \check{K})$ the *Legendre transform* of (B, K).

It is clear that if one knows two out of the three of B, $\check{B}$ and g, one knows the last one also.

EXAMPLE 6.10. Suppose $B = M_{\mathbb{R}}/\Gamma$ as in Example 6.7, (1). Then one choice of potential function $K : M_{\mathbb{R}} \to \mathbb{R}$ is simply a convex quadratic function, determined by an inner product $\langle \cdot, \cdot \rangle$ on $M_{\mathbb{R}}$, with $K(x) = \frac{1}{2}\langle x, x\rangle$. Note that for $\gamma \in \Gamma$,

$$\begin{aligned} K(x+\gamma) &= \frac{1}{2}\langle x+\gamma, x+\gamma\rangle \\ &= \frac{1}{2}\langle x, x\rangle + \langle x, \gamma\rangle + \frac{1}{2}\langle \gamma, \gamma\rangle \\ &= K(x) + \langle x, \gamma\rangle + \frac{1}{2}\langle \gamma, \gamma\rangle, \end{aligned}$$

so $\alpha(\gamma)(x) = \langle x, \gamma\rangle + \frac{1}{2}\langle \gamma, \gamma\rangle$ and $d\alpha(\gamma)$ is the element of $N_{\mathbb{R}}$ corresponding to the functional $\langle \gamma, \cdot\rangle$ on $M_{\mathbb{R}}$. In other words, the inner product allows us to identify $M_{\mathbb{R}}$ with $N_{\mathbb{R}}$; in fact, $\check{\delta} : M_{\mathbb{R}} \to N_{\mathbb{R}}$ is easily seen to be this identification. This identifies $\Gamma \subseteq M_{\mathbb{R}}$ with

$$\check{\Gamma} = \{d\alpha(\gamma) \in N_{\mathbb{R}} | \gamma \in \Gamma\}.$$

EXERCISE 6.11. Show that in Example 6.7, (2), there is no convex function $K : M_{\mathbb{R}} \to \mathbb{R}$ with the property that $K - K \circ \Psi_\gamma = \alpha(\gamma)$ for some affine linear $\alpha(\gamma)$, for all $\gamma \in \Gamma$. Thus $B = M_{\mathbb{R}}/\Gamma$ cannot be a moduli space of special Lagrangian tori on some Calabi-Yau manifold.

6.2. The semi-flat SYZ picture

We shall now use the structures on the base of a special Lagrangian fibration detailed in the previous section to describe a simple form of mirror symmetry. See [**331**] for more details on semi-flat mirror symmetry.

6.2.1. The basic version. We can now use the structures discussed in §6.1 to define a toy version of mirror symmetry. Fix throughout this section an affine manifold B, and assume that all transition maps are in $M_{\mathbb{R}} \rtimes \mathrm{GL}(n, \mathbb{Z})$ (rather than $\mathrm{Aff}(M_{\mathbb{R}})$).

Given this data, let $y_1, \dots, y_n$ be local affine coordinates; then one obtains a family of lattices in TB, generated by $\partial/\partial y_1, \dots, \partial/\partial y_n$. This is well-defined globally because of the integrality assumption: a change of coordinates will produce a different basis for the same lattice, related by an element of $\mathrm{GL}(n, \mathbb{Z})$. This defines a local system $\Lambda \subseteq TB$. Similarly, locally $dy_1, \dots, dy_n$ generate a local system $\check{\Lambda} \subseteq T^*B$. We will also write $\Lambda_{\mathbb{R}} = \Lambda \otimes_{\mathbb{Z}} \mathbb{R}$ and $\check{\Lambda}_{\mathbb{R}} = \check{\Lambda} \otimes_{\mathbb{Z}} \mathbb{R}$. Again these are local systems contained in TB and T^*B, but now we allow real linear combinations of $\partial/\partial y_1, \dots, \partial/\partial y_n$ or $dy_1, \dots, dy_n$ as local sections.

We can now define two torus bundles:

$$X(B) := TB/\Lambda$$

is a T^n-bundle over B, as is

$$\check{X}(B) := T^*B/\check{\Lambda}.$$

We write

$$f : X(B) \to B$$

and

$$\check{f} : \check{X}(B) \to B,$$

and we say these are dual torus bundles.

These bundles $X(B)$ and $\check{X}(B)$ come along with some additional structures. First, $\check{X}(B)$ is naturally a symplectic manifold: the canonical symplectic form on T^*B descends to a symplectic form ω on $\check{X}(B)$. Second, $X(B)$ carries a complex structure. Locally, this can be described in terms of holomorphic coordinates. Let $U \subseteq B$ be an open set with affine coordinates $y_1, \ldots, y_n$, so TU has coordinate functions $y_1, \ldots, y_n$, $x_1 = dy_1, \ldots, x_n = dy_n$. Then

$$q_j = e^{2\pi\sqrt{-1}(x_j + \sqrt{-1}y_j)}$$

gives a system of holomorphic coordinates on TU/Λ.

EXERCISE 6.12. Check to see how the coordinates q_j transform under an integral affine change of coordinates y_j. Observe that this change of coordinates is holomorphic.

Completing this exercise shows that these coordinates give a well-defined complex structure on $X(B)$. Note that this can be described in a coordinate-free manner as follows. The differential of the developing map takes $T\tilde{B} = \tilde{B} \times M_{\mathbb{R}}$ to $TM_{\mathbb{R}} = M_{\mathbb{R}} \times M_{\mathbb{R}}$, which allows us to pull back the natural complex structure on $M_{\mathbb{R}} \times M_{\mathbb{R}} = M_{\mathbb{R}} \oplus \sqrt{-1}M_{\mathbb{R}} = M_{\mathbb{R}} \otimes \mathbb{C}$. The complex structure described by the above coordinates is just induced by the pullback of the canonical complex structure on $M_{\mathbb{R}} \otimes \mathbb{C}$.

Note that in addition $X(B)$ comes along with a natural local holomorphic n-form Ω given by

$$\frac{dq_1 \wedge \cdots \wedge dq_n}{q_1 \cdots q_n}.$$

This is not always globally well-defined, although $\Omega \wedge \bar{\Omega}$ is, as we see in the following exercise.

EXERCISE 6.13. Check that Ω is preserved by a change of coordinates $y_1, \ldots, y_n$ in $M_{\mathbb{R}} \rtimes \mathrm{SL}_n(\mathbb{Z})$, and is only preserved up to sign by a change of coordinates in $M_{\mathbb{R}} \rtimes \mathrm{GL}(n, \mathbb{Z})$. Thus, if the holonomy representation is contained in $M_{\mathbb{R}} \rtimes \mathrm{SL}_n(\mathbb{Z})$, we obtain a global holomorphic n-form Ω.

Now suppose in addition we have a metric g of Hessian form on B, with potential function $K : \tilde{B} \to \mathbb{R}$. Then in fact both $X(B)$ and $\check{X}(B)$ become Kähler manifolds:

PROPOSITION 6.14. *$K \circ f$ is a (multi-valued) Kähler potential on $X(B)$, defining a Kähler form $\omega = 2\sqrt{-1}\partial\bar{\partial}(K \circ f)$. This metric is Ricci-flat if and only if K satisfies the real Monge-Ampère equation*

$$\det \frac{\partial^2 K}{\partial y_i \partial y_j} = \text{constant}.$$

PROOF. Working locally with affine coordinates (y_i) and multi-valued complex coordinates $z_i = \frac{1}{2\pi\sqrt{-1}} \log q_i = x_i + \sqrt{-1}y_i$, we compute $\omega = 2\sqrt{-1}\partial\bar{\partial}(K \circ f) = \frac{\sqrt{-1}}{2} \sum \frac{\partial^2 K}{\partial y_i \partial y_j} dz_i \wedge d\bar{z}_j$ which is clearly positive. Furthermore, ω^n is proportional to $\Omega \wedge \bar{\Omega}$ if and only if $\det(\partial^2 K/\partial y_i \partial y_j)$ is constant. □

We write this Kähler manifold as $X(B, K)$.

Dually we have

PROPOSITION 6.15. *In local canonical coordinates $y_i, \check{x}_i$ on T^*B, the functions $z_i = \check{x}_i + \sqrt{-1}\partial K/\partial y_i$ on T^*B induce a well-defined complex structure on $\check{X}(B)$, with respect to which the canonical sympletic form ω is the Kähler form of a metric. Furthermore this metric is Ricci-flat if and only if K satisfies the real Monge-Ampère equation*

$$\det \frac{\partial^2 K}{\partial y_i \partial y_j} = \text{constant}.$$

PROOF. As in Exercise 6.12, it is easy to see that an affine linear change in the coordinates y_i (and hence an appropriate change in the coordinates $\check{x}_i$) results in a linear change of the coordinates z_i, so they induce a well-defined complex structure invariant under $\check{x}_i \mapsto \check{x}_i + 1$, and hence a complex structure on $\check{X}(B)$. So one computes that

$$\omega = \sum d\check{x}_i \wedge dy_i = \frac{\sqrt{-1}}{2} \sum g^{ij} dz_i \wedge d\bar{z}_j$$

where $g_{ij} = \partial^2 K/\partial y_i \partial y_j$. Then the metric is Ricci-flat if and only if $\det(g^{ij}) = \text{constant}$, if and only if $\det(g_{ij}) = \text{constant}$. □

As before, we call this Kähler manifold $\check{X}(B, K)$, and now observe

PROPOSITION 6.16. *There is a canonical isomorphism*

$$X(B, K) \cong \check{X}(\check{B}, \check{K})$$

of Kähler manifolds, where $(\check{B}, \check{K})$ is the Legendre transform of (B, K).

PROOF. Of course $B = \check{B}$ as manifolds, but they carry different affine structures. In addition, the metrics g induced by K and $\check{K}$ coincide by Proposition 6.4. Now identify TB and $T^*B = T^*\check{B}$ using this metric, so in local coordinates (y_i), $\partial/\partial y_i$ is identified with $\sum_j g_{ij} dy_j$. But $d\check{y}_i = \sum \frac{\partial^2 K}{\partial y_i \partial y_j} dy_j = \sum_j g_{ij} dy_j$, so $\partial/\partial y_i$ and $d\check{y}_i$ are identified. Thus this identification descends to a canonical identification of $X(B)$ and $\check{X}(\check{B})$.

We just need to check that this identification gives an isomorphism of Kähler manifolds. But the complex coordinate $\check{z}_i = \check{x}_i + \sqrt{-1}\partial\check{K}/\partial\check{y}_i$ on $T^*\check{B}$ is identified with the coordinate $z_i = x_i + \sqrt{-1}y_i$ on TB under this identification, so the complex structures agree. Finally, the Kähler forms are

$$\tfrac{\sqrt{-1}}{2} \sum g_{ij} dz_i \wedge d\bar{z}_j \text{ and } \tfrac{\sqrt{-1}}{2} \sum \check{g}^{ij} dz_i \wedge d\bar{z}_j$$

respectively, where $\breve{g}_{ij} = g(\partial/\partial\breve{y}_i, \partial/\partial\breve{y}_j)$. But

$$\begin{aligned}\delta_{ik} = \sum_j g_{ij}g^{jk} &= \sum_j g^{jk} g\,(\partial/\partial y_i, \partial/\partial y_j) \\ &= \sum_j g^{jk} g(\sum_k g_{ik}\partial/\partial\breve{y}_k, \sum_l g_{jl}\partial/\partial\breve{y}_l) \\ &= \sum g^{jk} g_{ik} g_{jl} \breve{g}_{kl} \\ &= \sum g_{il}\breve{g}_{lk},\end{aligned}$$

so $\breve{g}^{ij} = g_{ij}$.

Thus the two Kähler forms also agree. □

We can now state more explicitly the simplest form of mirror symmetry.

DEFINITION 6.17. If B is an affine manifold (with transition functions in $\mathbb{R}^n \rtimes \mathrm{GL}(n, \mathbb{Z})$) then we say $X(B)$ and $\check{X}(B)$ are *SYZ dual.* If, in addition, we have a convex function $K : \tilde{B} \to \mathbb{R}$, then we say

$$X(B, K) \cong \check{X}(\check{B}, \check{K})$$

and

$$\check{X}(B, K) \cong X(\ddot{B}, \ddot{K})$$

are SYZ dual.

In the former case, this is a duality between complex and symplectic manifolds, and in the latter between Kähler manifolds. We view either case as a simple version of mirror symmetry.

6.2.2. Semi-flat differential forms. In this section we will discuss forms on both $X(B)$ and $\check{X}(B)$, and their interplay, so it will be useful to work locally with affine coordinates $y_1, \ldots, y_n$ on B, and corresponding coordinates $x_1, \ldots, x_n$ on the tangent bundle and $\check{x}_1, \ldots, \check{x}_n$ on the cotangent bundle. In addition, we will assume in this section that the transition maps of B are contained in $\mathbb{R}^n \rtimes \mathrm{SL}_n(\mathbb{Z})$, so that $X(B)$ carries a nowhere vanishing holomorphic n-form.

DEFINITION 6.18. A *semi-flat differential form* of type (p, q) on $X(B)$ (or $\check{X}(B)$) is a $(p+q)$-form written locally on B as

$$\sum_{\substack{\#I=p \\ \#J=q}} \alpha_{IJ}(y) dy_I \wedge dx_J$$

(or

$$\sum_{\substack{\#I=p \\ \#J=q}} \alpha_{IJ}(y) dy_I \wedge d\check{x}_J$$

on $\check{X}(B)$), where I and J are multi-index sets and α_{IJ} are functions on B and $\#I$, $\#J$ are the number of elements in I, J. Note the type is independent of the choice of affine coordinates. Denote by $S^{p,q}$ (respectively $\check{S}^{p,q}$) the space of semi-flat forms on $X(B)$ (respectively on $\check{X}(B)$) of type (p,q).

For example, the symplectic form ω on $\check{X}(B)$ is a semi-flat form of type $(1,1)$.

A crucial point is that there is a natural isomorphism

$$\mu : S^{p,q} \xrightarrow{\cong} \check{S}^{p,n-q}$$

which we shall now define. First, let us focus on one torus $T^n \cong V/\Lambda$, where V is a vector space and Λ is a lattice. Then $H^p(V/\Lambda, \mathbb{R})$ is naturally identified with the space of constant p-forms on the torus, namely $\bigwedge^p V^\vee$. On the other hand, the dual torus is $V^\vee/\Lambda^\vee$, where $\Lambda^\vee \subseteq V^\vee$ is the lattice of linear functionals taking integer values on Λ. Then $H^p(V^\vee/\Lambda^\vee, \mathbb{R}) \cong \bigwedge^p V$. Also, there is a natural pairing

$$\bigwedge^p V^\vee \times \bigwedge^{n-p} V^\vee \to \bigwedge^n V^\vee$$

given by wedge product. Hence, after choosing an isomorphism $\bigwedge^n V^\vee \cong \mathbb{R}$ (which is equivalent to choosing a non-zero n-form), we obtain an identification of $\bigwedge^{n-p} V^\vee$ with $\bigwedge^p V$. Explicitly, given coordinates $x_1, \ldots, x_n$ on V, we can write an element of $\bigwedge^p V$ as

$$\sum_{\#J=p} a_J \frac{\partial}{\partial x_J},$$

where $J = \{j_1, \ldots, j_p\}$ is a multi-index set,

$$\frac{\partial}{\partial x_J} = \frac{\partial}{\partial x_{j_1}} \wedge \cdots \wedge \frac{\partial}{\partial x_{j_p}}.$$

If the chosen n-form is $dx_1 \wedge \cdots \wedge dx_n$, then we can write the corresponding element of $\bigwedge^{n-p} V^\vee$ as the contraction

$$\iota\left(\sum a_J \frac{\partial}{\partial x_J}\right) dx_1 \wedge \cdots \wedge dx_n.$$

Here we use the convention that the contraction of a polyvector $v_1 \wedge \cdots \wedge v_p$ with a q-form α is the $(q-p)$-form $\iota(v_1 \wedge \cdots \wedge v_p)\alpha$ which takes the value $\alpha(v_1, \ldots, v_p, w_1, \ldots, w_q)$ on the vectors $w_1, \ldots, w_q$. (This is mostly a question of making sign conventions explicit; this is tedious, but unfortunately very important.)

Now look at the global picture. Because of the assumption on the transition maps of B, the manifold $X(B)$ carries a global holomorphic n-form Ω, and locally

$$\Omega = \bigwedge_{i=1}^{n} (dx_i + \sqrt{-1} dy_i).$$

Thus on $X(B)$, we can restrict Ω to any fiber and obtain a canonical choice of n-form, $dx_1 \wedge \cdots \wedge dx_n$. We can then define the isomorphism

$$\mu : S^{p,q} \to \check{S}^{p,n-q}$$

locally by

$$\mu\left(\sum a_{IJ}(y) dy_I \wedge dx_J\right)$$
$$= (-1)^{q(q-1)/2+n(n-1)/2} \sum a_{IJ}(y) dy_I \wedge \iota(\partial/\partial\check{x}_J) d\check{x}_1 \wedge \cdots \wedge d\check{x}_n.$$

EXERCISE 6.19. (1) Check that μ is well-defined, independently of the choice of affine coordinates.
(2) Show that μ takes closed forms to closed forms.

We end this section with a discussion of the de Rham cohomology of local systems.

Let E be a local system on a manifold M with fiber $\mathbb{R}^n$. Let $\mathcal{E} = E \otimes C^\infty(M)$; here $C^\infty(M)$ denotes the sheaf of C^∞ functions on M. Then $\mathcal{E}$ is the sheaf of sections of a vector bundle on M. Locally, if $U \subseteq M$ is an open set on which $E|_U$ is trivial, with a basis of sections $s_1, \ldots, s_n \in \Gamma(U, E)$, then

$$\Gamma(U, \mathcal{E}) = \left\{\sum f_i s_i | f_i : U \to \mathbb{R} \text{ is } C^\infty\right\}.$$

Now $\mathcal{E}$ has a connection $\nabla : \mathcal{E} \to T^*M \otimes \mathcal{E}$ determined by E. Locally this takes the form $\nabla(\sum f_i s_i) = df_i \otimes s_i$. This connection is flat since $\nabla s_i = 0$, $i = 1, \ldots, n$. Thus we obtain a sequence of maps

$$\mathcal{E} \xrightarrow{\nabla} T^*M \otimes \mathcal{E} \xrightarrow{\nabla} \bigwedge^2 T^*M \otimes \mathcal{E} \xrightarrow{\nabla} \cdots$$

with $\nabla^2 = 0$ since the connection is flat. In other words, ∇ defines a complex. Furthermore, $\ker(\mathcal{E} \xrightarrow{\nabla} T^*M \otimes \mathcal{E}) = E$, so this complex is a resolution of E. This is a generalization of the usual de Rham complex $\bigwedge^0 T^*M \xrightarrow{d} \bigwedge^1 T^*M \xrightarrow{d} \cdots$, which is a resolution of the constant sheaf $\mathbb{R}$. Furthermore, each sheaf $\bigwedge^p T^*M \otimes \mathcal{E}$ is what is known as a fine sheaf (see e.g. [**94**]), and as such $H^q(M, \bigwedge^p T^*M \otimes \mathcal{E}) = 0$ for all $q > 0$. From standard homological arguments it follows that

$$H^p(M, E) = \frac{\ker(\nabla : \Gamma(M, \bigwedge^p T^*M \otimes \mathcal{E}) \to \Gamma(M, \bigwedge^{p+1} T^*M \otimes \mathcal{E}))}{\operatorname{im}(\nabla : \Gamma(M, \bigwedge^{p-1} T^*M \otimes \mathcal{E}) \to \Gamma(M, \bigwedge^p T^*M \otimes \mathcal{E}))}.$$

This is the de Rham realization of $H^p(M, E)$. This cohomology group can also be calculated via Čech cohomology (see §4.5.1), which will also prove useful.

Now consider the local system $\bigwedge^q \check{\Lambda}_{\mathbb{R}}$ on the affine manifold B. A given fiber $\bigwedge^q \check{\Lambda}_{\mathbb{R},b}$ can be viewed as the space of q-forms on the torus $\Lambda_{\mathbb{R},b}/\Lambda_b$. Under this identification, with affine coordinates $y_1, \ldots, y_n$ on an affine chart U, a local basis of sections can be identified with $\{dx_{i_1} \wedge \cdots \wedge dx_{i_p} | 1 \le i_1 < \cdots < i_p \le n\}$. Let $\mathcal{E}_q = \bigwedge^q \check{\Lambda} \otimes C^\infty(B)$. Then we see that the space of semi-flat differential forms on $f^{-1}(U)$ of type (p,q) can be identified with $\bigwedge^p T^*B \otimes \mathcal{E}_q$, identifying $\sum \alpha_{IJ} dy_I \wedge dx_J$ with $\sum \alpha_{IJ} dy_I \otimes dx_J$. Furthermore, the connection $\nabla : \bigwedge^p T^*B \otimes \mathcal{E}_q \to \bigwedge^{p+1} T^*B \otimes \mathcal{E}_q$ coincides with the differential $d : S^{p,q} \to S^{p+1,q}$. Thus we have shown

PROPOSITION 6.20. *There are natural maps*

$$\{\text{closed forms in } S^{p,q}\} \to H^p\left(B, \bigwedge^q \check{\Lambda}_{\mathbb{R}}\right)$$

and dually

$$\{\text{closed forms in } \check{S}^{p,q}\} \to H^p\left(B, \bigwedge^q \Lambda_{\mathbb{R}}\right).$$

It is useful also to explain how, given a de Rham representative for an element of $H^p(B, \bigwedge^q \Lambda_{\mathbb{R}})$, we can associate a Čech representative. To do so, begin with a closed semi-flat form α of type (p,q). In addition, choose an open covering $\{U_i\}_{i\in I}$ of B such that $U_{i_0} \cap \cdots \cap U_{i_p}$ is contractible for all $i_0, \ldots, i_p \in I$. Thus, over each U_i, there is a semi-flat $(p-1,q)$-form α_i^1 such that $d\alpha_i^1 = \alpha$ on $f^{-1}(U_i)$. (Indeed, $\alpha = \sum_{I,J} \alpha_{IJ}(y) dy_I \wedge dx_J$ on U_i and $d\alpha = 0$ implies $d\left(\sum_I \alpha_{IJ} dy_I\right) = 0$, and hence by contractibility of U_i, $\sum_I \alpha_{IJ} dy_I$ is exact.) Now on $U_i \cap U_j$, α_i^1 and α_j^1 don't necessarily agree, but $d(\alpha_j^1 - \alpha_i^1) = 0$. Let $\beta_{ij}^1 = \alpha_j^1 - \alpha_i^1$ on $U_i \cap U_j$. By the same argument we can find a semi-flat $(p-2,q)$-form α_{ij}^2 such that $d\alpha_{ij}^2 = \alpha_j^1 - \alpha_i^1$. Then for i_0, i_1, i_2,

$$\begin{aligned} d(\alpha_{i_0 i_1}^2 + \alpha_{i_1 i_2}^2 - \alpha_{i_0 i_2}^2) &= \alpha_{i_1}^1 - \alpha_{i_0}^1 + \alpha_{i_2}^1 - \alpha_{i_1}^1 - \alpha_{i_2}^1 + \alpha_{i_0}^1 \\ &= 0 \end{aligned}$$

on $U_{i_0} \cap U_{i_1} \cap U_{i_2}$. We set $\beta_{i_0 i_1 i_2}^2 = \alpha_{i_0 i_1}^2 + \alpha_{i_1 i_2}^2 - \alpha_{i_0 i_2}^2$. We continue this process inductively until we obtain closed semi-flat $(0,q)$-forms $\beta_{i_0 \cdots i_p}^p$ on $U_{i_0} \cap \cdots \cap U_{i_p}$. Since these are closed, they are in fact flat sections of $\bigwedge^q \check{\Lambda}_{\mathbb{R}}$, and define a Čech p-cocycle for $\bigwedge^q \Lambda_{\mathbb{R}}$.

EXERCISE 6.21. Work out the details of this inductive procedure as follows. Let

$$\begin{aligned} Z^{p,q} &= \ker(\nabla : \bigwedge^p T^*B \otimes \bigwedge^q \check{\Lambda}_{\mathbb{R}} \to \bigwedge^{p+1} T^*B \otimes \bigwedge^q \check{\Lambda}_{\mathbb{R}}) \\ &= \operatorname{im}(\nabla : \bigwedge^{p-1} T^*B \otimes \bigwedge^q \check{\Lambda}_{\mathbb{R}} \to \bigwedge^p T^*B \otimes \bigwedge^q \check{\Lambda}_{\mathbb{R}}) \end{aligned}$$

(the latter equality for $p > 0$). This gives exact sequences of sheaves

$$0 \to Z^{p,q} \to \bigwedge^p T^*B \otimes \bigwedge^q \check{\Lambda}_{\mathbb{R}} \to Z^{p+1,q} \to 0$$

with $Z^{0,q} = \bigwedge^q \check{\Lambda}_{\mathbb{R}}$. Start with a closed semi-flat (p_0,q)-form α, i.e., a section of $Z^{p_0,q}$, and use the Čech coboundary map to consider Čech representatives for the images of the coboundary maps coming from this exact sequence for $p = p_0 - 1, p_0 - 2, \ldots$:

$$\alpha \in H^0(B, Z^{p_0,q}) \to H^1(B, Z^{p_0-1,q}) \to \cdots$$
$$\to H^{p_0}(B, Z^{0,q}) = H^{p_0}(B, \textstyle\bigwedge^q \check{\Lambda}_{\mathbb{R}}).$$

Describe these Čech representatives, and show that they coincide with the β^i's described above.

6.2.3. SYZ with a B-field. Mirror symmetry is not complete without consideration of the B-field. Traditionally the B-field should take values in $H^2(X, \mathbb{R}/\mathbb{Z})$. However, for the moment we will consider the B-field in a smaller space, and take up the general case in §6.2.5.

Let B be an affine manifold with transition functions in $\mathbb{R}^n \rtimes \mathrm{GL}(n, \mathbb{Z})$ as usual. We have the local system Λ on B, as well as $\Lambda_{\mathbb{R}} = \Lambda \otimes_{\mathbb{Z}} \mathbb{R}$. The local system $\Lambda_{\mathbb{R}}$ is just the sheaf of sections of TB flat with respect to the induced flat connection on B.

Consider $f : X(B) \to B$. If $U \subseteq B$ is an open set, then $\Gamma(U, \Lambda_{\mathbb{R}}/\Lambda)$ acts holomorphically on $f^{-1}(U)$. Indeed, if $s \in \Gamma(U, \Lambda_{\mathbb{R}}/\Lambda)$, then s acts on $X(U) = TU/\Lambda|_U$ by fiberwise translation acting on the fiber TU_b/Λ_b by $x \mapsto x + s(b)$. This action is holomorphic: in holomorphic coordinates q_i, the action is

$$(q_1, \ldots, q_n) \mapsto (e^{2\pi\sqrt{-1}\theta_1} q_1, \ldots, e^{2\pi\sqrt{-1}\theta_n} q_n)$$

for a section $\sum \theta_i \partial/\partial y_i$ of $\Lambda_{\mathbb{R}}/\Lambda$, with $\theta_1, \ldots, \theta_n$ constants. Furthermore, if K is a potential function for a metric on B, then this action preserves the induced Kähler metric.

Thus, given a Čech 1-cocycle (b_{ij}) for $\Lambda_{\mathbb{R}}/\Lambda$, i.e., $b_{ij} \in \Gamma(U_i \cap U_j, \Lambda_{\mathbb{R}}/\Lambda)$ for some open covering $\{U_i\}$ of B, and $b_{ij} + b_{jk} = b_{ik}$ on $U_i \cap U_j \cap U_k$, we can construct a new complex manifold $X(B, \mathbf{B})$ or a new Kähler manifold $X(B, \mathbf{B}, K)$ as follows. Glue together the manifolds $X(U_i)$ (or $X(U_i, K)$) along $X(U_i \cap U_j)$ via the identifications

$$X(U_i) \supseteq X(U_i \cap U_j) \xrightarrow{b_{ij}} X(U_i \cap U_j) \subseteq X(U_j)$$

where b_{ij} is acting by translation. The Čech cocycle condition guarantees compatibility of these gluings and produces a complex manifold $X(B, \mathbf{B})$ or, in the presence of a potential function K, a Kähler manifold $X(B, \mathbf{B}, K)$.

Thus the full semi-flat SYZ picture is as follows. The data consists of an affine manifold B with potential K and B-fields $\mathbf{B} \in H^1(B, \Lambda_{\mathbb{R}}/\Lambda)$, $\check{\mathbf{B}} \in H^1(B, \check{\Lambda}_{\mathbb{R}}/\check{\Lambda})$. Now as we saw in the proof of Proposition 6.16, the local system $\check{\Lambda}$ defined using the affine structure on B is the same as the local system Λ defined using the affine structure on $\check{B}$, and so we say the pair

$$(X(B, \mathbf{B}, K), \check{\mathbf{B}})$$

is SYZ dual to

$$(X(\check{B},\check{\mathbf{B}},\check{K}),\mathbf{B}).$$

This may not look like the traditional form of the B-field giving a complexified Kähler class, so we will give another version of this construction shortly which will make this transparent.

EXAMPLE 6.22. We will now see all these concepts in the case of elliptic curves. Let $B = \mathbb{R}/\tau_2\mathbb{Z}$ for some real number $\tau_2 > 0$, with the affine structure on B induced by the natural one on $\mathbb{R}$. Then $X(B) = \mathbb{R}/\tau_2\mathbb{Z}\times\mathbb{R}/\mathbb{Z}$, and with coordinates y on $\mathbb{R}/\tau_2\mathbb{Z}$ and x on $\mathbb{R}/\mathbb{Z}$, the complex structure is given by the complex coordinate $z = x + \sqrt{-1}y$. Equivalently, $X(B)$ is the elliptic curve $\mathbb{C}/\langle 1, \sqrt{-1}\tau_2\rangle$, where $\langle 1, \sqrt{-1}\tau_2\rangle$ denotes the lattice in $\mathbb{C}$ generated by these two numbers.

To compute $H^1(B, \Lambda_{\mathbb{R}}/\Lambda)$, note that Λ is the constant sheaf $\mathbb{Z}$, so

$$H^1(B, \Lambda_{\mathbb{R}}/\Lambda) = H^1(S^1, \mathbb{R}/\mathbb{Z}) = \mathbb{R}/\mathbb{Z}.$$

We leave it as an exercise to the reader to show that if $\mathbf{B} \in H^1(B, \mathbb{R}/\mathbb{Z})$ is represented by $\tau_1 \in \mathbb{R}$, then $X(B, \mathbf{B})$ is isomorphic to the elliptic curve $\mathbb{C}/\langle 1, \tau\rangle$ where $\tau = \tau_1 + \sqrt{-1}\tau_2$. Of course, replacing τ_1 with $\tau_1 + 1$ has no affect on the lattice.

On the other hand, $\check{X}(B) = \mathbb{R}/\tau_2\mathbb{Z} \times \mathbb{R}/\mathbb{Z}$ is a symplectic manifold with symplectic form $dx \wedge dy$; this has volume τ_2.

If we allow a potential, we can take K to be any quadratic function in y. Since the metric only determines K up to an affine linear function, we can assume $K = \frac{\check{\tau}_2}{\tau_2}\frac{y^2}{2}$, so that $\check{B} = \mathbb{R}/\check{\tau}_2\mathbb{Z}$, and $\check{K} = \frac{\tau_2}{\check{\tau}_2}\frac{\check{y}^2}{2}$. Then the Kähler metric defined by $K \circ f$, using the complex coordinate z on $\mathbb{C}/\langle 1, \tau_1 + \sqrt{-1}\tau_2\rangle$, is $\omega = \frac{\sqrt{-1}}{2}\frac{\check{\tau}_2}{\tau_2}dz \wedge d\bar{z}$, which has volume $\check{\tau}_2$. Then with B-fields $\mathbf{B} = \tau_1$ and $\check{\mathbf{B}} = \check{\tau}_1$, $(X(B,\tau_1,K),\check{\tau}_1)$ is SYZ dual to $(X(\check{B},\check{\tau}_1,\check{K}),\tau_1)$. This is saying that the elliptic curve $\mathbb{C}/\langle 1, \tau_1 + \sqrt{-1}\tau_2\rangle$ with complexified Kähler class $\check{\tau}_1 + \sqrt{-1}\check{\tau}_2$ is mirror to the elliptic curve $\mathbb{C}/\langle 1, \check{\tau}_1 + \sqrt{-1}\check{\tau}_2\rangle$ with complexified Kähler class $\tau_1 + \sqrt{-1}\tau_2$.

EXERCISE 6.23. Consider, in Example 6.7, (2), the lattice $\Gamma \subseteq G$ generated by T_1 and T_2 with

$$\begin{aligned} T_1(m_1, m_2) &= (m_1 + 1, m_2) \\ T_2(m_1, m_2) &= (m_1 + em_2 + e(e-1)/2, m_2 + e) \end{aligned}$$

where $e > 0$ is an integer. Let $B = \mathbb{R}^2/\Gamma$. Show that $H^1(B, \Lambda_{\mathbb{R}}/\Lambda) = (\mathbb{R}/\mathbb{Z})^2 \oplus (\mathbb{Z}/e\mathbb{Z})$. Hint: Λ is a local system with monodromy about generators of $H_1(B,\mathbb{Z})$ corresponding to T_1 and T_2 being $\left(\begin{smallmatrix}1&0\\0&1\end{smallmatrix}\right)$ and $\left(\begin{smallmatrix}1&e\\0&1\end{smallmatrix}\right)$ respectively. The cohomology of the local system $\Lambda_{\mathbb{R}}/\Lambda$ is the same as the group cohomology $H^1(\mathbb{Z}^2, (\mathbb{R}/\mathbb{Z})^2)$, where $\mathbb{Z}^2 = H_1(B,\mathbb{Z})$ acts via the above matrices. Then compute this group cohomology using a standard method (see e.g. [**68**]).

There is a different way to think about the B-field more directly in terms of differential forms, and this approach will be extended to generalised complex structures in §6.2.5. We first recall some observations of Hitchin [**232**].

PROPOSITION 6.24. *Let X be a real $2n$-dimensional manifold. If Ω is a complex-valued C^∞ n-form on X satisfying the three properties*

(1) $d\Omega = 0$;

(2) Ω is locally decomposable (i.e., can be written locally as $\theta_1 \wedge \cdots \wedge \theta_n$, where $\theta_1, \ldots, \theta_n$ are 1-forms);

(3) $(-1)^{n(n-1)/2}(\sqrt{-1}/2)^n \Omega \wedge \bar{\Omega} > 0$ everywhere on X,

then Ω determines a complex structure on X for which Ω is a holomorphic n-form.

PROOF. Ω defines at each point an n-dimensional subspace $\Omega_X^{1,0}$ of $T^*X \otimes \mathbb{C}$, by taking a decomposition $\Omega = \theta_1 \wedge \cdots \wedge \theta_n$ and taking the subspace of $T^*X \otimes \mathbb{C}$ spanned by $\theta_1, \ldots, \theta_n$. It is a standard fact about exterior algebra that this gives a well-defined subspace which is in fact independent of the particular choice of decomposition. Condition (3) implies that this subspace is in fact complementary to its complex conjugate, and we obtain a splitting

$$T^*X \otimes_{\mathbb{R}} \mathbb{C} = \Omega_X^{1,0} \oplus \Omega_X^{0,1}.$$

We can then define an almost complex structure J by taking these two subspaces to be the $\pm\sqrt{-1}$ eigenspaces of J respectively.

To show that J is an integrable complex structure, we use the first condition. Writing $\Omega = \theta_1 \wedge \cdots \wedge \theta_n$, note that $0 = \Omega \wedge \theta_i$, so

$$\begin{aligned} 0 = d(\Omega \wedge \theta_i) &= d\Omega \wedge \theta_i \pm \Omega \wedge d\theta_i \\ &= \pm \Omega \wedge d\theta_i, \end{aligned}$$

Now we recall the Newlander-Nirenberg theorem. If $\theta_1, \ldots, \theta_n$ are a basis of $(1,0)$-forms with respect to J, then we can write

$$d\theta_i = \sum A_i^{jk} \theta_j \wedge \theta_k + \sum B_i^{jk} \theta_j \wedge \bar{\theta}_k + \sum C_i^{jk} \bar{\theta}_j \wedge \bar{\theta}_k.$$

Then the Newlander-Nirenberg theorem says that J is integrable if and only if $C_i^{jk} = 0$ for all i, j, k. But this is implied in our case by $0 = \Omega \wedge d\theta_i$, so J is integrable. □

For future reference, we also point out the data we need to specify a Ricci-flat Kähler manifold:

PROPOSITION 6.25. *Let X be a real $2n$-dimensional manifold. Suppose ω is a symplectic form on X and Ω is a complex-valued n-form on X such that*

(1) Ω satisfies the conditions of Proposition 6.24;

(2) ω is a positive $(1,1)$-form in the complex structure of Proposition 6.24;

(3) $(-1)^{n(n-1)/2}(\sqrt{-1}/2)^n\Omega\wedge\bar{\Omega}=\omega^n/n!$.

Then Ω induces a complex structure on X such that ω is a Kähler form on X whose corresponding metric is Ricci-flat.

PROOF. This is obvious, Ricci-flatness being equivalent to the holomorphic n-form being of constant length. □

Assume now that the transition maps of B lie in $\mathbb{R}^n \rtimes \mathrm{SL}_n(\mathbb{Z})$ as in §6.2.2. Instead of viewing the B-field as an element of $H^1(B,\Lambda_{\mathbb{R}}/\Lambda)$, we shall define it to be a semi-flat closed form $\mathbf{B}$ of type $(1,1)$ on $\check{X}(B)$. In particular, this represents an element of $H^1(B,\Lambda_{\mathbb{R}})$, and hence an element of $H^1(B,\Lambda_{\mathbb{R}}/\Lambda)$. Note that it is not necessarily the case that every element of $H^1(B,\Lambda_{\mathbb{R}}/\Lambda)$ is in the image of $H^1(B,\Lambda_{\mathbb{R}})$: for example, $H^2(B,\Lambda)$ may have torsion. For a B-field $\mathbf{B}$ not in the image of $H^1(B,\Lambda_{\mathbb{R}})$, $X(B,\mathbf{B})$ does not even coincide topologically with $X(B)$, and the construction which follows does not apply.

Consider

$$e^{\mathbf{B}+\sqrt{-1}\omega}\in\bigoplus_{p=0}^{n}\check{S}^{p,p},$$

$$e^{\mathbf{B}+\sqrt{-1}\omega}=1+\mathbf{B}+\sqrt{-1}\omega+\frac{(\mathbf{B}+\sqrt{-1}\omega)\wedge(\mathbf{B}+\sqrt{-1}\omega)}{2}+\cdots.$$

This can be considered formally or can be viewed as a generalised Calabi-Yau structure in the sense of Hitchin, as will be discussed in §6.2.5.

Then $\mu^{-1}(e^{\mathbf{B}+\sqrt{-1}\omega})\in\bigoplus_{p=0}^{n}S^{p,n-p}$ is in fact an n-form.

PROPOSITION 6.26. *$\mu^{-1}(e^{\mathbf{B}+\sqrt{-1}\omega})$ is the holomorphic n-form with respect to a complex structure on $X(B)$, and with this complex structure $X(B)$ is isomorphic to $X(B,\mathbf{B})$. Here we view $\mathbf{B}$ as defining an element of $H^1(B,\Lambda_{\mathbb{R}}/\Lambda)$ using Proposition 6.20.*

PROOF. We need to show that $\mu^{-1}(e^{\mathbf{B}+\sqrt{-1}\omega})$ satisfies the conditions of Proposition 6.24. It is clear it is closed, since each component of $e^{\mathbf{B}+\sqrt{-1}\omega}$ is closed and μ^{-1} takes closed forms to closed forms. On the other hand, a careful check with sign conventions shows that if $\mathbf{B}=\sum b_{ij}dy_i\wedge d\check{x}_j$ and of course $\omega=\sum dy_i\wedge d\check{x}_i$, then

$$\mu^{-1}(e^{\mathbf{B}+\sqrt{-1}\omega})=\bigwedge_{i=1}^{n}\left(dx_i+\sum_{j=1}^{n}(b_{ji}+\sqrt{-1}\delta_{ji})dy_j\right).$$

Thus $\mu^{-1}(e^{\mathbf{B}+\sqrt{-1}\omega})$ satisfies the hypotheses of Proposition 6.24, so we obtain a complex structure on the torus bundle $X(B)$. We need to compare it with the complex structure on $X(B,\mathbf{B})$, but to do so we first need to write down a Čech representative for $\mathbf{B}$. Fortunately, we have seen how to do this at the end of the previous section. Choose an open covering $\{U_i\}$ of B, and on U_i, with coordinates y_i, $\mathbf{B}=\sum b_{jk}dy_j\wedge d\check{x}_k$. We can then find functions

c_i^k such that $dc_i^k = \sum b_{jk} dy_j$. Here $(c_i^1, \ldots, c_i^n)$ should be viewed as a section c_i of TB, namely $c_i = \sum c_i^j \partial/\partial y_j$, and the difference $\beta_{ij} = c_j - c_i$ is a flat section of TB, i.e., a section of $\Lambda_{\mathbb{R}}$ over $U_i \cap U_j$, and (β_{ij}) is the Čech 1-cocycle representing $\mathbf{B}$. Then consider $f^{-1}(U_i)$. The standard complex structure is induced by the holomorphic volume form

$$\Omega = \bigwedge_{j=1}^{n} (dx_j + \sqrt{-1} dy_j).$$

If $T_{c_i} : f^{-1}(U_i) \to f^{-1}(U_i)$ denotes fiberwise translation by c_i, i.e.,

$$T_{c_i}(y_1, \ldots, y_n, x_1, \ldots, x_n) = (y_1, \ldots, y_n, x_1 + c_i^1, \ldots, x_n + c_i^n),$$

then

$$\begin{aligned} T_{c_i}^*(\Omega) &= \bigwedge_{j=1}^{n} (dx_j + dc_i^j + \sqrt{-1} dy_j) \\ &= \bigwedge_{j=1}^{n} \left(dx_j + \sum_{k=1}^{n} (b_{kj} + \sqrt{-1}\delta_{kj}) dy_k \right). \end{aligned}$$

Thus if we write $f : X(B) \to B$ for the torus bundle with the standard complex structure and $f' : X(B)' \to B$ for the torus bundle with the complex structure induced by $\mu^{-1}(e^{\mathbf{B}+\sqrt{-1}\omega})$, we see that there are local isomorphisms $T_{c_i} : (f')^{-1}(U_i) \to f^{-1}(U_i)$. Over $U_i \cap U_j$, we obtain a commutative diagram

$$\begin{array}{ccc} (f')^{-1}(U_i \cap U_j) & \xrightarrow{T_{c_i}} & f^{-1}(U_i \cap U_j) \\ \big\downarrow = & & \big\downarrow T_{c_j - c_i} = T_{\beta_{ij}} \\ (f')^{-1}(U_i \cap U_j) & \xrightarrow{T_{c_j}} & f^{-1}(U_i \cap U_j) \end{array}$$

so we obtain an isomorphism $X(B)' \cong X(B, (\beta_{ij}))$, as desired. □

EXAMPLE 6.27. Returning to the example of an elliptic curve, let $B = \mathbb{R}/\tau_2\mathbb{Z}$. We take $\Omega = dx + (\sqrt{-1} + \tau_1/\tau_2) dy$ on $X(B)$, $\mathbf{B} + \sqrt{-1}\omega = (\sqrt{-1} + \tau_1/\tau_2) dy \wedge d\check{x}$ on $\check{X}(B)$. If we change coordinates to $y' = y/\tau_2$, so that y' is periodic with period 1, these take the form

$$\Omega = dx + (\tau_1 + \sqrt{-1}\tau_2) dy', \quad \mathbf{B} + \sqrt{-1}\omega = (\tau_1 + \sqrt{-1}\tau_2) dy' \wedge d\check{x}$$

on $(\mathbb{R}/\mathbb{Z})^2$ and

$$\mu^{-1}(e^{\mathbf{B}+i\omega}) = \Omega.$$

6.2.4. Families of complex manifolds. We now show how to obtain an entire non-trivial family of complex structures from an affine manifold B, given some additional hypotheses. This is important as it helps identify the notion of large complex structure limit.

Assume B is an integral affine manifold. Then we actually get more out of our constructions. First, let $B' = B \times \mathbb{R}_{>0}$, and give B' an integral affine structure by specifying the developing map $\delta' : \tilde{B}' = \tilde{B} \times \mathbb{R}_{>0} \to M_{\mathbb{R}} \oplus \mathbb{R}$ by

$\delta'(b,r) = (r\delta(b), r)$. Note that if $\gamma \in \pi_1(B) = \pi_1(B')$, and $\rho(\gamma) \in \mathrm{Aff}(M)$ such that $\rho(\gamma) \circ \delta \circ \Psi_\gamma = \delta$, then we can write $\rho(\gamma)(m) = \mathbf{A}m + \mathbf{b}$ for $\mathbf{A} \in \mathrm{GL}(M), \mathbf{b} \in M$. Let Ψ'_γ be the deck transformation of $\tilde{B}'$ induced by γ, and let $\rho'(\gamma)(m,r) = (\mathbf{A}m + r\mathbf{b}, r)$. Then

$$\begin{aligned} \rho'(\gamma) \circ \delta' \circ \Psi'_\gamma(b,r) &= \rho'(\gamma) \circ \delta'(\Psi_\gamma(b), r) \\ &= \rho'(\gamma) \circ (r\delta\Psi_\gamma(b), r) \\ &= (r\mathbf{A}\delta\Psi_\gamma(b) + r\mathbf{b}, r) \\ &= (r\rho(\gamma)\delta\Psi_\gamma(b), r) \\ &= (r\delta(b), r) \\ &= \delta'(b,r). \end{aligned}$$

Thus the holonomy representation for δ' is given by $\rho' : \pi_1(B) \to \mathrm{GL}(M \oplus \mathbb{Z})$, as $\rho'(\gamma)$ is always linear. If $\mathbf{b}$ were not in M, but only in $M_\mathbb{R}$, $\mathbf{B}'$ would not be integral, and then we would not be able to talk about $X(B')$. However, as we can, we have an $(n+1)$-dimensional complex manifold $X(B')$, where $n = \dim M_\mathbb{R}$. We also have $X(\mathbb{R}_{>0})$, with $\mathbb{R}_{>0} \hookrightarrow \mathbb{R}$ the inclusion giving the affine structure on $X(\mathbb{R}_{>0})$. The coordinate $q = \exp(2\pi\sqrt{-1}(x + \sqrt{-1}y))$ identifies $X(\mathbb{R}_{>0})$ with the punctured unit disk $D^* \subseteq \mathbb{C}$. Meanwhile, the projection $B' \to \mathbb{R}_{\geq 0}$ induces a map on tangent bundles, and because the projection is affine linear, this induces a map $f : X(B') \to D^*$.

One can view this family over a punctured disk as a large complex structure limit degeneration. (Lacking of course is a fiber over 0, and ideally one would like to have a degenerate Calabi-Yau over 0, but that is a story for another day: see [**205, 206, 207, 208**].)

EXERCISE 6.28. Show that if $B = \mathbb{R}/n\mathbb{Z}$, n a positive integer, then $f : X(B') \to D^*$ is a family of elliptic curves over the punctured disk D^* with $f^{-1}(q) \cong \mathbb{C}/\langle 1, (n/2\pi\sqrt{-1}) \log q \rangle$.

This construction can be generalised as follows. We do not assume yet that B has an affine structure. Suppose we are given an $\mathbb{R}$-vector space $V_\mathbb{R}$ of continuous functions from $\tilde{B}$ to $M_\mathbb{R}$ which satisfies the property that there exists a representation $\tilde{\rho} : \pi_1(B) \to \mathrm{GL}(M)$ such that for each $\delta \in V_\mathbb{R}$, there exists a representation $\rho_\delta : \pi_1(B) \to \mathrm{Aff}(M_\mathbb{R})$ such that $\tilde{\rho}$ is the composition of ρ_δ with the projection $\mathrm{Aff}(M_\mathbb{R}) \to \mathrm{GL}(M_\mathbb{R})$, and furthermore, $\rho_\delta(\gamma) \circ \delta \circ \Psi_\gamma = \delta$. Suppose furthermore that $V_\mathbb{R}$ contains a lattice V such that whenever $\delta \in V$, ρ_δ has image in $\mathrm{Aff}(M)$. Finally, let $\mathcal{V} \subseteq V_\mathbb{R}$ be the subset

$$\mathcal{V} := \{\delta \in V_\mathbb{R} \,|\, \delta \text{ is an immersion}\},$$

which we assume to be non-empty and open. Each $\delta \in \mathcal{V}$ determines an affine structure on B. Also, $\mathcal{V}$ inherits an affine structure from $V_\mathbb{R}$, and if we view V as giving an integral structure on $V_\mathbb{R}$, we can then talk about $X(\mathcal{V})$.

Now put an affine structure on $B' = B \times \mathcal{V}$ via the developing map $\delta' : \tilde{B} \times \mathcal{V} \to M_{\mathbb{R}} \times V_{\mathbb{R}}$ defined by $\delta'(b, \delta) = (\delta(b), \delta)$. The holonomy map of δ' can be calculated as before: write for $\gamma \in \pi_1(B)$

$$\rho_\delta(\gamma)(m) = \tilde{\rho}(\gamma)(m) + \mathrm{Trans}_\gamma(\delta)$$

where $\mathrm{Trans}_\gamma(\delta)$ is the translational part of $\rho_\delta(\gamma)$. In particular, the map $\mathrm{Trans}_\gamma : V_{\mathbb{R}} \to M_{\mathbb{R}}$ is linear. Then setting

$$\rho'(\gamma)(m, \delta) = (\tilde{\rho}(\gamma)(m) + \mathrm{Trans}_\gamma(\delta), \delta),$$

one checks easily as before that

$$\rho'(\gamma) \circ \delta' \circ \Psi'_\gamma = \delta'.$$

Note that since $\mathrm{Trans}_\gamma(\delta) \in M$ whenever $\delta \in V$, $\rho'(\gamma) \in \mathrm{GL}(M \oplus V)$. So again the affine manifold structure induced on B' is integral, and we obtain the complex manifolds $X(B')$, $X(\mathcal{V})$, and a holomorphic map $f : X(B') \to X(\mathcal{V})$.

Note that $X(\mathcal{V}) = (V_{\mathbb{R}} + \sqrt{-1}\mathcal{V})/V$, which is a tube domain if $\mathcal{V}$ is a cone.

We would like to explicitly describe the fibers of the map $f : X(B') \to X(\mathcal{V})$. We will see shortly that the fibers are of the form $X(B, \mathbf{B})$ for various choices of affine structure on B and choices of B-field. To understand how the B-field is determined, we need

CONSTRUCTION 6.29. Consider $\delta_1 + \sqrt{-1}\delta_2 \in V_{\mathbb{R}} + \sqrt{-1}\mathcal{V}$. By assumption $\delta_2 : \tilde{B} \to M_{\mathbb{R}}$ is an immersion and hence induces an affine structure on B, yielding sheaves Λ and $\Lambda_{\mathbb{R}}$ on B. Now given this, we claim that δ_1 induces a natural class $[\delta_1] \in H^1(B, \Lambda_{\mathbb{R}})$. To see this, let $\pi : \tilde{B} \to B$ be the universal cover, and let $U \subseteq B$ be any sufficiently small open set so that $\pi^{-1}(U)$ is the disjoint union of open sets of the form $\Psi_\gamma(V)$, $\gamma \in \pi_1(B)$, for $V \subseteq \tilde{B}$ some open set. Identifying U with V, the immersion $\delta_2|_V : U \to M_{\mathbb{R}}$ induces an identification δ_{2*} of TU with $\delta_2(U) \times M_{\mathbb{R}}$. Thus the graph of $\delta_1|_V : U \to M_{\mathbb{R}}$ yields a section of TU.

If we replace V with $V' = \Psi_\gamma(V)$, then we obtain, for $i = 1, 2$, $\delta'_i : U \to M_{\mathbb{R}}$ with $\delta'_i = \delta_i \circ \Psi_\gamma = \rho_i(\gamma)^{-1} \circ \delta_i$, where $\rho_i = \rho_{\delta_i}$. In addition, $\delta'_{2*} : TU \to \delta'_2(U) \times M_{\mathbb{R}}$ is then given by $\delta'_{2*} = (\rho_2(\gamma)^{-1} \times \tilde{\rho}_2(\gamma)^{-1}) \circ \delta_{2*}$ (as $\tilde{\rho}_2(\gamma)^{-1}$ is the differential of $\rho_2(\gamma)^{-1}$). Thus we have a commutative diagram

$$\begin{array}{ccc} TU & \xrightarrow{\delta_{2*}} & \delta_2(U) \times M_{\mathbb{R}} \\ & \searrow^{\delta'_{2*}} & \uparrow{\scriptstyle \rho_2(\gamma) \times \tilde{\rho}_2(\gamma)} \\ & & \delta'_2(U) \times M_{\mathbb{R}} \end{array}$$

and graphs

$$\begin{aligned}\{(\delta_2(b),\delta_1(b))\,|\,b\in U\} &\subseteq \delta_2(U)\times M_{\mathbb{R}}\\ \{(\delta_2'(b),\delta_1'(b))\,|\,b\in U\} &= \{(\rho_2(\gamma)^{-1}\delta_2(b),\rho_1(\gamma)^{-1}\delta_1(b))\,|\,b\in U\}\\ &\subseteq \delta_2'(U)\times M_{\mathbb{R}}.\end{aligned}$$

The image of the latter in $\delta_2(U)\times M_{\mathbb{R}}$ is then

$$\{(\delta_2(b),\tilde{\rho}_2(\gamma)\rho_1(\gamma)^{-1}\delta_1(b))\,|\,b\in U\}=\{(\delta_2(b),\delta_1(b)-\mathrm{Trans}_\gamma(\delta_1))\,|\,b\in U\}.$$

Thus the two sections differ only by $\mathrm{Trans}_\gamma(\delta_1)$, which can be viewed as an element of $\Gamma(U,\Lambda_{\mathbb{R}})$.

Now if we choose an open covering $\{U_i\}$ of B of sets of the type just considered, and choices of sets $V_i\subseteq\tilde{B}$ over U_i, we obtain sections $\Gamma_i\in\Gamma(U_i,TB)$ which are graphs of δ_1 under the identifications induced by δ_2 on V_i. In addition, $\Gamma_j-\Gamma_i$ is an element of $\Gamma(U_i\cap U_j,\Lambda_{\mathbb{R}})$. This gives a Čech representative of the class $[\delta_1]\in H^1(B,\Lambda_{\mathbb{R}})$.

Note that if $\mathrm{Trans}_\gamma(\delta_1)\in M$ for all γ, then $[\delta_1]\in H^1(B,\Lambda)$. This is the case if $\delta_1\in V$ rather than $V_{\mathbb{R}}$. Thus if δ_1 is only defined modulo V, then $[\delta_1]$ is defined in $H^1(B,\Lambda_{\mathbb{R}})/H^1(B,\Lambda)$, or in $H^1(B,\Lambda_{\mathbb{R}}/\Lambda)$. □

REMARK 6.30. If $\delta_1=\delta_2=\delta$, then the class $[\delta]\in H^1(B,\Lambda_{\mathbb{R}})$ is known as the *radiance obstruction* of the affine structure on B induced by δ. See §1 of [**206**] or [**184**] for details. As the symplectic form ω on $\check{X}(B)$ is a semi-flat 2-form of type $(1,1)$, it represents a class in $H^1(B,\Lambda_{\mathbb{R}})$. This coincides with $[\delta]$. We leave the details to the reader.

PROPOSITION 6.31. *Let $\delta_1+\sqrt{-1}\delta_2$ represent a point in $X(\mathcal{V})=(V_{\mathbb{R}}+\sqrt{-1}\mathcal{V})/V$ so that δ_2 induces an affine structure on B. Then with $\mathbf{B}=[\delta_1]\in H^1(B,\Lambda_{\mathbb{R}}/\Lambda)$, we have an isomorphism*

$$f^{-1}(\delta_1+\sqrt{-1}\delta_2)\cong X(B,\mathbf{B}).$$

PROOF. We have $B'=B\times\mathcal{V}$, and an integral affine structure induced by δ' on B'. This yields a local system $\Lambda'\subseteq TB'$. On the other hand, $B\times\{\delta_2\}\subseteq B'$ itself is an affine manifold with affine structure given by δ_2, and this yields a local system $\Lambda\subseteq T(B\times\{\delta_2\})$. It is clear that for $b\in B\times\{\delta_2\}$,

$$\Lambda_b=\{v\in\Lambda_b'\,|\,v\text{ is tangent to }B\times\{\delta_2\}\}.$$

Now the map $f:X(B')\to X(\mathcal{V})$ is induced by the differential $p_*:TB'\to T\mathcal{V}=\sqrt{-1}\mathcal{V}\times V_{\mathbb{R}}$ of the projection $B'\to\mathcal{V}$. Thus, for example,

$$p_*^{-1}(\sqrt{-1}\delta_2,0)=T(B\times\{\delta_2\}),$$

and thus

$$f^{-1}(\sqrt{-1}\delta_2)=T(B\times\{\delta_2\})/\Lambda=X(B).$$

More generally, $TB'|_{B\times\{\delta_2\}}=T(B\times\{\delta_2\})\times V_{\mathbb{R}}$, and $p_*^{-1}(\sqrt{-1}\delta_2,\delta_1)=T(B\times\{\delta_2\})\times\{\delta_1\}$. Clearly this latter space is just a translation of the subbundle $T(B \times \{\delta_2\})$ in TB', but not as a complex manifold. To identify the complex structure on $T(B \times \{\delta_2\}) \times \{\delta_1\}$, we need to study $\Lambda_{B\times\{\delta_2\}}$ more carefully.

To this end, consider first an open set $U \subseteq B$ small enough so that $\pi^{-1}(U) = \coprod \Psi_\gamma(V)$ as in Construction 6.29, giving $\delta_2 : U \to M_{\mathbb{R}}$, an immersion. Let us first use δ_2 to identify U with $\delta_2(U) \subseteq M_{\mathbb{R}}$, so that δ_2 is actually the identity, and $U \times \mathcal{V}$ carries an affine structure induced by δ'. Let $b \in U$. Then $T(U \times \mathcal{V})_{(b,\delta_2)} = M_{\mathbb{R}} \oplus V_{\mathbb{R}}$, and we have a differential $\delta'_* : T(U \times \mathcal{V})_{(b,\delta_2)} \to M_{\mathbb{R}} \oplus V_{\mathbb{R}}$ which we compute as follows. If $(m, \delta) \in M_{\mathbb{R}} \oplus V_{\mathbb{R}}$, then let $\gamma : (-\epsilon, \epsilon) \to U \times \mathcal{V}$ be given by $\gamma(t) = (b, \delta_2) + t(m, \delta)$. Then

$$\begin{aligned} \delta'_*(m,\delta) &= \frac{d}{dt}(\delta' \circ \gamma(t))\Big|_{t=0} \\ &= \frac{d}{dt}(\delta'(b+tm, \delta_2 + t\delta))\Big| \\ &= \frac{d}{dt}(\delta_2(b+tm) + t\delta(b+tm), \delta_2 + t\delta)\Big|_{t=0} \end{aligned}$$

Recalling that δ_2 is taken to be the identity, we obtain

$$\delta'_*(m,\delta) = (m + \delta(b), \delta).$$

Thus in particular, $\delta'_*(-\delta(b), \delta) = (0, \delta)$, and the section

$$b \mapsto (-\delta(b), \delta)$$

of

$$T(U \times \mathcal{V})|_{U\times\{\delta_2\}}$$

is a section of $\Lambda'_{\mathbb{R}}|_{U\times\{\delta_2\}}$. So translation by the section $b \mapsto (-\delta_1(b), \delta_1)$ induces a biholomorphic map between $T(U \times \{\delta_2\}) \times \{0\} \subseteq T(U \times \mathcal{V})|_{U\times\{\delta_2\}}$ with $T(U \times \{\delta_2\}) \times \{\delta_1\}$.

Now cover B with open sets $\{U_i\}$ as in Construction 6.29. Making choices as before, we obtain graphs $\Gamma_i \subseteq T(U_i \times \{\delta_2\})$ of δ_1. Then what we have shown is that the maps $\psi_i : T(U_i \times \{\delta_2\}) \times \{0\} \to T(U_i \times \{\delta_2\}) \times \{\delta_1\}$, given by translating by the section $(-\Gamma_i, \delta_1)$ of $T(U_i \times \mathcal{V})$, are biholomorphic.

Hence we obtain a commutative diagram

$$\begin{array}{ccc}
T(U_i \times \{\delta_2\}) \times \{0\} & & \\
\uparrow & & \\
T((U_i \cap U_j) \times \{\delta_2\}) \times \{0\} & \xrightarrow{(-\Gamma_i, \delta_1)} & T((U_i \cap U_j) \times \{\delta_2\}) \times \{\delta_1\} \\
\Big\downarrow {\scriptstyle (\Gamma_j - \Gamma_i, 0)} & & \Big\downarrow {\scriptstyle =} \\
T((U_i \cap U_j) \times \{\delta_2\}) \times \{0\} & \xrightarrow[(-\Gamma_j, \delta_1)]{} & T((U_i \cap U_j) \times \{\delta_2\}) \times \{\delta_1\} \\
\downarrow & & \\
T(U_j \times \{\delta_2\}) \times \{0\} & &
\end{array}$$

and we see $p_*^{-1}(\sqrt{-1}\delta_2, \delta_1)$ is obtained by regluing TB using the cocycle

$$(U_{ij}, \Gamma_j - \Gamma_i) = [\delta_1].$$

Thus $f^{-1}(\delta_1 + \sqrt{-1}\delta_2)$ is obtained by regluing $X(B)$ via

$$\mathbf{B} = [\delta_1] \in H^1(B, \Lambda_{\mathbb{R}}/\Lambda).$$

□

The family $f : X(B') \to X(\mathcal{V})$ is not topologically trivial. To see precisely how this works, choose $\delta_2 \in \mathcal{V}$ and $\delta_1 \in V$, and consider $S^1 = (\mathbb{R}\delta_1/\mathbb{Z}\delta_1) + \sqrt{-1}\delta_2 \subseteq X(\mathcal{V})$. Fix $0 + \sqrt{-1}\delta_2$ as a basepoint of this circle. We will define a continuous family of diffeomorphisms, for $0 \leq r \leq 1$,

$$\phi_r : f^{-1}(\sqrt{-1}\delta_2) \to f^{-1}(r\delta_1 + \sqrt{-1}\delta_2).$$

Since

$$f^{-1}(\sqrt{-1}\delta_2) = f^{-1}(\delta_1 + \sqrt{-1}\delta_2),$$

ϕ_1 will give an automorphism of $f^{-1}(\sqrt{-1}\delta_2)$, called the *monodromy diffeomorphism*. If this is not isotopic to the identity, then the family $f^{-1}(S^1) \to S^1$ is non-trivial. More precisely, it can be obtained topologically from $[0,1] \times f^{-1}(\sqrt{-1}\delta_2)$ by identifying $\{0\} \times f^{-1}(\sqrt{-1}\delta_2)$ and $\{1\} \times f^{-1}(\sqrt{-1}\delta_2)$ via ϕ_1.

To define ϕ_r, simply look on the level of tangent bundles:

$$p_* : T(B \times \{\delta_2\}) \times V_{\mathbb{R}} \to V_{\mathbb{R}},$$

and we have $\tilde{\phi}_r : p_*^{-1}(0) \to p_*^{-1}(\delta_1)$ by adding $(0, \delta_1)$ to each point $(x, c) \in p_*^{-1}(0)$. However, we have seen locally in the proof of Proposition 6.31 that after dividing out by Λ', $p_*^{-1}(0)$ and $p_*^{-1}(\delta_1)$ are identified by adding the local section of $TB'|_{U \times \{\delta_2\}}$ defined by $b \mapsto (-\delta_1(b), \delta_1)$. Thus ϕ_1 can be locally described on $f^{-1}(\sqrt{-1}\delta_2) = TB/\Lambda$ by fiberwise translation $TB_b \ni y \mapsto y + \delta_1(b)$. Now on different choices of open sets, the sections $b \mapsto \delta_1(b)$ differ by elements of Λ, as $\delta_1 \in V$ rather than just $V_{\mathbb{R}}$. Thus $b \mapsto \delta_1(b) \bmod \Lambda$

defines a section of $X(B,0) \to B$, and ϕ_1 is given by fiberwise translation by this section.

Thus we have shown

PROPOSITION 6.32. *The monodromy transformation $\phi_1 : X(B) \to X(B)$ induced by a loop of the form $(\mathbb{R}\delta_1/\mathbb{Z}\delta_1) \times \sqrt{-1}\delta_2$ in $X(\mathcal{V})$ is given by translation of $X(B)$ by a section of $X(B)$ determined by δ_1.*

6.2.5. Generalized complex structures. We now address a mystery which arose in our discussion of the B-field. Traditionally, for a Calabi-Yau manifold X, the B-field takes values in $H^2(X,\mathbb{R})/H^2(X,\mathbb{Z})$ or $H^2(X,\mathbb{R}/\mathbb{Z})$. However, in §6.2.3 we have taken the B-field to live in $H^1(B,\Lambda_{\mathbb{R}}/\Lambda)$, or taken the B-field to be represented by a semi-flat 2-form of type $(1,1)$. In general, this does not give all elements of $H^2(X,\mathbb{R}/\mathbb{Z})$. For example, if B is a two-dimensional torus with the standard affine structure, then $H^1(B,\Lambda_{\mathbb{R}}/\Lambda) = (\mathbb{R}/\mathbb{Z})^4$ while $H^2(X(B),\mathbb{R}/\mathbb{Z}) = (\mathbb{R}/\mathbb{Z})^6$. The discrepancy in dimensions is more apparent when one takes into consideration that the moduli space of complex structures on a two-dimensional complex torus is four-dimensional. So there are additional choices of B-field whose meaning is not immediately clear.

The resolution to this puzzle, as first suggested by Kapustin [**284**] and developed further by Ben-Bassat [**38, 39**] and Chiantese, Gmeiner and Jeschek [**96**], is to enlarge the moduli spaces involved by considering Hitchin's notion [**236**] of *generalized complex structure*. We explain this concept here.

We begin with a generalization of the Lie bracket. Let M be a real manifold, and let $X+\xi, Y+\eta$ be C^∞ sections of $TM \oplus \bigwedge^p T^*M$. Then we define the *Courant bracket* by

$$[X+\xi, Y+\eta] = [X,Y] + \mathcal{L}_X\eta - \mathcal{L}_Y\xi - \frac{1}{2}d(\iota(X)\eta - \iota(Y)\xi).$$

Here $\mathcal{L}$ denotes Lie derivative. We will be interested particularly in the case $p = 1$. We will make use of a natural inner product of signature (n,n) on $TM \oplus T^*M$ given by

$$\langle X+\xi, Y+\eta\rangle = \frac{1}{2}(\xi(Y) + \eta(X)).$$

We then have

DEFINITION 6.33. A *generalized complex structure* on a manifold M of even dimension is a subbundle $E \subseteq (TM \oplus T^*M) \otimes \mathbb{C}$ such that

(1) $(TM \oplus T^*M) \otimes \mathbb{C} = E \oplus \bar{E}$.
(2) The space of sections of E is closed under Courant bracket.
(3) E is isotropic with respect to the inner product on $TM \oplus TM^*$.

The third condition can be rephrased as follows. View E and $\bar{E}$ as the $\pm\sqrt{-1}$ eigenspaces for an operator $\mathcal{J} : TM \oplus T^*M \to TM \oplus T^*M$

with $\mathcal{J}^2 = -1$. The third condition is equivalent to $\mathcal{J}$ being orthogonal with respect to the indefinite inner product. Indeed, if $\mathcal{J}$ is orthogonal and $X+\xi \in E$, then $\langle X+\xi, X+\xi\rangle = \langle \mathcal{J}(X+\xi), \mathcal{J}(X+\xi)\rangle = -\langle X+\xi, X+\xi\rangle$, and the converse is similarly easily checked.

EXAMPLES 6.34. (1) If M has an almost complex structure J, we can take $E = T^{0,1}M \oplus \Omega^{1,0}M$, where we take the vector fields of type $(0,1)$ and 1-forms of type $(1,0)$. Conditions (1) and (3) of the definition of generalized complex structure are then obvious, and condition (2) boils down, after using the Cartan formula for Lie derivative and the fact that $(1,0)$-forms are zero on $(0,1)$ vector fields, to

$$[X,Y] + \iota(X)d\eta - \iota(Y)d\xi \in E$$

whenever $X+\xi, Y+\eta \in E$. This is clearly equivalent to the two conditions: $T^{0,1}M$ being closed under Lie bracket, and the differential of a $(1,0)$ form being of type $(2,0)+(1,1)$. However, these two conditions are both equivalent to the integrability of the almost complex structure J, by the Newlander-Nirenberg Theorem. Thus condition (2) of the definition of generalized complex structure is equivalent to the integrability of J. Note that

$$\mathcal{J} = \begin{pmatrix} -J & 0 \\ 0 & J^* \end{pmatrix}.$$

(2) Let M be an even-dimensional manifold with a non-degenerate two-form ω. Then ω induces an identification $\omega : TM \to T^*M$ by $\omega(X) = \iota(X)\omega$. Take

$$\mathcal{J} = \begin{pmatrix} 0 & \omega^{-1} \\ -\omega & 0 \end{pmatrix}.$$

The $+\sqrt{-1}$ eigenspace E is

$$E = \{X + \sqrt{-1}\omega(X) | X \in TM \otimes \mathbb{C}\}.$$

Again conditions (1) and (3) of Definition 6.33 are obvious, while for (2), the argument given in the previous example shows that E being closed under Courant bracket is equivalent to ω being closed, i.e., M is a symplectic manifold.

(3) If $\mathbf{B}$ is a closed 2-form, then we obtain an automorphism $\mathbf{B}$ of $TM \oplus T^*M$ given by

$$\mathbf{B} : X + \xi \mapsto X + \xi + \iota(X)\mathbf{B}.$$

This satisfies $\mathbf{B}[X+\xi, Y+\eta] = [\mathbf{B}(X+\xi), \mathbf{B}(Y+\eta)]$; indeed, via repeated use of the Cartan formula,

$$\begin{aligned}
[\mathbf{B}(X+\xi), \mathbf{B}(Y+\eta)] &= [X+\xi, Y+\eta] + [X, \iota(Y)(\mathbf{B})] + [\iota(X)(\mathbf{B}), Y] \\
&= [X+\xi, Y+\eta] + \mathcal{L}_X \iota(Y)\mathbf{B} - \mathcal{L}_Y \iota(X)\mathbf{B} \\
&\quad + d(\iota(Y)\iota(X)\mathbf{B}) \\
&= [X+\xi, Y+\eta] + \mathcal{L}_X \iota(Y)\mathbf{B} - d(\iota(Y)\iota(X)\mathbf{B}) \\
&\quad - \iota(Y)d(\iota(X)\mathbf{B}) + d(\iota(Y)\iota(X)\mathbf{B}) \\
&= [X+\xi, Y+\eta] + \mathcal{L}_X \iota(Y)\mathbf{B} - \iota(Y)\mathcal{L}_X\mathbf{B} \\
&\quad + \iota(Y)\iota(X)d\mathbf{B} \\
&= [X+\xi, Y+\eta] + \iota([X,Y])\mathbf{B} + \iota(Y)\iota(X)d\mathbf{B} \\
&= \mathbf{B}[X+\xi, Y+\eta].
\end{aligned}$$

The next to last equality is a property of Lie derivatives; see, e.g., [**319**], Proposition 5.3, page 140. The last line follows from $\mathbf{B}$ being closed.

In addition, the inner product on $TM \oplus T^*M$ is clearly invariant under the action of $\mathbf{B}$. So given a generalized complex structure E, $\mathbf{B}(E)$ is also a generalized complex structure. This is twisting with a B-field.

If we have a notion of generalized complex structure, we should also have the notion of a *generalized Calabi-Yau manifold.* We take our cue from Proposition 6.24, which demonstrates that a complex structure on a Calabi-Yau manifold is determined completely by the holomorphic n-form. We will replace this form with a more general form, called a *pure spinor.* We consider the positive and negative spinor bundles on M: if $\dim M = n$, then

$$\begin{aligned}
S^+ &= \left(\bigwedge^{\text{even}} T^*M\right) \otimes \left(\bigwedge^n TM\right)^{1/2} \\
S^- &= \left(\bigwedge^{\text{odd}} T^*M\right) \otimes \left(\bigwedge^n TM\right)^{1/2}
\end{aligned}$$

Denoting by $S^\pm$ either S^+ or S^-, define $\sigma : S^\pm \to S^\pm$ by

$$\sigma(\varphi) = (-1)^{p(p-1)/2}\varphi$$

for φ a form of pure degree p. We then have a natural bilinear form

$$\langle \varphi, \psi \rangle = (\sigma(\varphi) \wedge \psi)_{\text{top}}.$$

Here we take the top degree part, getting a section of $\bigwedge^n T^*M \otimes \bigwedge^n TM$ and then take the trace to get a number. Equivalently, one could remove the square roots of $\bigwedge^n TM$ from the definition of $S^\pm$ and then obtain values of the pairing in $\bigwedge^n T^*M$, which we shall do below.

$TM \oplus T^*M$ acts on $S^\pm$ by Clifford multiplication:

$$(X+\xi)\cdot\varphi = \iota(X)\varphi + \xi \wedge \varphi.$$

A spinor $\varphi \in S^{\pm} \otimes \mathbb{C}$ is *pure* if the subbundle

$$E_{\varphi} = \{X + \xi \in (TM \oplus T^*M) \otimes \mathbb{C} | (X + \xi) \cdot \varphi = 0\}$$

is a maximally isotropic subbundle of $(TM \oplus T^*M) \otimes \mathbb{C}$.

We will need

LEMMA 6.35. *$E_{\varphi} \cap E_{\psi} = 0$ if and only if $\langle \varphi, \psi \rangle \neq 0$.*

PROOF. See [**88**], III 2.4. □

DEFINITION 6.36. A *generalized Calabi-Yau manifold* is a manifold M of even dimension with a closed form $\varphi \in S^{\pm} \otimes \mathbb{C}$ which is a pure spinor and has $\langle \varphi, \bar{\varphi} \rangle$ everywhere non-zero.

EXAMPLES 6.37. (1) If M is a Calabi-Yau manifold with nowhere vanishing holomorphic n-form Ω, then Ω is a pure spinor: $(X + \xi) \cdot \Omega = 0$ if and only if X is a vector field of type $(0,1)$ and ξ is a form of type $(1,0)$. So E_{Ω} is the generalized complex structure associated to the complex structure on M. We see the conditions of the definition of generalized Calabi-Yau manifold correspond to the conditions of Proposition 6.24.

(2) If $\varphi = \exp(\sqrt{-1}\omega)$ with ω a symplectic form on M, then $(X+\xi) \cdot \varphi = 0$ if and only if $\sqrt{-1}\iota(X)\omega = \xi$. Thus E_{φ} is as in Example 6.34.

(3) A B-field $\mathbf{B}$ acts on spinors by $\varphi \mapsto \exp(\mathbf{B}) \wedge \varphi$. Now

$$\begin{aligned}
(\mathbf{B}^{-1}(X + \xi)) \cdot (\exp(\mathbf{B}) \wedge \varphi) &= (X + \xi - \iota(X)\mathbf{B}) \cdot (\exp(\mathbf{B}) \wedge \varphi) \\
&= \iota(X)(\exp(\mathbf{B}) \wedge \varphi) \\
&\quad + \xi \wedge \exp(\mathbf{B}) \wedge \varphi - (\iota(X)\mathbf{B}) \wedge \exp(\mathbf{B}) \wedge \varphi \\
&= (\iota(X)\mathbf{B}) \wedge \exp(\mathbf{B}) \wedge \varphi + \exp(\mathbf{B}) \wedge \iota(X)\varphi \\
&\quad + \xi \wedge \exp(\mathbf{B}) \wedge \varphi - (\iota(X)\mathbf{B}) \wedge \exp(\mathbf{B}) \wedge \varphi \\
&= \exp(\mathbf{B}) \wedge ((X + \xi) \cdot \varphi).
\end{aligned}$$

Thus if φ is pure, so is $\exp(\mathbf{B}) \wedge \varphi$. In addition, if φ defines a generalized Calabi-Yau structure then

$$\langle \exp(\mathbf{B}) \wedge \varphi, \exp(\mathbf{B}) \wedge \bar{\varphi} \rangle = \langle \varphi, \bar{\varphi} \rangle$$

is nowhere zero. Thus $\exp(\mathbf{B}) \wedge \varphi$ defines a generalized Calabi-Yau structure, corresponding to twisting the associated generalized complex structure by $-\mathbf{B}$.

The important point here is that a generalized Calabi-Yau structure gives a generalized complex structure:

THEOREM 6.38. *If φ gives a generalized Calabi-Yau structure, then E_{φ} is a generalized complex structure.*

PROOF. We already know E_φ is maximally isotropic, so the third condition for generalized complex structure is satisfied. On the other hand, since $\langle\varphi,\bar\varphi\rangle$ is nowhere zero,

$$0 = E_\varphi \cap E_{\bar\varphi} = E_\varphi \cap \bar{E}_\varphi$$

by Lemma 6.35. This gives the first condition. So we need to show E_φ is closed under the Courant bracket. Suppose $X+\xi, Y+\eta \in E_\varphi$. Then

$$\iota(X)\varphi + \xi\wedge\varphi = 0 = \iota(Y)\varphi + \eta\wedge\varphi.$$

Then using $d\varphi = 0$ and the Cartan formula,

$$\begin{aligned}
\iota([X,Y])\varphi &= \mathcal{L}_X(\iota(Y)\varphi) - \iota(Y)\mathcal{L}_X\varphi\\
&= -\mathcal{L}_X(\eta\wedge\varphi) - \iota(Y)d(\iota(X)\varphi)\\
&= -(\mathcal{L}_X\eta)\wedge\varphi - \eta\wedge\mathcal{L}_X\varphi + \iota(Y)d(\xi\wedge\varphi)\\
&= -(\mathcal{L}_X\eta)\wedge\varphi - \eta\wedge d(\iota(X)\varphi) + \iota(Y)((d\xi)\wedge\varphi)\\
&= -(\mathcal{L}_X\eta)\wedge\varphi + \eta\wedge d(\xi\wedge\varphi) + \iota(Y)((d\xi)\wedge\varphi)\\
&= -(\mathcal{L}_X\eta)\wedge\varphi + \eta\wedge d(\xi)\wedge\varphi + (\iota(Y)d\xi)\wedge\varphi + (d\xi)\wedge\iota(Y)\varphi\\
&= -(\mathcal{L}_X\eta)\wedge\varphi + \eta\wedge d(\xi)\wedge\varphi + (\iota(Y)d\xi)\wedge\varphi - (d\xi)\wedge\eta\wedge\varphi\\
&= -(\mathcal{L}_X\eta)\wedge\varphi + (\iota(Y)d\xi)\wedge\varphi
\end{aligned}$$

so, by skew symmetry,

$$\begin{aligned}
\iota([X,Y])\varphi &= \frac{1}{2}(\iota([X,Y])\varphi - \iota([Y,X])\varphi)\\
&= \frac{1}{2}(-(\mathcal{L}_X\eta)\wedge\varphi + (\iota(Y)d\xi)\wedge\varphi + (\mathcal{L}_Y\xi)\wedge\varphi - (\iota(X)d\eta)\wedge\varphi)\\
&= [\iota(Y)d\xi + \frac{1}{2}d(\iota(Y)\xi) - \iota(X)d\eta - \frac{1}{2}d(\iota(X)\eta)]\wedge\varphi\\
&= [\mathcal{L}_Y\xi - \mathcal{L}_X\eta - \frac{1}{2}(d(\iota(Y)\xi - \iota(X)\eta))]\wedge\varphi.
\end{aligned}$$

However, from the definition of Courant bracket, this implies $[X+\xi, Y+\eta]\cdot\varphi = 0$, as desired. □

Now we come back to mirror symmetry. We return to our basic setup, in which B is an affine manifold with transition maps in $\mathbb{R}^n \rtimes \mathrm{SL}_n(\mathbb{Z})$. Then we can talk about a spinor φ being semi-flat, as long as it is a sum of semi-flat forms of various types. Thus we obtain semi-flat generalized Calabi-Yau structures on $X(B)$ and $\check{X}(B)$. In particular, using the definition of μ given in §6.2.2, we have that

THEOREM 6.39. *If φ is a semi-flat pure spinor defining a generalized Calabi-Yau structure on $X(B)$, then $\mu(\varphi)$ is a semi-flat pure spinor defining a generalized Calabi-Yau structure on $\check{X}(B)$.*

PROOF. We need to understand how the mirror transform acts on elements of $T \oplus T^*$. Let $S_{1,0}$ and $S_{0,1}$ denote the sheaves of *semi-flat vector fields* of type $(1,0)$ and $(0,1)$ respectively on $X(B)$. Here $S_{1,0}$ is given by vector fields of the form $\sum_{i=1}^n a_i(y)\partial/\partial y_i$, while $S_{0,1}$ is given by vector fields of the form $\sum_{i=1}^n a_i(y)\partial/\partial x_i$. Again, these sheaves do not depend on the choice of local affine coordinates. Similarly, we define $\check{S}_{1,0}$ and $\check{S}_{0,1}$ as sheaves of semi-flat vector fields of types $(1,0)$ and $(0,1)$ on $\check{X}(B)$, consisting of vector fields locally of the form $\sum_{i=1}^n a_i(y)\partial/\partial y_i$ and $\sum_{i=1}^n a_i(y)\partial/\partial \check{x}_i$ respectively. We continue to use the notation $S^{p,q}$ of §6.2.2 for semi-flat forms of type (p,q). Observe that

$$
\begin{array}{rlrl}
S^{1,0} & \cong T^*B & S^{0,1} & \cong \check{\Lambda} \otimes C^\infty(B) \\
\check{S}^{1,0} & \cong T^*B & \check{S}^{0,1} & \cong \Lambda \otimes C^\infty(B) \\
S_{1,0} & \cong TB & S_{0,1} & \cong \Lambda \otimes C^\infty(B) \\
\check{S}_{1,0} & \cong TB & \check{S}_{0,1} & \cong \check{\Lambda} \otimes C^\infty(B)
\end{array}
$$

Thus there is a natural isomorphism

$$\tilde{\mu} : S_{1,0} \oplus S_{0,1} \oplus S^{1,0} \oplus S^{0,1} \to \check{S}_{1,0} \oplus \check{S}_{0,1} \oplus \check{S}^{1,0} \oplus \check{S}^{0,1}$$

which identifies $S_{0,1}$ with $\check{S}^{0,1}$ and $S^{0,1}$ with $\check{S}_{0,1}$. Explicitly, in local coordinates,

$$\tilde{\mu}(dy_i) = dy_i, \quad \tilde{\mu}(\partial/\partial y_i) = \partial/\partial y_i \quad \tilde{\mu}(dx_i) = \partial/\partial \check{x}_i, \quad \tilde{\mu}(\partial/\partial x_i) = d\check{x}_i.$$

If $\varphi \in S^{p,q}$, we note that for $X + \xi \in S_{1,0} \oplus S_{0,1} \oplus S^{1,0} \oplus S^{0,1}$,

$$\tilde{\mu}(X+\xi) \cdot \mu(\varphi) = \mu((X+\xi)\cdot\varphi). \tag{6.1}$$

Indeed, we only need to check for $X + \xi$ of the form $\partial/\partial y_i$, $\partial/\partial x_i$, dy_i, and dx_i. For $X + \xi = \partial/\partial y_i$, the claim is obvious, since μ only acts on the part of φ involving dx_i's. The same is true for $X + \xi = dy_i$. If $J \subseteq \{1, \dots, n\}$ is an index set with $j \notin J$ and $\#J = q$, then $\tilde{\mu}(\partial/\partial x_j) \cdot \mu(dy_I \wedge dx_J) = \pm d\check{x}_j \wedge dy_I \wedge \iota(\partial/\partial \check{x}_J) d\check{x}_1 \wedge \cdots \wedge d\check{x}_n = 0$, and $(\partial/\partial x_j) \cdot (dy_I \wedge dx_J) = 0$. On the other hand, if $j \notin J$, $\#J = q - 1$, then

$$
\begin{aligned}
& \tilde{\mu}(\partial/\partial x_j) \cdot \mu(dy_I \wedge dx_j \wedge dx_J) \\
= \;& d\check{x}_j \wedge (-1)^{q(q-1)/2+n(n-1)/2} dy_I \wedge \iota(\partial/\partial \check{x}_j \wedge \partial/\partial \check{x}_J) d\check{x}_1 \wedge \cdots \wedge d\check{x}_n \\
= \;& (-1)^{q(q-1)/2+n(n-1)/2} (-1)^p (-1)^{q-1} dy_I \wedge \iota(\partial/\partial \check{x}_J) d\check{x}_1 \wedge \cdots \wedge d\check{x}_n \\
= \;& (-1)^{(q-1)(q-2)/2+n(n-1)/2} (-1)^p dy_I \wedge \iota(\partial/\partial \check{x}_J) d\check{x}_1 \wedge \cdots \wedge d\check{x}_n
\end{aligned}
$$

while

$$
\begin{aligned}
& \mu(\partial/\partial x_j \cdot (dy_I \wedge dx_j \wedge dx_J)) \\
= \;& \mu((-1)^p dy_I \wedge dx_J) \\
= \;& (-1)^{(q-1)(q-2)/2+n(n-1)/2} (-1)^p dy_I \wedge \iota(\partial/\partial \check{x}_J) d\check{x}_1 \wedge \cdots \wedge d\check{x}_n.
\end{aligned}
$$

Similarly, if $j \in J$, then $\tilde{\mu}(dx_j)\cdot\mu(dy_I\wedge dx_J) = (\partial/\partial\check{x}_j)\cdot(\pm dy_I\wedge\iota(\partial/\partial\check{x}_J)d\check{x}_1\wedge\cdots\wedge d\check{x}_n) = 0 = \mu(dx_j\cdot dy_I\wedge dx_J)$. On the other hand, if $j\notin J$, then

$$\begin{aligned}
&\tilde{\mu}(dx_j)\cdot\mu(dy_I\wedge dx_J)\\
=\;& (\partial/\partial\check{x}_j)\cdot((-1)^{q(q-1)/2+n(n-1)/2}dy_I\wedge\iota(\partial/\partial\check{x}_J)d\check{x}_1\wedge\cdots\wedge d\check{x}_n)\\
=\;& (-1)^{q(q-1)/2+n(n-1)/2}(-1)^p dy_I\wedge\iota(\partial/\partial\check{x}_J\wedge\partial/\partial\check{x}_j)d\check{x}_1\wedge\cdots\wedge d\check{x}_n
\end{aligned}$$

while

$$\begin{aligned}
&\mu((dx_j)\cdot(dy_I\wedge dx_J))\\
=\;& \mu((-1)^{p+q}dy_I\wedge dx_J\wedge dx_j)\\
=\;& (-1)^{q(q+1)/2+n(n-1)/2}(-1)^{p+q}dy_I\wedge\iota(\partial/\partial\check{x}_J\wedge\partial/\partial\check{x}_j)d\check{x}_1\wedge\cdots\wedge d\check{x}_n\\
=\;& (-1)^{q(q-1)/2+n(n-1)/2}(-1)^{p}dy_I\wedge\iota(\partial/\partial\check{x}_J\wedge\partial/\partial\check{x}_j)d\check{x}_1\wedge\cdots\wedge d\check{x}_n.
\end{aligned}$$

This demonstrates that $\tilde{\mu}$ and μ respect Clifford multiplication. On the other hand, it is clear that $\tilde{\mu}$ respects the indefinite inner product on semi-flat sections of $T\oplus T^*$.

Because φ is semi-flat, $E_\varphi\subseteq TX(B)\oplus T^*X(B)$ is generated by semi-flat sections. Thus E_φ is maximal isotropic if and only if $E_{\mu(\varphi)}$ is maximal isotropic. By Exercise 6.19, φ is closed if and only if $\mu(\varphi)$ is closed.

Finally, we check that the inner product on the space of semi-flat spinors is preserved up to sign by μ. Let $dy_I\wedge dx_J$ be of type (p,q), $dy_{I'}\wedge dx_{J'}$ of type $(n-p,n-q)$. Then $\langle dy_I\wedge dx_J, dy_{I'}\wedge dx_{J'}\rangle = 0$ unless $I'=\{1,\ldots,n\}\setminus I$ and $J'=\{1,\ldots,n\}\setminus J$. If this is the case, then

$$\begin{aligned}
\langle dy_I\wedge dx_J, dy_{I'}\wedge dx_{J'}\rangle &= (-1)^{(p+q)(p+q-1)/2}dy_I\wedge dx_J\wedge dy_{I'}\wedge dx_{J'}\\
&= (-1)^{(p+q)(p+q-1)/2+q(n-p)}dy_I\wedge dy_{I'}\wedge dx_J\wedge dx_{J'},
\end{aligned}$$

while

$$\begin{aligned}
&\langle\mu(dy_I\wedge dx_J),\mu(dy_{I'}\wedge dx_{J'})\rangle\\
=\;& (-1)^{q(q-1)/2+(n-q)(n-q-1)/2}\\
&\cdot\langle dy_I\wedge(-1)^{q(q+1)/2+\sum_{j\in J}j}d\check{x}_{J'}, dy_{I'}\wedge(-1)^{(n-q)(n-q+1)/2+\sum_{j\in J'}j}d\check{x}_J\rangle\\
=\;& (-1)^{n(n-1)/2+(p+n-q)(p+n-q-1)/2}dy_I\wedge d\check{x}_{J'}\wedge dy_{I'}\wedge d\check{x}_J\\
=\;& (-1)^{n(n-1)/2+(p+n-q)(p+n-q-1)/2+(n-q)(n-p+q)}dy_I\wedge dy_J\wedge d\check{x}_J\wedge d\check{x}_{J'}.
\end{aligned}$$

One can check that the signs in front of the two expressions differ by a factor of $(-1)^n$. Of course, this gives $2n$-forms on two different manifolds; to compare them, we need to obtain numbers by choosing nowhere vanishing $2n$-forms on $X(B)$ and $\check{X}(B)$. But a canonical choice is $dy_1\wedge\cdots\wedge dy_n\wedge dx_1\wedge\cdots\wedge dx_n$ and $dy_1\wedge\cdots\wedge dy_n\wedge d\check{x}_1\wedge\cdots\wedge d\check{x}_n$ respectively. (These forms are proportional to $\Omega\wedge\bar{\Omega}$ and ω^n respectively, where Ω is the canonical holomorphic n-form on $X(B)$ and ω the canonical symplectic form on $\check{X}(B)$.) In any event, we see that

$$\langle\varphi,\psi\rangle = (-1)^n\langle\mu(\varphi),\mu(\psi)\rangle \tag{6.2}$$

for semi-flat spinors φ, ψ, and hence if $\langle \varphi, \bar{\varphi} \rangle \neq 0$, then $\langle \mu(\varphi), \mu(\bar{\varphi}) \rangle \neq 0$. □

Now of course we are often interested in Calabi-Yau manifolds with metrics, so a natural question is: what is the generalization of a Kähler manifold, or a Ricci-flat metric on a generalized Calabi-Yau manifold. A solution to this was given in Gualtieri's thesis [**212**].

DEFINITION 6.40. A *generalized Kähler structure* on a real manifold M is a pair $(\mathcal{J}_1, \mathcal{J}_2)$ of commuting generalized complex structures such that $G = -\mathcal{J}_1 \mathcal{J}_2$ defines a positive definite metric on $TM \oplus T^*M$ via $G(X + \xi, Y + \eta) = \langle G(X + \xi), Y + \eta \rangle$.

EXAMPLE 6.41. Let (g, J, ω) be a usual Kähler structure on a manifold M, with g a Riemannian metric, J the (integrable) almost complex structure and ω the Kähler form (so $\omega(\cdot, J\cdot) = g(\cdot, \cdot)$). Then both J and ω define generalized complex structures, via

$$\mathcal{J}_1 = \begin{pmatrix} -J & 0 \\ 0 & J^* \end{pmatrix}, \quad \mathcal{J}_2 = \begin{pmatrix} 0 & \omega^{-1} \\ -\omega & 0 \end{pmatrix}.$$

Note that

$$G = -\mathcal{J}_1 \mathcal{J}_2 = -\mathcal{J}_2 \mathcal{J}_1 = \begin{pmatrix} 0 & g^{-1} \\ g & 0 \end{pmatrix}$$

where $g : TM \to T^*M$ is the identification induced by g. So $G(X + \xi, Y + \eta) = \langle g^{-1}(\xi) + g(X), Y + \eta \rangle = \frac{1}{2}(g^{-1}(\xi, \eta) + g(X, Y))$.

EXERCISE 6.42. Any generalized Kähler structure can be twisted by a B-field, by twisting both $\mathcal{J}_1$ and $\mathcal{J}_2$. Write down formulas for the twisting by a B-field of the standard notion of Kähler structure discussed above.

Now we can define generalized Calabi-Yau metrics:

DEFINITION 6.43. A *generalized Calabi-Yau metric* is defined by a generalized Kähler structure $(\mathcal{J}_1, \mathcal{J}_2)$ coming from pure spinors φ_1 and φ_2 with

$$\langle \varphi_1, \bar{\varphi}_1 \rangle = C \langle \varphi_2, \bar{\varphi}_2 \rangle$$

for some non-zero constant C.

EXAMPLE 6.44. In Example 6.41, if the manifold M is a Calabi-Yau manifold, then Ricci-flatness is equivalent to ω^n being proportional to $\Omega \wedge \bar{\Omega}$. But this is the same condition as that given in Definition 6.43.

Mirror symmetry now operates on generalized Calabi-Yau metrics:

THEOREM 6.45. *Let φ_1, φ_2 be pure semi-flat spinors determining a generalized Calabi-Yau metric on $X(B)$. Then $\mu(\varphi_1), \mu(\varphi_2)$ determine a generalized Calabi-Yau metric on $\check{X}(B)$.*

PROOF. By Theorem 6.39, $\mu(\varphi_1)$ and $\mu(\varphi_2)$ both define generalized Calabi-Yau structures on $\check{X}(B)$. Furthermore, by (6.2), the condition

$$\langle \mu(\varphi_1), \mu(\bar{\varphi}_1)\rangle = C\langle \mu(\varphi_2), \mu(\bar{\varphi}_2)\rangle$$

holds, and we just need to show that $\mu(\varphi_1)$ and $\mu(\varphi_2)$ define a generalized Kähler structure. Denote by $\check{\mathcal{J}}_1$ and $\check{\mathcal{J}}_2$ the almost complex structures on $T\check{X}(B)\oplus T^*\check{X}(B)$ defined by $\mu(\varphi_1)$ and $\mu(\varphi_2)$. Then on $\check{S}_{1,0}\oplus\check{S}_{0,1}\oplus\check{S}^{1,0}\oplus\check{S}^{0,1}$,

$$\check{\mathcal{J}}_i = \tilde{\mu}\mathcal{J}_i\tilde{\mu}^{-1},$$

as the space of semi-flat sections of $E_{\mu(\varphi_i)}$ is just the image under $\tilde{\mu}$ of the space of semi-flat sections of E_{φ_i}, by (6.1). Then

$$-\check{\mathcal{J}}_1\check{\mathcal{J}}_2 = -\tilde{\mu}\mathcal{J}_1\mathcal{J}_2\tilde{\mu}^{-1},$$

so if $-\mathcal{J}_1\mathcal{J}_2$ defines a positive definite metric, so does $-\check{\mathcal{J}}_1\check{\mathcal{J}}_2$. □

EXAMPLE 6.46. Suppose $\dim_{\mathbb{R}} M = 2$. If φ is an odd pure spinor on M, the generalized Calabi-Yau conditions are $d\varphi = 0$ and $\varphi\wedge\bar{\varphi}$ is nowhere zero. In this case φ defines an ordinary complex structure.

If φ is an even spinor, $\varphi = c+\beta$, where c is a constant and β is a 2-form. Then

$$\langle c+\beta, \bar{c}+\bar{\beta}\rangle = c\bar{\beta}-\bar{c}\beta \neq 0$$

implies $c\neq 0$, $\operatorname{Im}(\beta/c)\neq 0$, so $\beta/c = \mathbf{B}+\sqrt{-1}\omega$, where $\mathbf{B}$ is an arbitrary real closed 2-form and ω is a symplectic form. Thus

$$\varphi = c\exp(\mathbf{B}+\sqrt{-1}\omega).$$

EXERCISE 6.47. Check that if $\dim_{\mathbb{R}} M = 2$, then (φ_1,φ_2) define a generalized Calabi-Yau metric if and only if one spinor is even, one is odd, and the even spinor is of the form $c\exp(\mathbf{B}+\sqrt{-1}\omega)$ with ω a Kähler form in the complex structure determined by the other spinor.

EXAMPLE 6.48. Suppose $\dim_{\mathbb{R}} M = 4$. First suppose φ is an odd pure spinor defining a generalized Calabi-Yau structure, with $\varphi = \beta+\gamma$ with β a 1-form and γ a 3-form. Consider first the condition that E_φ is maximal isotropic. If $X+\xi\in E_\varphi$, then

$$0 = (X+\xi)\cdot(\beta+\gamma) = \iota(X)\beta + (\xi\wedge\beta+\iota(X)\gamma)+\xi\wedge\gamma.$$

This shows $X\in\ker\beta$, so X is constrained to live in a three-dimensional space at each point of M. On the other hand, $\xi\wedge\beta+\iota(X)\gamma = 0$ implies ξ is determined by $\iota(X)\gamma$ up to adding a multiple of β. Thus E_φ can only be four-dimensional if $X+\xi+c\beta\in E_\varphi$ whenever $X+\xi\in E_\varphi$. However this happens only if $\beta\wedge\gamma = 0$, i.e., $\gamma = \beta\wedge\nu$ for some two-form ν. On the other hand, if $\varphi = \beta\wedge(1+\nu)$ and $X\in\ker\beta$, then

$$(X+\xi)\cdot\beta\wedge(1+\nu) = \xi\wedge\beta\wedge(1+\nu) - \beta\wedge\iota(X)\nu,$$

and this is zero if $\xi = -\iota(X)\nu + c\beta$. Thus

$$E_\varphi = \{(X, -\iota(X)\nu + c\beta) | X \in \ker\beta, c \in \mathbb{C}\}$$

is maximal isotropic. Now

$$\langle \beta + \gamma, \bar\beta + \bar\gamma \rangle = \beta \wedge \bar\gamma + \bar\beta \wedge \gamma = \beta \wedge \bar\beta \wedge (\bar\nu - \nu).$$

Thus the condition that this is nowhere zero and $d\varphi = 0$ tells us that, locally, there is a function $f : M \to \mathbb{C}$ with $df = \beta$ defining a fibration locally on M, with $\operatorname{Im}\nu$ defining a symplectic structure on the fibers and $\operatorname{Re}\nu$ a B-field on the fibers. So these generalized complex structures are a type of hybrid mix of complex and symplectic structures.

EXERCISE 6.49. Determine when two odd spinors φ_1 and φ_2 of this form yield a generalized Calabi-Yau metric.

Next consider the even pure spinor case, with $\varphi = c + \beta + \gamma$ with c a constant, β a two-form, and γ a four-form. First suppose $c \neq 0$. By rescaling, we can assume that $c = 1$. Then, if $X + \xi \in E_\varphi$, we have

$$0 = (\xi + \iota(X)\beta) + (\xi \wedge \beta + \iota(X)\gamma),$$

so $\xi = -\iota(X)\beta$ is determined by X, and then

$$0 = -(\iota(X)\beta) \wedge \beta + \iota(X)\gamma$$

or

$$0 = \iota(X)(-\beta \wedge \beta/2 + \gamma).$$

In order for E_φ to be maximal, this must hold for all X, and thus we must have

$$\gamma = \beta \wedge \beta/2,$$

so

$$\varphi = \exp(\beta).$$

Furthermore,

$$0 \neq \langle 1 + \beta + \beta \wedge \beta/2, 1 + \bar\beta + \bar\beta \wedge \bar\beta/2 \rangle = \frac{\bar\beta \wedge \bar\beta}{2} - \beta \wedge \bar\beta + \frac{\beta \wedge \beta}{2} = \frac{1}{2}(\bar\beta - \beta)^2.$$

Thus $\beta = \mathbf{B} + \sqrt{-1}\omega$, where ω is symplectic. So φ induces a symplectic manifold twisted by a B-field.

If $c = 0$, then

$$\dim_{\mathbb{C}}\{X \in T_b M \otimes \mathbb{C} \,|\, \iota(X)\beta = 0\} \leq 2$$

with equality if and only if β is decomposable, or equivalently, $\beta \wedge \beta = 0$, and similarly

$$\dim_{\mathbb{C}}\{\xi \in T_b^* M \otimes \mathbb{C} \,|\, \xi \wedge \beta = 0\} \leq 2,$$

with equality if and only if β is decomposable. As $(X + \xi) \cdot \varphi = 0$ if and only if $\iota(X)\beta = 0$ and $\iota(X)\gamma = -\xi \wedge \beta$, it follows that E_φ can be maximal if and only if $\beta \wedge \beta = 0$. In addition, $\langle \varphi, \bar\varphi \rangle = -\beta \wedge \bar\beta \neq 0$, and this tells us that β

is the holomorphic two-form of a complex structure. In particular, we can write $\varphi = \beta + \beta \wedge \mathbf{B}$, where $\mathbf{B}$ is a not necessarily closed, arbitrary two-form. Since only the $(0,2)$ part of $\mathbf{B}$ with respect to the complex structure induced by β is relevant, we can assume $\mathbf{B} = f\bar{\beta}$ for some function f, so that $\varphi = \beta + f\beta \wedge \bar{\beta}$.

EXERCISE 6.50. (1) Let φ_1, φ_2 define a generalized Calabi-Yau metric on a manifold M of dimension $2n$. Then show that $\dim(E_{\varphi_1} \cap E_{\varphi_2}) = \dim(E_{\varphi_1} \cap \bar{E}_{\varphi_2}) = n$. Thus in particular, $\langle \varphi_1, \varphi_2 \rangle = \langle \varphi_1, \bar{\varphi}_2 \rangle = 0$. [Hint: Show that the ± 1 eigenspaces $G^{\pm}$ of G acting on $TM \oplus T^*M$ are both $2n$-dimensional. Then express $G^{\pm} \otimes \mathbb{C}$ in terms of E_{φ_i} and $E_{\bar{\varphi}_i}$.]

(2) Conversely, if $n = 2$, show that if φ_1, φ_2 define generalized Calabi-Yau structures, then together they define a generalized Calabi-Yau metric if and only if $\langle \varphi_1, \varphi_2 \rangle = \langle \varphi_1, \bar{\varphi}_2 \rangle = 0$ and $\langle \varphi_1, \bar{\varphi}_1 \rangle = C \langle \varphi_2, \bar{\varphi}_2 \rangle$ for $C > 0$ a constant. [Hint: use the fact that for two pure even spinors φ and ψ, $\dim(E_\varphi \cap E_\psi)$ is even.]

EXAMPLE 6.51. We can finally understand the answer to the question posed at the beginning of the section. If B is an affine manifold with transition maps in $\mathbb{R}^n \rtimes SL_n(\mathbb{Z})$, we now allow the B-field to be any closed semi-flat 2-form. Then a generalized Calabi-Yau structure given by a twisted symplectic form $\exp(\mathbf{B} + \sqrt{-1}\omega)$ on $X(B)$ defines a mirror generalized Calabi-Yau structure $\mu(\exp(\mathbf{B} + \sqrt{-1}\omega))$ on $\check{X}(B)$. What we saw in Proposition 6.26 is that if $\mathbf{B}$ and ω are semi-flat of type $(1,1)$, then $\mu(\exp(\mathbf{B} + \sqrt{-1}\omega))$ defines an ordinary complex structure on $\check{X}(B)$. Otherwise, one only obtains a generalized complex structure. By the above analysis in the real four-dimensional case, we see that if $\dim_{\mathbb{R}} B = 2$, then we obtain either a twisted complex structure or a twisted symplectic structure. In fact, in this two-dimensional case, one sees that if $\mathbf{B} + \sqrt{-1}\omega$ contains a part which is of type $(0,2)$, then $\mu(\exp(\mathbf{B} + \sqrt{-1}\omega))$ has a type $(0,0)$ component, and hence must yield a twisted symplectic structure as its mirror. Otherwise, the mirror is a twisted complex structure. Of course, the situation could be more complicated in higher dimensions.

6.3. A- and B-branes in the semi-flat case

We now return to the original duality between A- and B-branes which motivated the SYZ conjecture, as discussed in §1.1.5. By restricting our attention to a particularly simple class of branes, we will get an exact isomorphism between moduli spaces of A- and B-branes on semi-flat mirror pairs.

6.3.1. B-branes. Let us now fix, as in §6.2.3, an affine manifold B with transition maps in $\mathbb{R}^n \rtimes \mathrm{GL}(n,\mathbb{Z})$, and a B-field $\mathbf{B} \in H^1(B, \Lambda_{\mathbb{R}}/\Lambda)$. This gives rise to a complex manifold $X(B, \mathbf{B})$. Let us consider B-branes on $X(B,\mathbf{B})$ which are relatively simple, given by pairs $(Y,\mathcal{L})$. Here Y is a complex submanifold of $X(B,\mathbf{B})$ such that $f|_Y : Y \to f(Y) \subseteq B$ is a fibration in d-dimensional tori over a d-dimensional submanifold of B. In addition, $\mathcal{L}$ is a line bundle on Y of a certain sort.

Focus first locally, where $U \subseteq B$ is an open set identified with a subset of $M_{\mathbb{R}}$, where as usual $M = \mathbb{Z}^n$ and $M_{\mathbb{R}} = M \otimes_{\mathbb{Z}} \mathbb{R}$. Assume the B-field is trivial on this open subset. Set $X = f^{-1}(U) = U \times \sqrt{-1}(M_{\mathbb{R}}/M)$. Assume $Y \subseteq X$ is a complex manifold as above, fibering in tori over a submanifold of U. Looking at a point $x \in Y$, $f(x) = b$, the tangent space T_xY must be a complex subspace of T_xX. We can thus split $T_xY = V \oplus JV$, where J is the almost complex structure on X and V is the vertical tangent space of Y, i.e., the tangent space of the fiber $Y_b = X_b \cap Y$ at x. Then $f_*(JV)$ is the tangent space to $f(Y)$ at b. On the other hand, $f_* : J(T_x(X_b)) \to T_bB$ is an isomorphism, so the vector space JV is completely determined by $T_b(f(Y))$, independent of $x \in Y_b$. Thus V itself is independent of $x \in Y_b$, so Y_b is simply a linear torus. Then $Y_b = (V/V \cap M) + g(b)$, where $g : f(Y) \to M_{\mathbb{R}}$ is a function giving the translation of the torus away from the origin of $M_{\mathbb{R}}$.

There is a constraint on V: it must be a subspace of $M_{\mathbb{R}}$ defined over $\mathbb{Q}$ so that $V/V \cap M$ is a torus. Thus as $b \in f(Y)$ varies, because $V \subseteq M_{\mathbb{R}}$ must vary continuously, it in fact must be constant. Thus $f(Y)$ spans a rational affine subspace of $M_{\mathbb{R}}$.

Note that we can also restrict the nature of the function g: in order for the tangent space at a point $x \in Y$ to be of the form $V \oplus JV$, g must be constant modulo V, as can be easily checked.

REMARK 6.52. In terms of equations, the complex submanifolds of the restricted sort being considered here can be written in terms of the complex coordinates of §6.2.1 via monomial equations

$$a_1 \prod q_i^{b_{i1}} = \cdots = a_m \prod q_i^{b_{im}} = 1,$$

with (b_{ij}) an integer matrix. Applying $-\frac{1}{2\pi}\log|\cdot|$ to these equations gives the affine linear equations defining the image of this complex submanifold in U.

We now globalize these observations to $X(B,\mathbf{B})$.

DEFINITION 6.53. If B is an affine manifold then an *affine subspace* of B is a submanifold which is locally given as the zero locus of an affine linear function $M_{\mathbb{R}} \to \mathbb{R}^m$ for some m. If the holonomy of B is contained in $M_{\mathbb{R}} \rtimes \mathrm{GL}(n,\mathbb{Z})$, the affine subspace is said to be *rational* if the differential of this affine linear map is defined over $\mathbb{Q}$.

PROPOSITION 6.54. *Let B be an affine manifold with holonomy contained in $M_{\mathbb{R}} \rtimes \mathrm{GL}(n,\mathbb{Z})$, and let $\mathbf{B} \in H^1(B,\Lambda_{\mathbb{R}}/\Lambda)$ be represented by a Čech 1-cocycle (b_{ij}) on an open covering $\{U_i\}$ of B. Then giving a* *holomorphic submanifold* $Y \subseteq X(B, \mathbf{B})$ *fibering in* d*-dimensional tori over a* d*-dimensional* $f(Y) \subseteq B$ *is equivalent to the following data: (1) a* d*-dimensional rational affine submanifold* $B' \subseteq B$ *with* $\Lambda' \subseteq \Lambda|_{B'}$ *the local system of horizontal integral tangent vectors tangent to* B', $\Lambda'_{\mathbb{R}} := \Lambda' \otimes \mathbb{R}$, *and (2) elements* $g_i \in \Gamma(U_i \cap B', \Lambda_{\mathbb{R}}|_{B'})$ *such that* $g_j - g_i = b_{ij}|_{U_i \cap U_j \cap B'}$ mod $\Gamma(U_i \cap U_j \cap B', \Lambda'_{\mathbb{R}} + \Lambda|_{B'})$.

PROOF. Suppose we are given such data. Write $g : X(B) \to B$ and $f : X(B, \mathbf{B}) \to B$. Then over U_i we have naturally a complex submanifold $X(B' \cap U_i) = T(B' \cap U_i)/\Lambda'|_{B' \cap U_i} \subseteq g^{-1}(U_i)$ and we then translate by g_i inside $g^{-1}(U_i)$ to get a complex submanifold $g_i + X(B' \cap U_i)$. Now we obtain $X(B, \mathbf{B})$ by gluing $g^{-1}(U_i)$ and $g^{-1}(U_j)$ via translation by b_{ij}. Consider a point $x \in g^{-1}(U_i)$ with $g(x) = b \in U_i \cap U_j$ and $x \in g_i(b) + X(B' \cap U_i)$. Then $x = g_i(b) + x'$ is identified with $x + b_{ij} = g_i(b) + b_{ij} + x' = g_j(b) + x'$ mod $\Lambda'_{\mathbb{R},b} + \Lambda_b$. But x' plus anything in $\Lambda'_{\mathbb{R},b} + \Lambda_b$ is in $X(B' \cap U_j)$, so $x + b_{ij} \in g_j + X(B' \cap U_j)$. Thus the gluing maps identify $g_i + X(B' \cap U_i)$ with $g_j + X(B' \cap U_j)$ over $U_i \cap U_j$.

The converse follows from the local discussion at the beginning of this section. □

EXERCISE 6.55. Let $Y \subseteq X(B, \mathbf{B})$ be determined by data $B' \subseteq B$, (g_i). Then show that there is an isomorphism $Y \cong X(B', \mathbf{B}')$, where $\mathbf{B}' \in H^1(B', \Lambda'_{\mathbb{R}}/\Lambda')$ is represented by the Čech 1-cocycle $(b_{ij} + g_i - g_j)_{ij}$, with

$$\begin{aligned} b_{ij} + g_i - g_j &\in \Gamma(U_i \cap U_j \cap B', (\Lambda|_{B'} + \Lambda'_{\mathbb{R}})/\Lambda|_{B'}) \\ &= \Gamma(U_i \cap U_j \cap B', \Lambda'_{\mathbb{R}}/(\Lambda|_{B'} \cap \Lambda'_{\mathbb{R}})) \\ &= \Gamma(U_i \cap U_j \cap B', \Lambda'_{\mathbb{R}}/\Lambda'). \end{aligned}$$

Note that we have an exact sequence

$$0 \to \Lambda'_{\mathbb{R}}/\Lambda' \to (\Lambda_{\mathbb{R}}/\Lambda)|_{B'} \to \Lambda_{\mathbb{R}}|_{B'}/(\Lambda|_{B'} + \Lambda'_{\mathbb{R}}) \to 0, \tag{6.3}$$

and given the B-field $\mathbf{B} = (b_{ij})$, the existence of sections $g_i \in \Gamma(U_i \cap B', \Lambda_{\mathbb{R}}|_{B'})$ with $b_{ij} + g_i - g_j \in \Gamma(U_i \cap U_j \cap B', \Lambda|_{B'} + \Lambda'_{\mathbb{R}})$ is equivalent to the image of $\mathbf{B}|_{B'} \in H^1(B', (\Lambda_{\mathbb{R}}/\Lambda)|_{B'})$ in $H^1(B', \Lambda_{\mathbb{R}}|_{B'}/(\Lambda|_{B'} + \Lambda'_{\mathbb{R}}))$ being zero. Show that all holomorphic submanifolds $Y \subseteq X(B, \mathbf{B})$ fibering in tori over B' are in one-to-one correspondence with Čech 0-cochains (g_i) for $\Lambda_{\mathbb{R}}|_{B'}/(\Lambda|_{B'} + \Lambda'_{\mathbb{R}})$ satisfying $g_j - g_i = b_{ij} \mod \Lambda|_{B'} + \Lambda'_{\mathbb{R}}$. Show that $H^0(B', \Lambda_{\mathbb{R}}|_{B'}/(\Lambda|_{B'} + \Lambda'_{\mathbb{R}}))$ acts freely and transitively on the set of such choices. (We say this set of choices is a *torsor* over $H^0(B', \Lambda_{\mathbb{R}}|_{B'}/(\Lambda|_{B'} + \Lambda'_{\mathbb{R}}))$). □

REMARK 6.56. We may also consider the case where the fibers of $Y \to f(Y)$ are disjoint unions of tori. This can be accomplished by allowing $B' \to B$ to be an unramified covering of its image, rather than an embedding.

EXAMPLE 6.57. If $B = M_{\mathbb{R}}/\Gamma$ for a lattice $\Gamma \subseteq M_{\mathbb{R}}$, note first that if Γ is general, then B contains no compact rational affine subspaces. (This happens when there is no sublattice of Γ spanning a rational subspace of $M_{\mathbb{R}}$). But even if there is a compact rational affine subspace B' of B, then for a general choice of $\mathbf{B}$, $\mathbf{B}$ does not restrict to zero in $H^1(B', \Lambda_{\mathbb{R}}|_{B'}/(\Lambda|_{B'} + \Lambda'_{\mathbb{R}}))$, as can be easily checked. This reflects the fact that a general complex torus does not contain any complex subtorus. □

To first approximation, we might wish to view a B-brane as more than just a complex submanifold, but rather a complex submanifold along with a complex vector bundle. In a further simplification, we assume this vector bundle is a line bundle. Because the complex submanifolds we are considering are always of the form $X(B', \mathbf{B}')$ by Exercise 6.55, we will first describe line bundles on $X(B, \mathbf{B})$.

We will in fact describe only certain line bundles on $X(B, \mathbf{B})$. Suppose we have an open covering $\{U_i\}$ of B, say with coordinate charts $\psi_i : U_i \to M_{\mathbb{R}}$. We can construct a complex line bundle on $X(B, \mathbf{B})$ by taking the line bundle to be trivial on $f^{-1}(U_i)$, with transition functions $s_{ij} : f^{-1}(U_i \cap U_j) \to \mathbb{C}^\times$. In other words, we identify

$$f^{-1}(U_i) \times \mathbb{C} \supseteq f^{-1}(U_i \cap U_j) \times \mathbb{C} \to f^{-1}(U_i \cap U_j) \times \mathbb{C} \subseteq f^{-1}(U_j) \times \mathbb{C}$$

using the map $(x, c) \mapsto (x, s_{ij}(x) \cdot c)$. Now we will only consider certain invertible functions on $f^{-1}(U_i \cap U_j)$. If we use coordinate charts $\psi_i : U_i \to M_{\mathbb{R}}$ to identify $f^{-1}(U_i \cap U_j)$ with $\psi_i(U_i \cap U_j) \times \sqrt{-1}(M_{\mathbb{R}}/M) \subseteq M \otimes \mathbb{C}^\times$, then for $n \in N = \mathrm{Hom}_{\mathbb{Z}}(M, \mathbb{Z})$ define q^n to be the character $M \otimes \mathbb{C}^\times \to \mathbb{C}^\times$ determined by n. This is a holomorphic function. The functions aq^n for $n \in N$ are invertible functions on $M \otimes \mathbb{C}^\times$, and hence on $f^{-1}(U_i \cap U_j)$. Of course, there may be many other invertible functions; locally any function on $f^{-1}(U_i \cap U_j)$ has a Laurent expansion as an infinite sum of such terms, but we will not allow these more complicated functions. The space of functions of the form aq^n, which we will call *monomial functions*, forms a sheaf $\mathcal{P}_{B,\mathbf{B}}$ on B, i.e.

$$\mathcal{P}_{B,\mathbf{B}}(U) = \{\text{monomial functions on } f^{-1}(U)\}.$$

The group of line bundles we are interested in is classified by the cohomology group $H^1(B, \mathcal{P}_{B,\mathbf{B}})$.

This motivates the following definition:

DEFINITION 6.58. A *semi-flat B-brane* on $X(B, \mathbf{B})$ is a holomorphic submanifold Y of $X(B, \mathbf{B})$ fibering in tori over a rational affine submanifold B' of B, along with a line bundle $\mathcal{L}$ on Y, whose transition functions are monomial functions on Y.

To classify semi-flat B-branes, let's describe $\mathcal{P}_{B,\mathbf{B}}$ in a more intrinsic fashion. We have a subsheaf $\mathbb{C}^\times \subseteq \mathcal{P}_{B,\mathbf{B}}$ consisting of the constant functions. On the other hand, if $b \in B$, then we can identify some small neighbourhood U of b canonically up to translation with a neighbourhood U' in $\Lambda_{\mathbb{R},b}$, so

$f^{-1}(U) \cong U' \times \sqrt{-1}(\Lambda_{\mathbb{R},b}/\Lambda_b)$ and $\mathcal{P}_{B,\mathbf{B}}(U) \cong \mathbb{C}^\times \times \check{\Lambda}_b = \mathbb{C}^\times \times \check{\Lambda}(U)$. Thus we have an exact sequence

$$1 \to \mathbb{C}^\times \to \mathcal{P}_{B,\mathbf{B}} \to \check{\Lambda} \to 0.$$

This sequence does not in general split: this is already clear locally in that the choice of U' affects the identification of $\mathcal{P}_{B,\mathbf{B}}(U) \cong \mathbb{C}^\times \times \check{\Lambda}(U)$. Furthermore, the extension even depends on the B-field. More precisely,

PROPOSITION 6.59. *The extension class of*

$$1 \to \mathbb{C}^\times \to \mathcal{P}_{B,\mathbf{B}} \to \check{\Lambda} \to 0$$

in $\operatorname{Ext}^1_B(\check{\Lambda}, \mathbb{C}^\times) \cong H^1(B, \Lambda \otimes \mathbb{C}^\times)$ *is represented by*

$$\exp(2\pi\sqrt{-1}(\mathbf{B} + \sqrt{-1}[\delta])),$$

where $\mathbf{B} + \sqrt{-1}[\delta] \in H^1(B, \Lambda \otimes \mathbb{C})$ *is the "complexified radiance obstruction of* B*,"* $[\delta]$ *being the radiance obstruction of* B *(see Remark 6.30).*

PROOF. Let $\{U_i\}$ be an open cover with each U_i contractible, and

$$\{(U_i \cap U_j, b_{ij})\}$$

a Čech representative for the B-field. Let $\psi_i : U_i \to M_{\mathbb{R}}$ be coordinate charts. As observed above, this chart yields an isomorphism of $\mathcal{P}_{B,\mathbf{B}}|_{U_i}$ with the constant sheaf $\mathbb{C}^\times \times N$, giving us local splittings $\alpha_i : \check{\Lambda}|_{U_i} \to \mathcal{P}_{B,\mathbf{B}}|_{U_i}$. The extension class, as an element of $H^1(B, \Lambda \otimes \mathbb{C}^\times)$, is represented by the Čech one-cocycle given by the difference of these splittings, i.e., by $\{(U_i \cap U_j, \alpha_j/\alpha_i)\}$ with α_j/α_i a map $\check{\Lambda}|_{U_i\cap U_j} \to \mathbb{C}^\times$, which can be viewed as a section of $\Lambda \otimes \mathbb{C}^\times$ over $U_i \cap U_j$. (Here we are writing the sheaf of groups $\mathcal{P}_{B,\mathbf{B}}$ multiplicatively.) Now let us compare α_i and α_j directly. To do this, we need to understand how $X(B, \mathbf{B})$ is constructed by gluing together pieces of the form $X(\psi_i(U_i))$. In particular, $X(\psi_i(U_i \cap U_j))$ is glued to $X(\psi_j(U_i \cap U_j))$ by twisting the canonical identification with b_{ij}. The canonical identification is given as follows. The transition map $\psi_j \circ \psi_i^{-1} \in M_{\mathbb{R}} \rtimes \mathrm{GL}(M)$ induces a map on tangent bundles $(\psi_j \circ \psi_i^{-1})_* : T(\psi_i(U_i \cap U_j)) \to T(\psi_j(U_i \cap U_j))$, which induces the canonical identification $X(\psi_i(U_i \cap U_j)) \to X(\psi_j(U_i \cap U_j))$. This is then composed with fiberwise addition by b_{ij} to obtain the identification

$$\phi_{ij} : X(\psi_i(U_i \cap U_j)) \to X(\psi_j(U_i \cap U_j)).$$

Explicitly, using the inclusions of $X(\psi_i(U_i \cap U_j))$ and $X(\psi_j(U_i \cap U_j))$ in $M_{\mathbb{R}} \times M_{\mathbb{R}}/M = M \otimes_{\mathbb{Z}} \mathbb{C}^\times$, ϕ_{ij} acts by $\psi_j \circ \psi_i^{-1} \in \operatorname{Aff}(M_{\mathbb{R}})$ on $M_{\mathbb{R}}$ and by

$$m \mapsto \operatorname{Lin}(\psi_j \circ \psi_i^{-1})(m) + b_{ij}$$

on $M_{\mathbb{R}}/M$, where $\operatorname{Lin} : \operatorname{Aff}(M_{\mathbb{R}}) \to \mathrm{GL}(M_{\mathbb{R}})$ is the projection. Alternatively, writing an element $m \in M \otimes_{\mathbb{Z}} \mathbb{C}^\times$ as $\exp(2\pi\sqrt{-1}(m_1 + \sqrt{-1}m_2))$ for $m_1, m_2 \in$

$M_{\mathbb{R}}$,

$$\begin{aligned}\phi_{ij}(m) &= \phi_{ij}(\exp(2\pi\sqrt{-1}(m_1+\sqrt{-1}m_2))) \\ &= \exp(2\pi\sqrt{-1}(\mathrm{Lin}(\psi_j\circ\psi_i^{-1})(m_1)+b_{ij}+\sqrt{-1}(\psi_j\circ\psi_i^{-1}(m_2)))) \\ &= \exp(2\pi\sqrt{-1}(\mathrm{Lin}(\psi_j\circ\psi_i^{-1})(m_1+\sqrt{-1}m_2))) \\ &\quad\cdot\exp(2\pi\sqrt{-1}(b_{ij}+\sqrt{-1}\,\mathrm{Trans}(\psi_j\circ\psi_i^{-1}))) \\ &= \mathrm{Lin}(\psi_j\circ\psi_i^{-1})(m)\exp(2\pi\sqrt{-1}(b_{ij}+\sqrt{-1}\,\mathrm{Trans}(\psi_j\circ\psi_i^{-1})))\end{aligned}$$

where $\mathrm{Trans}:\mathrm{Aff}(M_{\mathbb{R}})\to M_{\mathbb{R}}$ is the projection onto the linear part. (Note: Trans is not a homomorphism.) Then for $n\in N$, $m\in M\otimes_{\mathbb{Z}}\mathbb{C}^\times$,

$$\begin{aligned}&(q^n\circ\phi_{ij})(m) \\ =\;& q^n(\mathrm{Lin}(\psi_j\circ\psi_i^{-1})(m)\exp(2\pi\sqrt{-1}(b_{ij}+\sqrt{-1}\,\mathrm{Trans}(\psi_j\circ\psi_i^{-1})))) \\ =\;& q^{{}^t\mathrm{Lin}(\psi_j\circ\psi_i^{-1})(n)}(m)q^n(\exp(2\pi\sqrt{-1}(b_{ij}+\sqrt{-1}\,\mathrm{Trans}(\psi_j\circ\psi_i^{-1})))) \\ =\;& q^{{}^t\mathrm{Lin}(\psi_j\circ\psi_i^{-1})(n)}(m)\exp(2\pi\sqrt{-1}\langle n, b_{ij}+\sqrt{-1}\,\mathrm{Trans}(\psi_j\circ\psi_i^{-1})\rangle).\end{aligned}$$

Thus

$$q^n\circ\phi_{ij}=\exp(2\pi\sqrt{-1}\langle n,b_{ij}+\sqrt{-1}\,\mathrm{Trans}(\psi_j\circ\psi_i^{-1})\rangle)q^{{}^t\mathrm{Lin}(\psi_j\circ\psi_i^{-1})(n)}.$$

So we see that $\alpha_j/\alpha_i=\exp(2\pi\sqrt{-1}(b_{ij}+\sqrt{-1}\,\mathrm{Trans}(\psi_j\circ\psi_i^{-1})))$ is a section of $\Lambda\otimes\mathbb{C}^\times$ over $U_i\cap U_j$. The result then follows from the description of the radiance obstruction in Construction 6.29 and Remark 6.30. □

It is a standard fact that the connecting homomorphisms for the long exact sequence of cohomology are given by cup product with the extension class. This gives:

PROPOSITION 6.60. *An element*

$$\bar{\alpha}\in H^1(B,\check{\Lambda})$$

lifts to an element

$$\alpha\in H^1(B,\mathcal{P}_{B,\mathbf{B}})$$

in the long exact sequence

$$H^1(B,\mathbb{C}^\times)\to H^1(B,\mathcal{P}_{B,\mathbf{B}})\to H^1(B,\check{\Lambda})\to H^2(B,\mathbb{C}^\times)$$

if and only if $\bar{\alpha}\cdot\exp(2\pi\sqrt{-1}(B+\sqrt{-1}[\delta]))=1$ *with respect to the natural cup product*

$$H^1(B,\check{\Lambda})\times H^1(B,\Lambda\otimes\mathbb{C}^\times)\to H^2(B,\mathbb{C}^\times).$$

EXAMPLE 6.61. Let $B=M_{\mathbb{R}}/\Gamma$ for a lattice $\Gamma\subseteq M_{\mathbb{R}}$, and let $\check{\Gamma}\subseteq N_{\mathbb{R}}$ be the dual lattice. Then Λ (and $\check{\Lambda}$) is the constant local system. The exact sequence

$$1\to\mathbb{C}^\times\to\mathcal{P}_{B,\mathbf{B}}\to\check{\Lambda}\to 0$$

gives a long exact cohomology sequence

$$1 \to \mathbb{C}^\times \to H^0(B, \mathcal{P}_{B,\mathbf{B}}) \to N \to \check{\Gamma} \otimes_{\mathbb{Z}} \mathbb{C}^\times \to H^1(B, \mathcal{P}_{B,\mathbf{B}})$$
$$\to \check{\Gamma} \otimes_{\mathbb{Z}} N \to \bigwedge^2 \check{\Gamma} \otimes_{\mathbb{Z}} \mathbb{C}^\times.$$

The connecting map $\check{\Gamma} \otimes_{\mathbb{Z}} N \to \bigwedge^2 \check{\Gamma} \otimes_{\mathbb{Z}} \mathbb{C}^\times$ is given by cup product with

$$\exp(2\pi\sqrt{-1}(\mathbf{B} + \sqrt{-1}[\delta])) \in H^1(B, \Lambda \otimes \mathbb{C}^\times) = \check{\Gamma} \otimes_{\mathbb{Z}} M \otimes_{\mathbb{Z}} \mathbb{C}^\times.$$

The B-field can take any value in $H^1(B, \Lambda_{\mathbb{R}}) = \check{\Gamma} \otimes_{\mathbb{Z}} M_{\mathbb{R}}$, and the radiance obstruction is given by the canonical inclusion $\Gamma \to M_{\mathbb{R}}$ as an element of $\mathrm{Hom}(\Gamma, M_{\mathbb{R}}) = \check{\Gamma} \otimes_{\mathbb{Z}} M_{\mathbb{R}}$. (Exercise!) So we see for a general choice of B-field, the map $H^1(B, \mathcal{P}_{B,\mathbf{B}}) \to H^1(B, \check{\Lambda})$ is in fact zero.

Note also that $H^0(B, \mathcal{P}_{B,\mathbf{B}}) = \mathbb{C}^\times$, since the only global holomorphic functions on the compact space $X(B, \mathbf{B})$ are constants. So we obtain in any event an exact sequence

$$1 \to \frac{\check{\Gamma} \otimes_{\mathbb{Z}} \mathbb{C}^\times}{N} \to H^1(B, \mathcal{P}_{B,\mathbf{B}}) \to \check{\Gamma} \otimes_{\mathbb{Z}} N$$

with the first map being an isomorphism for a general choice of $\mathbf{B} + \sqrt{-1}[\delta]$.

EXERCISE 6.62. Describe the quotient $(\check{\Gamma} \otimes_{\mathbb{Z}} \mathbb{C}^\times)/N$ as a complex manifold of the form $X(B', \mathbf{B}')$ for some affine torus B' and B-field $\mathbf{B}'$. Note in particular that the inclusion of N in $\check{\Gamma} \otimes_{\mathbb{Z}} \mathbb{C}^\times$ depends on the initial choice of data.

Putting this all together, we obtain the following classification of semi-flat B-branes:

THEOREM 6.63. *Suppose we are given a rational affine submanifold B' of B and a class $[\mathcal{L}] \in H^1(B', \check{\Lambda}')$, where $\check{\Lambda}'$ is the dual of Λ'. Then there exists a semi-flat B-brane $(Y, \mathcal{L})$ fibering over B' isomorphic to $X(B', \mathbf{B}')$ for some $\mathbf{B}' \in H^1(B', \Lambda'_{\mathbb{R}}/\Lambda')$ with line bundle $\mathcal{L} \in H^1(B', \mathcal{P}_{B',\mathbf{B}'})$ mapping to $[\mathcal{L}] \in H^1(B', \check{\Lambda}')$ if and only if*

(1) $\mathbf{B}'$ is a lift of $\mathbf{B}|_{B'}$ to $H^1(B', \Lambda'_{\mathbb{R}}/\Lambda')$ via the exact sequence (6.3).

(2)

$$[\mathcal{L}] \cdot \exp(2\pi\sqrt{-1}(\mathbf{B}' + \sqrt{-1}[\delta_{B'}])) = 1$$

under the natural cup product

$$H^1(B', \check{\Lambda}') \times H^1(B', \Lambda' \otimes \mathbb{C}^\times) \to H^2(B', \mathbb{C}^\times).$$

Here $[\delta_{B'}]$ denotes the class of the radiance obstruction on B'.

Furthermore, Y is uniquely specified by a Čech 0-cochain (g_i) for

$$\Lambda_{\mathbb{R}}|_{B'}/(\Lambda|_{B'} + \Lambda'_{\mathbb{R}})$$

with $(b_{ij}+\tilde{g}_i-\tilde{g}_j)_{ij}$ *representing* $\mathbf{B}'$ *for any choice of lifts* $\tilde{g}_i$ *of* g_i *to* $(\Lambda_{\mathbb{R}}/\Lambda)|_{B'}$. *The set of choices of* Y *for a given* $\mathbf{B}'$ *is a torsor over*

$$\operatorname{im}(H^0(B',(\Lambda_{\mathbb{R}}/\Lambda)|_{B'})\to H^0(B',\Lambda_{\mathbb{R}}|_{B'}/(\Lambda|_{B'}+\Lambda'_{\mathbb{R}}))).$$

Also, $\mathcal{L}$ *is uniquely determined up to tensoring with a line bundle in*

$$H^1(B',\mathbb{C}^\times)/H^0(B',\check{\Lambda}_{B'})\subseteq H^1(B',\mathcal{P}_{B',\mathbf{B}'}).$$

PROOF. This is a combination of Proposition 6.54, Exercise 6.55, and Proposition 6.60. □

We can construct a wider range of vector bundles as follows. Let $B'\to B$ be a covering map. Then the affine structure on B induces an affine structure on B', and pulling back the B-field $\mathbf{B}$ to $\mathbf{B}'\in H^1(B',\Lambda'_{\mathbb{R}}/\Lambda')$, we obtain $X(B',\mathbf{B}')$, and a natural map $\pi: X(B',\mathbf{B}')\to X(B,\mathbf{B})$, which is a map of complex manifolds. Given a line bundle $\mathcal{L}$ on $X(B',\mathbf{B}')$, obtained via the previous construction, then $\pi_*\mathcal{L}$ is a vector bundle on $X(B,\mathbf{B})$, of rank equal to the degree of the map π (the number of points in $\pi^{-1}(x)$ for any $x\in X(B,\mathbf{B})$). In addition, new vector bundles can be built as extensions of previously constructed vector bundles. In this manner, one can construct all vector bundles on elliptic curves: see for example [**398**]. However, we will stick to line bundles to keep the discussion in the next section reasonably simple.

6.3.2. A-branes. Next we consider A-branes. We will again consider a restricted class: firstly we only consider Lagrangian, rather than coisotropic, branes. Secondly, we will not insist on *special* Lagrangian branes. Indeed, in the previous section, we fixed the complex structure on $X(B,\mathbf{B})$ and considered holomorphic submanifolds with line bundles; we ignored that part of the condition on branes coming from the choice of complexified Kähler form. Likewise, on this side we will work with the symplectic manifold $\check{X}(B)$, along with a choice of B-field $\mathbf{B}$. For a pair $(L,\mathcal{L})$ where $L\subseteq\check{X}(B)$ is Lagrangian, and $\mathcal{L}$ is a $U(1)$-bundle on L with a connection $\nabla_{\mathcal{L}}$, we wish to know what the required condition for $(L,\mathcal{L})$ to be an A-brane. If there were no B-field, we would simply require the connection to be flat, i.e., the curvature two-form F of the connection would be zero. However, it follows from the discussion in §3.5.2.7 that the presence of the B-field gives the condition that in fact $\mathbf{B}|_L+F=0$. Because of slightly different conventions in this chapter, we actually require the curvature two-form F to satisfy $F=2\pi\sqrt{-1}\mathbf{B}|_L$. (This is a change of sign, and in this chapter we view the connection as being given by a connection one-form with values in the Lie algebra of $U(1)$.) Here we have represented the B-field $\mathbf{B}$ by a 2-form. Now if $F=2\pi\sqrt{-1}\mathbf{B}|_L$, then $\mathbf{B}|_L$ represents an integral cohomology class of L. This is then a necessary condition for a Lagrangian cycle L to support an A-brane structure. If $\mathbf{B}|_L$ is integral, then there is a $U(1)$-bundle with curvature $2\pi\sqrt{-1}\mathbf{B}|_L$, and modulo gauge equivalence, the set of such bundles is given by $H^1(L,\mathbb{R})/H^1(L,\mathbb{Z})$.

DEFINITION 6.64. Let $B' \subseteq B$ be a rational affine subspace. We define $L(B') \subseteq \check{X}(B)$ by

$$L(B') := (TB')^{\perp}/(\check{\Lambda}|_{B'} \cap (TB')^{\perp}),$$

where $(TB')^{\perp}$ is the subbundle of $(T^*B)|_{B'}$ which annihilates $TB' \subseteq TB|_{B'}$.

Given an open cover $\{U_i\}$ of B' and sections g_i of $T^*B|_{U_i}$ such that $g_j - g_i \in \Gamma(U_i \cap U_j, \check{\Lambda}|_{B'} + (TB')^{\perp})$, we define $L(B', \{(U_i, g_i)\}) \subseteq \check{X}(B)$ by

$$L(B', \{(U_i, g_i)\}) := \bigcup_i (L(U_i) + g_i).$$

The condition on $g_j - g_i$ guarantees that $L(U_i) + g_i$ and $L(U_j) + g_j$ agree on the overlap.

Note that the condition on $\{(U_i, g_i)\}$ tells us that this data in fact determines a section

$$\begin{aligned} g &\in \Gamma(B', T^*B|_{B'}/(\check{\Lambda}|_{B'} + (TB')^{\perp}) \\ &\cong \Gamma(B', T^*B'/\check{\Lambda}'), \end{aligned}$$

where $\check{\Lambda}'$ is the subsheaf of flat integral sections of T^*B'. Thus we can write

$$L(B', g) := L(B', \{(U_i, g_i)\}).$$

We obtain, via the exact sequence

$$0 \to \check{\Lambda}' \to T^*B' \to T^*B'/\check{\Lambda}' \to 0,$$

the *class* of g, which is the image $[g]$ of g in $H^1(B', \check{\Lambda}')$ under the connecting homomorphism. □

In general, there are too many choices of g with a given class $[g]$ for which $L(B', g)$ is Lagrangian. Lagrangian submanifolds are not sufficiently rigid for our purposes, and in general have infinite dimensional moduli spaces. The full definition of A-brane given in §1.1.4 rigidifies A-branes by insisting they be special Lagrangian. However, we have not specified a complex structure on $\check{X}(B)$, so here we will adopt a different approach, standard in Floer theory, by considering Lagrangian submanifolds up to Hamiltonian isotopy.

EXERCISE 6.65. Given $\check{f} : \check{X}(B) \to B$ and a C^∞-function φ on B, show that the time-one Hamiltonian flow associated to the Hamiltonian vector field H_φ, defined by the equation $\iota(H_\varphi)\omega = d(\varphi \circ \check{f})$, is given by fiberwise translation by the 1-form $d\varphi$.

This exercise motivates us to consider $L(B', g)$ and $L(B', g')$ to be *equivalent* if $g - g'$ is the image of an exact 1-form on B'.

By analogy with Definition 6.58, we define:

DEFINITION 6.66. A *semi-flat A-brane* on $\check{X}(B)$ is a triple $(L, \mathcal{L}, \nabla_{\mathcal{L}})$ where L is a Lagrangian submanifold of $\check{X}(B)$ fibering in linear tori over a rational affine submanifold B' of B, along with a $U(1)$-bundle $\mathcal{L}$ on L with connection $\nabla_{\mathcal{L}}$ whose curvature 2-form satisfies $\mathbf{F} = 2\pi\sqrt{-1}\mathbf{B}|_L$.

We say $(L_1, \mathcal{L}_1, \nabla_{\mathcal{L}_1})$ and $(L_2, \mathcal{L}_2, \nabla_{\mathcal{L}_2})$ are *equivalent* if there is an exact 1-form α on B' such that $T_\alpha(L_1) = L_2$, where T_α denotes fiberwise translation by α; $T_\alpha^*(\mathcal{L}_2) \cong \mathcal{L}_1$, and the pull-back of $\nabla_{\mathcal{L}_2}$ is gauge equivalent to $\nabla_{\mathcal{L}_1}$, i.e., differs from $\nabla_{\mathcal{L}_1}$ by the differential of a function $L_1 \to U(1)$.

We now wish to classify semi-flat A-branes up to equivalence.

LEMMA 6.67. *Let $B' \subseteq B$ be a rational affine subspace covered by coordinate charts $\{U_i\}$ of B, and suppose $y_1, \ldots, y_n$ are affine coordinates on U_i chosen so that B' is given by $y_{p+1} = \cdots = y_n = 0$. Then we have coordinates*

$$(y_1, \ldots, y_p, \check{x}_{p+1}, \ldots, \check{x}_n)$$

on $L(B', g)$ and coordinates

$$(y_1, \ldots, y_n, \check{x}_1, \ldots, \check{x}_n)$$

on $\check{X}(B)$. Let $g \in \Gamma(B', T^*B'/\check{\Lambda}')$ be locally represented as a section of $T^*B|_{B'}$, given in these coordinates by $g = \sum_{j=1}^n g^j dy_j$. Then we can parameterize $L(B', g)$ by

$$\begin{aligned}&(y_1, \ldots, y_p, \check{x}_{p+1}, \ldots, \check{x}_n)\\ &\qquad\mapsto (y_1, \ldots, y_p, 0, \ldots, 0, g^1, \ldots, g^p, \check{x}_{p+1} + g^{p+1}, \ldots, \check{x}_n + g^n).\end{aligned}$$

Suppose $\mathbf{B} = \sum_{i,j} b_{ij} dy_i \wedge d\check{x}_j$. Then

$$(\mathbf{B} + \sqrt{-1}\omega)|_{L(B',g)} = \sum_{i=1}^{p} \sum_{j=p+1}^{n} b_{ij} dy_i \wedge d\check{x}_j + \sum_{i=1}^{p}\sum_{j=1}^{n} (b_{ij} + \sqrt{-1}\delta_{ij}) dy_i \wedge dg^j.$$

PROOF. Pulling back

$$\mathbf{B} + \sqrt{-1}\omega = \sum_{i,j=1}^{n} (b_{ij} + \sqrt{-1}\delta_{ij}) dy_i \wedge d\check{x}_j$$

under the parameterization gives

$$\begin{aligned}&\sum_{i=1}^{p} \sum_{j=p+1}^{n} (b_{ij} + \sqrt{-1}\delta_{ij}) dy_i \wedge d\check{x}_j + \sum_{i=1}^{p}\sum_{j=1}^{n} (b_{ij} + \sqrt{-1}\delta_{ij}) dy_i \wedge dg^j\\ &= \sum_{i=1}^{p} \sum_{j=p+1}^{n} b_{ij} dy_i \wedge d\check{x}_j + \sum_{i=1}^{p}\sum_{j=1}^{n} (b_{ij} + \sqrt{-1}\delta_{ij}) dy_i \wedge dg^j\end{aligned}$$

as desired. ∎

REMARK 6.68. In what follows, we will talk about semi-flat differential forms on $L(B', g)$. As in §6.2.2, these are forms constant on fibers of $L(B', g)$. However, unlike in §6.2.2, we cannot split these forms into forms of type (p, q) globally. Indeed, in the notation of the lemma, a form on $L(B', g)$ locally of the form $dy_I \wedge d\check{x}_J$ for $I \subseteq \{1, \ldots, p\}$, $J \subseteq \{p+1, \ldots, n\}$ changes appearance with a change of coordinates and a choice of representative for g: there might be additional terms with fewer $d\check{x}_i$'s and more dy_j's. Thus we can speak of forms of type (p, q) modulo forms of type $(p+1, q-1) + (p+2, q-2) + \cdots$. For example, the calculation of the Lemma shows that $\mathbf{B}|_{L(B',g)}$ is a form of type $(1, 1)$ modulo forms of type $(2, 0)$, the $(2, 0)$ part not being well-defined.

One other point which may be confusing in what follows is that we will be talking about two different types of restrictions of $\mathbf{B}$. We can view $\mathbf{B}$ as a 2-form on $\check{X}(B)$, in which case we can restrict it to submanifolds to obtain a 2-form on a submanifold. Or $\mathbf{B}$ can be viewed as an element of $H^1(B, \Lambda_{\mathbb{R}})$, and as such, can be restricted to B' to obtain a class in $H^1(B', \Lambda_{\mathbb{R}}|_{B'})$. We write the former restriction as $\mathbf{B}|_M$ for a submanifold $M \subseteq \check{X}(B)$, and the latter restriction by $\mathbf{B}|_{B'}$.

PROPOSITION 6.69. *Given a rational affine manifold $B' \subseteq B$ and a class*

$$[g] \in H^1(B', \check{\Lambda}'),$$

there exists a representative $g \in H^0(B', T^*B'/\check{\Lambda}')$ such that $L(B', g)$ is Lagrangian if and only if the image of $[g]$ in $H^2(B', \mathbb{R})$ is zero under the connecting homomorphism in the long exact sequence associated to

$$0 \longrightarrow \mathbb{R} \longrightarrow \mathrm{Aff}_{\mathbb{Z}}(B', \mathbb{R}) \xrightarrow{d} \check{\Lambda}' \longrightarrow 0,$$

where $\mathrm{Aff}_{\mathbb{Z}}(B', \mathbb{R})$ denotes the sheaf of affine linear functions on B' with integral slope and d is exterior derivative. Furthermore, the set of choices of Lagrangians of the form $L(B', g)$ modulo equivalence is equal to the set of all liftings of $[g]$ to $H^1(B', \mathrm{Aff}_{\mathbb{Z}}(B', \mathbb{R}))$.

PROOF. Locally, writing $g = \sum_{j=1}^n g^j dy_j$, we see from the Lemma that $L(B', g)$ is Lagrangian if and only if $\sum_{j=1}^p dg^j \wedge dy_j = 0$, or equivalently, $dg = 0$ if we interpret g as a section of T^*B'. Now writing $(T^*B')_{\text{closed}}$ for the subsheaf of T^*B' consisting of closed 1-forms, we see that $L(B', g)$ is Lagrangian if and only if $g \in H^0(B', T^*B'/\check{\Lambda}')$ is a section of $H^0(B', (T^*B')_{\text{closed}}/\check{\Lambda}')$. There is an exact sequence

$$0 \to \check{\Lambda}' \to (T^*B')_{\text{closed}} \to (T^*B')_{\text{closed}}/\check{\Lambda}' \to 0,$$

yielding a long exact sequence

$$H^0(B', (T^*B')_{\text{closed}}/\check{\Lambda}') \to H^1(B', \check{\Lambda}') \to H^1(B', (T^*B')_{\text{closed}})$$

so $[g] \in H^1(B', \check{\Lambda}')$ can be lifted to $g \in H^0(B', (T^*B')_{\text{closed}}/\check{\Lambda}')$ if and only if the image of $[g]$ in $H^1(B', (T^*B')_{\text{closed}})$ is zero. However, there is an exact sequence

$$0 \longrightarrow \mathbb{R} \longrightarrow C^\infty(B') \xrightarrow{d} (T^*B')_{\text{closed}} \longrightarrow 0$$

(where $C^\infty(B')$ denotes the sheaf of C^∞ functions on B'), showing

$$H^1(B', (T^*B')_{\text{closed}}) \cong H^2(B', \mathbb{R}).$$

Note we have a commutative diagram of exact sequences

$$\begin{array}{ccccccccc}
 & & & & 0 & & 0 & & \\
 & & & & \downarrow & & \downarrow & & \\
0 & \longrightarrow & \mathbb{R} & \longrightarrow & \operatorname{Aff}_{\mathbb{Z}}(B', \mathbb{R}) & \xrightarrow{d} & \check{\Lambda}' & \longrightarrow & 0 \\
 & & \downarrow{=} & & \downarrow & & \downarrow & & \\
0 & \longrightarrow & \mathbb{R} & \longrightarrow & C^\infty(B') & \xrightarrow{d} & (T^*B')_{\text{closed}} & \longrightarrow & 0 \\
 & & \downarrow & & \downarrow & & \downarrow & & \\
 & & 0 & \longrightarrow & C^\infty(B') / \operatorname{Aff}_{\mathbb{Z}}(B', \mathbb{R}) & \xrightarrow{\cong} & (T^*B')_{\text{closed}} / \check{\Lambda}' & \longrightarrow & 0 \\
 & & & & \downarrow & & \downarrow & & \\
 & & & & 0 & & 0 & &
\end{array}$$

Thus $[g]$ lifts to a section g yielding a Lagrangian $L(B', g)$ if and only if the image of $[g]$ in $H^2(B', \mathbb{R})$ is zero under the connecting homomorphism of the exact sequence in the first row. Furthermore, as $H^1(B', C^\infty(B')) = 0$, if $[g]$ lifts to $\bar{g} \in H^1(B', \operatorname{Aff}_{\mathbb{Z}}(B', \mathbb{R}))$, it in turn lifts to an element

$$g \in H^0(B', C^\infty(B') / \operatorname{Aff}_{\mathbb{Z}}(B', \mathbb{R})) \cong H^0(B', (T^*B')_{\text{closed}} / \check{\Lambda}'),$$

and any such lifting makes $L(B', g)$ Lagrangian. Two such liftings g, g' are equivalent if $g - g'$ comes from $H^0(B', C^\infty(B'))$, and so we see that the lifting $\bar{g}$ of $[g]$ specifies an equivalence class of Lagrangians $L(B', g)$. $\square$

COROLLARY 6.70. *Given a rational affine submanifold $B' \subseteq B$ and a class*

$$[g] \in H^1(B', \check{\Lambda}'),$$

there exists a representative g of $[g]$ such that $L(B', g)$ is Lagrangian if and only if the cup product satisfies

$$[g] \cdot [\delta_{B'}] = 0$$

under the natural cup product

$$H^1(B', \check{\Lambda}') \times H^1(B', \Lambda'_{\mathbb{R}}) \to H^2(B', \mathbb{R})$$

where $[\delta_{B'}] \in H^1(B', \Lambda'_{\mathbb{R}})$ is the radiance obstruction of B'.

PROOF. One can show, as in the proof of Proposition 6.59, that the extension class of the exact sequence

$$0 \to \mathbb{R} \to \mathrm{Aff}_{\mathbb{Z}}(B', \mathbb{R}) \to \check{\Lambda}' \to 0$$

is the radiance obstruction $[\delta_{B'}]$ of B'. From this one sees that the image of $[g] \in H^1(B', \check{\Lambda}')$ in $H^2(B', \mathbb{R})$ is precisely the cup product of $[g]$ with the radiance obstruction $[\delta_{B'}]$, exactly as in the proof of Proposition 6.60. (See for example [**206**], Proposition 1.12.) $\square$

To understand the existence of a $U(1)$-bundle with the correct properties, observe that we need the restriction of the 2-form $\mathbf{B}$ to $L = L(B', g)$ to represent an integral class in $H^2(L, \mathbb{Z})$. Once this is the case, there is a suitable $U(1)$-bundle with connection on L. Given one such choice, the set of all such $U(1)$-bundles, modulo gauge equivalence, satisfying the curvature condition is parameterized by $H^1(L, \mathbb{R}/\mathbb{Z})$.

It is somewhat difficult, given the technology we have at hand, to analyze this integrality condition, so instead let us understand the simpler condition that $\mathbf{B}|_{L(B',g)}$ represents $0 \in H^2(L(B', g), \mathbb{R})$, i.e., there is a globally defined 1-form $\mathbf{A}$ on $L(B', g)$ with $d\mathbf{A} = \mathbf{B}|_{L(B',g)}$. A choice of such an $\mathbf{A}$ of course can be viewed as a connection 1-form on $L(B', g)$, hence a connection with curvature $2\pi\sqrt{-1}\mathbf{B}|_{L(B',g)}$. We wish to classify such $\mathbf{A}$'s, up to exact forms that just represent a change of gauge.

THEOREM 6.71. *$\mathbf{B}|_{L(B',g)}$ is exact on $L(B', g)$ if and only if there exists a lift of $\mathbf{B}|_{B'} \in H^1(B', \Lambda_{\mathbb{R}}|_{B'})$ to $\mathbf{B}' \in H^1(B', \Lambda'_{\mathbb{R}})$ in the exact sequence*

$$H^1(B', \Lambda'_{\mathbb{R}}) \to H^1(B', \Lambda_{\mathbb{R}}|_{B'}) \to H^1(B', \Lambda_{\mathbb{R}}|_{B'}/\Lambda'_{\mathbb{R}})$$

such that

$$[g] \cdot [\mathbf{B}'] = 0$$

under the natural cup product

$$H^1(B', \check{\Lambda}') \times H^1(B', \Lambda'_{\mathbb{R}}) \to H^2(B', \mathbb{R}).$$

Furthermore, if there is such a $\mathbf{B}'$, there is a one-to-one correspondence between 1-forms $\mathbf{A}$ on $L(B', g)$, modulo exact forms, with $d\mathbf{A} = \mathbf{B}|_{L(B',g)}$, and the data

(1) *A choice of lift $\mathbf{B}'$ of $\mathbf{B}|_{B'}$ and $\alpha \in \Gamma(B', C^\infty(B') \otimes (\Lambda_{\mathbb{R}}|_{B'}/\Lambda'_{\mathbb{R}}))$ such that for any lift $\tilde{\alpha} \in \Gamma(B', C^\infty(B') \otimes \Lambda_{\mathbb{R}}|_{B'})$ of α, $\mathbf{B}|_{B'} - \nabla\tilde{\alpha}$ represents $\mathbf{B}'$. Here ∇ is the natural connection on $C^\infty(B') \otimes \Lambda_{\mathbb{R}}|_{B'}$ whose flat sections are sections of $\Lambda_{\mathbb{R}}|_{B'}$. For a given $\mathbf{B}'$ the set of such α is a torsor over*

$$\mathrm{im}(H^0(B', \Lambda_{\mathbb{R}}|_{B'}) \to H^0(B', \Lambda_{\mathbb{R}}|_{B'}/\Lambda'_{\mathbb{R}})).$$

(2) A choice of lift of $[g] \in H^1(B', \breve{\Lambda}')$ *to* $H^1(B', \mathcal{P}^{\mathbb{R}}_{B',\mathbf{B}'})$*, where* $\mathcal{P}^{\mathbb{R}}_{B',\mathbf{B}'}$ *is the sheaf defined as the extension*

$$0 \to \mathbb{R} \to \mathcal{P}^{\mathbb{R}}_{B',\mathbf{B}'} \to \breve{\Lambda}' \to 0$$

with extension class $\mathbf{B}' \in H^1(B', \Lambda'_{\mathbb{R}})$*.*

PROOF. We saw from Lemma 6.67 that $\mathbf{B}|_{L(B',g)}$ can be locally written as a sum of semi-flat differential forms of type $(1,1)$ and $(2,0)$. The part of type $(2,0)$ is not well-defined, as a change in representative for g (replacing g with another section of $T^*B|_{B'}$ congruent to g modulo $\breve{\Lambda}|_{B'} + (TB')^{\perp}$) will change the part of type $(2,0)$. Nevertheless, the $(1,1)$ part is independent of choices, and gives a well-defined element β of $H^1(B', \Lambda_{\mathbb{R}}|_{B'}/\Lambda'_{\mathbb{R}})$, as in §6.2.2 associated to the local system $\Lambda_{\mathbb{R}}|_{B'}/\Lambda'_{\mathbb{R}}$. Note that β is the image of $\mathbf{B}|_{B'} \in H^1(B', \Lambda_{\mathbb{R}}|_{B'})$ under the natural map

$$H^1(B', \Lambda_{\mathbb{R}}|_{B'}) \to H^1(B', \Lambda_{\mathbb{R}}|_{B'}/\Lambda'_{\mathbb{R}}).$$

Now from the exact sequence

$$0 \to \Lambda'_{\mathbb{R}} \to \Lambda_{\mathbb{R}}|_{B'} \to \Lambda_{\mathbb{R}}|_{B'}/\Lambda'_{\mathbb{R}} \to 0,$$

we see that $\mathbf{B}|_{B'}$ lifts to an element $\mathbf{B}' \in H^1(B', \Lambda'_{\mathbb{R}})$ if and only if $\beta \in H^1(B', \Lambda_{\mathbb{R}}|_{B'}/\Lambda'_{\mathbb{R}})$ is zero, which in turn is equivalent to the existence of a section

$$\alpha \in \Gamma(B', C^{\infty}(B') \otimes (\Lambda_{\mathbb{R}}|_{B'}/\Lambda'_{\mathbb{R}}))$$

with $\nabla\alpha = \beta$, where

$$\nabla : C^{\infty}(B') \otimes (\Lambda_{\mathbb{R}}|_{B'}/\Lambda'_{\mathbb{R}}) \to T^*B' \otimes (\Lambda_{\mathbb{R}}|_{B'}/\Lambda'_{\mathbb{R}})$$

is the connection defined in §6.2.2. Note that if $\mathbf{B}|_{L(B',g)} = d\mathbf{A}$, then $\mathbf{A}$ is cohomologous to a semi-flat form of type $(0,1)$ plus $(1,0)$, and the $(0,1)$ part determines an α as above with $\nabla\alpha = \beta$. Thus if $\mathbf{B}|_{L(B',g)}$ is exact, $\mathbf{B}|_{B'}$ can be lifted to $H^1(B', \Lambda'_{\mathbb{R}})$.

In any event, a lifting $\mathbf{B}'$ is determined by α, by choosing any lifting of

$$\alpha \in \Gamma(B', C^{\infty}(B') \otimes (\Lambda_{\mathbb{R}}|_{B'}/\Lambda'_{\mathbb{R}}))$$

to a section $\tilde{\alpha} \in \Gamma(B', C^{\infty}(B') \otimes \Lambda_{\mathbb{R}}|_{B'})$, and then $\mathbf{B}' = \mathbf{B}|_{B'} - \nabla\tilde{\alpha}$ defines a section of $\Lambda'_{\mathbb{R}} \otimes T^*B'$, determining an element $\mathbf{B}' \in H^1(B', \Lambda'_{\mathbb{R}})$. Note that any two lifts $\tilde{\alpha}_1, \tilde{\alpha}_2$ of α satisfy $\tilde{\alpha}_1 - \tilde{\alpha}_2 \in \Gamma(B', C^{\infty}(B') \otimes \Lambda'_{\mathbb{R}})$ so $\nabla(\tilde{\alpha}_1 - \tilde{\alpha}_2)$ represents the zero cohomology class in $H^1(B', \Lambda'_{\mathbb{R}})$. Thus the lifts $\mathbf{B}'_1$ and $\mathbf{B}'_2$ of $\mathbf{B}|_{B'}$ determined by $\tilde{\alpha}_1$ and $\tilde{\alpha}_2$ are cohomologous, so the cohomology class of $\mathbf{B}'$ is completely determined by α. On the other hand, changing α by adding an

$$\alpha' \in \Gamma(B', C^{\infty}(B') \otimes (\Lambda_{\mathbb{R}}|_{B'}/\Lambda'_{\mathbb{R}}))$$

with $\nabla\alpha' = 0$ changes the class of $\mathbf{B}'$ by the image of $\alpha' \in H^0(B', \Lambda_{\mathbb{R}}|_{B'}/\Lambda'_{\mathbb{R}})$ in $H^1(B', \Lambda'_{\mathbb{R}})$.

Explicitly, in local coordinates, if $\tilde{\alpha} = \sum_{j=1}^n \tilde{\alpha}_j d\check{x}_j$,

$$\mathbf{B}' = \sum_{i=1}^{p}\sum_{j=1}^{n}\left(b_{ij} - \frac{\partial\tilde{\alpha}_j}{\partial y_i}\right) dy_i \wedge d\check{x}_j = \sum_{i,j=1}^{p}(b_{ij} - \frac{\partial\tilde{\alpha}_j}{\partial y_i})dy_i \wedge d\check{x}_j, \tag{6.4}$$

(the latter equality follows from the assumption that $\mathbf{B}|_{B'} - \nabla\tilde{\alpha}$ is a section of $\Lambda'_{\mathbb{R}} \otimes T^*B'$). Now $\tilde{\alpha}$ can be viewed as a semi-flat $(0,1)$-form on $\check{f}^{-1}(B')$ and then $\mathbf{B}'$ can be viewed as a semi-flat form on $\check{f}^{-1}(B')$, namely the form $\mathbf{B}|_{\check{f}^{-1}(B')} - d\tilde{\alpha}$, which is of course homologous to $\mathbf{B}|_{\check{f}^{-1}(B')}$. Restricting this further to $L(B',g)$ gives, by the Lemma, with g locally representable by $\sum_{j=1}^p g^j dy_j$, the semi-flat 2-form of type $(2,0)$

$$\mathbf{B}|_{L(B',g)} - d\tilde{\alpha}|_{L(B',g)} = \sum_{i,j=1}^{p}(b_{ij} - \frac{\partial\tilde{\alpha}_j}{\partial y_i})dy_i \wedge dg^j,$$

which is now globally well-defined on B'. If this form is exact, i.e., $\mathbf{B}|_{L(B',g)} - d\tilde{\alpha}|_{L(B',g)} = d\tilde{\alpha}'$, then we see that $\mathbf{A} = \tilde{\alpha}|_{L(B',g)} + \tilde{\alpha}'$ satisfies $\mathbf{B}|_{L(B',g)} = d\mathbf{A}$. Conversely, if there is such an $\mathbf{A}$, then clearly $\mathbf{B}|_{L(B',g)} - d\tilde{\alpha}|_{L(B',g)}$ is exact.

Now let's understand the cohomology class of $\mathbf{B}|_{L(B',g)} - d\tilde{\alpha}|_{L(B',g)}$. The class $[g] \in H^1(B', \check{\Lambda}')$ has a de Rham representation as follows. Locally g is represented as a section of $T^*B' = C^\infty(B') \otimes \check{\Lambda}'$, $g = \sum_{i=1}^p g^i dy_i = \sum_{i=1}^p g^i \partial/\partial\check{x}_i$ (thinking of elements of $\check{\Lambda}'_{\mathbb{R},b}$ as tangent vectors to the fibers of $\check{X}(B') \to B'$). Two representations g, g' differ by $(g')^i = g^i + n_i$, for $n_i \in \mathbb{Z}$. Applying the natural connection (see §6.2.2) $\nabla : C^\infty(B') \otimes \check{\Lambda}' \to T^*B' \otimes \check{\Lambda}'$ locally yields $\nabla g = \sum_{i=1}^p dg^i \otimes \partial/\partial\check{x}_i$, and this no longer depends on the choice of representative, yielding a section of $T^*B' \otimes \check{\Lambda}'$ over B'. Then $[g]$ is easily seen to be represented by this section of $T^*B' \otimes \check{\Lambda}'$. The cup product

$$H^1(B', \check{\Lambda}') \times H^1(B', \Lambda'_{\mathbb{R}}) \to H^2(B', \mathbb{R})$$

is locally represented by

$$\left(\sum_{i,j} \alpha_{ij} dy_i \otimes \frac{\partial}{\partial\check{x}_j}\right) \cdot \left(\sum_{k,l} \beta_{kl} dy_k \wedge d\check{x}_l\right) \mapsto \sum_{i,j,k,l} \alpha_{ij}\beta_{kl}\delta_{jl} dy_i \wedge dy_k.$$

So the cup product of $[g]$ with $\mathbf{B}'$ is locally represented by

$$\sum_{i,j=1}^{p}(b_{ij} - \frac{\partial\tilde{\alpha}_j}{\partial y_i})dg^j \wedge dy_i,$$

precisely the form $-\mathbf{B}|_{L(U,g)} + d\tilde{\alpha}|_{L(U,g)}$. This shows that $\mathbf{B}|_{L(B',g)}$ is exact if and only if the image of $\mathbf{B}|_{B'}$ in $H^1(B', \Lambda_{\mathbb{R}}|_{B'}/\Lambda'_{\mathbb{R}})$ is zero, and the cup product $[g] \cdot \mathbf{B}' = 0$ in $H^2(B', \mathbb{R})$ for some lift $\mathbf{B}'$ of $\mathbf{B}|_{B'}$.

There only remains the description of 1-forms $\mathbf{A}$ on $L(B',g)$, modulo exact forms, with $d\mathbf{A} = \mathbf{B}|_{L(B',g)}$. It is not difficult to see that if $d\mathbf{A} =$

$\mathbf{B}|_{L(B',g)}$, then because $\mathbf{B}$ is semi-flat, $\mathbf{A}$ is cohomologous to a semi-flat 1-form on $L(B',g)$. So we can assume that $\mathbf{A}$ is semi-flat.

First, as we saw, choosing the $(0,1)$ part of $\mathbf{A}$ is equivalent to choosing an

$$\alpha \in \Gamma(B', C^\infty(B') \otimes (\Lambda_{\mathbb{R}}|_{B'}/\Lambda'_{\mathbb{R}}))$$

with $\nabla\alpha = \beta$. The section α can be viewed as specifying a semi-flat form of type $(0,1)$ modulo forms of type $(1,0)$ on $L(B',g)$, with $d\alpha = \mathbf{B}|_{L(B',g)}$ modulo semi-flat forms of type $(2,0)$. Choosing a lift $\tilde{\alpha}$ of α is just choosing a particular representative for the form, and in particular gives the choice of $\mathbf{B}'$.

Given the choice of $\tilde{\alpha}$, we now wish to show that the choice of semi-flat $\tilde{\alpha}'$, up to exact forms, with

$$d\tilde{\alpha}' = \mathbf{B}|_{L(B',g)} - d\tilde{\alpha}|_{L(B',g)}, \tag{6.5}$$

is in one-to-one correspondence with liftings of $[g]$ to $H^1(B', \mathcal{P}^{\mathbb{R}}_{B',\mathbf{B}'})$. Indeed, the extension $\mathcal{P}^{\mathbb{R}}_{B',\mathbf{B}'}$ can be expressed as follows. The vector bundle $C^\infty(B') \otimes \mathcal{P}^{\mathbb{R}}_{B',\mathbf{B}'}$ necessarily splits as $C^\infty(B') \oplus (C^\infty(B') \otimes \check{\Lambda}')$, carrying a flat connection $\nabla_{\mathcal{P}}$ defined by

$$\nabla_{\mathcal{P}}(f,\beta) = (df - \mathbf{B}'(\beta), \nabla_{\check{\Lambda}'}(\beta)).$$

Here $\nabla_{\check{\Lambda}'}$ is the flat connection on $C^\infty(B') \otimes \check{\Lambda}'$ whose flat sections are the sections of $\check{\Lambda}'_{\mathbb{R}}$, and $\mathbf{B}' \in T^*B' \otimes \Lambda'_{\mathbb{R}} = \mathrm{Hom}_{C^\infty(B')}(\check{\Lambda}' \otimes C^\infty(B'), T^*B')$ evaluates on a section of $C^\infty(B') \otimes \check{\Lambda}'$ to give a 1-form on B'. It is an elementary exercise to check that if we define $\mathcal{P}^{\mathbb{R}}_{B',\mathbf{B}'}$ to be the sheaf of those sections of $C^\infty(B') \oplus (C^\infty(B') \otimes \check{\Lambda}')$ which are flat with respect to this connection, and with projection to the second component being integral flat sections of $\check{\Lambda}' \otimes C^\infty(B')$, then the extension class of $\mathcal{P}^{\mathbb{R}}_{B',\mathbf{B}'}$ is indeed $\mathbf{B}'$. This can be done using the Čech representation of the extension class used in Proposition 6.59.

Now we are given the class $[g] \in H^1(B', \check{\Lambda}')$, represented by $\tilde{g} \in T^*B' \otimes \check{\Lambda}'$ given locally by $\tilde{g} = \sum_{i=1}^p dg^i \otimes \partial/\partial\check{x}_i$, with $\nabla_{\check{\Lambda}'}\tilde{g} = 0$. Giving a lifting of this to an element of $H^1(B', \mathcal{P}^{\mathbb{R}}_{B',\mathbf{B}'})$ means giving a pair $(\tilde{\alpha}', \tilde{g}) \in T^*B' \oplus (T^*B' \otimes \check{\Lambda}')$ with

$$0 = \nabla_{\mathcal{P}}(\tilde{\alpha}', \tilde{g}) = (d\tilde{\alpha}' - \mathbf{B}'(\tilde{g}), \nabla_{\check{\Lambda}'}\tilde{g}) = (d\tilde{\alpha}' - \mathbf{B}'(\tilde{g}), 0).$$

But locally

$$\mathbf{B}'(\tilde{g}) = \sum_{i,j=1}^{p} (b_{ij} - \frac{\partial\tilde{\alpha}_j}{\partial y_i}) dy_i \wedge dg^j,$$

so $\nabla_{\mathcal{P}}(\tilde{\alpha}', g) = 0$ is precisely the statement (6.5). So a choice of lifting is equivalent to a choice of $\tilde{\alpha}'$, modulo exact 1-forms. This completes the result. $\square$

We will now state, without proof, the full-strength result we need to compare semi-flat A-branes with the semi-flat B-branes of the previous section. We need to understand when $\mathbf{B}|_{L(B',g)}$ is integral rather than just exact. We omit a proof since an argument requires a deeper understanding of Čech representatives for cohomology classes than we wish to go into here.

THEOREM 6.72. *$\mathbf{B}|_{L(B',g)}$ is integral on $L(B',g)$ if and only if there exists a lift of $\mathbf{B}|_{B'} \in H^1(B', (\Lambda_{\mathbb{R}}/\Lambda)|_{B'})$ to $\mathbf{B}' \in H^1(B', \Lambda'_{\mathbb{R}}/\Lambda')$ in the exact sequence*

$$H^1(B', \Lambda'_{\mathbb{R}}/\Lambda') \to H^1(B', (\Lambda_{\mathbb{R}}/\Lambda)|_{B'}) \to H^1(B', \Lambda_{\mathbb{R}}|_{B'}/(\Lambda|_{B'} + \Lambda'_{\mathbb{R}}))$$

such that

$$[g] \cdot [\mathbf{B}'] = 0$$

under the natural cup product

$$H^1(B', \check{\Lambda}') \times H^1(B', \Lambda'_{\mathbb{R}}/\Lambda') \to H^2(B', \mathbb{R}/\mathbb{Z}).$$

Furthermore, if $\mathbf{B}|_{L(B',g)}$ is integral, and $\mathcal{L}$ is a $U(1)$-bundle on $L(B',g)$ with first Chern class $\mathbf{B}|_{L(B',g)}$, there is a one-to-one correspondence between semi-flat connections on $\mathcal{L}$ with curvature $2\pi\sqrt{-1}\mathbf{B}|_{L(B',g)}$, modulo gauge equivalence, and the data

(1) A choice of lift $\mathbf{B}'$ of $\mathbf{B}|_{B'}$ and a Čech 0-cochain (g_i) for

$$\Lambda_{\mathbb{R}}|_{B'}/(\Lambda|_{B'} + \Lambda'_{\mathbb{R}})$$

with $(b_{ij} + \tilde{g}_i - \tilde{g}_j)_{ij}$ representing $\mathbf{B}'$ for any choice of lifts $\tilde{g}_i$ of g_i to $(\Lambda_{\mathbb{R}}/\Lambda)|_{B'}$. The set of choices of (g_i) for a given $\mathbf{B}'$ is a torsor over

$$\operatorname{im}\left(H^0(B', (\Lambda_{\mathbb{R}}/\Lambda)|_{B'}) \to H^0(B', \Lambda_{\mathbb{R}}|_{B'}/(\Lambda|_{B'} + \Lambda'_{\mathbb{R}}))\right).$$

(2) A choice of lift of $[g] \in H^1(B, \check{\Lambda}')$ to $H^1(B', \mathcal{P}^{\mathbb{R}/\mathbb{Z}}_{B',\mathbf{B}'})$, where $\mathcal{P}^{\mathbb{R}/\mathbb{Z}}_{B',\mathbf{B}'}$ is the sheaf defined as the extension

$$0 \to \mathbb{R}/\mathbb{Z} \to \mathcal{P}^{\mathbb{R}/\mathbb{Z}}_{B',\mathbf{B}'} \to \check{\Lambda}' \to 0$$

with extension class $\mathbf{B}' \in H^1(B', \Lambda'_{\mathbb{R}}/\Lambda')$.

Comparing this result with that of Theorem 6.63, one sees that the moduli spaces of semi-flat A-branes and B-branes coincide. There is one subtlety we should remark on: there is no canonical isomorphism. Indeed, to obtain this isomorphism we are choosing non-canonical identifications of torsors. The set of line bundles with a given Chern class is a *torsor* over the group of line bundles with Chern class zero: i.e. the group of line bundles of Chern class zero acts freely and transitively on the set of line bundles of a given Chern class via tensor product, but there is no canonical bijection between line bundles of Chern class zero and line bundles of a given Chern class. Such an identification is achieved after choosing some line bundle of the given Chern class.

Similarly, the group $H^1(X,\mathbb{R}/\mathbb{Z})$ acts on the space of $U(1)$ connections modulo gauge equivalence on a line bundle $\mathcal{L}$ on X, but the space of such connections is not canonically isomorphic to $H^1(X,\mathbb{R}/\mathbb{Z})$. One obtains an isomorphism only after choosing one such connection. Again the space of such connections is a torsor over $H^1(X,\mathbb{R}/\mathbb{Z})$. This noncanonicity on both sides means that as phrased here, we have not exhibited canonical isomorphisms between moduli of semi-flat A- and B-branes.

EXAMPLE 6.73. Consider the case when $B=\mathbb{R}/\tau_2\mathbb{Z}$, $\mathbf{B}=\tau_1 dx\wedge dy$, so that $X(B,\mathbf{B})\cong\mathbb{C}/\langle 1,\tau\rangle$, an elliptic curve with periods 1 and $\tau=\tau_1+\sqrt{-1}\tau_2$, and take $B'=B$. Then the set of semi-flat B-branes fibering over B' is just the Picard group of $X(B,\mathbf{B})$.

On the other side, $\check{X}(B)$ is just a symplectic torus, and any section of the S^1 fibration $\check{X}(B)\to B$ is Lagrangian. We can choose a class $[g]\in H^1(B,\check{\Lambda})=\mathbb{Z}$. Thinking of $\check{X}(B)=\mathbb{R}^2/\langle(1,0),(0,\tau_2)\rangle$, any section is Hamiltonian isotopic to a section given by a line $x=[g]y/\tau_2+b$ of slope $[g]/\tau_2$, where $[g]$ is viewed as an integer, and $b\in\mathbb{R}$ is defined modulo $\mathbb{Z}$. Note that from the exact sequence of Proposition 6.69, $H^1(B,\mathrm{Aff}_{\mathbb{Z}}(B,\mathbb{R}))$ fits into an exact sequence

$$0\to\mathbb{R}/\mathbb{Z}\to H^1(B,\mathrm{Aff}_{\mathbb{Z}}(B,\mathbb{R}))\to\mathbb{Z}\to 0.$$

In fact, the pair $(b,[g])$ defines an element in $H^1(B,\mathrm{Aff}_{\mathbb{Z}}(B,\mathbb{R}))$. Of course, in this case $\mathbf{B}|_{L(B,g)}=0$, and the space of flat $U(1)$-bundles modulo gauge equivalence on $L(B,g)$ is just $H^1(L(B,g),\mathbb{R}/\mathbb{Z})=\mathbb{R}/\mathbb{Z}$. Thus, in the correspondence between A- and B-branes, line bundles of degree d on $X(B,\mathbf{B})$ correspond to lines of slope d/τ_2 on $\check{X}(B)$, and the isomorphism class of the B-brane then determines the translational part of $L(B,g)$ and the connection on $L(B,g)$.

6.4. Compactifications

6.4.1. Affine manifolds with singularities. While the semi-flat picture described in the previous sections may be simple and gives an elegant explanation for many aspects of mirror symmetry, it does not actually give a particularly satisfying range of examples, essentially just giving mirror symmetry for complex tori. The reason, of course, is that torus bundles have a rather limited topology; in particular, their topological Euler characteristic is always zero. Hence we can't explain mirror symmetry even for K3 surfaces using the semi-flat picture. To get around this, we need to include the possibility of singular fibers.

To explore this, we will start with some more interesting examples of affine manifolds than previously discussed.

Let $M=\mathbb{Z}^n$, $M_{\mathbb{R}}=M\otimes_{\mathbb{Z}}\mathbb{R}$, $N=\mathrm{Hom}(M,\mathbb{Z})$, $M_{\mathbb{R}}=N\otimes_{\mathbb{Z}}\mathbb{R}$ as usual. Let $\Xi\subseteq M_{\mathbb{R}}$ be any n-dimensional convex polytope containing $0\in M_{\mathbb{R}}$ in its interior. Choose further a polyhedral decomposition, $\mathcal{P}$ of Ξ, by which we mean the following: $\mathcal{P}$ is a collection of closed polytopes contained in Ξ covering Ξ such that

- if $\sigma \in \mathcal{P}$, every face of σ is in $\mathcal{P}$;
- if $\sigma_1, \sigma_2 \in \mathcal{P}$, then $\sigma_1 \cap \sigma_2$ is a face of σ_1 and σ_2.

Recall that if $\sigma \subset M_{\mathbb{R}}$ is a polytope, then the barycenter $\mathrm{Bar}(\sigma)$ of σ is the average of the vertices of σ. The first barycentric subdivision of σ is then the triangulation of σ consisting of all simplices spanned by barycenters of ascending chains of faces of σ. Thus, given a polyhedral decomposition $\mathcal{P}$, we can define the first barycentric subdivision $\mathrm{Bar}(\mathcal{P})$ of $\mathcal{P}$ to be the triangulation consisting of all simplices in the first barycentric subdivisions of all $\sigma \in \mathcal{P}$.

Now put $B = \partial\Xi$, so that B is an $(n-1)$-dimensional sphere, and define $\Delta \subseteq B$ to be the union of all simplices of $\mathrm{Bar}(\mathcal{P})$ contained in $(n-2)$-dimensional faces of Ξ and which do not contain a vertex of $\mathcal{P}$.

EXAMPLE 6.74. (1) Take $\Xi \subseteq \mathbb{R}^3$ to be the convex hull of

$$P_0 = (-1,-1,-1),\ P_1 = (3,-1,-1),\ P_2 = (-1,3,-1),\ P_3 = (-1,-1,3).$$

We will not choose $\mathcal{P}$ itself, but only dictate what the restriction of $\mathcal{P}$ to each dimension $n-2$ face (in this case edge) is. In other words, we describe for each $n-2$ face σ the decomposition

$$\mathcal{P}_\sigma = \{\tau \cap \sigma | \tau \in \mathcal{P} \text{ such that } \tau \cap \sigma \neq \emptyset\}.$$

So, choose in this example the decomposition along each edge to be the decomposition into four intervals with integral vertices, as in Figure 1.

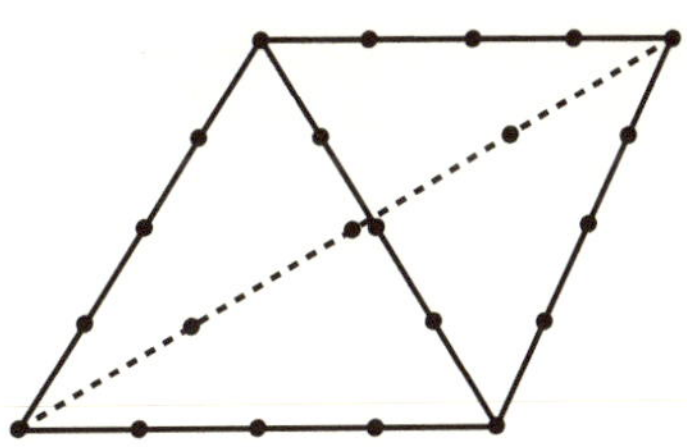

FIGURE 1

The set Δ is just the set of barycenters of these intervals: this gives 24 points.

(2) One dimension up, take Ξ to be the convex hull of

$$\begin{aligned} P_0 &= (-1,-1,-1,-1),\\ P_1 &= (4,-1,-1,-1),\\ P_2 &= (-1,4,-1,-1),\\ P_3 &= (-1,-1,4,-1),\\ P_4 &= (-1,-1,-1,4). \end{aligned}$$

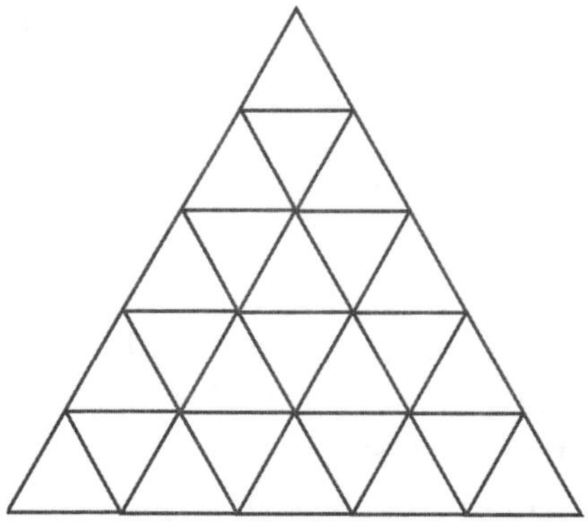

FIGURE 2

Take $\mathcal{P}$ restricted to each two-face to look like Figure 2, so that Δ on this face looks like Figure 3.

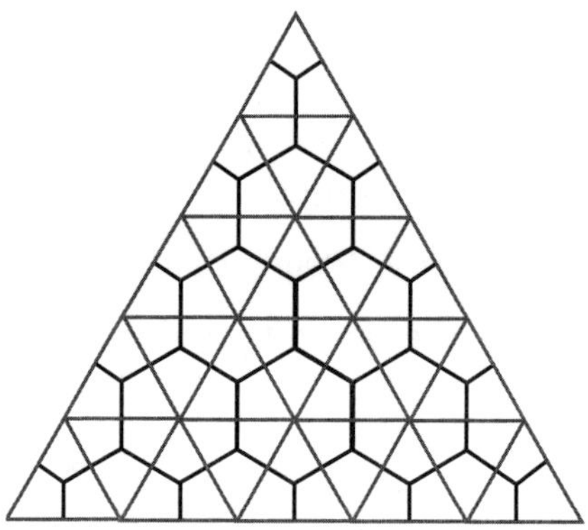

FIGURE 3

Continuing with the construction, we will now construct an affine structure on $B_0 := B \setminus \Delta$. To do so, for each $(n-1)$-dimensional face σ of Ξ, let $W_\sigma = \mathrm{Int}(\sigma)$. Secondly, for every vertex v of $\mathcal{P}$ (i.e., $\{v\} \in \mathcal{P}$) contained in a $(n-2)$-dimensional face of σ, define an open set

$$W_v = \bigcup_{\substack{\sigma \in \mathrm{Bar}(\mathcal{P}) \\ v \in \sigma \subseteq B}} \mathrm{Int}(\sigma).$$

EXERCISE 6.75. Check that

$$\{W_\sigma | \sigma \text{ is an } (n-1)\text{-dimensional face of } \Xi\} \cup \{W_v | v \text{ is a vertex of } \mathcal{P}\}$$

forms an open covering of B_0, with $W_{\sigma_1} \cap W_{\sigma_2} \neq \emptyset$ if and only if $\sigma_1 = \sigma_2$, and $W_{v_1} \cap W_{v_2} \neq \emptyset$ if and only if $v_1 = v_2$.

Now we define affine coordinate charts on W_σ and W_v. Of course σ is contained in an affine hyperplane $\mathbb{A}_\sigma$ in $M_\mathbb{R}$, and hence we obtain a canonical embedding

$$\psi_\sigma : \mathrm{Int}(\sigma) \to \mathbb{A}_\sigma.$$

Secondly, consider the projection

$$\psi_v : W_v \subseteq M_\mathbb{R} \to M_\mathbb{R}/v\mathbb{R}.$$

EXERCISE 6.76. Check that ψ_v embeds W_v topologically in $M_\mathbb{R}/v\mathbb{R}$.

Clearly the transition maps $\psi_v \circ \psi_\sigma^{-1}$ are affine transformations. Thus we obtain an affine structure on B_0.

DEFINITION 6.77. An *affine manifold with singularities* is a topological manifold B along with a set $\Delta \subseteq B$ which is locally a finite union of locally closed submanifolds of codimension at least 2. Furthermore $B_0 = B \setminus \Delta$ has the structure of an affine manifold. An affine manifold with singularities is *integral* if the affine structure on B_0 is integral. We always denote by $i : B_0 \to B$ the inclusion map.

In the example at hand, we indeed only have an affine structure on B_0, and Δ is the singular locus.

In order to study the nature of the singularities of the affine structure, it is useful to study the monodromy of the affine structure. Given a loop $\gamma \in \pi_1(B_0, b)$ based at a point b, we can parallel transport vectors in $\Lambda_{\mathbb{R},b} = TB_b$ along the loop γ using the flat connection on TB. This gives a linear map $T_\gamma : \Lambda_{\mathbb{R},b} \to \Lambda_{\mathbb{R},b}$, which we call the *monodromy* about γ. Thus we obtain a representation $\pi_1(B_0, b) \to \mathrm{Aut}(\Lambda_{\mathbb{R},b})$ given by $\gamma \mapsto T_\gamma^{-1}$, the *monodromy representation* of the local system $\Lambda_\mathbb{R}$.

EXERCISE 6.78. Show that T_γ is the linear part of $\rho(\gamma^{-1})$, where $\rho : \pi_1(B_0, b) \to \mathrm{Aff}(\Lambda_{\mathbb{R},b})$ is the holonomy representation of Definition 6.6.

Of course, if γ is a small loop about the discriminant locus Δ, and T_γ is non-trivial, then it isn't possible to extend the affine structure across Δ near γ. Thus monodromy can detect singularities.

PROPOSITION 6.79. *Let γ be a loop in B_0 which is based at a vertex v_1 of $\mathcal{P}$ contained in a face τ of Ξ, passing into an $(n-1)$-dimensional face σ_1 of Ξ containing τ, through a vertex v_2 of $\mathcal{P}$ contained in τ, into an $(n-1)$-dimensional face σ_2 of Ξ containing τ, and then back to v_1 (see Figure 4). Let $w_1, w_2 \in N_\mathbb{R}$ be the unique elements such that w_i takes the value 1 on σ_i. Let $T_\gamma : \Lambda_{\mathbb{R},v_1} \to \Lambda_{\mathbb{R},v_1}$ denote parallel transport around γ. Note that $\Lambda_{\mathbb{R},v_1}$ can be identified as a vector space with $M_\mathbb{R}/v_1\mathbb{R}$ via the differential of the coordinate chart $\psi_{v_1} : W_{v_1} \to M_\mathbb{R}/v_1\mathbb{R}$. Under this identification, T_γ is given by*

$$T_\gamma(m) = m + \langle w_1 - w_2, m\rangle(v_2 - v_1) \mod v_1.$$

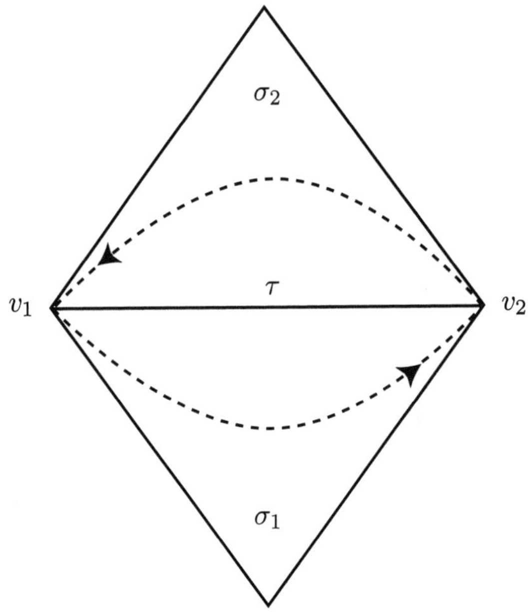

FIGURE 4

PROOF. The tangent space of σ_i is canonically identified with the space $w_i^\perp \subseteq M_{\mathbb{R}}$. Parallel transport of tangent vectors from W_{σ_i} into W_{v_j} is given by the isomorphism $w_i^\perp \to M_{\mathbb{R}}/v_j\mathbb{R}$ induced by the projection. Then T_γ is obtained by going around the following diagram clockwise:

$$\begin{array}{ccc} M_{\mathbb{R}}/v_1\mathbb{R} & \xleftarrow{\psi_1} & w_1^\perp \\ \uparrow{\scriptstyle \psi_4} & & \downarrow{\scriptstyle \psi_2} \\ w_2^\perp & \xrightarrow{\psi_3} & M_{\mathbb{R}}/v_2\mathbb{R} \end{array}$$

Thus, if $m \in M_{\mathbb{R}}$ represents an element of $M_{\mathbb{R}}/v_1\mathbb{R}$, then

$$\begin{aligned} T_\gamma(m) &= \psi_4(\psi_3^{-1}(\psi_2(\psi_1^{-1}(m)))) \\ &= \psi_4(\psi_3^{-1}(\psi_2(m - \langle w_1, m\rangle v_1))) \\ &= \psi_4(\psi_3^{-1}(m - \langle w_1, m\rangle v_1 \mod v_2)) \\ &= \psi_4(m - \langle w_1, m\rangle v_1 - \langle w_2, m - \langle w_1, m\rangle v_1\rangle v_2) \\ &= m - \langle w_2, m\rangle v_2 + \langle w_1, m\rangle v_2 \mod v_1 \\ &= m + \langle w_1 - w_2, m\rangle (v_2 - v_1) \mod v_1. \end{aligned}$$

□

Notice that for general Ξ, this transformation need not be in $\mathrm{GL}_{n-1}(\mathbb{Z})$. However, there is an important case when it is.

DEFINITION 6.80. We say an integral lattice polytope Ξ is *reflexive* if 0 is in the interior of Ξ and the affine hyperplane spanned by any $(n-1)$-dimensional face of Ξ is given by the affine linear equation $\langle y, \cdot\rangle = 1$ for some $y \in N$.

PROPOSITION 6.81. *If Ξ is a reflexive polytope, then B_0 is integral if it is constructed as above from a polyhedral decomposition $\mathcal{P}$ of Ξ with only integral vertices. Thus, in this case, B is an integral affine manifold with singularities.*

PROOF. If v is a vertex, σ an $(n-1)$-dimensional face of Ξ containing v, then there exists $w \in N$ such that w takes the value 1 on σ. Then the transition map $\psi_v \circ \psi_\sigma^{-1}$ is the projection $w^\perp \to M_{\mathbb{R}}/v\mathbb{R}$, which clearly restricts to an integral isomorphism

$$w^\perp \cap M \to M/v\mathbb{Z}.$$

□

Since there is a huge number of reflexive polytopes (in four dimensions, Kreuzer and Skarke obtained 473,800,776 reflexive polytopes [**315**]), this supplies a rich collection of examples of non-compact integral affine manifolds. There is in fact a more general construction which allows one to generate affine manifolds from reflexive polytopes and some additional data, so in fact one obtains an infinite number of examples. See Haase and Zharkov [**216**], and Gross [**203**].

Of course, the condition of reflexivity has not been pulled out of thin air! It is worth at this point reminding the reader (see MS1, §7.10 or Batyrev [**32**] for more details) of the fundamental role reflexive polytopes play in the most basic mirror symmetry construction. A reflexive polytope Ξ defines a Gorenstein projective toric variety $(\mathbb{P}_\Xi, \mathcal{O}_{\mathbb{P}_\Xi}(1))$. A hyperplane section of $\mathbb{P}_\Xi$ is a Calabi-Yau variety (usually singular, but one can find a maximal partial crepant projective resolution, which resolves singularities in codimension less than 4). Mirror pairs of Calabi-Yau varieties then can be found as hyperplane sections of $\mathbb{P}_\Xi$ and $\mathbb{P}_{\Xi^*}$, where Ξ^* is the dual, or polar, polytope of Ξ defined as

$$\Xi^* = \{n \in N_{\mathbb{R}} | \langle n, m\rangle \geq -1 \text{ for all } m \in \Xi\}.$$

The basic philosophy that the reader should keep in mind in what follows is that the integral affine manifold B produced from a reflexive polytope Ξ should yield both the hyperplane section of $\mathbb{P}_\Xi$ and its mirror. More precisely, B comes with an open subset $B_0 \subseteq B$ carrying an integral affine structure. This gives rise to $X(B_0)$ and $\check{X}(B_0)$. The philosophy is that $X(B_0)$ should be compactifiable to a hyperplane section of $\mathbb{P}_{\Xi^*}$ and $\check{X}(B_0)$ should be compactifiable to a hyperplane section of $\mathbb{P}_\Xi$. However, these are only topological statements. See Gross and Siebert [**204**] for details of this, or Gross [**202**].

EXAMPLE 6.82. The polytopes given in Example 6.74 are reflexive polytopes. Let's explore the monodromy in these examples.

(1) Choose any two adjacent vertices along an edge. Any choice will give the same results, so take $v_1 = (-1,-1,-1)$, $v_2 = (0,-1,-1)$. Then for the two-dimensional faces σ_1 and σ_2 containing both v_1 and v_2, we can take $w_1 = (0,0,-1)$ and $w_2 = (0,-1,0)$. Thus

$$T_\gamma(m) = m + \langle (0,1,-1), m\rangle (1,0,0) \mod v_1.$$

If we take $(1,0,0) \mod v_1$ and $(0,1,0) \mod v_1$ to form a basis for $M/v_1\mathbb{Z}$, the matrix for T_γ is $\begin{pmatrix} 1 & 1 \\ 0 & 1 \end{pmatrix}$.

(2) In this case, it is interesting to study the behaviour of monodromy in a neighbourhood of each vertex of the discriminant locus Δ. There are two distinct types of vertices: the ones contained in the interior of two-dimensional faces, and the ones contained in the interior of one-dimensional faces. Let's look at each of these in detail. First, consider the triangle with vertices $v_1 = (-1,-1,-1,-1)$, $v_2 = (0,-1,-1,-1)$, $v_3 = (-1,0,-1,-1)$. This triangle is a two-dimensional cell of $\mathcal{P}$, and is contained in two three-faces of Ξ, σ_1 and σ_2, with $w_1 = (0,0,-1,0)$ and $w_2 = (0,0,0,-1)$. Let γ_i be the loop involving v_i, v_{i+1}, σ_1 and σ_2 as in Proposition 6.79, with indices taken modulo 3. Then Proposition 6.79 yields

$$\begin{aligned} T_{\gamma_1}(m) &= m + \langle (0,0,-1,1), m\rangle (1,0,0,0) \mod v_1 \\ T_{\gamma_2}(m) &= m + \langle (0,0,-1,1), m\rangle (-1,1,0,0) \mod v_2 \\ T_{\gamma_3}(m) &= m + \langle (0,0,-1,1), m\rangle (0,-1,0,0) \mod v_3. \end{aligned}$$

Now $T_{\gamma_i} : M_\mathbb{R}/v_i\mathbb{R} \to M_\mathbb{R}/v_i\mathbb{R}$, so to compare these three monodromy transformations, it is convenient to parallel transport the tangent spaces $\Lambda_{\mathbb{R},v_i}$ to the tangent space in the interior, say, of σ_1, which is identified canonically with $w_1^\perp$. Take as basis of $w_1^\perp$ the vectors $e_1 = (1,0,0,0)$, $e_2 = (0,-1,0,0)$ and $e_3 = (0,0,0,1)$. In this basis, T_{γ_1}, T_{γ_2} and T_{γ_3} have matrices

$$\begin{pmatrix} 1 & 0 & 1 \\ 0 & 1 & 0 \\ 0 & 0 & 1 \end{pmatrix}, \begin{pmatrix} 1 & 0 & -1 \\ 0 & 1 & -1 \\ 0 & 0 & 1 \end{pmatrix} \text{ and } \begin{pmatrix} 1 & 0 & 0 \\ 0 & 1 & 1 \\ 0 & 0 & 1 \end{pmatrix}.$$

Note $T_{\gamma_3} \circ T_{\gamma_2} \circ T_{\gamma_1}$ is the identity, as one should expect.

Now let's look to see what's happening near a vertex of Δ on an edge. Take, say, $v_1 = (-1,-1,-1,-1)$, $v_2 = (0,-1,-1,-1)$. These are the end-points of an interval contained in an edge, and are contained in three three-faces σ_1, σ_2 and σ_3 of Ξ, with $w_1 = (0,0,-1,0)$, $w_2 = (0,0,0,-1)$ and $w_3 = (0,-1,0,0)$. If γ_i' is the loop involving v_1, v_2, w_i and w_{i+1} as in

Proposition 6.79, then

$$\begin{aligned} T_{\gamma_1'}(m) &= m + \langle(0,0,-1,1),m\rangle(1,0,0,0) \mod v_1 \\ T_{\gamma_2'}(m) &= m + \langle(0,1,0,-1),m\rangle(1,0,0,0) \mod v_1 \\ T_{\gamma_3'}(m) &= m + \langle(0,-1,1,0),m\rangle(1,0,0,0) \mod v_1. \end{aligned}$$

In a basis $e_1' = (1,0,0,0)$, $e_2' = (0,1,0,0)$ and $e_3' = (0,0,-1,0)$ for $M_{\mathbb{R}}/v_1\mathbb{R}$, these monodromy transformations have matrices

$$\begin{pmatrix} 1 & 0 & 1 \\ 0 & 1 & 0 \\ 0 & 0 & 1 \end{pmatrix}, \begin{pmatrix} 1 & 1 & 0 \\ 0 & 1 & 0 \\ 0 & 0 & 1 \end{pmatrix} \text{ and } \begin{pmatrix} 1 & -1 & -1 \\ 0 & 1 & 0 \\ 0 & 0 & 1 \end{pmatrix}.$$

This is fundamentally different behaviour. In the first case, there is a two-dimensional subspace of $w_1^\perp$ left invariant by the monodromy transformations, but in the second case there is only a one-dimensional subspace. These points will have quite different properties. Note, however, that the monodromy about a simple loop around one of the edges of Δ always looks the same; after a change of basis, it is always of the form $\left(\begin{smallmatrix} 1 & 1 & 0 \\ 0 & 1 & 0 \\ 0 & 0 & 1 \end{smallmatrix}\right)$, which looks very much like the two-dimensional case. □

EXERCISE 6.83. Let $\tau \in \mathcal{P}$, $\tau \subseteq B$. Show that for any two vertices v_1 and v_2 of τ and any $(n-1)$-dimensional faces σ_1 and σ_2 of Ξ containing τ determining a loop γ, the monodromy transformation $T_\gamma : M_{\mathbb{R}}/v_1\mathbb{R} \to M_{\mathbb{R}}/v_1\mathbb{R}$ leaves invariant the tangent space to the affine space spanned by τ. This puts the behaviour in the above example in context.

6.4.2. Local models for compactifications. If B is an integral affine manifold with singularities, we obtain dual torus bundles $X(B_0) \to B_0$ and $\check{X}(B_0) \to B_0$. (If we wish, we may decorate these as in §6.2.1 and §6.2.3, but this will not be particularly relevant for the following discussion.) Here $X(B_0)$ is a complex manifold and $\check{X}(B_0)$ is a symplectic manifold. We can then ask whether it is possible to compactify $X(B_0)$ or $X(\check{B}_0)$ in the following sense: If B is compact, is there an $X(B)$ which is compact and a commutative diagram

$$\begin{array}{ccc} X(B_0) & \hookrightarrow & X(B) \\ \downarrow & & \downarrow \\ B_0 & \hookrightarrow & B \end{array}$$

(and one can ask for a similar diagram for $\check{X}(B)$). The easiest form of this question is a topological one, demanding only that $X(B)$ or $\check{X}(B)$ be a topological compactification of $X(B_0)$ or $\check{X}(B_0)$ respectively. However, one could refine this question and demand that $X(B)$ be a compactification of $X(B_0)$ as a complex manifold, and that $\check{X}(B)$ be a compactification of $\check{X}(B_0)$ as a symplectic manifold.

REMARK 6.84. Here is a very important point. In cases of interest, we don't expect compactifications of $X(B_0)$ as a complex manifold, but we do expect compactifications of $\check{X}(B_0)$ as a symplectic manifold. In dimension three, Castaño-Bernard and Matessi [**90**] have obtained results for symplectic compactifications. We shall see later on in §7.3.7 that complex structures need to be deformed before compactification. It is widely believed that it is this difference between the symplectic (A-model) and complex (B-model) sides which produces instanton corrections. This will be explored more later. □

Just as in Definition 6.17, we can say that if $X(B)$ and $\check{X}(B)$ are constructed, then they are *SYZ dual* to each other. The Strominger-Yau-Zaslow conjecture then suggests that SYZ dual manifolds are mirror dual.

We will discuss these compactifications in great detail in dimensions two and three. One of the great benefits of studying this in dimension three is that some basic phenomena of mirror symmetry become transparent, such as the change of sign of Euler characteristic, at a purely local (for the base B) level. We shall see this at the end of this section. This section will cover local models for these compactifications. We will first study the two-dimensional case, as in Example 6.74, (1), and then study the three-dimensional cases occurring in Example 6.74, (2) around the two different sorts of vertices.

The basic ingredient is the *nut*[1]. For our purposes, we can view this as follows. Let $\overline{X} = \mathbb{C}^2$, $X = \mathbb{C}^2 \setminus \{(0,0)\}$, $\overline{Y} = \mathbb{R}^3$ and $Y = \mathbb{R}^3 \setminus \{(0,0,0)\}$. Consider the map $p : X \to Y$ given by

$$p(z_1, z_2) = (2\,\mathrm{Re}(z_1 z_2), 2\,\mathrm{Im}(z_1 z_2), |z_1|^2 - |z_2|^2). \tag{6.6}$$

This expresses X as an S^1-bundle over Y, with fibers being orbits of the S^1 action

$$(z_1, z_2) \mapsto (e^{\sqrt{-1}\theta} z_1, e^{-\sqrt{-1}\theta} z_2),$$

and we can compute the first Chern class of this bundle as follows. Choose a connection 1-form on the bundle; for this, we may take

$$\theta = \sqrt{-1}\,\mathrm{Im}(\bar{z}_1 dz_1 - \bar{z}_2 dz_2)/(|z_1|^2 + |z_2|^2).$$

Then the first Chern class of this bundle is represented by the curvature 2-form

$$\frac{d\theta}{2\pi\sqrt{-1}} = \frac{-(u_1 du_2 \wedge du_3 + u_2 du_3 \wedge du_1 + u_3 du_1 \wedge du_2)}{4\pi(u_1^2 + u_2^2 + u_3^2)^{3/2}},$$

where u_1, u_2, u_3 are coordinates on $\mathbb{R}^3$. To determine the cohomology class represented by this 2-form, one simply integrates this form over the unit

[1]See §7.1.4 for discussion of the terminology *nuts* and *bolts*.

two-sphere in $\mathbb{R}^3$, and we find

$$\int_{S^2} \frac{d\theta}{2\pi\sqrt{-1}} = 1.$$

Thus the first Chern class of the bundle is $1 \in \mathbb{Z} \cong H^2(Y,\mathbb{Z})$. Here $\overline{X}$ is the nut, and $\bar{p} : \overline{X} \to \overline{Y}$, given by the same formula as p, is a partial compactification of the S^1-bundle $X \to Y$. In general, it is used as follows.

Let $\overline{Y}$ be any three-manifold (not necessarily compact), and let $Y \subseteq \overline{Y}$ be an open set with $\overline{Y} \setminus Y$ a discrete set of points. Suppose we are given an S^1-bundle $q : X \to Y$ with the property that for every point $y \in \overline{Y} \setminus Y$, there is an open neighbourhood U_y of y in $\overline{Y}$ such that $U_y \setminus \{y\} \cong \mathbb{R}^3 \setminus \{0\}$, and such that $q^{-1}(U_y \setminus \{y\}) \to U_y \setminus \{y\}$ is an S^1-bundle with $c_1 = \pm 1$. Then we can patch the nut into $q^{-1}(U_y \setminus \{y\})$: there is a commutative diagram

$$\begin{array}{ccc} q^{-1}(U_y \setminus \{y\}) & \hookrightarrow & \mathbb{C}^2 \\ \downarrow{\scriptstyle q} & & \downarrow{\scriptstyle \bar{p}} \\ U_y \setminus \{y\} & \hookrightarrow & \mathbb{R}^3 \end{array}$$

Hence we can glue in $\mathbb{C}^2$, and obtain $\bar{q} : \overline{X} \to \overline{Y}$.

The following exercise shows that the resulting space $\overline{X}$ does not depend on the specific gluing.

EXERCISE 6.85. Let $Y, \overline{Y}$ and $q : X \to Y$ be as above. Show that there is a unique topology on the set $\overline{X} = X \cup (\overline{Y} \setminus Y)$ extending the topology on X such that there is a commutative diagram of continuous maps

$$\begin{array}{ccc} X & \hookrightarrow & \overline{X} \\ \downarrow{\scriptstyle q} & & \downarrow{\scriptstyle \bar{q}} \\ Y & \hookrightarrow & \overline{Y} \end{array}$$

where $\bar{q}$ is proper (the inverse image of a compact set is compact) and $\bar{q}|_{\overline{Y}\setminus Y}$ is the identity on $\overline{Y} \setminus Y$. Furthermore, $\overline{X}$ is a manifold.

We can now construct the key two-dimensional example.

CONSTRUCTION 6.86. Let $\overline{Y} = S^1 \times D$, where D is a two-dimensional open disk with a chosen point $0 \in D$. Choose also a point $p \in S^1$, and let

$$Y = \overline{Y} \setminus \{(p, 0)\}.$$

Now we can calculate $H^2(Y, \mathbb{Z})$, so we can identify Chern classes of S^1-bundles over Y. Notice that $H^2(\overline{Y}, \mathbb{Z}) = 0$. Then by the relative cohomology long exact sequence, we have

$$0 = H^2(\overline{Y}, \mathbb{Z}) \to H^2(Y, \mathbb{Z}) \to H^3(\overline{Y}, Y, \mathbb{Z}) \to H^3(\overline{Y}, \mathbb{Z}) = 0.$$

We can calculate $H^3(\overline{Y}, Y, \mathbb{Z})$ as follows. There is a neighbourhood U of $(p, 0)$ homeomorphic to $\mathbb{R}^3$ such that $U \setminus \{(p, 0)\}$ is homeomorphic to $\mathbb{R}^3 \setminus \{0\}$. By excision,

$$H^3(\overline{Y}, Y, \mathbb{Z}) \cong H^3(U, U \setminus \{(p,0)\}, \mathbb{Z}).$$

However, the relative cohomology long exact sequence then yields

$$H^3(U, U \setminus \{(p,0)\}, \mathbb{Z}) \cong H^2(U \setminus \{(p,0)\}, \mathbb{Z}) \cong \mathbb{Z},$$

since $\mathbb{R}^3 \setminus \{0\}$ retracts onto S^2. Thus we conclude that $H^2(Y, \mathbb{Z}) \cong \mathbb{Z}$. Furthermore, we take an S^1-bundle $q : X \to Y$ with $c_1 = 1$, so we can add a nut and get $\bar{q} : \overline{X} \to \overline{Y}$. Now compose $\bar{q}$ with the projection $\overline{Y} \to D$, to obtain $f : \overline{X} \to D$.

Let's look at the fibers of f. If $x \in D$, $x \neq 0$, then $f^{-1}(x)$ is an S^1-bundle over $S^1 \times \{x\}$, i.e., a two-torus. But $f^{-1}(0)$ looks like a T^2 with one circle pinched to a point, see Figure 5. The shaded disk in Figure 5 is

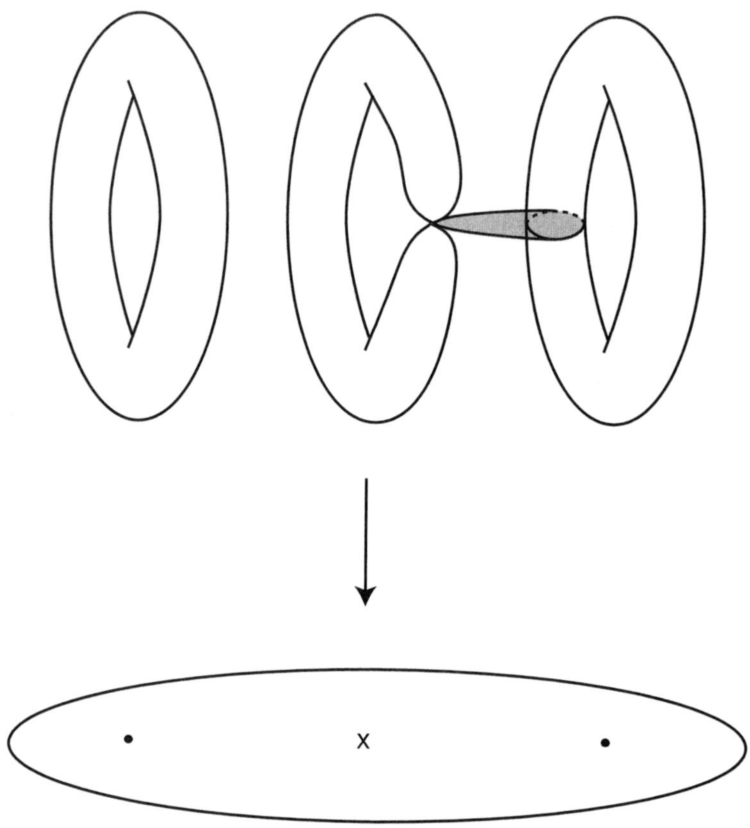

FIGURE 5

obtained as follows. Take a closed line segment in D with an endpoint at 0 and an endpoint at $b \in D$. Lift this line segment to a line segment in Y with an endpoint at $(p, 0)$. Then take the inverse image of this line segment in $\overline{X}$ to obtain a disk whose boundary is an S^1 contained in $f^{-1}(b)$. This S^1

gives a well-defined cycle in $H_1(f^{-1}(b),\mathbb{Z})$, and is usually called a *vanishing cycle*. In $\overline{X}$ it is homologous to zero, hence the terminology.

It is really worth understanding at this point the role of monodromy here. The torus bundle $f : f^{-1}(D\setminus\{0\}) \to D\setminus\{0\}$ is *not* the trivial one. If $b \in D\setminus\{0\}$ is a basepoint, we obtain a monodromy transformation

$$T : H_1(f^{-1}(b),\mathbb{Z}) \to H_1(f^{-1}(b),\mathbb{Z})$$

as follows. Let $\gamma : [0,1] \to D\setminus\{0\}$ be a simple counterclockwise loop based at b. Pull back the fibration $f : \overline{X} \to D$ via γ to obtain a family $X' \to [0,1]$. Since $[0,1]$ is contractible, we can trivialize, i.e., there is an isomorphism $\psi : X' \cong [0,1] \times T^2$ compatible with the map $X' \to [0,1]$. Using the fact that $X'_0 = X'_1$, we obtain a diffeomorphism φ of $T^2 \cong X'_0$, which is the composition

$$X'_0 \xrightarrow{\psi} \{0\} \times T^2 = \{1\} \times T^2 \xrightarrow{\psi^{-1}} X'_1.$$

If $S^1 \subseteq D\setminus\{0\}$ is the image of γ, then to recover the fibration $f^{-1}(S^1) \to S^1$ topologically, one glues $\{0\} \times T^2$ to $\{1\} \times T^2$ via φ. The diffeomorphism φ is often called a monodromy diffeomorphism. Since the trivialization is not unique, the monodromy diffeomorphism is only determined up to isotopy. However, φ does induce a unique map on homology $\varphi_* : H_1(f^{-1}(b),\mathbb{Z}) \to H_1(f^{-1}(b),\mathbb{Z})$, and this is the monodromy transformation. One can also view this simply as the map which takes a cycle in $f^{-1}(b)$ and follows it continuously as we proceed around the loop. We don't necessarily get back to the cycle we started with.

We can see this very explicitly in this example. Consider a circle $C \subseteq D\setminus\{0\}$ wrapping around 0 once, and let $Z = f^{-1}(C)$. By construction, Z is an S^1-bundle over $T^2 = C \times S^1 \subseteq Y$ with c_1 being the restriction of $1 \in H^2(Y,\mathbb{Z})$ to $H^2(C \times S^1,\mathbb{Z})$. The restriction map is easily seen to be the identity. By pulling back this S^1-bundle to $C \times \mathbb{R}$ via the covering $C \times \mathbb{R} \to C \times S^1$, this S^1-bundle becomes trivial, i.e., is the projection $C \times \mathbb{R} \times S^1 \to C \times \mathbb{R}$. We can recover the S^1-bundle $Z \to C \times S^1$ by dividing $C \times \mathbb{R} \times S^1$ by the $\mathbb{Z}$-action

$$(e^{2\pi\sqrt{-1}\theta}, \rho, e^{2\pi\sqrt{-1}\tau}) \mapsto (e^{2\pi\sqrt{-1}\theta}, \rho+1, e^{2\pi\sqrt{-1}(\theta+\tau)}).$$

(Here we identify C and S^1 with $U(1)$). It is then easy to see that the projection

$$(C \times \mathbb{R} \times S^1)/\mathbb{Z} \mapsto C \times S^1$$

is an S^1-bundle with $c_1 = 1$. (Exercise!) Now consider the map $\gamma : [0,1] \to C \subseteq D$ parameterizing the loop C as above based at a point $y \in D$, and pull back $f : X \to D$ by γ to get $X' \to [0,1]$. By the above description of Z, we see $X' \cong [0,1] \times S^1 \times S^1$, explicitly via the isomorphism

$$([0,1] \times \mathbb{R} \times S^1)/\mathbb{Z} \to [0,1] \times S^1 \times S^1$$

given by

$$(\theta, \rho, e^{2\pi\sqrt{-1}\tau}) \mapsto (\theta, e^{2\pi\sqrt{-1}\rho}, e^{2\pi\sqrt{-1}(\tau-\theta\rho)})$$

so that the monodromy diffeomorphism is given by

$$(e^{2\pi\sqrt{-1}\rho}, e^{2\pi\sqrt{-1}\tau}) \mapsto (e^{2\pi\sqrt{-1}\rho}, e^{2\pi\sqrt{-1}(\rho+\tau)}).$$

This is an example of a *Dehn twist*, which can be thought of as follows. Take a torus, and cut along a circle, as pictured. Give the resulting cylinder a 360 degree twist, and then reglue, as depicted in Figure 6. If e_1 and

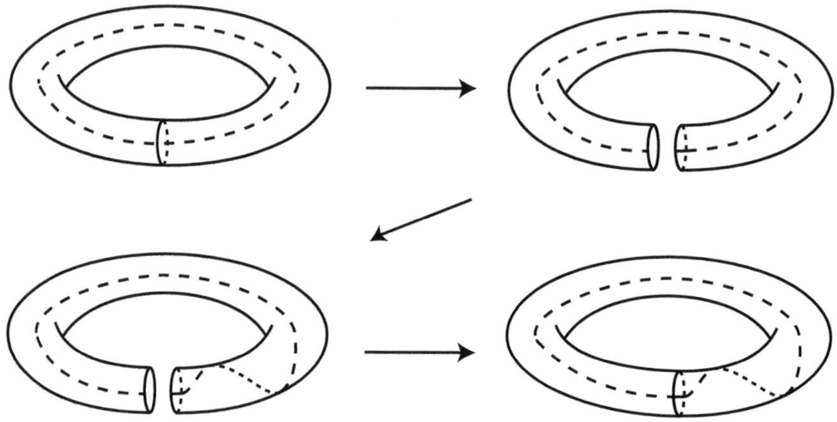

FIGURE 6

e_2 are the generators of $H_1(f^{-1}(y), \mathbb{Z})$ given by ρ = constant and τ = constant respectively, then the corresponding transformation on homology $T : H_1(f^{-1}(y), \mathbb{Z}) \to H_1(f^{-1}(y), \mathbb{Z})$ is given by the matrix $\left(\begin{smallmatrix} 1 & 1 \\ 0 & 1 \end{smallmatrix}\right)$ in this basis. Note that the cycle left invariant by monodromy here is the vanishing cycle.

CONSTRUCTION 6.87. There is another construction of the T^2-fibration of Construction 6.86 which generalizes well to higher dimension. Consider $\overline{X} \subseteq \mathbb{C}^2 \times \mathbb{C}^*$, with coordinates z_1, z_2 on $\mathbb{C}^2$ and w on $\mathbb{C}^*$, with $\overline{X}$ defined by the equation $z_1z_2 = w - 1$, and define $f : \overline{X} \to \mathbb{R}^2$ by $f(z_1, z_2, w) = (|z_1|^2 - |z_2|^2, \ln |w|)$. This factors through $\bar{q} : \overline{X} \to \overline{Y} = \mathbb{R} \times \mathbb{C}^* \cong D \times S^1$ defined by $\bar{q}(z_1, z_2, w) = (|z_1|^2 - |z_2|^2, w)$. It is easy to check that if $X = \overline{X} \setminus \{(0, 0, 1)\}$, $Y = \overline{Y} \setminus \{(0, 1)\}$, then $q = \bar{q}|_X : X \to Y$ is an S^1-bundle (with S^1 acting on the fibers by $(z_1, z_2, w) \mapsto (e^{i\theta}z_1, e^{-i\theta}z_2, w)$. It is clear that $\overline{X}$ is the compactification of X by a nut, and that the fibration $f : \overline{X} \to \mathbb{R}^2 \cong D$ is then topologically equivalent to the one constructed in Construction 6.86.

We will next examine the local models necessary for three-dimensional compactifications. It is in the three-dimensional case that we really see mirror symmetry in this topological picture.

CONSTRUCTION 6.88. Let $f : \overline{X} \to D$ be the T^2-fibration of Construction 6.86. Then we obtain a T^3-fibration $\overline{X} \times S^1 \times (0, 1) \to D \times (0, 1)$ given by $(x, p, r) \mapsto (f(x), r)$. The singular fibers, which occur over $\{0\} \times (0,1)$, are of the form $f^{-1}(0) \times S^1$, where $f^{-1}(0)$ is a pinched torus. A simple loop about $\Delta = \{0\} \times (0,1)$ induces a monodromy transformation $\begin{pmatrix} 1 & 1 & 0 \\ 0 & 1 & 0 \\ 0 & 0 & 1 \end{pmatrix}$ in a suitable basis. Alternatively, we can describe $\overline{X} \subseteq \mathbb{C}^2 \times (\mathbb{C}^*)^2$ with coordinates z_1, z_2 on $\mathbb{C}^2$ and w_1, w_2 on $(\mathbb{C}^*)^2$ with $\overline{X}$ defined by the equation $z_1 z_2 = w_1 - 1$ and $\overline{X} \to \mathbb{R}^3$ defined by $(|z_1|^2 - |z_2|^2, \ln|w_1|, \ln|w_2|)$ as in Construction 6.87.

CONSTRUCTION 6.89. We can generalize the nut to higher dimensions as follows. Let $\overline{Y}$ be a manifold, $Y \subseteq \overline{Y}$ be an open subset with $\overline{Y} \setminus Y = S$, a submanifold of codimension three. Consider an S^1-bundle $q : X \to Y$. Suppose that locally near S the first Chern class of this S^1-bundle is 1. What we mean by this is as follows. For each point $y \in S$, there is an open neighbourhood U of y in $\overline{Y}$ homeomorphic to $B^{n-3} \times \mathbb{R}^3$, where B^{n-3} is an open $(n-3)$-dimensional ball, such that $S \cap U = B^{n-3} \times \{0\}$. Then we require that the restriction of the S^1-bundle to $U \setminus (S \cap U)$ have first Chern class $1 \in H^2(B^{n-3} \times (\mathbb{R}^3 \setminus \{0\}), \mathbb{Z}) = \mathbb{Z}$. Given this there exists a unique topology on $\overline{X} = X \coprod S$ extending the topology on X such that $\overline{X}$ is a topological manifold and there is a commutative diagram of continuous maps

$$\begin{array}{ccc} X & \hookrightarrow & \overline{X} \\ \downarrow{\scriptstyle q} & & \downarrow{\scriptstyle \bar{q}} \\ Y & \hookrightarrow & \overline{Y} \end{array}$$

with $\bar{q}$ proper and $\bar{q}|_S$ the identity. To see this, note that the topology on $\overline{X}$ with basis consisting of open sets in X and inverse images of open sets in $\overline{Y}$ under $\bar{q}$ is easily seen to be the unique topology in which $\bar{q}$ is proper. Now if $y \in S$ is a point, let $U \subseteq \overline{Y}$ be the open neighbourhood of y as above, with $U \cong B^{n-3} \times \mathbb{R}^3$. Then the restriction of q to $\{y'\} \times (\mathbb{R}^3 - \{0\})$ for any point $q \in U \cap S$ is an S^1-bundle with first Chern class ± 1. Thus by using the nut we have a commutative diagram

$$\begin{array}{ccc} q^{-1}(U - S) & \hookrightarrow & B^{n-3} \times \mathbb{C}^2 \\ \downarrow{\scriptstyle q} & & \downarrow{\scriptstyle \bar{q}'} \\ U - S & \hookrightarrow & B^{n-3} \times \mathbb{R}^3 \end{array}$$

where $\bar{q}'$ is the identity on the first factor and the map p of (6.6) on the second factor. Thus, by uniqueness of the compactification, $\bar{q}'$ coincides with $\bar{q}^{-1}(U) \to U$, and in particular, $\overline{X}$ is a manifold.

CONSTRUCTION 6.90. Let B be a three-dimensional ball, and let $\Delta \subseteq B$ with $\Delta = l_1 \cup l_2 \cup l_3 \cup \{0\}$, where l_1, l_2 and l_3 are (open) line segments as depicted in Figure 7. Let $\overline{Y} = B \times (\mathbb{R}^2/\mathbb{Z}^2)$. We will describe a specific surface $S \subseteq \overline{Y}$. Let C_1 be the circle $(1,0)\mathbb{R}/(1,0)\mathbb{Z} \subseteq \mathbb{R}^2/\mathbb{Z}^2$, and let C_2 be the circle $(0,1)\mathbb{R}/(0,1)\mathbb{Z} \subseteq \mathbb{R}^2/\mathbb{Z}^2$. Let $S_1 = \bar{l}_1 \times C_1$, $S_2 = \bar{l}_2 \times C_2$ be cylinders

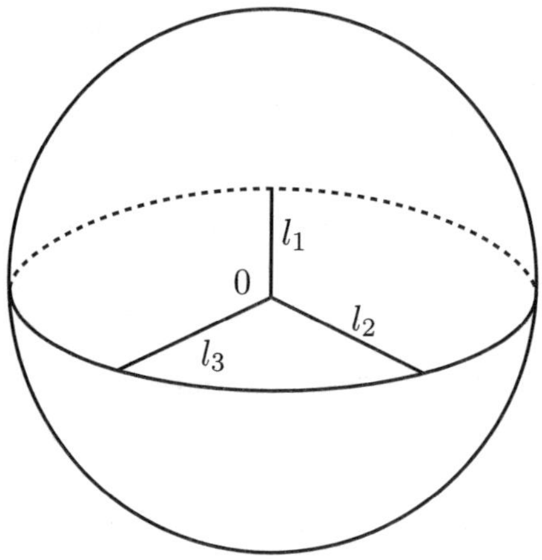

FIGURE 7

contained in $\overline{Y}$ sitting over the closures of $\bar{l}_1$ and $\bar{l}_2$ of l_1 and l_2. Construct S_3 as follows. Choose a homotopy between the circle $C_3 = (1,-1)\mathbb{R}/(1,-1)\mathbb{Z}$ and $C_1 - C_2$ which is sufficiently nice, i.e., take some continuous map

$$H : S^1 \times [0,1] \to \mathbb{R}^2/\mathbb{Z}^2$$

such that $H(S^1 \times \{1\}) = C_3$ and $H(S^1 \times \{0\}) = C_1 \cup C_2$, and sufficiently nice means that if we take a parameterization $\alpha : [0,1) \to \bar{l}_3$ with $\alpha(0) = 0$, then

$$S_3 = \{(\alpha(t), H(\theta, t)) \in \overline{Y} | \theta \in S^1, 0 \leq t < 1\} \subseteq \overline{Y}$$

is a smooth surface away from $\{0\} \times (\mathbb{R}^2/\mathbb{Z}^2)$. Set $S = S_1 \cup S_2 \cup S_3$. These three surfaces fit together to give a pair of pants fibering over Δ as depicted in Figure 8. Note that there is no way to make S a smooth surface, so this will be a purely topological construction.

Now put $Y = \overline{Y} \setminus S$. We can try to calculate $H^2(Y,\mathbb{Z})$ as follows. We have the relative cohomology long exact sequence

$$H^2(\overline{Y}, Y, \mathbb{Z}) \to H^2(\overline{Y}, \mathbb{Z}) \to H^2(Y, \mathbb{Z}) \to H^3(\overline{Y}, Y, \mathbb{Z}) \to H^3(\overline{Y}, \mathbb{Z}).$$

Now $\overline{Y}$ retracts onto $T^2 = \mathbb{R}^2/\mathbb{Z}^2$, so $H^2(\overline{Y},\mathbb{Z}) = H^2(T^2,\mathbb{Z})$ and $H^3(\overline{Y},\mathbb{Z}) = 0$. Furthermore, by Alexander duality, $H^p(\overline{Y}, Y, \mathbb{Z}) = H^{p-3}(S,\mathbb{Z})$, so

$$H^2(\overline{Y}, Y, \mathbb{Z}) = 0$$

and

$$H^3(\overline{Y}, Y, \mathbb{Z}) = H^0(S, \mathbb{Z}) = \mathbb{Z}.$$

This gives a short exact sequence

$$0 \to H^2(\overline{Y}, \mathbb{Z}) \to H^2(Y, \mathbb{Z}) \to \mathbb{Z} \to 0.$$

Finally, the restriction map $H^2(Y,\mathbb{Z}) \to H^2(\{b\} \times (\mathbb{R}^2/\mathbb{Z}^2), \mathbb{Z})$ for $b \in B \setminus \Delta$ splits the above exact sequence, so we get a canonical isomorphism $H^2(Y,\mathbb{Z}) \cong H^2(\overline{Y},\mathbb{Z}) \oplus \mathbb{Z}$. Under this identification, consider the element

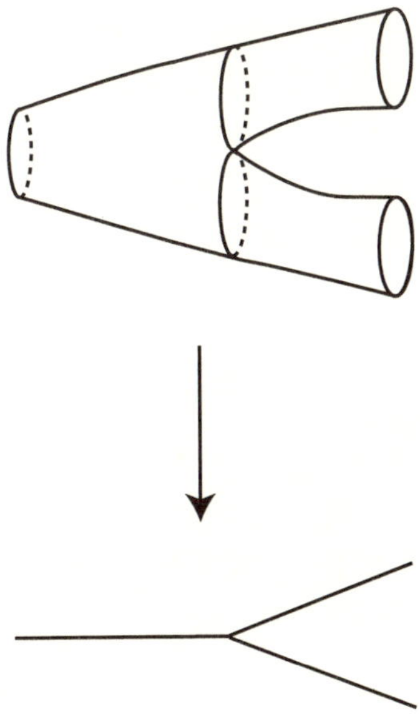

FIGURE 8

c_1 of $H^2(Y, \mathbb{Z})$ given by $(0, 1)$. In fact, if $y \in S$ and U is a small neighbourhood of y as in Construction 6.89, then the restriction of c_1 to $U \setminus S$ is $1 \in H^2(U \setminus S, \mathbb{Z}) = \mathbb{Z}$. Then if $q : X \to Y$ is an S^1-bundle with first Chern class c_1, q satisfies the conditions of Construction 6.89 and we can compactify to get a six-manifold $\overline{X}$ and a map $\bar{q} : \overline{X} \to \overline{Y}$. Composing with the projection to B gives a map $f : \overline{X} \to B$.

What have we achieved? Because c_1 restricts to zero on each $\{b\} \times T^2$ for $b \in B \setminus \Delta$, $f^{-1}(b)$ is a trivial S^1-bundle over T^2, i.e., a T^3. Thus f is a T^3-fibration. Furthermore, if $b \in l_i$ for some i, then $f^{-1}(b)$ is of the form $F \times S^1$, where F is a pinched torus. Finally, $f^{-1}(0)$ is a more degenerate fiber, with singular locus a figure eight, as $(\{0\} \times (\mathbb{R}^2/\mathbb{Z}^2)) \cap S$ is a figure eight. Note that the topological Euler characteristic of this singular fiber is -1. As a result, we call this fiber a *negative fiber.*

EXERCISE 6.91. In Construction 6.90, compute the monodromy of the torus bundle $f^{-1}(B \setminus \Delta) \to B \setminus \Delta$. Compare this with the monodromy of the affine structure calculated in Example 6.82, (2). Which sort of vertex does this example correspond to?

CONSTRUCTION 6.92. Let $\overline{X} \subseteq \mathbb{C}^3 \times \mathbb{C}^*$ with coordinates z_1, z_2, z_3 on $\mathbb{C}^3$ and coordinate w on $\mathbb{C}^*$, with $\overline{X}$ defined by the equation

$$z_1 z_2 z_3 = w - 1.$$

Consider the map $f : \overline{X} \to \mathbb{R}^3$ defined by

$$f(z_1, z_2, z_3) = (|z_1|^2 - |z_2|^2, |z_1|^2 - |z_3|^2, \log |w|).$$

We will now analyze this map. First, note that it factors as

$$\overline{X} \xrightarrow{\bar{q}} \overline{Y} \xrightarrow{g} \mathbb{R}^3$$

with $\overline{Y} = \mathbb{R}^2 \times \mathbb{C}^*$,

$$\bar{q}(z_1, z_2, z_3) = (|z_1|^2 - |z_2|^2, |z_1|^2 - |z_3|^2, w)$$

and

$$g(x_1, x_2, w) = (x_1, x_2, \log |w|).$$

The map g allows us to identify $\overline{Y}$ with $\mathbb{R}^3 \times S^1$. Meanwhile, $\bar{q}$ is invariant under the T^2-action

$$(z_1, z_2, z_3, w) \mapsto (e^{\sqrt{-1}\theta_1} z_1, e^{\sqrt{-1}\theta_2} z_2, e^{-\sqrt{-1}(\theta_1+\theta_2)} z_3, w).$$

It is easy to see that each fiber of $\bar{q}$ is a single orbit of this T^2-action. However, there are some degenerate orbits: let

$$S = \{(z_1, z_2, z_3, w) \in \overline{X} | \text{two of } z_1, z_2, z_3 \text{ are zero and } w = 1\}.$$

Then if $y \in S \backslash \{(0,0,0,1)\}$, the T^2-orbit of y is a circle, and if $y = (0,0,0,1)$, the T^2-orbit of y is a point. Thus, in particular, if $X = \overline{X} \setminus S$, $Y = \overline{Y} \setminus \bar{q}(S)$, then $q : X \to Y$, the restriction of $\bar{q}$ to X, is a principal T^2-bundle. Note that

$$\begin{aligned} \bar{q}(S) = & \quad \{(0, x_2, 1) \in \overline{Y} | x_2 \leq 0\} \\ & \cup \ \{(x_1, 0, 1) \in \overline{Y} | x_1 \leq 0\} \\ & \cup \ \{(x_1, x_2, 1) \in \overline{Y} | x_1 = x_2 \geq 0\}. \end{aligned}$$

Put $\Delta = f(S)$. We can write $\Delta = l_1 \cup l_2 \cup l_3 \cup \{(0,0,0)\}$, with

$$\begin{aligned} l_1 &= \{(0, x_2, 0) \in \mathbb{R}^3 | x_2 < 0\}, \\ l_2 &= \{(x_1, 0, 0) \in \mathbb{R}^3 | x_1 < 0\}, \\ l_3 &= \{(x_1, x_2, 0) \in \mathbb{R}^3 | x_1 = x_2 > 0\}. \end{aligned}$$

For a point $y \in \mathbb{R}^3 \setminus \Delta$, $g^{-1}(y) = S^1$, so $f^{-1}(y)$ is a principal T^2-bundle over S^1, i.e., a three-torus. Thus f is a T^3-fibration, with degenerate fibers over Δ. We wish to analyze the behaviour near the general singular fiber, i.e., over l_i, $i = 1, 2, 3$, more closely. So let U_i be an open set of $\mathbb{R}^3$, homeomorphic to $l_i \times D$, where D is a two-dimensional disk, such that $l_i \subseteq U_i$, $(0,0,0) \notin U_i$, and $U_i \cap l_j = \emptyset$ for $i \neq j$. See Figure 9. We can study $f^{-1}(U_i) \to U_i$. Note that

$$f^{-1}(U_i) \cap S = \begin{cases} \{(z_1, z_2, z_3) | z_1 = z_2 = 0, z_3 \neq 0\} & i = 1 \\ \{(z_1, z_2, z_3) | z_1 = z_3 = 0, z_2 \neq 0\} & i = 2 \\ \{(z_1, z_2, z_3) | z_2 = z_3 = 0, z_1 \neq 0\} & i = 3 \end{cases}$$

Take, for example, $i = 1$. Then we can factor the quotient

$$\bar{q} : f^{-1}(U_1) \to f^{-1}(U_1)/T^2 = g^{-1}(U_1) = U_1 \times S^1$$

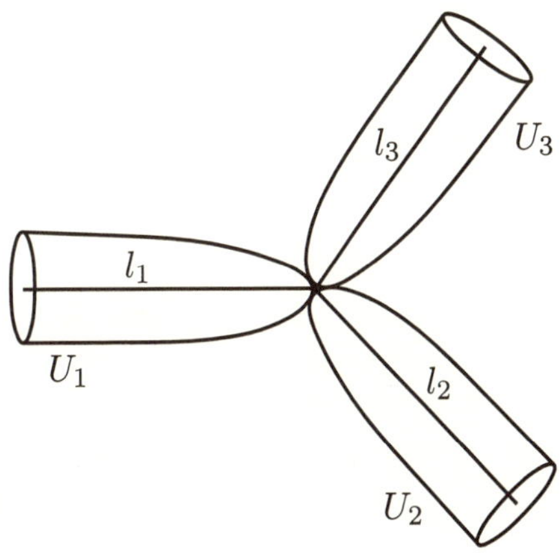

FIGURE 9

by first dividing $f^{-1}(U_1)$ out by the S^1-action

$$(z_1, z_2, z_3) \mapsto (e^{i\theta}z_1, e^{-i\theta}z_2, z_3),$$

giving

$$f^{-1}(U_1) \xrightarrow{\bar{q}_1} f^{-1}(U_1)/S^1 \xrightarrow{\bar{q}_2} U_1 \times S^1.$$

Now $\bar{q}_2$ is a (necessarily trivial) S^1-bundle, as the residual action of T^2 on $f^{-1}(U_1)/S^1$ now acts freely. Thus $f^{-1}(U_1)/S^1 \cong U_1 \times S^1 \times S^1 = U_1 \times T^2$, and $\bar{q}_1 : f^{-1}(U_1) \to U_1 \times T^2$ is an S^1-bundle degenerating over the surface $\bar{q}_1(S \cap f^{-1}(U_1))$ contained in $U_1 \times T^2$: explicitly, this is the surface

$$l_1 \times \{\text{point}\} \times S^1 \subseteq U_1 \times S^1 \times S^1.$$

We obtain a similar picture over U_2 and U_3.

EXERCISE 6.93. Using the above description, show that $f^{-1}(U_i) \to U_i$ coincides topologically with the fibration of Construction 6.88. Using this, compute the monodromy of the T^3-bundle $f^{-1}(\mathbb{R}^3 \setminus \Delta) \to \mathbb{R}^3 \setminus \Delta$ about loops around l_1, l_2 and l_3. Choosing a suitable basis, show this monodromy is transpose inverse to the monodromy of the fibration of Construction 6.90.

Finally, we note that $f^{-1}(0,0,0)$ looks like a T^3 with a T^2 pinched to a point, as in Figure 10. This is a fiber of Euler characteristic $+1$. As a result, we call this a *positive fiber.*

We now see a simple local form of SYZ duality. Positive and negative fibers are SYZ dual to each other. By this, we mean that in Constructions 6.90 and 6.92 we have constructed fibrations $X(B) \to B$ and $\check{X}(B) \to B$, where B is a neighbourhood of one of the trivalent vertices of, say, Example 6.74, (2). These fibrations are SYZ dual, and one particular consequence of this duality is that the Euler characteristic of the singular fibers are

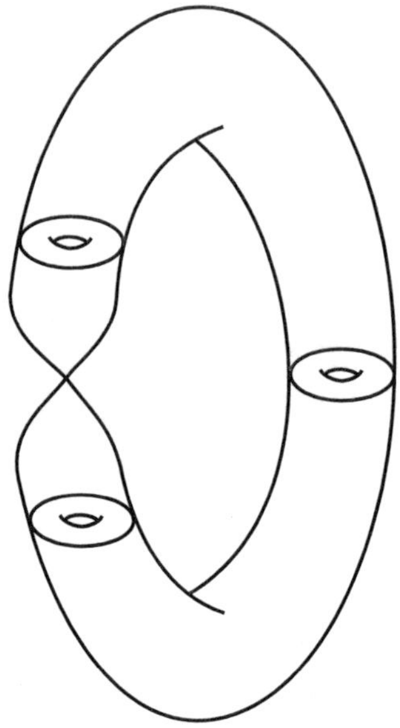

FIGURE 10

opposite in sign. *We view this as a local manifestation of mirror symmetry.* In particular, this observation will explain the global change in sign of the Euler characteristic of SYZ dual torus fibrations.

6.4.3. Compactifications. We can use the local models described in the previous section to topologically compactify two and three dimensional torus fibrations with the right sort of monodromy. We will not give quite as many details as we did in the previous section, as these are largely technical, and so relegate these details to the exercises.

First, let B be a two-dimensional integral affine manifold with singularities. Suppose that for any simple loop γ about a singular point of B based at a point b near the singular point, the monodromy transformation $T_\gamma : \Lambda_b \to \Lambda_b$ takes the form $\left(\begin{smallmatrix} 1 & 1 \\ 0 & 1 \end{smallmatrix}\right)$ in some basis. This is the case, for example, in Example 6.82, (1). Now consider the torus bundle $f_0 : X(B_0) \to B_0$. In general, $\Lambda_b \cong H_1(f_0^{-1}(b), \mathbb{Z})$ and T_γ coincides with the monodromy transformation $H_1(f_0^{-1}(b), \mathbb{Z}) \to H_1(f_0^{-1}(b), \mathbb{Z})$. Thus the restriction on the monodromy of the affine manifold guarantees that the local monodromy of the torus bundle near the singular points coincides with that given in Construction 6.86.

Construction 6.86 gave us a T^2-fibration $\overline{X} \to D$. There $\overline{X}$ and D were open manifolds. By replacing D with a closed disk $\overline{D}$ contained inside D and replacing $\overline{X}$ with the inverse image of $\overline{D}$, we obtain a T^2-fibration $\bar{f} : \overline{X} \to \overline{D}$ of manifolds with boundary, with one singular fiber. Then $\bar{f} : \partial\overline{X} \to \partial\overline{D}$ is a T^2-bundle over a circle, which we have already described explicitly. If we identify $\overline{D}$ with a small closed disk $\overline{D'} \subseteq B$ containing one singular point $b_0 \in B$, we obtain a T^2-bundle $f_0^{-1}(\partial\overline{D'}) \to \partial\overline{D'}$.

EXERCISE 6.94. Describe the T^2-bundle $\bar{f} : \partial\overline{X} \to \partial\overline{D}$ as $\mathcal{E}/\Gamma \to \partial\overline{D}$ where $\mathcal{E}$ is a (necessarily trivial) rank 2 (real) vector bundle on $\partial\overline{D}$ and Γ is a family of lattices in $\mathcal{E}$. Of course the T^2-bundle $f_0^{-1}(\partial\overline{D'}) \to \partial\overline{D'}$ is of the same form by construction, i.e., $TB_0|_{\partial\overline{D'}}/\Lambda|_{\partial\overline{D'}} \to \partial\overline{D'}$. Then show that because the local systems $\Lambda|_{\partial\overline{D'}}$ and Γ have the same monodromy, there is a homeomorphism $\partial\overline{X} \to f_0^{-1}(\partial\overline{D'})$ preserving the fiber structure and is linear on each fiber.

As a result of this exercise, we can glue $\bar{f} : \overline{X} \to \overline{D}$ to $f_0^{-1}(B_0 \setminus (D' \cap B_0)) \to B_0 \setminus (D' \cap B_0)$ along their common boundary using this homeomorphism. Doing this gluing at each singular point produces a compactified fibration $X(B) \to B$ as a topological fibration. This procedure can be done in such a way as to produce a differentiable manifold $X(B)$ so that $X(B) \to B$ is C^∞.

This procedure can just as well be carried out for $\check{X}(B_0) \to B_0$, since the conjugacy class of the monodromy transformations doesn't change under dualizing. In fact, in this case the compactification can be carried out in the symplectic category (see [**435, 90**]) but we will not give details here.

Moving on to the three-dimensional case, we again must make an assumption on the monodromy of the fibration. We will give a rather ad hoc definition for the restriction. This is a special case of a more general definition given in [**206**], §1.5.

DEFINITION 6.95. Let B be a three-dimensional integral affine manifold with singularities. We say B is *simple* if

(1) The monodromy of Λ about each edge of the graph Δ takes the form $\begin{pmatrix} 1&1&0\\0&1&0\\0&0&1\end{pmatrix}$ in a suitable basis, and
(2) each vertex of Δ is trivalent, and if $b \in B_0$ is a point near a vertex b_0 of Δ, $\gamma_1, \gamma_2, \gamma_3$ simple loops based at b around the three edges of Δ coming out of b_0, with $\gamma_1, \gamma_2, \gamma_3$ oriented so that $\gamma_1\gamma_2\gamma_3 = 1$ in $\pi_1(B_0, b)$, then $T_{\gamma_i} : \Lambda_b \to \Lambda_b$, $i = 1, 2, 3$ take the form in a suitable basis

$$\begin{pmatrix} 1&0&1\\0&1&0\\0&0&1\end{pmatrix}, \begin{pmatrix} 1&0&-1\\0&1&-1\\0&0&1\end{pmatrix} \text{ and } \begin{pmatrix} 1&0&0\\0&1&1\\0&0&1\end{pmatrix}.$$

or

$$\begin{pmatrix} 1&0&1\\0&1&0\\0&0&1\end{pmatrix}, \begin{pmatrix} 1&1&0\\0&1&0\\0&0&1\end{pmatrix} \text{ and } \begin{pmatrix} 1&-1&-1\\0&1&0\\0&0&1\end{pmatrix}.$$

Our goal is to compactify $X(B_0)$ or $\check{X}(B_0)$ for B a simple compact integral affine manifold with singularities.

Let's start with a special case. Assume that a connected component C of Δ has no trivalent vertices. In this case C will be a circle, and we can show how to extend $f : X(B_0) \to B_0$ across C. Let B' be a tubular neighbourhood of C in B, so that $B' \cap \Delta = C$, and $\partial \overline{B}'$ is a two-torus as depicted in Figure 11. We can restrict the T^3-bundle $f : X(B_0) \to B_0$ to $\partial \overline{B}'$, and ask what the

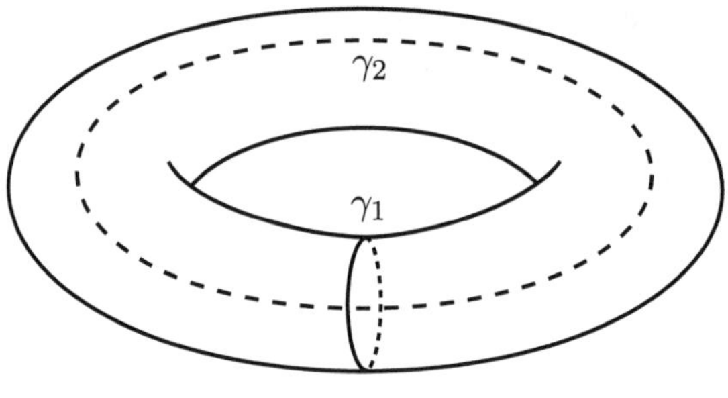

FIGURE 11

monodromy of this torus bundle about the generators γ_1 and γ_2 of $\pi_1(\partial \overline{B}', b)$ can be, where $b \in \partial \overline{B}'$ is a base point. By assumption, there is a basis e_1, e_2, e_3 of $H_1(f^{-1}(b), \mathbb{Z})$ such that $T_{\gamma_1} = \begin{pmatrix} 1&1&0\\0&1&0\\0&0&1 \end{pmatrix}$. Now since $\pi_1(\partial \overline{B}', b)$ is abelian, T_{γ_1} and T_{γ_2} must commute. If we write $T_{\gamma_2} = (a_{ij})_{1 \le i,j \le 3}$, then this commutativity ensures that

$$a_{21} = a_{23} = a_{31} = 0 \text{ and } a_{11} = a_{22}$$

so

$$T_{\gamma_2} = \begin{pmatrix} a_{11} & a_{12} & a_{13} \\ 0 & a_{11} & 0 \\ 0 & a_{32} & a_{33} \end{pmatrix}.$$

Since $\det T_{\gamma_2} = \pm 1$, we must have $a_{11} = \pm 1$ and $a_{33} = \pm 1$. The group $G \subset GL_3(\mathbb{Z})$ of matrices commuting with T_{γ_1} is then generated by the matrices

$$\begin{pmatrix} 1 & a & b \\ 0 & 1 & 0 \\ 0 & 0 & 1 \end{pmatrix}, a, b \in \mathbb{Z}, \begin{pmatrix} 1 & 0 & 0 \\ 0 & 1 & 0 \\ 0 & 1 & 1 \end{pmatrix}, \begin{pmatrix} 1 & 0 & 0 \\ 0 & 1 & 0 \\ 0 & 0 & -1 \end{pmatrix}, \begin{pmatrix} -1 & 0 & 0 \\ 0 & -1 & 0 \\ 0 & 0 & 1 \end{pmatrix}.$$

Now let $\bar{h} : \overline{X} \to \overline{D}$ be the fibration of Construction 6.86 over a closed disk $\overline{D}$ just as we considered in the two-dimensional case. Then $\dim \overline{X} = 4$ and $\bar{h}$ has one singular fiber, a pinched torus. We will construct a torus fibration over $\overline{B} := S^1 \times \overline{D}$ by identifying $\{0\} \times \overline{X} \times S^1$ and $\{1\} \times \overline{X} \times S^1$ in $[0,1] \times \overline{X} \times S^1$ via a homeomorphism $\varphi : \overline{X} \times S^1 \to \overline{X} \times S^1$ fitting into a commutative diagram

$$\begin{array}{ccc} \overline{X} \times S^1 & \longrightarrow & \overline{X} \times S^1 \\ \downarrow & & \downarrow \\ \overline{D} & \xrightarrow{=} & \overline{D} \end{array}$$

where the vertical arrows are projection to $\overline{X}$ followed by $\bar{h}$. Making this identification will yield a six-manifold $\overline{Z}$ with a fibration $\bar{f} : \overline{Z} \to \overline{B}$, with singular fibers over $S^1 \times \{0\}$. We obtain in this way a torus fibration whose monodromy can be described as follows. Recall that by the construction of $\overline{X}$, there is a map $\overline{X} \to \overline{Y} = S^1 \times \overline{D}$ with an S^1-action on fibers (with one fixed point of this action). Fix a point $b \in \overline{D} \setminus \{0\}$, and let e_1, e_2 be generators of $H_1(\bar{h}^{-1}(b), \mathbb{Z})$ chosen so that e_1 is an orbit of this S^1-action and monodromy about 0 takes the form $\left(\begin{smallmatrix} 1 & 1 \\ 0 & 1 \end{smallmatrix}\right)$. If we then identify b with a point in the base $\overline{B}$ of the fibration $\bar{f}$, given by $b \in \{0\} \times (\overline{D} \setminus \{0\}) \subseteq S^1 \times (\overline{D} \setminus \{0\})$, then we can take a basis of $H_1(\bar{f}^{-1}(b), \mathbb{Z}) = H_1(\bar{h}^{-1}(b) \times S^1, \mathbb{Z})$ given by $e_1 = e_1 \times \{\mathrm{pt}\}$, $e_2 = e_2 \times \{\mathrm{pt}\}$, $e_3 = \{\mathrm{pt}\} \times S^1$. Then monodromy about the loop γ_1 takes the form $T_{\gamma_1} = \left(\begin{smallmatrix} 1 & 1 & 0 \\ 0 & 1 & 0 \\ 0 & 0 & 1 \end{smallmatrix}\right)$ as desired. As $\varphi : \{0\} \times \overline{X} \times S^1 \to \{1\} \times \overline{X} \times S^1$ has been chosen to preserve fibers, φ acts on $\bar{f}^{-1}(b)$, and $T_{\gamma_2} = \varphi_* : H_1(\bar{f}^{-1}(b), \mathbb{Z}) \to H_1(\bar{f}^{-1}(b), \mathbb{Z})$. Thus, in order to realize a particular T_{γ_2}, we need to choose a diffeomorphism φ realizing this transformation on fibers. We show how to do this for each generator of the group G.

$\begin{pmatrix} 1 & a & b \\ 0 & 1 & 0 \\ 0 & 0 & 1 \end{pmatrix}$: We have the S^1-fibration $\bar{q} : \overline{X} \to \overline{Y} = S^1 \times \overline{D}$, hence an induced map

$$\overline{X} \times S^1 \xrightarrow{\bar{q} \times \mathrm{id}} (S^1 \times \overline{D}) \times S^1.$$

Identifying S^1 with the unit circle in $\mathbb{C}$, there is a map $m : S^1 \times S^1 \to S^1$ given by $(z_1, z_2) \mapsto z_1^a z_2^b$. Let $\pi : S^1 \times \overline{D} \times S^1 \to S^1 \times S^1$ be the projection. Then remembering that S^1 acts on $\overline{X}$, hence on $\overline{X} \times S^1$ with trivial action on the second factor, we can define

$$\varphi(x) = (m \circ \pi \circ (\bar{q} \times \mathrm{id})(x)) \cdot x.$$

It is easy to see that φ preserves fibers of $\overline{X} \times S^1 \to \overline{D}$, and the action on each fiber induces the matrix $\left(\begin{smallmatrix} 1 & a & b \\ 0 & 1 & 0 \\ 0 & 0 & 1 \end{smallmatrix}\right)$ on homology.

$\begin{pmatrix} 1 & 0 & 0 \\ 0 & 1 & 0 \\ 0 & 1 & 1 \end{pmatrix}$: We act on $(S^1 \times \overline{D}) \times S^1$ by $(z_1, b, z_2) \mapsto (z_1, b, z_1 z_2)$. Note that $\overline{X} \times S^1 \to (S^1 \times \overline{D}) \times S^1$ restricts to an S^1-bundle over $((S^1 \times \overline{D}) \setminus \{(p, 0)\}) \times S^1$, and it is easy to check that this preserves the Chern class of this S^1-bundle. Hence this homeomorphism lifts to a homeomorphism of $\overline{X} \times S^1$.

$\begin{pmatrix} 1 & 0 & 0 \\ 0 & 1 & 0 \\ 0 & 0 & -1 \end{pmatrix}$: Just act on $\overline{X} \times S^1$ by $(x, z) \mapsto (x, \bar{z})$.

$\begin{pmatrix} -1 & 0 & 0 \\ 0 & -1 & 0 \\ 0 & 0 & 1 \end{pmatrix}$: Think of each fiber of $\overline{X} \to \overline{D}$ as an abelian group, and act by negation. Alternatively, using the description of $\overline{X} \to \overline{D}$ in Construction 6.87, we can act on $\overline{X}$ by $(z_1, z_2, w) \mapsto (\bar{z}_1, \bar{z}_2, \bar{w})$.

Thus we are able to construct a T^3-fibration $\bar{f} : \overline{Z} \to \overline{B}$ with any desired monodromy, given the form of T_{γ_1}. We may then glue this in to $X(B_0)$.

EXERCISE 6.96. (Details of gluing). Identify the closure $\overline{B}'$ of the tubular neighbourhood of $C \cong S^1$ with $\overline{B}$. Describe the T^3-bundle $\bar{f} : \bar{f}^{-1}(\partial \overline{B}) \to \partial \overline{B}$ as $\mathcal{E}/\Gamma \to \partial \overline{B}$, where $\mathcal{E}$ is a trivial rank 3 vector bundle on $\partial \overline{B}$ and Γ is a family of lattices in $\mathcal{E}$. Show as in Exercise 6.94 that using this, the fibrations $\bar{f}$ and $f : X(B_0 \setminus B') \to B_0 \setminus B'$ can be glued along their common boundary.

REMARK 6.97. We should take some additional care in the gluing with regards to the existence of a section. Of course, $X(B_0) \to B_0$ has a section given by the zero section of TB_0. We would like to ensure that the compactification $X(B) \to B$ we construct also has a section. We can do this by finding a section of each of our local models and ensuring that when we glue, the zero section of $X(B_0) \to B_0$ and the chosen section of the local model can be identified.

For example, first consider the T^2 fibration $\overline{X} \to \overline{D}$ considered above. We have the S^1 fibration $\bar{q} : \overline{X} \to \overline{Y} = S^1 \times \overline{D}$, degenerating over $(p, 0) \in S^1 \times \overline{D}$. Choose a point $p' \in S^1$, $p' \neq p$. Then $\bar{q}^{-1}(\{p'\} \times \overline{D}) \to \{p'\} \times \overline{D}$ is an S^1-bundle over a disk, hence necessarily trivial, so we can choose a section of this S^1-bundle, defining a section

$$\sigma : \overline{D} \to \overline{X}$$

given by the composition

$$\overline{D} \xrightarrow{\cong} \{p'\} \times \overline{D} \to \bar{q}^{-1}(\{p'\} \times \overline{D}) \subseteq \overline{X},$$

where the second arrow is the chosen section of the S^1-bundle. We leave it to the reader to check that in Exercise 6.94 we can identify the T^2-bundle $\partial \overline{X} \to \partial \overline{D}$ with $\mathcal{E}/\Gamma \to \partial \overline{D}$ in such a way that $\sigma|_{\partial \overline{D}}$ is identified with the zero section. This enables us to perform the gluing of Exercise 6.94 so that the sections match, hence yielding a compactification with a section.

In the three-dimensional case we have just considered, we can follow the same procedure as long as $\bar{f} : \overline{Z} \to S^1 \times D$ constructed above always has a section. We leave it to the reader to check this is the case for each choice of T_{γ_2} above, and that the gluing of Exercise 6.96 can then be performed in such a way that sections glue.

We are now ready to complete the three-dimensional compactification. First, if Δ has any connected components without vertices, we extend the fibre bundle $X(B_0) \to B_0$ across these circles as above.

We proceed in two steps to complete the compactification. Let B' be a tubular neighbourhood of the remaining components of Δ. Let l be a line segment contained in Δ joining two vertices, and let $l' \subseteq l$ be a smaller closed interval contained in l. Choose a closed tubular neighbourhood $\overline{U} \cong \overline{D} \times l'$

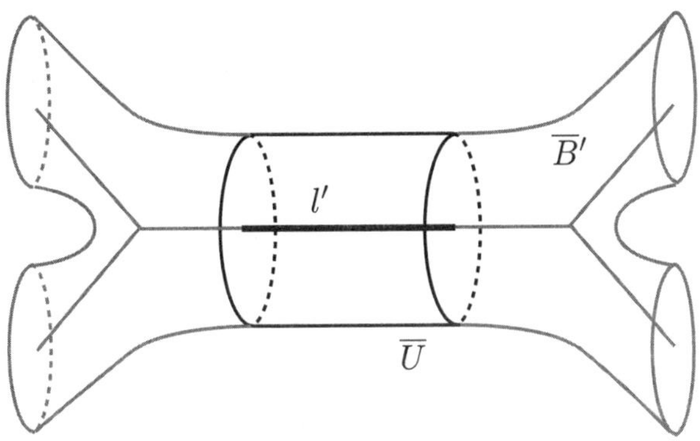

FIGURE 12

of l' as depicted in Figure 12, so that $\partial\overline{D} \times l' \subseteq \partial\overline{B}'$. Then as before, we can glue in a fibration $\overline{X} \times S^1 \times l' \to \overline{U}$. We do this for each such leg.

We note that as in Remark 6.97, this gluing can be done in such a way that the zero section of $X(B\backslash B') \to B\backslash B'$ matches up with a chosen section of $\overline{X} \times S^1 \times l' \to \overline{U}$.

There is another compatibility we need to preserve. Both $\overline{X} \times S^1 \times l'$ and $X(\overline{U} \cap (B \setminus B'))$ have natural $T^2 = S^1 \times S^1$ actions: For $\overline{X} \times S^1 \times l'$, the first factor S^1 acts on $\overline{X}$ by construction, and the second factor acts on the S^1 factor. On the other hand, if the monodromy around the leg is given in a basis e_1, e_2, e_3 by the matrix $\left(\begin{smallmatrix} 1 & 1 & 0 \\ 0 & 1 & 0 \\ 0 & 0 & 1 \end{smallmatrix}\right)$, then locally on $\overline{U} \cap (B \setminus B')$, e_1 and e_3 give well-defined sections of Λ, and $T^2 = (\mathbb{R}e_1/\mathbb{Z}e_1) \times (\mathbb{R}e_3/\mathbb{Z}e_3)$ acts naturally by translation on $X(\overline{U} \cap (B \setminus B'))$. We demand that we perform the gluing so that these two actions match up. It is not difficult to see that this can be done.

Having performed this gluing, we obtain a fibration $f : X(B_0') \to B_0'$, where $B_0' \subseteq B$ is the union of $B\backslash B'$ and the sets $\overline{U}$ occurring above. We now have left a number of neighbourhoods of trivalent vertices, as depicted in Figure 13. Let U be an open neighbourhood of a trivalent vertex as depicted, and let $V = U \cap (B \setminus B_0')$ so that we have a fibration $f : X(U \setminus V) \to U \setminus V$.

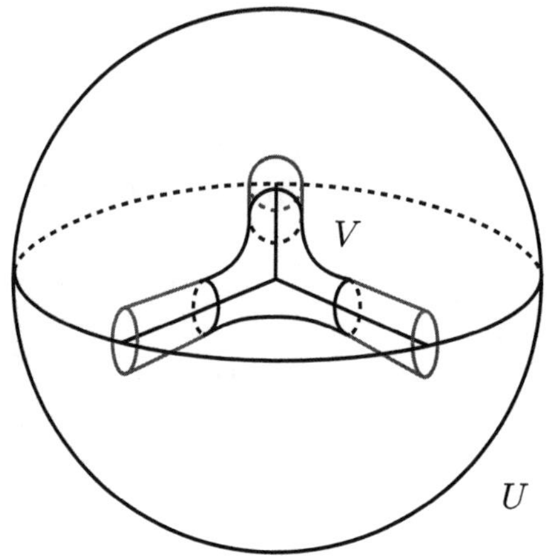

FIGURE 13

The trivalent vertex is either a positive or negative vertex in the sense of §6.4.2.

Consider first the case of a negative vertex. The monodromy of Λ near the vertex has one invariant: in the basis e_1, e_2, e_3 chosen in Definition 6.95, (2), e_1 is invariant. Then $S^1 = \mathbb{R}e_1/\mathbb{Z}e_1$ acts on $X(U \setminus \Delta)$ by translation, and because we have insisted that the gluing preserves this action, we in fact get an action of S^1 on $f^{-1}(\partial\overline{V})$.

EXERCISE 6.98. Show that $f^{-1}(\partial\overline{V})/S^1 \cong T^2 \times \partial\overline{V}$. Using this, show that $f^{-1}(\partial\overline{V}) \to \partial\overline{V}$ coincides with the map $f^{-1}(\partial\overline{B}) \to \partial\overline{B}$ of Construction 6.90, and thus this fibration can be glued in. Note this can be done so that a choice of section is glued to the zero section of $f : X(B_0') \to B_0'$, and so that the S^1 actions match, so that the S^1 action extends to the compactification.

Similarly, in the case of the positive vertex, there is a rank 2 monodromy invariant subgroup of Λ, hence a T^2-action. This time $f^{-1}(\partial\overline{V})/T^2 \cong S^1 \times \partial\overline{V}$, and one can check as before that the positive vertex fibration of Construction 6.92 can be glued in. As always, this can be done in a way which preserves the section and the local T^2-action.

Having finished the compactification by following the above steps, we obtain a torus fibration $f : X(B) \to B$ with the properties

(1) $f|_{f^{-1}(B\setminus B')}$ coincides with $X(B \setminus B') \to B \setminus B'$.
(2) f has a section, extending the zero-section of $X(B \setminus B') \to B \setminus B'$.
(3) For any open set $U \subseteq B$, there is a fiberwise torus action on $f^{-1}(U) \to U$ of the torus $\Gamma(U \setminus \Delta, \Lambda_{\mathbb{R}}/\Lambda)$.

This follows from the fact that we have preserved the section and action locally.

REMARK 6.99. This action allows us to perform a "topological twisting" on $f : X(B) \to B$. Indeed, given an element $\mathbf{B} \in H^1(B, i_*(\Lambda_{\mathbb{R}}/\Lambda))$, represented as a Čech 1-cocycle (b_{ij}) with $b_{ij} \in \Gamma((U_i \cap U_j) \setminus \Delta, \Lambda_{\mathbb{R}}/\Lambda)$, we can reglue the fibration $f : X(B) \to B$ to obtain $X(B, \mathbf{B}) \to B$ as we did in §6.2.3. However, this now occurs at a purely topological level. We may obtain a different topological manifold if the new fibration $X(B, \mathbf{B}) \to B$ does not have a section.

EXAMPLE 6.100. Consider B as given in Example 6.74. In each example, we can now construct compactifications $X(B) \to B$ of $X(B_0) \to B_0$ and $\check{X}(B) \to B$ of $\check{X}(B_0) \to B_0$. As mentioned at the end of §6.4.1, we anticipate that $X(B)$ should, topologically, be homeomorphic to a (resolution of a) hyperplane section of $\mathbb{P}_{\Xi^*}$ and $\check{X}(B)$ should be homeomorphic to a (resolution of a) hyperplane section of $\mathbb{P}_{\Xi}$. This indeed happens in the examples given there.

Specifically, one can show that in the two-dimensional example, we obtain a topological K3 surface. In the three-dimensional example, it was the main result of Gross [**202**] that $X(B)$ is homeomorphic to the mirror quintic and $\check{X}(B)$ is homeomorphic to the quintic.

For more general choices of reflexive polytopes, in any dimension, it is possible to show some similar results; this has been explored in Gross [**203**] and Gross and Siebert [**204**]. We omit details of these statements.

6.4.4. Cohomology of SYZ fibrations. Let B be an affine manifold with singularities, $i : B_0 \hookrightarrow B$ the inclusion. Suppose we have topological compactifications of $X(B_0)$ and $\check{X}(B_0)$, which we call $X(B)$ and $\check{X}(B)$ respectively, along with torus fibrations $f : X(B) \to B$, $\check{f} : \check{X}(B) \to B$ extending $f_0 : X(B_0) \to B_0$ and $\check{f}_0 : \check{X}(B_0) \to B_0$. Assuming we are in dimensions two and three, such compactifications exist given suitable restrictions on the monodromy, i.e., B simple. In addition, if B is simple, then in dimension three, the Euler characteristics of $X(B)$ and $\check{X}(B)$ are of opposite sign. Indeed, only a finite number of fibers of $X(B) \to B$ and $\check{X}(B) \to B$ have non-zero Euler characteristic, and thus the Euler characteristic of $X(B)$ or $\check{X}(B)$ is just the sum of the Euler characteristics of these individual fibers. Thus

$$\begin{aligned}\chi(X(B)) = &\big(\#\text{ of positive fibers of } X(B) \to B\big)\\ &- \big(\#\text{ of negative fibers of } X(B) \to B\big).\end{aligned}$$

But the SYZ dualizing process interchanges positive and negative fibers, so clearly $\chi(X(B)) = -\chi(\check{X}(B))$.

However, mirror symmetry demands at the very least an interchange of Hodge numbers between mirror pairs. Why should we expect this to occur? We will explain this here in dimension three, but first we give the general setup.

Given $f : X(B) \to B$, we can consider the sheaf $R^q f_* \mathbb{Q}$. This is the sheaf associated with the presheaf $U \mapsto H^q(f^{-1}(U), \mathbb{Q})$ on B. Similarly, we have the sheaf $R^q f_{0*}\mathbb{Q}$ on B_0. Because f_0 is a torus bundle, the latter sheaf is a local system, i.e., locally isomorphic to a constant sheaf with coefficients in the cohomology of a torus. Of course, $(R^q f_*\mathbb{Q})|_{B_0} = R^q f_{0*}\mathbb{Q}$, but $R^q f_*\mathbb{Q}$ generally fails to be a local system along $\Delta = B \setminus B_0$. However, it is always true that the stalks of $R^p f_*\mathbb{Q}$ are given by

$$(R^p f_*\mathbb{Q})_b \cong H^p(f^{-1}(b), \mathbb{Q}).$$

So if $f^{-1}(b)$ is singular, we might expect this cohomology group to be different from that of a non-singular torus.

DEFINITION 6.101. We say $f : X \to B$ is $\mathbb{Q}$-*simple* if

$$i_* R^p f_{0*}\mathbb{Q} = R^p f_*\mathbb{Q}$$

for all p. More generally, if G is any abelian group, we say f is G-*simple* if

$$i_* R^p f_{0*} G = R^p f_* G.$$

Note by definition, if $U \subseteq B$ is open, then

$$\Gamma(U, i_* R^p f_{0*}\mathbb{Q}) = \Gamma(U \cap B_0, R^p f_{0*}\mathbb{Q}).$$

Thus simplicity essentially requires that if $b \in B \setminus B_0$, then

$$H^p(f^{-1}(b), \mathbb{Q}) \cong \Gamma(U \cap B_0, R^p f_{0*}\mathbb{Q})$$

for U a sufficiently small open neighbourhood of b.

This is quite a powerful assumption. Let's see how this helps us. First consider the cohomology of a single torus V/Γ, where V is an n-dimensional vector space and Γ a rank n lattice. Then

$$\begin{aligned} H_1(V/\Gamma, \mathbb{Z}) &\cong \Gamma \\ H^1(V/\Gamma, \mathbb{Z}) &\cong \Gamma^\vee \\ H^p(V/\Gamma, \mathbb{Z}) &\cong \bigwedge^p \Gamma^\vee. \end{aligned}$$

On the other hand, viewing $V^\vee/\Gamma^\vee$ as the dual torus, where $\Gamma^\vee$ is identified with

$$\{v \in V^\vee | v(\Gamma) \subseteq \mathbb{Z}\},$$

we see $H^p(V^\vee/\Gamma^\vee, \mathbb{Z}) \cong \bigwedge^p \Gamma$. If we choose a generator of $\bigwedge^n \Gamma^\vee \cong \mathbb{Z}$, then the natural pairing

$$\bigwedge^p \Gamma^\vee \otimes \bigwedge^{n-p} \Gamma^\vee \to \bigwedge^n \Gamma^\vee \xrightarrow{\cong} \mathbb{Z}$$

induces an isomorphism

$$\bigwedge^p \Gamma^\vee \cong \bigwedge^{n-p} \Gamma.$$

Thus, up to a choice of generator of $\bigwedge^n \Gamma^\vee$, there is a canonical isomorphism

$$H^p(V/\Gamma, \mathbb{Z}) \cong H^{n-p}(V^\vee/\Gamma^\vee, \mathbb{Z}).$$

Applying this to $f_0 : X(B_0) \to B_0$ and $\check{f}_0 : \check{X}(B_0) \to B_0$, keeping in mind that $X(B_0) = TB_0/\Lambda$ and $\check{X}(B_0) = T^*B_0/\check{\Lambda}$, we have

$$R^1 f_{0*}\mathbb{Z} = \check{\Lambda} \text{ and } R^1 \check{f}_{0*}\mathbb{Z} = \Lambda.$$

Thus

$$R^p f_{0*}\mathbb{Z} = \bigwedge^p \check{\Lambda} \text{ and } R^p \check{f}_{0*}\mathbb{Z} = \bigwedge^p \Lambda.$$

Tensoring with $\mathbb{Q}$ gives

$$R^p f_{0*}\mathbb{Q} = \bigwedge^p \check{\Lambda}_{\mathbb{Q}} \text{ and } R^p \check{f}_{0*}\mathbb{Q} = \bigwedge^p \Lambda_{\mathbb{Q}}$$

where $\Lambda_{\mathbb{Q}} = \Lambda \otimes_{\mathbb{Z}} \mathbb{Q}$ and $\check{\Lambda}_{\mathbb{Q}} = \check{\Lambda} \otimes_{\mathbb{Z}} \mathbb{Q}$. Now assuming that $\bigwedge^n \Lambda \cong \mathbb{Z}$ (which is the case if the transition maps of B are in $\mathbb{R}^n \rtimes \mathrm{SL}_n(\mathbb{Z})$, as in Exercise 6.13), we then have a natural pairing as in that section

$$\bigwedge^p \Lambda \times \bigwedge^{n-p} \Lambda \to \bigwedge^n \Lambda \cong \mathbb{Z},$$

which yields an isomorphism

$$\bigwedge^p \check{\Lambda} \cong \bigwedge^{n-p} \Lambda.$$

Thus

$$R^p f_{0*}\mathbb{Q} \cong R^{n-p} \check{f}_{0*}\mathbb{Q},$$

and hence

$$i_* R^p f_{0*}\mathbb{Q} \cong i_* R^{n-p} \check{f}_{0*}\mathbb{Q}.$$

$\mathbb{Q}$-simplicity now tells us that

$$R^p f_*\mathbb{Q} \cong R^{n-p} \check{f}_*\mathbb{Q}. \tag{6.7}$$

This will be immensely useful for determining the relationship between the cohomology of $X(B)$ and $\check{X}(B)$ using the Leray spectral sequence.

We have used the word "simple", an overused word in mathematics, twice. Let us see that the simplicity of Definition 6.95 implies the $\mathbb{Q}$-simplicity of Definition 6.101.

PROPOSITION 6.102. *If B is three-dimensional and simple, then the compactifications $f : X(B) \to B$ and $\check{f} : \check{X}(B) \to B$ of f_0 and $\check{f}_0$ are $\mathbb{Q}$-simple.*

PROOF. Let $b \in \Delta$. We will work with f; the story for $\check{f}$ is the same. It is enough to show, for a sufficiently small neighbourhood U of b for which $\Gamma(U, R^p f_*\mathbb{Q}) \cong H^p(f^{-1}(b), \mathbb{Q})$, that the restriction map

$$\Gamma(U, R^p f_*\mathbb{Q}) \to \Gamma(U \setminus \Delta, R^p f_*\mathbb{Q}) = \Gamma(U, i_* R^p f_{0*}\mathbb{Q})$$

is an isomorphism. Now $\Gamma(U, i_* R^p f_{0*}\mathbb{Q})$ can be identified with the space of monodromy invariant elements of $H^p(f^{-1}(b_0), \mathbb{Q})$, where $b_0 \in U \setminus \Delta$. In particular, we just need to show that the restriction of cohomology classes

$$H^p(f^{-1}(U), \mathbb{Q}) \to H^p(f^{-1}(b_0), \mathbb{Q})$$

gives an isomorphism onto the monodromy invariant subspace W of the cohomology group $H^p(f^{-1}(b_0),\mathbb{Q})$, or dually the map

$$H_p(f^{-1}(b_0),\mathbb{Q}) \to H_p(f^{-1}(U),\mathbb{Q})$$

induces an isomorphism between $H_p(f^{-1}(b_0),\mathbb{Q})/W^\perp$ and $H_p(f^{-1}(U),\mathbb{Q})$. The latter can be identified with $H_p(f^{-1}(b),\mathbb{Q})$ via a deformation retract of $f^{-1}(U)$ onto $f^{-1}(b)$. So this statement can then be checked individually in the three different possibilities for $f^{-1}(b)$: the general singular fiber constructed in Construction 6.88, the negative fiber of Construction 6.90, and the positive fiber of Construction 6.92. Let us consider the negative case and leave the other cases to the reader. First, this is trivial for $p = 0, 3$. Now $f^{-1}(b)$ is an S^1-fibration over a T^2, with the S^1's degenerating to points over a figure eight, and it is easy to see that $H_1(f^{-1}(b),\mathbb{Q}) = \mathbb{Q}^2$ is generated by circles which map to the two generators of $H_1(T^2,\mathbb{Q})$. On the other hand, you will have computed the monodromy of this fibration in Exercise 6.91, and found that monodromy acting on $H_1(f^{-1}(b_0),\mathbb{Z})$ coming from going around the loops around the legs of Δ is given by the matrices

$$\begin{pmatrix}1&0&1\\0&1&0\\0&0&1\end{pmatrix},\ \begin{pmatrix}1&1&0\\0&1&0\\0&0&1\end{pmatrix},\ \text{and}\ \begin{pmatrix}1&-1&-1\\0&1&0\\0&0&1\end{pmatrix},$$

with a basis e_1, e_2, e_3 with e_1 the class of an orbit of the S^1 action on $f^{-1}(U)$ used to construct $f^{-1}(U)$, and where e_2, e_3 are a choice of deformations of the two generators of $H_1(f^{-1}(b),\mathbb{Z})$ described above to $f^{-1}(b_0)$. The fact that there is not a unique choice of such a deformation is encoded in the monodromy. But from this description of monodromy, $W^\perp$ is generated by e_1, so we get the desired isomorphism. A similar argument works for H_2, with $H_2(f^{-1}(b),\mathbb{Q})$ generated by a T^2 which is a section of the projection $f^{-1}(b) \to T^2$. We leave the remaining details to the reader. □

We will now explain in detail what the Leray spectral sequence is and what it does for us. For $f : X \to B$ any continuous map, the Leray spectral sequence is

$$E_2^{p,q} = H^p(B, R^q f_*\mathbb{Q}) \Rightarrow E_\infty^n = H^n(X,\mathbb{Q}).$$

This notation contains a lot of information. First, we can put the groups $H^p(B, R^q f_*\mathbb{Q})$ in an array, forming the so-called E_2 term of the spectral sequence.

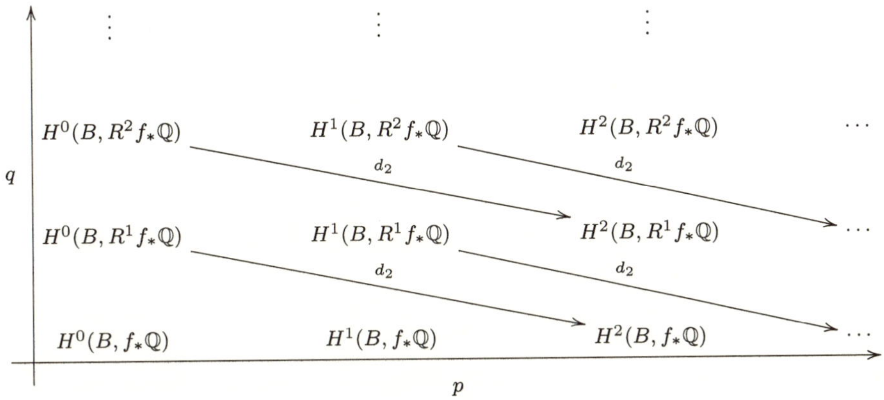

This comes with additional data of maps

$$d_2 : H^p(B, R^q f_*\mathbb{Q}) \to H^{p+2}(B, R^{q-1} f_*\mathbb{Q}).$$

This map is zero if p or $q-1$ is negative. But there is more: the composition of two of these maps is always zero, and if we set

$$E_3^{p,q} = \frac{\ker(d_2 : H^p(B, R^q f_*\mathbb{Q}) \to H^{p+2}(B, R^{q-1} f_*\mathbb{Q}))}{\operatorname{im}(d_2 : H^{p-2}(B, R^{q+1} f_*\mathbb{Q}) \to H^p(B, R^q f_*\mathbb{Q}))},$$

then we also obtain maps

$$d_3 : E_3^{p,q} \to E_3^{p+3,q-2}.$$

(We do not explain where any of these maps come from; this is part of the magic of spectral sequences. If you would like to understand where the machinery of spectral sequences comes from, consult any text on homological algebra, e.g. [**53**].) We can carry on like this, and at the nth step we have groups $E_n^{p,q}$ along with maps

$$d_n : E_n^{p,q} \to E_n^{p+n,q-n+1}$$

and

$$E_{n+1}^{p,q} = \frac{\ker(d_n : E_n^{p,q} \to E_n^{p+n,q-n+1})}{\operatorname{im}(d_n : E_n^{p-n,q+n-1} \to E_n^{p,q})}.$$

Since $E_n^{p,q} = 0$ for $p < 0$ or $q < 0$, it follows that for a given p, q there is a sufficiently large n such that $E_n^{p,q} = E_{n+1}^{p,q} = \cdots$. If we set $E_\infty^{p,q}$ to be this fixed group, then the magic of spectral sequences says the following: for each n, there is a filtration

$$H^n(X, \mathbb{Q}) = F^0 \supseteq \cdots \supseteq F^n \supseteq 0$$

such that

$$F^p / F^{p+1} \cong E_\infty^{p,n-p}.$$

In other words, the Leray spectral sequence should be viewed as a machine that takes as input the cohomology groups of certain sheaves on B, and after turning a crank, produces the graded pieces of a filtration of the cohomology of X.

The general difficulty is that we do not know what these maps d_n are, and even the groups we begin with may be hard to compute.

Let's see what this gives us for the maps $f : X(B) \to B$, $\check{f} : \check{X}(B) \to B$ as above, with $\dim B = 3$. First, $R^0 f_{0*}\mathbb{Q} = f_{0*}\mathbb{Q} = \mathbb{Q}$, as $f^{-1}(U)$ is connected for any open set U. We assume that $R^3 f_{0*}\mathbb{Q} \cong \mathbb{Q}$ also: this is just the assumption $\bigwedge^3 \check{\Lambda} \cong \mathbb{Z}$, which is equivalent to the holonomy of B_0 being contained in $\mathbb{R}^3 \rtimes \mathrm{SL}_3(\mathbb{Z})$. This allows us to choose a global section of $R^3 f_{0*}\mathbb{Q}$. By $\mathbb{Q}$-simplicity,

$$R^p f_*\mathbb{Q} = i_* R^p f_{0*}\mathbb{Q} = i_*\mathbb{Q} = \mathbb{Q}$$

for $p = 0, 3$. (Note that $i_*\mathbb{Q} = \mathbb{Q}$ follows from the fact that $B \setminus B_0$ is codimension two!) Second, if we further assume that B is simply connected ($B = S^3$), then we know $H^p(B, R^q f_*\mathbb{Q})$ for $q = 0, 3$. This gives us the E_2 terms in the Leray spectral sequence for f and $\check{f}$:

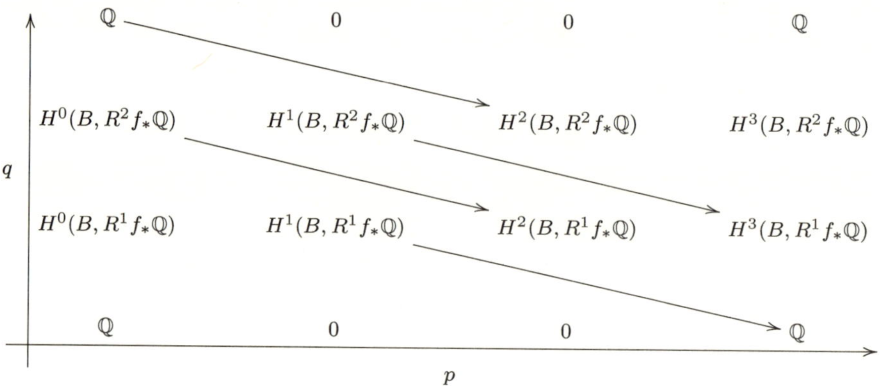

and

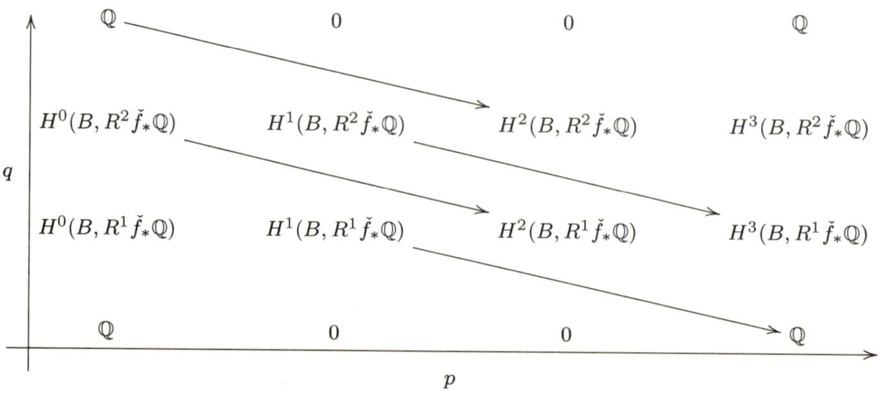

These two arrays have a symmetry, i.e., there are natural isomorphisms coming from (6.7)

$$H^p(B, R^q f_*\mathbb{Q}) \cong H^p(B, R^{3-q} \check{f}_*\mathbb{Q})$$

which reverse the order of the rows. This is analogous to the symmetry of the Hodge diamond for mirror pairs of Calabi-Yau 3-folds.

Now consider in the first diagram the $E_2^{0,1}$ term, $H^0(B, R^1 f_*\mathbb{Q})$. The two maps with domain or range $E_2^{0,1}$ are both zero, so $E_2^{0,1} = E_\infty^{0,1}$. Also, $E_\infty^{1,0} = 0$. Thus there exists a filtration of $H^1(X(B), \mathbb{Q}) = F^0 \supseteq F^1$ with $F^0/F^1 = E_2^{0,1}$ and $F^1 = E_2^{1,0} = 0$.

Suppose $H^1(X(B), \mathbb{Q}) = 0$. (This is the case if, say, $X(B)$ has finite fundamental group, which is usually the case for interesting Calabi-Yau threefolds. This includes all hypersurfaces in toric varieties.) Then we conclude $H^0(B, R^1 f_*\mathbb{Q}) = 0$. A similar argument shows that, since $H^5(X(B), \mathbb{Q}) = 0$ by Poincaré duality, we also have $H^3(B, R^2 f_*\mathbb{Q}) = 0$.

If furthermore $H^1(\check{X}(B), \mathbb{Q}) = 0$, then similarly

$$H^0(B, R^1 \check{f}_*\mathbb{Q}) = H^3(B, R^2 \check{f}_*\mathbb{Q}) = 0.$$

Combining this with the symmetry between the groups for f and $\check{f}$, we see that E_2 terms for f and $\check{f}$ now look like

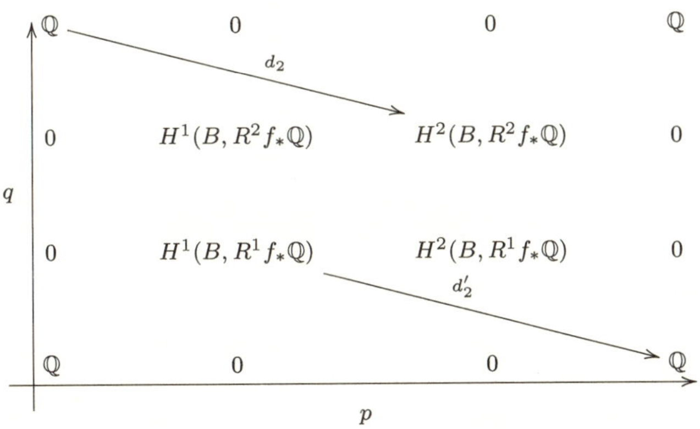

and

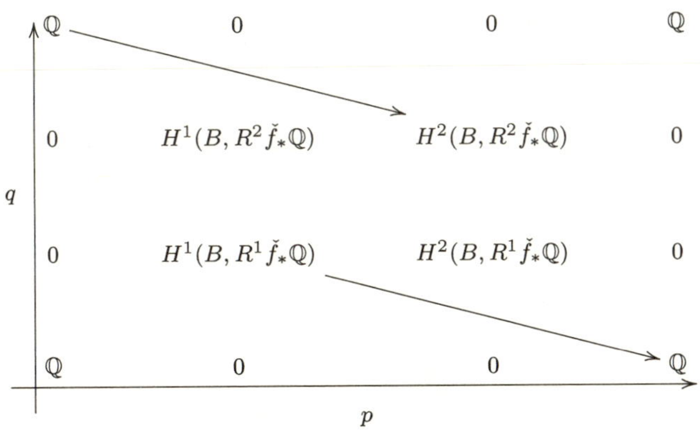

The picture is almost complete. In the first spectral sequence there are at most two non-zero maps

$$d_2 : H^0(B, R^3 f_*\mathbb{Q}) \to H^2(B, R^2 f_*\mathbb{Q})$$

and

$$d_2' : H^1(B, R^1 f_* \mathbb{Q}) \to H^3(B, f_* \mathbb{Q}).$$

To show these are zero, we introduce one more assumption: suppose f has a section $\sigma : B \to X(B)$. In §6.4.3, we in fact constructed the compactification $f : X(B) \to B$ to have a section. There may be other torus fibrations which are compactifications of twists of $X(B_0)$, but we won't consider these in this argument. Now note that $E_\infty^{3,0}$ is a quotient of $E_2^{3,0} = H^3(B, \mathbb{Q})$ and in addition $E_\infty^{3,0} \subseteq H^3(X, \mathbb{Q})$. A general fact about the Leray spectral sequence is that the composition

$$E_2^{3,0} \to E_\infty^{3,0} \hookrightarrow H^3(X, \mathbb{Q})$$

is the pullback

$$f^* : H^3(B, \mathbb{Q}) \to H^3(X, \mathbb{Q}).$$

But the composition

$$H^3(B, \mathbb{Q}) \xrightarrow{f^*} H^3(X, \mathbb{Q}) \xrightarrow{\sigma^*} H^3(B, \mathbb{Q})$$

is the identity since $f \circ \sigma$ is the identity, so f^* is injective. Thus $d_2' = 0$.

Similarly, $E_\infty^{0,3}$ is a subgroup of $E_2^{0,3}$, and the composed map

$$H^3(X, \mathbb{Q}) \to E_\infty^{0,3} \to E_2^{0,3} = H^0(X, R^3 f_* \mathbb{Q})$$

is the natural map as follows: $R^3 f_* \mathbb{Q}$ is the sheaf associated to the presheaf

$$U \mapsto H^3(f^{-1}(U), \mathbb{Q}),$$

so in particular there is a natural map $H^3(f^{-1}(B), \mathbb{Q}) \to \Gamma(B, R^3 f_* \mathbb{Q})$. Clearly the Poincaré dual class of $\sigma(B)$ in $H^3(X, \mathbb{Q})$ restricts to a non-zero element of $H^0(B, R^3 f_{0*} \mathbb{Q})$. Thus $d_2 = 0$ also. So all the maps in E_2 are zero, and there is no room for maps in E_n for $n > 2$. Thus $E_2^{p,q} = E_\infty^{p,q}$. In such a case we say the spectral sequence *degenerates* at the E_2 term.

Putting this all together, we see that with the various assumptions on f and $\check{f}$ we have isomorphisms

$$\begin{aligned} H^0(B, \mathbb{Q}) &\cong H^0(X(B), \mathbb{Q}) \\ H^1(B, R^1 f_* \mathbb{Q}) &\cong H^2(X(B), \mathbb{Q}) \\ H^2(B, R^2 f_* \mathbb{Q}) &\cong H^4(X(B), \mathbb{Q}) \\ H^3(B, R^3 f_* \mathbb{Q}) &\cong H^6(X(B), \mathbb{Q}), \end{aligned}$$

and similar statements for $\check{f} : \check{X}(B) \to B$. In addition, we have a filtration

$$H^3(X(B), \mathbb{Q}) = F^0 \supseteq F^1 \supseteq F^2 \supseteq F^3 \supseteq 0$$

with

$$\begin{aligned} F^0/F^1 &\cong H^0(B, R^3 f_*\mathbb{Q}) \\ F^1/F^2 &\cong H^1(B, R^2 f_*\mathbb{Q}) \\ F^2/F^3 &\cong H^2(B, R^1 f_*\mathbb{Q}) \\ F^3 &\cong H^3(B, f_*\mathbb{Q}). \end{aligned}$$

Now $\dim H^2(\check{X}(B), \mathbb{Q}) = \dim H^4(\check{X}(B), \mathbb{Q})$ by Poincaré duality, so

$$\dim F^1/F^2 = \dim F^2/F^3.$$

Putting this all together, we get

THEOREM 6.103. *Let B be a three-dimensional simple affine manifold with singularities with holonomy in $\mathbb{R}^3 \rtimes \mathrm{SL}_3(\mathbb{Z})$, and let $f : X(B) \to B$ and $\check{f} : \check{X}(B) \to B$ be the compactification of $f_0 : X(B_0) \to B_0$ and $\check{f}_0 : \check{X}(B_0) \to B_0$ with section given in §6.4.3. Suppose furthermore that $H^1(X(B), \mathbb{Q}) = H^1(\check{X}(B), \mathbb{Q}) = 0$. Then*

$$\dim H^{2*}(X(B), \mathbb{Q}) = \dim H^3(\check{X}(B), \mathbb{Q})$$

and

$$\dim H^{2*}(\check{X}(B), \mathbb{Q}) = \dim H^3(X(B), \mathbb{Q}).$$

Here H^{2*} denotes even cohomology. In particular, if $X(B)$ and $\check{X}(B)$ have complex structures making them Calabi-Yau 3-folds, then

$$\begin{aligned} h^{1,1}(X(B)) &= h^{1,2}(\check{X}(B)) \\ h^{1,2}(X(B)) &= h^{1,1}(\check{X}(B)). \end{aligned}$$

The Leray spectral sequence makes sense with $\mathbb{Q}$ replaced by $\mathbb{Z}$. In this case, the situation is subtler because torsion can appear in various entries of the spectral sequence. This is quite an interesting phenomenon, and is explained in [**201**], §3.

This analysis of the cohomology of $X(B)$ and $\check{X}(B)$ raises an interesting question: what is the filtration on H^3 provided by the Leray spectral sequence? Looking at the monodromy weight filtration in families of compactifications provides an answer. We will outline this here, without going into much detail.

Assume B_0 in fact carries an *integral* affine structure, so that we can construct an integral affine structure on $B_0' = B_0 \times \mathbb{R}_{>0}$ as in §6.2.4, and thus a family $X(B_0') \to X(\mathbb{R}_{>0}) = D^*$. We saw in Proposition 6.32 that the monodromy of this family was described by translation by a section $T_\delta : X(B_0) \to X(B_0)$, where the section is determined by the affine structure, i.e., is essentially the graph of the developing map $\delta : \tilde{B}_0 \to M_{\mathbb{R}}$. It can be shown that this translation can be extended, after replacing the section with a homotopic one, to a map $T_\delta : X(B) \to X(B)$.

This motivates the following theorem. Suppose we are in the situation of Theorem 6.103 and we are given a section $\sigma : B \to X(B)$ inducing an automorphism $T_\sigma : X(B) \to X(B)$. This always involves the choice of a zero-section $\sigma_0 : B \to X(B)$. Now consider the cohomology class $[\sigma - \sigma_0] \in H^3(X(B), \mathbb{Q})$ Poincaré dual to $\sigma - \sigma_0$. Because $[\sigma - \sigma_0]$ restricted to any fiber of $X(B) \to B$ is trivial, it follows that $[\sigma - \sigma_0] \in F^1$ and hence modulo F^2 defines a class $D_\sigma \in H^1(B, R^2 f_* \mathbb{Q}) \cong H^1(B, R^1 \check{f}_* \mathbb{Q}) \cong H^2(\check{X}(B), \mathbb{Q})$.

THEOREM 6.104. *Let B, $f : X(B) \to B$, $\check{f} : \check{X}(B) \to B$ satisfy the hypotheses of Theorem 6.103, and suppose $\sigma : B \to X(B)$ is a section inducing a homeomorphism*

$$T_\sigma : X(B) \to X(B)$$

given by

$$T_\sigma(v) = v + \sigma(f_0(v))$$

on $X(B_0)$. Then

$$T_\sigma^* : H^3(X(B), \mathbb{Q}) \to H^3(X(B), \mathbb{Q})$$

satisfies

$$(T_\sigma^* - I)(F^i) \subseteq F^{i+1},$$

inducing maps

$$T_\sigma^* - I : F^i/F^{i+1} = H^i(B, R^{3-i} f_* \mathbb{Q}) \to F^{i+1}/F^{i+2} = H^{i+1}(B, R^{2-i} f_* \mathbb{Q}).$$

Furthermore, the diagram

$$\begin{array}{ccc} H^i(B, R^{3-i} f_* \mathbb{Q}) & \overset{T_\sigma^* - I}{\longrightarrow} & H^{i+1}(B, R^{2-i} f_* \mathbb{Q}) \\ \downarrow \cong & & \downarrow \cong \\ H^{2i}(\check{X}(B), \mathbb{Q}) & \overset{\cup D_\sigma}{\longrightarrow} & H^{2i+2}(\check{X}(B), \mathbb{Q}) \end{array}$$

is commutative up to sign.

PROOF. See [**200**], Theorem 4.1. □

Using this theorem, one is able to show that the filtration $F^\bullet$ coincides with the monodromy weight filtration on $H^3(X(B), \mathbb{Q})$ induced by T_δ. Furthermore, one can use the commutative diagram of the theorem to show that the $(1,1)$ and $(1,2)$-Yukawa couplings of mirrors agree to zeroth order; i.e., the topological $(1,1)$-coupling on $\check{X}(B)$ agrees with that predicted by the $(1,2)$-couplings on the complex moduli space of $\check{X}(B)$. See [**200**] for more details.

CHAPTER 7

Metric Aspects of Calabi-Yau Manifolds

For the most part, we have ignored issues that arise from metrics. In Chapter 2 we dealt with topological, rather than conformal, field theories. In Chapters 4 and 5, we dealt mostly with questions of D-branes which led us to considering the derived category of coherent sheaves on varieties; this has nothing to do with the metric (until we begin to consider stability conditions). On the mirror side, we considered the Fukaya category, for which only the symplectic form, not the metric, is important. In the discussion of the Strominger-Yau-Zaslow conjecture, the metric begins to play a role, since without a metric we can't define special Lagrangian submanifolds. However, in Chapter 6 we quickly abandoned the metric in order to give some topological versions of SYZ, the full version of SYZ involving special Lagrangian tori being too hard.

In this chapter, we wish to take the metric more seriously, with a goal of understanding the SYZ conjecture in a deeper way. Many of the results we will talk about are quite difficult, so this chapter will have the flavour of a survey, explaining for the most part results without proofs.

We begin by discussing examples of Ricci-flat metrics. Most of the examples given will be metrics on complex manifolds, where it is particularly easy to state the meaning of Ricci-flatness. If X is a Kähler manifold with Kähler form ω, and Ω is a nowhere vanishing holomorphic n-form on X, then

$$\omega^n = e^f \cdot (-1)^{m(m-1)/2} \frac{\sqrt{-1}^m m!}{2^m} \Omega \wedge \bar{\Omega}$$

for some function $f : X \to \mathbb{R}$, and the Ricci curvature form ρ is given by

$$\rho = -\sqrt{-1}\partial\bar{\partial} f.$$

The metric is Ricci-flat if $\rho = 0$. In the case when X is compact, this is equivalent to f being constant. Even if there is no global holomorphic n-form, the Ricci curvature form can be defined locally in this manner.

If X is compact, Ricci-flat metrics are known to exist by results of Yau [**475**], but little is known about these metrics. However, in non-compact examples with lots of symmetry explicit Ricci-flat metrics can be written down. We will survey some of the examples appearing in the literature. We will also discuss some non-Kähler examples in less detail: in particular, some examples of G_2-holonomy which are playing an increasingly important role in the physics literature.

After finishing with this discussion of examples, we shall review work of Dominic Joyce on constructing examples of special Lagrangian submanifolds. This work produced surprising examples of special Lagrangian fibrations which had a radically different behaviour than expected. In particular, the discriminant locus of these fibrations was codimension one, as opposed to the nice codimension two discriminant loci studied for topological torus fibrations in Chapter 6. In particular, one startling consequence of this work is that under dualizing, the discriminant locus is likely to change. Thus the strong form of SYZ duality envisaged in Chapter 6, where non-singular torus fibers are dualized and singular fibers are changed as explained in that chapter, cannot hold. This requires a reevaluation and reformulation of the SYZ conjecture. Essentially, the physical argument of Strominger, Yau and Zaslow should only hold at the large complex structure limit, and as one moves away from this limit, one expects to see some inaccuracies in the conjecture. It is possible to make this more precise mathematically by studying the behaviour of Ricci-flat metrics near large complex structure limit points. So finally, we survey the work of Gross and Wilson [**210**] on large complex structure limits of K3 surfaces. This ultimately leads to a recasting of the SYZ conjecture which is perhaps within reach of modern-day techniques.

7.1. Examples of Ricci-flat metrics and various ansätze

7.1.1. Some basic examples. If X is a compact Kähler manifold with vanishing first Chern class, then Yau's theorem [**475**] ensures that for any choice of Kähler *class*, there exists a unique Ricci-flat Kähler metric whose associated $(1,1)$-form ω is in the given class; often, we confuse the metric and its associated $(1,1)$-form by referring to the metric as ω. The problem reduces to solving the complex Monge-Ampère equation on X. Given a Kähler metric ω on X whose Ricci-form is $i\partial\bar{\partial}f$, where f is a function on X normalized so that $\int_X e^f\omega^n = \int_X \omega^n$, we need to find a function ϕ (which we can normalize by the condition $\int_X \phi\omega^n = 0$) such that $\omega + i\partial\bar{\partial}\phi$ is the Kähler form of some metric, and

$$(\omega + \sqrt{-1}\partial\bar{\partial}\phi)^n = e^f\omega^n.$$

Then the metric $\omega + \sqrt{-1}\partial\bar{\partial}\phi$ is Ricci-flat.

The uniqueness of this metric dates back to Calabi, and is proved for instance by a simple integration by parts argument. The existence is proved by Yau using the method of continuity: one solves for functions ϕ_t ($t \in [0,1]$), normalized by the condition $\int_X \phi_t\omega^n = 0$, such that

(1) $\omega + \sqrt{-1}\partial\bar{\partial}\phi_t$ is the Kähler form of some metric, and

(2) $(\omega + \sqrt{-1}\partial\bar{\partial}\phi_t)^n = e^{tf}\frac{\int_X \omega^n}{\int_X e^{tf}\omega^n}\,\omega^n.$

So $t = 0$ has solution $\phi_0 = 0$, whilst $t = 1$ is the equation we want to solve. Let

$$A = \{t \in [0,1] \ : \ \exists\, C^{k,\alpha} \text{ solution } \phi_t\},$$

where $k \geq 2$ and $0 < \alpha < 1$. Here $C^{k,\alpha}$ denotes functions whose kth derivative is Hölder continuous with exponent α. A real-valued function f is Hölder continuous with exponent α if there exists a constant C such that $|f(x) - f(y)| \leq C\|x - y\|^\alpha$ for all x, y in the domain of f. For $\alpha = 1$, this is Lipschitz continuity. The hypothesis is typical of regularity required to solve PDEs. Yau's proof works by showing that A is both open and closed, and hence $A = [0,1]$. The openness part of this follows from standard techniques from analysis such as Hodge Theory and the Inverse Function Theorem on Banach spaces; the fact that A is closed follows using highly non-trivial a priori bounds (independent of t) for the solutions of the family of equations above.

Although we know therefore that these metrics exist, there is not however a single explicit (non-flat) example known in the compact case, and any such metric will be very difficult to find in practice because of the absence of continuous families of isometries. As we shall see later in the chapter, we can however hope, in appropriate circumstances, to find good approximations to such metrics, obtained by gluing techniques. These approximations will depend on knowing the explicit form of Ricci-flat Kähler metrics on certain *non-compact* manifolds. For non-compact manifolds, the metric will no longer be unique, but we may well be able to impose certain symmetries, and in that way reduce the highly complicated non-linear partial differential equation given above to ordinary differential equations that we can solve reasonably explicitly.

Whilst we will not use this in the sequel, we also remark that there has been recent progress on approximating Ricci-flat metrics *numerically*. Using ideas of Donaldson in [**130**], Douglas, Karp, Lukic and Reinbacher in [**140**] have calculated good numerical approximations to Ricci-flat metrics on hypersurfaces.

EXAMPLE 7.1. *The Eguchi-Hanson metric.* Let $X = \mathbb{C}^2/\{\pm 1\}$, the quadric cone in $\mathbb{C}^3$, with resolution $T^*\mathbb{CP}^1$, diffeomorphic to the real tangent bundle of S^2. On $\mathbb{C}^2$ we set $u = |z_1|^2 + |z_2|^2$ and look for Ricci-flat Kähler metrics with potential functions of the form $f(u)$. Solving, we obtain explicit solutions (see [**307**], page 293)

$$f_a(u) = u\big(1 + \frac{a^2}{u^2}\big)^{\frac{1}{2}} + a \log \frac{u}{(u^2 + a^2)^{\frac{1}{2}} + a}$$

for $a > 0$. The Kähler form $i\partial\bar{\partial} f_a$ can be shown to extend over the zero section E of $T^*\mathbb{CP}^1$, giving the Eguchi-Hanson metric, an SO(3)-invariant Ricci-flat ALE (asymptotically locally Euclidean) Kähler metric, with Kähler class of the form $-2a[E]$. The 2-sphere E has curvature a^{-1} and volume $4\pi a$. On any compact set in the complement of E, the Eguchi-Hanson metric converges rapidly to the orbifold flat metric as $a \to 0$ [**307**].

This relatively simple example is hyperkähler, and is of importance because it defines a Ricci-flat Kähler metric on both the smoothing and resolution of a simple node in complex dimension 2. For instance, if we have an orbifold metric on a compact K3 surface X with nodes, by taking the value of a to be small and gluing in appropriate copies of the Eguchi-Hanson metric, we may obtain good approximations to a Ricci-flat metric on a resolution or smoothing (see [**307**] for more details).

This example is also an example of a *cohomogeneity one* metric. This means that there is a group G acting on the space by isometries, and the general orbit of the group is codimension one. In the above example, $G = \mathrm{SO}(3)$. As we shall see, there is usually one orbit which is smaller dimensional, giving a picture like that of Figure 1. In the Eguchi-Hanson metric, the general orbit is $S^3/\{\pm 1\}$, while the special orbit is an S^2, the zero-section of the cotangent bundle.

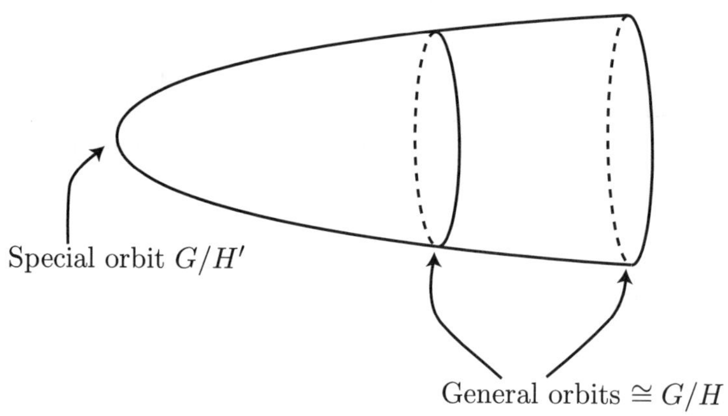

FIGURE 1

The Eguchi-Hanson metric may be generalized in various ways. In dimension two, by use of the hyperkähler quotient construction [**238**], Kronheimer [**316**] produced ALE gravitational instantons, which are hyperkähler metrics on the resolution of an arbitrary rational double point in dimension 2; these metrics had already been constructed for the A_k singularities (via the Gibbons-Hawking construction, as multi-instanton metrics). There is also a natural generalization to higher dimensions; these are the generalized Eguchi-Hanson metrics [**79, 177, 274**], which are complete metrics with holonomy $\mathrm{SU}(k)$ living on the kth power of the Hopf bundle over $\mathbb{CP}^{k-1}$, or equivalently on the crepant resolution of the quotient singularity $\mathbb{C}^k/C_k$, where $C_k = \langle \zeta \rangle$ denotes the group of kth roots of unity acting diagonally.

The metric can be expressed in terms of a Kähler potential depending only on $\sum |z_i|^2$ [**274**]. This metric was written down for $k = 3$ in a different context in [**340**]. These metrics are ALE, and are the only explicit examples known of ALE metrics with holonomy $\mathrm{SU}(k)$. They are of cohomogeneity one with respect to $U(k)$ acting on the bundle in the natural way; their codimension one orbits are $U(k)/(U(k-1) \times C_k) = S^{2k-1}/C_k$, and they have one singular (i.e., higher codimension) orbit $\mathbb{CP}^{k-1}$; see below for further discussion of cohomogeneity one metrics. These metrics are in fact a special case of metrics with holonomy contained in $\mathrm{SU}(k)$ defined on the canonical bundle of any Kähler-Einstein manifold with positive curvature (see [**79**] and Theorem 8.1 of [**408**]).

In a similar vein to the above $\mathrm{SU}(k)$-holonomy metrics, there are then the well-known hyperkähler metrics due to Calabi.

EXAMPLE 7.2. *Calabi hyperkähler metrics.*

These are complete hyperkähler metrics living on the holomorphic cotangent bundle $T^*\mathbb{CP}^k$ of complex projective space, also of cohomogeneity one, with isometry group $U(k+1)$ acting on codimension one orbits of the form

$$U(k+1)/U(k-1) \times U(1),$$

and with one singular orbit $\mathbb{CP}^k$ (see [**79**]). We endow $\mathbb{CP}^k$ with a multiple of the Fubini-Study metric, with constant holomorphic sectional curvature 2κ. This induces a Hermitian metric on the holomorphic cotangent bundle $T^*\mathbb{CP}^k$. Working locally over an affine open set $U \subset \mathbb{P}^k$, Calabi then writes down an explicit Kähler potential on $T^*\mathbb{P}^k|_U$, and checks that this gives a well-defined global metric on the holomorphic cotangent space, which is hyperkähler. This metric restricts to the given one on $\mathbb{P}^k$, and the hyperkähler triple of 2-forms consists of the Kähler form of the metric and the real and imaginary parts of the natural complex symplectic form on the holomorphic cotangent bundle. In fact, the hyperkähler metric is uniquely determined by these properties, if one imposes the additional condition of being invariant under the natural action of $U(1)$ on the fibers of $T^*\mathbb{CP}^k$ (see [**146, 455**]). In the case of $k = 1$, the Kähler potential of the Calabi metric is just the Kähler potential of the Eguchi-Hanson metric with parameter a, scaled by a factor $a/2$, where $a > 0$ is given by $\kappa = a^{-2}$ — the restriction of the metric to $\mathbb{P}^1$ therefore has curvature $2a^{-2} = 2\kappa$, as required. For arbitrary k, these metrics are of cohomogeneity one, with isometry group $U(k+1)$ acting on codimension one orbits of the form $U(k+1)/U(k-1) \times U(1)$, and with one singular orbit $\mathbb{CP}^k$.

We can however also consider the case of nodes in higher dimensions, and this leads us to the Stenzel metrics.

EXAMPLE 7.3. *The Stenzel and related metrics.* Another relevant generalization of Eguchi-Hanson to higher dimensions is the following: given a quadric cone in $\mathbb{C}^{n+1}$ with equation

$$z_0^2 + z_1^2 + \cdots + z_n^2 = 0$$

we can set $\rho = |z_0|^2 + |z_1|^2 + \cdots + |z_n|^2$ and look for a Ricci-flat Kähler potential of the form $f(\rho)$ — demanding the Ricci form be zero, we get $f(\rho) = \rho^{(n-1)/n}$. For $n = 2$, this is just the flat orbifold metric. For $n = 3$, the resulting Kähler Ricci-flat metric may be reinterpreted as a conical metric over the Einstein manifold $T^{1,1}$; this latter manifold is topologically $S^3 \times S^2$, but the metric is not the product Einstein metric [**84**]. Similarly, we can solve for a Ricci-flat Kähler potential of the form $f(\rho)$ on the smoothing (a non-singular quadric Q) with equation

$$z_0^2 + z_1^2 + \cdots + z_n^2 = \epsilon.$$

By writing $z_j = x_j + iy_j$ and taking real and imaginary parts of the equation for Q, we see that Q is diffeomorphic to the tangent bundle (or equivalently the cotangent bundle, identified with the tangent bundle via a metric) of the n-sphere S^n. The equations reduce to an ODE

$$\rho(f')^n + f''(f')^{n-1}(\rho^2 - |\epsilon|^2) = c > 0,$$

which may be solved explicitly. These metrics are the Stenzel metrics on $Q \cong T(S^n)$. As $\epsilon \to 0$, we recover the conifold metric. They are Ricci-flat Kähler metrics on the smoothing of the node, which in the case $n = 2$ coincides with the Eguchi-Hanson metric. In the case $n = 3$, the metric was first written down by Candelas and de la Ossa [**84**], and is asymptotically conical with base the Einstein manifold $T^{1,1}$. For $n = 3$, a corresponding metric on the small resolution of the node is also known explicitly [**387**], and is asymptotically conical with base the Einstein manifold $T^{1,1}/C_2$.

A natural generalization of the Stenzel metrics exists on the tangent bundle of any rank one compact symmetric space, and these are described in [**388, 328**]. An elegant derivation of these metrics is also contained in [**213**]. All these metrics are complete. The existence of complete, Ricci-flat Kähler metrics on complexifications of higher rank symmetric spaces is proved in [**28**].

As well as the specific examples we have mentioned above, there are various forms of ansatz, where we assume a certain symmetry on a class of examples, and find the resulting differential equations for solutions where the metric is for instance Ricci-flat, Einstein, or has a specified holonomy group. We will not usually be able to write down analytically a general solution, and perhaps not even be able to find sporadic solutions. Many papers in the physics literature are devoted to studying the solutions to ansätze numerically. In the case of hyperkähler metrics in dimension 4, we do have an ansatz, found by Gibbons and Hawking, which can be solved analytically for the general solution, and which will be highly relevant in the following sections.

7.1.2. The Gibbons-Hawking ansatz for dimension 4 hyperkähler metrics. In this ansatz, we describe the construction of equivariant hyperkähler metrics on S^1-bundles over open subsets of $\mathbb{R}^3$. The description of the Gibbons-Hawking ansatz we give below is based on the work of Gibbons and Hawking, Hitchin, and others.

Let $U \subseteq \mathbb{R}^3$ be an open set with the Euclidean metric, with coordinates u_1, u_2, u_3. Let $\pi : X \to U$ be a principal S^1-bundle, with S^1-action $S^1 \times X \to X$ written as $(e^{\sqrt{-1}t}, x) \mapsto e^{\sqrt{-1}t} \cdot x$. Let θ be a connection 1-form on X, i.e., a $\mathfrak{u}(1) = \sqrt{-1}\mathbb{R}$-valued 1-form, invariant under the S^1-action and such that $\theta(\partial/\partial t) = \sqrt{-1}$. The curvature of the connection θ is $d\theta = \pi^* F$ for a 2-form F on U, and $\sqrt{-1}F/2\pi$ represents the first Chern class of the bundle (see [**94**], Appendix). Suppose V is a positive real function on U satisfying $*dV = F/2\pi\sqrt{-1}$. Let

$$\begin{aligned} \omega_1 &= du_1 \wedge \theta/2\pi\sqrt{-1} + V du_2 \wedge du_3 \\ \omega_2 &= du_2 \wedge \theta/2\pi\sqrt{-1} + V du_3 \wedge du_1 \\ \omega_3 &= du_3 \wedge \theta/2\pi\sqrt{-1} + V du_1 \wedge du_2. \end{aligned}$$

Then $\omega_1^2 = \omega_2^2 = \omega_3^2$ is nowhere zero, and $\omega_i \wedge \omega_j = 0$, for $i \neq j$. Furthermore, $*dV = F/2\pi\sqrt{-1}$ implies that $d\omega_i = 0$ for all i, since for instance

$$d\omega_1 = -du_1 \wedge d\theta/2\pi\sqrt{-1} + dV \wedge du_2 \wedge du_3 = -du_1 \wedge *dV + dV \wedge du_2 \wedge du_3 = 0.$$

Therefore $\omega_1, \omega_2, \omega_3$ define a hyperkähler metric on X. Note that V is harmonic, since $dF = 0$ implies that $*d * dV = 0$.

Let θ_0 denote the real 1-form $\theta/2\pi\sqrt{-1}$, and observe that

$$-\omega_1 - \sqrt{-1}\omega_2 = (\theta_0 - \sqrt{-1}V du_3) \wedge (du_1 + \sqrt{-1} du_2).$$

By taking this to be the (holomorphic) 2-form Ω on X, by Proposition 6.24, this determines an integrable (since Ω is closed) almost complex structure on X, where $du_1 + \sqrt{-1}du_2$ and $\theta_0 - \sqrt{-1}V du_3$ span the holomorphic cotangent space inside the complexified cotangent space. It follows that the (integrable) almost complex structure J on the cotangent space is given by

$$J(du_1) = -du_2, \quad J(du_3) = -V^{-1}\theta_0.$$

Thus, if we consider the Kähler form $\omega = \omega_3$ as an alternating tensor, and use the relation that if g is the Riemannian metric, then $g(\zeta, \xi) = \omega(\zeta, J\xi)$, we obtain an expression for the metric

$$ds^2 = V d\mathbf{u} \cdot d\mathbf{u} + V^{-1}\theta_0^2.$$

Usually, we shall in fact start from a positive harmonic function V on U such that $- * dV$ represents the Chern class of the bundle. Then we can always find a connection 1-form θ with $d\theta/2\pi\sqrt{-1} = *dV$, such a θ being uniquely determined up to pullbacks of closed 1-forms from U, and hence we obtain hyperkähler metrics as above.

REMARK 7.4. We will need to calculate later some information about the curvature of this metric. To do this, we can work locally, and therefore take the orthonormal moving coframe given by $V^{1/2}du_1, V^{1/2}du_2, V^{1/2}du_3$ and $V^{-1/2}\theta_0$. We can moreover write the connection form locally as

$$\theta_0 = \frac{dt}{2\pi} + A_1du_1 + A_2du_2 + A_3du_3,$$

where $\nabla V = \nabla \times \mathbf{A}$. To calculate the curvature, we may then apply Cartan's method. We obtain, after some serious calculations [**210**], that the norm squared of the curvature is

$$\|R\|^2 = 12V^{-6}|\nabla V|^4 + V^{-4}\Delta(|\nabla V|^2) - 6V^{-5}(\nabla V)\cdot(\nabla(|\nabla V|^2)).$$

Using the fact that V is harmonic, we then recover, again after some more calculations [**210**], the compact formula given in equation (32) of [**375**] that

$$\|R\|^2 = \frac{1}{2}V^{-1}\Delta\Delta(V^{-1}).$$

EXAMPLE 7.5. In §6.4.2, we considered the map $p : X = \mathbb{C}^2 \setminus \{(0,0)\} \to \mathbb{R}^3 \setminus \{(0,0,0)\}$, given by

$$p(z_1, z_2) = (2\ \mathrm{Re}(z_1z_2), 2\ \mathrm{Im}(z_1z_2), |z_1|^2 - |z_2|^2).$$

This map exhibits X as an S^1-bundle over $\mathbb{R}^3 \setminus \{(0,0,0)\}$, with Chern class ± 1. The action of S^1 on X is given by $e^{\sqrt{-1}t}\cdot(z_1, z_2) = (e^{\sqrt{-1}t}z_1, e^{-\sqrt{-1}t}z_2)$. Note also that if we compose p with projection onto the first two factors, we obtain the map sending (z_1, z_2) to $2z_1z_2$, holomorphic with respect to the standard complex structures.

We now choose a positive harmonic function V on $\mathbb{R}^3 \setminus \{(0,0,0)\}$ such that, with S^2 the unit sphere in $\mathbb{R}^3$,

$$-\int_{S^2} *dV = \int_{S^2} \sqrt{-1}\,F/2\pi = \pm 1,$$

i.e., the Chern number is correct. The particular examples of such V we consider are

$$V = e + \frac{1}{4\pi|\mathbf{u}|} = e + \frac{1}{4\pi\sqrt{u_1^2+u_2^2+u_3^2}},$$

where $e \geq 0$. The integral

$$\int_{S^2} *d\left(\frac{1}{4\pi\sqrt{u_1^2+u_2^2+u_3^2}}\right)$$

is easily seen to be ± 1 (depending on the orientation of the sphere).

We take the connection form to be

$$\theta = \sqrt{-1}\ \mathrm{Im}(\bar{z}_1dz_1 - \bar{z}_2dz_2)/(|z_1|^2 + |z_2|^2).$$

Then

$$d\theta/2\pi\sqrt{-1} = \frac{-(u_1du_2\wedge du_3 + u_2du_3\wedge du_1 + u_3du_1\wedge du_2)}{4\pi(u_1^2+u_2^2+u_3^2)^{3/2}} = *dV$$

as required. We therefore obtain hyperkähler metrics on X. For all $e \geq 0$, it can be shown that these extend to metrics on $\mathbb{C}^2$, or in the language of §6.4.3, they extend over the *nut*. In fact, such metrics are ALF (asymptotically locally flat), approaching a flat metric when $|\mathbf{u}| \to \infty$, whilst being periodic in t. When $e = 1$, the metric obtained is the Taub-NUT metric, and when $e = 0$, it is just a flat metric on $\mathbb{C}^2$. To prove the assertions for $e = 0$, straightforward calculations show that, with $z_j = x_j + \sqrt{-1}y_j$,

$$\begin{aligned}
\omega_1 &= \frac{1}{\pi}(dx_2 \wedge dy_1 - dx_1 \wedge dy_2), \\
\omega_2 &= \frac{1}{\pi}(dx_1 \wedge dx_2 - dy_1 \wedge dy_2), \\
\omega_3 &= \frac{1}{\pi}(dx_1 \wedge dy_1 + dx_2 \wedge dy_2).
\end{aligned}$$

So $\omega_1, \omega_2, \omega_3$ extend to $\mathbb{C}^2$, and yield a flat metric, as claimed.

The calculations being local, we can extend this example to the case of multi-instanton metrics, inserting nuts over any finite set of points in $\mathbb{R}^3$. This means we take V of the form $e + \frac{1}{4\pi}\sum_j \frac{1}{r_j}$, where r_j denotes the Euclidean distance in $\mathbb{R}^3$ to the jth point of the subset. Taking $e = 0$, for two points we recover the Eguchi-Hanson metric yet again, and for k distinct points we obtain the Gibbons-Hawking gravitational multi-instantons; if we take the points to be collinear, a simple argument shows that this yields a complete Ricci-flat Kähler metric on the resolution of an A_{k-1} rational double point singularity [**231**]. As noted above, these were generalized later to give metrics on the resolution of any rational double point singularity [**316**]. Taking $e = 1$ above, we obtain instead the multi-Taub-NUT metrics; these are not ALE.

7.1.3. S^1-invariant Ricci-flat metrics on elliptic fibrations. One application of the Gibbons-Hawking ansatz is the construction of Ricci-flat metrics on (local) elliptic fibrations. We do this as follows.

In the Gibbons-Hawking ansatz, we can consider the case when $U = B \times \mathbb{R}$, with B a contractible open subset of $\mathbb{R}^2$ — in particular, the S^1-bundle X over U is topologically trivial. Set $y_1 = u_1$, $y_2 = u_2$, so then $y = y_1 + \sqrt{-1}y_2$ is a complex coordinate on B. The hyperkähler structure on X gives rise to a complex structure on X, under which the function y may be seen to be holomorphic, i.e., the map $X \to B$ is holomorphic [**210**]. If we fix a holomorphic section of the fibration, then we can integrate up the holomorphic form $\theta_0 - \sqrt{-1}V du$ to obtain a holomorphic coordinate $x = x_1 + \sqrt{-1}x_2$ on the universal cover $\tilde{X}$ of X such that the holomorphic 2-form on X is just $dx \wedge dy$. This enables us to identify $\tilde{X}$ with T^*B, with x, y then being holomorphic canonical coordinates on T^*B; that is, the point (x, y) corresponds to the 1-form $xdy \in T^*B_y$, where y also denotes the point in B with coordinate y. Moreover, in these coordinates, the S^1-action on X

yields an $\mathbb{R}$-action on T^*B which is just translation on the coordinate x_1, and so X is isomorphic to $(T^*B)/\mathbb{Z}$. For details of this see Construction 2.6 in [**210**]. Conversely, suppose we are given such an S^1-invariant Ricci-flat Kähler metric on $(T^*B)/\mathbb{Z}$ of the above type, namely that with respect to canonical holomorphic coordinates x, y on T^*B, the metric is independent of x_1. It is an easy verification [**210**] that this does indeed arise from the Gibbons-Hawking ansatz.

We shall be particularly interested in the metrics described above when V and θ are themselves periodic in $u = u_3$. The hyperkähler metric descends to one on the corresponding S^1-fibration over $Y = B \times S^1$ if and only if the three 2-forms $\omega_1, \omega_2, \omega_3$ are invariant under changing u by a period, which in turn is saying that the periodicity in u is independent of $y = u_1 + \sqrt{-1}u_2$. We shall now change notation and denote this $S^1 \times S^1$ fibration over B by X (the universal cover $\tilde{X}$ being the same as before). Since the restriction of the Kähler form ω_3 to a fiber X_y is just $du \wedge \theta_0 = du \wedge dt/2\pi$, the volume of any fiber is just the periodicity in u. With the complex coordinate y on B, the map $f : X \to B$ is a holomorphic map to the open subset B of $\mathbb{C}$, whose fibers are elliptic curves. Moreover, in this case, we also have that V and θ are periodic in u, with the period in u being constant, namely the volume of the elliptic fibers of $f : X \to B$.

Let us now apply the Gibbons-Hawking ansatz to the manifold X constructed in Construction 6.86 where $\overline{Y} = D \times S^1$, D a disc in $\mathbb{C}$ centred at the origin, and $S^1 = \mathbb{R}/\epsilon\mathbb{Z}$ for some $\epsilon > 0$. If V and θ, defined on $D \times \mathbb{R} \setminus \{0\} \times \epsilon\mathbb{Z}$, are chosen to be periodic in $u = u_3$ with period ϵ, then we obtain a hyperkähler metric on X. If moreover V is chosen to have local behaviour of the form $\frac{1}{4\pi r}$ (modulo a harmonic function) at the origin, and hence at all the points $(0, 0, n\epsilon)$, then the metric will extend to a hyperkähler metric on the relative compactification $\overline{X} \to \overline{Y}$. The family $\overline{X} \to D$ is holomorphic, and may be identified as a Kodaira degeneration of Type I_1, with singular fiber a nodal elliptic curve (topologically just a torus with one circle pinched to a point, as described in Construction 6.86); the volume of any fiber is the constant ϵ. This is the basic idea behind the construction of the Ooguri-Vafa metric, which we describe in §7.3.4. We shall in fact want to start from a fixed Kodaira degeneration of Type I_1, and thus a given complex structure on $\overline{X}$, and shall therefore need to choose the constant in V rather carefully to ensure that the hyperkähler metric we define does correspond to the given complex structure. See §7.3.4 for further details.

7.1.4. Other holonomies. We briefly survey some of the other constructions of Ricci-flat metrics.

Using twistor techniques, the Gibbons-Hawking ansatz was generalized in [**389**] to investigate $4n$-dimensional hyperkähler metrics admitting an action of the n-torus T^n. The hyperkähler case was studied further in [**171**], where the authors studied $4n$-dimensional hyperkähler manifolds X which admit a tri-holomorphic free action of the n-torus T^n. It turns out that the metric may be written in coordinates adapted to the torus action, in a form similar to the Gibbons-Hawking ansatz in dimension 4, and such that the non-linear Einstein equations reduce to a set of linear equations (essentially saying that certain functions on Euclidean 3-space are harmonic). In the case of 8-manifolds ($n = 2$), the solutions can be described geometrically, in terms of arrangements of 3-dimensional linear subspaces in Euclidean 6-space [**171**]. Another important method for the construction of hyperkähler metrics is via the hyperkähler quotient construction of [**238**]; for a discussion of hyperkähler metrics obtained in this way, the reader is referred to [**46**]. The hyperkähler quotient construction yields many examples starting from a flat quaternionic vector space $\mathbb{H}^n \cong \mathbb{R}^{4n}$ with an action by a subgroup G of $\mathrm{Sp}(n)$. In particular, in the case when G is a torus, the paper cited determines which hyperkähler metrics arise in this way.

There are in fact many explicit examples known of metrics on non-compact manifolds with $\mathrm{SU}(n)$ or $\mathrm{Sp}(2n)$ holonomy, many to be found only in the physics literature. The other holonomy groups automatically yielding Ricci-flat metrics are the special holonomy groups G_2 in dimension 7 and $\mathrm{Spin}(7)$ in dimension 8. If one is looking for compact examples, there are again only existence statements available for the metrics (Chapters 11-15 of [**274, 314**]); to give explicit examples of such metrics, one needs to look for non-compact examples. Until fairly recently only three explicit examples of *complete* metrics (in dimension 7) with G_2-holonomy and one explicit example (in dimension 8) with $\mathrm{Spin}(7)$-holonomy were known [**76, 178**] — non-complete examples due to Bryant were known before these. The three G_2-holonomy examples are asymptotically conical and live on the bundle of self-dual two-forms over S^4, the bundle of self-dual two-forms over $\mathbb{CP}^2$, and the spin bundle of S^3 (topologically $\mathbb{R}^4 \times S^3$), respectively. The metrics are of cohomogeneity one with respect to the Lie groups $\mathrm{SO}(5)$, $\mathrm{SU}(3)$ and $\mathrm{SU}(2) \times \mathrm{SU}(2) \times \mathrm{SU}(2)$ respectively. Recall that a cohomogeneity-one metric has a Lie group acting via isometries, with general (principal) orbits of real codimension one. A classification theorem for irreducible hyperkähler metrics in dimensions $4n > 4$, of cohomogeneity one with respect to a compact simple Lie group, was proved in [**107**]. In particular, if the metric is complete, then X is the holomorphic cotangent bundle of projective n-space $T^*\mathbb{CP}^n$, and the metric is the Calabi hyperkähler metric, described above.

The principal orbits in the above three examples with G_2-holonomy are $\mathbb{CP}^3$, $\mathrm{SU}(3)/T^2$ and $S^3 \times S^3$, respectively. The examples are complete, with orbit space being $\mathbb{R}_+$, and have one singular (i.e., higher codimension) orbit, which the physicists call a *bolt* (a nut being just a bolt which is a single point). In the three examples, the bolt is just the base space of the bundle, that is S^4, $\mathbb{CP}^2$ and S^3 respectively. In fact these examples may be regarded as deformations of conical metrics on the cone over the principal orbit, with the vertex of the cone replaced by a bolt. The last of these examples is of particular interest in that its principal orbit is $S^3 \times S^3$, the case of importance for physicists, because of its relation to conifold transitions.

These holonomy-G_2 examples are all examples in which a Lie group G acts with low codimension orbits. This is a general feature of explicit examples of Einstein metrics. The simplest case of such a situation would be when there is a single orbit of a group action, in which case the metric manifold is homogeneous. For metrics on homogeneous manifolds, the Einstein condition may be expressed purely algebraically. Moreover, all homogeneous Ricci-flat manifolds are flat [**44**], and so no interesting metrics occur. The next case to consider is that of cohomogeneity one with respect to G, i.e., the orbits of G are codimension one in general. Here, the Einstein condition reduces to a system of non-linear ordinary differential equations in one variable, namely the parameter on the orbit space. In the Ricci-flat case, the theorem of Cheeger-Gromoll [**92**] implies that the manifold has at most one end. Since we are in the non-compact case, the orbit space is $\mathbb{R}_+$ and there is just one singular orbit. Geometrically, if the principal orbit is of the form G/K, the singular orbit (the bolt) is G/H for some subgroup $H \supset K$; if G is compact, a necessary and sufficient condition for the space to be a smooth manifold is that H/K is diffeomorphic to a sphere [**369**]. In many examples, this is impossible because of the form of the group G, and so any metric constructed will not be complete. Summing up therefore, if we apply some ansatz, such as the Hitchin ansatz described below, to find families of cohomogeneity one metrics with holonomy say G_2 or Spin(7), and we ask about completeness of the metric, the first issue is that of regularity at the bolt, and the second issue is whether the solution may be integrated out to infinity. The latter condition will in general cut down the number of parameters allowed.

Various ansätze have been produced recently for finding cohomogeneity one G_2-holonomy or Spin(7)-holonomy metrics, and further examples of such metrics have appeared in the physics literature. Most of these may however be described in terms of a very recent ansatz due to Hitchin [**235**], which is both of theoretical and practical importance. In particular, it has enabled the construction of new cohomogeneity one metrics with special holonomy G_2 and Spin(7). A partial classification of cohomogeneity one metrics with G_2-holonomy may be found in [**99**].

7.1.5. The Hitchin ansatz for special holonomy metrics.

An account of this ansatz may be found in the volume of Proceedings of the Clay School 2002 [**237**], as well as in the original paper [**235**]; we shall therefore restrict ourselves to explaining the main features. The starting point is the concept of stable forms; a p-form on a manifold X of real dimension n is said to be *stable* if, on each tangent space $V = T_P X$, it lies in an open orbit of $\bigwedge^p V^*$ under the action of $\mathrm{GL}(n)$. If $p = 1$, then stable is the same as everywhere non-zero; if $p = 2$ and $n = 2m$ or $2m+1$, then stable is the same as everywhere of rank $2m$. Of more interest is the case of stable 3-forms, which can only occur in dimensions 6, 7 and 8. Observing the fact that an open orbit in $\bigwedge^p V^*$ implies by duality an open orbit in $\bigwedge^{n-p} V^*$, and hence the existence of stable $(n-p)$-forms, we have now accounted for all possible stable forms. No other cases occur.

For $n = 6$, there is an open orbit of $\bigwedge^3 V^*$ for which the stabilizer is $\mathrm{SL}(3,\mathbb{C})$ (and an open orbit where it is $\mathrm{SL}(3,\mathbb{R}) \times \mathrm{SL}(3,\mathbb{R})$), and we shall be interested in the stable forms with stabilizer (at each point) being the former. Similarly, for $n = 7$, we want the stabilizer to be G_2, and for $n = 8$ we want stabilizer $\mathrm{SU}(3)$; again, both these corresponding to open orbits in $\bigwedge^3 V^*$. From now on, we shall include in our definition that these are the stabilizers of *stable* 3-forms.

Let us now restrict ourselves to the most interesting cases, namely $p = 3$. Corresponding to any given stable p-form ρ, there is a volume form $\phi(\rho)$, homogeneous in ρ of degree $\frac{n}{p}$, which may be written down in terms of ρ — these volume forms are explicitly written down in the Appendix to [**235**]. There is also a canonically defined $(n-p)$-form $\hat{\rho}$, which has the property that $\hat{\rho}\wedge\rho = \frac{n}{p}\phi(\rho)$. For $n = 6$ and $p = 3$, $\hat{\rho}$ is determined by the property that $\Omega = \rho + i\hat{\rho}$ is a complex $(3,0)$-form preserved by $\mathrm{SL}(3,\mathbb{C})$ [**233**]. When $n = 7$, the stable 3-form ρ determines a metric. Explicitly, we can define this as follows: take the bilinear form on the tangent space $b(u,v) = -\frac{1}{6}\iota(u)\rho\wedge\iota(v)\rho\wedge\rho$; this defines a linear map $K_\rho : V \to V^* \otimes \bigwedge^7 V^*$; the volume form ϕ is $(\det K_\rho)^{\frac{1}{9}}$, and the metric at the point is then $g(u,v) = b_\rho(u,v)(\det K_\rho)^{-\frac{1}{9}}$. The metric determines the Hodge $*$-operation, and $\hat{\rho} = *\rho$ [**233**]. There is a similar story for $n = 8$, with $\hat{\rho} = - * \rho$.

Hitchin then considers, for a fixed cohomology class, the volume functional $\Phi(\rho) = \int_X \phi(\rho)$, which is invariant under the action of $\mathrm{Diff}(X)$, and he shows that the critical points ρ of this functional correspond to solutions of the equations $d\rho = 0 = d\hat{\rho}$. In the case $n = 6$, this says that the complex $(3,0)$-form Ω is closed, and defines the structure on X of a complex threefold with trivial canonical bundle. In the case $n = 7$, we observed above that the 3-form ρ determines the metric. If ρ is a closed stable form representing the given class, the condition for the metric to have G_2-holonomy is known to be $d * \rho = 0$ ([**147**] or [**274**], Proposition 10.1.3), and this is just the statement that ρ is a critical point for Φ. The case $n = 8$ remains something of a mystery in that no examples are known.

A related list of metrics on compact simply connected manifolds M correspond in a natural way to special holonomy metrics on the cone X over M, i.e., the warped product $M \times \mathbb{R}_+$. The two dimensions here of most interest to us will be $n = 6$, where the metrics have been traditionally called *nearly Kähler,* and correspond to G_2-metrics on the cone, and $n = 7$ where they are known as manifolds with *weak holonomy* G_2, and correspond to Spin(7) metrics on the cone (to get holonomy on the cone of precisely the type claimed, one needs to exclude a few special cases [**29**]). The basic idea of [**235**] is that these metrics may be obtained by finding critical points of a functional subject to certain constraints.

Let us concentrate now on the case of nearly Kähler metrics in six dimensions (and the consequent G_2-metrics in seven dimensions). We refer to [**235, 237**] for details. We can specify a symplectic form on M either by giving a closed stable 2-form ω, or a stable 4-form $\omega^2/2$ which is critical for the volume functional; the latter turns out to be a more useful approach. We shall now denote the volume functional (previously denoted Φ) on stable 3-forms by V and on stable 4-forms by W. We choose cohomology classes $\mathcal{A} \in H^3(M, \mathbb{R})$ and $\mathcal{B} \in H^4(M, \mathbb{R})$. We think of these as infinite dimensional affine spaces of all closed forms representing these cohomology classes. Then $\mathcal{A} \times \mathcal{B}$ is a product of affine spaces, with the tangent space at any point being the product of the exact 3-forms and the exact 4-forms. Given an exact 3-form $\rho = dF$ and an exact 4-form $\sigma = d\beta$, we define

$$\langle \rho, \sigma \rangle = \int_M F \wedge \sigma = -\int_M \rho \wedge \beta,$$

and hence a formal symplectic structure on $\mathcal{A} \times \mathcal{B}$ given by the formula $\langle \rho_1, \sigma_2 \rangle - \langle \rho_2, \sigma_1 \rangle$.

Suppose now we are given stable forms ρ and $\sigma = \omega^2/2$ as above (though not necessarily exact) on M. We therefore have an almost complex structure determined by ρ on M with a nowhere vanishing $(3,0)$-form $\Omega = \rho + i\hat{\rho}$, integrable if and only if $d\rho = 0 = d\hat{\rho}$. This almost complex structure together with the 2-form ω determines a Hermitian form on M, and we say that the pair (ρ, σ) is of *positive type* if this form is a metric. For the case when (ρ, σ) is a pair of *exact* stable forms of positive type, saying that they form a critical point of $3V(\rho) + 8W(\sigma)$ (the particular constants here are not crucial, since we can always rescale the metrics) subject to the constraint $\langle \rho, \sigma \rangle$ being constant is equivalent to the corresponding metric being nearly Kähler — one checks directly that for some constant λ (where in fact $d\hat{\rho} = -4\lambda\sigma$ and $d\omega = 3\lambda\rho$), the 3-form $\frac{r^2}{\lambda} dr \wedge \omega + r^3 \rho$ on $M \times \mathbb{R}_+$ is stable, and both closed and co-closed, and hence defines a G_2-holonomy metric. The fact that the metric on M is nearly Kähler follows from [**29**].

Returning to the general case, we shall wish to impose two compatibility conditions between the stable forms ρ and σ. The first of these conditions is that $\rho \wedge \omega = 0$, which should be interpreted as saying that ω is of type $(1,1)$ with respect to the almost complex structure induced by ρ (or alternatively that ρ is primitive with respect to ω). The second is that we have proportionality of volume forms, i.e., $\phi(\rho) = c\phi(\sigma)$ for some constant c. This says (assuming that the pair is of positive type) that $\Omega = \rho + i\hat{\rho}$ has constant length with respect to the metric. In the example given above with exact stable forms, these conditions in fact come out for free, and so do not need to be imposed. If ρ and σ were both critical with respect to their volume functionals, we would get an SU(3)-holonomy metric on M.

The theorem from [**235, 237**] concerning existence of G_2-holonomy metrics is the following: Suppose we have $(\rho(t), \sigma(t)) \in \mathcal{A} \times \mathcal{B}$ as above but depending on a parameter t, a pair of stable forms of positive type (with $\sigma(t) = \omega(t)^2/2$), which we assume evolve via the Hamiltonian flow of the functional $H(\rho, \sigma) = V(\rho) - 2W(\sigma)$ with respect to the symplectic form on $\mathcal{A} \times \mathcal{B}$ defined above. This latter condition reduces to the two equations

$$\partial\rho/\partial t = d\omega, \quad \partial\sigma/\partial t = \omega \wedge \partial\omega/\partial t = -d\hat{\rho}.$$

If for time $t = t_0$ the forms satisfy the compatibility conditions $\rho \wedge \omega = 0$ and $\phi(\rho) = 2\phi(\sigma)$, it is shown in [**235**] that these continue to hold for all t (therefore in particular the solutions satisfy the Hamiltonian constraint $H = 0$). Moreover, the 3-form $\varphi = \omega \wedge dt + \rho$ then defines a G_2-holonomy metric on the 7-manifold $M \times (a, b)$ for some open interval $t_0 \in (a, b) \subset \mathbb{R}$; the first of the above two equations corresponds to the condition $d\varphi = 0$, and the second to $d{*}\varphi = 0$. In the case when the classes $\mathcal{A}$ and $\mathcal{B}$ are trivial we have that the fibers are nearly Kähler.

For practical purposes, we need to reduce down to finite dimensional spaces of forms, and one can do this by imposing symmetries. The first interesting case, and one for which calculations are possible, will be that of cohomogeneity one metrics. It should be noted that [**99**] shows that the conditions of G_2-holonomy cohomogeneity one under a compact, connected Lie group severely restricts the principal orbits that can occur, which (up to finite quotients) will be S^6, $\mathbb{CP}^3$, $\mathrm{SU}(3)/T^2$, $S^3 \times S^3$, $S^3 \times T^3$ or T^6. All the known explicit examples of *complete* metrics in dimension 7 with holonomy precisely G_2 are of cohomogeneity one and have principal orbits $S^3 \times S^3$, $\mathrm{SU}(3)/(U(1) \times U(1))$ or $\mathbb{CP}^3$.

Suppose for instance, we take $M = S^3 \times S^3$, and act on the left by the natural action of $G = \mathrm{SU}(2) \times \mathrm{SU}(2)$. We can consider the forms on M which are invariant under G and anti-invariant under the action of C_2 switching the two factors — this latter condition means that the 3-forms change sign whilst the 4-forms do not. We will then use cohomology classes $\mathcal{A}$ and $\mathcal{B}$ as above, except only consider them as vector spaces of those forms representing the cohomology classes which are invariant under G and anti-invariant under C_2. Since $H^4(M, \mathbb{Z}) = 0$, the class $\mathcal{B}$ is necessarily trivial; for $\mathcal{A}$ we shall take the class represented by the difference of the pullbacks of generators of $H^3(S^3, \mathbb{Z}) = \mathbb{Z}$ from the two factors. Both $\mathcal{A}$ and $\mathcal{B}$ are three-dimensional (affine) spaces. The six first order equations derived from the Hamiltonian flow condition are written down in [**235**], and are equivalent to those derived in [**57**]. By imposing further symmetries, these may be solved [**57**] for an explicit one-parameter family of G_2-metrics, which in the most symmetric case corresponds to the previous example of [**76, 178**] with principal orbit $S^3 \times S^3$. The metrics are in general no longer asymptotically conical, but for sufficiently positive and negative values of the parameter, they are complete [**57**]. By taking a limit of their system of equations, the authors also recover the SU(3) deformed conifold metric of [**84**] (the Stenzel metric for $n = 3$). Related results were also obtained in [**105**]. In [**56**] and [**105**], a rather general system of equations for cohomogeneity one metrics with G_2-holonomy and principal orbits $S^3 \times S^3$ was written down, and in [**105**] three types of solutions were identified, providing a G_2-unification of the deformed conifold and resolved conifold SU(3)-holonomy metrics described above; these families of solutions yield regular and complete metrics for appropriate values of the parameters. It might be remarked that [**57**] works directly from the condition for G_2-holonomy in terms of the 3-form being closed and co-closed, but setting these calculations in the context of the Hitchin ansatz as in [**235**] clarifies the approach.

The Hitchin ansatz was used in [**215**] (see also [**237**]) to produce explicit examples of cohomogeneity one G_2-holonomy metrics with principal orbits $S^3 \times T^3$. The Hamiltonian flow condition reduces to two first order differential equations, which can be solved explicitly. As we remarked above, no explicit examples of complete cohomogeneity one G_2-holonomy metrics with principal orbits $S^3 \times T^3$ are known, and in particular these examples turn out not to be complete. In fact, we cannot insert a bolt and retain smoothness.

The same ansatz was applied again in [**97**] (see also [**237**]) to produce further examples of cohomogeneity one G_2-holonomy metrics with principal orbits $S^3 \times S^3$. Here, they take forms invariant under the natural action of $\mathrm{SU}(2) \times \mathrm{SU}(2)$, but instead of taking the class $\mathcal{A}$ to correspond to the difference of the pullbacks from the two factors of the generator for $H^3(S^3, \mathbb{Z})$ (i.e., corresponding to the pair of integers $(1, -1)$), they take a more general class $\mathcal{A}$ corresponding to a pair of arbitrary integers (m, n). The forms are no longer assumed anti-invariant under the C_2-action; there is however a natural closed 3-form in $\mathcal{A}$ given by the appropriate combination of invariant 3-forms pulled back from the two factors of S^3, and the authors consider only the forms which differ from this one by exact forms which are anti-invariant under the action of C_2. Thus the affine spaces of forms being considered are still three-dimensional, and the obvious slight generalization of the Hitchin ansatz remains valid. The Hamiltonian flow condition reduces to two first order differential equations [**97, 237**], which are then solved explicitly [**97**]. Again, one could have defined the 3-form directly, and then checked the closed and co-closed condition directly, but arguing via the Hitchin ansatz is more transparent.

The equations derived in [**97**] in fact provide a generalization of the previously found examples of G_2-holonomy metrics, of cohomogeneity one under $\mathrm{SU}(2) \times \mathrm{SU}(2)$ with principal orbit $S^3 \times S^3$. Moreover, the first order system of equations from **[215]** is recovered as a limit of their equations, and the metrics from **[215]** are obtainable from their solutions. The authors then argue that the only *complete non-singular* solutions with $S^3 \times T^3$ principal orbits that are produced as limits in this way are: flat $\mathbb{R}^4 \times T^3$, and the direct product of Eguchi-Hanson with T^3, both of which clearly have degenerate holonomy. In particular, they deduce that the examples in **[215]** are not complete. These and many of the other examples arising in **[97]** fail to be regular at the bolt.

A similar theory exists for finding weak holonomy G_2 metrics in dimension seven, and therefore non-compact Spin(7)-holonomy manifolds in dimension eight **[235, 237]**. On the compact 7-manifold M, one looks for an appropriate 4-form ρ. We briefly describe the basic ideas that are involved. Any given cohomology class $\mathcal{A}$ in $H^4(M, \mathbb{R})$ is an affine space, on which there is a naturally defined (indefinite) metric, the corresponding quadratic form on the exact 4-forms sending $d\gamma$ to $\int_M \gamma \wedge d\gamma$. One then chooses $\rho(t)$ to be a family of closed stable 4-forms evolving via the gradient flow of the volume functional $\Phi(\rho)$. The 4-form $\varphi = dt \wedge *\rho + \rho$ is shown then to satisfy the conditions for defining a Spin(7)-holonomy metric on $X = M \times (a, b)$ for some open interval $(a, b) \subset \mathbb{R}$. By restricting to cohomogeneity one metrics, this theory has also provided a practical ansatz for finding holonomy Spin(7) metrics. The examples found by this method include new examples which are complete **[105]**. The reader is referred to the references in **[97, 105]** for a good selection from the recent literature on this subject.

In addition, in **[120]**, the authors relate the Hitchin functionals in six and seven dimensions to topological string theory and topological M-theory.

7.2. Examples of special Lagrangian submanifolds

7.2.1. The difficulties in the mathematics of special Lagrangian submanifolds. We now return to the original formulation of the SYZ conjecture, as discussed in Chapter 6. The reader may have noticed that in Chapter 6, after discussing the moduli space of special Lagrangian submanifolds, we then proceeded to forget about the special Lagrangian condition. The reason for this is that the greatest progress has been made with simpler forms of the SYZ conjecture, in which the moduli space, as an affine manifold, plays the most important role. Indeed, we will argue in this chapter that this is actually a reasonable point of view. But first, we would like to explain some of the difficulties that arise in proving the SYZ conjecture.

The first difficulty arises in trying to find examples of special Lagrangian submanifolds. As we know from §7.1, no one has ever written down an explicit Ricci-flat metric on a compact Calabi-Yau manifold. Hence we cannot hope to solve directly the special Lagrangian equations $\omega|_M = 0, \operatorname{Im}\Omega|_M =$

0. One might hope this may not be necessary, that there might be some other trick for finding them.

Recall that in §1.1.4 we defined the more general notion of a calibration and a calibrated submanifold. A closed p-form ϕ on a Riemannian manifold X is a calibration if $\phi|_V \leq \mathrm{Vol}(V)$ for V any p-dimensional subspace of a tangent space of X. A p-dimensional submanifold M of X is *calibrated* by ϕ if $\phi|_M = \mathrm{Vol}(M)$.

There are tricks for constructing other types of calibrated submanifolds. For example, if X is an n-dimensional Kähler manifold with Kähler form ω, then for any $p \leq n$, $\omega^p/p!$ is a calibration. Wirtinger's theorem tells us that if $M \subseteq X$ is a $2p$-real dimensional submanifold, then

$$\left.\frac{\omega^p}{p!}\right|_M = \mathrm{Vol}(M)$$

if and only if M is a complex submanifold. On the other hand, as we know, in the algebraic case complex submanifolds are algebraic and are defined by polynomial equations. So it is very easy to write down $\omega^p/p!$-calibrated submanifolds.

Unfortunately no one has thought up a general trick like this in the special Lagrangian case. There is only one situation where it is easy to write down examples. If X is a Calabi-Yau defined over $\mathbb{R}$, i.e., X has an anti-holomorphic involution $\iota : X \to X$, then the fixed locus of ι, i.e., the set of real points of X, is always special Lagrangian. Indeed, let ω and Ω be the Kähler form and holomorphic n-form respectively. Then as Ω is of type $(n,0)$, $\iota^*\Omega$ is of type $(0,n)$ and thus $\iota^*\Omega = \pm\overline{\Omega}$, while $\iota^*\omega$ remains a form of type $(1,1)$ which is now negative. Hence $-\iota^*\omega$ is positive and still defines a Ricci-flat metric, so $-\iota^*\omega = \omega$ by uniqueness of Ricci-flat metrics. From this, it follows that if $M = \{x \in X | \iota(x) = x\}$, then $\omega|_M = 0$ and either $\mathrm{Re}\,\Omega|_M = 0$ or $\mathrm{Im}\,\Omega|_M = 0$. In any event, M is special Lagrangian.

This method has been applied in [**306**] and [**75**] to obtain examples of special Lagrangian tori and other manifolds in compact Calabi-Yau three-folds, including the quintic. Unfortunately, this is a rather special situation and is unlikely to lead to a general method of finding tori.

The only other approach to constructing examples which has met with some success is in situations where one can approximate Ricci-flat metrics reasonably accurately. This method has been applied by Peng Lu [**340**] and A. Kovalev in unpublished work. To illustrate the simplest example, from [**340**], let E be the unique elliptic curve with periods 1 and $\xi = \exp(2\pi\sqrt{-1}/3)$, so that it has an automorphism of order 3. Then $\mathbb{Z}_3$ acts diagonally on $E \times E \times E$, and this action has $3^3 = 27$ fixed points of order 3. Thus $X = E \times E \times E/\mathbb{Z}_3$ has 27 singularities locally of the form $\mathbb{C}^3/\mathbb{Z}_3$, with action given by $(z_1, z_2, z_3) \mapsto (\xi z_1, \xi z_2, \xi z_3)$. There is a resolution of singularities $\pi : \tilde{X} \to X$ obtained by blowing up the 27 points.

Now $E \times E \times E$ carries a flat metric, so X carries a flat orbifold metric. We can glue in the three-dimensional generalized Eguchi-Hanson metric mentioned in §7.1.1, a metric on the total space of $\mathscr{O}_{\mathbb{P}^2}(-3)$. This latter space is a resolution of $\mathbb{C}^3/\mathbb{Z}_3$ with exceptional locus $\mathbb{P}^2$. We can take the volume of the exceptional locus as a parameter. These metrics are glued to neighbourhoods of the exceptional divisors of π, and so we obtain an approximately Ricci-flat metric. It will be Ricci-flat, and in fact flat, outside a compact neighbourhood of the exceptional locus, and be Ricci-flat, coinciding with the generalized Eguchi-Hanson metric, in a small neighbourhood of the exceptional locus. In the region we glue, the metric fails to be Ricci-flat. However, using techniques adapted by R. Kobayashi [**307**] from the proof of Yau's theorem on the existence of Ricci-flat metrics [**475**], one can show that this metric is close to a genuine Ricci-flat metric, as long as the volume of the exceptional divisors is close to zero. Then one shows that a special Lagrangian submanifold on X avoiding the singular points will deform to a special Lagrangian submanifold on $\tilde{X}$.

This is in fact the most encouraging method of constructing examples in a more general setting. In particular, in §7.3, we will explain how such an approach may lead to the construction of the "correct" special Lagrangian tori. For the moment, we will say little more about this.

So the first difficulty is constructing any special Lagrangian submanifold. There are others. Suppose that we have constructed a special Lagrangian torus M in an n-dimensional Calabi-Yau manifold X. McLean's deformation theory results (§6.1.1) tell us that the torus deforms in an n-dimensional family. However, the next problem is that we don't even know if this family foliates X in a neighbourhood of M. Indeed, recall from §6.1.1 that infinitesimal deformations of M in X correspond to harmonic 1-forms on M: given a normal vector field ν to M determining an infinitesimal deformation, the corresponding 1-form is $\iota(\nu)\omega$ on M. Note that if there were a one-parameter family of special Lagrangian deformations of M intersecting M, then the corresponding normal field to M would have zeroes. In addition ν has zeroes if and only if $\iota(\nu)\omega$ has zeroes. There are examples of special Lagrangian tori in non-compact Calabi-Yau manifolds with harmonic one-forms having zeroes (see [**353**]). Thus, as M deforms, it is possible that it intersects itself. There is no way to rule this possibility out and guarantee that deformations of M foliate X locally. We note that this is not the case in dimension 2, where harmonic 1-forms on tori never have zeroes.

The next problem is that McLean's result is a local one. We may not be able to deform M so that its deformations fill up all of X. One reason this might happen is demonstrated by the following picture: suppose there is a subset $Z \subseteq X$ fibering over an n-dimensional closed ball B, say $Z \to B$, with fibers being special Lagrangian, but all fibers over the boundary of B

being singular. Then, as McLean's result says nothing about singular special Lagrangians, there is no guarantee that we can continue to deform.

To overcome this problem, one can try to study how singularities develop as special Lagrangian submanifolds deform. We shall see some work in this direction in §7.2.5. The most general context in which to study this, however, is geometric measure theory [**145**]. This is far too technical a subject to go into here, but for an excellent introduction, see [**364**]. Geometric measure theory replaces submanifolds with the notion of rectifiable currents. A p-current is a functional on the space of p-forms, and a rectifiable current is, roughly speaking, one obtained by integrating p-forms over a reasonably behaved p-dimensional set, with multiplicities. The theory is designed to study questions of volume minimization. Suppose we want to find a p-dimensional rectifiable current minimizing volume in a given homology class. Then we take a sequence of rectifiable currents whose volumes converge to the infimum of volumes of all such currents representing that class. Then an important result of geometric measure theory, Federer's compactness theorem, says that this sequence has a convergent subsequence in a suitable topology on the space of rectifiable p-currents, and the volume of the limit is the limit of the volumes. Thus one easily obtains volume minimizing rectifiable currents. One is forced to consider currents instead of subsets because it is difficult to describe a topology for the space of subsets in which such convergence statements would work.

As special Lagrangian submanifolds are already volume minimizing, one consequence of this theory is that limits of special Lagrangian submanifolds exist, and are volume minimizing (and special Lagrangian) rectifiable currents. So now the problem is: what do special Lagrangian rectifiable currents look like?

Unfortunately, there has been no progress on this question. The only result known is Almgren's famous regularity result, which states that volume minimizing currents are actually manifolds (with multiplicity) off a set of (Hausdorff) codimension two. The proof was published posthumously, and is 955 pages in print [**8**]. It is already a challenging problem to determine if there is a shorter proof of Almgren's result for special Lagrangian currents. But far stronger information is required before the deformation theory of special Lagrangian currents is understood. Vast technical challenges remain, and it seems highly unlikely that a proof of the strong form of the SYZ conjecture will be discovered in the near future. On the other hand, much progress in understanding special Lagrangian singularities has been made, and will be surveyed in this section. In addition, in §7.3.6, we will give a slightly weaker form of the SYZ conjecture which seems much more likely to be provable, and is probably sufficient for most purposes.

7.2.2. Special Lagrangian submanifolds with torus invariance. Most of what we know about special Lagrangian submanifolds comes from certain sorts of examples. These are easiest to describe when there is a great deal of symmetry. Various cases have been considered in the literature; here we will focus on examples invariant under a torus action.

Fix a triple (X, ω, Ω), where X is an n-dimensional complex manifold, ω is a Kähler form, and Ω is a nowhere vanishing holomorphic n-form. We do not insist on the Kähler metric being Ricci-flat, for reasons we shall see in a moment. Dominic Joyce coined the term *almost Calabi-Yau manifold* for this data.

Now suppose we have a Hamiltonian action of a torus $T := T^m$ acting on X, preserving both the forms ω and Ω. This group action is induced by a moment map $\mu : X \to \mathbf{t}^*$, where $\mathbf{t}^*$ is the dual of the Lie algebra of T. A vector $\xi \in \mathbf{t}$ acts infinitesimally on X via the Hamiltonian vector field $\rho(\xi)$ defined by

$$\iota(\rho(\xi))\omega = d(\langle \mu, \xi \rangle),$$

where $\langle \mu, \xi \rangle$ denotes the function $X \to \mathbb{R}$ defined by $x \mapsto \langle \mu(x), \xi \rangle$.

A model example to keep in mind is the action of T^2 on $\mathbb{C}^3$ given by $(z_1, z_2, z_3) \mapsto (e^{\sqrt{-1}(\theta_1+\theta_2)} z_1, e^{-\sqrt{-1}\theta_1} z_2, e^{-\sqrt{-1}\theta_2} z_3)$ with $(e^{\sqrt{-1}\theta_1}, e^{\sqrt{-1}\theta_2}) \in T^2$. This is just the action of the diagonal subgroup of $\mathrm{SU}(3)$ on $\mathbb{C}^3$; we need $\mathrm{SU}(3)$ instead of $U(3)$ to preserve the holomorphic 3-form $dz_1 \wedge dz_2 \wedge dz_3$. The moment map is $\mu(z_1, z_2, z_3) = (|z_1|^2 - |z_2|^2, |z_1|^2 - |z_3|^2)$, assuming ω is the standard symplectic form.

Given such an action, one obtains the symplectic quotient $X' := X//T = \mu^{-1}(p)/T$, for some choice of $p \in \mathbf{t}^*$; this comes with an induced symplectic form ω'. Also, if $X_1, \ldots, X_m$ are the vector fields on X generating the $T = T^m$ action, i.e., a choice of basis of $\mathbf{t}$, let $\Omega' = \iota(X_1, \ldots, X_m)\Omega$, the contraction of Ω by the vector fields $X_1, \ldots, X_m$. One can check that this descends to a form Ω' on X'. Thus we obtain an almost Calabi-Yau manifold (X', ω', Ω'), as long as p is not a critical value of μ. Note that even if (X, ω, Ω) was Calabi-Yau instead of just almost Calabi-Yau, there is no reason for (X', ω', Ω') to be Calabi-Yau.

The following easy result was observed by a number of people in different forms (see [**227, 185, 199, 266**]) and can be proved as an exercise:

PROPOSITION 7.6. *Suppose $M' \subseteq X'$ is special Lagrangian, i.e., $\omega'|_{M'} = 0$, $\operatorname{Im}\Omega'|_{M'} = 0$. Then the inverse image M of M' in $\mu^{-1}(p) \subseteq X$ is special Lagrangian in X.*

EXAMPLE 7.7. Given the T^2-action mentioned above on $\mathbb{C}^3$, it is easy to check that $\mathbb{C}^3//T^2$ is a complex curve, with single complex coordinate $w = z_1z_2z_3$, and that $\Omega' = dw$. Any real curve in X' is Lagrangian, so the only special Lagrangian submanifolds of X' are given by $\operatorname{Im} w = \text{constant}$. Thus we obtain special Lagrangian submanifolds of $\mathbb{C}^3$ given by the equations

$$|z_1|^2 - |z_2|^2 = c_1, \quad |z_1|^2 - |z_3|^2 = c_2, \quad \operatorname{Im} z_1z_2z_3 = c_3.$$

So the fibers of the map $f : \mathbb{C}^3 \to \mathbb{R}^3$ given by

$$f(z_1, z_2, z_3) = (|z_1|^2 - |z_2|^2, |z_1|^2 - |z_3|^2, \operatorname{Im} z_1 z_2 z_3)$$

are special Lagrangian. This was the first example of a special Lagrangian fibration, given in the original paper of Harvey and Lawson [**226**]. (Compare with Construction 6.92).

7.2.3. T^{n-1}-invariant (cohomogeneity 1) examples. There are numerous examples of non-compact almost Calabi-Yau manifolds coming from toric geometry, coming as resolutions or smoothings of toric varieties. See [**199**] for more details of the examples covered in this section. Here we will explain the general setup for resolutions of Gorenstein toric singularities, and give a few examples of these as well as a smoothing example. We assume familiarity with basic toric geometry (see for example MS1, Chapter 7).

Let $M \cong \mathbb{Z}^n$, $N = \operatorname{Hom}_{\mathbb{Z}}(M, \mathbb{Z})$, $M_{\mathbb{R}} := M \otimes_{\mathbb{Z}} \mathbb{R}$, $N_{\mathbb{R}} := N \otimes_{\mathbb{Z}} \mathbb{R}$ as usual, and let $\sigma \subseteq M_{\mathbb{R}}$ be a strongly convex rational polyhedral cone. We will assume that there exists an $n_0 \in N$ and generators $m_1, \ldots, m_s \in M$ of the cone σ such that $\langle n_0, m_i \rangle = 1$ for all i. Furthermore, if $m \in \sigma \cap (M \setminus \{0\})$, then $\langle n_0, m \rangle \geq 1$. If this is the case, we say the cone is *Gorenstein*, because the affine toric variety Y_σ defined by the cone σ is a Gorenstein variety. In the affine toric case, this means the canonical class is in fact trivial, so this is a local version of a Calabi-Yau manifold.

Let P be the convex hull of $m_1, \ldots, m_s$, i.e., $P = \{m \in \sigma | \langle n_0, m \rangle = 1\}$. We suppose further that we are given a subdivision of P using only integral vertices, with Σ the subdivision of σ induced by the subdivision of P. The rational polyhedral fan Σ defines a toric variety Y_Σ, along with a morphism $Y_\Sigma \to Y_\sigma$ which is a crepant partial resolution, i.e., $K_{Y_\Sigma} = 0$.

EXAMPLE 7.8. (1) Take σ to be generated by $(1, 0, 1)$, $(0, 1, 1)$ and $(-1, -1, 1)$. Then $n_0 = (0, 0, 1)$, P is the convex hull of the three generators, and can be subdivided by adding the point $(0, 0, 1)$, as in Figure 2. In this case Y_σ can be identified with $\mathbb{C}^3/\mathbb{Z}_3$, with the $\mathbb{Z}_3$ action being diagonal multiplication by third roots of unity, while Y_Σ is the blow-up of Y_σ at the singular point, and is isomorphic to the total space of $\mathscr{O}_{\mathbb{CP}^2}(-3)$.

(2) Take σ to be generated by $(0, 0, 1)$, $(1, 0, 1)$, $(0, 1, 1)$ and $(1, 1, 1)$. There are two possible subdivisions, as depicted in Figure 3. Y_σ is an ordinary double point $(xy - zw = 0)$ and the two subdivisions correspond to the two small resolutions of the ordinary double point.

We can use Proposition 7.6 to construct special Lagrangian fibrations on Y_Σ. We need a holomorphic n-form and a Kähler form on Y_Σ, so we discuss this first. Choose a basis $e_1, \ldots, e_n$ of M with dual basis $e_1^*, \ldots, e_n^*$ chosen so that $n_0 = e_1^* + \cdots + e_n^*$. Then $e_1^*, \ldots, e_n^*$ correspond to coordinates $z_1, \ldots, z_n$ on $M \otimes \mathbb{C}^*$. (In fact if $e_1^*, \ldots, e_n^* \in \sigma^\vee$, then $z_1, \ldots, z_n$ extend to

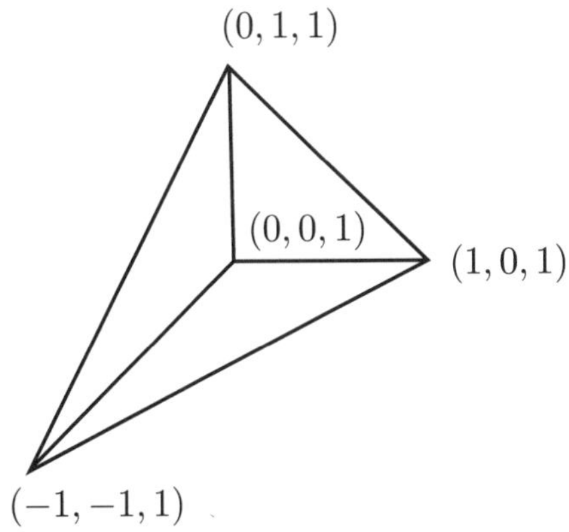

FIGURE 2

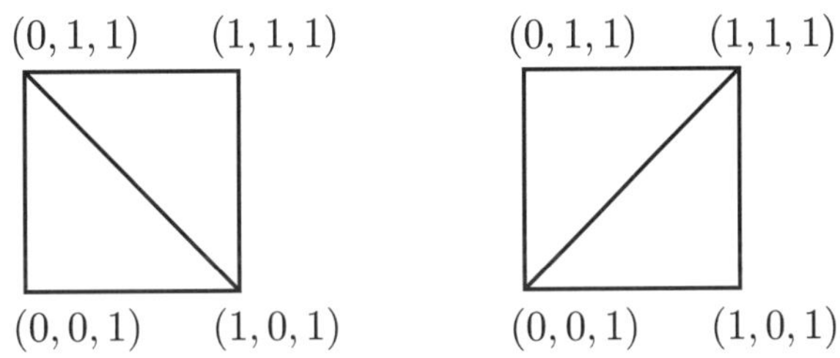

FIGURE 3

regular functions on Y_σ, and hence to Y_Σ. This can always be arranged.) Then we can take $\Omega = dz_1 \wedge \cdots \wedge dz_n$.

EXERCISE 7.9. Check that Ω is independent, up to sign, of the choice of basis. Furthermore, show that Ω is in fact nowhere vanishing on Y_Σ. (Hint: Check that Ω has no zeroes generically along any codimension one orbit of $\mathbb{C}^* \otimes M$ acting on Y_Σ.)

Note that Ω is preserved by the T^{n-1}-action given by $(\theta_1, \ldots, \theta_{n-1})$ taking $(z_1, \ldots, z_n)$ to $(e^{\sqrt{-1}\theta_1} z_1, \ldots, e^{\sqrt{-1}\theta_{n-1}} z_{n-1}, e^{-\sqrt{-1}(\theta_1 + \cdots + \theta_{n-1})} z_n)$.

For the Kähler form ω, we choose a metric invariant under the same action. In situations where Ricci-flat metrics are known to exist on Y_Σ (e.g., quotient singularities, when results of Joyce [**259, 260**] apply) or the conifold case (Example 7.8, (2)), the invariance of the metric follows from uniqueness results for the metrics. For example, in Example 7.8, (1) there is a unique ALE Ricci-flat metric for a fixed volume of the exceptional $\mathbb{CP}^2$, and hence it must be T^2-invariant. Similarly, in Example 7.8, (2), there is a unique asymptotically conical metric for a given volume of the exceptional $\mathbb{CP}^1$. Both these metrics were mentioned in §7.1.1.

Given the T^{n-1}-invariant triple $(Y_\Sigma, \omega, \Omega)$, we can now apply Proposition 7.6. We have a moment map $\mu : Y_\Sigma \to \mathbb{R}^{n-1}$. Given $p \in \mathbb{R}^{n-1}$, $Y'_p :=$

$\mu^{-1}(p)/T^{n-1} = Y_\Sigma//T^{n-1}$ is a Riemann surface. Because $w = \prod_{i=1}^n z_i$, the monomial defined by n_0, is invariant under the $(\mathbb{C}^*)^{n-1}$ action, w defines a complex coordinate on Y_p' and $\Omega' = dw$, as can be easily checked. As Y_p' is two-real-dimensional, any one-dimensional submanifold is automatically Lagrangian, and special Lagrangian submanifolds are then given by the equation $\operatorname{Im} w =$ constant. Thus the fibers of the map $f : Y_\Sigma \to \mathbb{R}^n$ given by $f(y) = (\mu(y), \operatorname{Im} w)$ are special Lagrangian. Note that this is a generalization of Example 7.7.

It is interesting to analyze these fibrations in detail. Unless one knows a bit more about μ, it is difficult to do so. However, one can describe the fibers of $f : Y_\Sigma \to \mathbb{R}^n$ in case the moment map μ is proper. This will be the case, say, if the metric is ALE. We omit the details of the derivation, which may be found in [**199**], Theorem 2.2. The main point is that if μ is proper, one can show that the map $q : Y_\Sigma \to \mathbb{R}^{n-1} \times \mathbb{C}$ given by $q = (\mu, w)$ is in fact the quotient map for the action of T^{n-1} on Y_Σ. Thus for $x \in \mathbb{R}^n$, $f^{-1}(x)$ is a union of T^{n-1}-orbits fibering over a line. If all these orbits are $(n-1)$-dimensional, then $f^{-1}(x) = T^{n-1} \times \mathbb{R}$, while if one of these orbits drops dimension, then $f^{-1}(x)$ is a singular fiber. Thus the discriminant locus of f is precisely the image under f of degenerate T^{n-1}-orbits, and this can be easily understood. Y_Σ is a union of $M \otimes \mathbb{C}^*$-orbits, and these orbits are in one-to-one correspondence with the cones of Σ. If τ is such a cone, then it is a standard fact of toric geometry that the corresponding orbit is fixed by the subtorus $(\mathbb{R}\tau \cap M) \otimes \mathbb{C}^*$. (Here $\mathbb{R}\tau$ is the real vector space spanned by τ.) If this stabilizer intersects the torus T^{n-1}, then the T^{n-1}-orbits are degenerate on this toric stratum. But the original description of our T^{n-1} shows this T^{n-1} is the compact part of the algebraic torus $n_0^\perp \otimes \mathbb{C}^*$. Thus, if $n_0^\perp \cap \mathbb{R}\tau \neq \{0\}$, the toric stratum of $\mathbb{R}\tau$ is a union of degenerate T^{n-1}-orbits. However, given that all one-dimensional cones of Σ are generated by elements of P, and $\langle n_0, P \rangle = 1$, it follows that $n_0^\perp \cap \mathbb{R}\tau \neq \{0\}$ if and only if $\dim \tau \geq 2$. Thus the discriminant locus of f is the image of all codimension ≥ 2 toric strata of Y_Σ. Now the function w is the monomial defined by n_0, and as n_0 is non-zero on each ray in Σ, w must in fact vanish on each codimension ≥ 1 stratum. Thus the discriminant locus is contained in $\mathbb{R}^{n-1} \times \{0\}$, and is in fact the image of the codimension ≥ 2 strata under the moment map μ, or topologically, is the quotient of the codimension ≥ 2 strata under the T^{n-1}-action. This is easy to describe. For example, in Example 7.8, (1), the discriminant locus is as in Figure 4, while in Example 7.8, (2), depending on the choice of small resolution, we get a picture as in Figure 5. Note the lengths of the interior lines will be determined by the Kähler class of ω.

One gets a similar picture whenever an n-dimensional almost Calabi-Yau manifolds has a T^{n-1}-action. However, there is another nice case derived

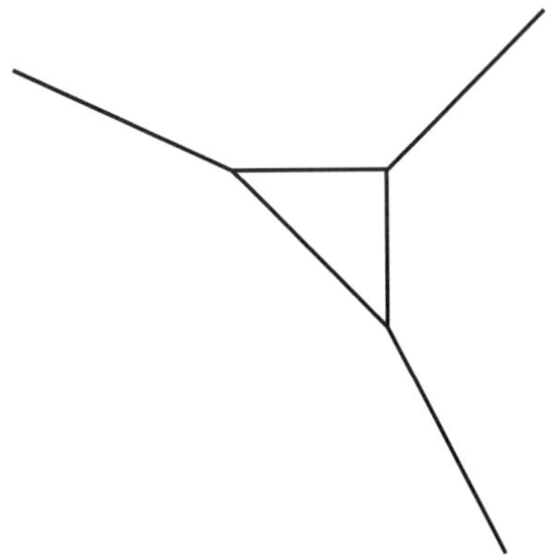

FIGURE 4

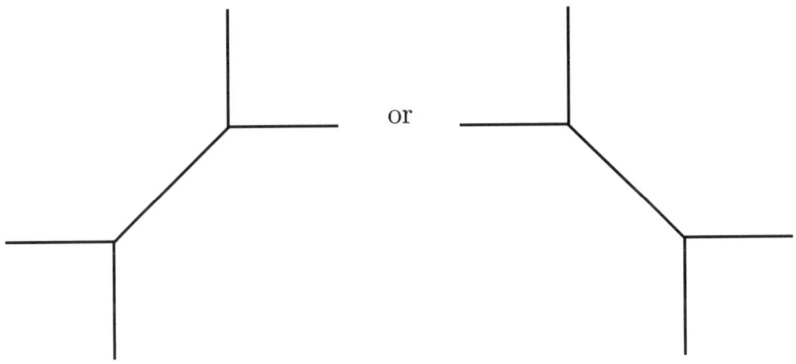

FIGURE 5

from toric geometry, namely smoothings of toric singularities. We will describe the simplest case here, namely the smoothing of the ordinary double point.

Consider the affine quadric Q_t, given by $xy - zw = t$ in $\mathbb{C}^4$, where t is a fixed number. Then Q_t has a T^2-action given by

$$(x, y, z, w) \mapsto (e^{\sqrt{-1}\theta_1}x, e^{-\sqrt{-1}\theta_1}y, e^{\sqrt{-1}\theta_2}z, e^{-\sqrt{-1}\theta_2}w).$$

In addition, Q_t carries a holomorphic 3-form given by

$$\Omega = -\frac{dx \wedge dy \wedge dz}{z} = \frac{dx \wedge dy \wedge dw}{w} = -\frac{dy \wedge dz \wedge dw}{y} = \frac{dx \wedge dz \wedge dw}{x}.$$

The equalities follow from $xdy + ydx = zdw + wdz$ on Q_t. Then Ω is clearly invariant under the given T^2-action.

The affine quadric carries a Ricci-flat metric (the Stenzel metric, Example 7.3), and this metric is also invariant under the T^2-action. Thus we can apply Proposition 7.6 again. The function xy is invariant on Q_t, so descends to a complex coordinate on $\mathbb{C} = \mu^{-1}(p)/T^2$ for $p \in \mathbb{R}^2$. Furthermore, $\Omega' = d(xy)$, as can be checked by calculating $\iota(X_1, X_2)\Omega$, where X_1

and X_2 are the vector fields generating the T^2-action. Thus $f = (\mu, \mathrm{Im}(xy))$ yields a special Lagrangian fibration. As before, the general fiber of f is $\mathbb{R} \times T^2$, and the discriminant locus is given by the image of the union of codimension ≥ 2 orbits of the T^2-action on Q_t. But the locus of points with non-trivial stabilizer on Q_t are the two hyperbolae $x = y = 0, zw = -t$ and $z = w = 0, xy = t$. The function xy takes the value 0 and t respectively on these two hyperbolae, and the image under the moment map of these hyperbolae are straight lines, non-parallel, in $\mathbb{R}^2$. Thus the discriminant locus consists of two lines, at heights 0 and $\mathrm{Im}\, t$. Note that if $\mathrm{Im}\, t = 0$, the lines cross. We can also let $t \to 0$, and Q_0 is singular. We can then take the small resolution, and view the procedures we have seen here as three different ways of resolving the ordinary double point; see Figure 6.

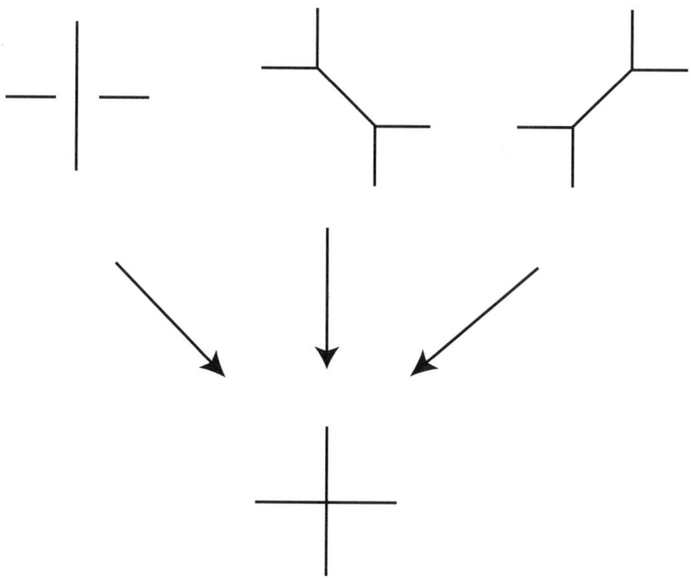

FIGURE 6

7.2.4. S^1-invariant examples in $\mathbb{C}^3$. It is too difficult to consider S^1-actions on general Calabi-Yau manifolds; instead, we will work with the S^1-action on $\mathbb{C}^3$ given by

$$(z_1, z_2, z_3) \mapsto (e^{\sqrt{-1}\theta} z_1, e^{-\sqrt{-1}\theta} z_2, z_3),$$

and consider S^1-invariant special Lagrangians. This is already very illuminating, as we begin to get very different behaviour than we saw in the T^{n-1} case. This situation has been studied extensively by Dominic Joyce in a sequence of papers [**266, 267, 268**].

The moment map of the S^1-action with respect to the standard symplectic form $\omega = \frac{\sqrt{-1}}{2} \sum_{i=1}^{3} dz_i \wedge d\bar{z}_i$ is $\mu(z_1, z_2, z_3) = |z_1|^2 - |z_2|^2$. Let's compute the reduced symplectic and holomorphic forms on $\mu^{-1}(2a)/S^1$. (The factor of 2 is a convenient normalization.) The vector field generating the S^1-action is

$$X = 2\sqrt{-1}(\bar{z}_1\partial_{\bar{z}_1} - z_1\partial_{z_1} - \bar{z}_2\partial_{\bar{z}_2} + z_2\partial_{z_2}).$$

Thus

$$\begin{aligned} \iota(X)\Omega &= -2\sqrt{-1}(z_1 dz_2 \wedge dz_3 + z_2 dz_1 \wedge dz_3) \\ &= -2\sqrt{-1}d(z_1 z_2) \wedge dz_3. \end{aligned}$$

Clearly $w = z_1 z_2$ and z_3 are the invariant functions determining the complex structure on $\mu^{-1}(2a)/S^1$, so Ω', up to a constant, is $dw \wedge dz_3$.

On the other hand, ω' on $\mu^{-1}(2a)/S^1$ will depend on the value of a. To compute this, we proceed as follows. We have the map $\pi : \mathbb{C}^3 \to \mathbb{C}^2$ given by $\pi(z_1, z_2, z_3) = (z_1 z_2, z_3) = (w, z_3)$. Under this map, the push-forwards of tangent vectors $\partial/\partial z_i$ at the point with coordinates (z_1, z_2, z_3) are computed as $\pi_*(\partial/\partial z_1) = z_2(\partial/\partial w)$, $\pi_*(\partial/\partial z_2) = z_1(\partial/\partial w)$, and $\pi_*(\partial/\partial z_3) = \partial/\partial z_3$. Now to compute $\omega'(v_1, v_2)$, for two tangent vectors at a point (w, z_3), we need to lift this point to a point p in $\mu^{-1}(2a)$, and then lift v_1 and v_2 to tangent vectors to $\mu^{-1}(2a)$ at the point p, say $\tilde{v}_1, \tilde{v}_2$. Then $\omega'(v_1, v_2) = \omega(\tilde{v}_1, \tilde{v}_2)$. Explicitly, we can lift (w, z_3) by using the S^1-action to make z_1 real and positive, and then

$$p = \left(\sqrt{\frac{2a + \sqrt{4a^2 + 4|w|^2}}{2}}, \sqrt{\frac{-2a + \sqrt{4a^2 + 4|w|^2}}{2}}\frac{w}{|w|}, z_3\right) = (z_1, z_2, z_3)$$

will do. The tangent vector $\partial/\partial z_3$ lifts to $\partial/\partial z_3$ at p, being tangent to $\mu^{-1}(2a)$, while $\partial/\partial w$ lifts to some linear combination $\alpha\partial/\partial z_1 + \beta\partial/\partial z_2$. The condition that this be tangent to $\mu^{-1}(a)$ is that $(\alpha\partial/\partial z_1 + \beta\partial/\partial z_2)(|z_1|^2 - |z_2|^2) = 0$, i.e., $\alpha\bar{z}_1 - \beta\bar{z}_2 = 0$, while the condition that $\pi_*(\alpha\partial/\partial z_1 + \beta\partial/\partial z_2) = \partial/\partial w$ is that $\alpha z_2 + \beta z_1 = 1$. Solving these two equations for α and β, we find

$$\alpha = \frac{\bar{z}_2}{|z_1|^2 + |z_2|^2}, \quad \beta = \frac{\bar{z}_1}{|z_1|^2 + |z_2|^2}.$$

Thus

$$\begin{aligned} \omega'(\partial/\partial w, \partial/\partial \bar{w}) &= \omega(\alpha\partial/\partial z_1 + \beta\partial/\partial z_2, \bar{\alpha}\partial/\partial \bar{z}_1 + \bar{\beta}\partial/\partial \bar{z}_2) \\ &= \frac{\sqrt{-1}}{2}(|\alpha|^2 + |\beta|^2) \\ &= \frac{\sqrt{-1}}{2}\left(\frac{1}{|z_1|^2 + |z_2|^2}\right). \end{aligned}$$

Thus we see

$$\begin{aligned}\omega' &= \frac{\sqrt{-1}}{2}\left(\frac{1}{2\sqrt{a^2+|w|^2}}dw\wedge d\bar{w}+dz_3\wedge d\bar{z}_3\right),\\ \Omega' &= -2\sqrt{-1}dw\wedge dz_3.\end{aligned}$$

Note that for $a=0$, ω' is in fact singular along the locus $w=0$. This is an important feature: critical points of the moment map yield singularities in the reduced symplectic form.

To look for special Lagrangian submanifolds in $\mathbb{C}^2$ with respect to these two forms, we choose an $\mathbb{R}^2$ in $\mathbb{C}^2$ and try to find special Lagrangian submanifolds which are graphs of functions over $\mathbb{R}^2$. For example, we can take coordinates x and y on $\mathbb{R}^2$, and write

$$M' = \{(w,z_3)\in\mathbb{C}^2\,|\,w=v(x,y)+\sqrt{-1}y,\,z_3=x+\sqrt{-1}u(x,y),\ x,y\in S\subseteq\mathbb{R}^2\}$$

where S is a domain on $\mathbb{R}^2$. In this case M' is special Lagrangian with respect to the forms ω',Ω' if

$$\begin{aligned}0 &= \operatorname{Im}\sqrt{-1}\,d(v+\sqrt{-1}y)\wedge d(x+\sqrt{-1}u)\\ &= dv\wedge dx-dy\wedge du\\ &= \left(\frac{\partial v}{\partial y}-\frac{\partial u}{\partial x}\right)dy\wedge dx\end{aligned}$$

and

$$\begin{aligned}0 &= \frac{1}{2\sqrt{a^2+|w|^2}}\left(d(v+\sqrt{-1}y)\wedge d(v-\sqrt{-1}y)\right)\\ &\quad +d(x+\sqrt{-1}u)\wedge d(x-\sqrt{-1}u)\\ &= \frac{1}{2\sqrt{a^2+|w|^2}}\left(-2\sqrt{-1}dv\wedge dy\right)-2\sqrt{-1}dx\wedge du\\ &= 2\sqrt{-1}\left(\frac{1}{2\sqrt{a^2+|w|^2}}\frac{\partial v}{\partial x}+\frac{\partial u}{\partial y}\right)dy\wedge dx.\end{aligned}$$

Thus u and v must satisfy the equations

$$\frac{\partial v}{\partial y} = \frac{\partial u}{\partial x} \tag{7.1}$$

$$\frac{\partial v}{\partial x}+2\sqrt{a^2+y^2+v^2}\frac{\partial u}{\partial y} = 0. \tag{7.2}$$

These equations can be viewed as a perturbation of the Cauchy-Riemann equations, $\partial v/\partial y=\partial u/\partial x$, $\partial v/\partial x=-\partial u/\partial y$. The first equation allows us to find, if S is a simply connected equation in $\mathbb{R}^2$, a potential function

$f : S \to \mathbb{R}$ such that $\partial f/\partial x = v$ and $\partial f/\partial y = u$. Then f satisfies the single equation

$$\frac{\partial^2 f}{\partial x^2} + 2\sqrt{a^2 + y^2 + (\partial f/\partial x)^2}\frac{\partial^2 f}{\partial y^2} = 0. \tag{7.3}$$

This is a second order, quasi-linear elliptic equation, though it is degenerate when $a = 0$ and $y = 0$ as the coefficient of $\partial^2 f/\partial y^2$ may become zero.

Here are some sample solutions. (See Joyce [**266**], §5.)

EXAMPLE 7.10. (1) An explicit global solution to Equations (7.1) and (7.2) with $a = 0$ is

$$\begin{aligned} u(x,y) &= y\tanh x \\ v(x,y) &= \frac{1}{2}y^2\operatorname{sech}^2 x - \frac{1}{2}\cosh^2 x. \end{aligned}$$

Note that when $y = 0$, $v(x,y) \neq 0$. This means that, on the surface $M' \subseteq \mathbb{C}^2$ defined by this u and v, w is never zero. Thus the special Lagrangian submanifold $M \subseteq \mathbb{C}^3$ is disjoint from the fixed locus $z_1 = z_2 = 0$ of the S^1-action. Thus we see that M is $S^1 \times \mathbb{R}^2$, and is non-singular.

(2) This example is a familiar one: we write it in a more useful fashion than giving u and v: take, for $a \geq 0$,

$$M_a = \left\{ (z_1, z_2, z_3) \in \mathbb{C}^3 \,\middle|\, \begin{array}{l} |z_1|^2 - |z_2|^2 = 2a, |z_2|^2 = |z_3|^2, \\ \operatorname{Im}(z_1z_2z_3) = 0, \operatorname{Re}(z_1z_2z_3) \geq 0 \end{array} \right\}.$$

This M_a is S^1-invariant (though it is in fact also T^2-invariant). M_a is contained in a fiber of the map of Example 7.7, but because of the inequality, it is not a whole fiber of that map. However, M_a is in fact $S^1 \times \mathbb{R}^2$ again for $a > 0$, so has no singularities. Indeed, given $|z_3|^2$, we know for a point in M_a, $|w|^2 = (|z_3|^2 + 2a)|z_3|^2$. However, $\operatorname{Im}(wz_3) = 0$, $\operatorname{Re}(wz_3) \geq 0$ then fixes an S^1 of choices for w and z_3, unless $w = z_3 = 0$. Thus the set of S^1-orbits is parametrized by an $\mathbb{R}^2$ (described in polar coordinates). When $a = 0$, we get one-half of the double cone over a two-torus $|z_1|^2 = |z_2|^2 = |z_3|^2$, $\operatorname{Im}(z_1z_2z_3) = 0$. Intuitively, an S^1-orbit in M_a has shrunk down to the zero-dimensional orbit $z_1 = z_2 = z_3 = 0$, creating the singularity.

This example can be expressed in terms of Equations (7.1) and (7.2): $\operatorname{Im}(wz_3) = 0$ becomes $xy + uv = 0$, so $v = -xy/u$, while $(|z_3|^2 + 2a)|z_3|^2 = |w|^2$ becomes

$$\begin{aligned} (x^2 + u^2 + 2a)(x^2 + u^2) &= y^2 + v^2 \\ &= y^2 + x^2y^2/u^2. \end{aligned} \tag{7.4}$$

This becomes the cubic equation in u^2

$$u^6 + u^4(2x^2 + 2a) + u^2(-y^2 + (x^2 + 2a)x^2) - x^2y^2 = 0.$$

We need to check that for every x and y, there is a unique positive solution of this equation for u^2. There are four cases:

(i) $x \neq 0$, $y \neq 0$, and there are three real solutions $u^2 = \gamma_1, \gamma_2, \gamma_3$. Then $\gamma_1\gamma_2\gamma_3 = x^2y^2 > 0$, while $\gamma_1 + \gamma_2 + \gamma_3 = -(2x^2 + 2a) < 0$. Thus two of the solutions are negative and one is positive.

(ii) $x \neq 0$, $y \neq 0$, and there is one real solution γ and two complex conjugate solutions $\delta, \bar{\delta}$. Then $\gamma|\delta|^2 > 0$ as in case (i), so $\gamma > 0$ is the unique positive real solution.

(iii) $y = 0$. Then the equation is

$$u^2(u^2 + x^2 + 2a)(u^2 + x^2) = 0.$$

Then the only non-negative root is 0 (possibly with some multiplicity if $x = 0$).

(iv) $x = 0$ but $y \neq 0$. Then the equation is

$$u^2(u^4 + 2au^2 - y^2) = 0.$$

This has the root $u^2 = 0$ and one positive root, since the second factor is negative when $u^2 = 0$. However if $u^2 = 0$ and $x^2 = 0$, then by (7.4), $y^2 + v^2 = 0$, i.e., $y = v = 0$. So there is only one allowable positive root under the assumption that $y \neq 0$.

This gives the existence of a function u, and v is determined by u. Hence we get solutions to Equations (7.1), (7.2) which are clearly already quite complicated, but shows that there exist interesting non-trivial solutions. One can check that this solution is smooth for $a > 0$, and smooth except at $x = y = 0$ when $a = 0$, where it is only continuous [**266**], §5.

In general, given a solution u, v to (7.1), (7.2) for some a, we will obtain, for $a \neq 0$, a non-singular special Lagrangian manifold, fibered in S^1's over a domain in $\mathbb{R}^2$. However, if $a = 0$, we have singularities where the three-dimensional special Lagrangian manifold intersects the union of zero-dimensional orbits, i.e., the locus where $z_1 = z_2 = 0$. Since $|z_1|^2 = |z_2|^2$ in $\mu^{-1}(0)$, $z_1 = z_2 = 0$ is equivalent to $w = 0$, or $y = v = 0$. Thus if v is zero at a point where y is zero, we obtain a singularity. In Example 7.10, (2), we saw that this only occurs when $x = y = 0$, which corresponds to the vertex of the cone.

Before discussing Equations (7.1), (7.2) in greater detail, let's use the above example to give a very important counterexample of Joyce.

EXAMPLE 7.11. (Joyce, [**261**]) First, rewrite Example 7.10, (2) as

$$M_a = \left\{ (z_1, z_2, z_3) \in \mathbb{C}^3 \,\middle|\, \begin{matrix} |z_1|^2 - a = |z_2|^2 + a = |z_3|^2 + |a|, \\ \mathrm{Im}(z_1z_2z_3) = 0, \mathrm{Re}(z_1z_2z_3) \geq 0 \end{matrix} \right\}.$$

One can then check easily that for $a < 0$, this is also special Lagrangian. For $a \geq 0$, M_a is one-half of the fiber $f^{-1}(2a, 2a, 0)$, where f is the map of Example 7.7, while for $a \leq 0$, it is one-half of the fiber $f^{-1}(2a, 0, 0)$. Now introduce a translation by c in z_3: for each $c \in \mathbb{C}$,

$$M_{a,c} = \{(z_1, z_2, z_3) \in \mathbb{C}^3 | (z_1, z_2, z_3 - c) \in M_a\}.$$

This gives a three-parameter family of special Lagrangian submanifolds of $\mathbb{C}^3$. In fact, it yields a special Lagrangian fibration of $\mathbb{C}^3$, given by $F(z_1, z_2, z_3) = (a, b)$ with $2a = |z_1|^2 - |z_2|^2$ and

$$b = \begin{cases} z_3 & a = z_1 = z_2 = 0 \\ z_3 - \bar{z}_1\bar{z}_2/|z_1| & a \geq 0, z_1 \neq 0 \\ z_3 - \bar{z}_1\bar{z}_2/|z_2| & a < 0. \end{cases}$$

Indeed, let us check that $F^{-1}(a, c) = M_{a,c}$ for $a \geq 0$; $a < 0$ follows similarly. Note that $(z_1, z_2, z_3) \in M_{a,c}$ if and only if $|z_1|^2 - |z_2|^2 = 2a, |z_2|^2 = |z_3 - c|^2$, and $z_1 z_2(z_3 - c) \in [0, \infty)$. Thus $z_1 z_2(z_3 - c) = |z_1||z_2||z_3 - c| = |z_1||z_2|^2$; so as long as $z_1 z_2 \neq 0$, we can divide by $z_1 z_2$ and obtain $c = z_3 - |z_1||z_2|^2/(z_1 z_2) = z_3 - \bar{z}_1\bar{z}_2/|z_1|$. On the other hand, if $z_1 z_2 = 0$, then either $a = 0$, in which case $z_1 = z_2 = 0$ and $z_3 = c$, or $a > 0$ and $z_2 = z_3 - c = 0$, $z_1 \neq 0$, and $z_3 - \bar{z}_1\bar{z}_2/|z_1| = c$ again. Thus $M_{a,c} \subseteq F^{-1}(a, c)$. The reverse argument shows equality.

Thus $F^{-1}(a, c)$ is homeomorphic to $\mathbb{R}^2 \times S^1$ when $a \neq 0$, but is singular, i.e., a cone over a T^2, when $a = 0$. Thus the discriminant locus of F is $\{0\} \times \mathbb{C}$, i.e., is codimension one!

This example was a crucial one for the understanding of the SYZ conjecture. A codimension one discriminant locus provides a very different picture than the nice, codimension two discriminant locus we saw both in Chapter 6 and in the T^2-invariant fibrations we saw in the previous section. Once one knew that such codimension one behaviour was possible, certain aspects of the full version of the SYZ conjecture needed to be rethought. We will return to these philosophical issues in the next section.

First, let us return to the behaviour of Equations (7.1) and (7.2). We will survey the results of Joyce's analysis of these equations, which demonstrates the existence of many interesting solutions. It is easier to study the equation (7.3), with $u = \partial f/\partial y$ and $v = \partial f/\partial x$. Joyce solved the Dirichlet problem for (7.3):

THEOREM 7.12. *Suppose S is a strictly convex domain in $\mathbb{R}^2$ invariant under $(x, y) \mapsto (x, -y)$ and let $k \geq 0$ be an integer, $\alpha \in (0, 1)$. Let $a \in \mathbb{R}$, $\phi \in C^{k+3,\alpha}(\partial S)$. If $a \neq 0$ then there is a unique $f \in C^{k+3,\alpha}(S)$ with $f = \phi$ on ∂S satisfying equation (7.3). If $a = 0$ there is a unique $f \in C^1(S)$ with $f = \phi$ on ∂S, and such that f is twice weakly differentiable and satisfies (7.3) with weak derivatives.*

PROOF. [**266**], Theorem 7.6, and [**267**], Theorems 9.20, 9.21. □

While the proof of this result for $a \neq 0$ only requires some fairly standard techniques from the theory of quasi-linear elliptic PDEs, the case $a = 0$, when the equation can be degenerate, is technically very difficult and we give no details here.

This theorem yields lots of examples of special Lagrangian submanifolds (with boundary) of $\mathbb{C}^3$, homeomorphic to $S \times S^1$ if $a \neq 0$ and possibly singular if $a = 0$, with singularities occurring when $v = y = 0$, i.e., $\partial f/\partial x = y = 0$. For "generic" solutions, the singularities are cones over T^2 as seen in Example 7.10, (2).

Now at first sight, it would seem to be difficult to understand the geometry of these examples just by knowing the values of the potential f on ∂S, as opposed to the values of u and v. However, Joyce has also found a way to get more explicit information about families of solutions. The idea is as follows. Suppose we have a domain $S \subseteq \mathbb{R}^2$, and two solutions (u_1, v_1) and (u_2, v_2) to Equations (7.1) and (7.2) for a given value of a. We would like to understand the intersection of the two special Lagrangian submanifolds M_1 and M_2 of $\mathbb{C}^3$ defined by these two solutions. Of course, when $(u_1, v_1) = (u_2, v_2)$, the two submanifolds M_1', M_2' of $\mathbb{C}^2$ defined as the graphs of these solutions intersect. Then M_1 and M_2 intersect along a circle over that point of intersection (unless $a = v = y = 0$, in which case M_1 and M_2 intersect in a point.) Thus we try to count the number of intersection points of M_1' and M_2' by counting the number of zeroes of $(u_1, v_1) - (u_2, v_2)$. This can be done by analogy with a standard result of complex analysis. Recall that the *winding number* of a closed curve $\gamma : S^1 \to \mathbb{C} \setminus \{0\}$ is defined to be $\frac{1}{2\pi\sqrt{-1}} \int_\gamma \frac{dz}{z}$. Then if $S \subseteq \mathbb{C}$ is a domain with $f : S \to \mathbb{C}$ a holomorphic function, non-zero on ∂S, then $f|_{\partial S} : \partial S \to \mathbb{C} \setminus \{0\}$ is a closed curve, and its winding number is equal to the number of zeroes of f in the interior of S, counted with multiplicity. Because Equations (7.1), (7.2) are a perturbed version of the Cauchy-Riemann equations, we might hope a similar result to be true, which is indeed the case. We first need to define the multiplicity of a zero.

DEFINITION 7.13. Let $a \in \mathbb{R}$, and let $(u_1, v_1), (u_2, v_2) : S \to \mathbb{R}^2$ be two solutions of (7.1), (7.2). We assume they are in $C^1(S)$ if $a \neq 0$ or $C^0(S)$ if $a = 0$. A zero (b, c) of $(u_1, v_1) - (u_2, v_2)$ is called *singular* if $a = c = 0$ and $v_1(b, 0) = v_2(b, 0)$. (Thus a singular zero corresponds to M_1 and M_2 intersecting at a singular point.)

Let $(b, c) \in \operatorname{Int}(S)$ be an isolated zero of $(u_1, v_1) - (u_2, v_2)$. Then the *multiplicity* of (b, c) is the winding number of $(u_1, v_1) - (u_2, v_2)$ around the circle $\gamma : S^1 \to S$ defined by $\gamma(\theta) = (b + \epsilon \cos\theta, c + \epsilon \sin\theta)$, where ϵ is chosen sufficiently small so that (b, c) is the only zero inside γ.

Joyce proves ([**266**], §6, [**268**], §7)

THEOREM 7.14. *In the situation of Definition 7.13,*

(1) the multiplicity of any isolated zero in $\operatorname{Int}(S)$ *is a positive integer.*
(2) Either $(u_1, v_1) \equiv (u_2, v_2)$ *or there are at most countably many zeroes of* $(u_1, v_1) - (u_2, v_2)$ *in* $\operatorname{Int}(S)$*, all isolated.*

This analysis allows Joyce to prove the following analogue of a standard complex analysis result.

THEOREM 7.15. *In the situation of Definition 7.13, suppose* $(u_1, v_1) \neq (u_2, v_2)$ *at every point of* ∂S. *Then* $(u_1, v_1) - (u_2, v_2)$ *has finitely many zeroes in* S, *all isolated, say with multiplicities* $k_1, \ldots, k_n$, *and the winding number about zero of* $(u_1, v_1) - (u_2, v_2)$ *along* ∂S *is* $\sum_{i=1}^n k_i$.

PROOF. See [**266**], Theorem 6.7 in the case $a \neq 0$ and [**268**], Theorem 7.7 in the case $a = 0$. □

There is a similar result which is more useful in the context of the solution to the Dirichlet problem, Theorem 7.12:

THEOREM 7.16. *Under the same hypotheses as Theorem 7.12, suppose we are given two different boundary conditions* $\phi_1, \phi_2 \in C^{k+3,\alpha}(\partial S)$ *giving two solutions* f_1, f_2 *yielding* (u_1, v_1), (u_2, v_2). *Suppose* $\phi_1 - \phi_2$ *has exactly* l *local maxima and* l *local minima on* ∂S. *Then* $(u_1, v_1) - (u_2, v_2)$ *has finitely many zeroes in* $S \setminus \partial S$, *all isolated, say with multiplicities* $k_1, \ldots, k_n$. *Then* $\sum_{i=1}^n k_i \leq l - 1$.

PROOF. See [**266**], Theorem 7.11 in the case $a \neq 0$ and [**268**], Theorem 7.10 in the case $a = 0$. □

Thus in particular, if $l = 1$ and M_1, M_2 are the special Lagrangian submanifolds of $\mathbb{C}^3$ defined by (u_1, v_1) and (u_2, v_2) respectively, we are guaranteed that $M_1 \cap M_2 = \emptyset$.

One can use this to obtain examples of special Lagrangian fibrations. Take S a strictly convex domain in $\mathbb{R}^2$ invariant under $(x, y) \mapsto (x, -y)$, and let U be an open subset of $\mathbb{R}^3$. Suppose we are given a continuous map $\Phi : U \to C^{3,\alpha}(\partial S)$ such that if $(a, b, c) \neq (a, b', c')$ then $\Phi(a, b, c) - \Phi(a, b', c')$ has only one local maximum and one local minimum. Then for $\beta \in U$, let $f_\beta \in C^{3,\alpha}(S)$ be the solution to the Dirichlet problem for (7.3) with $f_\beta|_{\partial S} = \Phi(\beta)$ and the value of a being the first component of β. Then f_β defines a special Lagrangian submanifold $M_\beta \subseteq \mathbb{C}^3$, and $M_\beta \cap M_{\beta'} = \emptyset$. One can show that M_β varies continuously with β, so we get a special Lagrangian fibration on some subset of $\mathbb{C}^3$.

For example, one can fix some $\phi \in C^{3,\alpha}(\partial S)$ and take $\Phi(a, b, c) = \phi + bx + cy$.

7.2.5. Special Lagrangian submanifolds with conical singularities. We have now seen how in three dimensions the "generic" singularity of a $U(1)$-invariant special Lagrangian submanifold looks like the cone over a T^2. To understand better the geometry of special Lagrangian submanifolds, it is therefore useful to study the deformation theory of special Lagrangian submanifolds with mild singularities. Recall from §6.1.1 that if X is a Calabi-Yau manifold, and $M \subseteq X$ is a compact non-singular special

Lagrangian submanifold, then the moduli space of special Lagrangian deformations of M inside X is smooth at the point corresponding to M and has tangent space $H^1(M, \mathbb{R})$, and more canonically, the tangent space is the space of harmonic 1-forms on M.

We begin by studying cones more generally, working in $\mathbb{C}^m$ with $m \geq 3$. A singular special Lagrangian submanifold $C \subseteq \mathbb{C}^m$ is called a *cone* if $C = tC := \{tx | x \in C\}$ for all $t > 0$. A cone has a link $\Sigma = S^{2m-1} \cap C$, where S^{2m-1} is the unit sphere in $\mathbb{C}^m$. The dimension of the link is $m - 1$.

Write $C' = C \setminus \{0\}$. Then C' is the image of the map $(0, \infty) \times \Sigma \to \mathbb{C}^m$ given by $(r, s) \mapsto rs \in \mathbb{C}^m$. Under this identification, we can write the metric on C' as $dr^2 + r^2 g_\Sigma$, where g_Σ is the metric on Σ.

To motivate slightly the formula we are about to give, we note that we can compute the Laplacian Δ on C' in terms of the Laplacian Δ_Σ on Σ: for a function $f : \Sigma \to \mathbb{R}$,

$$\Delta(r^\alpha f) = r^{\alpha-2}(\Delta_\Sigma f - \alpha(\alpha + m - 2)f). \tag{7.5}$$

This can be done via the explicit formula for the Laplacian, in a metric g_{ij} with respect to local coordinates x_i,

$$\Delta(f) = -\frac{1}{\sqrt{\det(g_{ij})}} \sum_{k,l} \frac{\partial}{\partial x_k} \left(\sqrt{\det(g_{ij})} g^{kl} \frac{\partial}{\partial x_l} f \right).$$

Thus whenever $\alpha(\alpha + m - 2)$ is an eigenvalue of Δ_Σ, with eigenfunction f, $r^\alpha f$ is a harmonic function on C'. So, let

$$\mathcal{D}_\Sigma = \{\alpha \in \mathbb{R} \,|\, \alpha(\alpha + m - 2) \text{ is an eigenvalue of } \Delta_\Sigma\}.$$

As Σ is compact, $\mathcal{D}_\Sigma$ is a discrete set. In addition, each eigenvalue comes with a multiplicity, the dimension of the space of eigenfunctions of that eigenvalue, and we write $m_\Sigma : \mathcal{D}_\Sigma \to \mathbb{N}$ for the function giving the multiplicity. The significance of $\mathcal{D}_\Sigma$, roughly, can be explained as follows. There is a theory of elliptic partial differential equations on manifolds with ends developed by Lockhart and McOwen [**337**] and Melrose [**357**]. If one considers the space of harmonic functions on C' with asymptotic behaviour near $r = 0$ like $r^\alpha f$ for f a function on Σ, then the dimension of this space depends on α. In fact, the dimension of this space will change when α is an element of $\mathcal{D}_\Sigma$. (More details will be given in a few pages.)

Finally, let G be the subgroup of $\mathrm{SU}(m)$ preserving C. Joyce defines the *stability index* of the cone C to be

$$\text{s-ind}(C) = -b_0(\Sigma) - m^2 - 2m + 1 + \dim G + \sum_{\alpha \in \mathcal{D}_\Sigma \cap [0,2]} m_\Sigma(\alpha).$$

Here b_0 denotes the 0th Betti number of Σ, i.e., the number of connected components of Σ.

We say C is *stable* if $\text{s-ind}(C) = 0$.

There are many known examples of special Lagrangian cones, the first being the cone over the torus

$$C = \{(z_1, \dots, z_m) \in \mathbb{C}^m \mid \operatorname{Im}(z_1 \cdots z_m) = 0, \operatorname{Re}(z_1 \cdots z_m) > 0, |z_1| = \cdots = |z_m|\}.$$

The link Σ is the torus

$$T^{m-1} = \left\{ \frac{1}{\sqrt{m}} \left(e^{\sqrt{-1}\theta_1}, \dots, e^{\sqrt{-1}\theta_{m-1}}, e^{-\sqrt{-1}(\theta_1 + \cdots + \theta_{m-1})} \right) \middle| \theta_1, \dots, \theta_{m-1} \in \mathbb{R} \right\} \subseteq \mathbb{C}^m.$$

The induced metric on the torus $T^{m-1} = \mathbb{R}^{m-1}/2\pi\mathbb{Z}^{m-1}$ is calculated as follows: the tangent vector $\partial/\partial\theta_i$ on T^{m-1} maps to the tangent vector

$$\frac{1}{\sqrt{m}} \Big(-\sin\theta_i \frac{\partial}{\partial x_i} + \cos\theta_i \frac{\partial}{\partial y_i} + \sin(\theta_1 + \cdots + \theta_{m-1}) \frac{\partial}{\partial x_m} - \cos(\theta_1 + \cdots + \theta_{m-1}) \frac{\partial}{\partial y_m} \Big)$$

in $\mathbb{C}^m$, with $z_i = x_i + \sqrt{-1}y_i$ coordinates on $\mathbb{C}^m$. Thus

$$g_{ij} = g(\partial/\partial\theta_i, \partial/\partial\theta_j) = \frac{1}{m}(\delta_{ij} + 1)$$

and

$$g^{ij} = m\delta_{ij} - 1.$$

The eigenfunctions of Δ on this flat torus are of course of the form

$$f = e^{\sqrt{-1}(n_1\theta_1 + \cdots + n_{m-1}\theta_{m-1})}$$

for $(n_1, \dots, n_{m-1}) \in \mathbb{Z}^{m-1}$, and

$$\Delta f = \sum_{k,l} g^{kl} n_k n_l f,$$

hence the eigenvalue of f is

$$m \sum_{i=1}^{m-1} n_i^2 - \sum_{i,j}^{m-1} n_i n_j.$$

From this one can tediously calculate that

$$\sum_{\alpha \in \mathcal{D}_\Sigma \cap [0,2]} m_\Sigma(\alpha) = 13, 27, 51, 93, \dots$$

for $m = 3, 4, 5, 6, \dots$. On the other hand, the Lie group preserving C is $U(1)^{m-1}$. Thus

$$\text{s-ind}(C) = 0, 6, 20, 50, \dots$$

and in fact C is stable only in the case $m = 3$. This is the only known example of a stable cone.

Next we define the notion of special Lagrangian submanifolds with conical singularities. Let X be a Calabi-Yau manifold, $M \subseteq X$ a special Lagrangian submanifold with isolated singularities. Let $x \in M$ be a singular point, B_R a ball of radius R in $\mathbb{C}^m$ carrying the standard symplectic form, $D : B_R \to X$ a symplectomorphic embedding of B_R with $D(0) = x$. In other words, D is a Darboux neighbourhood of x. Now suppose C is a special Lagrangian cone in $\mathbb{C}^m$ with link Σ. Let $\varphi : \Sigma \times (0, R) \to B_R$ be the embedding $\varphi(y, r) = ry$, and suppose there is a smooth $\varphi' : \Sigma \times (0, R') \to B_{R'}$ such that $D \circ \varphi' : \Sigma \times (0, R') \to D(B_R)$ is a diffeomorphism onto $(D(B_R) \setminus \{x\}) \cap M$. Finally, let $2 < \mu < 3$ with $(2, \mu] \cap \mathcal{D}_\Sigma = \emptyset$. Then we say M has a *conical singular point* at x with rate μ and cone C if

$$|\varphi - \varphi'| = O(r^{\mu-1})$$

and

$$|\nabla(\varphi - \varphi')| = O(r^{\mu-2})$$

as $r \to 0$. Here the absolute value and the Levi-Civita connection are computed in the cone metric on $\Sigma \times (0, R)$. Essentially this is just saying that M is close to the cone C near x in a suitable metric. We note that μ is chosen to be less than 3, because a change in the Darboux coordinates will result in a change of $|\varphi - \varphi'|$ by $O(r^2)$. The constraint on μ involving $\mathcal{D}_\Sigma$ is designed to make the deformation theory work below.

Joyce proves a number of results about special Lagrangian submanifolds with conical singularities. We summarize the most important ones here.

The first is a generalization of McLean's deformation theory results. Given $M \subseteq X$ a compact special Lagrangian submanifold with only conical singular points $x_1, \ldots, x_n$ with cones $C_1, \ldots, C_n$, we can try to deform M without destroying the singular points. We say a special Lagrangian $M' \subseteq X$ is a *deformation* of M if it has singular points $x'_1, \ldots, x'_n$ with cones $C'_1, \ldots, C'_n$, there is a homeomorphism $\iota : M \to M'$ with $\iota(x_i) = x'_i$ and which is a diffeomorphism away from the singular points, and the inclusions $M \hookrightarrow X$ and $M \xrightarrow{\iota} M' \hookrightarrow X$ are isotopic as continuous maps, so there is a continuous family of maps $M \to X$ interpolating between these two inclusions.

Let $\mathcal{M}_{SL}(M)$ denote the set of deformations of M; it is possible to put a topology on this space. Then Joyce gives in [**263**] the following description of $\mathcal{M}_{SL}(M)$ near M. Let $M^o = M \setminus \{x_1, \ldots, x_n\}$. There is a map $H^1_c(M^o, \mathbb{R}) \to H^1(M^o, \mathbb{R})$ from de Rham cohomology with compactly supported forms to ordinary de Rham cohomology. Call the image I_M. Then there is a space O_M, the *obstruction space*, of dimension $\sum_{i=1}^n \text{s-ind}(C_i)$, and a C^∞ map $\Phi : U \to O_M$ for U some open neighbourhood of 0 in I_M, such that $\Phi^{-1}(0)$ is homeomorphic to an open neighbourhood of M in $\mathcal{M}_{SL}(M)$.

To understand roughly how the obstructions enter, which didn't exist in the non-singular case, we give a brief sketch of the proof.

One first proves a Lagrangian neighbourhood theorem, showing that there is some tubular neighbourhood of M^o in X which is symplectomorphic to an open neighbourhood of M^o embedded as the zero section in the cotangent bundle of M^o. Any small deformation of M can then be identified, away from the singularities of the deformation, with the graph of a section of T^*M^o. Since we are looking for a Lagrangian section, this is in fact the graph of a closed 1-form α, which then defines a cohomology class $[\alpha] \in H^1(M^o, \mathbb{R})$. Of course, α must satisfy some additional conditions because it comes from a small deformation of M, and this translates into a decay condition on α near the singularities. In turn, this forces α to in fact be exact in neighbourhoods of the singular points. This shows that $[\alpha]$ is in fact in the image of $H^1_c(M^o, \mathbb{R})$. So choose a vector space I_M of compactly supported 1-forms representing the image of $H^1_c(M^o, \mathbb{R})$ in $H^1(M^o, \mathbb{R})$, and choose $\beta \in I_M$ representing α. So $\alpha = \beta + df$ for some C^∞ function f on M^o. The fact that M^o is special Lagrangian can then be expressed as a second-order non-linear PDE. The precise form depends on the complex structure, but its linearization at $\beta = f = 0$ is $d^*(\beta + df) = 0$, or $\Delta f = -d^*\beta$. Thus one needs to understand the cokernel of the Laplacian on M^o. Now we use the theory of PDEs on manifolds with cylindrical ends developed by Lockhart and McOwen [**337**]. One looks at function spaces $L^p_{k,\alpha}(M^o)$, which denotes the Sobolev space of functions whose first k derivatives are in L^p and whose behaviour near a conical point x_i with cone C_i and link Σ_i is $r^\alpha g$, with g a function on Σ_i. Then by the formula (7.5), we see that Δ takes $L^p_{k,\mu}(M^o)$ to $L^p_{k-2,\mu-2}(M^o)$. Then given $\mu \in (2,3)$ and $\mathcal{D}_{\Sigma_i} \cap (2, \mu] = \emptyset$, [**337**] tells us that the cokernel of $\Delta : L^p_{k,\mu}(M^o) \to L^p_{k-2,\mu-2}(M^o)$ is of dimension $\sum_{i=1}^n \sum_{\alpha \in \mathcal{D}_{\Sigma_i} \cap [0,2]} m_{\Sigma_i}(\alpha)$. This is not quite $\sum_{i=1}^n \text{s-ind}(C_i)$, but by allowing the singular points to move, and allowing f to have asymptotic behaviour of the form constant$+r^\mu g$, one obtains the correct cokernel, which is essentially the obstruction space O_M.

In short, it is the eigenfunctions for Δ_{Σ_i} which ultimately cause the obstructions.

Note that if the singularities of M are *stable*, then in fact $\mathcal{M}_{SL}(M)$ is smooth. This is the case in three dimensions, when the cone singularities involved are just the cones over T^2 described in Example 7.10.

We would actually like to know more about the moduli space of special Lagrangian submanifolds, and in particular, given a *singular* special Lagrangian submanifold $M \subseteq X$, we would like to know whether or not M can be deformed to a *non-singular* special Lagrangian submanifold. If M has only conical singularities, we can again turn to results of Joyce in [**264, 265**]. We first need to study asymptotically conical special Lagrangian submanifolds of $\mathbb{C}^m$.

Let $C \subseteq \mathbb{C}^m$ be a special Lagrangian cone, with only an isolated singularity at 0, with link $\Sigma = C \cap S^{2m-1}$, $\iota : \Sigma \times (0, \infty) \to \mathbb{C}^m$ the embedding $\iota(\sigma, r) = r\sigma$ as usual. If $M \subseteq \mathbb{C}^m$ is a special Lagrangian submanifold of $\mathbb{C}^m$, we say M is *asymptotically conical with rate* $\lambda < 2$ *and cone* C if there exists a compact subset $K \subseteq M$ and a diffeomorphism $\varphi : \Sigma \times (T, \infty) \to M \setminus K$ for some $T > 0$ such that

$$\begin{aligned} |\varphi - \iota| &= O(r^{\lambda-1}) \\ |\nabla(\varphi - \iota)| &= O(r^{\lambda-2}) \end{aligned}$$

as $r \to \infty$. Here the gradient and absolute value are computed in the cone metric on $\Sigma \times (T, \infty)$.

EXAMPLE 7.17. There are a number of examples, but the only ones we will use here are asymptotic to the cone over a torus, and are the obvious generalization of Example 7.10, (2). Set $\mathbf{a} = (a_1, \ldots, a_m)$, with $a_i \geq 0$ for all i and exactly two of the a_i's zero and the rest positive. Set

$$M^{\mathbf{a}} = \left\{ (z_1, \ldots, z_m) \in \mathbb{C}^m \,\middle|\, \begin{array}{c} \mathrm{Re}(z_1 \cdots z_m) \geq 0, \mathrm{Im}(z_1 \cdots z_m) = 0, \\ |z_1|^2 - a_1 = \cdots = |z_m|^2 - a_m \end{array} \right\}.$$

As in Example 7.10, (2), $M^{\mathbf{a}}$ is in fact a manifold without boundary, and is asymptotic to the cone $M^{\mathbf{0}}$. The rate is in fact 0.

Note that $M^{\mathbf{a}}$ is homeomorphic to $T^{m-2} \times \mathbb{R}^2$. Indeed, without loss of generality, we can take $a_{m-1} = a_m = 0$. Then $|z_i|^2 > 0$ for $1 \leq i \leq m-2$, and so the T^{m-2}-action on $M^{\mathbf{a}}$ given by

$$(z_1, \ldots, z_m) \mapsto (e^{\sqrt{-1}\theta_1} z_1, \ldots, e^{\sqrt{-1}\theta_{m-2}} z_{m-2}, e^{-\sqrt{-1}(\theta_1 + \cdots + \theta_{m-2})} z_{m-1}, z_m)$$

is free. The quotient of $M^{\mathbf{a}}$ by this T^{m-2} action is $\mathbb{C} = \mathbb{R}^2$ with complex coordinate z_m.

Thus, for each of the $\binom{m}{2}$ choices of two zero a_i's, there is an $(m-2)$-dimensional family of asymptotically conical special Lagrangian submanifolds with cone $M^{\mathbf{0}}$. In particular, this gives $\binom{m}{2}$ different smoothings of $M^{\mathbf{0}}$!

Given M an asymptotically conical special Lagrangian submanifold with rate $\lambda < 2$ and cone C, we define $\mathcal{M}^{\lambda}_{SL}(M)$ to be the moduli space of asymptotically conical special Lagrangian submanifolds M' with rate λ and cone C, isotopic to M as an asymptotically conical submanifold of $\mathbb{C}^m$.

Note that if M is rate λ, it is also of rate λ' for any $\lambda' \in [\lambda, 2)$, and it is possible for $\mathcal{M}^{\lambda}_{SL}(M)$ to depend on λ.

Marshall [**349**] and Pacini [**385**] proved the following (see also [**262**], Theorem 6.8).

THEOREM 7.18. *Let M be an asymptotically conical special Lagrangian submanifold in $\mathbb{C}^m$ with cone C and rate $\lambda < 2$, $\Sigma = C \cap S^{2m-1}$ and $\mathcal{D}_\Sigma$, m_Σ be the data obtained from Δ_Σ. Then*

(1) If $\lambda \in (0,2) \setminus \mathcal{D}_\Sigma$, then $\mathcal{M}^\lambda_{SL}(M)$ is a manifold with

$$\dim \mathcal{M}^\lambda_{SL}(M) = b^1(M) - b^0(M) + \sum_{\alpha \in \mathcal{D}_\Sigma \cap [0,\lambda]} m_\Sigma(\alpha).$$

(2) If $\lambda \in (2-m, 0)$, then $\mathcal{M}^\lambda_{SL}(M)$ is a manifold whose dimension is $b^{m-1}(M)$.

Here of course $b^i(M) = \dim_{\mathbb{R}} H^i(M, \mathbb{R})$ is the ith Betti number. Note that if $0 < \lambda < \inf(D_\Sigma \cap (0,\infty))$ then the only contribution to the summation in the first part is from $m_\Sigma(0)$, which is the dimension of the space of harmonic functions on Σ. This of course is just the number of connected components of Σ, as Σ is compact, and this gives $b^0(\Sigma)$.

EXAMPLE 7.19. In Example 7.17, if we take $0 < \lambda < \inf(\mathcal{D}_\Sigma \cap (0,\infty))$, then as $b^0(M^{\mathbf{a}}) = b^0(\Sigma) = 1$, we have $\dim M^\lambda_{SL}(M^{\mathbf{a}}) = b^1(T^{m-2} \times \mathbb{R}^2) = m-2$. This is the same as the dimension of the space of possible choices for $\mathbf{a}$.

To discuss the smoothing result which will be of use to us, we need to define one cohomological invariant. Given $M \subseteq \mathbb{C}^m$ asymptotically conical special Lagrangian, the symplectic form ω of course restricts to zero on M. As such, it represents an element of the *relative* de Rham cohomology group $H^2(\mathbb{C}^m, M, \mathbb{R})$, which is calculated as closed 2-forms on $\mathbb{C}^m$ vanishing on M modulo differentials of closed 1-forms on $\mathbb{C}^m$ vanishing on M. But we have the relative cohomology exact sequence

$$0 = H^1(\mathbb{C}^m, \mathbb{R}) \to H^1(M, \mathbb{R}) \to H^2(\mathbb{C}^m, M, \mathbb{R}) \to H^2(\mathbb{C}^m, \mathbb{R}) = 0$$

showing $H^1(M,\mathbb{R}) \cong H^2(\mathbb{C}^m, M, \mathbb{R})$. On the other hand, as there is a compact set $K \subseteq M$ such that $M \setminus K \cong \Sigma \times (T, \infty)$, there is a a natural restriction map $H^1(M,\mathbb{R}) \to H^1(\Sigma, \mathbb{R})$. Thus ω defines a class in $H^1(M,\mathbb{R})$ and we denote the image of this class in $H^1(\Sigma,\mathbb{R})$ by $Y(M)$.

EXAMPLE 7.20. In Example 7.17, $\Sigma \cong T^{m-1}$, so $H^1(\Sigma, \mathbb{R}) = \mathbb{R}^{m-1}$. To calculate $Y(M)$, we can proceed as follows, using the dual sequence of relative homology

$$0 = H_2(\mathbb{C}^m, \mathbb{R}) \to H_2(\mathbb{C}^m, M, \mathbb{R}) \to H_1(M, \mathbb{R}) \to H_1(\mathbb{C}^m, \mathbb{R}) = 0.$$

Thus a cohomology class in $H^1(M,\mathbb{R})$ coming from $[\omega] \in H^2(\mathbb{C}^m, M, \mathbb{R})$ can be calculated by integrating ω over surfaces with boundary on M.

Without loss of generality, let's do this for the smoothing $M^{\mathbf{a}}$ in Example 7.17 given by $a_1, \ldots, a_{m-2} > 0$, $a_{m-1} = a_m = 0$. Then we can take Σ to be given, for some $r > 0$, by

$$\Sigma = \{(e^{\sqrt{-1}\theta_1}\sqrt{r+a_1}, \ \ldots, \ e^{\sqrt{-1}\theta_{m-1}}\sqrt{r+a_{m-1}}, \quad e^{-\sqrt{-1}(\theta_1+\ldots+\theta_{m-1})}\sqrt{r})\},$$

and generators of $H_1(\Sigma, \mathbb{R})$ are loops, for $1 \le i \le m-1$,

$$\gamma_i = \{(\sqrt{r+a_1}, \ldots, e^{\sqrt{-1}\theta}\sqrt{r+a_i}, \ \ldots, \ \sqrt{r+a_{m-1}}, \quad e^{-\sqrt{-1}\theta}\sqrt{r})\},$$

bounded by the image of

$$\varphi_i : D = \{z \in \mathbb{C} | |z| \le 1\} \to \mathbb{C}^m$$

given by

$$\varphi_i(z) = (\sqrt{r+a_1}, \ldots, z\sqrt{r+a_i}, \ldots, \sqrt{r+a_{m-1}}, \quad \bar{z}\sqrt{r}).$$

Then

$$\begin{aligned} \int_{\varphi_i(D)} \omega &= \int_D \varphi_i^*(\omega) \\ &= \frac{\sqrt{-1}}{2} \int_D a_i dz \wedge d\bar{z} \\ &= \pi a_i. \end{aligned}$$

So $Y(M^{\mathbf{a}}) = (\pi a_1, \ldots, \pi a_{m-2}, 0)$.

Note that from the explicit description of $M^{\mathbf{a}}$ in Example 7.17, $Y(M^{\mathbf{a}})$ is in the image of $H^1(M^{\mathbf{a}}, \mathbb{R}) \to H^1(\Sigma, \mathbb{R})$, this image being $(m-2)$-dimensional.

We can now state the version of Joyce's smoothing result most relevant for our discussion.

THEOREM 7.21. *Let X be a Calabi-Yau m-fold with $2 < m < 6$, $M \subseteq X$ a compact special Lagrangian submanifold with conical singularities at $x_1, \ldots, x_n$ and cones $C_1, \ldots, C_n$. Let $M_1, \ldots, M_n$ be asymptotically conical special Lagrangian submanifolds in $\mathbb{C}^m$ with cones $C_1, \ldots, C_n$ and rates $\lambda_1, \ldots, \lambda_n$, with $\lambda_i \le 0$ for each i. Suppose $M^o = M \setminus \{x_1, \ldots, x_n\}$ is connected, and that there exists a class $\rho \in H^1(M^o, \mathbb{R})$ whose image under the natural map $H^1(M^o, \mathbb{R}) \to \bigoplus_{i=1}^n H^1(\Sigma_i, \mathbb{R})$ is $(Y(M_1), \ldots, Y(M_n))$. Then there exists an $\epsilon > 0$ and a smooth family $\{\tilde{M}^t | t \in (0, \epsilon]\}$ of compact, non-singular special Lagrangian submanifolds such that $\tilde{M}^t \to M$ in the sense of currents.*

The basic idea is as follows. For any $t > 0$, tM_i is also asymptotically conical: we always have at least this one-parameter family of deformations of M_i. As $t \to 0$, $tM_i \to C_i$ in the sense of currents. One replaces a neighbourhood of each singularity $x_i \in M$, which looks like C_i, with tM_i for t small. This gluing procedure gives a submanifold which is a smoothing of M, but is not special Lagrangian. However, it is *close* to being special Lagrangian, and delicate analysis is necessary in order to show that for small enough t one can deform this to a genuine special Lagrangian submanifold. However, there is a cohomological obstruction, given by the $Y(M_i)$'s. We note that the natural map $H^1(M^o, \mathbb{R}) \to H^1(\Sigma_i, \mathbb{R})$ comes from the fact that if U is a small neighbourhood of $x_i \in M$, then $U \cap M^o = (0, r) \times \Sigma_i$, and $H^1(M^o, \mathbb{R}) \to H^1(\Sigma_i, \mathbb{R})$ is just the restriction map.

This completes our survey of Joyce's work on conical special Lagrangian submanifolds. In the next section, we will apply this to speculation about the true nature of special Lagrangian fibrations in three dimensions.

7.2.6. Special Lagrangian T^3-fibrations. As we have already remarked, Joyce's examples show that we need to dismiss the idea that special Lagrangian fibrations have a codimension two discriminant locus in general, and thus the pictures of topological duality we saw in Chapter 6 cannot be expected to hold. Here we shall follow Joyce's speculations as to the behaviour of special Lagrangian fibrations which we might hope to see in nature. Of course, we still don't know a single example of a special Lagrangian fibration on a compact Calabi-Yau threefold with full SU(3)-holonomy, so one should keep in mind that these fibrations may not even exist. However, Joyce's work yields many local examples which support the basic picture we will give here. These ideas are all due to Joyce.

The basic idea is that we expect the discriminant locus to be an "amoeba". The original notion of amoeba, due to Gelfand, Kapranov and Zelevinsky [**173**], is the image of a holomorphic curve in $(\mathbb{C}^*)^2$ under the map $(z_1, z_2) \to (\log|z_1|, \log|z_2|)$. These sets tend to look like amoebas.

Let's consider a topological example. Consider the surface $S \subseteq (\mathbb{C}^*)^2$ defined by the equation $z_1 + z_2 + 1 = 0$. Define the map $\log : (\mathbb{C}^*)^2 \to \mathbb{R}^2$ by $\log(z_1, z_2) = (\log|z_1|, \log|z_2|)$. Then $\log(S)$ can be seen as follows. First consider the image of S under the absolute value map (i.e., don't take logs). This image is depicted in Figure 7. The line segment $r_1 + r_2 = 1$

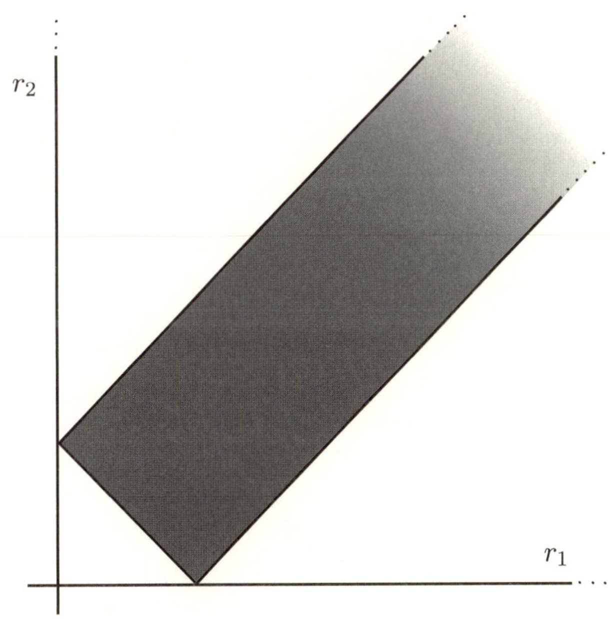

FIGURE 7

with $r_1, r_2 \geq 0$ is the image of $\{(-a, a-1) | 0 < a < 1\} \subseteq S$; the ray $r_2 = r_1 + 1$ with $r_1 \geq 0$ is the image of $\{(-a, a-1) | a < 0\} \subseteq S$; and the ray $r_1 = r_2 + 1$ is the image of $\{(-a, a-1) | a > 1\} \subseteq S$. The map $S \to |S|$ is one-to-one on the boundary of $|S|$ and two-to-one in the interior, with (z_1, z_2) and $(\bar{z}_1, \bar{z}_2)$ mapping to the same point in $|S|$. Taking the logarithm of this picture, we obtain the amoeba of S, $\log(S)$ as depicted in Figure 8. Now consider $S = S \times \{0\} \subseteq \overline{Y} = (\mathbb{C}^*)^2 \times \mathbb{R} = T^2 \times \mathbb{R}^3$. We

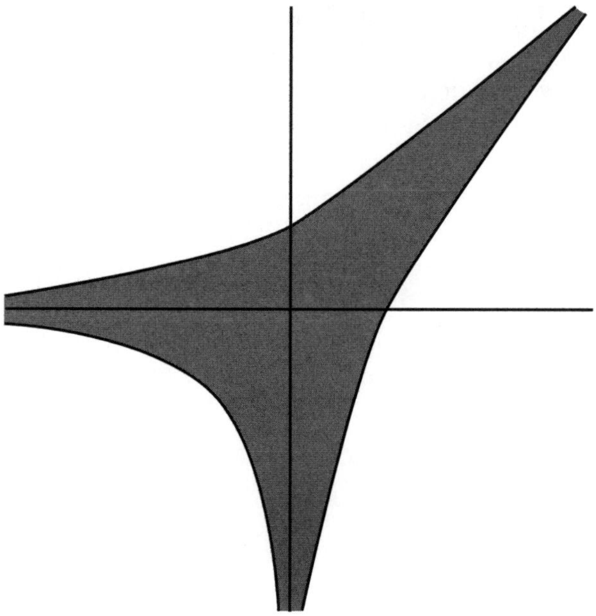

FIGURE 8

can then use Construction 6.89 to obtain a six-dimensional space $\overline{X}$, with a map $\pi : \overline{X} \to \overline{Y}$, an S^1-bundle over $\overline{Y} \setminus S$ degenerating over S, so that $\pi^{-1}(S) \xrightarrow{\cong} S$. We then have a T^3-fibration on $\overline{X}$, $f : \overline{X} \to \mathbb{R}^3$, by composing π with the map $(\log, \mathrm{id}) : (\mathbb{C}^*)^2 \times \mathbb{R} \to \mathbb{R}^3 = B$. Clearly the discriminant locus of f is $\log(S) \times \{0\}$. If b is in the interior of $\log(S) \times \{0\}$, then $f^{-1}(b)$ is obtained topologically by contracting two circles $\{p_1\} \times S^1$ and $\{p_2\} \times S^1$ on $T^3 = T^2 \times S^1$ to points. These are the familiar conical singularities we have seen in the special Lagrangian situation. However, we do not currently know how to deform this fibration to a special Lagrangian one.

If $b \in \partial(\log(S) \times \{0\})$, then $f^{-1}(b)$ has a slightly more complicated singularity, but only one. We shall ignore these fibers and instead examine how the "generic" singular fiber fits in with the results of the previous section. In particular, for b in the interior of $\log(S) \times \{0\}$, locally this discriminant locus splits B into two regions, and these regions represent two different possible smoothings of $f^{-1}(b)$. Let's see how this behaviour is possible using Theorem 7.21.

Assume now that $f : X \to B$ is a special Lagrangian fibration with topology given by the above example, with discriminant locus Δ being an amoeba. Let $b \in \mathrm{Int}(\Delta)$, and set $M = f^{-1}(b)$. Set $M^o = M \backslash \{x_1, x_2\}$, where x_1, x_2 are the two conical singularities of M. Suppose that the tangent cones to these two conical singularities, C_1 and C_2, are both cones of the form M^0 considered in Example 7.17. Then the links of these cones, Σ_1 and Σ_2, are T^2's, and one expects that topologically these can be described as follows. Note that $M^o \cong (T^2 \setminus \{y_1, y_2\}) \times S^1$ where y_1, y_2 are two points in T^2. We assume that the link Σ_i takes the form $\gamma_i \times S^1$, where γ_i is a simple loop around y_i. If these assumptions hold, then to see how M can be smoothed using Theorem 7.21, we consider the restriction maps in cohomology

$$H^1(M^o, \mathbb{R}) \to H^1(\Sigma_1, \mathbb{R}) \oplus H^1(\Sigma_2, \mathbb{R}).$$

It is not difficult to see that the image of this map is two-dimensional. Indeed, if we write a basis e_1^i, e_2^i of $H^1(\Sigma_i, \mathbb{R})$ where e_1^i is Poincaré dual to $[\gamma_i] \times \mathrm{pt}$ and e_2^i is Poincaré dual to $\mathrm{pt} \times S^1$, it is not difficult to see the image of the restriction map is spanned by $\{(e_1^1, e_1^2)\}$ and $\{(e_2^1, -e_2^2)\}$. Now our particular model of a topological fibration we've just given is not special Lagrangian, so in particular we don't know exactly how the tangent cones to M at x_1 and x_2 are sitting inside $\mathbb{C}^3$, and thus we can't compare them directly with an asymptotically conical smoothing. So to make a plausibility argument, choose new bases f_1^i, f_2^i of $H^1(\Sigma_i, \mathbb{R})$ so that if $M_i^{(a,0,0)}$, $M_i^{(0,a,0)}$ and $M_i^{(0,0,a)}$ are the three possible smoothings of the two singular tangent cones at the singular points x_1, x_2 of M, given by Example 7.17, then $Y(M_i^{(a,0,0)}) = \pi a f_1^i$, $Y(M_i^{(0,a,0)}) = \pi a f_2^i$, and $Y(M_i^{(0,0,a)}) = -\pi a(f_1^i + f_2^i)$. (It can be checked that this is the correct behaviour as in Example 7.20.)

Suppose that in this new basis, the image of the restriction map is spanned by the pairs $(f_1^1, r f_2^2)$ and $(r f_2^1, f_1^2)$ for $r > 0$, $r \neq 1$. Then, by Theorem 7.21, there are two possible ways of smoothing M, either by gluing in $M_1^{(a,0,0)}$ and $M_2^{(0,ra,0)}$ at the singular points x_1 and x_2 respectively, or by gluing in $M_1^{(0,ra,0)}$ and $M_2^{(a,0,0)}$ at x_1 and x_2 respectively. This could correspond to deforming M to a fiber over a point on one side of the discriminant locus of f or the other side. See [**262**], §10.3 for more details and other possibilities. (For example, the condition that $r \neq 1$ above is included to ensure there aren't three possible smoothings. For other possibilities of the image of the restriction map in terms of the bases $\{f_i^j\}$, there may be none or one smoothing.)

This at least gives a plausibility argument for the existence of a special Lagrangian fibration of the topological type given by f. To date, no such fibrations have been constructed, however.

The initial example of Joyce (Example 7.11) giving a special Lagrangian fibration with codimension one discriminant and singular fibers with cone over T^2 singularities, along with Joyce's much deeper analysis of $U(1)$-invariant special Lagrangians and conical singularities outlined above, forced significant rethinking of the Strominger-Yau-Zaslow picture. Instead of the nice codimension two discriminant locus hoped for, one is forced to confront a codimension one discriminant locus in special Lagrangian fibrations. This leads inevitably to the conclusion that a "strong form" of the Strominger-Yau-Zaslow conjecture cannot hold. In particular, one is forced to conclude that if $f : X \to B$ and $\check{f} : \check{X} \to B$ are dual special Lagrangian fibrations, then their discriminant loci cannot coincide. Thus one cannot hope for a fiberwise definition of the dualizing process, and one needs to refine the concept of dualizing fibrations. This will be taken up again in §7.3.6. Here, let us explain why the discriminant locus must change under dualizing.

The key lies in the behaviour of the positive and negative vertices of §6.4.2. As we saw in the explicit construction of positive and negative vertices, in the positive case the critical locus of the local model of the fibration is a union of three holomorphic curves, while in the negative case the critical locus is a pair of pants. In a "generic" special Lagrangian fibration, we expect the critical locus to remain roughly the same, but its image in the base B will be fattened out. In the negative case, this image will be an amoeba as in Figure 8. In the case of the positive vertex, the critical locus, at least locally, consists of a union of three holomorphic curves, so that we expect the discriminant locus to be the union of three different amoebas. Figure 9 shows the new discriminant locus for these two cases, as predicted by Joyce. Now, under dualizing, positive and negative vertices are interchanged. Thus

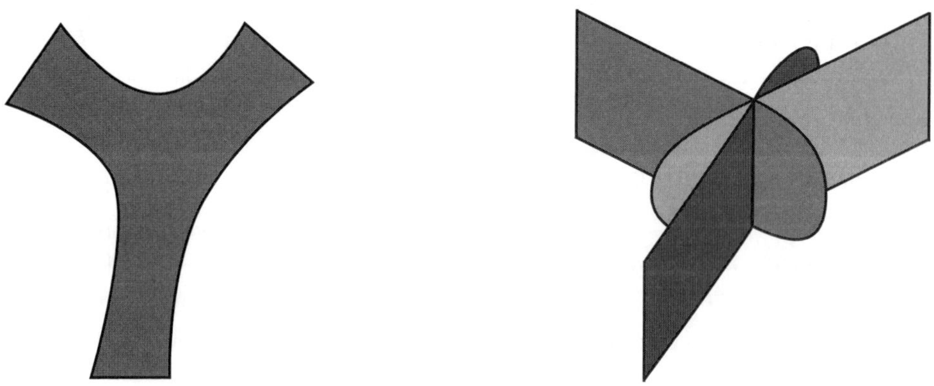

FIGURE 9. The fattenings of the negative and positive vertices. The left-hand picture is the negative vertex, with the fattening of the two-dimensional discriminant locus being a planar amoeba. The right-hand picture gives amoeba-like thickenings of each of the three legs of the discriminant locus, each in its own plane, the planes intersecting transversally along a line.

the discriminant locus must change.

This is all quite speculative, of course, and underlying this is the assumption that the discriminant loci are just fattenings of the graphs considered in Chapter 6. However, it is clear that a new notion of dualizing is necessary to cover this eventuality. In §7.3.6, we shall explain a revised version of the SYZ conjecture which will take care of these problems.

7.3. Large complex structure limits

7.3.1. Approximations to Ricci-flat metrics on compact manifolds. We commented in §7.1 that no explicit (non-flat) examples of Ricci-flat metrics are known on compact manifolds. In certain limiting cases, one can however write down very accurate approximations to the Ricci-flat metrics. To illustrate this, we first look at the case of Kummer surfaces, as studied by R. Kobayashi [**307**].

EXAMPLE 7.22. Let X be the Kummer surface of a complex torus Y, that is, the minimal resolution of $\overline{X} = Y/\{\pm 1\}$. Here $\{\pm 1\} \cong \mathbb{Z}_2$ acts on Y by -1 acting as negation. Then $\overline{X}$ has 16 nodes and X has (-2)-curves $E_1, \ldots, E_{16}$. We consider Kähler classes on X of the form

$$\kappa - \sum_{i=1}^{16} a_i e_i,$$

where e_i represents the class of E_i, κ is the pullback of an orbifold Kähler class on $\overline{X}$ and the a_i are small positive numbers. Kobayashi constructs an approximately Ricci-flat metric as follows.

Locally at least analytically each node is of the form $\mathbb{C}^2/\{\pm 1\}$, the quadric cone in $\mathbb{C}^3$, with resolution $T^*\mathbb{CP}^1$, diffeomorphic to the real tangent bundle of S^2. In Example 7.1 we described explicitly the Eguchi-Hanson metric on this resolution, an SO(3)-invariant Ricci-flat ALE Kähler metric, with Kähler class of the form $-2a[E]$. In this metric, the 2-sphere E has curvature a^{-1} and volume $4\pi a$.

On any compact set in the complement of E, the Eguchi-Hanson metric converges to the orbifold flat metric as $a \to 0$. On our original K3 orbifold $\overline{X}$, for each node we have a neighbourhood analytically of the form $B/\{\pm 1\}$, for some ball B around the origin in $\mathbb{C}^2$. Taking a fixed annulus in B, we can glue together the Kähler potentials corresponding to the flat metric outside the annulus and the Eguchi-Hanson metric (parameter a_i) inside the annulus (on the resolution), thus obtaining a metric ω_a in the appropriate Kähler class on the Kummer surface X, which is almost Ricci-flat for small $a = (a_1, \ldots, a_{16})$, and whose curvature only becomes large near the exceptional curves.

We now run the Yau program, obtaining a function u_a and a Ricci-flat metric

$$\tilde{\omega}_a = \omega_a + i\partial\bar{\partial}u_a,$$

i.e., if Ω denotes the holomorphic 2-form, then $(i/2)^2\Omega \wedge \bar{\Omega}$ is a constant multiple of the volume form $\tilde{\omega}_a^2/2$.

Moreover, an analysis of Yau's proof in [**307**] yields explicit bounds for the various norms of u_a (cf. §7.3.5 below):

(I) C^0-bound $||u_a||_{C^0} \leq C|a|^2$.

(II) C^2-bound, most conveniently written (assuming that as $a \to 0$, all the a_i's stay within a constant multiple of each other) as

$$(1 - C|a|^{1/2})\omega_a \leq \tilde{\omega}_a \leq (1 + C|a|^{1/2})\omega_a.$$

With appropriate scaling, there are also C^i-bounds for $i = 2, 3, 4$ in a neighbourhood of each (-2)-curve.

(III) $C^{k,\alpha}$-bounds

$$||u_a||_{C^{k,\alpha}} \leq C|a|^2$$

on any fixed relatively compact subdomain in the complement of the (-2)-curves. (Recall the definition of $C^{k,\alpha}$ from §7.1.1.) Here one applies a general result (Theorem 17.14 in [**179**]) to get from the C^2-bound to a $C^{2,\alpha}$-bound, and then standard arguments using bootstrapping and Schauder estimates for the stronger statements.

REMARK 7.23. For arbitrary rational double point singularities, one can glue in Kronheimer's ALE gravitational instantons [**316**] instead of the Eguchi-Hanson metric, and the same argument works [**307**].

For degenerations of an arbitrary K3 surface to one with rational double point singularities, Kobayashi repeats the argument, taking as background metric the orbifold Ricci-flat Kähler metric (which exists by Yau's theorem). Since this is only Euclidean locally up to third order, one does the gluing not on fixed annuli but on annuli at distance approximately say $a^{3/8}$.

The bounds obtained in the more general case are slightly worse — for instance in (I) and (III) one replaces $|a|^2$ by $|a|^{13/8}$. Note that here, unlike the Kummer case, we only have an explicit approximation to the Calabi-Yau metric near the (-2)-curves.

The above description of the approximately Ricci-flat metric, as one degenerates the Kähler structure on the K3 surface to one with rational double points, enables us to deduce that the metric limit (in the sense of Gromov-Hausdorff, see §7.3.6 below) of the Ricci-flat metrics is just the orbifold Ricci-flat metric. Limits of Ricci-flat metrics with bounded volume and diameter on a K3 surface were studied in general by Anderson [**13**], who showed that they converged to an orbifold metric. In the case when the diameter is unbounded, Anderson showed that collapsing must occur, which means that, if we rescale the metrics to have bounded diameters, the Gromov-Hausdorff limit will have strictly smaller dimension. For Kähler degenerations of K3 surfaces, there are only two possibilities, namely the orbifold degeneration described above, and the elliptic fibration case studied below in §7.3.3 to §7.3.6. This latter case will turn out to be related (via hyperkähler rotation) to large complex structure limits of K3 surfaces.

7.3.2. Large complex structure limits for K3 surfaces. The definition of large complex structure limit points for Calabi-Yau manifolds includes the condition that they are boundary points of complex moduli with maximally unipotent monodromy. The definition for Calabi-Yau threefolds in [**365**] includes an integrality condition. In this section however, we shall only study the case of K3 surfaces; writing things in terms of periods, we then obtain concrete descriptions of such limit points and the explicit formulae for mirror symmetry. The description we give here arose originally from the work of Pinkham, Nikulin and Dolgachev, and details may found in [**126**].

Let L denote the K3 lattice, and consider the moduli space of K3 surfaces X together with an isomorphism of $H^2(X,\mathbb{Z})$ with L. A large complex structure limit point for such marked K3 surfaces is a boundary point of maximally unipotent monodromy, which in turn gives rise to a primitive isotropic ($E^2 = 0$) vector $E \in L$ generating the piece W_0 in the monodromy weight filtration on $H^2(X,\mathbb{Q})$ — compare also Theorem 6.104. Starting from the class E, we produce an involution acting on the moduli space of triples $(X, \mathbf{B}+\sqrt{-1}\omega, \Omega)$ where X is a marked K3 surface, Ω is the class of a holomorphic 2-form on X, $\omega \in E^\perp \otimes \mathbb{R}$ a Kähler class on X, and the B-field $\mathbf{B}$ lies in $E^\perp/E \otimes \mathbb{R}$. In addition, Ω is normalized so that $\operatorname{Im}\Omega \in E^\perp \otimes \mathbb{R}$ and $\omega^2 = (\operatorname{Re}\Omega)^2 = (\operatorname{Im}\Omega)^2$.

Consider the period domain (of complex dimension 20)

$$D = \{[\Omega] \in \mathbb{P}(L_\mathbb{C}) \,|\, [\Omega]^2 = 0,\ [\Omega]\cdot[\bar{\Omega}] > 0\} \cong O(3,19)/O(2)\times O(1,19).$$

The projection map

$$q : \mathbb{P}(L_\mathbb{C}) \setminus [E] \to \mathbb{P}(L_\mathbb{C}/\langle E\rangle),$$

restricted to D, identifies the subset of D given by

$$\{[\Omega] \in D \,|\, \Omega \cdot E \neq 0\}$$

with an open subset of the affine space

$$\{v \in L_\mathbb{C}/\langle E\rangle \,|\, v\cdot E = 1\}.$$

Normalizing Ω so that $\operatorname{Im}\Omega \cdot E = 0$, and choosing a class $\sigma_0 \in L$ with $\sigma_0^2 = -2$ and $\sigma_0 \cdot E = 1$ (so that $\langle E, \sigma_0\rangle$ is a hyperbolic lattice), we obtain an identification of $q([\Omega])$ as a point of $E^\perp/\langle E\rangle$ by subtracting σ_0 from $[\Omega]$. Modulo multiples of E, this determines a point $\check{\mathbf{B}} + \sqrt{-1}\check{\omega}$ of $L_\mathbb{C}$. Explicitly, we have

$$\check{\mathbf{B}} \equiv (E\cdot\operatorname{Re}\Omega)^{-1}\operatorname{Re}\Omega - \sigma_0 \mod E$$

and

$$\check{\omega} \equiv (E\cdot\operatorname{Re}\Omega)^{-1}\operatorname{Im}\Omega \mod E. \tag{7.6}$$

The required involution then interchanges $(X, \mathbf{B} + \sqrt{-1}\omega, \Omega)$ with $(\check{X}, \check{\mathbf{B}} + \sqrt{-1}\check{\omega}, \check{\Omega})$, where

$$\check{\Omega} \equiv (E \cdot \operatorname{Re}\Omega)^{-1}(\sigma_0 + \mathbf{B} + \sqrt{-1}\omega) \qquad \operatorname{mod} E \tag{7.7}$$

and $\check{X}$ denotes a marked K3 surface with period $\check{\Omega}$. Note here that the actual classes of $\check{\Omega}$ and $\check{\omega}$ are determined completely by the relations $(\operatorname{Re}\check{\Omega})^2 = (\operatorname{Im}\check{\Omega})^2 = \check{\omega}^2$ and $\check{\omega} \cdot (\operatorname{Re}\check{\Omega}) = \check{\omega} \cdot (\operatorname{Im}\check{\Omega}) = (\operatorname{Re}\check{\Omega}) \cdot (\operatorname{Im}\check{\Omega}) = 0$.

We will argue that this formula coincides with the SYZ picture of mirror symmetry. The key point is that it is easy to construct special Lagrangian torus fibrations on K3 surfaces, using the so-called *hyperkähler trick*. On a K3 surface with a Ricci-flat, hence hyperkähler, metric, there is an S^2's worth of complex structures compatible with the metric. On a hyperkähler manifold, there are three complex structures I, J and K compatible with the metric satisfying $I^2 = J^2 = K^2 = -1$ and $IJ = K$, $JK = I$ and $KI = J$. The general complex structure compatible with the metric is $aI + bJ + cK$ with $a^2 + b^2 + c^2 = 1$. In terms of forms, if in complex structure I the Kähler form is ω and the holomorphic two-form is Ω, normalized so that $\omega^2 = (\operatorname{Re}\Omega)^2 = (\operatorname{Im}\Omega)^2$, then in the K complex structure, the holomorphic two-form is $\Omega_K = \operatorname{Im}\Omega + \sqrt{-1}\omega$ and the Kähler form is $\omega_K = \operatorname{Re}\Omega$. Since we can always first multiply Ω by an arbitrary phase factor $e^{\sqrt{-1}\theta}$, we can normalize Ω so that $\operatorname{Im}\Omega \cdot E = 0$. Then the fact that $\Omega_K \cdot E = 0$ implies that E is the class of an effective divisor with $E^2 = 0$, and for generic choice of data this will be the fiber of an elliptic fibration $f : X_K \to \mathbb{CP}^1$. Since Ω_K restricts to zero on a fiber of this map, $f : X_I \to \mathbb{CP}^1$ is a fibration with general fiber a special Lagrangian torus. We then take σ_0 to be the C^∞-section of f corresponding to the zero section of the Jacobian fibration of $X_K \to \mathbb{CP}^1$; explicitly, this says [**210**], page 482, that the complex structure determined by $\Omega' = \Omega_K - f^*\sigma_0^*\Omega_K$ is just that of the Jacobian fibration. (Recall that the Jacobian fibration of an elliptic fibration is one whose fibers are isomorphic to the original fibration, but which has a section.) Denote the class of this section also by σ_0.

The interpretation of mirror symmetry in terms of dualizing the special Lagrangian fibration $f : X \to \mathbb{CP}^1$ coincides with the explicit formulae derived above, as determined by the classes E and σ_0, under the assumption that all fibers of f are irreducible. This latter claim is checked in §4 of [**209**] under the assumption that σ_0 is a special Lagrangian section of f (in other words $X_K \to \mathbb{CP}^1$ is a Jacobian fibration, i.e., has a holomorphic section), and follows in the general case from the discussion in §1 of [**210**]. If $f : X \to B$ denotes the special Lagrangian fibration constructed above, where $B = S^2$, the Leray spectral sequence

$$H^i(S^2, R^j f_*\mathbb{Q}) \Rightarrow H^n(X, \mathbb{Q})$$

degenerates at the E_2 term, and the non-zero terms are shown in the following diagram:

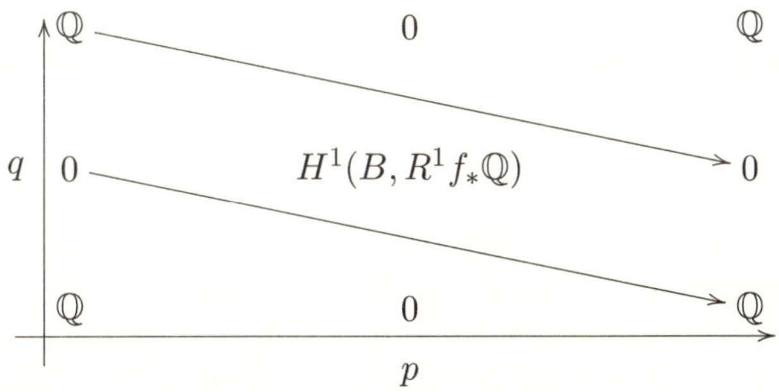

(Proposition 4.1 of [**209**]). This in turn yields a filtration

$$0 \subseteq F_0 \subseteq F_1 \subseteq F_2 = H^2(X, \mathbb{Q}),$$

where $F_1 = E^{\perp}$, such that

$$\begin{aligned} F_0 &\cong H^2(S^2, f_*\mathbb{Q}) \cong \mathbb{Q}E \\ F_1/F_0 &\cong H^1(S^2, R^1 f_*\mathbb{Q}) \\ F_2/F_1 &\cong H^0(S^2, R^2 f_*\mathbb{Q}) \cong \mathbb{Q}\sigma_0. \end{aligned}$$

The filtration in fact coincides with the weight filtration determined by the large complex structure limit point. Now the description of mirror symmetry given in §6.2.3 suggests that at the cohomological level, the mirror involution should interchange Ω and $\exp(\mathbf{B}+\sqrt{-1}\omega)$. We can see what this interchange does on the level of the filtration just described as follows. First identify the dual of $X \to S^2$ with the original fibration via Poincaré duality. If we normalize Ω further so that $\operatorname{Re}\Omega \cdot E = 1$ (that is, the form Ω restricts to the volume form on the fiber), then we have $\operatorname{Im}\Omega \in F_1 \otimes_{\mathbb{Q}} \mathbb{R}$ and $\operatorname{Re}\Omega \in \sigma_0 + F_1 \otimes_{\mathbb{Q}} \mathbb{R}$. Thus $\Omega - \sigma_0 \in F_1 \otimes_{\mathbb{Q}} \mathbb{C}$, and the image of this class in $(F_1/F_0) \otimes_{\mathbb{Q}} \mathbb{C}$ should coincide with $\check{\mathbf{B}} + \sqrt{-1}\check{\omega}$. Conversely, we may consider $\mathbf{B} + i\omega$ as an element of $F_1/F_0 \otimes_{\mathbb{Q}} \mathbb{C}$ and lift it to the unique class α in $F_1 \otimes_{\mathbb{Q}} \mathbb{C}$ with the property that $\alpha + \sigma_0 = \check{\Omega}$ satisfies $\check{\Omega}^2 = 0$. This is precisely the description of mirror symmetry we gave in (7.6) and (7.7). For a more detailed explanation of this dualizing process at the level of forms, see [**201**].

Given the above explicit formula for the mirror map, we shall now simply define the large complex structure limit of $\check{X}$ to be the mirror to the large Kähler structure limit of X. In the latter limit, we keep the complex structure on X fixed but allow the Kähler form to go to infinity. More precisely, if $\{\mathbf{B}_l + \sqrt{-1}t_l\omega\}$ is a sequence of complexified Kähler forms on X with $t_l > 0$, $t_l \to \infty$, then we say the sequence $\{\mathbf{B}_l + \sqrt{-1}t_l\omega\}$ is approaching the large Kähler limit in the complexified Kähler moduli space of X.

DEFINITION 7.24. For each l, let $\check{X}_l$ be the K3 surface given by the data $(\check{X}_l, \check{\mathbf{B}}_l + i\check{\omega}_l, \check{\Omega}_l)$ mirror to $(X, \mathbf{B}_l + \sqrt{-1}t_l\omega, t_l\Omega)$. The sequence of surfaces $\{\check{X}_l\}$ is said to *approach a large complex structure limit point.*

We are in fact cheating slightly here, by only approaching the large Kähler limit along a ray. The more general approach might be to allow a more general sequence of Kähler forms. However, this is more difficult to deal with because the Jacobian fibration which arises below will be varying, and this obscures our main objectives. With the above definition, we have formulae for $\check{\Omega}_l$ and $\check{\omega}_l$ in terms of t_l, Ω, ω and $\mathbf{B}_l$.

The Kähler class $\check{\omega}_l$ is represented by a Ricci-flat metric $\check{g}_l$, and we would like to understand the behaviour of this metric as $t_l \to \infty$. It is convenient to perform a hyperkähler rotation, i.e., $\check{g}_l$ is also a Kähler metric on the K3 surface $\check{X}_{l,K}$ with

$$\begin{aligned}\check{\Omega}_{l,K} &= \operatorname{Im}\check{\Omega}_l + \sqrt{-1}\check{\omega}_l\\ \check{\omega}_{l,K} &= \operatorname{Re}\check{\Omega}_l.\end{aligned}$$

This equality holds on the level of forms. Explicitly, in cohomology, we have (page 481 of [**210**]) that

$$\begin{aligned}\check{\Omega}_{l,K} &= (E\cdot\operatorname{Re}\Omega)^{-1}(\omega+\sqrt{-1}\operatorname{Im}\Omega - ((\omega+\sqrt{-1}\operatorname{Im}\Omega)\cdot(\sigma_0+\mathbf{B}_l))E)\\ \check{\omega}_{l,K} &= (t_lE\cdot\operatorname{Re}\Omega)^{-1}(\sigma_0+\mathbf{B}_l) \mod E.\end{aligned}$$

Assuming a general choice of data, E will represent the class of a fiber of an elliptic fibration $f_l : \check{X}_{l,K} \to \mathbb{CP}^1$. This elliptic fibration coincides with a special Lagrangian T^2-fibration on $\check{X}_l$. Note that the area of the fiber of f_l with respect to the metric $\check{g}_l$ is $(t_lE\cdot\operatorname{Re}\Omega)^{-1}$, which goes to zero as $t_l \to \infty$.

Now $\check{\Omega}_{l,K}$ depends on l, but these classes only differ by the pullback of a class from $\mathbb{CP}^1$. This in fact tells us that the elliptic K3 surfaces $\check{X}_{l,K}$ are closely related; in particular, all these elliptic surfaces have the same Jacobian $\check{J}_K$, which is the unique elliptic K3 surface with a holomorphic section with complex structure induced by $\check{\Omega}_{l,K} + f_l^*\alpha$ for some 2-form α on $\mathbb{CP}^1$.

This then leads us to the following question:

QUESTION 7.25. *Let $j : J \to \mathbb{CP}^1$ be an elliptic K3 surface with a section, and let $f_l : X_l \to \mathbb{CP}^1$ be a sequence of elliptic K3 surfaces with Jacobian $j : J \to \mathbb{CP}^1$. Let ω_l be a Ricci-flat Kähler metric on X_l with* $\mathrm{Vol}(X_l)$ *independent of l. Let $\epsilon_l = Area_{\omega_l}(f_l^{-1}(y))$ for any point $y \in \mathbb{CP}^1$, and suppose $\epsilon_l \to 0$ as $l \to \infty$. Describe the behaviour of the metric ω_l as $l \to \infty$.*

In [**210**], this question is solved in the case that the map j has 24 Kodaira type I_1 fibers (true for the generic K3 elliptic fibration). From now on, we make this assumption on the singular fibers. We describe below how to construct very accurate approximations to the metric. For simplicity, we shall deal with the case when all the elliptic K3 surfaces $f_l : X_l \to \mathbb{CP}^1$ coincide with the Jacobian fibration — in the above formulation, this corresponds to taking special values of the B-fields $\mathbf{B}_l$. The general case follows similarly, with only minor modifications to the arguments [**210**].

7.3.3. The metric away from the singular fibers. From now on, let $f : X \to \mathbb{CP}^1$ denote a fixed elliptic K3 surface with holomorphic section σ_0 and 24 singular fibers of type I_1. Let Ω denote a holomorphic 2-form on X and ω a Kähler form with $[\omega]^2 = [\mathrm{Re}\,\Omega]^2 = [\mathrm{Im}\,\Omega]^2$. We set $\epsilon = \int_{X_b} \omega$ to be the area of the fibers. If g denotes the corresponding Ricci-flat metric on X, we are interested in the behaviour of g when ϵ becomes small. It turns out that this metric behaves very differently near the singular fibers as opposed to elsewhere. We describe in this section what will turn out to be a good approximation to the metric g away from the singular fibers, the so-called semi-flat metric. To describe these metrics explicitly, we shall need a more canonical description of our fibration, together with a canonical system of coordinates.

We denote by $\mathcal{T}^*_{\mathbb{CP}^1}$ the holomorphic cotangent bundle on the base; given an open set U of $\mathbb{CP}^1$ and a smooth section $\alpha \in \Gamma(U, \mathcal{T}^*_{\mathbb{CP}^1})$, there is a natural action $F_\alpha : f^{-1}(U) \to f^{-1}(U)$ defined as follows. First assume that α is compactly supported on U, and so the 1-form $f^*(\alpha)$ is compactly supported on $f^{-1}(U)$. Using the holomorphic symplectic form Ω on X, we obtain a unique vector field V, tangent to the fibers of f, such that $\iota(V)\Omega = f^*(\alpha)$, and hence, by integration, a flow $\phi_t : f^{-1}(U) \to f^{-1}(U)$ for all t. The required map F_α is defined to be ϕ_1. Assuming furthermore that α is closed, an easy check verifies that F_α preserves the form Ω and acts fiberwise. One now observes that $V|_{f^{-1}(b)}$ depends only on the value of α at $b \in U$, and so for any smooth section $\alpha \in \Gamma(U, \mathcal{T}^*_{\mathbb{CP}^1})$ we still have a fiberwise action F_α defined. This recipe then defines a map $\pi : \mathcal{T}^*_{\mathbb{CP}^1} \to X$, taking $(P, \alpha) \in \mathcal{T}^*_{\mathbb{CP}^1}$ to $F_\alpha(\sigma_0(P))$, with the zero section corresponding to the given holomorphic section σ_0. This map is analysed in §2 and §7 of [**201**]; in our case, the image of π is just the complement X^0 in X of the critical locus of f, and the kernel of π is a (degenerating) family of lattices in the fibers (varying holomorphically). This kernel may be identified as $R^1 f_* \mathbb{Z} \subset \mathcal{T}^*_{\mathbb{CP}^1}$; here we are making the canonical identification $R^1 f_* \mathcal{O}_X \cong \omega_{\mathbb{CP}^1}$, valid for elliptic K3 surfaces, where $\omega_{\mathbb{CP}^1}$ denotes the canonical sheaf of the base, namely the sheaf of holomorphic sections of $\mathcal{T}^*_{\mathbb{CP}^1}$. Thus X^0 has been identified as a quotient of the holomorphic cotangent bundle by $\Lambda \cong R^1 f_* \mathbb{Z}$. Locally over $\mathbb{CP}^1$, we have holomorphic canonical coordinates on $\mathcal{T}^*_{\mathbb{CP}^1}$ denoted by (x, y), where y is a local holomorphic coordinate on the base and (x, y) corresponds to $x dy$ in $\mathcal{T}^*_{\mathbb{CP}^1}$, the zero section being given locally by $x = 0$. Moreover, there is a canonical holomorphic symplectic form on the holomorphic cotangent bundle, defined locally by $d(x dy) = dx \wedge dy$. We observe that $\pi^*(\Omega)$ is a nowhere vanishing holomorphic 2-form on $\mathcal{T}^*_{\mathbb{CP}^1}$, from which it follows easily that it agrees with the canonical holomorphic symplectic form up to a constant. When $f : X \to \mathbb{CP}^1$ is not a Jacobian fibration, there is still a similar theory, but the above description needs slight modifications involving pullbacks of 2-forms from the base ([**201**], §7).

Returning now to our elliptic K3 surface with given holomorphic section σ_0, we can therefore describe the fibration away from the singular fibers as $f_0 : X_0 \to B_0$, where B_0 is an open subset of $\mathbb{C}$ and $X_0 = \mathcal{T}^*_{B_0}/\Lambda$, for Λ a holomorphically varying family of lattices in $\mathcal{T}^*_{B_0}$. Moreover, we may take local holomorphic coordinates (x, y) on X_0 corresponding to the holomorphic canonical coordinates on $\mathcal{T}^*_{B_0}$, where $y = y_1 + \sqrt{-1}y_2$ and $x = x_1 + \sqrt{-1}x_2$. The holomorphic 2-form on X_0 will be induced from $\Omega = dx \wedge dy$ on $\mathcal{T}^*_{B_0}$. Given our original fibration $f : X \to \mathbb{CP}^1$, and B_0 an open subset of $\mathbb{CP}^1$ disjoint from the discriminant locus Δ (24 points), it is these coordinates that we use over B_0. We now express the Kähler form ω locally in terms of these coordinates:

$$\begin{aligned}\omega &= \frac{\sqrt{-1}}{2}W(dx \wedge d\bar{x} + \bar{b}dx \wedge d\bar{y} + bdy \wedge d\bar{x} + (W^{-2} + |b|^2)dy \wedge d\bar{y}) \\ &= \frac{\sqrt{-1}}{2}(W(dx + bdy) \wedge \overline{(dx + bdy)} + W^{-1}dy \wedge d\bar{y}).\end{aligned}$$

Here the functions W and b are defined by the above expression, and the coefficient of $dy \wedge d\bar{y}$ is chosen to ensure the normalization $\omega^2 = (\operatorname{Im}\Omega)^2$. The function W is positive real-valued and the function b is complex-valued. The characterization of Ricci-flat metrics in terms of forms (Proposition 6.25) then ensures that the associated metric is Ricci-flat so long as it is Kähler. The Kähler condition is now $d\omega = 0$. This equation can be written as

$$\begin{aligned}\partial_y W &= \partial_x(Wb) \\ \partial_y(W\bar{b}) &= \partial_x(W(W^{-2} + |b|^2)).\end{aligned}$$

Note that expanding the second equation gives

$$W\partial_y\bar{b} + \bar{b}\partial_y W = -W^{-2}\partial_x W + (\partial_x W)|b|^2 + W(b\partial_x\bar{b} + \bar{b}\partial_x b).$$

Using the first equation to replace $\partial_y W$ and simplifying gives the above two equations being equivalent to

$$(K1) \qquad (\partial_y - b\partial_x)\bar{b} = -W^{-3}\partial_x W$$

$$(K2) \qquad (\partial_y - b\partial_x)W = W\partial_x b.$$

In this set-up, the obvious metrics to investigate are the semi-flat ones, as described in §6.2.1. In terms of the above description, the metric is semi-flat if the function W is constant on the fibers. With the above coordinates, let $\tau_1(y), \tau_2(y)$ be two holomorphic functions on B_0 such that $\tau_1(y)dy, \tau_2(y)dy$

generate a lattice $\Lambda(y) \subseteq \mathcal{T}^*_{B_0,y}$ for each $y \in B_0$, giving us the holomorphically varying family of lattices $\Lambda \subseteq \mathcal{T}^*_{B_0} = B_0 \times \mathbb{C}$. Typically, we may allow τ_1 and τ_2 to be multi-valued. Assuming without loss of generality that $\mathrm{Im}(\bar{\tau}_1\tau_2) > 0$, then a Ricci-flat metric on $X_0 = (B_0 \times \mathbb{C})/\Lambda$ is given by the data

$$\begin{aligned} W &= \frac{\epsilon}{\mathrm{Im}(\bar{\tau}_1\tau_2)} \\ b &= -\frac{W}{\epsilon}[\mathrm{Im}(\tau_2\bar{x})\partial_y\tau_1 + \mathrm{Im}(\bar{\tau}_1 x)\partial_y\tau_2]. \end{aligned}$$

It is easy to check that these satisfy the equations (K1) and (K2). This metric, a priori defined on $\mathcal{T}^*_B$, descends to a metric on X, and the area of a fiber of $f : X \to B$ is ϵ. We call this metric on X *the standard semi-flat metric*, with Kähler form ω_{SF}. This metric was first considered in [**196**].

It may be checked explicitly that this metric is independent of the particular choice of generators for Λ, so that multi-valuedness of τ_1 and τ_2 does not cause a problem. Furthermore, the metric is independent of the choice of the coordinate y (keeping in mind that a change of the coordinate y necessitates a change of the canonical coordinate x, and hence the functions τ_1, τ_2). This may also be seen as follows: The inclusion $R^1 f_{0*}\mathbb{Z} \cong \Lambda \subseteq \mathcal{T}^*_{B_0}$ allows one to identify $(R^1 f_{0*}\mathbb{R}) \otimes C^\infty(B_0)$ with the underlying C^∞ vector bundle $\mathcal{T}^*_{B_0}$, along with the Gauss-Manin connection ∇_{GM} on $\mathcal{T}^*_{B_0}$, the flat connection whose flat sections are sections of $R^1 f_{0*}\mathbb{R}$. The standard semi-flat metric is the unique semi-flat Ricci-flat Kähler metric satisfying the conditions

(1) The area of each fiber is ϵ;
(2) $\omega^2_{SF} = (\mathrm{Re}\,\Omega)^2 = (\mathrm{Im}\,\Omega)^2$;
(3) The orthogonal complement of each vertical tangent space is the horizontal tangent space of ∇_{GM} at that point.

The reader should be aware however that if $T_\sigma : X_0 \to X_0$ denotes translation by a holomorphic section σ, then $T^*_\sigma\omega_{SF}$ may give rise to a different semi-flat metric, satisfying conditions (1) and (2) but not (3). However, if σ is not only holomorphic but a flat section with respect to the Gauss-Manin connection (so that $\sigma(y) = a_1\tau_1(y) + a_2\tau_2(y)$ for constants a_1, a_2) then T_σ is an isometry and $T^*_\sigma\omega_{SF} = \omega_{SF}$, $T^*_\sigma\Omega = \Omega$.

It will also be useful to have the Kähler potential for the metric. This is a function φ such that $\omega = \frac{\sqrt{-1}}{2}\partial\bar{\partial}\varphi$. Let ϕ_1 and ϕ_2 be anti-derivatives of τ_1 and τ_2 respectively. Then we can take

$$\varphi = \frac{\epsilon}{\mathrm{Im}(\bar{\tau}_1\tau_2)}\left(-\frac{\bar{x}^2}{2}\frac{\tau_1}{\bar{\tau}_1} + |x|^2 - \frac{x^2}{2}\frac{\bar{\tau}_1}{\tau_1}\right) + \frac{\sqrt{-1}}{2\epsilon}(\phi_1\bar{\phi}_2 - \bar{\phi}_1\phi_2).$$

This is well-defined on subsets $\mathcal{T}^*_U \subseteq \mathcal{T}^*_B$ for U simply connected, but not on $\mathcal{T}^*_B/\Lambda$.

In summary, the semi-flat metrics defined above are Ricci-flat, the fibers have area ϵ and diameter of the form $O(\epsilon^{\frac{1}{2}})$, and the curvature tensor R may be calculated (see below) to have C^0-norm $\|R\| = O(\epsilon)$. Further details may be found in [**210**].

7.3.4. The metric in a neighbourhood of a singular fiber. Away from the singular fibers, we saw in the previous section that suitable coordinates for calculations were the complex action-angle coordinates. In a neighbourhood of a singular fiber (recall that the singular fibers are assumed to be of Type I_1), these coordinates will not be suitable, the appropriate coordinates now being those arising from the Gibbons-Hawking ansatz described in §7.1.2.

The aim of this section is to describe a certain hyperkähler metric on a neighbourhood of each singular fiber in our elliptically fibered K3 surface, and to derive various estimates associated with this metric. If the fibers are assumed to have volume ϵ, then away from the singular fiber this metric will decay very rapidly, for small ϵ, to a semi-flat metric. Since the singular fibers were assumed to be of Kodaira type I_1, locally around each singular fiber, one of the periods is invariant under monodromy (and in fact, by an appropriate choice of holomorphic coordinate y on the base, may be taken to be constant, value 1), whilst the other period will be multivalued and tend to infinity. The metric we define will be an S^1-invariant metric (as described in §7.1.3) on the smooth part of the fibration, and will be most conveniently described in the Gibbons-Hawking coordinates. This metric was first written down (in a slightly different form) by Ooguri and Vafa [**383**], (see also [**416**]) and so will be referred to as the *Ooguri-Vafa* metric.

Suppose $D_r \subset \mathbb{C}$ is the disc of radius $r < 1$, with center the origin, and $f : \overline{X} \to D_r$ an elliptic fibration, with singular fiber over the origin of type I_1. Let $\overline{Y} = D_r \times \mathbb{R}/\epsilon\mathbb{Z}$ and $Y = (D_r \times \mathbb{R} - \{0\} \times \epsilon\mathbb{Z})/\epsilon\mathbb{Z}$. It is straightforward to check that there is an induced map $\bar{\pi} : \overline{X} \to \overline{Y}$ of C^∞ manifolds, which restricts to an S^1-bundle $\pi : X \to Y$ with Chern class ± 1, the sign dependent on the choice of orientation for the fiber. The reader will not be surprised to learn that $\bar{\pi} : \overline{X} \to \overline{Y}$ is precisely the topological compactification of $\pi : X \to Y$ described in Construction 6.86, with the singularity of the I_1-fiber of the elliptic fibration corresponding to the insertion of a nut over the point $(\{0\} \times \epsilon\mathbb{Z})/\epsilon\mathbb{Z}$. The plan now is to define a hyperkähler metric on X via the Gibbons-Hawking ansatz applied to $\pi : X \to Y$, and then check that it extends to a hyperkähler metric on $\overline{X}$. We will however need to take care to ensure that the hyperkähler metric we define in this way is compatible with the given complex structure on $\overline{X}$.

We apply the Gibbons-Hawking ansatz as described in §7.1.2 and §7.1.3, with $U = D \times \mathbb{R} \setminus \{0\} \times \epsilon\mathbb{Z}$. We denote by y_1, y_2 the coordinates u_1, u_2 on the unit disc $D \subset \mathbb{C}$, and by $u = u_3$ the coordinate on $\mathbb{R}$. We want to write down a function V, harmonic on U, periodic in u with period ϵ, and with singularities of the correct type at the points $\{0\} \times \epsilon\mathbb{Z}$. For instance, around zero, V should behave like a harmonic function plus a term $\frac{1}{4\pi|\mathbf{x}|}$, from which it will follow that the resulting hyperkähler metric on X extends to $\overline{X}$ — this is essentially just the calculation performed in Example 7.5 yielding the Taub-NUT metric. We are led therefore to take $V = V_0 + f(y_1, y_2)$, where f is a harmonic function in y_1, y_2 on D, and

$$V_0 = \frac{1}{4\pi} \sum_{n=-\infty}^{\infty} \left(\frac{1}{\sqrt{(u+n\epsilon)^2 + y_1^2 + y_2^2}} - a_{|n|} \right),$$

where $a_n = \frac{1}{n\epsilon}$ $(n > 0)$, thus ensuring appropriate convergence, and a_0 is chosen appropriately to ensure that the periods do not change as we change ϵ, and so we are defining metrics on a fixed elliptic fibration.

Straightforward complex analysis shows that V_0 as defined above does converge to a harmonic function on U. The value of a_0 that we take will be

$$a_0 = 2(-\gamma + \log(2\epsilon))/\epsilon,$$

where γ is Euler's constant. For this choice of a_0, we can show the following (cf. Lemma 3.1 of [**210**]).

LEMMA 7.26. *With the notation as above, V_0 has an expansion, valid when $|y| \neq 0$,*

$$V_0 = -\frac{1}{4\pi\epsilon} \log |y|^2 + \frac{1}{2\pi\epsilon} \sum_{\substack{m=-\infty \\ m \neq 0}}^{m=\infty} e^{2\pi\sqrt{-1}mu/\epsilon} K_0(2\pi|my|/\epsilon)$$

where $y = y_1 + \sqrt{-1}y_2$ and K_0 is the modified Bessel function. From this expansion, it follows that there exists a constant C such that for any $0 < r_0 < 1$, there exists an $\epsilon_0 > 0$ such that for all $\epsilon < \epsilon_0$, $|y| > r_0$,

$$\left| V_0 + \frac{1}{4\pi\epsilon} \log |y|^2 \right| \leq \frac{C}{\epsilon} e^{-2\pi|y|/\epsilon}.$$

Thus, on a fixed annulus in D, we have $\epsilon V_0 \sim -\frac{1}{2\pi} \log r$ as $\epsilon \to 0$, where $r^2 = y_1^2 + y_2^2$. Moreover, if $r \leq 1$, and f is a harmonic function on the disc D_r of radius r such that $f(y) - \frac{1}{4\pi} \log |y|^2 > 0$ for $|y| \leq r$, then there exists an ϵ_0 such that for all $\epsilon < \epsilon_0$,

$$V_0 + f(y)/\epsilon > 0$$

in $D_r \times \mathbb{R}$.

Suppose now $D_r \subseteq \mathbb{C}$ is a disc of radius $r < 1$ centered on the origin, and let $h(y) = f(y_1, y_2) + \sqrt{-1}g(y_1, y_2)$ be a holomorphic function on D_r, so that

$$-\frac{1}{4\pi} \log |y|^2 + f(y_1, y_2) > 0$$

on D_r. Let V_0 be the harmonic function on Y defined in Lemma 7.26, and $V = V_0 + f(y_1, y_2)/\epsilon$, with ϵ chosen small enough so that $V > 0$ on Y.

PROPOSITION 7.27. *There exists a connection 1-form θ on X such that $d\theta/2\pi\sqrt{-1} = *dV$, thus defining a hyperkähler metric on X via the Gibbons-Hawking ansatz. This metric extends to a hyperkähler metric on $\overline{X}$, and yields a holomorphic elliptic fibration $\overline{X} \to D_r$ with periods 1 and*

$$\frac{1}{2\pi\sqrt{-1}} \log y + \sqrt{-1}h(y) + C,$$

for some real constant C. By appropriate choice of θ, this constant C may be taken to be zero.

We justify briefly the statements concerning the periods in Proposition 7.27. From the construction of our metric, it will be automatic that one of the periods is constant of value 1. The other period $\tau(y)$ will be given locally (for $y \neq 0$) as $\int_\gamma dx$, where x, y denote the canonical holomorphic coordinates, and γ is an S^1 in the fiber X_y mapping isomorphically to $\{y\} \times S^1 \subset Y$. From the discussion in §7.1.3 regarding the relationship between the canonical holomorphic coordinates and the Gibbons-Hawking coordinates, we may rewrite this as

$$\tau(y) = \int_\gamma \theta_0 - \sqrt{-1} \int_\gamma V du,$$

where as before θ_0 denotes the real 1-form $\theta/2\pi\sqrt{-1}$. Calculating the imaginary part of this, one finds that

$$\int_\gamma dx_2 = -\int_\gamma V du = \pm(\frac{1}{4\pi} \log |y|^2 - f(y_1, y_2)),$$

using the Fourier expansion for V_0 from Lemma 7.26. We choose the orientation of γ to obtain the choice of sign to be minus. Then $\int_\gamma dx_1$ is necessarily locally a harmonic conjugate of $-\frac{1}{4\pi} \log |y|^2 + f(y_1, y_2)$, and so the period of γ is

$$\frac{1}{2\pi\sqrt{-1}} \log y + \sqrt{-1}h(y) + C$$

for some real constant C. Now θ_0 may be modified by adding a term adu ($a \in \mathbb{R}$) without changing the fact that $d\theta_0 = *dV$. If θ_0 is changed in this way, we have

$$\begin{aligned} \int_\gamma dx_1 &= \int_\gamma \theta_0 + adu \\ &= a\epsilon + \int_\gamma \theta_0. \end{aligned}$$

We can therefore choose a suitably to obtain $C = 0$, and hence the period is as claimed.

In this way, we have ensured that the hyperkähler metrics we have defined are all compatible with the given fixed complex structure on $\overline{X}$. Moreover, over a fixed annulus in D, the asymptotic formula for V_0 in Lemma

7.26 may be reinterpreted as saying that the Ooguri-Vafa metrics tend to a semi-flat metric exponentially fast. One remarks here that with periods 1 and $\frac{1}{2\pi\sqrt{-1}}\log y + \sqrt{-1}h(y)$ as above, the semi-flat metric corresponds to taking V to be $\epsilon^{-1}(-\frac{1}{4\pi}\log|y|^2 + f(y)) = W^{-1}$, where $W = \epsilon/\operatorname{Im}(\tau)$ as in the previous section.

The explicit nature of the formulae which appear in the Ooguri-Vafa metrics, and in particular the Fourier expansion for V_0 given in Lemma 7.26, makes it possible to estimate accurately the diameters and curvatures. With the fibers having area ϵ as above, it is shown in Proposition 3.5 of [**210**] that the diameter of the singular fiber is $O((\epsilon\log\epsilon^{-1})^{1/2})$, whilst the diameter of a smooth fiber is simply $O(\epsilon^{1/2})$, in agreement with our calculations in the semi-flat case. Geometrically, this implies that if we rescale the metrics so that the area of the fibers is 1, then the diameter of the singular fiber is $O((\log\epsilon^{-1})^{1/2})$, which becomes arbitrary large as $\epsilon \to 0$.

To estimate the curvature R for the Ooguri-Vafa metrics, we appeal to the formula for $\|R\|^2$ we gave in §7.1.2 for any metric obtained via the Gibbons-Hawking ansatz. Again, using the Fourier expansion for V_0, a slightly intricate argument (Proposition 3.8 from [**210**]) yields the following estimate for $\|R\|$.

PROPOSITION 7.28. *Let $R(\epsilon)$ denote the curvature tensor of an Ooguri-Vafa metric as defined above, with fiber of area ϵ. Then there exist positive constants C, C' (independent of ϵ) such that, for all sufficiently small ϵ,*

$$C'\epsilon^{-1}\log(\epsilon^{-1})^{-2} < \|R(\epsilon)\|_{C^0} < C\epsilon^{-1}\log(\epsilon^{-1}),$$

where $\|\cdot\|_{C^0}$ denotes the usual C^0-norm.

Thus, for the Ooguri-Vafa metrics, the curvature blows up at the node on the singular fiber of the elliptic fibration. This again should be contrasted with its behaviour on the smooth fibers (or equivalently for the semi-flat metrics), where $\|R(\epsilon)\|_{C^0} = O(\epsilon)$. This latter fact may be proved either directly, or by writing the semi-flat metrics in terms of the Gibbons-Hawking ansatz, and then using again the formula from §7.1.2.

7.3.5. Accurate approximations to the Ricci-flat metrics. We now construct very accurate approximations (as $\epsilon = \epsilon_l \to 0$) to the Ricci-flat metrics appearing in Question 7.25. Let $f : X \to B = \mathbb{CP}^1$ denote the elliptic fibration as before, with holomorphic section σ_0, and let $\Delta = \{p_1, \ldots, p_{24}\}$ denote the discriminant locus of f, above which the fibers have I_1 singularities. Let $B_0 = B \setminus \Delta$ and $X_0 = f^{-1}(B_0)$. From §7.3.3, we then have the standard semi-flat metric ω_{SF} defined on X_0, with fibers having area ϵ. The dependence of ω_{SF} on ϵ should be born in mind in what follows. The recipe we describe constructs an approximately Ricci-flat metric on X by means of gluing in suitable twists of the Ooguri-Vafa metric in a neighbourhood of each singular fiber to this background metric.

Let us now concentrate on a fixed point $p \in \Delta$; we suppose that U is a contractible neighbourhood of p with $U \cap \Delta = \{p\}$, and set $U^* = U - \{p\}$. We denote $f^{-1}(U)$ by X_U and $f^{-1}(U^*)$ by X_{U^*}. In order to perform the gluing, we need the following fact, proved in Lemma 4.3 of [**210**].

LEMMA 7.29. *Let ω_{SF} denote (the Kähler form of) the standard semi-flat metric on X_0 with fibers of area ϵ, and ω a Kähler form on X_U, also with fibers of area ϵ. Then the class $[\omega_{SF} - \omega] = 0$ in $H^2(X_{U^*}, \mathbb{R})$, and furthermore, there exists a holomorphic section σ of $f : X_U \to U$ and a function φ on X_{U^*} such that*

$$\omega_{SF} - T_\sigma^* \omega = \sqrt{-1}\partial\bar{\partial}\varphi,$$

where T_σ is translation by the section σ.

Choosing now a holomorphic coordinate y in a neighbourhood of p, we can express the holomorphic periods of f as $\tau_1(y), \tau_2(y)$, where τ_1 is taken to be single-valued. In T_U^*, this coincides with the holomorphic differential $\tau_1(y)dy$. Locally, there exists a function $g(y)$ with $dg = \tau_1(y)dy$; since $\tau_1(p) \neq 0$, we can use g as a local holomorphic coordinate in a neighbourhood of p. Replacing y by g, we can then assume that $\tau_1(y) = 1$ and also that $y = 0$ at p. By the results of the previous section, we can then construct for all ϵ less than some ϵ_0 an Ooguri-Vafa metric ω_{OV} (with fiber area ϵ) on $f^{-1}(U)$, for some $U = \{y \mid |y| < r\}$, for some r which only depends on the period τ_2 and ϵ_0, but not ϵ. Fix $r_1 < r_2 < r$, and let $U_i = \{y \mid |y| < r_i\}$, with U_{r_1,r_2} the annulus $\{y \mid r_1 < |y| < r_2\}$.

This enables us to glue in the local metric ω_{OV} to the background metric ω_{SF} using an appropriate fixed C^∞ cut-off function $\psi(r)$ with $\psi(r^2) = 1$ for $r \leq r_1$ and $\psi(r^2) = 0$ for $r \geq r_2$. The point here is that by Lemma 7.29 there exists a holomorphic section σ of $f : X_U \to U$ and a function φ on X_{U^*} such that

$$\omega_{SF} - T_\sigma^* \omega_{OV} = \sqrt{-1}\partial\bar{\partial}\varphi$$

over U^*. By considering

$$\omega_{SF} - \sqrt{-1}\partial\bar{\partial}(\psi(|y|^2)\varphi),$$

we obtain a closed real $(1,1)$-form agreeing with $T_\sigma^*\omega_{OV}$ for $|y| \leq r_1$, and with ω_{SF} for $|y| \geq r_2$. We now perform such a gluing at each singular fiber, obtaining a global closed real $(1,1)$-form ω_{new}. For sufficiently small ϵ , it may be checked that ω_{new} is positive-definite, and therefore defines a Kähler metric on X. On the other hand, it is not clear that $\int_X \omega_{\text{new}}^2 = \int_X (\operatorname{Re}\Omega)^2$. Recall however that for small ϵ, the forms ω_{SF} and ω_{OV} differ over the annulus U_{r_1,r_2} by $O(e^{-C/\epsilon})$. By construction $\omega_{\text{new}}^2 = (\operatorname{Re}\Omega)^2$ outside of $f^{-1}(U_{r_1,r_2})$, and ω_{new}^2 and $(\operatorname{Re}\Omega)^2$ differ only by $O(e^{-C/\epsilon})$ on $f^{-1}(U_{r_1,r_2})$, so $\int_X \omega_{\text{new}}^2 - \int_X (\operatorname{Re}\Omega)^2 = O(e^{-C/\epsilon})$. Now noting that

$$([\omega_{\text{new}}] + aE)^2 = [\omega_{\text{new}}]^2 + 2a\epsilon,$$

we can find a two-form α on B supported on U_{r_1,r_2}, with C^k-norm $O(e^{-C/\epsilon})$ for each k, such that $\int_X(\omega_{\rm new}+f^*\alpha)^2=\int_X(\operatorname{Re}\Omega)^2$. Set $\omega_\epsilon=\omega_{\rm new}+f^*\alpha$. Because α is still small, ω_ϵ is still positive and defines the desired Kähler metric.

This glued metric is clearly Ricci-flat outside the gluing regions (the points above the chosen annuli around points of Δ). However, it is very close to being globally Ricci-flat, in that the Ricci form is at most $O(e^{-C/\epsilon})$. Moreover, if we set $F_\epsilon=\log(\frac{\Omega\wedge\bar\Omega/2}{\omega_\epsilon^2})$, then both F_ϵ and its Laplacian (with respect to the metric ω_ϵ) have C^0-norm at most $O(e^{-C/\epsilon})$. The calculations referred to in the previous two sections show that the metric defined by ω_ϵ has diameter behaving at worst like $O(\epsilon^{-1/2})$, and that the curvature tensor R has $\|R\|_{C^0}\to\infty$ as $\epsilon\to 0$, but behaving at worst like $O(\epsilon^{-1}\log(\epsilon^{-1}))$. For a fixed smooth fiber, $\|R\|\le C\epsilon$, for some constant C depending on the fiber.

We claim now that the metric constructed in this way is a very accurate approximation to the Ricci-flat metric — this we do by analysing Yau's proof for the existence of the Ricci-flat metric, in a similar way to what was done by Kobayashi for Kähler degenerations of K3 surfaces to surfaces with rational double point singularities, as mentioned in §7.3.1. The starting metric we take in running this program is of course the metric corresponding to ω_ϵ. Yau's proof yields a function u_ϵ on X such that

$$\begin{aligned}(\omega_\epsilon+i\partial\bar\partial u_\epsilon)^2 &= e^{F_\epsilon}\omega_\epsilon^2\\ \int_X u_\epsilon\omega_\epsilon^2 &= 0.\end{aligned}$$

Moreover, an analysis of the proof yields various bounds on u_ϵ, which confirm the claim being made that ω_ϵ is a very accurate approximation to the Ricci-flat metric, which we shall denote by $\tilde\omega_\epsilon$. This analysis is carried out in §5 of [**210**].

(I) The C^0-estimate we obtain is that, for some positive constants C_1, C_2 independent of ϵ, we have $\|u_\epsilon\|_\infty\le C_1\epsilon^{-5}e^{-C_2/\epsilon}$. The proof of this is via Sobolev inequalities. The above estimate on the diameter of the metric corresponding to ω_ϵ ensures that the Sobolev constant is bounded below by $O(\epsilon^5)$. The fact that F_ϵ has C^0-norm at most $O(e^{-C/\epsilon})$ then produces the estimate. Of course, for small ϵ, the exponential term is the significant term in this estimate, and so the C^0-norm tends to zero exponentially as $\epsilon\to 0$.

(II) The C^2-estimate we obtain is that, for some constant $C>1$ independent of ϵ,
$$C^{-1}\omega_\epsilon\le\tilde\omega_\epsilon\le C\omega_\epsilon$$
for all sufficiently small ϵ. This estimate follows via Yau's argument using the maximum principle, and needs all the various bounds for

the metric ω_ϵ we obtained above. This argument lies at the heart of Yau's proof and our estimates on u_ϵ. This estimate already implies that, with the Ricci-flat metrics $\tilde{\omega}_\epsilon$, the fibers of $f : X \to \mathbb{CP}^1$ collapse as $\epsilon \to 0$ (cf. §7.3.6 below).

(III) The $C^{k,\alpha}$-estimate (for $k \geq 2$ and $0 < \alpha < 1$) we obtain says that for any simply-connected open set $U \subset B$ with closure $\overline{U} \subset B_0 = B \backslash \Delta$, there exist positive constants C_1, C_2, depending on U and k but independent of ϵ, such that

$$\|u_\epsilon\|_{C^{k,\alpha}} \leq C_1 e^{-C_2/\epsilon},$$

for ϵ sufficiently small. Here one applies a general result (Theorem 17.14 in [**179**]) to get from the C^2-bound to a $C^{2,\alpha}$-bound, and then standard arguments using bootstrapping and Schauder estimates for the stronger statements.

The basic reason why the proofs of these estimates work is that the quantities which blow up as $\epsilon \to 0$, such as the diameter and the curvature, do so at worst polynomially in ϵ, whilst the quantities such as F_ϵ and the Ricci curvature which tend to zero do so exponentially fast, being bounded by $O(e^{-C/\epsilon})$. The analysis undertaken of the a priori estimates which come out of Yau's proof show that the latter behaviour will always dominate over the former.

In summary therefore, we have that ω_ϵ is globally within a constant multiple of the Ricci-flat metric $\tilde{\omega}_\epsilon$, and that over relatively compact subsets of $\mathbb{CP}^1 \setminus \Delta$ it is arbitrarily close to $\tilde{\omega}_\epsilon$ in any $C^{k,\alpha}$-norm, the error being $O(e^{-C/\epsilon})$ for ϵ sufficiently small. We shall see the geometric significance of these statements in the next section, where we describe the Gromov-Hausdorff convergence of metric spaces.

7.3.6. Gromov-Hausdorff convergence. In this section, we make precise the notion of *metric convergence* which has been alluded to earlier at various points. The results of the previous section will then imply an important result concerning complex degenerations of K3 surfaces to large complex structure limits, or equivalently concerning the Kähler degenerations of elliptic K3 surfaces studied above. With the proper normalization, this says that in the limit, the K3 surfaces in fact converge to 2-spheres, equipped with a standard metric. This in turn will lead us to a general conjecture concerning the metric limits of Calabi–Yau manifolds as the complex structure degenerates to large complex structure limit, which has provided motivation for recent work by a number of authors.

DEFINITION 7.30. Let (X, d_X), (Y, d_Y) be two compact metric spaces. Suppose there exist maps $f : X \to Y$ and $g : Y \to X$ (not necessarily continuous!) such that for all $x_1, x_2 \in X$,

$$|d_X(x_1, x_2) - d_Y(f(x_1), f(x_2))| < \epsilon$$

and for all $x \in X$,

$$d_X(x, g \circ f(x)) < \epsilon,$$

and the two symmetric properties for Y hold. Then we say that the Gromov-Hausdorff distance between X and Y is at most ϵ. The Gromov-Hausdorff distance $d_{GH}(X, Y)$ is the infimum of all such ϵ.

The Gromov-Hausdorff distance defines a topology on the set of compact metric spaces, and hence a notion of convergence. It follows from results of Gromov (see, e.g., [**391**], pg. 281, Cor. 1.11) that the class of compact Ricci-flat manifolds with diameter $\leq D$ is precompact. Moreover, we can find a subsequence converging (in the sense of Gromov-Hausdorff) to a compact length space. A length space is a metric space in which the distance between two points is the infimum of the lengths of paths joining them. In the compact case, this distance is achievable as the length of some curve joining the points. In particular, if we have a sequence of Calabi-Yau n-folds whose complex structure converges to a large complex structure limit point (or any other boundary point for that matter) and whose metrics have diameter bounded above, then there is a convergent subsequence. The basic question one now asks is: what is the limit? If the limit has real dimension less than $2n$, we say that *collapsing* occurs.

The results of the previous section provide an answer to this question for the degenerations of K3 surfaces studied there, but with the metrics normalized so as to have bounded diameter rather than bounded volume — this normalization is achieved by scaling all the metrics by ϵ. We state the result in the more general form, where we are not assuming that the elliptically fibered K3 surfaces necessarily have a holomorphic section.

THEOREM 7.31. *Let $j : J \to B$ be an elliptically fibered K3 surface with a section and singular fibers all of type I_1, and let $f_i : X_i \to B$ be a sequence of elliptically fibered K3 surfaces with Jacobian j. Let ω_i correspond to a Ricci-flat Kähler metric on X_i with ω_i^2 independent of i, and with $\int_{f_i^{-1}(b)} \omega_i = \epsilon_i \to 0$ as $i \to \infty$. Then the sequence of Riemannian manifolds $(X_i, \epsilon_i \omega_i)$ converges in the Gromov-Hausdorff sense to B, the metric on B being induced from a standard Riemannian metric on the complement of the discriminant locus. In local coordinates, the periods may be written as $\tau_1(y), \tau_2(y)$ with $\mathrm{Im}(\bar{\tau}_1 \tau_2) > 0$, and then the Riemannian metric is given by $\mathrm{Im}(\bar{\tau}_1 \tau_2)\, dy \otimes d\bar{y}$.*

It is important to point out that this theorem is not at all obvious— just because the areas of the fibers are going to zero doesn't mean they cannot do something perverse, such as stretch out but get thinner in a way that the area goes to zero, but the fibers remain "big".

In fact, to prove the theorem, we approximate the Ricci-flat metrics ω_i as in §7.3.5. Away from the singular fibers, we have very fast convergence to the semi-flat metric, and hence metric convergence to the appropriate open subset of B with the stated Riemannian metric. On a *fixed* neighbourhood of a singular fiber, we can only use the C^2-estimate, as given in (II), and this is insufficient to imply the claimed metric convergence. If however we also allow ourselves to shrink the neighbourhoods of the singular fibers, letting the radii of the chosen annuli in B tend suitably to zero as $i \to \infty$ (cf. the case of degenerations of an arbitrary K3 surface to one with rational double point singularities, as described in Remark 7.23), then it is a straightforward argument, using the C^2-estimate locally around the singular fibers, that the Gromov-Hausdorff limit is as claimed. The limit Riemannian metric on the complement B_0 of the discriminant locus in B is just a multiple of the McLean metric on B_0, as described in §6.1.1, corresponding to the special Lagrangian torus fibration $X_0 \to B_0$ obtained by hyperkähler rotation back to the complex structure I and equipped with the semi-flat metric. To see this, recall that for tangent vectors u, v at $b \in B_0$, the McLean metric is defined by

$$g(u,v) = -\int_{f^{-1}(b)} \iota(u)\omega_I \wedge \iota(v) \operatorname{Im} \Omega_I = -\int_{f^{-1}(b)} \iota(u) \operatorname{Im} \Omega_K \wedge \iota(v) \operatorname{Re} \Omega_K,$$

where in local coordinates $\Omega_K = dx \wedge dy$. Thus, an easy calculation verifies that

$$2\, g(\partial/\partial y, \partial/\partial \bar{y}) = \int_{f^{-1}(b)} dx_1 \wedge dx_2 = \operatorname{Im}(\bar{\tau}_1 \tau_2).$$

The above theorem leads us to a conjecture for Calabi–Yau n-folds, made in §6 of [**210**], and also independently by Kontsevich, Soibelman and Todorov.

CONJECTURE 7.32. *Let $\overline{\mathcal{M}}$ be a compactified moduli space of complex deformations of a simply-connected Calabi-Yau n-fold X with holonomy group* SU(n)*, and let $p \in \overline{\mathcal{M}}$ be a large complex structure limit point (see* [**365**] *for the precise Hodge-theoretic definition of this notion). Let (X_i, g_i) be a sequence of Calabi-Yau manifolds with Ricci-flat Kähler metric which are complex deformations of X, with the sequence $[X_i] \in \overline{\mathcal{M}}$ converging suitably to p, and $C_1 \geq Diam(X_i) \geq C_2 > 0$ for all i. Then a subsequence of (X_i, g_i) converges to a metric space (X_∞, d_∞), where X_∞ is homeomorphic to S^n and d_∞ is a metric on X_∞. Furthermore, $X_\infty \setminus \Delta$ carries an affine structure and d_∞ is induced by a Monge-Ampère (Riemannian) metric on $X_\infty \setminus \Delta$, with $\Delta \subseteq X_\infty$ some subset of codimension 2.*

The phrase *converging suitably to p* is discussed in §6 of [**210**]; one needs in particular to rule out cases where the limit would have real dimension strictly less than n. In the mirror, this could be caused for instance by approaching the large Kähler structure limit in such a way as to approach the boundary of the projectivised Kähler cone.

7.3.7. The future of the SYZ conjecture. The Strominger-Yau-Zaslow conjecture has now been around for more than ten years. As explained in this chapter, it now seems very unlikely that it will ever be proved in its original form. However, Conjecture 7.32 suggests a limiting version of the SYZ conjecture which seems likely to be more amenable to proof.

Let us start with a maximally unipotent degeneration of Calabi-Yau n-folds $\mathcal{X} \to D$, D a disk, and let $t_i \to 0 \in D$. Suppose for t_i sufficiently close to zero, there is a special Lagrangian T^n whose homology class is invariant under monodromy, or more specifically, generates the space W_0 of the monodromy weight filtration associated to the monodromy operator (this is where we expect to find fibers of a special Lagrangian fibration associated to a maximally unipotent degeneration). Let $B_{0,i}$ be the moduli space of deformations of this torus; every point of $B_{0,i}$ corresponds to a smooth special Lagrangian torus in $\mathcal{X}_{t_i}$. This manifold then comes equipped with the McLean metric and affine structures defined in §6.1.1. One can then compactify $B_{0,i} \subseteq B_i$ (probably by taking the closure of $B_{0,i}$ in the space of special Lagrangian currents; the details aren't important here). This gives a series of compact metric spaces (B_i, d_i) with the metric d_i induced by the McLean metric. If the McLean metric is normalized to keep the diameter of B_i constant independent of i, then we can hope that (B_i, d_i) converges to a compact metric space (B_∞, d_∞). Then here is the limiting form of SYZ:

CONJECTURE 7.33. *If $(\mathcal{X}_{t_i}, g_i)$ converges to (X_∞, g_∞) and (B_i, d_i) is non-empty for large i and converges to (B_∞, d_∞), then B_∞ and X_∞ are isometric up to scaling. Furthermore, there is a subspace $B_{\infty,0} \subseteq B_\infty$ with $\Delta := B_\infty \setminus B_{\infty,0}$ of Hausdorff codimension 2 in B_∞ such that $B_{\infty,0}$ is a Monge-Ampère manifold, with the Monge-Ampère metric inducing d_∞ on $B_{\infty,0}$.*

Essentially what this is saying is that as we approach the large complex structure limit, special Lagrangian tori fill out more and more of the Calabi-Yau manifold. However, there is no point in moduli where we can be sure to obtain a special Lagrangian fibration on the entire manifold.

The conjecture is worded in this way since it is much more likely to be provable than stronger forms. Here is an outline of an approach to proving this conjecture:

(1) Given the maximally unipotent degeneration $\mathcal{X} \to D$, guess what the limiting object $X_\infty = B_\infty$ is, but only as an *affine manifold with singularities*.
(2) Solve the affine Calabi conjecture, i.e., find a Monge-Ampère metric on B_∞. This will give a Ricci-flat metric on $X(B_{\infty,0})$.

(3) Find local models for Ricci-flat metrics near singular fibers of compactifications of $X(B_{\infty,0})$.
(4) Glue in these local models, and show that the resulting metric is close to an actual Ricci-flat metric, just as was done in §7.3.5. This will give a good description of the Ricci-flat metric on a Calabi-Yau manifold near the large complex structure limit.
(5) Since the metric away from the singular fibers is close to the semi-flat metric, one would expect that the special Lagrangian fibers in the semi-flat metric would deform to special Lagrangian tori with respect to the actual Ricci-flat metric. This would yield (open subsets of) the moduli spaces $B_{i,0}$ for t_i close to zero. This allows us to compare X_i and B_i and prove the limiting result.

The first step is largely understood; indeed, we have already seen examples in §6.4.1 of affine manifolds with singularities which conjecturally are the correct ones. Certainly, they are correct at a topological level. The second and third steps are the most challenging. There has been no success at proving the affine Calabi conjecture in dimension larger than 2, but see [**338, 339**] for some work in that direction. For the third step, there are no local models analogous to the Ooguri-Vafa metric near positive or negative vertices, but see [**476**] for some work in this direction. However, the expectation is that the fourth step will only need techniques similar to those sketched above from [**210**].

This leaves open the question of how we do mirror symmetry using this modified version of the SYZ conjecture. Essentially, we would follow these steps:

(1) We begin with a maximally unipotent degeneration of Calabi-Yau manifolds $\mathcal{X} \to D$, along with a choice of polarization. This gives us a Kähler class $[\omega_t] \in H^2(\mathcal{X}_t, \mathbb{R})$ for each $t \in D \setminus 0$, represented by ω_t the Kähler form of a Ricci-flat metric g_t.
(2) Identify the Gromov-Hausdorff limit of a sequence $(\mathcal{X}_{t_i}, r_i g_{t_i})$ with $t_i \to 0$, and r_i a scale factor which keeps the diameter of $\mathcal{X}_{t_i}$ constant. The limit will be, if the above conjectures work, an affine manifold with singularities B along with a Monge-Ampère metric.
(3) Perform a Legendre transform to obtain a new affine manifold with singularities $\check{B}$, though with the same metric.
(4) Try to construct a compactification of $X_\epsilon(\check{B}_0) := T\check{B}_0/\epsilon\Lambda$ for small $\epsilon > 0$ to obtain a complex manifold $X_\epsilon(\check{B})$. This will be the mirror manifold.

Actually, we need to elaborate on this last step a bit more. The problem is that while we expect that it should be possible in general to construct symplectic compactifications of the symplectic manifold $\check{X}(B_0)$ (and hence get the mirror as a symplectic manifold), we don't expect to be able to compactify $X_\epsilon(\check{B}_0)$ as a complex manifold. We have seen this explicitly in the description of the Ooguri-Vafa metric: one needed to perturb the semi-flat metric (which in particular gives a perturbed complex structure) before it could be extended across the singular fibre. Instead, the general expectation is that a "small" deformation of $X_\epsilon(\check{B}_0)$ is necessary before it can be compactified. One expects the deformation to be of size $O(e^{-C/\epsilon})$, measured in an appropriate way. Furthermore, this small deformation is critically important in mirror symmetry: *it is this small deformation which provides the B-model instanton corrections.*

Because of the importance of this last issue, it has already been studied by several authors: Fukaya in [**166**] has studied the problem directly using heuristic ideas, while Kontsevich and Soibelmann [**312**] have modified the problem of passing from an affine manifold to a complex manifold by instead producing a non-Archimedean space. We give this problem a name:

QUESTION 7.34 (The reconstruction problem). *Given an integral affine manifold with singularities B (see §6.4.1), construct a complex manifold $X_\epsilon(B)$ which is a compactification of a small deformation of $X_\epsilon(B_0)$.*

There has recently been much progress on this last problem in work of Gross and Siebert. Discussing this work is beyond the scope of the book, but briefly, the idea is that given an integral affine manifold with singularities B and some additional combinatorial data on B, one can construct a degeneration of Calabi-Yau manifolds. It is in fact possible to work on a purely algebro-geometric level, and obtain in this way a correspondence between the geometry of degenerations and the geometry of affine manifolds. This provides an algebro-geometric version of the SYZ conjecture, which holds out the promise of being far more powerful than the differential-geometric versions discussed here. In particular, it gives a direct link between rational curves and periods. We send the interested reader to [**205, 206, 203, 207, 208**].

CHAPTER 8

The Mathematics of Homological Mirror Symmetry

The goal of this chapter is to state a mathematically precise form of Kontsevich's homological mirror symmetry (HMS) conjecture. Primarily we shall explain the algebraic structures involved in some detail, namely A_∞-algebras and categories. By way of example, we shall give a (partial) proof in the simplest case, that of elliptic curves. Finally we briefly discuss Seidel's recent proof of HMS for the quartic K3 surface.

Roughly put, HMS states that if X and $\check{X}$ are a mirror pair of Calabi-Yau manifolds, then the category of A-branes on X is isomorphic to the category of B-branes on $\check{X}$. Mathematically, we have already seen in Chapter 5 that the category of B-branes is the derived category of coherent sheaves on $\check{X}$, $D^b(\check{X})$, defined rigorously in §4.5. We have discussed in §3.6 that the category of A-branes on X is the *Fukaya category* $\mathrm{Fuk}(X)$, whose objects are, roughly, Lagrangian submanifolds of X. So HMS should posit an equivalence of categories between $D^b(\check{X})$ and $\mathrm{Fuk}(X)$.

There is a basic problem with this equivalence. $\mathrm{Fuk}(X)$ is not actually a genuine category, but is something known as an A_∞-category. This means in particular that composition of morphisms is not precisely associative, but only associative "up to homotopy". On the other hand, $D^b(\check{X})$ is a triangulated category, while $\mathrm{Fuk}(X)$ is not triangulated. So some work understanding HMS is necessary before it becomes a precise statement.

To begin, we need to understand what an A_∞-category is, and this entails understanding the rather complex set of A_∞ relations.

8.1. A_∞-algebras and categories

8.1.1. Introduction: A_∞ spaces. The definition of A_∞-algebras and categories is a bit difficult to motivate, so we will begin by exploring the idea of A_∞ spaces, introduced by Stasheff in [**430**], and which have already been alluded to in §2.5. Suppose we have a topological space X and a "multiplication map" $m_2 : X \times X \to X$. This map may or may not be associative; imposing associativity is an extra condition. An A_∞ space imposes a weaker structure, which requires m_2 to be associative up to homotopy, along with "higher order" versions of this. Indeed, there are very standard situations where one has natural multiplication maps which are not associative, but obey the weaker conditions we will describe.

The standard example is when X is the loop space of another space M, i.e., if $m_0 \in M$ is a chosen base point,

$$X = \{x : [0,1] \to M \mid x \text{ continuous}, x(0) = x(1) = m_0\}.$$

Composition of loops is then defined, with

$$x_2x_1(t) = \begin{cases} x_2(2t) & 0 \le t \le 1/2 \\ x_1(2t-1) & 1/2 \le t \le 1. \end{cases}$$

However, this composition is not associative, but $x_3(x_2x_1)$ and $(x_1x_2)x_3$ are homotopic loops. This homotopy is depicted schematically by Figure 1. On

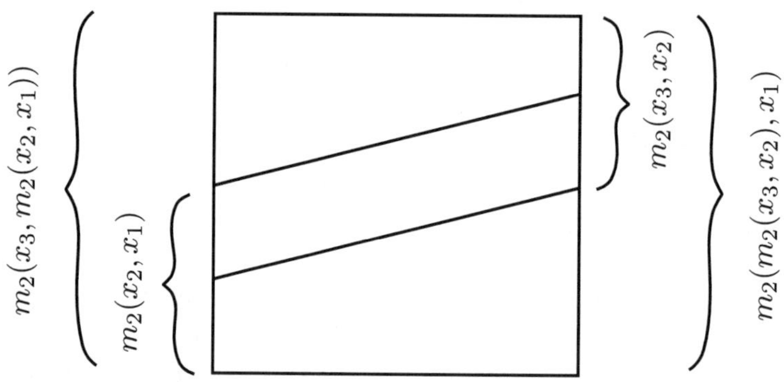

FIGURE 1

the left, we first traverse x_3 from time 0 to time $1/2$, then traverse x_2 from time $1/2$ to time $3/4$, and then x_1 from time $3/4$ to time 1. On the right, we first traverse x_3 from time 0 to time $1/4$, x_2 from time $1/4$ to time $1/2$, and then x_1 from time $1/2$ to time 1. By continuously deforming these times, we can homotop one of the loops to the other. This homotopy can be represented by a map

$$m_3 : [0,1] \times X \times X \times X \to X$$

such that $\{0\} \times X \times X \times X \to X$ is given by

$$(x_3, x_2, x_1) \mapsto m_2(x_3, m_2(x_2, x_1))$$

and $\{1\} \times X \times X \times X \to X$ is given by

$$(x_3, x_2, x_1) \mapsto m_2(m_2(x_3, x_2), x_1).$$

The question then arises: if we have four elements $x_1, \ldots, x_4$ of X, there are a number of different ways of putting brackets in their product, and these are related by the homotopies defined by m_3. Indeed, we can relate

$$((x_4x_3)x_2)x_1 \text{ and } x_4(x_3(x_2x_1))$$

in two different ways:

$$((x_4x_3)x_2)x_1 \sim (x_4x_3)(x_2x_1) \sim x_4(x_3(x_2x_1))$$

and

$$((x_4x_3)x_2)x_1 \sim (x_4(x_3x_2))x_1 \sim x_4((x_3x_2)x_1) \sim x_4(x_3(x_2x_1)).$$

Here each $\sim$ represents a homotopy given by m_3. Schematically, we can represent this by a polygon, which we will call $\overline{\mathcal{S}_4}$, with each vertex labelled by one of the ways of associating $x_4x_3x_2x_1$, and the edges represent homotopies between them; see Figure 2. In other words, the homotopies m_3 yield

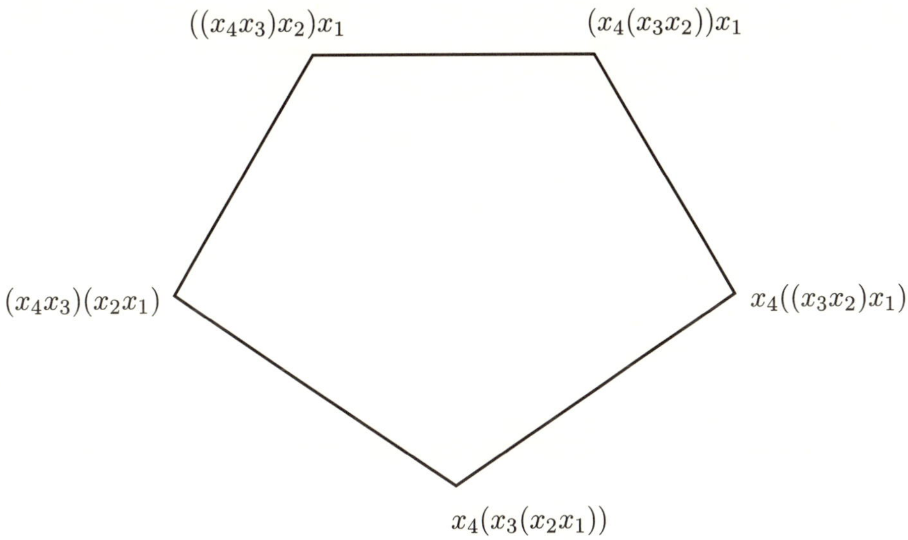

FIGURE 2

a map $\partial\overline{\mathcal{S}_4} \times X^4 \to X$ which is defined using appropriate combinations of m_2 and m_3 on each edge of the boundary of $\overline{\mathcal{S}_4}$. For example, restricting to the edge with vertices $((x_4x_3)x_2)x_1$ and $(x_4(x_3x_2))x_1$, this map is given by $(s, x_4, \dots, x_1) \mapsto m_2(m_3(s, x_4, x_3, x_2), x_1)$.

We can then impose a new condition on the structure exhibited so far, namely that this map extend across $\overline{\mathcal{S}_4}$, giving a map

$$m_4 : \overline{\mathcal{S}_4} \times X^4 \to X.$$

Let's see what happens now if we try to choose ways of associating the expression $x_5x_4x_3x_2x_1$. We represent the different choices by $\overline{\mathcal{S}_5}$, depicted in Figure 3.

Again, each vertex represents one way of associating the product and the edges represent homotopies between them. There are two sorts of two-dimensional faces: the pentagons clearly are copies of $\overline{\mathcal{S}_4}$, while the squares come from applying two associations in the two different orders. We think of such a face as $\overline{\mathcal{S}_3} \times \overline{\mathcal{S}_3}$, where $\overline{\mathcal{S}_3} = [0,1]$. Note that each face arises

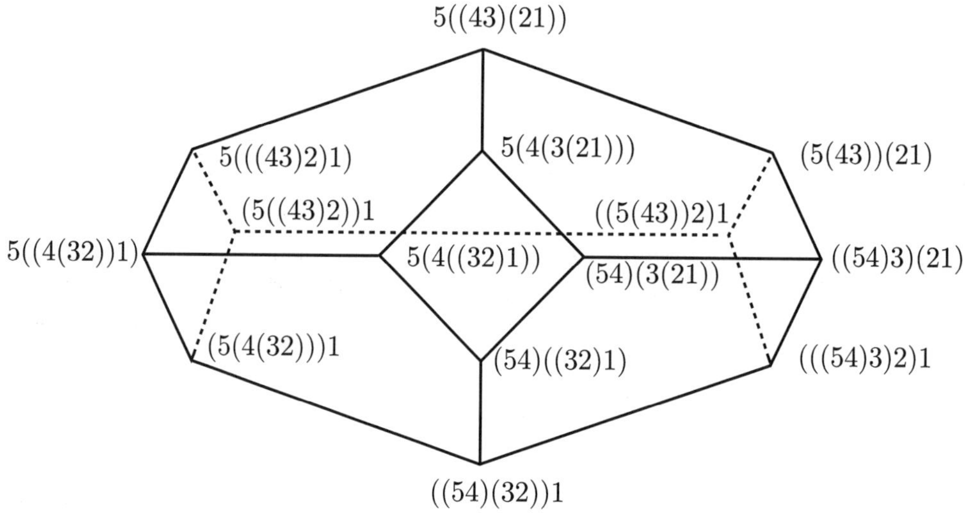

FIGURE 3

by choosing one set of brackets for $x_5 \cdots x_1$, excluding the trivial choices of $(x_5 \cdots x_1)$ and $x_5 \cdots (x_i) \cdots x_1$. For example, the upper left-hand front pentagon corresponds to ways of bracketing 5(4321), the left-hand square to ways of bracketing 5(432)1, and the right-hand upper front pentagon corresponds to ways of bracketing 534(21).

As before, this allows us to define a map

$$\partial \overline{\mathcal{S}_5} \times X^5 \to X.$$

For example, on the upper left-hand pentagon, this map takes the form

$$(s, x_5, \dots, x_1) \mapsto m_2(x_5, m_4(s, x_4, x_3, x_2, x_1))$$

while on the left-hand square, which we write as $\overline{\mathcal{S}_3} \times \overline{\mathcal{S}_3}$, we can take

$$((s,t), x_5, \dots, x_1) \mapsto m_3(s, x_5, m_3(t, x_4, x_3, x_2), x_1).$$

On the upper right-hand pentagon we have

$$(s, x_5, \dots, x_1) \mapsto m_4(s, x_5, x_4, x_3, m_2(x_2, x_1)).$$

Now since m_4 itself restricts to a composition of m_2 and m_3 on the boundary of $\overline{\mathcal{S}_4}$, one can check that these maps defined on each face are in fact compatible and hence yield a map

$$\partial \overline{\mathcal{S}_5} \times X^5 \to X.$$

The next "higher associativity" relation, if it exists, would be an extension of this map to a map

$$m_5 : \overline{\mathcal{S}_5} \times X^5 \to X.$$

Stasheff [**430**] constructed an infinite sequence of such "associahedra," $\overline{\mathcal{S}_2} = \{\text{pt}\}$, $\overline{\mathcal{S}_3} = [0, 1]$, $\overline{\mathcal{S}_4}, \dots$ such that the vertices of $\overline{\mathcal{S}_d}$ correspond to ways of bracketing completely $x_d \cdots x_1$, and the maximal proper faces of $\overline{\mathcal{S}_d}$ are in one-to-one correspondence with non-trivial ways of choosing a single set of brackets, the face corresponding to

$$x_d \cdots x_{p+q+1}(x_{p+q} \cdots x_{q+1})x_q \cdots x_1$$

being isomorphic to

$$\overline{\mathcal{S}_{d-p+1}} \times \overline{\mathcal{S}_p}.$$

One particular description of $\overline{\mathcal{S}_d}$ will be especially useful, namely as the moduli space of *ribbon trees*.

DEFINITION 8.1. A *(metric) ribbon tree* is a connected tree with a finite number of vertices and edges, with no bivalent vertices, with the additional data of a cyclic ordering of edges at each vertex and a length assigned to each edge in $(0, \infty]$.

Such a tree has, say, $d + 1$ external vertices, or leaves: these are the univalent vertices. We choose one of them and call it the outgoing vertex, and call the other external vertices incoming vertices. The cyclic ordering, along with this choice of outgoing vertex, allows us to label uniquely the incoming vertices $v_1, \ldots, v_d$ in a unique way compatible with the cyclic ordering. Think of the tree as oriented, with all edges pointing towards the outgoing vertex. We give an example in Figure 4, where the semi-circular arrows denote the cyclic ordering. If we forget the attached lengths, and all

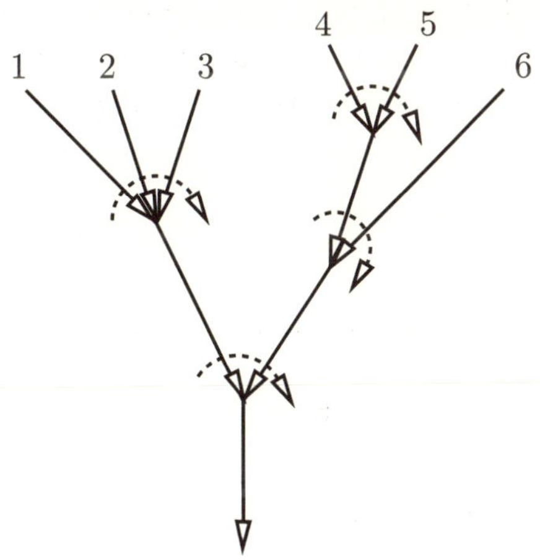

FIGURE 4

interior vertices are trivalent, then we can view the tree as telling us how to associate $x_d \cdots x_1$.

DEFINITION 8.2. $\overline{\mathcal{S}_d}$ is the moduli space of ribbon trees with $d+1$ external vertices, with each edge connected to an external vertex (an *external edge*) being of infinite length. This is a compactification of the moduli space $\mathcal{S}_d \subseteq \overline{\mathcal{S}_d}$ of ribbon trees such that all internal edges have finite length.

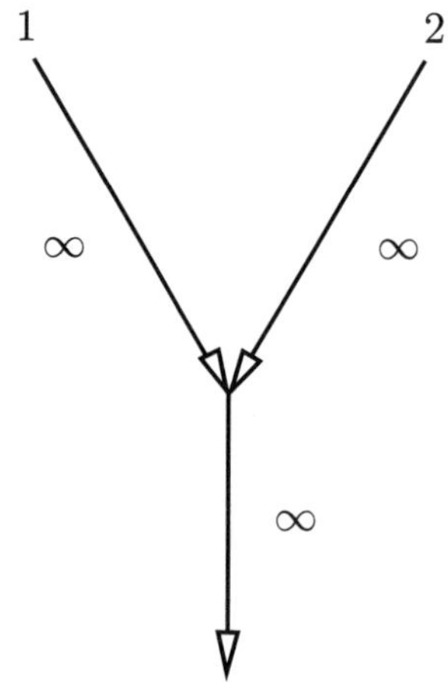

FIGURE 5

Clearly $\overline{\mathcal{S}}_2 = \{\text{pt}\}$ consists of the tree depicted in Figure 5. In general, a ribbon tree with $d+1$ external vertices and only trivalent interior vertices has $d-2$ interior edges, and each edge has a length in $(0,\infty]$. However, we can consider limiting cases where the length of an edge goes to 0, and think of this as contracting the corresponding edge. So any such trivalent tree with interior edge lengths in $[0,\infty]$ gives rise to a ribbon tree, and this then yields a cube $[0,\infty]^{d-2}$ inside $\overline{\mathcal{S}_d}$. Thus $\overline{\mathcal{S}_d}$ has a cubical subdivision with the cube indexed by combinatorial types of trivalent trees. For example, if $d=3$, we get the subdivision of $\overline{\mathcal{S}_3}$, an interval, as in Figure 6. For $d=4$ we get a

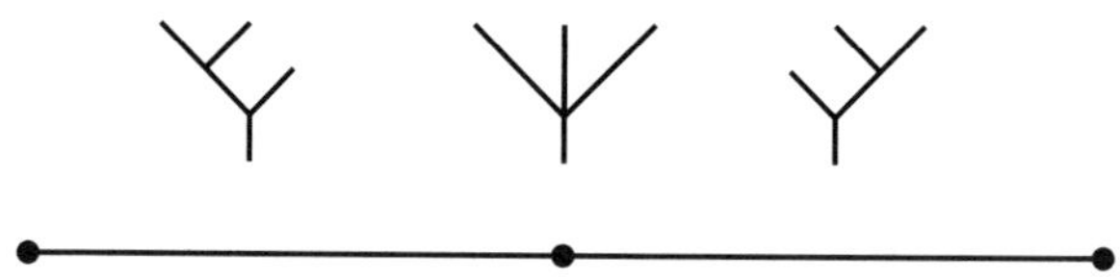

FIGURE 6

subdivision as depicted in Figure 7. The faces of $\overline{\mathcal{S}_d}$ now correspond to trees with edges with infinite length. In general, $\overline{\mathcal{S}_d}$ can be viewed as a polytope where each maximal proper face is given by a choice of one non-trivial set of brackets as before, i.e.,

$$x_d \cdots x_{p+q+1}(x_{p+q} \cdots x_{q+1})x_q \cdots x_1 \tag{8.1}$$

which corresponds to trees described by giving $(T,T') \in \overline{\mathcal{S}_{d-p+1}} \times \overline{\mathcal{S}_p}$: this gives a tree in $\overline{\mathcal{S}_d}$ by identifying the outgoing edge of T' with the $(q+1)$st incoming edge of T, and giving this edge infinite length, as depicted in Figure 8.

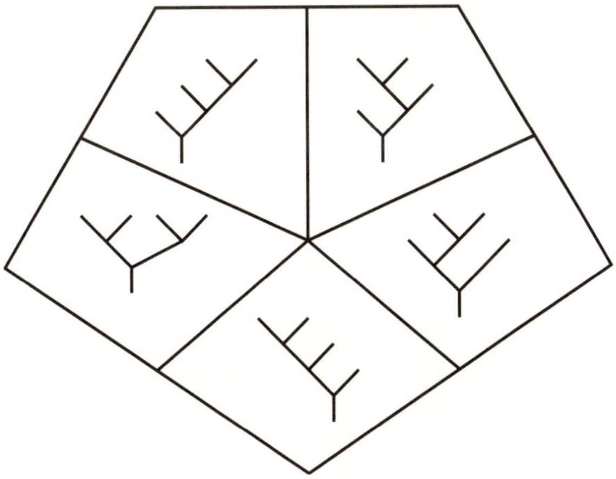

FIGURE 7

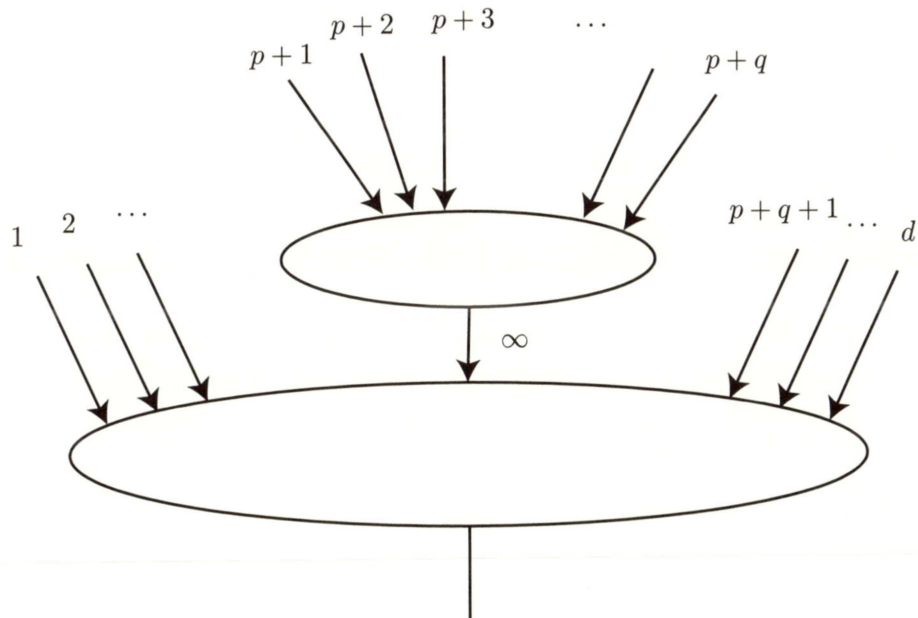

FIGURE 8

We can now give the general definition of an A_∞ space: this is a space X with a sequence of maps for $d \geq 2$,

$$m_d : \overline{\mathcal{S}_d} \times X^d \to X$$

along with compatabilities on the boundary of $\overline{\mathcal{S}_d}$ given by the combinatorial structure of the boundary generalizing the examples given above for small

d: when m_d is restricted to the boundary face determined by the bracketing (8.1), we obtain the map

$$\overline{\mathcal{S}_{d-p+1}} \times \overline{\mathcal{S}_p} \times X^d \to X$$

given by

$$(T, T', x_d, \ldots, x_1) = m_{d-p+1}(T, x_d, \ldots, x_{p+q+1}, \\ m_p(T', x_{p+q}, \ldots, x_{q+1}), x_q, \ldots, x_1).$$

REMARK 8.3. Another interpretation for $\overline{\mathcal{S}_d}$ is as a subset of the real locus of $\overline{\mathcal{M}}_{0,d+1}$, the moduli space of $(d+1)$-pointed stable curves of genus 0. The real locus of $\mathcal{M}_{0,d+1}$ consists of Riemann spheres with distinct marked points $x_1, \ldots, x_d, y$ on the equator. Then $\mathcal{S}_d$ can be identified with the subset where those points occur in cyclic order, and $\overline{\mathcal{S}_d}$ with the closure of this set. One can also view this moduli space as a moduli space of $(d+1)$-pointed disks with the marked points $x_1, \ldots, x_d, y$ appearing in cyclic order on the boundary of the disk: at the boundary $\overline{\mathcal{S}_d} \setminus \mathcal{S}_d$ disks bubble off, see, e.g., Figure 9 in the $d = 4$ case.

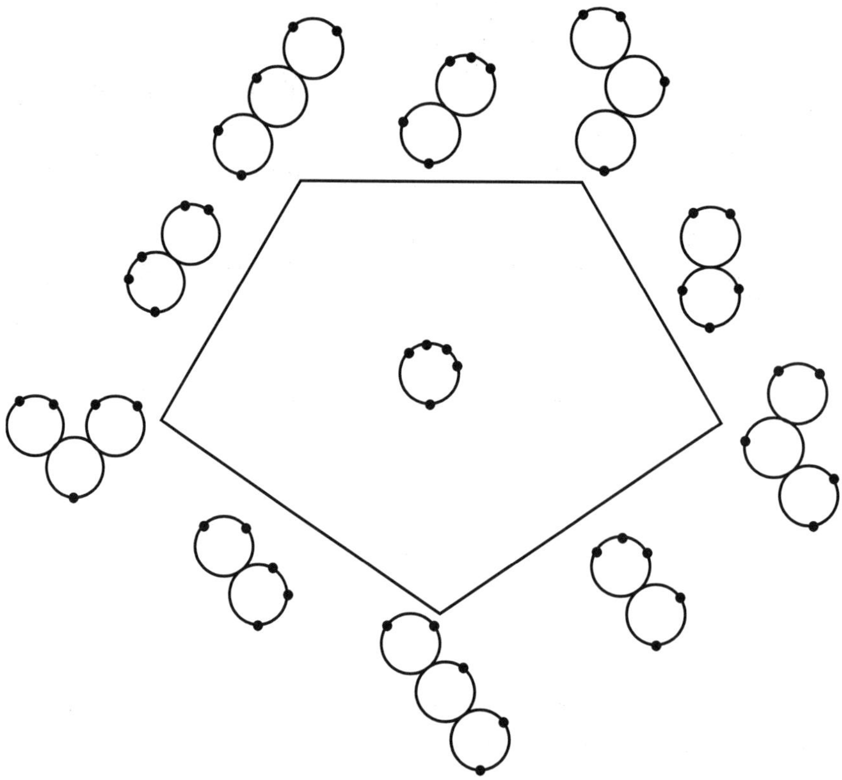

FIGURE 9. Compare with the trees in Figure 7.

8.1.2. A_∞-algebras and categories. We now define the algebraic and category theoretic analogues of the A_∞ spaces defined above. With the above description of A_∞ spaces, these definitions should look more natural. The basic idea is that an A_∞-algebra is a non-associative algebra with higher multiplications measuring the failure of associativity.

We now fix a ground field k in all that follows.

DEFINITION 8.4. A (non-unital) A_∞-algebra A is a $\mathbb{Z}$-graded k-vector space

$$A = \bigoplus_{p \in \mathbb{Z}} A^p$$

with graded k-linear maps, for $d \geq 1$,

$$m_d : A^{\otimes d} \to A$$

of degree $2 - d$ satisfying for each $d \geq 1$ the relation

(8.2)
$$\sum_{\substack{1 \leq p \leq d \\ 0 \leq q \leq d-p}} (-1)^{\deg a_1 + \cdots + \deg a_q - q} m_{d-p+1}(a_d, \ldots, a_{p+q+1}, m_p(a_{p+q}, \ldots, a_{q+1}), a_q, \ldots, a_1) = 0.$$

Let's try to unravel this definition for small d. First, $d = 1$. We have m_1, a map of degree 1, and (8.2) becomes

$$m_1(m_1(a_1)) = 0,$$

i.e., m_1 turns A into a complex.

The map m_2 is degree 0, i.e., $\deg m_2(a_2, a_1) = \deg a_2 + \deg a_1$. We should view m_2 as a multiplication map. Then for $d = 2$, the choices for (p, q) in (8.2) are $(1, 0)$, $(1, 1)$ and $(2, 0)$, and (8.2) becomes

$$m_2(a_2, m_1(a_1)) + (-1)^{\deg a_1 - 1} m_2(m_1(a_2), a_1) + m_1(m_2(a_2, a_1)) = 0. \tag{8.3}$$

Up to sign, this says m_1 is a graded derivation with respect to the multiplication operation. More precisely, if we set $\partial(a) = (-1)^{\deg a} m_1(a)$, $a_2 \cdot a_1 = (-1)^{\deg a_1} m_2(a_2, a_1)$, then (8.3) can be rewritten as

$$\partial(a_2 \cdot a_1) = (\partial a_2) \cdot a_1 + (-1)^{\deg a_2} a_2 \cdot (\partial a_1). \tag{8.4}$$

If multiplication were associative, this would yield what's known as a *differential graded algebra*. However, let's look at what (8.2) gives for $d = 3$. Here the possible pairs are

$$(p, q) = (1, 0), (1, 1), (1, 2), (2, 0), (2, 1), (3, 0),$$

and we get

(8.5)

$$\begin{aligned} m_3(a_3,a_2,m_1(a_1)) &+ (-1)^{\deg a_1-1}m_3(a_3,m_1(a_2),a_1) \\ &+(-1)^{\deg a_1+\deg a_2-2}m_3(m_1(a_3),a_2,a_1) \\ &+m_2(a_3,m_2(a_2,a_1))+(-1)^{\deg a_1-1}m_2(m_2(a_3,a_2),a_1) \\ &+m_1(m_3(a_3,a_2,a_1))=0. \end{aligned}$$

The fourth and fifth terms are

$$(-1)^{\deg a_2}a_3\cdot(a_2\cdot a_1)-(-1)^{\deg a_2}(a_3\cdot a_2)\cdot a_1,$$

so this expression tells us by how much multiplication fails to be associative. Note in particular that if $H^*(A)$ denotes the cohomology of A with respect to ∂ (or equivalently m_1), then (8.4) tells us that if $a_2,a_1\in A$ with $\partial a_2,\partial a_1=0$, and $[a_2],[a_1]$ are elements in $H^*(A)$ represented by a_2 and a_1, then the multiplication $[a_2]\cdot[a_1]=[a_2\cdot a_1]$ is well-defined. If $\partial a_i=0$ for $i=1,2,3$, then (8.5) gives

$$a_3\cdot(a_2\cdot a_1)-(a_3\cdot a_2)\cdot a_1=\pm\partial(m_3(a_3,a_2,a_1)),$$

so multiplication on $H^*(A)$ is associative.

For arbitrary d, it is clear that the A_∞ relation we obtain is a sum of two sorts of terms: the terms which arise for $1<p<d$ correspond to non-trivial choices of putting one set of brackets in $a_d\cdots a_1$, and hence these terms are in one-to-one correspondence with the codimension one faces of $\overline{S}_d$. Furthermore, the terms arising from $p=1$ and $p=d$ give a contribution

$$\pm m_d(a_d,\ldots,m_1(a_1))\pm\cdots\pm m_d(m_1(a_d),\cdots,a_1)\pm m_1(m_d(a_d,\ldots,a_1)).$$

This gives the relationship between m_d and the lower-order multiplications. This should be seen as analogous to the definition of A_∞ spaces, in which m_d gave an extension to $\overline{S}_d$ of a map on $\partial\overline{S}_d$ defined by lower-order multiplications.

We also wish to define morphisms of A_∞-algebras.

DEFINITION 8.5. A morphism of A_∞-algebras $f:A\to B$ is a family

$$f_d:A^{\otimes d}\to B$$

of graded maps of degree $1-d$ such that for each $d\geq 1$,

$$\begin{aligned} &\sum_r\sum_{s_1,\ldots,s_r} m_r^B(f_{s_r}(a_d,\ldots,a_{d-s_r+1}),\ldots,f_{s_1}(a_{s_1},\ldots,a_1)) \\ &\qquad=\sum_{p,q}(-1)^{\deg a_1+\cdots+\deg a_q-q}\cdot \\ &\qquad\qquad f_{d-p+1}(a_d,\ldots,a_{p+q+1},m_p^A(a_{p+q},\ldots a_{q+1}),a_q,\ldots,a_1), \end{aligned} \tag{8.6}$$

where the sum on the left is over all $r \geq 1$ and partitions $s_1 + \cdots + s_r = d$, and the sum on the right is over $1 \leq p \leq d$, $0 \leq q \leq d-p$.

A morphism is a *quasi-isomorphism* if f_1 is a quasi-isomorphism in the usual sense of complexes.

Again, let's try to unwind this definition. First, if $f_n = 0$ for $n \geq 2$, we just have the relation, for each d,

$$m_d^B(f_1(a_d), \ldots, f_1(a_1)) = f_1(m_d^A(a_d, \ldots, a_1)). \tag{8.7}$$

If this is the case, the morphism is called *strict*.

If the morphism is not strict, the idea is that the higher f_n's measure the failure of (8.7). In particular, for small d, (8.6) gives:

$$d = 1: \quad m_1^B(f_1(a_1)) = f_1(m_1^A(a_1)),$$

so f_1 is a morphism of chain complexes;

(8.8)

$$d = 2: m_2^B(f_1(a_2), f_1(a_1)) + m_1^B(f_2(a_2, a_1)) = \\ f_2(a_2, m_1^A(a_1)) + (-1)^{\deg a_1 - 1} f_2(m_1^A(a_2), a_1) + f_1(m_2^A(a_2, a_1)).$$

We can compose morphisms of A_∞-algebras $f_d : A^{\otimes d} \to B$ and $g_d : B^{\otimes d} \to C$ by

$$(g \circ f)_d(a_d, \ldots, a_1) = \sum_r \sum_{s_1, \ldots, s_r} g_r(f_{s_r}(a_d, \ldots, a_{d-s_r+1}), \ldots, f_{s_1}(a_{s_1}, \ldots, a_1)),$$

where $r \geq 1$ and $s_1 + \cdots + s_r = d$.

We now generalize the above concepts easily to categories.

DEFINITION 8.6. A (non-unital) A_∞-category $\mathcal{A}$ consists of a collection of objects $\operatorname{Ob}\mathcal{A}$, a $\mathbb{Z}$-graded k-vector space $\operatorname{Hom}_{\mathcal{A}}(X_0, X_1)$ for any $X_0, X_1 \in \operatorname{Ob}\mathcal{A}$, and for every $d \geq 1$, k-linear composition maps

$$m_d : \operatorname{Hom}_{\mathcal{A}}(X_{d-1}, X_d) \otimes \cdots \otimes \operatorname{Hom}_{\mathcal{A}}(X_0, X_1) \to \operatorname{Hom}_{\mathcal{A}}(X_0, X_d)$$

of degree $2 - d$, satisfying

$$\sum_{p,q} (-1)^{\deg a_1 + \cdots + \deg a_q - q} m_{d-p+1}(a_d, \ldots, a_{p+q+1}, \\ m_p(a_{p+q}, \ldots, a_{q+1}), a_q, \ldots, a_1) = 0.$$

An A_∞-functor between two A_∞-categories consists of a map

$$F : \operatorname{Ob}\mathcal{A} \to \operatorname{Ob}\mathcal{B}$$

and maps

$$F_d : \operatorname{Hom}_{\mathcal{A}}(X_{d-1}, X_d) \otimes \cdots \otimes \operatorname{Hom}_{\mathcal{A}}(X_0, X_1) \to \operatorname{Hom}_{\mathcal{B}}(F(X_0), F(X_d))$$

of degree $1-d$ satisfying

$$\sum_{r}\sum_{s_1,\ldots,s_r} m_r^{\mathcal{B}}(F_{s_r}(a_d,\ldots,a_{d-s_r+1}),\ldots,F_{s_1}(a_{s_1},\ldots,a_1))$$
$$=\sum_{p,q}(-1)^{\deg a_1+\cdots+\deg a_q-q}.$$
$$F_{d-p+1}(a_d,\ldots,a_{p+q+1},m_p^{\mathcal{A}}(a_{p+q},\ldots,a_{q+1}),a_q,\ldots,a_1),$$

with the indices as in Definition 8.5.

F is a *quasi-isomorphism* if

$$F_1:\mathrm{Hom}_{\mathcal{A}}(X_0,X_1)\to\mathrm{Hom}_{\mathcal{B}}(F(X_0),F(X_1))$$

is a quasi-isomorphism for each X_0, X_1.

F is *strict* if $F_d=0$ for $d>1$.

Note that given an A_∞-category $\mathcal{A}$, one obtains several other "categories": $H^*(\mathcal{A})$ is the "category" whose objects are $\mathrm{Ob}\,\mathcal{A}$ and whose morphisms are

$$\mathrm{Hom}_{H^*(\mathcal{A})}(X_0,X_1)=H^*(\mathrm{Hom}_{\mathcal{A}}(X_0,X_1)).$$

This is almost a genuine category, in that composition is now associative but there are not necessarily any identity morphisms. We also have the category $H^0(\mathcal{A})$, with objects $\mathrm{Ob}\,\mathcal{A}$ and morphisms $\mathrm{Hom}_{H^0(\mathcal{A})}(X_0,X_1)=H^0(\mathrm{Hom}_{\mathcal{A}}(X_0,X_1))$.

8.2. Examples and constructions

8.2.1. Coherent sheaves. Here we'll consider the easiest example of an A_∞-category to describe, and use it to motivate a general construction of A_∞ structures.

Let X be a non-singular algebraic variety over a field k, with an affine open cover $\mathcal{U}=\{U_i\}_{i\in I}$, where I is an ordered index set. Denote by $\mathrm{D}^b_\infty(X)$ the category whose objects are bounded complexes of locally free sheaves. To define morphisms, first define the complex $\mathcal{H}om^\bullet_{\mathcal{O}_X}(\mathcal{E}^\bullet,\mathcal{F}^\bullet)$, with

$$\mathcal{H}om^q_{\mathcal{O}_X}(\mathcal{E}^\bullet,\mathcal{F}^\bullet)=\bigoplus_m \mathcal{H}om_{\mathcal{O}_X}(\mathcal{E}^m,\mathcal{F}^{m+q})$$

graded maps of degree q, and

$$\delta:\mathcal{H}om^q_{\mathcal{O}_X}(\mathcal{E}^\bullet,\mathcal{F}^\bullet)\to\mathcal{H}om^{q+1}_{\mathcal{O}_X}(\mathcal{E}^\bullet,\mathcal{F}^\bullet)$$

taking $(f_m:\mathcal{E}^m\to\mathcal{F}^{m+q})_m$ to

$$(\delta(f))_m=d_{\mathcal{F}}\circ f_m-(-1)^q f_{m+1}\circ d_{\mathcal{E}}:\mathcal{E}^m\to\mathcal{F}^{m+q+1}.$$

Next, we take the total complex of the Čech complex of $\mathcal{H}om^\bullet_{\mathcal{O}_X}(\mathcal{E}^\bullet,\mathcal{F}^\bullet)$:

$$\mathrm{Hom}^n_{\mathrm{D}^b_\infty(X)}(\mathcal{E}^\bullet,\mathcal{F}^\bullet)=\bigoplus_{p+q=n}\check{C}^p(\mathcal{U},\mathcal{H}om^q_{\mathcal{O}_X}(\mathcal{E}^\bullet,\mathcal{F}^\bullet)).$$

Here

$$\check{C}^p(\mathcal{U},\mathcal{G}) = \bigoplus_{i_0<\cdots<i_p} \Gamma(U_{i_0\cdots i_p},\mathcal{G}),$$

with $U_{i_0\cdots i_p} = U_{i_0}\cap\cdots\cap U_{i_p}$. We can define a differential ∂ on the complex $\mathrm{Hom}^\bullet_{\mathrm{D}^b_\infty(X)}(\mathcal{E}^\bullet,\mathcal{F}^\bullet)$, via the total differential

$$\partial = d + (-1)^p\delta,$$

where $d : \check{C}^p(\mathcal{U},\mathcal{G}) \to \check{C}^{p+1}(\mathcal{U},\mathcal{G})$ is the usual Čech differential, defined by

$$(d(\alpha_{i_0\cdots i_p}))_{i_0\cdots i_{p+1}} = \sum_{j=0}^{p}(-1)^j\alpha_{i_0\cdots\hat{\imath}_j\cdots i_{p+1}}|_{U_{i_0\cdots i_{p+1}}}.$$

This complex is related to the usual notion of morphism in the derived category by

$$H^i(\mathrm{Hom}^\bullet_{\mathrm{D}^b_\infty(X)}(\mathcal{E}^\bullet,\mathcal{F}^\bullet)) = \mathrm{Ext}^i(\mathcal{E}^\bullet,\mathcal{F}^\bullet).$$

(See §4.5.) For the moment, we will avoid taking cohomology, and instead view $\mathrm{D}^b_\infty(X)$ as an A_∞-category, defining, with the sign conventions we have chosen,

$$m_1(f) = (-1)^{\deg f}\partial f,$$

and composition: with

$$\begin{aligned} f = (f_{n,i_0\cdots i_p}) &\in \check{C}^p(\mathcal{U},\mathcal{H}om^q_{\mathscr{O}_X}(\mathcal{E}^\bullet,\mathcal{F}^\bullet)) \\ g = (g_{n',j_0\cdots j_{p'}}) &\in \check{C}^{p'}(\mathcal{U},\mathcal{H}om^{q'}_{\mathscr{O}_X}(\mathcal{F}^\bullet,\mathcal{G}^\bullet)) \end{aligned}$$

the composition $g\circ f$ is given by

$$\begin{aligned} (g\circ f)_{n,i_0\cdots i_{p+p'}} = ((-1)^{q'p}g_{n+q,i_0\cdots i_{p'}}\circ f_{n,i_{p'}\cdots i_{p+p'}})|_{U_{i_0\cdots i_{p+p'}}} :& \\ \mathcal{E}^n|_{U_{i_0\cdots i_{p+p'}}} \to \mathcal{G}^{n+q+q'}|_{U_{i_0\cdots i_{p+p'}}}&. \end{aligned}$$

Then define

$$m_2(g,f) = (-1)^{\deg f}g\circ f.$$

One easily checks (8.3), or equivalently that (8.4) is satisfied. Furthermore, because composition of morphisms of sheaves is already associative, we can take $m_n = 0$ for $n\geq 3$, and we obtain an A_∞-category.

Of course, this is stronger than an A_∞-category: it is in fact a DG-category (differential graded), which means the higher multiplication maps are trivial. So why should this be what we want, as we have neither the correct morphisms nor an interesting A_∞ structure?

The answer comes from the following algebraic statement, originally due to Kadeishvili [**279**], and which we will present a version of in the category context, due to Kontsevich and Soibelman [**311**].

THEOREM 8.7. *Given an A_∞-algebra A, there is an A_∞-algebra structure on $H^*(A)$ with $m_1 = 0$ and a quasi-isomorphism of A_∞-algebras $i : H^*(A)\to A$.*

We will explain a more general construction for an arbitrary A_∞-category $\mathcal{A}$.

We will need to make a choice, for every pair $X, Y \in \mathrm{Ob}(\mathcal{A})$, of a projector, i.e., a chain map (of degree 0)

$$\Pi : \mathrm{Hom}_{\mathcal{A}}(X, Y) \to \mathrm{Hom}_{\mathcal{A}}(X, Y)$$

with

$$\Pi^2 = \Pi,$$

and a chain homotopy (of degree -1)

$$H : \mathrm{Hom}_{\mathcal{A}}(X, Y) \to \mathrm{Hom}_{\mathcal{A}}(X, Y)[-1]$$

satisfying

$$\Pi - 1 = m_1 H + H m_1.$$

This implies in particular that the image of Π is a complex homotopic to

$$\mathrm{Hom}_{\mathcal{A}}(X, Y);$$

in particular, the interesting case for us will be when the image of Π is in fact isomorphic to $H^*(\mathrm{Hom}_{\mathcal{A}}(X, Y))$, but the construction works in any event. We obtain a new A_∞-category, $\Pi\mathcal{A}$, with

$$\mathrm{Ob}(\Pi\mathcal{A}) = \mathrm{Ob}(\mathcal{A})$$

and

$$\mathrm{Hom}_{\Pi\mathcal{A}}(X, Y) = \Pi\big(\mathrm{Hom}_{\mathcal{A}}(X, Y)\big).$$

Then we set

(1) $m_1^{\Pi\mathcal{A}} = m_1^{\mathcal{A}}$ (which makes sense since Π is a chain map).
(2) $m_2^{\Pi\mathcal{A}} = \Pi \circ m_2^{\mathcal{A}}$.
(3) $m_d^{\Pi\mathcal{A}} = \sum_T m_{d,T}$, where T runs over all ribbon trees (with no lengths attached to the edges) with d inputs and 1 output. We define $m_{d,T}$ by attaching m_n to each interior vertex of valency $n+1$, H to each interior edge, and Π to the outgoing vertex, and define $m_{d,T}$ to be the obvious sequence of compositions coming from this diagram. For example, for T the tree depicted in Figure 10, we get

$$m_{d,T}(a_5, \ldots, a_1) = \Pi\big(m_2(a_5, H(m_3(a_4, a_3, H(m_2(a_2, a_1)))))\big).$$

Furthermore, we can define an A_∞ functor $i : \Pi\mathcal{A} \to \mathcal{A}$. Here i is the identity on the level of objects, i_1 is just the inclusion of chain complexes, and we define i_d similarly to m_d, writing

$$i_d = \sum_T i_{d,T},$$

where T runs over the same trees, and $i_{d,T}$ is defined by replacing the output Π with H.

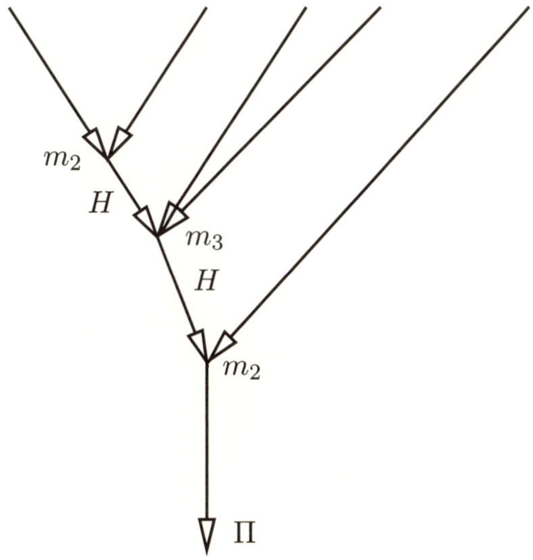

FIGURE 10

Let's see why this works. First, the relationship for m_2, remembering $m_1^{\Pi\mathcal{A}} = m_1$:

$$\begin{aligned} & m_2^{\Pi\mathcal{A}}(a_2, m_1(a_1)) + (-1)^{\deg a_1 - 1} m_2^{\Pi\mathcal{A}}(m_1(a_2), a_1) + m_1(m_2^{\Pi\mathcal{A}}(a_2, a_1)) \\ =& \Pi\big(m_2(a_2, m_1(a_1)) + (-1)^{\deg a_1 - 1} m_2(m_1(a_2), a_1) + m_1(\Pi(m_2(a_2, a_1)))\big) \\ =& 0, \end{aligned}$$

by the relation for m_2, and for the last term, the fact that Π commutes with m_1 and $\Pi^2 = \Pi$.

It is a good exercise to check that $m_d^{\Pi\mathcal{A}}$ for $d \geq 3$ satisfies the A_∞ relations, and that $\{i_d\}$ defines an A_∞ functor. We will sketch the argument here as given in [**311**], leaving to the reader the rather arduous task of checking the signs.

Define a new map

$$\hat{m}_d^{\Pi\mathcal{A}} : \mathrm{Hom}_{\Pi\mathcal{A}}(X_{d-1}, X_d) \otimes \cdots \otimes \mathrm{Hom}_{\Pi\mathcal{A}}(X_0, X_1) \to \mathrm{Hom}_{\Pi\mathcal{A}}(X_0, X_d)[3-n]$$

via

$$\hat{m}_d^{\Pi\mathcal{A}}(a_d, \ldots, a_1) = \sum_{(T,e)} \pm \hat{m}_{(T,e)}(a_d, \ldots, a_1),$$

where the sum runs over all trees T as before, along with a choice of edge of T. Then $\hat{m}_{(T,e)}$ is defined as before, except that if e is an internal edge, the label H on that edge is replaced by $Hm_1 + m_1H = \Pi - 1$, and if e is an external edge, an m_1 is added to that that edge, see Figure 11. A sign must be chosen for each tree which depends on e, but we leave it to the reader to determine the sign which makes the following argument work.

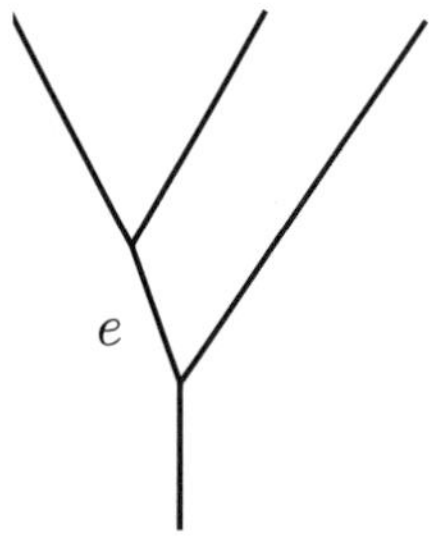

$$\Rightarrow \Pi m_2(a_3, (m_1 H + H m_1) m_2(a_2, a_1))$$

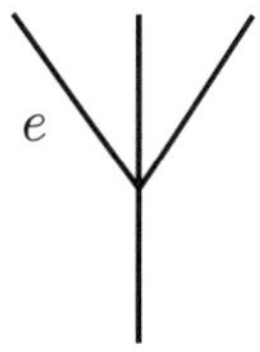

$$\Rightarrow \Pi m_3(a_3, a_2, m_1(a_1))$$

FIGURE 11

Now we calculate $\hat{m}_d^{\Pi\mathcal{A}}$ in two different ways. First we use the expression $\Pi - 1$. We then get three different sorts of contributions to $\hat{m}_d^{\Pi\mathcal{A}}$:

$$\hat{m}_d^{\Pi\mathcal{A}} = m_d^{\Pi\mathcal{A}, m_1} + m_d^{\Pi\mathcal{A}, \Pi} - m_d^{\Pi\mathcal{A}, 1}.$$

Here $m_d^{\Pi\mathcal{A}, m_1}$ consists of those contributions coming from pairs (T, e) with e an external edge. The two remaining terms are defined via a sum over pairs (T, e) for e an internal edge, with the H on the edge replaced with either Π or 1 in the two cases. The sign contribution is the same as in the definition of $\hat{m}_d^{\Pi\mathcal{A}}$.

On the other hand, we can use the expression $m_1 H + H m_1$ on each interior edge. Thus each pair (T, e) for e an internal edge produces two contributions. Now consider any internal vertex v. For every edge containing v, there is then exactly one term appearing in $\hat{m}_d^{\Pi\mathcal{A}}$ which has an m_1 on that edge adjacent to v as depicted in Figure 12. (Note that if an edge is external, there is no attached H.) We can then, for such a vertex, apply the A_∞ relation for $m_n^{\mathcal{A}}$ and rewrite this sum (with signs, which are in fact determined by the signs in the A_∞ relation for m_n) as a sum of terms of the form depicted in Figure 13. In this way, one sees that $\hat{m}_d^{\Pi\mathcal{A}}$ can be written as a sum of contributions from trees appearing in $m_d^{\Pi\mathcal{A}, 1}$; with care taken on the signs, one in fact sees that

$$\hat{m}_d^{\Pi\mathcal{A}} = -m_d^{\Pi\mathcal{A}, 1}.$$

Thus

$$m_d^{\Pi\mathcal{A}, m_1} + m_d^{\Pi\mathcal{A}, \Pi} = 0.$$

This turns out to be precisely the A_∞ relation for $m_d^{\Pi\mathcal{A}}$.

As an exercise, see how this works for m_3!

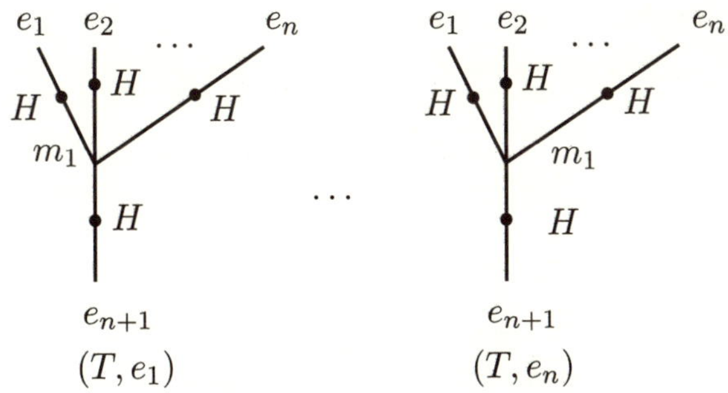

FIGURE 12

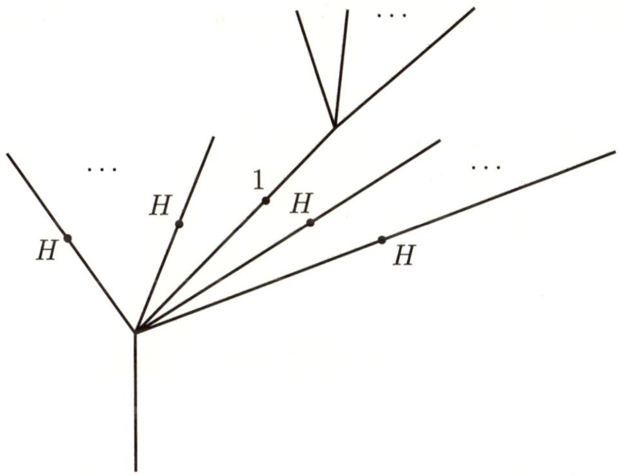

FIGURE 13

The argument that $\{i_d\}_{d\geq 1}$ form an A_∞ functor is similar. Let's check this for degrees 1 and 2, and again sketch the general case. For $d = 1$, we just need

$$m_1^{\mathcal{A}}(i_1(a_1)) = i_1(m_1^{\Pi\mathcal{A}}(a_1)),$$

which is obvious with i_1 the inclusion. For $d = 2$, observe that the definition of i_d gives

$$i_2(a_2, a_1) = Hm_2^{\mathcal{A}}(a_2, a_1).$$

Consider the expression

$$Hm_2^{\mathcal{A}}(a_2, m_1(a_1)) + (-1)^{\deg a_1 - 1} Hm_2^{\mathcal{A}}(m_1(a_2), a_1) + Hm_1(m_2^{\mathcal{A}}(a_2, a_1)).$$

Using the relationship for $m_2^{\mathcal{A}}$, this is in fact 0. However, $Hm_1 = -m_1H + \Pi - 1$, so we get, using the definition of i_2 for the first three terms, and the definition of $m_2^{\Pi A}$ for the fourth,

$$0 = i_2(a_2, m_1^{\Pi A}(a_1)) + (-1)^{\deg a_1 - 1} i_2(m_1^{\Pi A}(a_2), a_1) - m_1^A(i_2(a_2, a_1)) + i_1 m_2^{\Pi A}(a_2, a_1) - m_2^A(i_1(a_2), i_1(a_1)),$$

which reduces to (8.8).

For the general case, we proceed pretty much as before. Define

$$\hat{\imath}_d = \sum_{(T,e)} \pm \hat{\imath}_{(T,e)},$$

where as before the sum is over trees with d inputs and a choice of edge, and $\hat{\imath}_{(T,e)}$ is defined in the same way as $\hat{m}_{(T,e)}^{\Pi A}$, except that we replace the operator Π attached to the output vertex with H. The choice of signs is the same as in the definition of $\hat{m}_d^{\Pi A}$. Using $\Pi - 1$ attached to the interior edges, we can write

$$\hat{\imath}_d = i_d^{m_1} + i_d^{\Pi} - i_d^1,$$

where $i_d^{m_1}$, i_d^{Π} and i_d^1 are defined as precisely as in the m_d argument, again with H on the output vertex rather than Π. Using $m_1 H + H m_1$ instead of $\Pi - 1$ to calculate $\hat{\imath}_d$ gives as before

$$\hat{\imath}_d = -i_d^1,$$

and hence we have

$$i_d^{m_1} + i_d^{\Pi} = 0.$$

Now i_d^{Π} can be seen to produce an appropriately signed

$$\sum_{\substack{1<p<d \\ 0 \le q \le d-p}} \pm i_{d-p+1}(a_d, \ldots, a_{p+q+1}, m_p^{\Pi A}(a_{p+q}, \ldots, a_{q+1}), a_p, \ldots, a_1).$$

(Note the sum excludes the cases $p = 1, d$.) Those terms in $i_d^{m_1}$ with the selected edge being an incoming edge contribute the $p = 1$ terms:

$$\sum_{\substack{p=1 \\ 0 \le q \le d-p}} \pm i_{d-p+1}(a_d, \ldots, a_{q+2}, m_1^{\Pi A}(a_{q+1}), a_q, \ldots, a_1).$$

To get the remaining terms, consider the terms in $i_d^{m_1}$ with the selected edge being the outgoing edge: these give contributions $H m_1(\cdots)$. Using $H m_1 = -m_1 H + \Pi - 1$, one sees that the contribution from these terms in $i_d^{m_1}$ coming from $-m_1 H$ is $-m_1^A(i_d(a_d, \ldots, a_1))$, that coming from Π is $i_1 m_d^{\Pi A}(a_d, \ldots, a_1)$, and that coming from -1 is slightly more complicated: trees as depicted in Figure 14 contribute

$$-m_d^A(i_{s_r}(a_d, \ldots, a_{d-s_r+1}), \ldots, i_{s_1}(a_{s_1}, \ldots, a_1)).$$

This accounts for all terms in the relation for i_d, and again care with the signs shows the i_d relation.

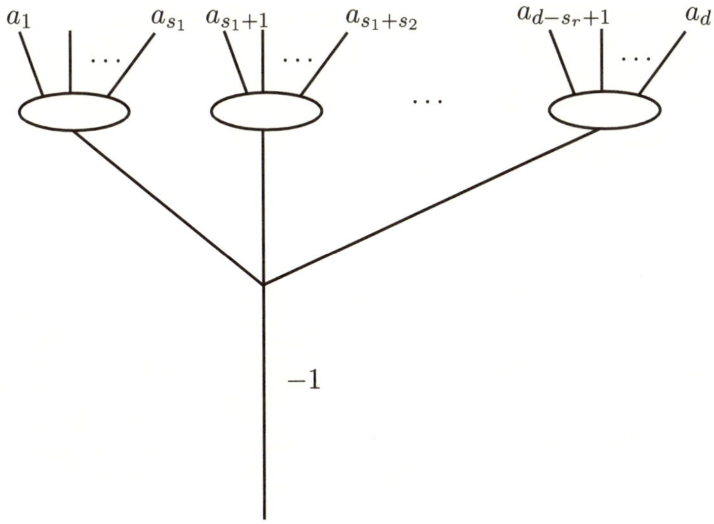

FIGURE 14

It is clear that $\{i_d\}_{d\geq 1}$ is a quasi-isomorphism of A_∞-categories, as this is just a condition on i_1, which is satisfied by the existence of the chain homotopy H.

We now return to the category $\mathrm{D}^b_\infty(X)$. The point of the above construction is that we can put an A_∞ structure on $H^*(\mathrm{D}^b_\infty(X))$. This can be done by choosing, for each $\mathcal{E}^\bullet$, $\mathcal{F}^\bullet$, an inclusion i of the complex $\mathrm{Ext}^\bullet(\mathcal{E}^\bullet, \mathcal{F}^\bullet)$ (the complex which is $H^i(\mathrm{Hom}^\bullet_{\mathrm{D}^b_\infty(X)}(\mathcal{E}^\bullet, \mathcal{F}^\bullet))$ in degree i with trivial boundary map) into $\mathrm{Hom}^\bullet_{\mathrm{D}^b_\infty(X)}(\mathcal{E}^\bullet, \mathcal{F}^\bullet)$ which is a quasi-isomorphism, along with choices of projector Π from $\mathrm{Hom}^\bullet_{\mathrm{D}^b_\infty(X)}(\mathcal{E}^\bullet, \mathcal{F}^\bullet)$ onto $i(\mathrm{Ext}^\bullet(\mathcal{E}^\bullet, \mathcal{F}^\bullet))$ and homotopy H. Then we obtain via the above construction an A_∞ structure on $H^*(\mathrm{D}^b_\infty(X))$. The latter category is equal to $\mathrm{D}^b(X)$ as an ordinary category, so this gives an A_∞ structure on $\mathrm{D}^b(X)$. The differentials m_1 are zero, but higher multiplication maps are non-trivial. In addition, this A_∞ structure on $\mathrm{D}^b(X)$ provides an A_∞-category quasi-isomorphic to $\mathrm{D}^b_\infty(X)$. Thus the A_∞ structure on $\mathrm{D}^b_\infty(X)$ should be viewed as highly non-trivial, even though the higher multiplication maps we initially defined are trivial.

EXAMPLE 8.8. Here we will give the simplest possible example of a non-trivial m_3 in $\mathrm{D}^b(\mathbb{P}^1)$, arising in

$$m_3 : \mathrm{Ext}^1(\mathcal{O}_{\mathbb{P}^1}(2), \mathcal{O}_{\mathbb{P}^1}) \otimes \mathrm{Hom}(\mathcal{O}_{\mathbb{P}^1}(1), \mathcal{O}_{\mathbb{P}^1}(2)) \otimes \mathrm{Hom}(\mathcal{O}_{\mathbb{P}^1}, \mathcal{O}_{\mathbb{P}^1}(1)) \longrightarrow \mathrm{Hom}(\mathcal{O}_{\mathbb{P}^1}, \mathcal{O}_{\mathbb{P}^1}).$$

To carry out the calculation, we will need Čech resolutions for

$$\begin{aligned}
&\mathcal{H}om(\mathscr{O}_{\mathbb{P}^1}, \mathscr{O}_{\mathbb{P}^1}(1)) = \mathcal{H}om(\mathscr{O}_{\mathbb{P}^1}(1), \mathscr{O}_{\mathbb{P}^1}(2)) = \mathscr{O}_{\mathbb{P}^1}(1)\\
&\mathcal{H}om(\mathscr{O}_{\mathbb{P}^1}, \mathscr{O}_{\mathbb{P}^1}(2)) = \mathscr{O}_{\mathbb{P}^1}(2)\\
&\mathcal{H}om(\mathscr{O}_{\mathbb{P}^1}(2), \mathscr{O}_{\mathbb{P}^1}) = \mathscr{O}_{\mathbb{P}^1}(-2)\\
&\mathcal{H}om(\mathscr{O}_{\mathbb{P}^1}(1), \mathscr{O}_{\mathbb{P}^1}) = \mathscr{O}_{\mathbb{P}^1}(-1)\\
&\mathcal{H}om(\mathscr{O}_{\mathbb{P}^1}, \mathscr{O}_{\mathbb{P}^1}) = \mathscr{O}_{\mathbb{P}^1}
\end{aligned}$$

We use the standard open cover U_0, U_1 of $\mathbb{P}^1$ with $U_0 = \mathbb{C}$, $U_1 = (\mathbb{C} \setminus \{0\}) \cup \{\infty\}$ with coordinates z, z^{-1} on U_0 and U_1 respectively. The Čech complexes for these five sheaves, along with our choices of H and Π when needed, are given below. H is only non-trivial from degree 1 to degree 0 in this situation.

$\mathscr{O}_{\mathbb{P}^1}(1)$	$\mathbb{C}[z] \oplus z\mathbb{C}[z^{-1}] \to \mathbb{C}[z, z^{-1}]$
m_1	$(f_0, f_1) \mapsto f_1 - f_0$
$\mathscr{O}_{\mathbb{P}^1}(2)$	$\mathbb{C}[z] \oplus z^2\mathbb{C}[z^{-1}] \to \mathbb{C}[z, z^{-1}]$
m_1	$(f_0, f_1) \mapsto f_1 - f_0$
Π_0	$(a_0 + a_1 z + a_2 z^2 + z^3 f_0(z), z^2 f_1(z^{-1}))$ $\mapsto (a_0 + a_1 z + a_2 z^2, a_0 + a_1 z + a_2 z^2)$
Π_1	0
H	$(z^3 f_0(z) + z^2 f_1(z^{-1})) \mapsto (z^3 f_0(z), -z^2 f_1(z^{-1}))$
$\mathscr{O}_{\mathbb{P}^1}(-2)$	$\mathbb{C}[z] \oplus z^{-2}\mathbb{C}[z^{-1}] \to \mathbb{C}[z, z^{-1}]$
m_1	$(f_0, f_1) \mapsto f_1 - f_0$
$\mathscr{O}_{\mathbb{P}^1}(-1)$	$\mathbb{C}[z] \oplus z^{-1}\mathbb{C}[z^{-1}] \to \mathbb{C}[z, z^{-1}]$
m_1	$(f_0, f_1) \mapsto f_1 - f_0$
Π	0
H	$(f_0(z) + z^{-1} f_1(z^{-1})) \mapsto (f_0(z), -z^{-1} f_1(z^{-1}))$
$\mathscr{O}_{\mathbb{P}^1}$	$\mathbb{C}[z] \oplus \mathbb{C}[z^{-1}] \to \mathbb{C}[z, z^{-1}]$
m_1	$(f_0, f_1) \mapsto f_1 - f_0$
Π_0	$(a_0 + z f_0(z), f_1(z^{-1})) \mapsto (a_0, a_0)$
Π_1	0
H	$(z f_0(z) + f_1(z^{-1})) \mapsto (z f_0(z), -f_1(z^{-1}))$

There are only three terms contributing to m_3 (see Figure 15), but the first is irrelevant since there is no m_3 at the Čech complex level. Furthermore,

$$\Pi m_2(a_3, H m_2(a_2, a_1)) = 0$$

since for $a_2 \in \mathrm{Hom}(\mathscr{O}_{\mathbb{P}^1}(1), \mathscr{O}_{\mathbb{P}^1}(2))$, $a_1 \in \mathrm{Hom}(\mathscr{O}_{\mathbb{P}^1}, \mathscr{O}_{\mathbb{P}^1}(1))$, $m_2(a_2, a_1)$ is degree 0 and H lives in degree 1. For $\Pi m_2(H m_2(a_3, a_2), a_1)$, we have the following table:

a_3	a_2	a_1	$m_2(a_3, a_2)$	$Hm_2(a_3, a_2)$	$\Pi m_2(Hm_2(a_3, a_2), a_1)$
z^{-1}	$(1,1)$	$(1,1)$	z^{-1}	$(0, -z^{-1})$	$(0,0)$
z^{-1}	(z,z)	$(1,1)$	1	$(1,0)$	$(1,1)$
z^{-1}	$(1,1)$	(z,z)	z^{-1}	$(0, -z^{-1})$	$(0,0)$
z^{-1}	(z,z)	(z,z)	1	$(1,0)$	$(1,1)$

The last column describes m_3; in particular, it is non-zero.

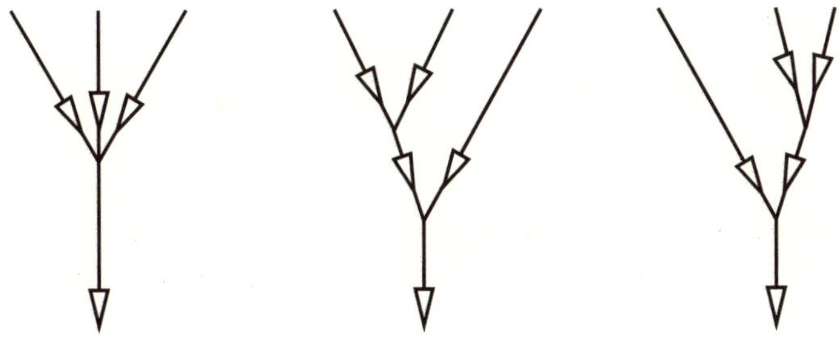

FIGURE 15

8.3. The Morse and Fukaya A_∞-categories

We now want to lead up to the definition of the Fukaya category, which has a fundamentally different flavour than what we considered previously. To warm up, we will give a simpler example, namely the Morse category, which has a similar flavour to the Fukaya category and is related to it in some situations.

8.3.1. The Morse category. Let us recall the basics of Morse theory (see for example §§3.4 and 10.5 in MS1). Throughout, let M be an n-dimensional compact oriented Riemannian manifold. A critical point p of a differentiable function $f : M \to \mathbb{R}$ is said to be *non-degenerate* if, in some local coordinates near p, the function f takes the form $f = x_1^2 + \cdots + x_k^2 - x_{k+1}^2 - \cdots - x_n^2$, in which case we say the critical point has *index* k and write

$$\deg p = k.$$

Given such a critical point, there are two manifolds, $W^s(p)$ and $W^u(p)$. Here $W^s(p)$ is the stable manifold, consisting of the points whose image under the *negative* gradient flow of f converges to p as $t \to +\infty$, while $W^u(p)$, the unstable manifold, consists of points whose image under the *positive* gradient flow of f converges to p as $t \to +\infty$. One checks easily from the local description of a critical point that

$$\deg p = \dim W^s(p).$$

DEFINITION 8.9. f is *Morse-Smale* if f has a finite number of critical points, all non-degenerate, and all stable and unstable submanifolds intersect each other transversally.

For any critical point p, $W^s(p)$ does not come with an intrinsic orientation, but we can consider the free abelian group o_p generated by the two possible orientations on $W^s(p)$, modulo the relations

$$[\Omega_p] + [-\Omega_p] = 0,$$

where $[\Omega_p]$, $[-\Omega_p]$ are the two opposite orientations on $W^s(p)$. As an abstract group, $o_p \cong \mathbb{Z}$.

We can now define the Morse complex

$$CM^i(f) = \bigoplus_{\substack{p\in Crit(f)\\ \deg(p)=i}} o_p.$$

To define a differential, which we shall write as m_1, we proceed as follows. Essentially m_1 counts the number of gradient flow lines connecting two critical points whose degrees differ by 1. Explicitly, first note that as M is oriented and $W^s(p)$ and $W^u(p)$ meet transversally at p and are of complementary dimension, an orientation on $W^s(p)$ induces an orientation on $W^u(p)$. If $\deg q = \deg p + 1$, then $\dim W^u(p) = n - \deg p$, $\dim W^s(q) = \deg p + 1$, so $W^u(p) \cap W^s(q)$ is one-dimensional, hence a union of line segments, the gradient flow trajectories of $-f$ from p to q. Furthermore, the orientations on $W^u(p)$ and $W^s(q)$ determine an orientation on $W^u(p) \cap W^s(q)$. The map f identifies these line segments with segments in $\mathbb{R}$, which carries a natural orientation. We can then take the difference $n(q,p)$ of the number of segments in $W^u(p) \cap W^s(q)$ where the orientation agrees with the orientation on $\mathbb{R}$ minus the number of segments where the orientations disagree. We then define

$$m_1 : CM^i(f) \to CM^{i+1}(f)$$

by

$$m_1([\Omega_p]) = \sum_q n(q,p)[\Omega_q].$$

Note that changing the orientations Ω_p or Ω_q changes the sign of the orientation on $W^u(p) \cap W^s(q)$, hence the sign of $n(q,p)$; thus m_1 is well-defined.

The map m_1 was already shown to satisfy $m_1^2 = 0$ in MS1, §10.5.4. Our goal here is to define an A_∞-category with higher multiplication maps, and $m_1^2 = 0$ will be subsumed within the A_∞ relations.

Let us now define the *Morse category*, Morse(M), whose objects will consist of all smooth real-valued functions on M. This was described by Fukaya in [**165**], but we follow Abouzaid's [**2**] exposition.

The objects of Morse(M) will consist of all smooth real-valued functions on M. However, morphism spaces and higher multiplication maps will not always exist, so what we are defining is not even an A_∞-category, but what we will call an A_∞-precategory, which we will make more precise in the next subsection.

If f_0, f_1 are two functions such that $f_1 - f_0$ is a Morse-Smale function on M, then we set

$$\mathrm{Hom}(f_0, f_1) = CM^\bullet(f_1 - f_0),$$

with m_1 as defined as above. Given $\{f_0, \dots, f_d\}$ satisfying some additional conditions to be made precise shortly, we will define higher multiplication maps

$$m_d : \mathrm{Hom}(f_{d-1}, f_d) \otimes \cdots \otimes \mathrm{Hom}(f_0, f_1) \to \mathrm{Hom}(f_0, f_d).$$

To do so, we proceed as follows.

DEFINITION 8.10. Given a sequence of functions $f_0, \dots, f_d$ and critical points $p_{i,i+1}$ of $f_{i+1} - f_i$ and $p_{0,d}$ of $f_d - f_0$, the moduli space of gradient trees

$$\mathcal{S}_d(p_{0,d}; p_{0,1}, \dots, p_{d-1,d})$$

is the space of maps $\phi : S \to M$, $S \in \mathcal{S}_d$ (the moduli space of ribbon trees) such that

(1) Label each edge e of S with a function f_e as follows. If e is an external incoming edge, labelled by $1 \le i \le d$, then $f_e = f_i - f_{i-1}$; otherwise, if e comes out of a vertex v, then f_e is the sum of all functions labelling the edges coming into v. Then the image of each edge e is a gradient line of f_e.
(2) The orientation of e is given by the gradient flow of $-f_e$. Also, if e is of length l, ϕ identifies the vector field given by a fixed parametrization of the edge by $[0, l]$ with the gradient vector field of $-f_e$.
(3) If e is an external edge, and $f_e = f_i - f_j$, then the image of ϕ converges to $p_{j,i}$ at the external vertex.

For a given sequence $f_0, \dots, f_d$, we can set

$$\mathcal{S}(f_0, \dots, f_d) = \coprod_{p_{0,d}, p_{0,1}, \dots, p_{d-1,d}} \mathcal{S}(p_{0,d}; p_{0,1}, \dots, p_{d-1,d}),$$

where the disjoint union is over all possible critical points.

This moduli space can be compactified by allowing the length of edges to go to ∞; in this case a gradient tree converges to a union of gradient trees, as depicted in Figure 16. This gives compactifications $\overline{\mathcal{S}}(f_0, \dots, f_d)$.

Let's explain how to construct $\mathcal{S}(p_{0,d}; p_{0,1}, \dots, p_{d-1,d})$. Write $p_i := p_{i-1,i}$, $q = p_{0,d}$, and let

$$\mathcal{E}(q; p_1, \dots, p_d) = W^s_{f_d - f_0}(q) \times \left(\prod_{i=1}^d W^u_{f_i - f_{i-1}}(p_i) \right) \times \mathcal{S}_d,$$

where the subscripts on the W's indicate which function we are taking the stable or unstable manifolds with respect to. For each $S \in \mathcal{S}_d$, pick a base

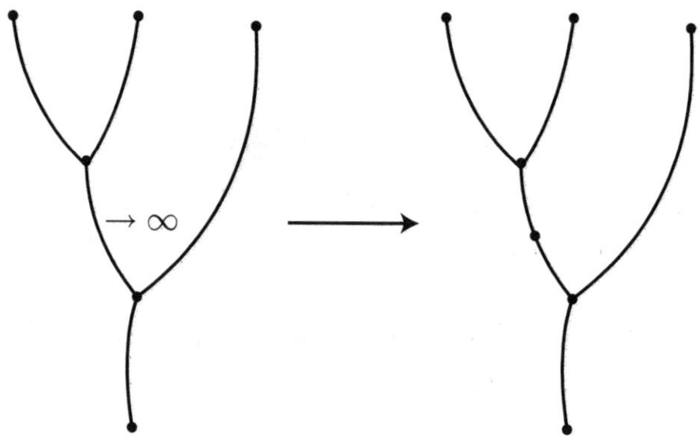

FIGURE 16

point $s \in S$ on the outgoing edge, some fixed distance from the last internal vertex. We can then, for a given S, define a map

$$\Phi_j^s : W^u_{f_j - f_{j-1}}(p_j) \to M$$

as follows. Let v_j be the internal vertex the jth incoming edge is attached to, and let A be the path in S from v_j to s. Now for a point $y \in W^u_{f_j - f_{j-1}}(p_j)$, take the image of y under the sequence of gradient flows of the functions on the edges traversed by A, with time for the gradient flow of $-f_e$ equal to the length of $e \cap A$. By taking these maps along with the inclusion map $W^s_{f_d - f_0}(q) \hookrightarrow M$, we obtain a $(d+1)$-component map

$$\Phi : \mathcal{E}(q; p_1, \ldots, p_d) \to M^{d+1}.$$

It is then clear that $\mathcal{S}(q; p_1, \ldots, p_d)$ is just the inverse image of the diagonal $\Delta \subseteq M^{d+1}$. We will say the collection $f_0, \ldots, f_d$ is *Morse-Smale* if each $f_j - f_i$ is Morse-Smale for $i < j$ and Φ is transversal with respect to Δ, i.e., $\Phi^{-1}(\Delta)$ is smooth of the expected dimension. Note that this expected dimension is

$$\begin{aligned} & \dim \mathcal{E}(q; p_1, \ldots, p_d) - \operatorname{codim}(\Delta / M^{d+1}) \\ = \; & \deg q + d \cdot \dim M - \sum \deg p_i + d - 2 - d \cdot \dim M \\ = \; & \deg q + d - 2 - \sum \deg p_i. \end{aligned}$$

Thus, in particular, $\overline{\mathcal{S}}(f_0, \ldots, f_d)$ is dimension zero when $2 - d + \sum \deg p_i = \deg q$.

Now there is one subtlety, which is that we need to orient $\overline{\mathcal{S}}(f_0, \ldots, f_d)$. This has to be done with great care so that the A_∞ relations will hold with the right sign convention; the only place this has been carried out in detail is in [**2**], Appendix B. We will omit these details.

When $2 - d + \sum \deg p_i = \deg q$, we set $n(q; p_1, \ldots, p_d)$ to be the signed number of gradient trees in $\mathcal{S}(q; p_1, \ldots, p_d)$, and then define m_d by the formula

$$m_d([\Omega_{p_d}] \otimes \cdots \otimes [\Omega_{p_1}]) = \sum_q n(q; p_1, \ldots, p_d)[\Omega_q].$$

Of course, the orientations $\Omega_{p_1}, \ldots, \Omega_{p_d}, \Omega_q$ influence the orientation of the moduli space $\mathcal{S}(q; p_1, \ldots, p_d)$ so that this is well-defined.

We have to explain why this definition of m_d satisfies the A_∞ relations. To see this, consider the compactification $\overline{\mathcal{S}}(q; p_1, \ldots, p_d)$ when

$$3 - d + \sum_d p_i = \deg q.$$

From this degree condition, the moduli space is one-dimensional. The boundary of this moduli space consists of contributions of the form

$$\mathcal{S}(q; p_1, \ldots, p_j, r, p_{i+j+1}, \ldots, p_d) \times \mathcal{S}(r; p_{j+1}, \ldots, p_{i+j})$$

for $1 \leq i \leq d$, $0 \leq j \leq d - i$, and where

$$\deg r = p - 2 + \sum_{k=j+1}^{i+j} \deg p_j.$$

Now consider a term appearing in the degree d A_∞ relation,

$$\begin{aligned} &m_{d-i+1}([\Omega_{p_d}], \ldots, [\Omega_{p_{i+j+1}}], m_i([\Omega_{p_{i+j}}], \ldots, [\Omega_{p_{j+1}}]), [\Omega_{p_j}], \ldots, [\Omega_{p_1}]) \\ &= \sum_r n(r; p_{j+1}, \ldots, p_{i+j}) m_{d-i+1}([\Omega_{p_d}], \ldots, [\Omega_{p_{i+j+1}}], [\Omega_r], [\Omega_{p_j}], \ldots, [\Omega_{p_1}]) \\ &= \sum_{q,r} n(r; p_{j+1}, \ldots, p_{i+j}) n(q; p_1, \ldots, p_j, r, p_{i+j+1}, \ldots, p_d)[\Omega_q]. \end{aligned}$$

Thus the contribution to the coefficient of $[\Omega_q]$ in the A_∞ relation coming from the critical points $p_1, \ldots, p_d$ is just a signed sum of the number of boundary points of $\overline{\mathcal{S}}(q; p_1, \ldots, p_d)$. However, the signed sum of the number of boundary points of a one-dimensional manifold with boundary is always zero, so if all orientations are chosen consistently, the A_∞ equations are satisfied. This choice of orientations has been described in [**2**].

8.3.2. A_∞-precategories. The construction of the previous subsection suffers from the defect that the morphism spaces are not defined between any two objects, and higher multiplications are subject to further constraints. Thus we have not even defined a genuine A_∞-category, but only what Kontsevich and Soibelman in [**311**] termed an A_∞-precategory, which we define here.

DEFINITION 8.11. A (non-unital) A_∞-precategory $\mathcal{A}$ consists of:

(1) A collection of objects $\mathrm{Ob}(\mathcal{A})$.

(2) For each $n \geq 2$, a collection of transversal sequences $\mathrm{Ob}^n_{\mathrm{tr}}(\mathcal{A}) \subseteq \mathrm{Ob}(\mathcal{A})^n$, i.e., a set of n-tuples of sequences.
(3) For $(X_0, X_1) \in \mathrm{Ob}^2_{\mathrm{tr}}(\mathcal{A})$, a $\mathbb{Z}$-graded chain complex $\mathrm{Hom}_{\mathcal{A}}(X_0, X_1)$.
(4) For $(X_0, \ldots, X_d) \in \mathrm{Ob}^{d+1}_{\mathrm{tr}}(\mathcal{A})$, a map

$$m_d : \mathrm{Hom}_{\mathcal{A}}(X_{d-1}, X_d) \otimes \cdots \otimes \mathrm{Hom}_{\mathcal{A}}(X_0, X_1) \to \mathrm{Hom}_{\mathcal{A}}(X_0, X_d)[2-d].$$

We require in addition:
(5) Every subsequence of a transversal sequence is transversal.
(6) The A_∞ relations are satisfied by the m_d's.

So Morse(M) is an example of an A_∞-precategory, where the set of transversal sequences are sequences $(f_0, \ldots, f_d)$ which are Morse-Smale and such that every subsequence is also Morse-Smale.

We will also want the notion of a quasi-equivalence of A_∞-precategories, first by giving the definition of a functor between A_∞-precategories:

DEFINITION 8.12. Let $\mathcal{A}$, $\mathcal{B}$ be A_∞-precategories. An A_∞-functor $F : \mathcal{A} \to \mathcal{B}$ consists of

(1) a map $F : \mathrm{Ob}(\mathcal{A}) \to \mathrm{Ob}(\mathcal{B})$ such that whenever $(X_1, \ldots, X_d) \in \mathrm{Ob}^d_{\mathrm{tr}}(\mathcal{A})$, $(F(X_1), \ldots, F(X_d)) \in \mathrm{Ob}^d_{\mathrm{tr}}(\mathcal{B})$.
(2) For each transversal sequence $(X_0, \ldots, X_d)$ of $\mathcal{A}$, maps

$$F_d : \mathrm{Hom}_{\mathcal{A}}(X_{d-1}, X_d) \otimes \cdots \otimes \mathrm{Hom}_{\mathcal{A}}(X_0, X_1) \to \mathrm{Hom}_{\mathcal{B}}(F(X_0), F(X_d))$$

satisfying the usual formula for an A_∞-functor.

DEFINITION 8.13. An A_∞-functor between A_∞-precategories $\mathcal{A}$, $\mathcal{B}$ is a *quasi-equivalence* if it induces an isomorphism at the level of $H^*(\mathcal{A})$, $H^*(\mathcal{B})$.

In many cases, such as we shall see in §8.4, it is often convenient to avoid defining things when various objects aren't transversal, and this formalism allows us to get away with this.

8.3.3. The Fukaya category. We will now describe the Fukaya category for a Calabi-Yau manifold X, equipped with a symplectic form ω and a nowhere vanishing holomorphic n-form Ω. We will discuss this in somewhat greater detail than was described earlier in the book (see §3.6) and in MS1. However, to attempt a completely technically correct definition is far beyond what can be accomplished here. For many more technical details, one can consult the book of Seidel [**421**], which we have followed closely.

We first describe the objects of the Fukaya category, Fuk(X). These will be quadruples $(L, \mathcal{L}, \tilde{\xi}_L, S_L)$, where:

- $L \subseteq X$ is an oriented Lagrangian submanifold.
- $\mathcal{L}$ is a $U(1)$-bundle with flat connection on L.
- There is always a function

$$\xi_L : L \to S^1$$

with

$$\xi_L = \frac{\Omega|_L}{|\Omega|_L|}.$$

In other words, at any point $x \in L$, we choose a positively oriented basis $\eta_1, \ldots, \eta_n$ of tangent vectors to L, and

$$\xi_L(x) = \frac{\Omega(\eta_1, \ldots, \eta_n)}{|\Omega(\eta_1, \ldots, \eta_n)|}.$$

One easily checks that ξ_L is independent of the choice of basis. Then $\tilde{\xi}_L$ is a *grading* of L, i.e., a choice of lift

$$\tilde{\xi}_L : L \to \mathbb{R}$$

with

$$\xi_L = e^{\pi\sqrt{-1}\tilde{\xi}_L}.$$

We note that not all Lagrangian submanifolds possess a grading: the function ξ_L determines a class in $H^1(L, \mathbb{Z})$ (pulling back the generator of $H^1(S^1, \mathbb{Z})$ via ξ_L) known as the *Maslov class* of L. See Figure 17. This must be zero in order for a grading to exist, and then it is not unique, but can be modified by adding any even integer. Note that changing the orientation of L requires shifting the grading by 1: this can be viewed as changing a brane to an anti-brane. Note that if L is in fact special Lagrangian, so that $\Omega|_L = e^{\sqrt{-1}\theta}\mathrm{Vol}(L)$ for some fixed angle θ, then ξ_L is the constant function $e^{\sqrt{-1}\theta}$. Thus special Lagrangian submanifolds always possess a grading.

- S_L is a spin structure on L. The obstruction to the existence of a spin structure on L is the second Steifel-Whitney class $w_2 \in H^2(L, \mathbb{Z}/2\mathbb{Z})$ of L; if $w_2 = 0$, then the spin structures are classified by $H^1(L, \mathbb{Z}/2\mathbb{Z})$. This choice will not play an important role in our exposition, but is in fact important for determining the orientations on the moduli spaces of J-holomorphic disks which determine the maps m_k. Since we will omit most discussions of these signs here, the reader need not worry too much about this.

In fact, a further constraint needs to be imposed on the Lagrangian: L should be *unobstructed* in the sense of [**164**]. This is an extremely delicate and complicated condition involving holomorphic disks with boundary in L. We shall discuss this more a bit later, albeit briefly, as it is immensely technical. Suffice it to say that it only becomes an issue when $\dim X \geq 3$.

Suppose we are given two objects of $\mathrm{Fuk}(X)$; for convenience we write these just as L_0 and L_1. Suppose that L_0 and L_1 intersect transversally. We wish to define $\mathrm{Hom}_{\mathrm{Fuk}(X)}(L_0, L_1)$. We will need several notions to define this even as a graded vector space, let alone as a complex.

First, we need to define the correct notion of index of an intersection point $x \in L_0 \cap L_1$. We do this as follows. Consider the symplectic vector

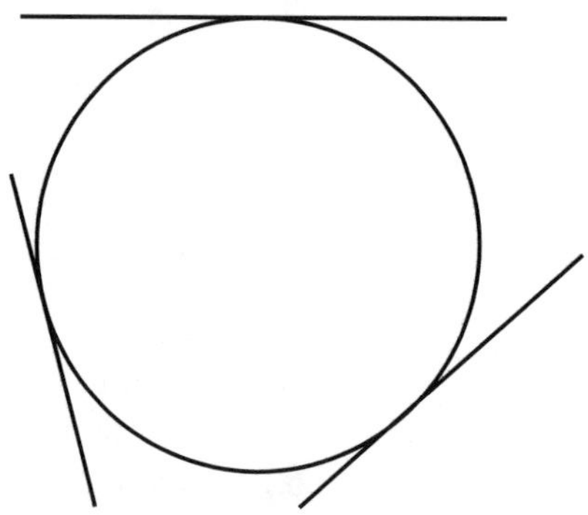

FIGURE 17. A contractible loop L in a surface X always has non-trivial Maslov class: in the example pictured, the angle the tangent line makes goes from 0 to 2π.

space $V := T_xX$. Denote by $\mathrm{Gr}(V)$ the Lagrangian Grassmannian, the space of all Lagrangian subspaces of V, so $\Lambda_i := T_xL_i$ for $i = 0, 1$ specifies two points in $\mathrm{Gr}(V)$. Note that by transversality $\Lambda_0 \cap \Lambda_1 = 0$. A path in $\mathrm{Gr}(V)$, Λ_t, $0 \leq t \leq 1$, connecting Λ_0 and Λ_1 is said to be *crossingless* if $\Lambda_0 \cap \Lambda_t = 0$ for $0 < t \leq 1$ and the *crossing form* at $t = 0$ is negative definite. One defines this quadratic form on Λ_0 first by choosing for small t an isomorphism $\phi_t : \Lambda_0 \to \Lambda_t$ with ϕ_0 the identity. Then the quadratic form is given by

$$v \mapsto \frac{d}{dt}\omega(v, \phi_t(v))\big|_{t=0}.$$

For example, we can always take $V = \mathbb{C}^n$ with the standard symplectic form, $\Lambda_0 = \mathbb{R}^n$, and suppose $\Lambda_1 = e^{\sqrt{-1}\pi c_1}\mathbb{R} \times \cdots \times e^{\sqrt{-1}\pi c_n}\mathbb{R}$ with $c_k \in (-1, 0]$. Then $\Lambda_t = e^{\sqrt{-1}\pi t c_1}\mathbb{R} \times \cdots \times e^{\sqrt{-1}\pi t c_n}\mathbb{R}$ is a crossingless path connecting Λ_0 and Λ_1. One can then always choose a function $\tilde{a} : [0, 1] \to \mathbb{R}$ such that

$$e^{2\pi\sqrt{-1}\tilde{a}(t)} = \left(\frac{\Omega|_{\Lambda_t}}{|\Omega|_{\Lambda_t}|}\right)^2.$$

Then we define the *absolute Maslov index* of the intersection point $x \in L_0 \cap L_1$ as

$$I(x) := (\tilde{\xi}_{L_1}(x) - \tilde{a}(1)) - (\tilde{\xi}_{L_0}(x) - \tilde{a}(0)).$$

This is in fact always in $\mathbb{Z}$. See Figure 18 for an example.

Second, for reasons of convergence, we cannot work over the ground field $\mathbb{C}$, but rather work over the *Novikov ring*, Λ_{nov}, which is defined to be the ring of formal power series of the form

$$\sum_{i \in \mathbb{Z}} a_i q^{\lambda_i}$$

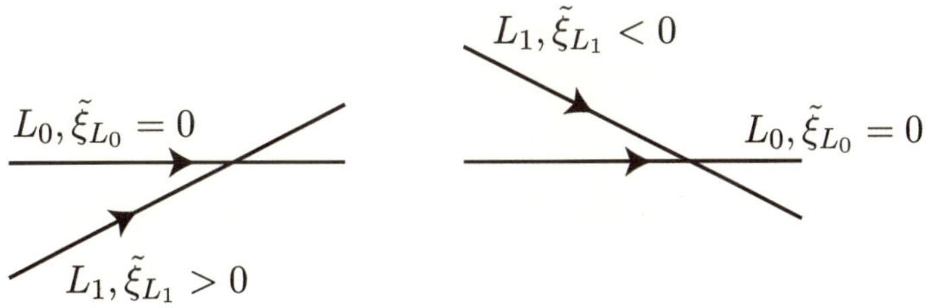

FIGURE 18. Two examples in $\mathbb{C}$. The arrows indicate the orientations, and the grading is a constant $\tilde{\xi} \in (-1, 1)$. In the first case, the index is 1, in the second 0.

where the coefficients $a_i \in \mathbb{Z}$ vanish for all sufficiently negative i, and λ_i are a sequence of real numbers satisfying

$$\lim_{i\to\infty} \lambda_i = \infty.$$

This latter condition guarantees that multiplication makes sense in this ring.

We can now define the Floer complex $\mathrm{Hom}(L_0, L_1)$ when L_0 intersects L_1 transversally. For now, we will restrict to the case that the local systems $\mathcal{L}_1$, $\mathcal{L}_2$ are trivial, and consider the general case later. We set

$$\mathrm{Hom}^i(L_0, L_1) = \bigoplus_{\substack{x\in L_0\cap L_1\\ I(x)=i}} [x] \cdot \Lambda_{\mathrm{nov}}.$$

We define the differential m_1 by using moduli spaces of pseudo-holomorphic strips. Take a family of almost complex structures $J := \{J_t\}_{0\le t\le 1}$ compatible with the symplectic form ω (i.e., so that $\omega(\cdot, J_t\cdot)$ defines a Riemannian metric on X). For $p, q \in L_0 \cap L_1$, let $\mathcal{M}(p, q, L_0, L_1, J)$ denote the moduli space of maps

$$u : \mathbb{R} \times [0,1] \to X$$

such that

$$\begin{gathered} u(\mathbb{R}\times\{0\}) \subseteq L_0, \quad u(\mathbb{R}\times\{1\}) \subseteq L_1 \\ \tfrac{\partial u}{\partial \tau} + J_t(u)\tfrac{\partial u}{\partial t} = 0 \\ \lim_{\tau\to-\infty} u(\tau, t) = p, \quad \lim_{\tau\to+\infty} u(\tau, t) = q. \end{gathered}$$

Here τ is the coordinate on $\mathbb{R}$ and t the coordinate on $[0,1]$. The second line should be viewed as the Cauchy-Riemann equations. For generic choice of J_t, one expects

$$\dim \mathcal{M}(p, q, L_0, L_1, J) = I(q) - I(p).$$

This is shown by computing the index of certain operators in a standard calculation in the theory of pseudo-holomorphic curves. We will not give any of these technical details here.

Note that $\mathbb{R}$ acts on this moduli space, just by acting by translation on the coordinate τ, replacing for $r \in \mathbb{R}$ a solution $u(\cdot, \cdot)$ with a solution

$u(\cdot + r, \cdot)$. The only time this $\mathbb{R}$-action is not free is when $p = q$ and u is a constant map. We obtain a quotient

$$\mathcal{M}(p, q, L_0, L_1, J)/\mathbb{R},$$

of dimension $I(q) - I(p) - 1$ (and which we take to be empty if $p = q$). It is possible to put an orientation on this space, having to do with our choices of spin structures on L_i, and hence when $I(q) = I(p) + 1$, $\mathcal{M}(p, q, L_0, L_1, J)/\mathbb{R}$ is a finite set, with each point u coming with a sign $s(u)$.

We use the above moduli spaces to define m_1 just as we did in the Morse category: namely, if $p, r \in L_0 \cap L_1$ with $I(r) = I(p) + 1$, then the coefficient of $[r]$ in $m_1([p])$ is

$$\sum_{u \in \mathcal{M}(p,r,L_0,L_1,J)/\mathbb{R}} (-1)^{s(u)} q^{\int u^*\omega} \in \Lambda_{\mathrm{nov}}.$$

If one wants to show that $m_1^2 = 0$, one would ideally like to proceed as in the Morse case, by considering the one-dimensional moduli spaces $\mathcal{M}(p, q, L_0, L_1, J)/\mathbb{R}$ when $I(q) = I(p) + 2$. Ideally, studying the boundary then gives $m_1^2 = 0$. However, unlike the Morse case, there are several different ways that the holomorphic strips can degenerate. They can degenerate to a union of strips connecting p to r and r to q, with $I(r) = I(p) + 1$, or a holomorphic disk can bubble off in such a way that its boundary is contained in L_0 or L_1. (See Figure 19.) (Those readers familiar with the usual situation for pseudo-holomorphic curves may also worry about spheres bubbling off, but this is a real codimension two phenomenon.) Now the first sort of degeneration is the sort which gives us information about m_1^2, but the second degeneration is novel. This causes possible obstructions to $m_1^2 = 0$. This

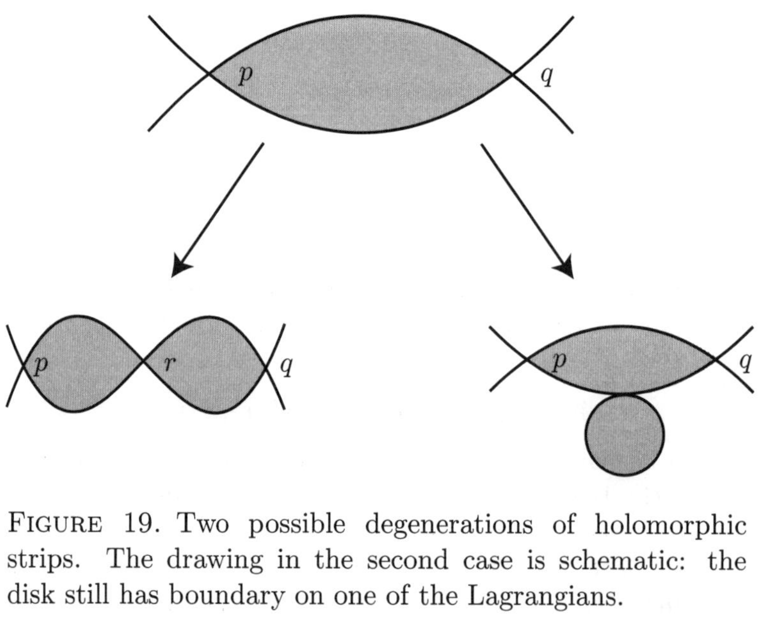

FIGURE 19. Two possible degenerations of holomorphic strips. The drawing in the second case is schematic: the disk still has boundary on one of the Lagrangians.

problem has been considered at great length in [**164**]. If the moduli space of holomorphic disks with boundary on Lagrangians L_0 and L_1 satisfies certain properties (hinted at in §3.6.2.3), then it is possible to modify the definition of m_1 in such a way that $m_1^2 = 0$. If a Lagrangian L does not satisfy the correct properties, then L is said to be *obstructed*, and we should not view it as an element of the Fukaya category. This is an additional restriction on objects in the Fukaya category.

In general, there might be many holomorphic disks with boundary on a Lagrangian— this naturally brings in the subject of open Gromov-Witten invariants. There is yet to be a wholly satisfactory mathematical theory of open Gromov-Witten invariants, but in many interesting cases there is: see [**297, 429**]. We shall not develop this topic here, but discuss briefly the expected dimension of the moduli space of such disks.

Given a holomorphic disk $u : D \to X$ with $u(\partial D) \subseteq L$, we can, for an arbitrary symplectic manifold X, define the Maslov class of u, $\mu(u)$. This is defined as follows: we can trivialize the bundle $u^*(TX)$ as $D \times \mathbb{C}^n$ as a symplectic vector bundle, with the standard symplectic structure on $\mathbb{C}^n$. This defines a map $\gamma : \partial D \to \mathrm{Gr}(\mathbb{C}^n)$, taking $x \in \partial D$ to $T_{\mu(x)}L \subseteq \mathbb{C}^n$ under this trivialization. The standard holomorphic n-form Ω on $\mathbb{C}^n$ defines as before a map $\mathrm{Gr}(\mathbb{C}^n) \to S^1$, taking the subspace spanned by $\eta_1, \ldots, \eta_n$ to $(\Omega(\eta_1, \ldots, \eta_n)/|\Omega(\eta_1, \ldots, \eta_n)|)^2$. The pull-back of the positive generator of $H^1(S^1, \mathbb{Z})$ to $H^1(\mathrm{Gr}(\mathbb{C}^n), \mathbb{Z})$ is the so-called *Maslov class* μ, and generates $H^1(\mathrm{Gr}(\mathbb{C}^n), \mathbb{Z})$. The Maslov class of u is then

$$\mu(u) := \int_{\partial D} \gamma^* \mu \in \mathbb{Z}.$$

If J is a general almost complex structure, let $\mathcal{M}(L, J)$ denote the space of J-holomorphic maps $u : D \to X$ with boundary in L. One expects the component of $\mathcal{M}(L, J)$ containing u to have dimension $\frac{1}{2} \dim_{\mathbb{R}} X + \mu(u)$ (modulo worries of multiple coverings). In our case, when L has a grading, in fact $\mu(u) = 0$. Remembering that $\mathrm{PSL}_2(\mathbb{R})$ acts on $\mathcal{M}(L, J)$ by reparametrization, we see that $\dim \mathcal{M}(L, J)/\mathrm{PSL}_2(\mathbb{R}) = \frac{1}{2} \dim_{\mathbb{R}} X - 3$. Thus we do not have to worry about this case for small dimension: only if the Calabi-Yau manifold X has complex dimension ≥ 3 does this become an issue.

From now on we will assume that we are in a situation where holomorphic disks with boundary on a Lagrangian do not appear. Of course, this excludes the case of greatest interest to us, the threefold case, but nevertheless we will continue with this assumption. We are then in a position to define

$$m_d : \mathrm{Hom}(L_{d-1}, L_d) \otimes \cdots \otimes \mathrm{Hom}(L_0, L_1) \to \mathrm{Hom}(L_0, L_d)[2-d].$$

For $p_{i-1,i} \in L_{i-1} \cap L_i$, $1 \leq i \leq d$, $p_{0,d} \in L_0 \cap L_d$, J some general almost complex structure, we define

$$\mathcal{M}(p_{0,d}; p_{0,1}, \ldots, p_{d-1,d}; L_0, \ldots, L_d)$$

to be the moduli space of maps $u : D \to X$ with D a disk having $d+1$ marked points $x_{0,1}, \ldots, x_{d-1,d}, x_{0,d} \in \partial D$, occurring in cyclic counterclockwise order, satisfying

$$\begin{aligned} u(x_{i-1,i}) &= p_{i-1,i} \quad 1 \leq i \leq d \\ u(x_{0,d}) &= p_{0,d} \\ u([x_{i-1,i}, x_{i,i+1}]) &\subseteq L_i \quad 1 \leq i \leq d \\ u([x_{d-1,d}, x_{0,d}]) &\subseteq L_d \\ u([x_{0,d}, x_{0,1}]) &\subseteq L_0 \end{aligned}$$

Here, for x, y two adjacent marked points on D, $[x, y]$ denotes the interval on ∂D connecting x and y. This is the moduli space of holomorphic polygons: see Figure 20. The expected dimension of this moduli space for a *given*

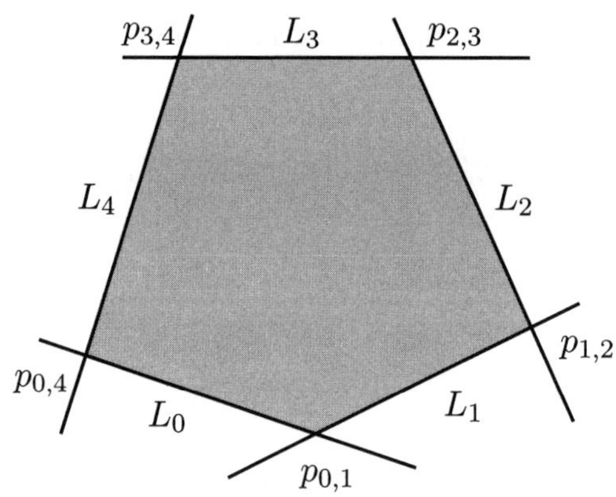

FIGURE 20

$(d+1)$-pointed marked disk is $I(p_{0,d}) - \sum_{i=1}^{d} I(p_{i-1,i})$. Since the moduli space of such $(d+1)$-pointed marked disks is the Stasheff associahedron $\mathcal{S}_d$, of dimension $d-2$, we get

$$\dim \mathcal{M}(p_{0,d}; p_{0,1}, \ldots, p_{d-1,d}; L_0, \ldots, L_d)/\mathrm{PSL}_2(\mathbb{R}) = I(p_{0,d}) + d - 2 - \sum_{i=1}^{d} I(p_{i-1,i}).$$

When this dimension is zero, the coefficient of $[p_{0,d}]$ in $m_d([p_{d-1,d}], \ldots, [p_{0,1}])$ is

$$\sum_{u \in \mathcal{M}(p_{0,d}; p_{0,1}, \ldots, p_{d-1,d}; L_0, \ldots, L_d)/\mathrm{PSL}_2(\mathbb{R})} (-1)^{s(u)} q^{\int u^* \omega} \in \Lambda_{\mathrm{nov}}.$$

Again s denotes the mysterious choice of sign, which is based on the choices of spin structures.

Assuming that there are no issues with holomorphic disks bubbling off, then these m_k's satisfy the A_∞ relations: this is shown as usual by analyzing the way the one-dimensional moduli spaces of disks are compactified.

We now explain how to modify the above formulae when the Lagrangians L_i have non-trivial flat $U(1)$ bundles $\mathcal{L}_i$. First of all, the Floer complex is more naturally written as

$$\mathrm{Hom}^i((L_0,\mathcal{L}_0),(L_1,\mathcal{L}_1)) = \bigoplus_{\substack{p\in L_0\cap L_1\\ I(p)=i}} \mathrm{Hom}(\mathcal{L}_{0,p},\mathcal{L}_{1,p})\otimes\Lambda_{\mathrm{nov}}.$$

So given $p\in L_0\cap L_1$, $t\in\mathrm{Hom}(\mathcal{L}_{0,p},\mathcal{L}_{1,p})$, we obtain an element

$$t_p\in\mathrm{Hom}^*((L_0,\mathcal{L}_0),(L_1,\mathcal{L}_1)).$$

Secondly, the m_d's are modified as follows. Given

$$u\in\mathcal{M}(p_{0,d};p_{0,1},\ldots,p_{d-1,d};L_0,\ldots,L_d),$$

parallel transport using the connection on L_i along $u([x_{i-1,i},x_{i,i+1}])$ for $1\le i\le d-1$ yields an identification $H_i^u:\mathcal{L}_{i,p_{i-1,i}}\cong\mathcal{L}_{i,p_{i,i+1}}$, while parallel transport along $u([x_{0,d},x_{0,1}])$ in L_0 and parallel transport along $u([x_{d-1,d},x_{0,d}])$ in L_d yields respectively

$$\begin{aligned} H_0^u:\mathcal{L}_{0,p_{0,d}} &\cong \mathcal{L}_{0,p_{0,1}}\\ H_d^u:\mathcal{L}_{d,p_{d-1,d}} &\cong \mathcal{L}_{d,p_{0,d}}.\end{aligned}$$

Then we define the coefficient of $m_d(t_{p_{d-1,d}},\ldots,t_{p_{0,1}})$ at the point $p_{0,d}$ to be

$$\sum(-1)^{s(u)}(H_d^u\circ t_{p_{d-1,d}}\circ H_{d-1}^u\circ t_{p_{d-2,d-1}}\circ\cdots\circ H_0^u)\otimes q^{\int u^*\omega}$$
$$\in\mathrm{Hom}(\mathcal{L}_{0,p_{0,d}},\mathcal{L}_{d,p_{0,d}})\otimes\Lambda_{\mathrm{nov}},$$

where the sum is over all $u\in\mathcal{M}(p_{0,d};p_{0,1},\ldots,p_{d-1,d};L_0,\ldots,L_d)/\mathrm{PSL}_2(\mathbb{R})$. The same formula applies for m_1 also, as well as for higher rank flat bundles.

So far these definitions only apply when $L_0,\ldots,L_k$ are pairwise transversal, which of course need not be the case in general. There are two ways around this. First, we could just define an A_∞-precategory, and then if we want to prove HMS we would show that it is quasi-equivalent as a precategory to the derived category of coherent sheaves of the mirror. The second approach is to prove that this construction is invariant under Hamiltonian deformation. This can be made use of in the following fashion, as is done in [**421**]: For every pair of Lagrangian submanifolds L_0, L_1, choose a family of Hamiltonian functions on X, i.e., $H_t: X\to\mathbb{R}$, $t\in[0,1]$, such that if ϕ^t denotes the time t Hamiltonian flow, then $\phi^1(L_0)$ intersects L_1 transversally. Then one can define $\mathrm{Hom}^*(L_0,L_1)$ as above but replacing L_0 with $\phi^1(L_0)$. This enables one to avoid transversality problems. In particular, for a single object L, we obtain $\mathrm{Hom}^*(L,L)$. One then shows that up to quasi-isomorphism, we get an A_∞-category independent of these choices: see [**421**] for details.

While we have been fairly vague in general, we would like to be more specific in the case that $\dim_{\mathbb{R}} X = 2$. For us, this will be the case that X is an elliptic curve, which we will treat in detail in the next section. For the moment, we will discuss the question of signs, as analyzed in [**421, 2**].

In the case of X being an elliptic curve, any one-dimensional submanifold is Lagrangian. Writing $X = \mathbb{C}/\Gamma$ for a lattice Γ, we can lift Lagrangian submanifolds to the universal cover $\tilde{X} = \mathbb{C}$. So given $L \subseteq X$ a circle, $\tilde{L} \subseteq \tilde{X}$ a lifting, $\tilde{L}$ is either a circle (if L represents a trivial element of $H_1(X, \mathbb{Z})$) or a line (diffeomorphic to $\mathbb{R}$). In the former case, there is no grading, as we saw in Figure 17. So we can assume $\tilde{L}$ is a line, with an orientation. In this case, we take the spin structure on L induced by the only spin structure on a disk D bounding the circle L. (This is the Neveu-Schwarz spin structure, see Example 2.3.) Figure 18 then gives us the indices of the intersection points.

Holomorphic disks arising in the calculation of the A_∞-category structure can now be taken to be ordinary holomorphic disks, i.e., polygons in $\mathbb{C}$, with boundary contained in $\tilde{L}_0 \cup \cdots \cup \tilde{L}_d$, see Figure 21. In that fig-

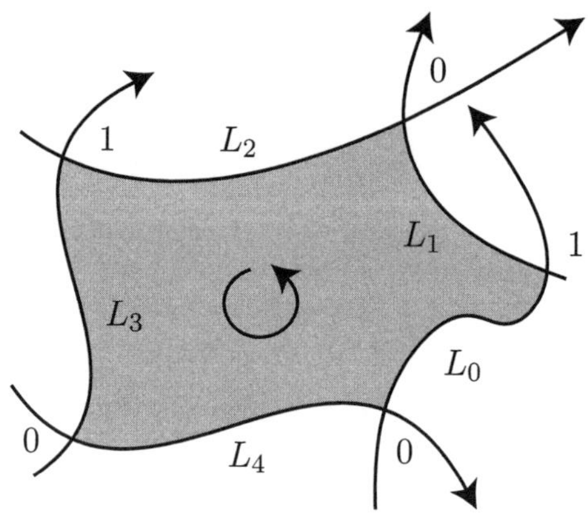

FIGURE 21. The standard orientation of the disk is indicated.

ure, we have given the indices of the intersection points $p_{0,1}, \ldots, p_{3,4}$ and $p_{0,4}$. This picture gives an element $u \in \mathcal{M}(p_{0,4}; p_{0,1}, \ldots, p_{3,4}; L_0, \ldots, L_d)$. Of course different choices of the liftings $\tilde{L}_i$ will give rise to different disks. The sign $(-1)^{s(u)}$ is determined as follows. For each index 1 point $p_{i,j}$ with $i < j$, we have a contribution to the sign $(-1)^{s(p_{i,j})}$, which is positive if and only if the natural orientation on the boundary of the disk coincides with the given orientation on L_j. For example, in Figure 21, the intersection points of index 1 contribute $+1$ from $p_{0,1}$ and -1 from $p_{2,3}$, giving a total sign of $(-1)(+1) = -1$. (In the special case when we are defining m_1 in

$\mathrm{Hom}(L_0, L_1)$, using strips, or equivalently bigons, the rule is that the sign is $+1$ if the given orientation of L_1 agrees with that induced as the boundary of the strip, and -1 if not.)

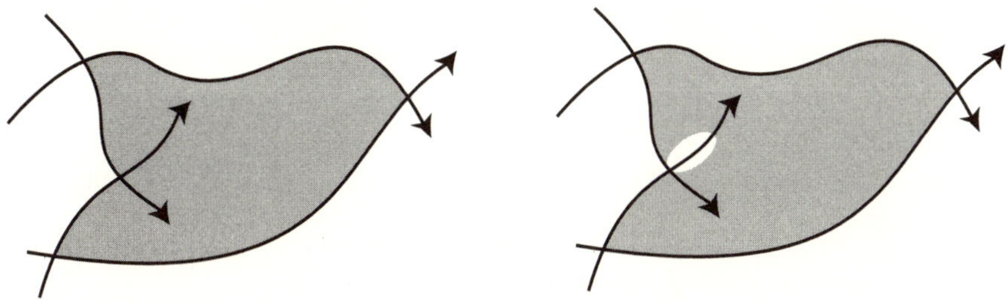

FIGURE 22

It is also worth noting here that disks in 0-dimensional moduli spaces always have convex corners at the intersection points. As we see in Figure 22, a concave corner of a disk would locally vary in a one-dimensional family: the boundary of the disk doubles back on itself in the second picture, and clearly the point at which it does so can vary.

We can carry this procedure out in a simple example. Take $\Gamma \subseteq \mathbb{C}$ to be a lattice $\langle 1, \tau \rangle$, and take $L = \mathbb{R}/\mathbb{Z} \subseteq X = \mathbb{C}/\Gamma$. A lifting $\tilde{L}$ of L is $\mathbb{R}$, and we wish to calculate $\mathrm{Hom}^*(L, L)$. Now according to the above discussion, we should take $L_0 = L$, L_1 a Hamiltonian perturbation of L. Viewing $\mathbb{C}/\mathbb{Z}$ as the cotangent bundle of the circle L, a possible Hamiltonian deformation of L is just the graph of an exact differential on L, e.g.

$$L_1 = \{(x, \sin(2\pi x)) | x \in \mathbb{R}\}.$$

Keeping in mind the periodicity, $L_0 \cap L_1 = \{p_0, p_1\}$, with $p_1 = (0,0)$, $p_0 = (1/2, 0)$, and p_1 is index 1, p_0 is index 0. Figure 23 shows the two holomorphic strips and the signs of their contributions are as listed. Thus $\mathrm{Hom}^*(L_0, L_1)$ is

$$[p_0]\Lambda_{\mathrm{nov}} \to [p_1]\Lambda_{\mathrm{nov}}$$

and then

$$m_1([p_0]) = (-q^{\int u_1^*\omega} + q^{\int u_2^*\omega})[p_1],$$

where $u_1 : D \to X$, $u_2 : D \to X$ correspond to the left and right-hand disks respectively. Noting that $\int u_1^*\omega = \int u_2^*\omega$ (a feature of any choice of similar Hamiltonian deformation), we see in fact that $m_1 = 0$ and we recover the cohomology of L with coefficients in Λ_{nov}, $H^*(L, \Lambda_{\mathrm{nov}})$, which is expected.

In the next section, we shall see some more complicated examples of polygons playing a role in the higher multiplication maps for the elliptic curve. But first, we shall have a final discussion of algebraic issues surrounding A_∞-categories.

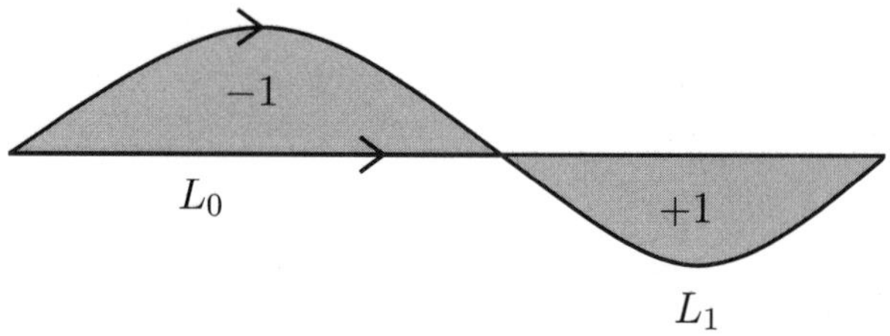

FIGURE 23

8.3.4. Identity morphisms and triangulated A_∞-categories. We have at least two remaining algebraic issues which arise when comparing the derived category of coherent sheaves on X with the Fukaya category of the mirror $\check{X}$ which would lead one to expect these cannot be isomorphic. The first problem is that $\mathrm{D}^b(X)$ has identity morphisms, while a priori $\mathrm{Fuk}(\check{X})$ does not. The second is that $\mathrm{D}^b(X)$ is a triangulated category, while a priori $\mathrm{Fuk}(\check{X})$ is not. The first problem turns out not to be a problem: in fact, contrary to expectations, $\mathrm{Fuk}(\check{X})$ is quasi-isomorphic to an A_∞-category with identity. The second problem is slightly more involved. Morally, a triangulated category requires some additional data over that of a category, namely a set of distinguished triangles satisfying the axioms given in §4.4.3. However, for A_∞-categories, being triangulated is merely a property. As we have defined it, $\mathrm{Fuk}(\check{X})$ is not known to be triangulated. However, there is an algebraic procedure, replacing $\mathrm{Fuk}(\check{X})$ with the *twisted* category $\mathrm{Tw}(\mathrm{Fuk}(\check{X}))$, which is a triangulated category containing $\mathrm{Fuk}(\check{X})$. This will take a little bit of effort.

Let us begin with the discussion of identities. Let $\mathcal{A}$ be a (non-unital) A_∞-category. Then there are several useful notions of identities. First, we say $\mathcal{A}$ is *cohomologically unital*, or *c-unital*, if the non-unital ordinary category $H^*(\mathcal{A})$ in fact has identities, i.e., $H^*(\mathcal{A})$ is actually an ordinary category.

On the other hand, we say $\mathcal{A}$ is *strictly unital* if for every object X there is a (necessarily unique) $e_X \in \mathrm{Hom}^0_{\mathcal{A}}(X, X)$ which satisfies

$$m_1(e_X) = 0$$

$$(-1)^{\deg a} m_2(e_{X_1}, a) = a = m_2(a, e_{X_0}) \text{ for } a \in \mathrm{Hom}_{\mathcal{A}}(X_0, X_1).$$

Any higher multiplication map involving e_X is zero.

The first point is that $\mathrm{Fuk}(X)$ is in fact c-unital. The reason for this is that in the Calabi-Yau case we are considering, for a single graded Lagrangian L, the Floer homology $H^*(\mathrm{Hom}_{\mathrm{Fuk}(X)}(L, L))$ is isomorphic to the ordinary cohomology group $H^*(L, \Lambda_{\mathrm{nov}})$, and of course $1 \in H^0(L, \Lambda_{\mathrm{nov}})$ is an identity. This fact was proved by Piunikhin, Salamon and Schwarz in [**392**] (in a slightly more special context). Seidel then proves in [**421**, Corollary 2.14] that any c-unital A_∞-category is quasi-isomorphic to a strictly unital

A_∞-category. We do not give the details here, but this demonstrates that we do not need to worry about this issue. As a result, from now on we will assume all our A_∞-categories are strictly unital.

We move on to the question of triangulated structures. This requires introducing a number of new concepts.

DEFINITION 8.14. Let $\mathcal{A}$ be an A_∞-category. A (right) A_∞-module $\mathcal{M}$ over $\mathcal{A}$ consists of a graded vector space $\mathcal{M}(X)$ for each $X \in \mathrm{Ob}(\mathcal{A})$, along with multiplication maps, for each $d \geq 1$, of the form

$$m_d^{\mathcal{M}} : \mathcal{M}(X_{d-1}) \otimes \mathrm{Hom}_{\mathcal{A}}(X_{d-2}, X_{d-1}) \otimes \cdots \otimes \mathrm{Hom}_{\mathcal{A}}(X_0, X_1) \to \mathcal{M}(X_0)[2-d]. \tag{8.9}$$

These should satisfy the relation

$$\begin{aligned}&\sum_n (-1)^{\deg a_1 + \cdots + \deg a_n - n} m_{n+1}^{\mathcal{M}}(m_{d-n}^{\mathcal{M}}(b, a_{d-1}, \ldots, a_{n+1}), \ldots, a_1) \\ &+ \sum_{m,n} (-1)^{\deg a_1 + \cdots + \deg a_n - n} \cdot \\ &\qquad m_{d-m+1}^{\mathcal{M}}(b, a_{d-1}, \ldots, a_{n+m+1}, m_m^{\mathcal{A}}(a_{n+m}, \ldots, a_{n+1}), \ldots, a_1) = 0.\end{aligned}$$

The second sum is over $n + m < d$. For example, for $d = 1$, this becomes $(m_1^{\mathcal{M}})^2 = 0$, i.e., $m_1^{\mathcal{M}}$ turns $\mathcal{M}(X)$ into a chain complex. For $d = 2$, this is

$$m_1^{\mathcal{M}}(m_2^{\mathcal{M}}(b, a_1)) + (-1)^{\deg a_1 - 1} m_2^{\mathcal{M}}(m_1^{\mathcal{M}}(b), a_1) + m_2^{\mathcal{M}}(b, m_1^{\mathcal{A}}(a)) = 0$$

and for $d = 3$, this is

$$\begin{aligned}m_1^{\mathcal{M}}(m_3^{\mathcal{M}}(b, a_2, a_1)) &+ (-1)^{\deg a_1 - 1} m_2^{\mathcal{M}}(m_2^{\mathcal{M}}(b, a_2), a_1) \\ &+ (-1)^{\deg a_1 + \deg a_2 - 2} m_3^{\mathcal{M}}(m_1^{\mathcal{M}}(b), a_2, a_1) \\ &+ m_3^{\mathcal{M}}(b, a_2, m_1^{\mathcal{A}}(a_1)) \\ &+ (-1)^{\deg a_1 - 1} m_3^{\mathcal{M}}(b, m_1^{\mathcal{A}}(a_2), a_1) + m_2^{\mathcal{M}}(b, m_2^{\mathcal{A}}(a_2, a_1)) \\ &= 0.\end{aligned}$$

As in the case of the original A_∞ relations, the first one says that $m_1^{\mathcal{M}}$ satisfies the Leibniz rule for module multiplication (i.e., $m_2^{\mathcal{M}}$), while $m_3^{\mathcal{M}}$ measures the failure of this module multiplication to be associative.

A_∞-modules over $\mathcal{A}$ themselves form an A_∞-category, $\mathbf{Mod}(\mathcal{A})$. We have to define the notion of morphisms of A_∞-modules, which we call pre-homomorphisms. If $\mathcal{M}_1$, $\mathcal{M}_2$ are two A_∞-modules, then a pre-homomorphism $t : \mathcal{M}_1 \to \mathcal{M}_2$ of degree $\deg t$ is given by a sequence of maps for $d \geq 1$

$$\begin{aligned}t_d : \mathcal{M}_1(X_{d-1}) \otimes \mathrm{Hom}_{\mathcal{A}}(X_{d-2}, X_{d-1}) &\otimes \cdots \otimes \mathrm{Hom}_{\mathcal{A}}(X_0, X_1) \\ &\to \mathcal{M}_2(X_0)[\deg t - d + 1].\end{aligned} \tag{8.10}$$

One then defines $m_1^{\mathbf{Mod}(\mathcal{A})}$ on the graded vector space of pre-homomorphisms $\mathrm{Hom}_{\mathbf{Mod}(\mathcal{A})}(\mathcal{M}_1, \mathcal{M}_2)$ by

$$\begin{aligned}(m_1^{\mathbf{Mod}(\mathcal{A})}&(t)_d)(b, a_{d-1}, \ldots, a_1) = \\ &\sum_n (-1)^* m_{n+1}^{\mathcal{M}_2}(t_{d-n}(b, a_{d-1}, \ldots, a_{n+1}), a_n, \ldots, a_1) \\ &+ \sum_n (-1)^* t_{n+1}(m_{d-n}^{\mathcal{M}_1}(b, a_{d-1}, \ldots, a_{n+1}), a_n, \ldots, a_1) \\ &+ \sum_{m,n} (-1)^* t_{d-m+1}(b, a_{d-1}, \ldots, m_n^{\mathcal{A}}(a_{n+m}, \ldots, a_{n+1}), a_n, \ldots, a_1).\end{aligned}$$

Here $*$ denotes the expression $\deg a_{n+1} + \cdots + \deg a_{d-1} + \deg b - d + n + 1$. Also, we need to define composition: given $t^1 : \mathcal{M}_0 \to \mathcal{M}_1$, $t^2 : \mathcal{M}_1 \to \mathcal{M}_2$,

$$\begin{aligned}(m_2^{\mathbf{Mod}(\mathcal{A})}&(t^2, t^1))_d(b, a_{d-1}, \ldots, a_1) \\ &= \sum_n (-1)^* t_{n+1}^2(t_{d-n}^1(b, a_{d-1}, \ldots, a_{n+1}), a_n, \ldots, a_1).\end{aligned}$$

The reader can expand these out for small degree to again get a feeling for what these things mean. Fortunately, in this case, all higher multiplication maps $m_d^{\mathbf{Mod}(\mathcal{A})}$, $d > 2$, vanish, so we have a differential graded category. We call t a homomorphism if $m_1^{\mathbf{Mod}(\mathcal{A})}(t) = 0$. This category is strictly unital, with $\mathrm{id} \in \mathrm{Hom}_{\mathbf{Mod}(\mathcal{A})}(\mathcal{M}, \mathcal{M})$ given by $\mathrm{id}_1(b) = (-1)^{\deg b} b$, $\mathrm{id}_d = 0$ for $d > 1$. An *isomorphism* is a homomorphism which is an isomorphism in $H^*(\mathbf{Mod}(\mathcal{A}))$.

We can now define something known as the *Yoneda embedding*, an A_∞-functor $\mathrm{Yon} : \mathcal{A} \to \mathbf{Mod}(\mathcal{A})$. This has a simple definition, namely for $Y \in \mathrm{Ob}(\mathcal{A})$, we set $\mathrm{Yon}(Y)$ to be the module $X \mapsto \mathrm{Hom}_{\mathcal{A}}(X, Y)$. Of course, the latter is already a chain complex, coming from $m_1^{\mathcal{A}}$, and to define the module structure, we need to provide maps as in (8.9). However, as in this case $\mathcal{M}(X_{d-1}) = \mathrm{Hom}_{\mathcal{A}}(X_{d-1}, Y)$ and $\mathcal{M}(X_0) = \mathrm{Hom}_{\mathcal{A}}(X_0, Y)$, these maps are just given by $m_d^{\mathcal{A}}$. So this gives the functor Yon as a map from objects of $\mathcal{A}$ to objects of $\mathbf{Mod}(\mathcal{A})$; however, we have to define the higher maps,

$$\begin{aligned}\mathrm{Yon}_d : \mathrm{Hom}_{\mathcal{A}}(X_{d-1}, X_d) \otimes \cdots &\otimes \mathrm{Hom}_{\mathcal{A}}(X_0, X_1) \to \\ &\mathrm{Hom}_{\mathbf{Mod}(\mathcal{A})}(\mathrm{Yon}(X_0), \mathrm{Yon}(X_d))[1-d].\end{aligned}$$

So given $a_1, \ldots, a_d$, we need to give a pre-homomorphism, i.e., a collection of maps

$$\begin{aligned}t_n : \mathrm{Yon}(X_0)(Y_{n-1}) \otimes \mathrm{Hom}_{\mathcal{A}}(Y_{n-2}, Y_{n-1}) &\otimes \cdots \otimes \mathrm{Hom}_{\mathcal{A}}(Y_0, Y_1) \to \\ &\mathrm{Yon}(X_d)(Y_0)[2 - d - n + \textstyle\sum \deg a_i],\end{aligned}$$

i.e.,

$$(\mathrm{Yon}_d)_n : \mathrm{Hom}_{\mathcal{A}}(Y_{n-1}, X_0) \otimes \mathrm{Hom}_{\mathcal{A}}(Y_{n-2}, Y_{n-1}) \otimes \cdots \otimes \mathrm{Hom}_{\mathcal{A}}(Y_0, Y_1) \to \mathrm{Hom}_{\mathcal{A}}(Y_0, X_d)[2 - d - n + \textstyle\sum \deg a_i].$$

There is an obvious choice, i.e.,

$$(\mathrm{Yon}_d(a_d, \ldots, a_1))_n(b, b_{n-1}, \ldots, b_1) = m^{\mathcal{A}}_{d+n}(a_d, \ldots, a_1, b, b_{n-1}, \ldots, b_1).$$

This defines an A_∞-functor Yon, as the reader can check.

In analogy with what happens in ordinary category theory, we say an object $Y \in \mathrm{Ob}(\mathcal{A})$ *quasi-represents* an A_∞-module $\mathcal{M}$ if there is an isomorphism $t : \mathrm{Yon}(Y) \to \mathcal{M}$. The advantage of using this setup is that it is easy to define algebraic operations in $\mathbf{Mod}(\mathcal{A})$, and these translate into concepts in $\mathcal{A}$ when the results of these operations are quasi-representable. Two examples are crucial for us.

First we have the shift functor: given an A_∞-module $\mathcal{M}$, we define $S\mathcal{M}$ by

$$(S\mathcal{M})(Y) = \mathcal{M}(Y)[1].$$

If $S(\mathrm{Yon}(Y))$ is quasi-representable by some $SY \in \mathrm{Ob}(\mathcal{A})$ for all $Y \in \mathrm{Ob}(\mathcal{A})$, then we obtain a shift functor $S : \mathcal{A} \to \mathcal{A}$. For example, in the Fukaya category, the shift functor can be realized just by adding 1 to the grading.

More important for us is the mapping cone construction. Let $Y_0, Y_1 \in \mathrm{Ob}(\mathcal{A})$, $c \in \mathrm{Hom}^0_{\mathcal{A}}(Y_0, Y_1)$ satisfying $m^{\mathcal{A}}_1(c) = 0$. Then we can define the *abstract mapping cone* of c to be the A_∞-module $Cone(c)$ defined by

$$Cone(c)(X) := \mathrm{Hom}_{\mathcal{A}}(X, Y_0)[1] \oplus \mathrm{Hom}_{\mathcal{A}}(X, Y_1),$$

$$m_d^{Cone(c)}((b_0, b_1), a_{d-1}, \ldots, a_1) = (m^{\mathcal{A}}_d(b_0, a_{d-1}, \ldots, a_1), m^{\mathcal{A}}_d(b_1, a_{d-1}, \ldots, a_1) + m^{\mathcal{A}}_{d+1}(c, b_0, a_{d-1}, \ldots, a_1)).$$

If there is an object of $\mathcal{A}$ quasi-representing $Cone(c)$, we denote it $\mathrm{Cone}(c)$. The module $Cone(c)$ comes along with canonical pre-homomorphisms

$$\begin{aligned} \iota &\in \mathrm{Hom}^0_{\mathbf{Mod}(\mathcal{A})}(\mathrm{Yon}(Y_1), Cone(c)), \\ \pi &\in \mathrm{Hom}^1_{\mathbf{Mod}(\mathcal{A})}(Cone(c), \mathrm{Yon}(Y_0)) \end{aligned}$$

given by

$$\iota_1(b_1) = (0, (-1)^{\deg b_1} b_1), \quad \pi_1(b_0, b_1) = (-1)^{\deg b_0 - 1} b_0,$$

with higher order parts to these pre-homomorphisms being zero. This gives us a "triangle" in $H^*(\mathbf{Mod}(\mathcal{A}))$:

(8.11)
$$\begin{array}{ccc} \mathrm{Yon}(Y_0) & \xrightarrow{[\mathrm{Yon}_1(c)]} & \mathrm{Yon}(Y_1) \\ & {\scriptstyle [\pi]}\nwarrow \quad \swarrow{\scriptstyle [\iota]} & \\ & Cone(c) & \end{array}$$

Here $[\cdot]$ denotes the cohomology class of the homomorphism in $H^*(\mathbf{Mod}(\mathcal{A}))$, which makes sense as $m_1^{\mathbf{Mod}(\mathcal{A})}$ applied to $\mathrm{Yon}_1(c)$, ι or π is zero, as can be checked by the reader. Remember that $[\pi]$ is degree one.

This now allows us to define the notion of an exact triangle in $\mathcal{A}$. It is a diagram in $H^*(\mathcal{A})$ of the form

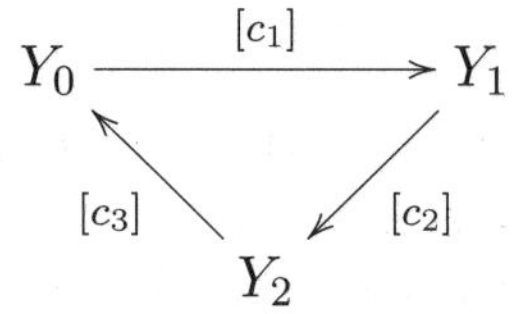

which becomes isomorphic to (8.11) under the Yoneda embedding, with $c_1 = c$.

DEFINITION 8.15. An A_∞-category $\mathcal{A}$ is *triangulated* if

(1) every morphism $[c_1]$ in $H^0(\mathcal{A})$ can be extended to an exact triangle;
(2) There is a shift functor $S : \mathcal{A} \to \mathcal{A}$.
(3) For every $Y \in \mathrm{Ob}(\mathcal{A})$, there is an object $\tilde{Y} \in \mathrm{Ob}(\mathcal{A})$ such that $S\tilde{Y} \cong Y$ in $H^0(\mathcal{A})$.

An example of a triangulated A_∞-category is $\mathrm{D}^b_\infty(X)$: the shift functor is the usual shift on complexes, $M \mapsto M[1]$, defined in §4.4.3, for M and objects of $\mathrm{D}^b_\infty(X)$. Furthermore, mapping cones in this category are defined as in §4.4.3.

The Fukaya category, on the other hand, is not known to be triangulated; at the very least, it is expected one would have to add immersed, non-embedded Lagrangians to the Fukaya category, and this presents technical difficulties in defining Floer homology. So we still need to fix this problem. The solution is the twist construction. Given an A_∞-category $\mathcal{A}$, we define a new A_∞-category, $\mathrm{Tw}(\mathcal{A})$, the category of *twisted complexes*, which is triangulated.

DEFINITION 8.16. Let $\mathcal{A}$ be an A_∞-category. A *twisted complex* is a sequence of objects $X_1, \ldots, X_n$ of $\mathcal{A}$ along with a strictly lower triangular matrix $\Delta = (\delta_{i,j})_{1 \le i,j \le n}$ of morphisms of degree 1, $\delta_{i,j} \in \mathrm{Hom}^1(X_j, X_i)$, such that

$$\sum_{k \ge 1} m_k(\Delta, \ldots, \Delta) = 0.$$

Here we now interpret m_k applied to matrices of morphisms using the usual matrix multiplication rule, so that, using the strictly lower triangular condition, this is equivalent to

$$\sum_{\substack{i_1 < \cdots < i_k \\ k \ge 1}} m_k(\delta_{i_k, i_{k-1}}, \ldots, \delta_{i_2, i_1}) = 0.$$

Note that this seemingly infinite sum therefore only has a finite number of non-zero terms. We write $X = (X_1, \ldots, X_n)$, $\Delta = \Delta^X$, so a twisted object is written (X, Δ^X).

The twisted category of $\mathcal{A}$ is the A_∞-category whose objects consist of twisted complexes of $\mathcal{A}$, with

$$\mathrm{Hom}_{\mathrm{Tw}(\mathcal{A})}((X, \Delta^X), (Y, \Delta^Y)) = \bigoplus_{i,j} \mathrm{Hom}_{\mathcal{A}}(X_i, Y_j).$$

If $(X^0, \Delta^{X^0}), \ldots, (X^d, \Delta^{X^d})$ is a sequence of twisted complexes, and

$$f^i := (f^i_{j,k}) \in \mathrm{Hom}_{\mathrm{Tw}(\mathcal{A})}((X^{i-1}, \Delta^{X^{i-1}}), (X^i, \Delta^{X^i})),$$

then the map $m_d^{\mathrm{Tw}(\mathcal{A})}$ is given by

$$m_d^{\mathrm{Tw}(\mathcal{A})}(f^d, \ldots, f^1) = \sum_{j_0, \ldots, j_d} m^{\mathcal{A}}_{d+j_0+\cdots+j_d}(\Delta^{X^d}, \ldots, \Delta^{X^d}, f^d, \Delta^{X^{d-1}}, \ldots, \Delta^{X^{d-1}}, f^{d-1}, \ldots, f^1, \Delta^{X^0}, \ldots, \Delta^{X_0}).$$

This requires some explanation. First, the sum is over all $j_0, \ldots, j_d \geq 0$. Secondly, Δ^{X_i} appears j_i times in this expression. Thirdly, the expression should be thought of as a matrix product, as before. We leave it to the reader to check that these m_d's satisfy the A_∞-relations!

The simplest example is direct sum, when $\Delta^X = 0$, so that the object $((X_1, \ldots, X_n), 0)$ represents $X_1 \oplus \cdots \oplus X_n$. A slightly less trivial example of a twisted complex is the mapping cone construction. We assume that the A_∞-category $\mathcal{A}$ has a shift functor S (as the Fukaya category does). Then given $c \in \mathrm{Hom}^0_{\mathcal{A}}(Y_0, Y_1)$, we get a twisted complex

$$C = \left((SY_0, Y_1), \begin{pmatrix} 0 & 0 \\ -S(c) & 0 \end{pmatrix}\right),$$

where the shift functor S is used to think of c as defining an element $S(c)$ of $\mathrm{Hom}^1_{\mathcal{A}}(SY_0, Y_1)$. Then note that there are canonical morphisms in

$$\mathrm{Hom}^0(Y_1, C) = \mathrm{Hom}^0(Y_1, SY_0) \oplus \mathrm{Hom}^0(Y_1, Y_1),$$

given by $(0, \mathrm{id}_{Y_1})$ and

$$\begin{aligned} \mathrm{Hom}^1(C, Y_0) &= \mathrm{Hom}^1(SY_0, Y_0) \oplus \mathrm{Hom}^1(Y_1, Y_0) \\ &= \mathrm{Hom}^0(Y_0, Y_0) \oplus \mathrm{Hom}^1(Y_1, Y_0) \end{aligned}$$

given by $(\mathrm{id}_{Y_0}, 0)$. Again, the reader can check that these maps give an exact triangle in $\mathrm{Tw}(\mathcal{A})$, i.e., C represents the mapping cone of c in $\mathrm{Tw}(\mathcal{A})$. In fact, generalizing this slightly, we see that $\mathrm{Tw}(\mathcal{A})$ is itself triangulated. Given $c \in \mathrm{Hom}^0_{\mathrm{Tw}(\mathcal{A})}((X^0, \Delta^{X^0}), (X^1, \Delta^{X^1}))$, we build $\mathrm{Cone}(c)$ as

$$C = \left((SX^0, X^1), \begin{pmatrix} \Delta^{X^0} & 0 \\ -S(c) & \Delta^{X^1} \end{pmatrix}\right).$$

We can think of $\mathrm{Tw}(\mathcal{A})$ as being the derived category of $\mathcal{A}$, and can write it as $\mathrm{D}^b(\mathcal{A})$.

Unfortunately, this is still not quite the correct category for us. The derived category $\mathrm{D}^b(X)$ of coherent sheaves on X has another property which $\mathrm{Fuk}(\check{X})$ or $\mathrm{D}^b(\mathrm{Fuk}(\check{X}))$ may not have. Namely, if $\mathcal{F}$ is an object in $\mathrm{D}^b(X)$, and $p : \mathcal{F} \to \mathcal{F}$ a morphism with $p^2 = p$, then one can think of the image of p as another object in $\mathrm{D}^b(X)$. Such a morphism p is called *idempotent*, and $\mathrm{D}^b(X)$ is *split-closed*, essentially saying that $\mathrm{D}^b(X)$ contains all images of idempotent morphisms.

In general, given a linear category, there is a way of constructing a larger category, the *Karoubi completion*, which contains all images of idempotent morphisms. This can be generalized to A_∞-categories, yielding from an A_∞-category $\mathcal{A}$ a split-closed category. We omit the details of this construction—the interested read can consult [**421**], §4. The construction has a similar flavour to the ones we have seen above. In particular, we denote by $\mathrm{D}^\pi(\mathcal{A})$ this process applied to $\mathrm{D}^b(\mathcal{A})$. This can be shown to be a split-closed triangulated A_∞-category. Once one does this, the hoped-for rigorous statement of Homological Mirror Symmetry becomes:

CONJECTURE 8.17. *Let X and $\check{X}$ be a mirror pair of Calabi-Yau manifolds. Then $\mathrm{D}^b_\infty(X)$ and $\mathrm{D}^\pi(\mathrm{Fuk}(\check{X}))$ are quasi-equivalent as A_∞-categories.*

8.4. The elliptic curve

We will now show a limited form of HMS for the elliptic curve. For simplicity, we will not try to compare the full derived and Fukaya categories, though what we do here constitutes most of the work. The case of the elliptic curve has already been studied in detail in [**397, 396**]; we will take a somewhat different approach here inspired by Abouzaid in [**2**] and developed in [**1**] which uses the Čech approach to defining the A_∞ structure on the derived category.

We will do this by building a degeneration of elliptic curves, a family $\mathcal{X} \to D$, where D can be viewed either as the unit disk with coordinate q or, in the category of schemes, $D = \operatorname{Spec} \mathbb{C}[[q]]$. This family will be a compactification of the family considered in §6.2.4. The advantage of working with such a family is that the parameter q then corresponds to the Kähler parameter q appearing in the Fukaya category of the mirror, i.e., the variable q in the Novikov ring. In particular, one should view either side as giving a family of categories parametrized by q. Furthermore, by working with an explicit degeneration, the methods introduced here fit well with the philosophy introduced by Gross and Siebert in [**205, 206**]. Although we do not elaborate on that philosophy here, it is helpful in this situation.

We will in fact build the degeneration $\mathcal{X} \to D$ along with a line bundle $\mathcal{L}$ which is ample (positive) when restricted to each fiber. We will then compute all higher multiplication maps involving powers, both positive and negative, of $\mathcal{L}$.

8.4.1. The degeneration: The Tate curve. The method of constructing our degeneration is a special case of a construction of Mumford for degenerations of abelian varieties [**373**]. This degeneration of elliptic curves is usually called the Tate curve. We start with the following data. Pick a degree d for the choice of line bundle, and consider a continuous piecewise linear function $\varphi : \mathbb{R} \to \mathbb{R}$ which has slope i on the interval $[i, i+1]$ for $i \in \mathbb{Z}$. We can choose this to take the value 0 on $[0,1]$ to be specific; in this case,

$$\varphi(x) = ix - \tfrac{i(i+1)}{2} \text{ for } x \in [i, i+1]$$

and φ satisfies a periodicity condition

$$\varphi(x+d) = \varphi(x) + d \cdot x + \frac{d(d-1)}{2}. \tag{8.12}$$

Let

$$\Delta = \{(x,y) \in \mathbb{R}^2 | y \geq \varphi(x)\}.$$

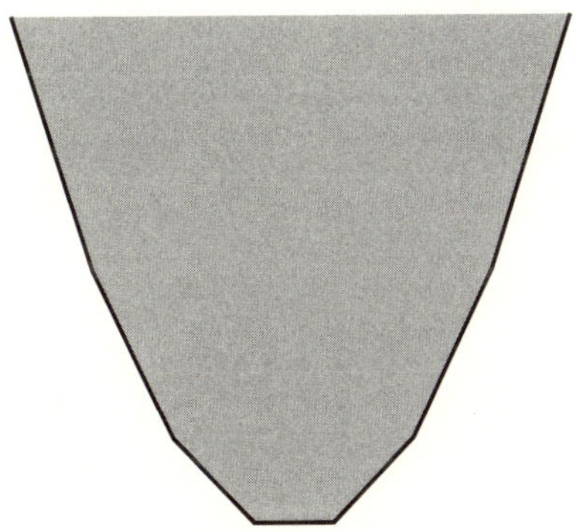

FIGURE 24. The graph of φ. The shaded area is Δ.

Just as a convex polytope determines a projective toric variety, an unbounded polytope can also determine a toric variety. This can be described via the *normal fan* of Δ. For each face $\sigma \subseteq \Delta$, define

$$\check{\sigma} = \{n \in (\mathbb{R}^2)^\vee : n|_\sigma = \text{constant}, \langle n, m\rangle \geq \langle n, m'\rangle \quad \forall m \in \Delta, m' \in \sigma\},$$

the *normal cone* to Δ at σ. We define the normal fan of Δ to be the collection of cones $\{\check{\sigma} | \sigma \subseteq \Delta \text{ a face}\}$. This is a fan living in $(\mathbb{R}^2)^\vee$. It is easy to see that if σ is the interval with endpoints $(i, \varphi(i))$ and $(i+1, \varphi(i+1))$, then $\check{\sigma}$ is the ray generated by $(-i, 1)$. Thus the normal fan Σ of Δ is as depicted in Figure 25. This violates a standard notion of fan in that it contains an infinite number of cones, but we should not be bothered by this: this fan still defines a toric variety X_Σ, covered by an *infinite* number of affine toric charts coming from the cones of Σ. Furthermore, $(0,1) \in \mathbb{R}^2$ takes the value 1 on the primitive generators of the rays of Σ, so if we denote by q the monomial corresponding to $(0,1)$, q is in fact a regular function on X_Σ, defining a map $g : X_\Sigma \to \mathbb{A}^1$.

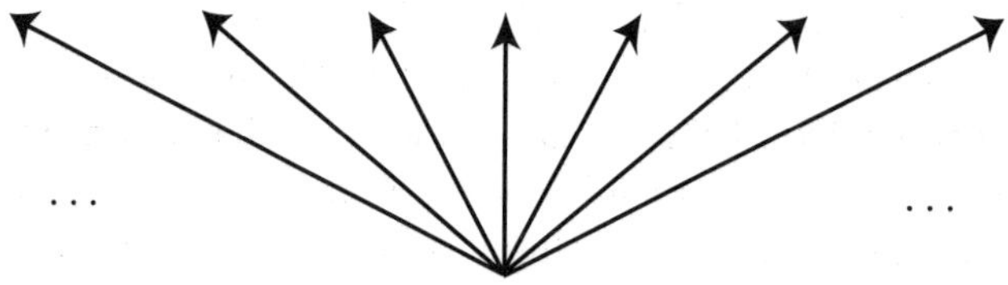

FIGURE 25. The normal fan of Δ.

Since q vanishes to order 1 on each toric divisor of X_Σ (corresponding to a ray of Σ) the fiber $g^{-1}(0)$ consists of an infinite chain of $\mathbb{P}^1$'s. On the other hand, it is not hard to see that any other fiber of g is a copy of $\mathbb{C}^*$.

Like any toric variety, X_Σ has an open cover indexed by the cones in the fan Σ, or equivalently, by the faces of Δ. In particular, we have an open cover $\{U_w | w \text{ is a vertex of } \Delta\}$, where $U_w = \operatorname{Spec} \mathbb{C}[\check{w}^\vee \cap \mathbb{Z}^2]$. This is the open subset of X_Σ corresponding to the cone $\check{w}$ of the fan Σ; alternatively $\check{w}^\vee$ can be viewed as just the tangent wedge to Δ at w, see Figure 26. Similarly, if σ is an edge of Δ, we obtain an open subset U_σ of X_Σ with $U_\sigma = \operatorname{Spec} \mathbb{C}[\check{\sigma}^\vee \cap \mathbb{Z}^2]$.

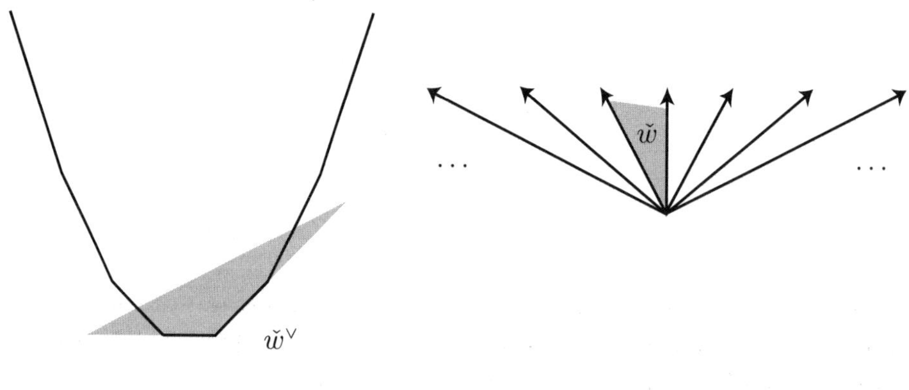

FIGURE 26

Now $\mathbb{Z}$ acts on X_Σ. We can describe this action by describing a $\mathbb{Z}$-action on the fan Σ. The generator of the action is given by the matrix $\left(\begin{smallmatrix} 1 & d \\ 0 & 1 \end{smallmatrix}\right)$ on $(\mathbb{R}^2)^\vee$: this takes a ray $\mathbb{R}_{\geq 0}(i,1)$ to $\mathbb{R}_{\geq 0}(i+d,1)$. This has the effect of acting on the big torus orbit $(\mathbb{C}^*)^2 \subseteq X_\Sigma$, with coordinates z, q corresponding to the standard basis $(1,0), (0,1)$ of $\mathbb{R}^2$, via

$$(z,q) \mapsto (zq^d, q).$$

Hence if we fix a non-zero value of q, the action on the $\mathbb{C}^*$ parametrized by z is $z \mapsto z \cdot q^d$. For $0 < |q| < 1$, the quotient of this fiber by the $\mathbb{Z}$-action is $\mathbb{C}^*/q^{\mathbb{Z}}$. This quotient coincides with

$$\mathbb{C}/\langle 1, \frac{1}{2\pi\sqrt{-1}} \log q\rangle;$$

note that $\exp(2\pi\sqrt{-1}\cdot)$ identifies $\mathbb{C}/\mathbb{Z}$ with $\mathbb{C}^*$. Clearly this is an elliptic curve as long as $|q| \neq 1$.

On the other hand, for $q = 0$, the action on the fiber shifts the infinite chain of $\mathbb{P}^1$'s d places, so the quotient by this $\mathbb{Z}$-action is a cycle of d rational curves, as depicted in Figure 27.

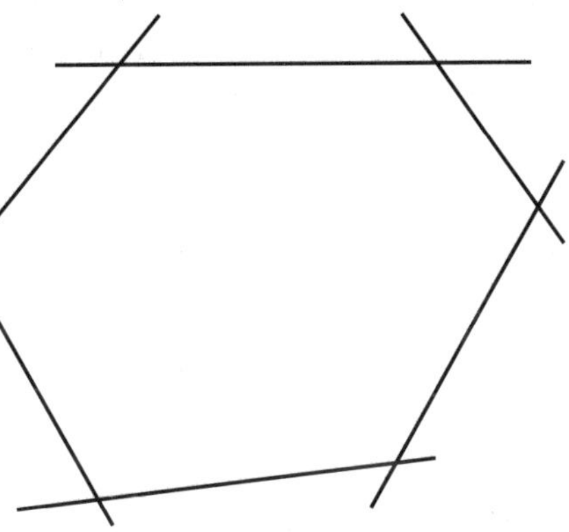

FIGURE 27

We can now take

$$\mathcal{X} = g^{-1}(D)/\mathbb{Z},$$

where D is the open unit disk. Then $g : \mathcal{X} \to D$, induced by q, is a degeneration of elliptic curves to a so-called Kodaira type I_d fiber, i.e., a cycle of d rational curves.

For our purposes, just as in the Fukaya category where we worked with the Novikov ring, a formal power series ring, it is more convenient for us to work with a formal version of this quotient. This is a rather technical point, but it allows us essentially to work entirely with the topology of the singular fiber, and also allows us to take the $\mathbb{Z}$ quotient in a more algebraic category.

Indeed, it is difficult to divide algebraic varieties by groups such as $\mathbb{Z}$. In this situation, we observe first of all that $g^{-1}(0)/\mathbb{Z}$ makes sense as an algebraic variety; as we saw, it is a cycle of d rational curves. On the other hand, let X_Σ^k denote the subscheme of X_Σ defined by the equation $q^{k+1} = 0$, so that X_Σ^k is a "kth order thickening" of $g^{-1}(0)$. Then in fact one can check (the details aren't so important for us) that $X_\Sigma^k/\mathbb{Z}$ also makes sense as a scheme, as a "kth order thickening" of $g^{-1}(0)/\mathbb{Z}$.

We can then construct the "formal scheme"

$$\hat{X}_\Sigma = \varprojlim X_\Sigma^k.$$

See [222], II §9, for the definition of a formal scheme. This is a ringed space whose underlying topological space is that of X_Σ^k (these coincide for all k), or equivalently, it has underlying topological space $g^{-1}(0)$. The sheaf of rings on $\hat{X}_\Sigma$ is given on an open set $U \subseteq g^{-1}(0)$ by

$$\Gamma(U, \mathcal{O}_{\hat{X}_\Sigma}) = \varprojlim \Gamma(U, \mathcal{O}_{X_\Sigma^k}).$$

(See, e.g., [25] for the definition of inverse limit and more details about completions of rings.) Essentially what this does is it allows us to consider functions in a "formal neighbourhood" of $g^{-1}(0)$ which are formal power series in q. For example, if U_w is the open set of X_Σ defined above, then $\hat{U}_w = g^{-1}(0) \cap U_w$ is topologically an open subset of $g^{-1}(0)$, and

$$\Gamma(\hat{U}_w, \mathcal{O}_{\hat{X}_\Sigma}) = \varprojlim \mathbb{C}[\breve{w}^\vee \cap \mathbb{Z}^2]/(q^n) =: \mathbb{C}[\breve{w}^\vee \cap \mathbb{Z}^2]\hat{\otimes}_{\mathbb{C}[q]}\mathbb{C}[[q]],$$

where the completed tensor product on the right is defined by the inverse limit. Then

$$\hat{\mathcal{X}} := \hat{X}_\Sigma/\mathbb{Z}$$

makes sense as a formal scheme also, and there is a map of formal schemes $\hat{g} : \hat{\mathcal{X}} \to \hat{\mathbb{A}}^1$ induced by q. Here $\hat{\mathbb{A}}^1$ is the ringed space consisting of the point 0 and the ring $\mathbb{C}[[q]]$. This should be viewed as a formal version of $g : \mathcal{X} \to D$.

We will actually calculate the desired multiplication maps on $\hat{\mathcal{X}}$; all answers will be formal power series in q. However, we first need to construct the desired line bundles on $\hat{\mathcal{X}}$.

To do so, we choose a relatively ample line bundle $\mathcal{L}$ on $\hat{\mathcal{X}}$. In fact, the choice of $\mathcal{L}$ is already present in Δ: a polytope determines not just a toric variety but a line bundle on it. In this case there is a line bundle $\mathcal{L} = \mathcal{O}_\Delta(1)$ on X_Σ which has a basis of sections which is in one-to-one correspondence with integral points of Δ. We can describe $\mathcal{L}$ via its transition maps in the usual way for line bundles on toric varieties, by saying that $\mathcal{L}|_{U_w}$ is trivial, identified naturally with $\mathcal{O}_{U_w} \cdot z^w$. Thus on $U_w \cap U_{w'}$, the transition map from $\mathcal{O}_{U_w}$ to $\mathcal{O}_{U_{w'}}$ is given by multiplication by $z^{w-w'}$. One sees easily that

$$\Delta = \bigcap_{w \text{ a vertex of } \Delta} \breve{w}^\vee + w$$

and any integral point of $m \in \Delta$ then gives a well-defined section z^m of $\mathcal{L}$. Note that if $\mathcal{L}$ is restricted to any irreducible component of $g^{-1}(0)$, we get the line bundle $\mathcal{O}_{\mathbb{P}^1}(1)$.

We want $\mathcal{L}$ to induce a line bundle $\hat{\mathcal{L}}$ on $\hat{\mathcal{X}}$. In order to do this, it is not enough to know how $\mathbb{Z}$ acts on $\mathcal{X}$, but how this $\mathbb{Z}$-action lifts to an action on $\mathcal{L}$. For this purpose, it is enough to describe a $\mathbb{Z}$-action on

$$\bigoplus_{n=0}^{\infty} H^0(X_\Sigma, \mathcal{L}^{\otimes n}).$$

This space of sections has a canonical basis given by the integral points of $C(\Delta) \subseteq \mathbb{R}^3$, where

$$C(\Delta) = \overline{\{(rm, r) \mid m \in \Delta, r \in [0, \infty)\}}.$$

Note that the integral points with third coordinate an integer n correspond to elements of a basis for $H^0(X_\Sigma, \mathcal{L}^{\otimes n})$. We want an action which satisfies the following properties:

(1) It is linear on $\mathbb{Z}^3 \subseteq \mathbb{R}^3$, as we want the action of $\mathbb{Z}$ on $\mathcal{L}$ to act by toric automorphisms of the total space of $\mathcal{L}$.
(2) This action preserves the third component, i.e., does not change the degree of a section.
(3) As the action on $\mathcal{L}$ should be compatible with the action on X_Σ, the action on $\mathbb{Z}^2 \times \{0\} \subseteq \mathbb{Z}^3$ should be generated by

$$\begin{pmatrix} 1 & 0 \\ d & 1 \end{pmatrix}$$

(the transpose of the action on $(\mathbb{R}^2)^\vee$).
(4) The action should identify $C(\Delta)$ with itself.

The first three conditions tell us that the generator of the action must be

$$\begin{pmatrix} 1 & 0 & * \\ d & 1 & * \\ 0 & 0 & 1 \end{pmatrix}$$

From the fourth condition, the action must preserve the boundary of $C(\Delta)$, and it follows from the periodicity condition (8.12) that the only choice for this generator is

$$T = \begin{pmatrix} 1 & 0 & d \\ d & 1 & \frac{d(d-1)}{2} \\ 0 & 0 & 1 \end{pmatrix}.$$

For a given $n \geq 0$, we will write this generator as an affine transformation

$$T_n(x, y) := (x + nd, xd + y + \frac{d(d-1)}{2}n),$$

the first two components of $T(x, y, n)$. This describes the action on sections of $\mathcal{L}^{\otimes n}$, a basis being represented by integral points of $n\Delta$.

As before, we can now divide $\mathcal{L}$ out by $\mathbb{Z}$, and this only makes sense on the formal neighbourhood $\hat{X}_\Sigma$ of $g^{-1}(0)$. We then get a line bundle $\hat{\mathcal{L}}$ on $\hat{\mathcal{X}}$.

The easiest way to describe what we have constructed is to write down a $\mathbb{C}[[q]]$-basis of sections for $H^0(\hat{\mathcal{X}}, \hat{\mathcal{L}}^{\otimes n})$. To do so, one writes down a basis of $\mathbb{Z}$-invariant sections of $H^0(X_\Sigma, \mathcal{L}^{\otimes n})$. This basis is as follows: pick any $p \in \frac{1}{n}\mathbb{Z}$. Then set

$$\vartheta_{n,p} := \sum_{s=-\infty}^{\infty} z^{T_n^s(np, n\varphi(p))}.$$

Note that on any given open set U_w, this makes sense formally, i.e.,

$$\vartheta_{n,p} \in (\mathbb{C}[\check{w}^\vee \cap \mathbb{Z}^2]\hat{\otimes}_{\mathbb{C}[q]}\mathbb{C}[[q]]) \cdot z^{nw}.$$

To see why this makes sense, one can check that for any given $k > 0$, all but a finite number of terms in the sum defining $\vartheta_{n,p}$ are zero in $(\mathbb{C}[\check{w}^\vee \cap \mathbb{Z}^2] \otimes_{\mathbb{C}[q]} \mathbb{C}[q]/(q^k)) \cdot z^{nw}$. The functions $\vartheta_{n,p}$ are known as *theta functions.* Note also that

$$\vartheta_{n,p} = \vartheta_{n,p+d}$$

as

$$T_n(np, n\varphi(p)) = (n(p+d), n\varphi(p+d)).$$

It is not difficult to see that the $\vartheta_{n,p}$'s generate the space of $\mathbb{Z}$-invariant sections of $\mathcal{L}^{\otimes n}|_{\hat{X}_\Sigma}$ as a $\mathbb{C}[[q]]$-module. Note however that we need to allow formal power series in q in order for there to exist $\mathbb{Z}$-invariant sections: $\mathcal{L}^{\otimes n}$ on X_Σ has no such sections. This is another way of seeing why we need to pass to the completion before we can take the quotient. Note more generally that if we have any $(x, y) \in \mathbb{Z}^2$ with $(x, y) \in n\Delta$, i.e., $y \geq n\varphi(x/n)$, then

$$\sum_{s=-\infty}^{\infty} z^{T_n^s(x,y)} = q^{y-n\varphi(x/n)}\vartheta_{n,x/n}. \tag{8.13}$$

We define

$$\mathrm{ord}_n(x, y) = y - n\varphi(x/n) \tag{8.14}$$

for any $n \in \mathbb{Z}$, $(x, y) \in \mathbb{Z}^2$. (8.13) is the motivation for this definition.

Note also that elements of this basis of $\mathbb{Z}$-invariant sections are in one-to-one correspondence with points on the integral affine manifold (see §6.1.2) $B = \mathbb{R}/d\mathbb{Z}$ with coordinates in $\frac{1}{n}\mathbb{Z}$, a point which we shall return to in a moment. We write this set of points as $B((1/n)\mathbb{Z})$.

At this point, a reader unhappy with the vague description of the quotient construction we have given may proceed as follows. Consider the $\mathbb{C}[[q]]$-algebra

$$\hat{R} = \bigoplus_{n=0}^{\infty} H^0(\hat{\mathcal{X}}, \hat{\mathcal{L}}^{\otimes n}).$$

One can define this simply by knowing the set of $\mathbb{C}[[q]]$-module generators,

$$\{1\} \cup \{\vartheta_{n,p} | p \in B((1/n)\mathbb{Z}), n \geq 1\},$$

and the multiplication map, i.e., the structure constants of the algebra, something which we will explain shortly. We can then consider the scheme $\mathrm{Proj}\,\hat{R}$, which is a scheme defined over $\mathrm{Spec}\,\mathbb{C}[[q]]$ (see for example [**222**], II,

§2). This is a genuine scheme, not a formal scheme. The fact that one can obtain a genuine scheme from a formal scheme in the presence of an ample line bundle is Grothendieck's existence theorem, see EGA III, 5.4.5, [**211**]. So the construction above can be viewed as a special case.

8.4.2. Multiplication of theta functions and triangles. Let us now turn to a geometric interpretation for theta functions and their multiplication. We have just seen an affine manifold B emerge from our description of the basis of theta functions. So it is natural to try to fit what we are doing into the context of Chapter 6.

We will in fact view a fiber of the map $\mathcal{X} \to D$ as $\check{X}(B)$, see §6.2.1. Indeed, the fiber comes along with the restriction of the line bundle $\mathcal{L}$, which is necessarily of degree d. On the other hand, $\check{X}(B)$ comes with a canonical symplectic form ω with $\int_{\check{X}(B)} \omega = d$, so ω represents $c_1(\mathcal{L})$.

The reader may object that this goes against the philosophy of §6.3: namely, one considers line bundles on $X(B)$ and Lagrangians on $\check{X}(B)$. This is indeed the case, but there is a reason for this switch. When we discuss the Fukaya category, we will be able to work, at least in the elliptic curve case, with a fixed almost complex structure, namely, the complex structure $X(B)$. It will be easy to describe holomorphic disks on $X(B)$. By choosing to work on $X(B)$ on the Fukaya side, we are making it easy to describe disks but difficult to describe Lagrangians. Similarly, by working on $\check{X}(B)$ for the derived category side, it is difficult to describe sheaves, but we get a canonically given cohomology class from the symplectic form. As a result, we restrict attention to line bundles with first Chern class represented by integral multiples of ω, and more specifically, powers of $\mathcal{L}$. On $X(B)$, we restrict attention to Lagrangians of a very special sort, namely sections which are induced by multiples of the developing map (§6.1.2), with trivial local system.

Let us be more explicit now and describe, in the spirit of §6.3, the Lagrangians on $X(B)$ which will be mirror dual to the line bundles $\mathcal{L}^{\otimes n}$. First, we need to specify a symplectic form on $X(B)$: by Proposition 6.14, this can be done by specifying a convex multi-valued function on B; the function $K = y^2/2$ will do nicely, with $\omega = dy \wedge dx$. Here we are briefly using the convention of Chapter 6 and writing y as the coordinate on B, and x the fiber coordinate on $X(B)$. Now any real curve contained in $X(B)$ is Lagrangian, so the only role ω plays is in calculating areas of holomorphic disks.

For any $n \in \mathbb{Z}$, let L_n be the image in $X(B)$ of the graph of $x = -ny$, oriented in the direction of increasing y. The minus sign has to do with the choice of sign conventions in defining the Fukaya category. Since x is a periodic coordinate with period 1, and y is a periodic coordinate with period d, this makes sense (see Figure 28). One first notices that if $f : X(B) \to B$

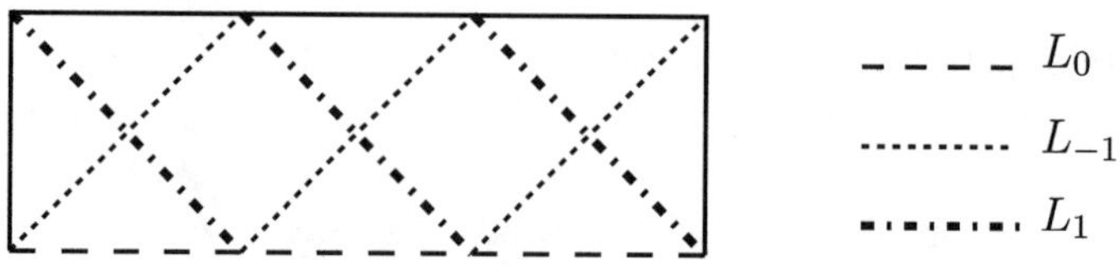

FIGURE 28

is the projection, then

$$f(L_{n_1} \cap L_{n_2}) = f(L_0 \cap L_{n_2 - n_1}) = B\left(\frac{1}{n_2 - n_1}\mathbb{Z}\right)$$

whenever $n_1 \neq n_2$. From §6.3, one expects that L_n should be mirror to a line bundle of degree nd (at least up to sign, but the sign here is imposed on us by conventions having to do with the Fukaya category— it is not particularly important), so the natural expectation is that $\hat{\mathcal{L}}^{\otimes n}$ is mirror to L_n. As a first step of understanding this duality, we have in fact already found a canonical isomorphism between the two sides at the level of cohomology.

Let us be more precise. Suppose $n_1 < n_2$. Then

$$\operatorname{Ext}^i(\hat{\mathcal{L}}^{\otimes n_1}, \hat{\mathcal{L}}^{\otimes n_2}) = H^i(\hat{\mathcal{X}}, \hat{\mathcal{L}}^{\otimes (n_2 - n_1)}) = \begin{cases} \bigoplus_{p \in B\left(\frac{1}{(n_2-n_1)}\mathbb{Z}\right)} [p]\mathbb{C}[[q]] & i = 0 \\ 0 & i = 1 \end{cases}$$

So in particular, the complex $\operatorname{Hom}_{\mathrm{D}^b_\infty(\hat{X})}(\hat{\mathcal{L}}^{\otimes n_1}, \hat{\mathcal{L}}^{\otimes n_2})$ is quasi-isomorphic to the complex

$$\cdots \to 0 \to \bigoplus_{p \in B\left(\frac{1}{(n_2-n_1)}\mathbb{Z}\right)} [p]\mathbb{C}[[q]] \to 0 \to \cdots . \tag{8.15}$$

On the other hand, if $n_1 > n_2$, then

$$\operatorname{Ext}^i(\hat{\mathcal{L}}^{\otimes n_1}, \hat{\mathcal{L}}^{\otimes n_2}) = H^i(\hat{\mathcal{X}}, \hat{\mathcal{L}}^{\otimes (n_2 - n_1)}) = \begin{cases} 0 & i = 0 \\ \bigoplus_{p \in B\left(\frac{1}{(n_2-n_1)}\mathbb{Z}\right)} [p]\mathbb{C}[[q]] & i = 1. \end{cases}$$

The last equality follows from Serre duality, but will be made more explicit later. Thus $\operatorname{Hom}_{\mathrm{D}^b_\infty(\hat{\mathcal{X}})}(\hat{\mathcal{L}}^{\otimes n_1}, \hat{\mathcal{L}}^{\otimes n_2})$ is quasi-isomorphic to the complex non-trivial only in degree 1

$$\cdots \to 0 \to \bigoplus_{p \in B\left(\frac{1}{(n_2-n_1)}\mathbb{Z}\right)} [p]\mathbb{C}[[q]] \to 0 \cdots . \tag{8.16}$$

We compare this to the Fukaya side of the picture. We need to choose gradings on the Lagrangians L_n. The L_n's are straight lines in the universal cover $\mathbb{R}^2$ of $\check{X}(B)$, so we can take the grading to be constant. As in Figure 18, we will always take the grading to be in the interval $(-1, 1)$, depending on the angle of the line. Having done so, it is then clear from Figure 18 that all points of $L_{n_1} \cap L_{n_2}$ are index 0 if $n_1 < n_2$ and index 1 if $n_1 > n_2$, and of course these intersection points are in one-to-one correspondence with $B(\frac{1}{(n_2-n_1)}\mathbb{Z})$. Thus the Floer complexes coincide with (8.15) or (8.16) in the two cases, after tensoring with Λ_{nov}.

We are now ready to compare multiplication maps. We will first do so for the compositions

$$\mathrm{Hom}(\hat{\mathcal{L}}^{\otimes n_0+n_1}, \hat{\mathcal{L}}^{\otimes n_0+n_1+n_2}) \otimes \mathrm{Hom}(\hat{\mathcal{L}}^{\otimes n_0}, \hat{\mathcal{L}}^{\otimes n_0+n_1}) \\ \to \mathrm{Hom}(\hat{\mathcal{L}}^{\otimes n_0}, \hat{\mathcal{L}}^{\otimes n_0+n_1+n_2}),$$

only for $n_1, n_2 > 0$, as we can use the explicit description of theta functions; the remaining cases will be dealt with once we introduce Čech complexes to compute the higher multiplication maps.

Without loss of generality, we can take $n_0 = 0$, in which case we want to compute the product

$$H^0(\hat{\mathcal{L}}^{\otimes n_1}) \otimes H^0(\hat{\mathcal{L}}^{\otimes n_2}) \to H^0(\hat{\mathcal{L}}^{\otimes (n_1+n_2)}).$$

Then for $p_1 \in B(\frac{1}{n_1}\mathbb{Z})$, $p_2 \in B(\frac{1}{n_2}\mathbb{Z})$,

$$\begin{aligned}
\vartheta_{n_1,p_1} \cdot \vartheta_{n_2,p_2} &= \left(\sum_{s_1=-\infty}^{\infty} z^{T_{n_1}^{s_1}(n_1 p_1, n_1\varphi(p_1))} \right) \left(\sum_{s_2=-\infty}^{\infty} z^{T_{n_2}^{s_2}(n_2 p_2, n_2\varphi(p_2))} \right) \\
&= \sum_{s_1,s_2=-\infty}^{\infty} z^{(n_1(p_1+s_1 d), n_1\varphi(p_1+s_1 d))} \cdot z^{(n_2(p_2+s_2 d), n_2\varphi(p_2+s_2 d))} \\
&= \sum_{s_1,s_2=-\infty}^{\infty} z^{((n_1+n_2)(\frac{n_1(p_1+s_1 d)+n_2(p_2+s_2 d)}{n_1+n_2}), n_1\varphi(p_1+s_1 d)+n_2\varphi(p_2+s_2 d))} \\
&= \sum_{s_1,s_2=-\infty}^{\infty} z^{((n_1+n_2)(\frac{n_1 p_1+n_2(p_2+(s_2-s_1)d)}{n_1+n_2}+s_1 d), (n_1+n_2)\varphi(\frac{n_1 p_1+n_2(p_2+(s_2-s_1)d)}{n_1+n_2}+s_1 d))} \\
&\quad \cdot q^{n_1\varphi(p_1+s_1 d)+n_2\varphi(p_2+s_2 d)-(n_1+n_2)\varphi(\frac{n_1 p_1+n_2(p_2+(s_2-s_1)d)}{n_1+n_2}+s_1 d)}.
\end{aligned}$$

Now using (8.12), one can check that

$$\begin{aligned}
&n_1\varphi(p_1+s_1 d) + n_2\varphi(p_2+s_2 d) \\
&\qquad - (n_1+n_2)\varphi\left(\frac{n_1 p_1 + n_2(p_2+(s_2-s_1)d)}{n_1+n_2} + s_1 d\right) \\
&\quad = n_1\varphi(p_1) + n_2\varphi(p_2+(s_2-s_1)d) - \\
&\qquad (n_1+n_2)\varphi\left(\frac{n_1 p_1 + n_2(p_2+(s_2-s_1)d)}{n_1+n_2}\right).
\end{aligned}$$

Define for $p_1 \in \frac{1}{n_1}\mathbb{Z}, p_2 \in \frac{1}{n_2}\mathbb{Z}$,

$$\deg(p_1, p_2) = n_1\varphi(p_1) + n_2\varphi(p_2) - (n_1+n_2)\varphi\left(\frac{n_1 p_1 + n_2 p_2}{n_1+n_2}\right).$$

By convexity of φ, this is always non-negative. We can then write

$$\vartheta_{n_1,p_1} \cdot \vartheta_{n_2,p_2} = \sum_{\alpha=-\infty}^{\infty} \sum_{s=-\infty}^{\infty} z^{T^s_{n_1+n_2}((n_1+n_2)(\frac{n_1p_1+n_2(p_2+\alpha d)}{n_1+n_2}),(n_1+n_2)\varphi(\frac{n_1p_1+n_2(p_2+\alpha d)}{n_1+n_2}))} q^{\deg(p_1,p_2+\alpha d)}$$

$$= \sum_{\alpha=-\infty}^{\infty} \vartheta_{n_1+n_2,\frac{n_1p_1+n_2(p_2+\alpha d)}{n_1+n_2}} q^{\deg(p_1,p_2+\alpha d)}.$$

This can be rewritten geometrically as follows. We are viewing p_1 and p_2 as points in $\mathbb{R}$, the universal cover of B; as such, they are only defined modulo d. Thus, we can pick liftings p_1 and $p_2+\alpha d$ for $\alpha \in \mathbb{Z}$, and the interval $[p_1, p_2+\alpha d]$ maps to an affine line segment on B joining p_1 and p_2. Thus the sum over all α can be viewed as a sum over affine line segments with endpoints p_1 mod $d\mathbb{Z}$ and p_2 mod $d\mathbb{Z}$. Note that such a line segment gives a contribution from the point of B which is a weighted average of the lifting of the endpoints.

REMARK 8.18. This is a generalization of a very simple phenomenon for projective toric varieties. If Δ is a compact lattice polytope defining a projective toric variety $(\mathbb{P}_\Delta, \mathcal{O}_{\mathbb{P}_\Delta}(1))$, then the points of $\Delta(\frac{1}{n}\mathbb{Z})$ form a basis for $H^0(\mathbb{P}_\Delta, \mathcal{O}_{\mathbb{P}_\Delta}(n))$. Multiplying two sections $m_1 \in \Delta(\frac{1}{n_1}\mathbb{Z})$, $m_2 \in \Delta(\frac{1}{n_2}\mathbb{Z})$ gives rise to a weighted average $\frac{n_1m_1+n_2m_2}{n_1+n_2} \in \Delta(\frac{1}{n_1+n_2}\mathbb{Z})$. □

We can now make contact with the Fukaya side of the story. Without loss of generality, we can take $n_0 = 0$ and consider the multiplication map

$$\mathrm{Hom}_{\mathrm{Fuk}(\check{X})}(L_{n_1}, L_{n_1+n_2}) \otimes \mathrm{Hom}_{\mathrm{Fuk}(\check{X})}(L_0, L_{n_1}) \to \mathrm{Hom}_{\mathrm{Fuk}(\check{X})}(L_0, L_{n_1+n_2}).$$

As we have seen earlier, p_1 and p_2 determine elements of $\mathrm{Hom}_{\mathrm{Fuk}(\check{X})}(L_0, L_{n_1})$ and $\mathrm{Hom}_{\mathrm{Fuk}(\check{X})}(L_{n_1}, L_{n_1+n_2})$ respectively. Given α, we obtain a lifting of L_0, L_{n_1} and $L_{n_1+n_2}$ to the universal cover $\mathbb{C}$ of $\check{X} = X(B) = \mathbb{C}/\langle d, \sqrt{-1}\rangle$ as depicted in Figure 29. The weighted average appears at the intersection point of $\tilde{L}_0 \cap \tilde{L}_{n_1+n_2}$, and so the contribution to the Floer product from the shaded triangle T, which is just a holomorphic disk, is just $q^{\int_T \omega}$, appearing with a positive sign according to §8.3.3. Now $\int_T \omega$ is the area of the triangle T, which as depicted in Figure 29, is

$$\frac{1}{2}\left(\frac{n_1p_1+n_2(p_2+\alpha d)}{n_1+n_2} - p_1\right) \cdot (n_1(p_2+\alpha d - p_1))$$

$$= \frac{1}{2}\frac{n_1n_2((p_2+\alpha d)-p_1)^2}{n_1+n_2}$$

$$= n_1\psi(p_1) + n_2\psi(p_2+\alpha d) - (n_1+n_2)\psi\left(\frac{n_1p_1+n_2(p_2+\alpha d)}{n_1+n_2}\right)$$

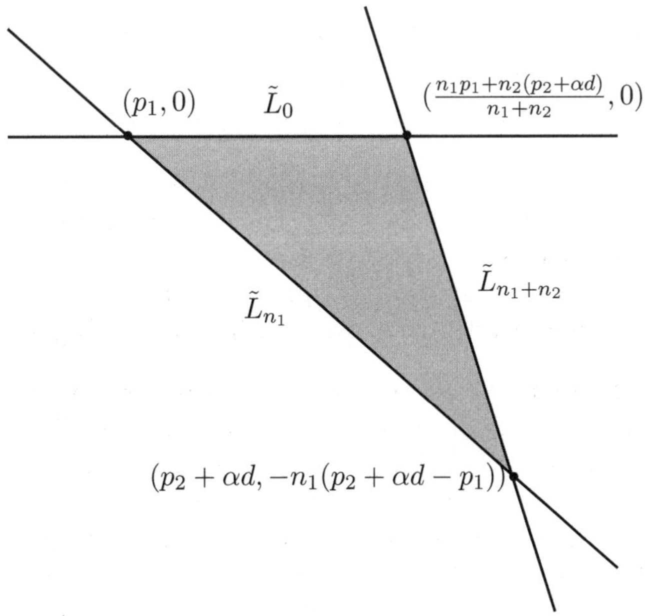

FIGURE 29

where

$$\psi(x) = \frac{x(x-1)}{2}, \tag{8.17}$$

as is easily checked. This is not the same thing as $\deg(p_1, p_2 + d\alpha)$, so at first glance we aren't getting the same contribution. However, setting

$$A(x) = \psi(x) - \varphi(x), \tag{8.18}$$

we note that $A(x + d) = A(x)$. We can then define a map for any n_1, n_2, $n_1 \neq n_2$,

$$\mathrm{Hom}_{\mathrm{D}^b(\hat{\mathcal{X}})}(\hat{\mathcal{L}}^{\otimes n_1}, \hat{\mathcal{L}}^{\otimes n_2}) \otimes_{\mathbb{C}[[q]]} \Lambda_{\mathrm{nov}} \to \mathrm{Hom}_{\mathrm{Fuk}(X(B))}(L_{n_1}, L_{n_2})$$

given, for $p \in B(\frac{1}{n_2-n_1}\mathbb{Z})$, by

$$[p] \mapsto [p]q^{-(n_2-n_1)A(p)}. \tag{8.19}$$

This fixes the problem, as in our situation for

$$H^0(\mathcal{L}^{\otimes n_1}) \otimes H^0(\mathcal{L}^{\otimes n_2}) \to H^0(\mathcal{L}^{\otimes n_1+n_2}),$$

the coefficient of $\left[\frac{n_1p_1+n_2(p_2+\alpha d)}{n_1+n_2}\right]$ in the product of $[p_1]$ and $[p_2]$ is given by $q^{\deg(p_1,p_2+\alpha d)}$. Transporting this product to

$$\mathrm{Hom}^0_{\mathrm{Fuk}(X(B))}(L_0, L_{n_1}) \otimes \mathrm{Hom}^0_{\mathrm{Fuk}(X(B))}(L_{n_1}, L_{n_1+n_2}) \\ \to \mathrm{Hom}^0_{\mathrm{Fuk}(X(B))}(L_0, L_{n_1+n_2})$$

under the identification (8.19) gives the coefficient of $\left[\frac{n_1p_1+n_2(p_2+\alpha d)}{n_1+n_2}\right]$ in the product of $[p_1]\otimes 1$ and $[p_2+\alpha d]\otimes 1$ as

$$q^{n_1A(p_1)}q^{n_2A(p_2)}q^{\deg(p_1,p_2+\alpha d)}q^{-(n_1+n_2)A((n_1p_1+n_2(p_2+\alpha d))/(n_1+n_2))} = q^{\mathrm{Area}(T)}.$$

Thus the multiplication of theta functions agrees with Floer multiplication.

We would now like to understand the higher products, and to do so, we need to get our hands dirty with the Čech description of $\mathrm{D}^b_\infty(\check{\mathcal{X}})$. However, we will first give a geometric description of the Fukaya side of the story which will mesh well with the above description of multiplication.

8.4.3. Tropical Morse trees. A key paper by Fukaya and Oh [**167**] related Floer homology of sections of the cotangent bundle of a manifold to the Morse homology of the manifold. The idea roughly is to approximate holomorphic disks by structures arising from gradient flow lines. This idea was exploited by Kontsevich and Soibelman in [**311**] to prove a version of homological mirror symmetry for abelian varieties. In the elliptic curve case, this step is particularly easy, and we shall use a variation on this idea, defining what we will call *tropical Morse trees*. In this exposition, we follow joint work of Abouzaid, Gross and Siebert in [**1**].

DEFINITION 8.19. Let B be an integral affine manifold. Given a sequence of distinct integers $n_0,\ldots,n_d\in\mathbb{Z}$ and any metric ribbon tree S we can label the edges e of S with integers n_e as follows. If e is an external incoming edge, attached to the ith external vertex, then $n_e = n_i - n_{i-1}$; otherwise, if e comes out of a vertex v, then n_e is the sum of all numbers labelling the edges coming into v. Then given in addition points

$$p_{i,i+1}\in B\left(\frac{1}{n_{i+1}-n_i}\mathbb{Z}\right)$$

and

$$p_{0,d}\in B\left(\frac{1}{n_d-n_0}\mathbb{Z}\right)$$

we define

$$\mathcal{S}^{\mathrm{trop}}_d(p_{0,d};p_{0,1},\ldots,p_{d-1,d})$$

to be the moduli space of *tropical Morse trees* on B, i.e., continuous maps $\phi: S\to B$ from a ribbon tree with a collection of *affine displacement vectors*, i.e., for each edge e of S, a section $\mathbf{v}_e\in\Gamma(e,(\phi|_e)^*TB)$, satisfying the following properties:

(1) If v is the ith external incoming vertex, then $\phi(v)=p_{i-1,i}$; if v is the external outgoing vertex, then $\phi(v)=p_{0,d}$.
(2) If e is an edge of S, then $\phi(e)$ is locally an affine line segment on B. (This line segment can have irrational slope). If e is an external edge, we also allow $\phi(e)$ to be a point.

(3) If v is an external vertex and e the unique edge of S containing v, then $\mathbf{v}_e(v) = 0$.

(4) For an edge e of S identified with $[0,1]$ with coordinate s, with the edge e oriented from 0 to 1, then $\mathbf{v}_e(s)$ is tangent to $\phi(e)$ at $\phi(s)$, pointing in the same direction as the orientation on $\phi(e)$ induced by that on e. Furthermore, using the affine structure to identify $(\phi|_e)^*TB$ with the trivial bundle on e, we have

$$\frac{d}{ds}\mathbf{v}_e(s) = n_e\phi_*(\partial/\partial s).$$

(5) If v is an internal vertex of S with incoming edges $e_1, \ldots, e_p$ and outgoing edge e_{out}, then

$$\mathbf{v}_{e_{\text{out}}}(v) = \sum_{i=1}^{p} \mathbf{v}_{e_i}(v).$$

(6) The length of an edge e in S (remember that each edge of a ribbon tree comes along with a length, the external edges having infinite length) coincides with

$$\frac{1}{n_e}\log\left(\frac{\mathbf{v}_e(1)}{\mathbf{v}_e(0)}\right).$$

Since $\mathbf{v}_e(0)$ and $\mathbf{v}_e(1)$ are proportional vectors pointing in the same direction, their quotient makes sense as a positive number. There is one special case: if e is an external edge that is contracted by ϕ, then $\mathbf{v}_e(0) = \mathbf{v}_e(1) = 0$, but we still take the length to be infinite.

REMARK 8.20. The meaning of (4) can be explained more clearly in the case that $B = \mathbb{R}^n$: it tells us that

$$\mathbf{v}_e(s) = \mathbf{v}_e(0) + n_e(\phi(s) - \phi(0)).$$

Item (6) in fact plays no significant role in this definition, other than giving us a map from $\mathcal{S}_d^{\text{trop}}(p_{0,d}; p_{0,1}, \ldots, p_{d-1,d})$ to $\mathcal{S}_d$. It will however play a more important role later on, when we need to construct a quasi-isomorphism of categories.

EXERCISE 8.21. One can understand the conditions of this definition by carrying out the following exercise, showing that the notion of tropical Morse tree is essentially just the Legendre transform of the notion of a gradient tree. Suppose B was obtained as the Legendre transform of $(\check{B}, \check{K})$, as in §6.1.2. We think of $\check{K}$ as a function on the universal cover $\tilde{B}$ of B. Given $n_0, \ldots, n_d \in \mathbb{Z}$, show that a tropical Morse tree on B coincides with a gradient tree on $\check{B}$ (with the metric induced by $\check{K}$) for the functions $(n_0\check{K}+l_0, \ldots, n_d\check{K}+l_d)$ for some affine linear functions $l_0, \ldots, l_d$ with integral slope on the universal cover of $\check{B}$. Furthermore, the length of time of the gradient flow on each edge is precisely that given in item (6) of the Definition.

EXAMPLE 8.22. The description of multiplication of theta functions already yields examples of tropical Morse trees. Let $B = \mathbb{R}/d\mathbb{Z}$, and let T be the ribbon tree with two incoming edges, and $n_1, n_2 > 0$. Then Figure 30 shows a tropical Morse tree representing the multiplication of $\vartheta_{n_1,p_{0,1}}$ and $\vartheta_{n_2,p_{1,2}}$ to get a contribution from $\vartheta_{n_3,p_{0,2}}$, with $n_3 = n_1 + n_2$. Note that the outgoing external edge must be contracted to a point, as otherwise the affine displacement vector cannot be zero at the outgoing vertex.

If on the other hand $n_1 > 0$, $n_2 < 0$ and $n_3 = n_1+n_2 < 0$, then Figure 31 gives another example of a tropical Morse tree. This time, by Condition (4) the incoming edge with negative weight must be contracted to a point. Note that, in both cases, we can view these maps of trees as factoring through the universal cover, so $p_{0,1}, p_{1,2}$ and $p_{0,2}$ can be any liftings of the points in B to $\mathbb{R}$.

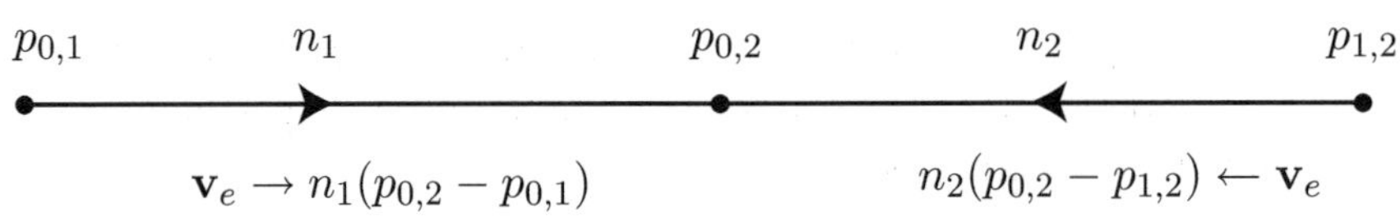

FIGURE 30. A tropical Morse tree: the requirement from (3) and (5) of the definition requires that $n_1(p_{0,2}-p_{0,1})+n_2(p_{0,2}-p_{1,2}) = 0$, or equivalently $p_{0,2} = (n_1p_{0,1} + n_2p_{1,2})/(n_1 + n_2)$.

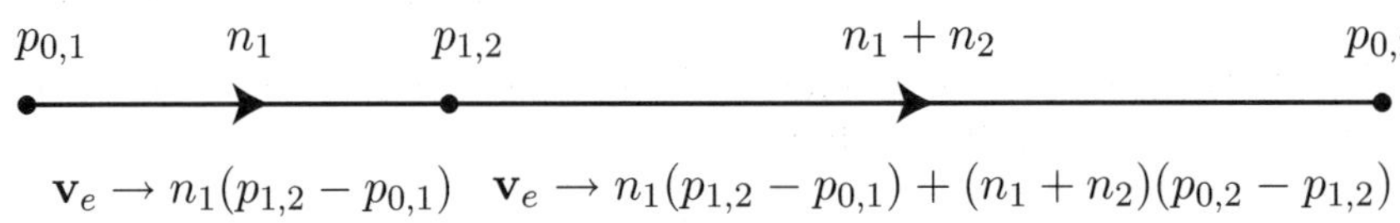

FIGURE 31. The condition that $\mathbf{v}_e$ is zero at $p_{0,2}$ again implies that $p_{0,2} = (n_1p_{0,1} + n_2p_{1,2})/(n_1 + n_2)$.

8.4.4. Holomorphic polygons and tropical Morse trees. We now explain how a tropical Morse tree gives rise to a union of polygons in $X(B)$. If $\dim B \geq 2$, this union need not be an actual holomorphic submanifold, as it is a piecewise linear object, but if $\dim B = 1$, it is always a holomorphic polygon.

The Lagrangians L_n of $X(B) \to B$ for $B = \mathbb{R}/d\mathbb{Z}$ can be generalized to any integral affine manifold B of dimension r: in local integral affine coordinates $y_1, \ldots, y_r$ on B and fiberwise coordinates $x_1, \ldots, x_r$ as in §6.2.1 on TB, one can define a section $\sigma_n : B \to X(B)$ via the local formula

$$\sigma_n(y_1, \ldots, y_r) = (y_1, \ldots, y_r, -ny_1, \ldots, -ny_r);$$

one checks easily that this is well-defined under integral affine changes of coordinates. We set

$$L_n = \sigma_n(B).$$

In this case we don't actually have a symplectic structure on $X(B)$, so it doesn't make sense to say L_n is Lagrangian.

Now we show how a tropical Morse tree $\phi : S \to B$ defines a piecewise linear disk. Any edge e of S is labelled by $n_e = n_j - n_i$ for some $j > i$. Consider the map

$$R_e : e \times [0,1] \to X(B)$$

defined by

$$R_e(s,t) = \sigma_{n_i}(\phi(s)) - t \cdot \mathbf{v}_e(s),$$

where $t \cdot \mathbf{v}_e(s)$ is viewed as a tangent vector at $\phi(s)$. Write the vertices of e as v_{in} and v_{out}. We note that

$$R_e(e \times \{0\}) \subseteq L_{n_i}$$

by definition, and if

$$R_e(v_{\text{in}} \times \{1\}) \subseteq L_{n_j},$$

i.e.,

$$-\mathbf{v}_e(v_{\text{in}}) = \sigma_{n_j}(\phi(v_{\text{in}})) - \sigma_{n_i}(\phi(v_{\text{in}})) \mod \Lambda,$$

then by Condition (4) of Definition 8.19, modulo Λ we have

$$\begin{aligned} -\frac{d}{ds}\mathbf{v}_e(s) &= -(n_j - n_i)(\phi|_e)_* \frac{\partial}{\partial s} \\ &= \frac{d}{ds}\big(\sigma_{n_j}(\phi(s)) - \sigma_{n_i}(\phi(s))\big) \end{aligned}$$

Thus $R_e(v_{\text{in}} \times \{1\}) \subseteq L_{n_j}$ implies $R_e(e \times \{1\}) \subseteq L_{n_j}$.

Next consider a vertex v of S. If v is an external incoming vertex v_i, then $\phi(v_i) \in B\left(\frac{1}{n_i - n_{i-1}}\mathbb{Z}\right)$ and so $\sigma_{n_i}(\phi(v_i)) = \sigma_{n_{i-1}}(\phi(v_i)) \in L_{n_{i-1}} \cap L_{n_i}$. In particular, $R_e(v_i \times \{1\}) \subseteq L_{n_i}$, so that

$$\begin{aligned} R_e(e \times \{0\}) &\subseteq L_{n_{i-1}}, \\ R_e(e \times \{1\}) &\subseteq L_{n_i} \end{aligned}$$

for e the unique edge attracted to v_i.

If v is any interior vertex with incoming edges $e_1, \ldots, e_p$, outgoing edge e_{out}, with e_j weighted by $n_{i_j} - n_{i_{j-1}}$, for $i_0 < \cdots < i_p$, then inductively we can assume

$$\begin{aligned} R_{e_j}(e_j \times \{0\}) &\subseteq L_{n_{i_{j-1}}}, \\ R_{e_j}(e_j \times \{1\}) &\subseteq L_{n_{i_j}}. \end{aligned}$$

Then (5) of Definition 8.19 tells us that in fact

$$\begin{aligned} R_{e_{\text{out}}}(v \times \{1\}) &= \sigma_{n_{i_0}}(\phi(v)) - \mathbf{v}_{e_{\text{out}}}(v) \\ &= \sigma_{n_{i_0}}(\phi(v)) - \sum \mathbf{v}_{e_j}(v) \\ &= \sigma_{n_{i_p}}(\phi(v)) \in L_{n_{i_p}}. \end{aligned}$$

Hence

$$\begin{aligned} R_{e_{\text{out}}}(e_{\text{out}} \times \{0\}) &\subseteq L_{n_{i_0}} \\ R_{e_{\text{out}}}(e_{\text{out}} \times \{1\}) &\subseteq L_{n_{i_p}}. \end{aligned}$$

Now at the vertex v, the rectangles $R_{e_i}(e_i \times [0,1])$, $1 \leq i \leq p$ and $R_{e_{\text{out}}}(e_{\text{out}} \times [0,1])$ do not necessarily fit together. However, there is a polygon P_v in $TB_{\phi(v)}$ whose edge vectors are $-\mathbf{v}_{e_1}(v), \ldots, -\mathbf{v}_{e_p}(v)$ and $-\mathbf{v}_{e_{\text{out}}}(v)$, as depicted in Figure 32.

By patching together the rectangles $e \times [0,1]$ and the polygonal regions P_v to create a fattening of the tree S, we obtain a topological disk, $\text{Fat}(S)$, see Figure 33. It is probably more intuitive to use triangles instead of rectangles for the external edges, as the edge of such a rectangle corresponding to the external vertex is always contracted. We also can ignore any contracted external edge. We obtain a continuous map $R : \text{Fat}(S) \to X(B)$ by patching together the maps R_e (we aren't too picky about the details here) and this gives a map from a disk into $X(B)$. Using the observations in §6.3, the image of R will be a union of holomorphic triangles, quadrilaterals and non-holomorphic polygons contained in fibers of $X(B) \to B$.

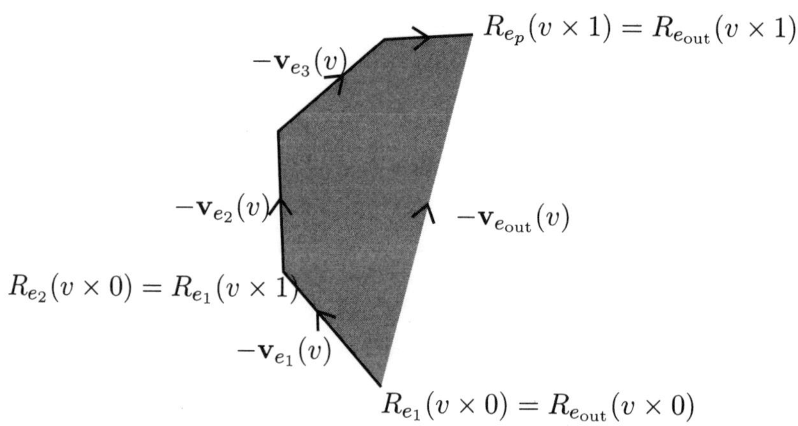

FIGURE 32

EXAMPLE 8.23. Applying the above construction on an elliptic curve always gives a genuine holomorphic polygon, exhibited as a union of triangles and quadrilaterals. There are no two-dimensional P_v's. For example, Figures

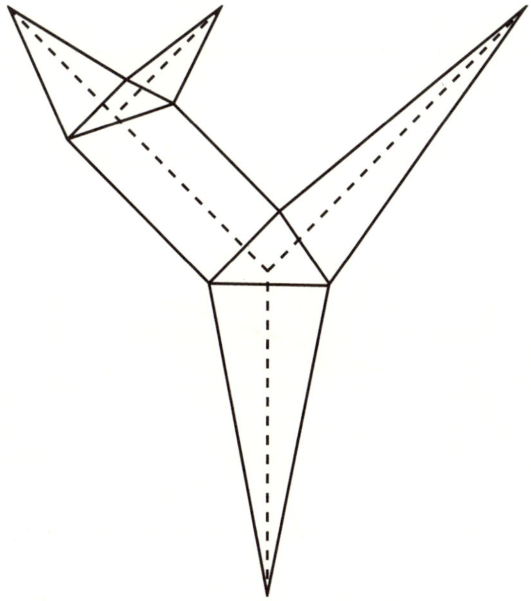

FIGURE 33

30 and 31 give triangles, the former corresponding to the triangle of Figure 29.

Conversely, in the elliptic curve case, given any holomorphic polygon bounded by Lagrangians $L_{n_{i_0}}, \ldots, L_{n_{i_d}}$, we can construct a tropical Morse tree which gives rise to it. We leave the details to the reader, providing some examples in Figures 34, 35.

8.4.5. Čech calculations. We can now state the main result of this section. We will consider the A_∞-precategories $\mathrm{D}(\hat{\mathcal{X}})$ and $\check{\mathrm{D}}(X(B))$ defined as follows. The objects of $\mathrm{D}(\hat{\mathcal{X}})$ are just arbitrary powers of the line bundle $\hat{\mathcal{L}}$ on $\hat{\mathcal{X}}$. A sequence of line bundle $(\hat{\mathcal{L}}^{\otimes n_1}, \ldots, \hat{\mathcal{L}}^{\otimes n_d})$ is transversal if $n_i \neq n_j$ for any $i \neq j$. Then $\mathrm{Hom}_{\mathrm{D}(\hat{\mathcal{X}})}(\hat{\mathcal{L}}^{\otimes n_1}, \hat{\mathcal{L}}^{\otimes n_2})$ will be a Čech complex computing the cohomology of $\hat{\mathcal{L}}^{\otimes(n_2-n_1)}$, so that $m_1^{\mathrm{D}(\hat{\mathcal{X}})}$ is given by the Čech differential and $m_2^{\mathrm{D}(\hat{\mathcal{X}})}$ is given by composition as in §8.2.1; all higher multiplication maps are zero.

The objects of $\check{\mathrm{D}}(X(B))$ will be the Lagrangians L_i, $i \in \mathbb{Z}$, oriented and graded as discussed in §8.3.3, with non-trivial spin structure. Morphisms and higher multiplication maps are given as usual as in §8.3.3, again with transversal sequences being $(L_{n_1}, \ldots, L_{n_d})$ with $n_i \neq n_j$ for any $i \neq j$.

Then the goal of this section is to prove

THEOREM 8.24. $\mathrm{D}(\hat{\mathcal{X}}) \otimes_{\mathbb{C}[[q]]} \Lambda_{\mathrm{nov}}$ *and* $\check{\mathrm{D}}(X(B))$ *are quasi-equivalent. Here the tensor product just means we tensor all spaces of morphisms by* Λ_{nov}.

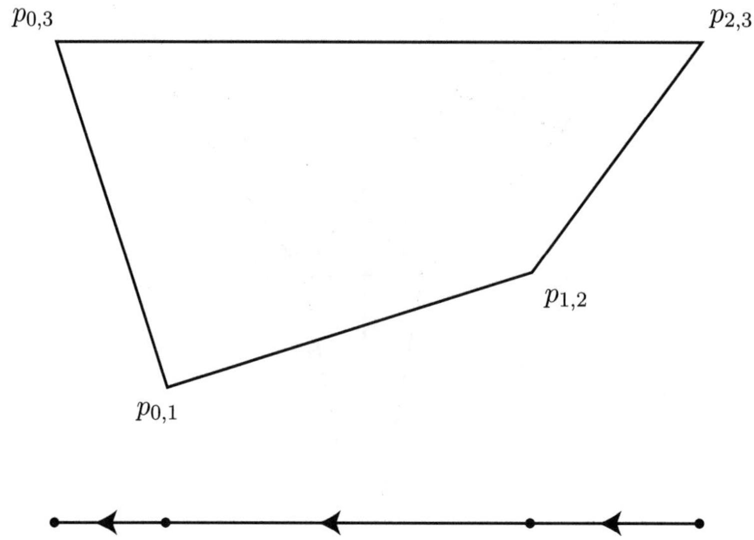

FIGURE 34. The external edges with vertices $p_{0,1}$ and $p_{1,2}$ are length zero.

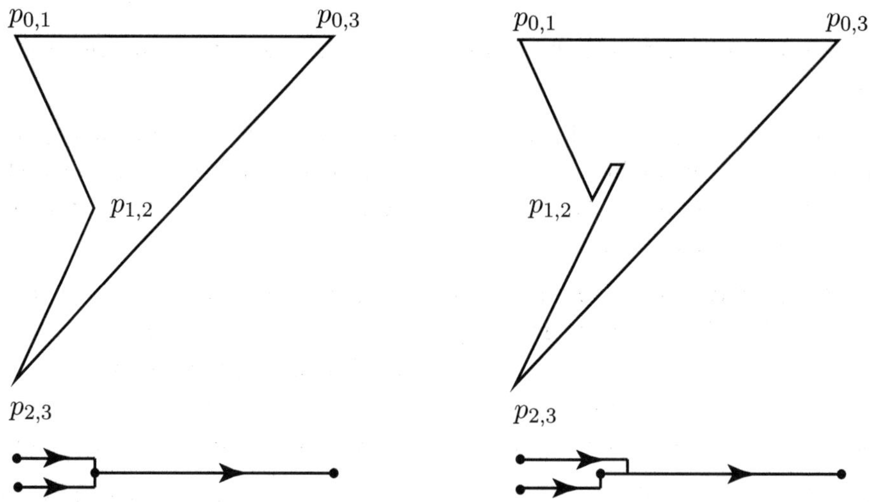

FIGURE 35. In the left-hand diagram, S has a quadrivalent vertex, with the edge coming from $p_{1,2}$ contracted. The Morse tree is drawn schematically since these edges are superimposed on B. This Morse tree is one element of a one-parameter family; the example on the right gives another element, with corresponding holomorphic polygon.

We will begin by describing our choice of Čech open covering, and hence the morphism spaces for $\mathrm{D}(\hat{\mathcal{X}})$.

We have seen that X_Σ has a natural open cover $\{U_w | w \text{ a vertex of } \sigma\}$. Each set U_w restricts to an open subset of $f^{-1}(0)$, and hence defines an open set $\hat{U}_w$ of $\hat{X}_\Sigma$; as remarked earlier, this is a ringed space with underlying topological space the same as the underlying topological space of $\operatorname{Spec} \mathbb{C}[\check{w}^\vee \cap \mathbb{Z}^2]/(q)$ but with structure sheaf associated to the ring

$$\varprojlim \mathbb{C}[\check{w}^\vee \cap \mathbb{Z}^2]/(q^n) = \mathbb{C}[\check{w}^\vee \cap \mathbb{Z}^2]\hat{\otimes}_{\mathbb{C}[q]}\mathbb{C}[[q]].$$

The open sets $\hat{U}_w$ descend to open sets of the quotient $\hat{\mathcal{X}} = \hat{X}_\Sigma/\mathbb{Z}$, so we have an open cover of $\hat{\mathcal{X}}$ indexed by a set of representatives of the vertices of Δ modulo the $\mathbb{Z}$-action, or equivalently, by elements of $B(\mathbb{Z})$. Furthermore, if w_1 and w_2 are the vertices of an edge σ of Δ, then $\hat{U}_{w_1} \cap \hat{U}_{w_2} = \hat{U}_\sigma$, while if w_1 and w_2 are not adjacent in $B(\mathbb{Z})$, then $\hat{U}_{w_1} \cap \hat{U}_{w_2} = \emptyset$. See Figure 36.

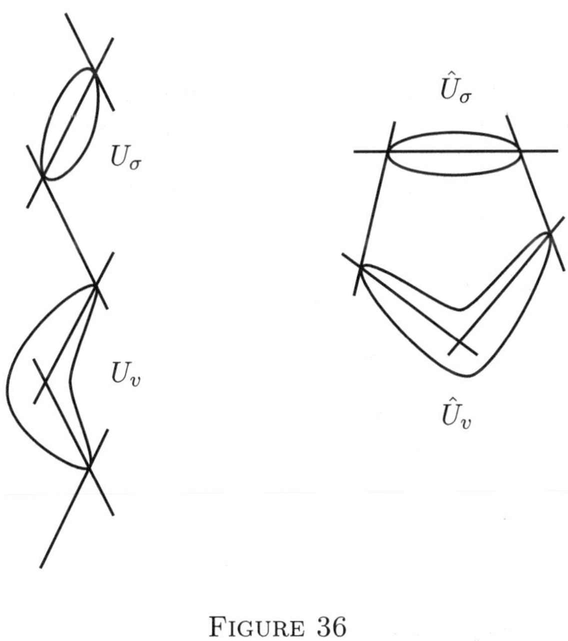

FIGURE 36

We now have to make some choices. We need to choose a specific set of representatives for these open sets, indexed by an ordered index set. We use the index set $I = \{0, \ldots, d-1\}$ with the standard ordering, and write

$$\hat{U}_i := \hat{U}_{w_i}$$

where

$$w_i := (i, \varphi(i)).$$

Hence $\hat{U}_{i,j} := \hat{U}_i \cap \hat{U}_j = \emptyset$ for $i < j$ unless $j = i + 1$ or $i = 0, j = d - 1$.

We describe the Čech complex as follows. Recall that $\hat{\mathcal{L}}^{\otimes n}|_{\hat{U}_w} = \mathscr{O}_{\hat{U}_w} \cdot z^{nw}$ canonically (this works for n positive or negative). Thus

$$\Gamma(\hat{U}_w, \hat{\mathcal{L}}^{\otimes n}) = \Gamma^n_w := (\mathbb{C}[\check{w}^\vee \cap \mathbb{Z}^2]\hat{\otimes}_{\mathbb{C}[q]}\mathbb{C}[[q]]) \cdot z^{nw}.$$

In particular, we have the identification

$$\Gamma(\hat{U}_i, \hat{\mathcal{L}}^{\otimes n}) = \Gamma^n_{w_i}. \tag{8.20}$$

If σ is an edge of Δ with endpoints w_i and w_j, then

$$\begin{aligned}\Gamma(\hat{U}_\sigma, \hat{\mathcal{L}}^{\otimes n}) = \Gamma(\hat{U}_{w_i} \cap \hat{U}_{w_j}, \hat{\mathcal{L}}^{\otimes n}) = \Gamma^n_\sigma \quad &:= \quad (\mathbb{C}[\check{\sigma}^\vee \cap \mathbb{Z}^2]\hat{\otimes}_{\mathbb{C}[q]}\mathbb{C}[[q]]) \cdot z^{nw_i} \\ &= \quad (\mathbb{C}[\check{\sigma}^\vee \cap \mathbb{Z}^2]\hat{\otimes}_{\mathbb{C}[q]}\mathbb{C}[[q]]) \cdot z^{nw_j}.\end{aligned}$$

Thus if σ_i is the interval with endpoints w_i and w_{i+1}, we have canonically for $0 \le i < d-1$

$$\Gamma(\hat{U}_{i,i+1}, \hat{\mathcal{L}}^{\otimes n}) = \Gamma^n_{\sigma_i}, \tag{8.21}$$

while

$$\Gamma(\hat{U}_{0,d-1}, \hat{\mathcal{L}}^{\otimes n}) = \Gamma^n_{\sigma_{d-1}}. \tag{8.22}$$

This creates a certain asymmetry we will have to deal with repeatedly in what follows. In particular, if $0 \le i < d-1$, the restriction maps

$$\begin{aligned}\Gamma(\hat{U}_i, \hat{\mathcal{L}}^{\otimes n}) \quad &\to \quad \Gamma(\hat{U}_{i,i+1}, \hat{\mathcal{L}}^{\otimes n}) \\ \Gamma(\hat{U}_{i+1}, \hat{\mathcal{L}}^{\otimes n}) \quad &\to \quad \Gamma(\hat{U}_{i,i+1}, \hat{\mathcal{L}}^{\otimes n})\end{aligned}$$

are the obvious inclusions $\Gamma^n_{w_i}, \Gamma^n_{w_{i+1}} \subseteq \Gamma^n_{\sigma_i}$, as is the restriction map

$$\Gamma(\hat{U}_{d-1}, \hat{\mathcal{L}}^{\otimes n}) \to \Gamma(\hat{U}_{0,d-1}, \hat{\mathcal{L}}^{\otimes n}).$$

On the other hand, the restriction map

$$\Gamma(\hat{U}_0, \hat{\mathcal{L}}^{\otimes n}) \to \Gamma(\hat{U}_{0,d-1}, \hat{\mathcal{L}}^{\otimes n})$$

is given by the composition

$$\Gamma^n_{w_0} \xrightarrow{T_n} \Gamma^n_{w_d} \subseteq \Gamma^n_{\sigma_{d-1}}.$$

The Čech complex is now

$$\begin{aligned}C^0(\hat{\mathcal{L}}^{\otimes n}) &= \bigoplus_{i \in I} \Gamma(\hat{U}_i, \hat{\mathcal{L}}^{\otimes n}) \\ &\to C^1(\hat{\mathcal{L}}^{\otimes n}) = \left(\bigoplus_{0 \le i < d-1} \Gamma(\hat{U}_{i,i+1}, \hat{\mathcal{L}}^{\otimes n}) \right) \oplus \Gamma(\hat{U}_{0,d-1}, \hat{\mathcal{L}}^{\otimes n}).\end{aligned}$$

We write an element of the Čech complex as $\sum_i (\tau_i, f_i)$, for

$$\tau_i \in \{w_i, \sigma_i | 0 \le i \le d-1\}$$

and $f_i \in \Gamma(\hat{U}_{\tau_i}, \mathcal{L}^{\otimes n})$. Then we have explicitly
(8.23)

$$m_1^{\mathrm{D}(\hat{\mathcal{X}})}\Big(\sum_{i=0}^{d-1}(w_i, z^{m_i})\Big) = \sum_{i=0}^{d-2}(\sigma_i, z^{m_{i+1}} - z^{m_i}) + (\sigma_{d-1}, z^{m_{d-1}} - z^{T_n(m_0)}).$$

On the other hand, we can describe $m_2^{\mathrm{D}(\hat{\mathcal{X}})}$ in all cases, with $0 \le j < d-1$ in the second and fourth lines:

(8.24)

$$m_2((w_j, z^{m_j}), (w_i, z^{m_i})) = \begin{cases} (w_i, z^{m_i+m_j}) & \text{if } i = j \\ 0 & \text{otherwise} \end{cases}$$

$$m_2((w_i, z^{m_i}), (\sigma_j, z^{m_j})) = \begin{cases} (\sigma_i, -z^{m_i+m_j}) & \text{if } i = j \\ 0 & \text{otherwise} \end{cases}$$

$$m_2((w_i, z^{m_i}), (\sigma_{d-1}, z^{m_j})) = \begin{cases} (\sigma_{d-1}, -z^{T_n(m_i)+m_j}) & \text{if } i = 0 \\ 0 & \text{otherwise} \end{cases}$$

$$m_2((\sigma_j, z^{m_j}), (w_i, z^{m_i})) = \begin{cases} (\sigma_j, z^{m_i+m_j}) & \text{if } i = j+1 \\ 0 & \text{otherwise} \end{cases}$$

$$m_2((\sigma_{d-1}, z^{m_j}), (w_i, z^{m_i})) = \begin{cases} (\sigma_{d-1}, z^{m_i+m_j}) & \text{if } i = d-1 \\ 0 & \text{otherwise} \end{cases}$$

Having described the structure of the spaces of morphisms in the A_∞-precategory $\mathrm{D}(\hat{\mathcal{X}})$, we next note that $\check{\mathrm{D}}(X(B))$ can be identified with the *tropical Morse category*:

DEFINITION 8.25. Let B be an integral affine manifold of dimension one. The *tropical Morse category* $\mathrm{TMC}(B)$ has as set of objects just the set of integers, transversal sequences are sequences $(n_1, \ldots, n_d)$ with $n_i \neq n_j$ for any $i \neq j$, $\mathrm{Hom}_{\mathrm{TMC}(B)}(n_1, n_2)$ is either

$$\bigoplus_{p \in B(\frac{1}{n_2-n_1}\mathbb{Z})} [p] \cdot \Lambda_{\mathrm{nov}} \to 0$$

if $n_2 > n_1$ or

$$0 \to \bigoplus_{p \in B(\frac{1}{n_2-n_1}\mathbb{Z})} [p] \cdot \Lambda_{\mathrm{nov}}$$

if $n_2 < n_1$. Higher multiplication maps are given by counting tropical Morse trees: given a transversal sequence $(n_0, \ldots, n_d)$, $p_{i-1,i} \in B(\frac{1}{n_i - n_{i-1}}\mathbb{Z})$, the contribution to the coefficient of $p_{0,d} \in B(\frac{1}{n_d - n_0}\mathbb{Z})$ in $m_d(p_{d-1,d}, \ldots, p_{0,1})$ is

$$\sum_\phi (-1)^{s(\phi)}(-1)^{\deg(\phi)}[p_{0,d}],$$

where ϕ runs over elements of $\mathcal{S}_d^{\rm trop}(p_{0,d};p_{0,1},\ldots,p_{d-1,d})$.

For the sign $s(\phi)$, for our purposes we can just assign the same sign that we would for the holomorphic disk defined by the tropical Morse tree; this works in two dimensions, and would obviously require something more delicate if we were working in higher dimensions.

Note that we have insisted in this definition that $\dim B = 1$. Indeed, one can make this definition in any dimension, but it is not at all clear that this defines an A_∞ (pre)-category in higher dimensions. The problem is that the moduli space of tropical Morse trees need not be the right dimension, and some virtual count or perturbation technique is necessary to turn the above description into an A_∞-category. The problems arise essentially because we are taking gradient flow lines with respect to very special functions, rather than general choices of functions. In one dimension, however, we don't need to worry about these issues, because as we saw in the previous section, tropical Morse trees are equivalent in dimension one to holomorphic disks in $X(B)$, and counting disks in two real dimensions does provide an A_∞ structure even without assuming the Lagrangians are perturbed generically, see, e.g., [**3**].

For the degree, lift $\phi : S \to B$ to the universal cover, $\tilde{\phi} : S \to \tilde{B} = \mathbb{R}$. Then set

$$\deg(\phi) = \sum_{i=1}^{d}(n_i - n_{i-1})\varphi(\tilde{\phi}(v_{i-1,i})) - (n_d - n_0)\varphi(\tilde{\phi}(v_{0,d})).$$

PROPOSITION 8.26. *$\check{\mathrm{D}}(X(B))$ is quasi-isomorphic to* $\mathrm{TMC}(B)$.

PROOF. Of course the object L_n of $\check{\mathrm{D}}(X(B))$ corresponds to the object n of $\mathrm{TMC}(B)$. We define a map

$$F_1 : \mathrm{Hom}_{\check{\mathrm{D}}(X(B))}(L_{n_0}, L_{n_1}) \to \mathrm{Hom}_{\mathrm{TMC}(B)}(n_0, n_1)$$

by

$$F_1([x]) = [x]q^{(n_1-n_0)A(x)} \tag{8.25}$$

where $A(x)$ is defined in (8.18). All higher maps F_d are taken to be zero. It is easy to check that

$$m_d^{\mathrm{TMC}(B)}(F_1(x_d),\ldots,F_1(x_1)) = F_1(m_d^{\check{\mathrm{D}}(X(B))}(x_d,\ldots,x_1)).$$

Indeed, these maps are defined by counting tropical Morse trees and holomorphic disks respectively, and we have seen that this is the same thing. Furthermore, the definition of the signs coincide, and the only question is comparing the areas of holomorphic disks with degrees of tropical Morse trees. Under (8.25), the degree of a tropical Morse tree is taken to the area of the corresponding disk, as follows from Lemma 8.27. □

LEMMA 8.27. *Let $\phi : S \to B$ be a tropical Morse tree on $B = \mathbb{R}/d\mathbb{Z}$, with incoming vertices $v_{i-1,i}$, outgoing vertex $v_{0,d}$, with a lifting $\tilde{\phi} : S \to \tilde{B} = \mathbb{R}$, $R : \mathrm{Fat}(S) \to X(B)$ as given in §8.4.4. Suppose ϕ is a point in a zero-dimensional moduli space of such trees. Then*

$$\int_{\mathrm{Fat}(S)} R^*\omega = \sum_{i=1}^{d} (n_i - n_{i-1})\psi(\tilde{\phi}(v_{i-1,i})) - (n_d - n_0)\psi(\tilde{\phi}(v_{0,d})),$$

where ψ is as defined in (8.17).

PROOF. The fact that ϕ is an element of a zero-dimensional moduli space implies that if we lift $R : \mathrm{Fat}(S) \to X(B)$ to the universal cover, $\tilde{R} : \mathrm{Fat}(S) \to \mathbb{C}$, then the image of $\mathrm{Fat}(S)$ is convex. Indeed, if not, then, as in Figure 22, one can vary this image in a one-parameter family of disks with boundary contained in the relevant Lagrangians. Thus, in particular, $\tilde{R}$ is one-to-one, and we can apply a standard formula for the area of a polygon in the plane with vertices with coordinates (y_i, x_i), $0 \le i \le d$: the area is

$$\frac{1}{2}\sum_{i=0}^{d-1}(y_i x_{i+1} - y_{i+1}x_i).$$

Now the edges of the polygon are line segments contained in lifts $\tilde{L}_0, \ldots, \tilde{L}_d$ of $L_0, \ldots, L_d$. Suppose the equation of $\tilde{L}_i$ is given by $x = -n_i y + a_i$ for some $a_i \in \mathbb{Z}$. If $\tilde{\phi}(v_{i-1,i}) = p_{i-1,i}$, then we know that $p_{i-1,i}$ is the first coordinate of the point $\tilde{L}_{i-1} \cap \tilde{L}_i$, so that

$$-n_{i-1}p_{i-1,i} + a_{i-1} = -n_i p_{i-1,i} + a_i.$$

Similarly, if $\tilde{\phi}(v_{0,d}) = p_{0,d}$, then this is the first coordinate of the point $\tilde{L}_0 \cap \tilde{L}_d$, so

$$-n_0 p_{0,d} + a_0 = -n_d p_{0,d} + a_d.$$

Without loss of generality, we can take $a_0 = 0$. Then one sees that for $1 \le i \le d$,

$$a_i = \sum_{j=1}^{i}(n_j - n_{j-1})p_{j-1,j},$$

and also

$$a_d = (n_d - n_0)p_{0,d}.$$

The area formula then gives (with $p_{d,d+1} := p_{0,d}$)

$$\begin{aligned}
&\frac{1}{2}\sum_{i=1}^{d}(p_{i-1,i}(-n_ip_{i,i+1}+a_i)-p_{i,i+1}(-n_ip_{i-1,i}+a_i))\\
&\quad=\frac{1}{2}\sum_{i=1}^{d}(p_{i-1,i}-p_{i,i+1})a_i\\
&\quad=\frac{1}{2}\sum_{i=1}^{d}\sum_{j=1}^{i}(n_j-n_{j-1})p_{j-1,j}(p_{i-1,i}-p_{i,i+1})\\
&\quad=\frac{1}{2}\sum_{j=1}^{d}(n_j-n_{j-1})p_{j-1,j}(p_{j-1,j}-p_{0,d})\\
&\quad=\sum_{j=1}^{d}\left((n_j-n_{j-1})\frac{p_{j-1,j}(p_{j-1,j}-1)}{2}-(n_j-n_{j-1})\frac{p_{j-1,j}(p_{0,d}-1)}{2}\right)\\
&\quad=\sum_{j=1}^{d}(n_j-n_{j-1})\psi(p_{j-1,j})-(n_d-n_0)\psi(p_{0,d}),
\end{aligned}$$

the last equality using the equality of the two different expressions for a_d. □

It is tempting to try a direct comparison of the A_∞ structure provided by the tropical Morse category and the A_∞ structure on $\mathrm{D}(\hat{\mathcal{X}})$ provided by the construction of §8.2.1. This would involve writing down projectors and homotopies for the Čech complexes. Unfortunately, this gives a description on the line bundle side which resembles to a certain extent the picture on the tropical Morse side, but not precisely. The problem is essentially that the discreteness of the Čech cover prevents us from seeing in detail tropical Morse trees doing interesting things inside line segments $[i, i+1]$ on B. Instead, it is easiest to write down an A_∞ functor in a more direct way which provides a quasi-isomorphism. Again, this is an adaptation of ideas from [**2**].

To begin, we will define a bit of extra structure. Let $\mathcal{P}$ be the polyhedral decomposition of B into line segments of length 1, i.e., $\mathcal{P}$ consists of the unit intervals $[i, i+1]$ and the vertices $\{i\}$. By abuse of notation, we write the image of the interval $[d-1, d]$ in B as $[0, d-1]$: probably $[d-1, 0]$ would make more sense but we want to keep the smaller index to the left. Pick for each $0 \le i < d$ a general point $\alpha_i \in (i, i+1)$. Then we can define the dual cell complex $\check{\mathcal{P}}$ consisting of the vertices $\{\alpha_i\}$ and the intervals $[\alpha_i, \alpha_{i+1}]$ for $0 \le i < d-1$ and $[\alpha_{d+1}, d+\alpha_0]$ (remember we are looking at these intervals modulo $d\mathbb{Z}$). Note there is an obvious one-to-one inclusion reversing correspondence between cells of $\mathcal{P}$ and cells of $\check{\mathcal{P}}$, which we write as $\sigma \mapsto \sigma^\vee$: e.g. $i^\vee = [\alpha_{i-1}, \alpha_i]$ for $0 < i \le d-1$.

For any small $\epsilon > 0$, we will produce a slight deformation $\check{\mathcal{P}}_\epsilon$ of $\check{\mathcal{P}}$: the one-dimensional cells are

$$\{[\alpha_i - \epsilon, \alpha_{i+1} - \epsilon] | 0 \leq i < d-2\} \cup \{[\alpha_{d-2} - \epsilon, \alpha_{d-1} + \epsilon], [\alpha_{d-1} + \epsilon, \alpha_0 + d - \epsilon]\}$$

(and of course the zero-dimensional cells are just the endpoints of these intervals). See Figure 37. Given a cell $\sigma \in \mathcal{P}$, we denote the interior of the dual cell in $\check{\mathcal{P}}_\epsilon$ by $\sigma_\epsilon^\vee$.

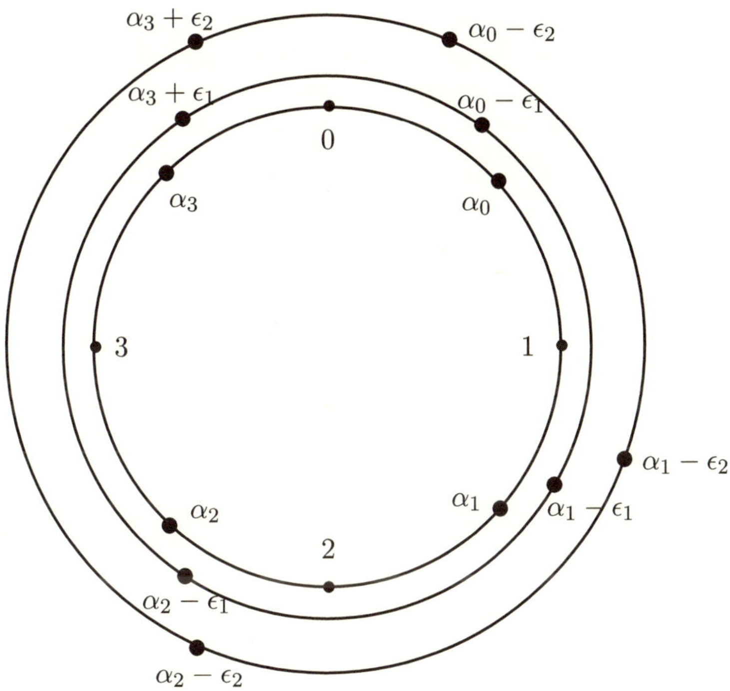

FIGURE 37. The inner circle shows both $\mathcal{P}$ and $\check{\mathcal{P}}$, the outer circles show $\check{\mathcal{P}}_{\epsilon_1}$ and $\check{\mathcal{P}}_{\epsilon_2}$ for $\epsilon_2 > \epsilon_1$

These decompositions have the feature that if $\epsilon_1 < \epsilon_2$, then for $0 \leq j < d-1$,

$$i_{\epsilon_1}^\vee \cap [j, j+1]_{\epsilon_2}^\vee \neq \emptyset \text{ if and only if } i = j$$

and

$$i_{\epsilon_1}^\vee \cap [0, d-1]_{\epsilon_2}^\vee \text{ if and only if } i = 0,$$

while for $0 \leq j < d-1$

$$[j, j+1]_{\epsilon_1}^\vee \cap i_{\epsilon_2}^\vee \neq \emptyset \text{ if and only if } i = j+1$$

and

$$[0, d-1]_{\epsilon_1}^\vee \cap i_{\epsilon_2}^\vee \text{ if and only if } i = d-1.$$

If we compare this with the formulae for $m_2^{\mathrm{D}(\check{\mathcal{X}})}$, (8.24), we see that this somewhat crazy set of polyhedral decompositions serves the purpose of enforcing the ordering of the index set for the Čech open cover when we perform the multiplication.

Next we define a variation on the notion of a tropical Morse tree on $B = \mathbb{R}/d\mathbb{Z}$.

DEFINITION 8.28. $\mathcal{T}_d$ is the moduli space of *shrubs*: these are ribbon trees with finite lengths assigned to all edges except the outgoing edge, (which still has infinite length), with the property that if v_{adj} is the unique vertex adjacent to the outgoing vertex, the distance from v_{adj} to any incoming vertex is independent of the incoming vertex. We call this condition the *length condition.*

$\mathcal{T}_1$ is just a point, corresponding to a line segment with infinite length, being the outgoing edge.

In what follows, let $\tilde{B} = \mathbb{R} \to B$ be the universal cover of B, and $\pi_1 : \mathbb{Z}^2 \to \mathbb{Z}$ the projection onto the first coordinate.

DEFINITION 8.29. Suppose we are given data

$$\begin{aligned}
&r \geq 1,\\
&n_0, \ldots, n_r \in \mathbb{Z},\\
&\tau_{i-1,i} \subsetneq B \text{ for } 1 \leq i \leq r,\\
&\tilde{\tau}_{i-1,i} \subset \tilde{B} \text{ a lift of } \tau_{i-1,i},\\
&m_{i-1,i} \in \mathbb{Z}^2 \text{ for } 1 \leq i \leq r,\\
&p_{0,r} \in B\left(\frac{1}{n_r - n_0}\mathbb{Z}\right).
\end{aligned}$$

Then we define

$$\mathcal{T}_r^{\text{trop}}(p_{0,r}; (\tau_{0,1}, m_{0,1}), \ldots, (\tau_{r-1,r}, m_{r-1,r}))$$

to be the moduli space of continuous maps $\phi : S \to B$ from a shrub $S \in \mathcal{T}_d$ along with a collection of affine displacement vectors $\mathbf{v}_e \in \Gamma(e, (\phi|_e)^* TB)$ on edges e of S, satisfying the following properties:

(1) If v_i is the ith incoming vertex, then $\phi(v_i) \in \tau_{i-1,i}$. If v_{out} is the external outgoing vertex, then $\phi(v_{\text{out}}) = p_{0,r}$.
(2) The same axiom as (2) of Definition 8.19.
(3) If e_i is the unique edge of S containing v_i, let $\tilde{v}_i \in \tilde{B}$ be the lift of $\phi(v_i)$ to $\tilde{\tau}_{i-1,i}$. Then

$$\mathbf{v}_{e_i}(v_i) = (n_i - n_{i-1}) \cdot \tilde{v}_i - \pi_1(m_{i-1,i}).$$

If e_{out} is the unique edge of S containing v_{out}, then $\mathbf{v}_{e_{\text{out}}}(v_{\text{out}}) = 0$.
(4) The same axiom as (4) of Definition 8.19.
(5) The same axiom as (5) of Definition 8.19.
(6) The same axiom as (6) of Definition 8.19.

REMARK 8.30. (1) The main difference between these trees and the previously defined tropical Morse trees is that the incoming vertices, instead of being points corresponding to intersections of Lagrangians, are specified to lie in certain cells, and the initial affine displacement vectors are not zero, but are specified by additional data of the lifts $\tilde{\tau}_{i-1,i}$ and $m_{i-1,i}$. Note that from the ith incoming vertex we head away from $\pi_1(m_{i-1,i})/(n_i - n_{i-1})$ if $n_i - n_{i-1} > 0$, while we head towards $\pi_1(m_{i-1,i})/(n_i - n_{i-1})$ if $n_i - n_{i-1} < 0$.

(2) Though we have not included the lifts $\tilde{\tau}_{i-1,i}$ in the notation, the moduli space does depend on these lifts. If $\tilde{\tau}_{i-1,i}$ is replaced by $\tilde{\tau}_{i-1,i} + d$, then $m_{i-1,i}$ needs to be replaced by $T_{n_i-n_{i-1}}(m_{i-1,i})$: check that making this change gives the same moduli space.

(3) By changing the lifts $\tilde{\tau}_{i-1,i}$ and the $m_{i-1,i}$ as in (2), for a given $\phi : S \to B$, one can find a lift $\tilde{\phi} : S \to \tilde{B}$ such that the image of the ith incoming vertex always lies in $\tilde{\tau}_{i-1,i}$. Then if an edge e of S is labelled by $n_j - n_{i-1}$, $j > i$, we in fact have $\mathbf{v}_e(p) = (n_j - n_{i-1})\tilde{\phi}(p) - \pi_1(\sum_{s=i}^{j} m_{s-1,s})$ for any $p \in e$. This is true initially by definition and can then be shown inductively. Since the affine displacement vector must be zero at the outgoing vertex v_{out}, we in fact see $\tilde{\phi}(v_{\text{out}}) = \pi_1(\sum_{i=1}^{r} m_{i-1,i})/(n_d - n_0)$. This is $p_{0,d} \mod d\mathbb{Z}$.

DEFINITION 8.31. Given

$$\phi \in \mathcal{T}_r^{\text{trop}}(p_{0,r}; (\tau_{0,1}, m_{0,1}), \ldots, (\tau_{r-1,r}, m_{r-1,r})),$$

let $\tilde{\phi} : S \to \tilde{B}$ be a lift as in Remark 8.30 (3) (which might entail changing the $\tilde{\tau}_{i-1,i}$'s and the $m_{i-1,i}$'s). Then we can define the *degree* of ϕ to be

$$\deg \phi = \text{ord}_{n_r - n_0}\left(\sum_{i=1}^{r} m_{i-1,i}\right).$$

PROPOSITION 8.32. *Let $\sigma_{0,1}, \ldots, \sigma_{d-1,d} \in \mathcal{P}$, $\epsilon_1, \ldots, \epsilon_d$ small, and*

$$\phi \in \mathcal{T}_d^{\text{trop}}(p_{0,d}; ((\sigma_{0,1})_{\epsilon_1}^{\vee}, m_{0,1}), \ldots, ((\sigma_{d-1,d})_{\epsilon_d}^{\vee}, m_{d-1,d})),$$

with

$$z^{m_{i-1,i}} \in \Gamma(\hat{U}_{\sigma_{i-1,i}}, \hat{\mathcal{L}}^{\otimes(n_i-n_{i-1})}).$$

for each i. Then

$$\deg \phi \geq 0.$$

PROOF. Let $\tilde{\phi} : S \to \tilde{B}$ be a lift of ϕ, as in Remark 8.30, (3). By the assumption on $z^{m_{i-1,i}}$, if the image of the ith incoming vertex of $\tilde{\phi}$ in $\tilde{B}$ is $\tilde{v}_i$, then $m_{i-1,i}$ lies above the straight line spanned by some edge of $(n_i - n_{i-1})\Delta$ which contains the point $((n_i - n_{i-1})\tilde{v}_i, (n_i - n_{i-1})\varphi(\tilde{v}_i))$. (Depending on $\tilde{v}_i$, there is either one or two such lines; in the latter case $m_{i-1,i}$ is contained in a wedge.) We will show inductively: *For any point $p \in S$ on some edge labelled by $n_e = n_j - n_{i-1}$, $\sum_{s=i}^{j} m_{s-1,s}$ lies above the line spanned by some edge of $(n_j - n_{i-1})\Delta$ which contains the point $((n_j - n_{i-1})\tilde{\phi}(p), (n_j - n_{i-1})\varphi(\tilde{\phi}(p)))$.*

First suppose this is true at the initial vertex v_{in} of an edge e of S. Then by Remark 8.30, (3), the affine displacement vector is

$$\mathbf{v}_e(v_{\mathrm{in}}) = (n_j - n_{i-1})\tilde{\phi}(v_{\mathrm{in}}) - \pi_1\left(\sum_{s=i}^{j} m_{s-1,s}\right),$$

which points away from

$$x := (n_j - n_{i-1})^{-1}\pi_1\left(\sum_{s=i}^{j} m_{s-1,s}\right)$$

if $n_j - n_{i-1} > 0$ and towards x if $n_j - n_{i-1} < 0$. In the former case, $\tilde{\phi}(e)$ must head away from x, so since the induction hypothesis holds at v_{in}, it is clear from Figure 38 that the induction hypothesis continues to hold at any $p \in e$. If instead $n_j - n_{i-1} < 0$, $\tilde{\phi}(e)$ heads towards x, but never passes it. From Figure 39, it is again clear that the induction hypothesis continues to hold at any $p \in e$.

Finally, to show that the induction hypothesis holds at v_{in}, note that $\sum_{s=i}^{j} m_{s-1,s}$ is the sum of the corresponding sums on the edges coming into v_{in}. Since the induction hypothesis holds at v_{in} for each of these edges, it holds for the edge e.

Now look at $\tilde{v}_{\mathrm{out}}$, the image under $\tilde{\phi}$ of the last vertex of S. Since

$$(n_d - n_0)\tilde{v}_{\mathrm{out}} = \pi_1\left(\sum_{i=1}^{d} m_{i-1,i}\right),$$

the induction hypothesis implies that $\sum_{i=1}^{d} m_{i-1,i}$ lies directly above

$$\big((n_d - n_0)\tilde{v}_{\mathrm{out}}, (n_d - n_0)\varphi(\tilde{v}_{\mathrm{out}})\big).$$

However, $\mathrm{ord}_{n_d - n_0}(\sum_{i=1}^{d} m_{i-1,i})$ is precisely the vertical distance between these two points, hence is non-negative. □

We can make a heuristic calculation of

$$\dim \mathcal{T}_r^{\mathrm{trop}}\big(p_{0,d}; (\tau_{0,1}, m_{0,1}), \ldots, (\tau_{r-1,r}, m_{r-1,r})\big).$$

Take an element of this space with trivalent vertices, and look at how we are allowed to vary it. It is not hard to see that before one imposes the length condition one can move the image of each interior vertex freely except for v_{adj}, whose image can be moved freely if and only if $\deg p_{0,r} = 1$. Furthermore, the image of the ith incoming vertex can be varied inside $\tau_{i,i+1}$. Moving these vertices changes the lengths of edges in a hopefully independent way, and imposing the length condition then imposes $r-1$ conditions. This gives a total dimension of

$$\begin{aligned}\#\text{ of interior vertices} - 1 + \deg p_{0,r} + \sum \dim \tau_{i,i+1} - (r-1)\\ = \deg p_{0,r} - 1 + \sum \dim \tau_{i,i+1}.\end{aligned}$$

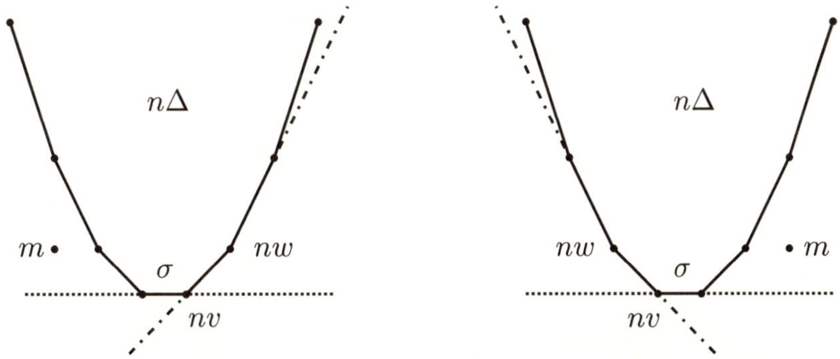

FIGURE 38. $n > 0$. The dashed lines depict Γ_σ^n and Γ_w^n. Note in the first figure that if $m \in \Gamma_\sigma^n$ appears to the left of nv, then also $m \in \Gamma_w^n$ with w to the right of v. In the second figure if $m \in \Gamma_\sigma^n$ appears to the right of nv, then also $m \in \Gamma_w^n$ with w to the left of v.

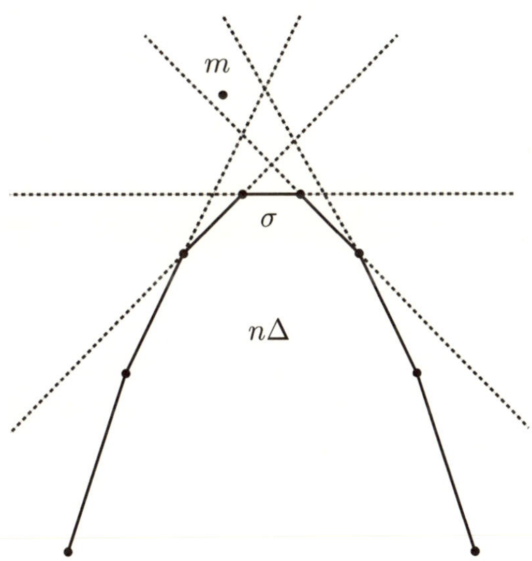

FIGURE 39. $n < 0$.

In fact, if $\dim \tau_{i-1,i} = 1$ for each i, this formula is always correct: once the positions of the first vertex and all internal vertices are fixed, the length condition determines the position of all other incoming vertices.

We will in fact be interested in moduli spaces

$$\mathcal{T}_d^{\mathrm{trop}} := \mathcal{T}_d^{\mathrm{trop}}\big(p_{0,d}; ((\sigma_{0,1})_{\epsilon_1}^\vee, m_{0,1}), \ldots, ((\sigma_{d-1,d})_{\epsilon_d}^\vee, m_{d-1,d})\big)$$

for any given sequence $\sigma_{i-1,i} \in \mathcal{P}$, $m_{i-1,i} \in \mathbb{Z}^2$, $i = 1, \ldots, d$, with $0 < \epsilon_1 < \cdots < \epsilon_d$. We will always use the convention that the lifts of the sets $(\sigma_{i-1,i})_{\epsilon_i}^\vee$

are one of the sets

$$\begin{aligned}(\alpha_j - \epsilon_i, \alpha_{j+1} - \epsilon_i), &\quad 0 \le j \le d-2\\ (\alpha_{d-2} - \epsilon_i, \alpha_{d-1} + \epsilon_i), &\\ (\alpha_{d-1} + \epsilon_i, \alpha_0 + d - \epsilon_i), &\\ \{\alpha_j - \epsilon_i\}, &\quad 0 \le j \le d-2\\ \{\alpha_{d-1} + \epsilon_i\}. &\end{aligned}$$

Now we can always find $0 < \epsilon_1 < \cdots < \epsilon_d$ such that $\mathcal{T}_d^{\mathrm{trop}}$ has the expected dimension. Indeed, let $I = \{i | \dim \sigma_{i-1,i}^{\vee} = 0\}$, and let $\mathcal{T}_d'$ be the above moduli space with $(\sigma_{i-1,i})_{\epsilon_i}^{\vee}$ replaced with an interval around it for each $i \in I$. We have a map

$$\mathrm{ev} : \mathcal{T}_d' \to B^{\#I}$$

given by evaluation of $\phi \in \mathcal{T}_d'$ on the incoming vertices indexed by I, and then

$$\mathcal{T}_d^{\mathrm{trop}} = \mathrm{ev}^{-1}\left(\prod_{i \in I} (\sigma_{i-1,i})_{\epsilon_i}^{\vee}\right).$$

Now as argued above, $\mathcal{T}_d'$ is the expected dimension, and either the image of ev is not of dimension equal to $\#I$, in which case $\mathcal{T}_d^{\mathrm{trop}}$ is empty for a general choice of the ϵ_i's, or else the dimension of the fibers is the expected dimension for sufficiently general choice of the ϵ_i's.

In fact, it is not difficult to see that one can do somewhat better when this moduli space is in fact of dimension zero: as we vary the ϵ_i's, the trees in the moduli space $\mathcal{T}_d^{\mathrm{trop}}$ will vary in a continuous way, although one may have jumping phenomena when boundary points of certain one-dimensional moduli spaces may appear. In particular, for any given choice of the data $(\sigma_{i-1,i}, m_{i-1,i})$ as before, one can find some ϵ such that all the moduli spaces $\mathcal{T}_d^{\mathrm{trop}}$ for any $0 < \epsilon_1 < \cdots < \epsilon_d < \epsilon$ are naturally isomorphic. We will use these moduli spaces to define our desired A_∞ functor. (Actually, there is a subtlety we wish to avoid in this exposition: the moduli spaces in question, even when zero-dimensional, may be infinite, and jumping phenomena can occur with ϵ arbitrarily small. However, the zero-dimensional moduli spaces are finite if we restrict to those ϕ with a given bound for their degree. This allows us to define the maps F_d below up to any given power of q, and hence allows us to define F_d over $\mathbb{C}[[q]]$. We will avoid discussing this subtlety below. For a less ad hoc solution to this problem, see [**2**].)

We define an A_∞ functor $F : \mathrm{D}(\hat{\mathcal{X}}) \to \mathrm{TMC}(B)$ as follows. On the level of objects, $F(\hat{\mathcal{L}}^{\otimes n}) = n$. For each $d \ge 1$ and transversal sequence $(\mathcal{L}^{\otimes n_0}, \ldots, \mathcal{L}^{\otimes n_d})$ define a map

$$F_d : C^*(\hat{\mathcal{L}}^{\otimes(n_d - n_{d-1})}) \otimes \cdots \otimes C^*(\hat{\mathcal{L}}^{\otimes(n_1 - n_0)}) \to \mathrm{Hom}^*_{\mathrm{TMC}(B)}(n_0, n_d)$$

as follows. Pick cells $\sigma_{0,1},\ldots,\sigma_{d-1,d} \in \mathcal{P}$, and $m_{i-1,i} \in \mathbb{Z}^2$ such that $z^{m_{i-1,i}} \in \Gamma(\hat{U}_{\sigma_{i-1,i}}, \hat{\mathcal{L}}^{\otimes(n_i-n_{i-1})})$ using identifications (8.20), (8.21) or (8.22). This defines an element

$$(\sigma_{d-1,d}, z^{m_{d-1,d}}) \otimes \cdots \otimes (\sigma_{0,1}, z^{m_{0,1}})$$

of

$$C^*(\hat{\mathcal{L}}^{\otimes(n_d-n_{d-1})}) \otimes \cdots \otimes C^*(\hat{\mathcal{L}}^{\otimes(n_1-n_0)}).$$

Given $p_{0,d} \in B\left(\frac{1}{n_d-n_0}\mathbb{Z}\right)$, the coefficient of $[p_{0,d}]$ in the image of this element in $\mathrm{Hom}^*_{\mathrm{TMC}(B)}(n_0, n_d)$ is

$$\begin{cases} \sum_\phi (-1)^{s'(\phi)} q^{\deg\phi}, & \text{if } \deg p_{0,d} - 1 + \sum_i \dim\sigma^\vee_{i-1,i} = 0 \\ 0 & \text{if } \deg p_{0,d} - 1 + \sum_i \dim\sigma^\vee_{i-1,i} \neq 0 \end{cases}$$

where the sum is over all ϕ in

$$\mathcal{T}^{\mathrm{trop}}_d(p_{0,d}; ((\sigma_{0,1})^\vee_{\epsilon_1}, m_{0,1}), \ldots, ((\sigma_{d-1,d})^\vee_{\epsilon_d}, m_{d-1,d})),$$

and $0 < \epsilon_1 < \cdots < \epsilon_d < \epsilon$ with ϵ chosen as in the previous paragraph so that this moduli space is independent of the particular choice of the ϵ_d's. We will omit the definition of the sign, $s'(\phi)$, here, as it is quite complicated. A hint of how to do this can be found in [**2**], Appendix B.

To understand the effect of F_d, let us first consider the case $d = 1$. Given a cell $\sigma \in \mathcal{P}$ and $z^m \in \Gamma(\hat{U}_\sigma, \hat{\mathcal{L}}^{\otimes(n_1-n_0)})$, there are two cases, $n_1 > n_0$ or $n_1 < n_0$. Consider the possible trees. It is easiest to think in terms of the lifts $\tilde{\phi} : S \to \tilde{B}$. In this case, these trees will just be line segments starting in $\sigma^\vee_\epsilon$ and heading either away or towards $\pi_1(m)/(n_1 - n_0)$. However, if $n_1 > n_0$, the affine displacement vector is increasing in length, and thus can never reach zero unless $\tilde{\phi}(v) = \pi_1(m)/(n_1 - n_0)$ for v the initial vertex. In this case we have a degenerate tree, with the single edge mapping to the point $\pi_1(m)/(n_1 - n_0)$. This case can only occur if $p_{0,1} = \pi_1(m)/(n_1 - n_0) \bmod d\mathbb{Z}$ and $p_{0,1} \in \sigma^\vee_\epsilon$, and this then is the only contribution. This makes a contribution of

$$q^{\mathrm{ord}_{n_1-n_0}(m)}[p_{0,1}].$$

The meaning of the exponent of q is that this is the maximum degree of divisibility of z^m in $\Gamma(\hat{U}_\sigma, \hat{\mathcal{L}}^{\otimes(n_1-n_0)})$ by q. Note this can only happen if $\dim\sigma = 0$, for if $\sigma^\vee_\epsilon$ is a point, ϵ will have been chosen sufficiently small so that $\sigma^\vee_\epsilon$ cannot be a rational point with denominator $n_1 - n_0$.

If $n_1 < n_0$, then the affine displacement vector is decreasing, and in fact reaches zero precisely when the line segment arrives at $\pi_1(m)/(n_1 - n_0)$. Thus there is always such a tree. The moduli space of such trees is of dimension $\dim\sigma^\vee_\epsilon$, so F_1 is zero on this Čech cochain if this dimension is 1, and otherwise $\dim\sigma = 1$ and we get the same formula for the contribution as in the case $n_1 > n_0$. However, we will define

$$s'(\phi) = \begin{cases} 1 & \sigma = [0, d-1], \\ 0 & \sigma \neq [0, d-1]. \end{cases}$$

The degree 1 relation for an A_∞ functor is

$$m_1^{\mathrm{TMC}(B)} \circ F_1 = F_1 \circ m_1^{\mathrm{D}(\hat{\mathcal{X}})}.$$

Of course, $m_1^{\mathrm{TMC}(B)} = 0$. Thus we only need to check $F_1 \circ m_1^{\mathrm{D}(\hat{\mathcal{X}})} = 0$ for $(\{i\}, z^m)$ as above, in the case $n_1 < n_0$. Then $m_1^{\mathrm{D}(\hat{\mathcal{X}})}$ consists of two terms, depending on i:

$$m_1^{\mathrm{D}(\hat{\mathcal{X}})}((\{i\}, z^m)) = \begin{cases} ([i-1, i], z^m) + ([i, i+1], -z^m) & 0 < i < d-1 \\ ([0, d-1], -z^{T_n(m)}) + ([0,1], -z^m) & i = 0 \\ ([d-2, d-1], z^m) + ([0, d-1], z^m) & i = d-1. \end{cases}$$

One then sees easily that applying F_1 to these possibilities always yields 0. It is also easy to see that F_1 induces a quasi-isomorphism of chain complexes, so that if F is an A_∞ functor, it is a quasi-isomorphism.

We sketch the argument that F is an A_∞ functor. Keeping in mind that in $\mathrm{D}(\hat{\mathcal{X}})$, only m_1 and m_2 are non-zero, we need to verify the following relation for each d:

$$\sum_r \sum_{s_1, \cdots, s_r} m_r^{\mathrm{TMC}(B)}(F_{s_r}(\cdots), \cdots, F_{s_1}(\cdots)) =$$

$$\sum_q (-1)^{\deg a_1 + \cdots + \deg a_q - q} F_d(a_d, \ldots, a_{q+2}, m_1^{\mathrm{D}(\hat{\mathcal{X}})}(a_{q+1}), a_q, \ldots, a_1)$$

$$+ \sum_q (-1)^{\deg a_1 + \cdots + \deg a_q - q} F_{d-1}(a_d, \ldots, a_{q+3}, m_2^{\mathrm{D}(\hat{\mathcal{X}})}(a_{q+2}, a_{q+1}), a_q, \ldots, a_1).$$

Fix $(n_0, \ldots, n_d)$, $a_i = (\sigma_{i-1,i}, z^{m_{i-1,i}})$, $i = 1, \ldots, d$, $p_{0,d} \in B\left(\frac{1}{n_d - n_0}\mathbb{Z}\right)$ for which we would like to prove this relation. Choose numbers $0 < \epsilon_1 < \cdots < \epsilon_d < \epsilon$ sufficiently small: how small will become clear in the argument that follows.

Now all coefficients of $[p_{0,d}]$ in the A_∞ functor relation are zero for degree reasons, unless the moduli space

$$\mathcal{T}_d^{\mathrm{trop}}(p_{0,d}; ((\sigma_{0,1})^\vee_{\epsilon_1}, m_{0,1}), \ldots, ((\sigma_{d-1,d})^\vee_{\epsilon_d}, m_{d-1,d}))$$

is one-dimensional. We consider the boundary of this moduli space. Naturally, the properly signed boundary will yield a total count of zero. Let us consider how an element of this moduli space can degenerate:

(1) The length of some edges can go to ∞. This means that the affine displacement vector becomes 0 on some edge other than the outgoing edge. Because the length from the vertex v_{adj} to any incoming vertex is the same,

this can only happen if every path from v_{adj} to an incoming vertex contains an edge whose length becomes infinite. It is easy to check that if the ϵ_i's are chosen sufficiently generally, then there is precisely one such edge along every such path. Thus one can view the limiting tree as decomposing into a number of trees of the type we consider in Definition 8.29 with their output vertices glued on to the input vertices of a tropical Morse tree. These degenerations contribute the first term.

(2) One of the incoming vertices moves to the boundary of the cell it lies on. Again, genericity of choices implies that this only happens for one such cell. These degenerations contribute to the second term.

(3) The length of an incoming edge can approach zero. In order for the lengths from v_{adj} back to any incoming vertex to be the same, at least two adjacent incoming edges must have length going to zero. Again, with generic choices, this only happens in pairs. So there is some i such that $\phi(v_i) = \phi(v_{i+1})$ for v_i, v_{i+1} the ith and $i+1$st input vertices, so we can view ϕ as a tree in

$$\mathcal{T}^{\mathrm{trop}}_{d-1}(p_{0,d}; ((\sigma_{0,1})^\vee_{\epsilon_1}, m_{0,1}), \ldots, ((\sigma_{i-1,i})^\vee_{\epsilon_i} \cap (\sigma_{i,i+1})^\vee_{\epsilon_{i+1}}, m_{i-1,i} + m_{i,i+1}), \\ \ldots, ((\sigma_{d-1,d})^\vee_{\epsilon_d}, m_{d-1,d})).$$

Indeed, if v' is the vertex adjacent to v_i and v_{i+1}, and e is the outgoing edge from v', we can choose a lift $\tilde{\phi} : S \to \tilde{B}$ with $\tilde{\phi}(v') = \tilde{\phi}(v_i) = \tilde{\phi}(v_{i+1})$ and

$$\mathbf{v}_e(v') = (n_{i+1} - n_{i-1})\tilde{\phi}(v') - \pi_1(m_{i-1,i} + m_{i,i+1}),$$

hence the appearance of the term $m_{i-1,i} + m_{i,i+1}$. Now if the intersection is empty, this of course never happens. If the intersection is non-empty but one of $(\sigma_{i-1,i})^\vee_{\epsilon_i}$ or $(\sigma_{i,i+1})^\vee_{\epsilon_{i+1}}$ is zero-dimensional, then the intersection is just this zero-dimensional cell, and this contribution corresponds precisely to m_2, as described by the last four cases of (8.24). Note that the condition $\epsilon_i < \epsilon_{i+1}$ enforces the ordering of the index set for the Čech open covering.

On the other hand, if the intersection is one-dimensional, then there are two possibilities. First we can have $\sigma_{i-1,i} \neq \sigma_{i,i+1}$, and the intersection is very small (of size roughly ϵ). Then ϵ can be assumed to be small enough so that in fact this moduli space is empty, and hence makes no contribution. On the other hand, if $\sigma_{i-1,i} = \sigma_{i,i+1}$, then $(\sigma_{i-1,i})^\vee_{\epsilon_i} \cap (\sigma_{i,i+1})^\vee_{\epsilon_{i+1}}$ is very close to $(\sigma_{i-1,i})^\vee_{\epsilon_i}$, so we can replace the intersection with this set without affecting the moduli space. Note that in this case $\dim \sigma_{i-1,i} = \dim \sigma_{i,i+1} = 0$, and gives the first case of (8.24). Thus we see these degenerations contribute to the third term of the A_∞ functor relation.

With a great deal of care on the signs, one finds that F is an A_∞ functor, proving Theorem 8.24.

To illustrate this proof, let us consider one possible case occurring in the relation involving F_2, which is

(8.26)

$$
\begin{aligned}
m_2^{\mathrm{TMC}}(F_1(\sigma_{1,2}, z^{m_{1,2}}), F_1(\sigma_{0,1}, z^{m_{0,1}}))
&= F_2((\sigma_{1,2}, z^{m_{1,2}}), m_1^{\mathrm{D}(\hat{\mathcal{X}})}(\sigma_{0,1}, z^{m_{0,1}})) \\
&\quad + (-1)^{\dim \sigma_{0,1}-1} F_2(m_1^{\mathrm{D}(\hat{\mathcal{X}})}(\sigma_{1,2}, z^{m_{1,2}}), (\sigma_{0,1}, z^{m_{0,1}})) \\
&\quad + F_1(m_2^{\mathrm{D}(\hat{\mathcal{X}})}((\sigma_{1,2}, z^{m_{1,2}}), (\sigma_{0,1}, z^{m_{0,1}}))).
\end{aligned}
$$

Consider the case where $n_1 - n_0$ and $n_2 - n_1$ are both positive and $\dim \sigma_{1,2} = \dim \sigma_{0,1} = 0$, so that the moduli space $\mathcal{T}_2^{\mathrm{trop}}$ is one-dimensional. Figure 40 shows some possibilities for how a tree may degenerate. The tree (1) shows

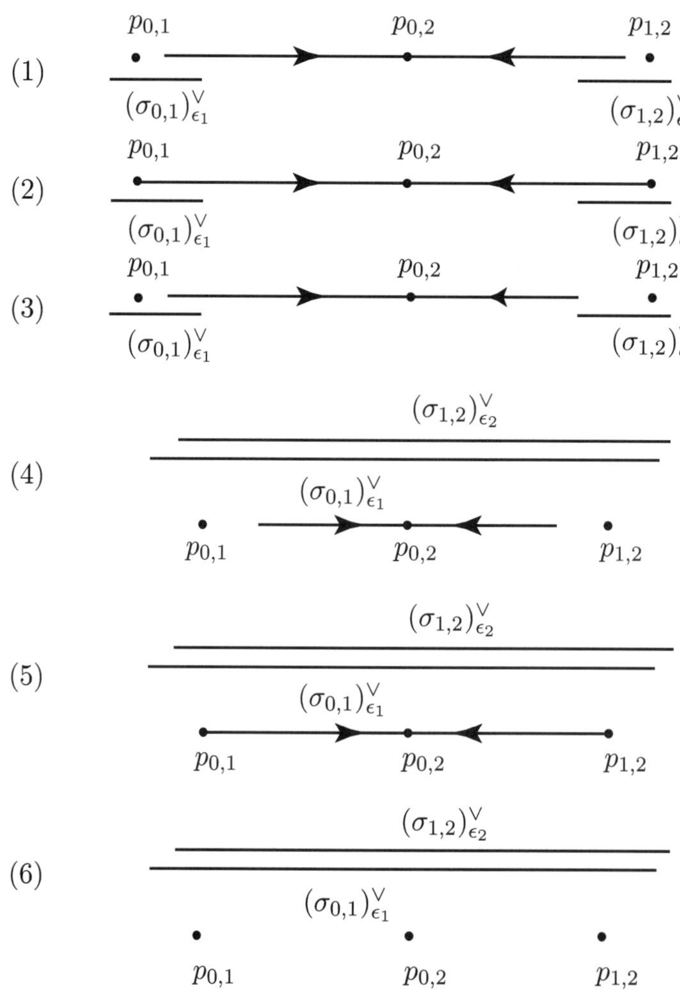

FIGURE 40

one particular case of a general tree in a one-parameter family, in which the initial points $\phi(v_{0,1})$ and $\phi(v_{1,2})$ appear in between $p_{0,1} := \pi_1(m_{0,1})/(n_1-n_0)$ and $p_{1,2} := \pi_1(m_{1,2})/(n_2-n_1)$. These points repel, and the two line segments meet at a point $p_{0,2}$. We can vary the two endpoints, but not in an independent fashion. As one endpoint approaches, say, $p_{0,1}$, the other endpoint must approach $p_{1,2}$, to obtain a degenerate case indicated by (2) in Figure 40. This now gives an ordinary tropical Morse tree, representing a contribution to the product $m_2^{\mathrm{TMC}}([p_{1,2}], [p_{0,1}])$, and $[p_{i-1,i}] = F_1\big((\sigma_{i-1,i}, m_{i-1,i})\big)$ for $i = 1, 2$. This gives a contribution to the left-hand side of the F_2 relation. If instead we move the endpoints away from $p_{0,1}$ and $p_{1,2}$, then at some point one endpoint lies on the boundary of the relevant cell, as in (3) of Figure 40. Depending on whether we reach the boundary of $(\sigma_{0,1})^\vee_{\epsilon_1}$ or $(\sigma_{1,2})^\vee_{\epsilon_2}$ first, we get a contribution to the first or second terms on the right-hand side of the F_2 relation.

Another example of possible behaviour occurs when $(\sigma_{0,1})^\vee_{\epsilon_1}$ and $(\sigma_{1,2})^\vee_{\epsilon_2}$ are not disjoint. In this case, the overlap is either a small set or a large set, the latter happening when $\sigma_{0,1} = \sigma_{1,2}$. In the latter case, if $p_{0,1}, p_{1,2} \in (\sigma_{0,1})^\vee_{\epsilon_1} \cap (\sigma_{1,2})^\vee_{\epsilon_2}$, we see that there are two possible degenerations for the family of trees indicated in (4) of Figure 40: either (5) the endpoints can move towards $p_{0,1}$ and $p_{1,2}$, again contributing to the left-hand side of (8.26), or (6) the endpoints could approach $p_{0,2}$, so that the lengths of these edges go to zero. This contributes to the last term of (8.26).

Of course there are many more possibilities, and some work has to be done to ensure that the signs work, but we omit this argument.

8.5. Seidel's result for the quartic K3 surface

In this last section, we will comment briefly on work of Paul Seidel [**420**] proving Homological Mirror Symmetry in one highly non-trivial case, that of the quartic K3 surface. To be precise, let X be a non-singular quartic surface in $\mathbb{P}^3$, equipped with a reasonable symplectic form, say the Kähler form of the Fubini-Study metric restricted to X. On the other hand, consider the mirror family, where we will be concerned about the complex structure. One takes the family defined by

$$q(x_0^4 + x_1^4 + x_2^4 + x_3^4) + x_0x_1x_2x_3 = 0 \tag{8.27}$$

in $\mathbb{P}^3 \times \operatorname{Spec} \mathbb{C}[[q]]$, and divides out by the group action

$$\Gamma_{16} := \{(a_0, a_1, a_2, a_3) | a_i \in \mathbb{C}, a_i^4 = 1, \prod a_i = 1\}/\{(a, a, a, a)|a^4 = 1\}.$$

This acts diagonally on $\mathbb{P}^3$, i.e., (a_0, a_1, a_2, a_3) takes

$$(x_0, x_1, x_2, x_3) \mapsto (a_0x_0, a_1x_1, a_2x_2, a_3x_3).$$

Then we define $\check{\mathcal{X}} \to \operatorname{Spec} \mathbb{C}[[q]]$ to be the quotient of the above family by Γ_{16}. The generic fiber $\hat{\mathcal{X}}_q$ is a K3 surface over the field $\mathbb{C}((q))$, and so we can consider the category $\mathrm{D}^b(\hat{\mathcal{X}}_q)$, the bounded derived category of coherent sheaves on $\hat{\mathcal{X}}_q$. We can tensor this category with Λ_{nov}, which contains the algebraic closure of $\mathbb{C}((q))$ (i.e., we tensor all the morphism spaces with Λ_{nov}). Then the statement of Seidel's result is, morally, as follows. (Legal disclaimer: There are some technical differences between the definitions given in this text and the definitions needed in Seidel's paper [**420**]. For precise statements, see [**420**].)

THEOREM 8.33. *There exists $\psi \in \mathrm{Aut}(\mathbb{C}[[q]])$ and an equivalence of triangulated A_∞-categories*

$$\mathrm{D}^\pi \,\mathrm{Fuk}(X) \cong \psi^* \,\mathrm{D}^b(\hat{\mathcal{X}}_q) \otimes \Lambda_{\mathrm{nov}}.$$

One should think of ψ as the mirror map, changing coordinates between the two q's; one would expect this to be non-trivial as q is not a canonical coordinate for the family $\hat{\mathcal{X}} \to \mathrm{Spec}\,\mathbb{C}[[q]]$. However, it is not known if this change of coordinates ψ really corresponds to the canonical coordinate.

Seidel's proof is very different than the one presented here for the elliptic curve, and is quite complicated. We can only content ourselves with an extremely brief sketch of the argument here.

The basic idea is to identify on both sides a small set of objects which split generate the respective categories. A set of objects of a category $\mathcal{A}$ *split generate* $\mathcal{A}$ if one can obtain any object from them by successive mapping cones, splitting off direct summands, and isomorphism. The categories generated by these objects can be studied via explicit methods and compared.

On the side of coherent sheaves, Beilinson's Theorem, Theorem 4.72, states that $\mathrm{D}^b(\mathbb{P}^3)$ is generated by a small set of locally free sheaves, namely $\Omega^i(i)[i]$, $0 \le i \le 3$. In fact, their restrictions to a quartic K3 surface split generate the bounded derived category of that surface. In our case, $\hat{\mathcal{X}}$ is described as a quotient of a family of K3 surfaces $\mathcal{X}$ by Γ_{16}, so instead of considering the derived category of the quotient, one can consider the Γ_{16}-equivariant derived category upstairs. Each split generator upstairs has 16 equivariant versions, and thus one obtains a set of 64 split generators for $\mathrm{D}^b(\hat{\mathcal{X}}_q)$.

On the other side, the Fermat family we considered (8.27), has, over $\mathrm{Spec}\,\mathbb{C}[q]$, 4 singular fibers each with 16 singular points. Each such singular point induces in some chosen general fiber in this family a Lagrangian sphere as vanishing cycle. This provides 64 Lagrangian spheres in this fiber, and chosen correctly, these yield the set of objects which are compared with those of the derived category. One shows that these 64 Lagrangians split generate $\mathrm{D}^\pi \,\mathrm{Fuk}(X)$. The most delicate part, which requires a great deal of computation, involves computing the subcategory generated by these objects; in particular, one has a number of multiplication maps to compute. Ultimately, this allows one to compare the two categories.

In fact, what one can show is that both categories are non-trivial deformations of the same category. Here, on both sides, q is the deformation parameter. By doing calculations with Hochschild cohomology (see §2.2.3 for some discussion of Hochschild cohomology), one can prove that there is only a one-parameter family of deformations. Thus the two families must coincide, but one cannot give the explicit mirror map.

Bibliography

[1] M. Abouzaid, M. Gross, and B. Siebert. Work in progress.

[2] Mohammed Abouzaid. *On homological mirror symmetry for toric varieties.* PhD thesis, University of Chicago, 2007.

[3] Mohammed Abouzaid. On the Fukaya categories of higher genus surfaces. *Adv. Math.*, 217(3):1192–1235, 2008.

[4] Ofer Aharony, Steven S. Gubser, Juan Maldacena, Hirosi Ooguri, and Yaron Oz. Large N field theories, string theory and gravity. *Phys. Rep.*, 323(3-4):183–386, 2000.

[5] Ofer Aharony, A. Hanany, Kenneth A. Intriligator, N. Seiberg, and M. J. Strassler. Aspects of $N = 2$ supersymmetric gauge theories in three dimensions. *Nucl. Phys.*, B499:67–99, 1997.

[6] Cecilia Albertsson, Ulf Lindström, and Maxim Zabzine. Superconformal boundary conditions for the WZW model. *J. High Energy Phys.*, (5):050, 13 pp. (electronic), 2003.

[7] A. Alexeevski and S. Natanzon. Noncommutative two-dimensional topological field theories and Hurwitz numbers for real algebraic curves. *Selecta Math. (N.S.)*, 12(3-4):307–377, 2006.

[8] Frederick J. Almgren, Jr. *Almgren's big regularity paper*, volume 1 of *World Scientific Monograph Series in Mathematics.* World Scientific Publishing Co. Inc., River Edge, NJ, 2000. Q-valued functions minimizing Dirichlet's integral and the regularity of area-minimizing rectifiable currents up to codimension 2, With a preface by Jean E. Taylor and Vladimir Scheffer.

[9] Orlando Alvarez. Theory of strings with boundaries: Fluctuations, topology, and quantum geometry. *Nucl. Phys.*, B216:125, 1983.

[10] L. Alvarez-Gaumé, D. Z. Freedman, and S. Mukhi. The background field method and the ultraviolet structure of the supersymmetric nonlinear σ-model. *Annals of Physics*, 134:85–109, 1981.

[11] Luis Alvarez-Gaumé, J. B. Bost, Gregory W. Moore, Philip C. Nelson, and Cumrun Vafa. Bosonization on higher genus riemann surfaces. *Commun. Math. Phys.*, 112:503, 1987.

[12] Luis Alvarez-Gaumé, Gregory W. Moore, Philip C. Nelson, Cumrun Vafa, and J. B. Bost. Bosonization in arbitrary genus. *Phys. Lett.*, B178:41–47, 1986.

[13] M. T. Anderson. The L^2 structure of moduli spaces of Einstein metrics on 4-manifolds. *Geom. Funct. Anal.*, 2(1):29–89, 1992.

[14] V. I. Arnold. Critical points of smooth functions, and their normal forms. *Uspehi Mat. Nauk*, 30(5(185)):3–65, 1975.

[15] Paul S. Aspinwall. A note on the equivalence of Vafa's and Douglas's picture of discrete torsion. *J. High Energy Phys.*, (12):Paper 29, 7, 2000.

[16] Paul S. Aspinwall. A point's point of view of stringy geometry. *J. High Energy Phys.*, (1):002, 15, 2003.

[17] Paul S. Aspinwall. D-branes on Calabi-Yau manifolds. In *Progress in string theory*, pages 1–152. World Sci. Publ., Hackensack, NJ, 2005.

[18] Paul S. Aspinwall and Michael R. Douglas. D-brane stability and monodromy. *J. High Energy Phys.*, (5):no. 31, 35, 2002.

[19] Paul S. Aspinwall, Brian R. Greene, and David R. Morrison. Measuring small distances in $N = 2$ sigma models. *Nuclear Phys. B*, 420(1-2):184–242, 1994.

[20] Paul S. Aspinwall, Brian R. Greene, and David R. Morrison. Calabi-Yau moduli space, mirror manifolds and spacetime topology change in string theory. In *Mirror symmetry, II*, volume 1 of *AMS/IP Stud. Adv. Math.*, pages 213–279. Amer. Math. Soc., Providence, RI, 1997.

[21] Paul S. Aspinwall, R. Paul Horja, and Robert L. Karp. Massless D-branes on Calabi-Yau threefolds and monodromy. *Comm. Math. Phys.*, 259(1):45–69, 2005.

[22] Paul S. Aspinwall and Sheldon Katz. Computation of superpotentials for D-branes. *Comm. Math. Phys.*, 264(1):227–253, 2006.

[23] Paul S. Aspinwall and Albion Lawrence. Derived categories and zero-brane stability. *J. High Energy Phys.*, (8):Paper 4, 27, 2001.

[24] Paul S. Aspinwall and David R. Morrison. Topological field theory and rational curves. *Comm. Math. Phys.*, 151(2):245–262, 1993.

[25] M. F. Atiyah and I. G. Macdonald. *Introduction to commutative algebra*. Addison-Wesley Publishing Co., Reading, Mass.-London-Don Mills, Ont., 1969.

[26] Michael Atiyah and Graeme Segal. Twisted K-theory. *Ukr. Mat. Visn.*, 1(3):287–330, 2004.

[27] Maurice Auslander, Idun Reiten, and Sverre O. Smalø. *Representation theory of Artin algebras*, volume 36 of *Cambridge Studies in Advanced Mathematics*. Cambridge University Press, Cambridge, 1997. Corrected reprint of the 1995 original.

[28] H. Azad and R. Kobayashi. Ricci-flat metrics on symmetric varieties. 1995.

[29] Christian Bär. Real Killing spinors and holonomy. *Comm. Math. Phys.*, 154(3):509–521, 1993.

[30] Sergey Barannikov and Maxim Kontsevich. Frobenius manifolds and formality of Lie algebras of polyvector fields. *Internat. Math. Res. Notices*, (4):201–215, 1998.

[31] Katrina Barron. A supergeometric interpretation of vertex operator superalgebras. *Internat. Math. Res. Notices*, (9):409–430, 1996.

[32] Victor V. Batyrev. Dual polyhedra and mirror symmetry for Calabi-Yau hypersurfaces in toric varieties. *J. Algebraic Geom.*, 3(3):493–535, 1994.

[33] Oliver Baues. Gluing affine 2-manifolds with polygons. *Geom. Dedicata*, 75(1):33–56, 1999.

[34] Katrin Becker, Melanie Becker, and Andrew Strominger. Fivebranes, membranes and non-perturbative string theory. *Nuclear Phys. B*, 456(1-2):130–152, 1995.

[35] K. Behrend and B. Fantechi. The intrinsic normal cone. *Invent. Math.*, 128(1):45–88, 1997.

[36] Roger E. Behrend, Paul A. Pearce, Valentina B. Petkova, and Jean-Bernard Zuber. Boundary conditions in rational conformal field theories. *Nucl. Phys.*, B570:525–589, 2000.

[37] A. A. Beilinson, J. Bernstein, and P. Deligne. Faisceaux pervers. *Astérisque*, 100:5–171, 1982.

[38] Oren Ben-Bassat. Mirror symmetry and generalized complex manifolds. I. The transform on vector bundles, spinors, and branes. *J. Geom. Phys.*, 56(4):533–558, 2006.

[39] Oren Ben-Bassat. Mirror symmetry and generalized complex manifolds. II. Integrability and the transform for torus bundles. *J. Geom. Phys.*, 56(7):1096–1115, 2006.

[40] David Berenstein and Michael R. Douglas. Seiberg duality for quiver gauge theories. 2002. arXiv:hep-th/0207027.

[41] Marcel Berger. *A Panoramic View of Riemannian Geometry*. Springer-Verlag, New York, 2003.

[42] Micha Berkooz, Michael R. Douglas, and Robert G. Leigh. Branes intersecting at angles. *Nuclear Phys. B*, 480(1-2):265–278, 1996.

[43] Nicole Berline, Ezra Getzler, and Michèle Vergne. *Heat kernels and Dirac operators*, volume 298 of *Grundlehren der Mathematischen Wissenschaften [Fundamental Principles of Mathematical Sciences]*. Springer-Verlag, Berlin, 1992.

[44] Arthur L. Besse. *Einstein manifolds*, volume 10 of *Ergebnisse der Mathematik und ihrer Grenzgebiete (3) [Results in Mathematics and Related Areas (3)]*. Springer-Verlag, Berlin, 1987.

[45] R. V. Bezrukavnikov and D. B. Kaledin. McKay equivalence for symplectic resolutions of quotient singularities. *Tr. Mat. Inst. Steklova*, 246(Algebr. Geom. Metody, Svyazi i Prilozh.):20–42, 2004.

[46] Roger Bielawski and Andrew S. Dancer. The geometry and topology of toric hyperkähler manifolds. *Comm. Anal. Geom.*, 8(4):727–760, 2000.

[47] Matthias Blau and George Thompson. Localization and diagonalization: a review of functional integral techniques for low-dimensional gauge theories and topological field theories. *J. Math. Phys.*, 36(5):2192–2236, 1995.

[48] T. S. Blyth. *Categories*. Longman, Harlow, 1986.

[49] Raf Bocklandt. Graded Calabi-Yau algebras of dimension 3. *J. Pure Appl. Algebra*, 212(1):14–32, 2008.

[50] A. I. Bondal and M. Kapranov. Framed triangulated categories. *Mat. Sb.*, 181(5):669–683, 1990.

[51] Alexei Bondal and Dmitri Orlov. Reconstruction of a variety from the derived category and groups of autoequivalences. *Compositio Math.*, 125(3):327–344, 2001.

[52] A. Borel, P.-P. Grivel, B. Kaup, A. Haefliger, B. Malgrange, and F. Ehlers. *Algebraic D-modules*, volume 2 of *Perspectives in Mathematics*. Academic Press Inc., Boston, MA, 1987.

[53] Raoul Bott and Loring W. Tu. *Differential forms in algebraic topology*, volume 82 of *Graduate Texts in Mathematics*. Springer-Verlag, New York, 1982.

[54] Wayne Boucher, Daniel Friedan, and Adrian Kent. Determinant formulae and unitarity for the $N = 2$ superconformal algebras in two dimensions or exact results on string compactification. *Phys. Lett. B*, 172(3-4):316–322, 1986.

[55] Peter Bouwknegt and Varghese Mathai. D-branes, B-fields and twisted K-theory. *J. High Energy Phys.*, (3):Paper 7, 11, 2000.

[56] Andreas Brandhuber. G_2 holonomy spaces from invariant three-forms. *Nuclear Phys. B*, 629(1-3):393–416, 2002.

[57] Andreas Brandhuber, Jaume Gomis, Steven S. Gubser, and Sergei Gukov. Gauge theory at large N and new G_2 holonomy metrics. *Nuclear Phys. B*, 611(1-3):179–204, 2001.

[58] Sheila Brenner and M. C. R. Butler. The equivalence of certain functors occurring in the representation theory of Artin algebras and species. *J. London Math. Soc. (2)*, 14(1):183–187, 1976.

[59] T. Bridgeland and V. Toledano-Laredo. Stability conditions and Stokes factors. 2008. arXiv:0801.3974.

[60] Tom Bridgeland. Equivalences of triangulated categories and Fourier-Mukai transforms. *Bull. London Math. Soc.*, 31(1):25–34, 1999.

[61] Tom Bridgeland. Flops and derived categories. *Invent. Math.*, 147(3):613–632, 2002.

[62] Tom Bridgeland. T-structures on some local Calabi-Yau varieties. *J. Algebra*, 289(2):453–483, 2005.

[63] Tom Bridgeland. Derived categories of coherent sheaves. In *International Congress of Mathematicians. Vol. II*, pages 563–582. Eur. Math. Soc., Zürich, 2006.

[64] Tom Bridgeland. Stability conditions on a non-compact Calabi-Yau threefold. *Comm. Math. Phys.*, 266(3):715–733, 2006.

[65] Tom Bridgeland. Stability conditions on triangulated categories. *Ann. of Math. (2)*, 166(2):317–345, 2007.

[66] Tom Bridgeland. Stability conditions on $K3$ surfaces. *Duke Math. J.*, 141(2):241–291, 2008.

[67] Tom Bridgeland, Alastair King, and Miles Reid. The McKay correspondence as an equivalence of derived categories. *J. Amer. Math. Soc.*, 14(3):535–554 (electronic), 2001.

[68] Kenneth S. Brown. *Cohomology of groups*, volume 87 of *Graduate Texts in Mathematics*. Springer-Verlag, New York, 1994.

[69] Ilka Brunner, Michael R. Douglas, Albion Lawrence, and Christian Römelsberger. D-branes on the quintic. *J. High Energy Phys.*, (8):Paper 15, 72, 2000.

[70] Ilka Brunner, Manfred Herbst, Wolfgang Lerche, and Bernhard Scheuner. Landau-Ginzburg realization of open string TFT. *J. High Energy Phys.*, (11):043, 28 pp. (electronic), 2006.

[71] Ilka Brunner and Kentaro Hori. Notes on orientifolds of rational conformal field theories. *J. High Energy Phys.*, (7):023, 59 pp. (electronic), 2004.

[72] Ilka Brunner and Kentaro Hori. Orientifolds and mirror symmetry. *J. High Energy Phys.*, 11:005, 2004.

[73] Ilka Brunner, Kentaro Hori, Kazuo Hosomichi, and Johannes Walcher. Orientifolds of Gepner models. *J. High Energy Phys.*, (2):001, 122 pp. (electronic), 2007.

[74] Ilka Brunner and Gregory Moore. 2D TFT on unoriented surfaces.

[75] Robert L. Bryant. Some examples of special Lagrangian tori. *Adv. Theor. Math. Phys.*, 3(1):83–90, 1999.

[76] Robert L. Bryant and Simon M. Salamon. On the construction of some complete metrics with exceptional holonomy. *Duke Math. J.*, 58(3):829–850, 1989.

[77] Jean-Luc Brylinski. *Loop spaces, characteristic classes and geometric quantization*, volume 107 of *Progress in Mathematics*. Birkhäuser Boston Inc., Boston, MA, 1993.

[78] Igor Burban and Bernd Kreussler. Derived categories of irreducible projective curves of arithmetic genus one. *Compos. Math.*, 142(5):1231–1262, 2006.

[79] E. Calabi. Métriques kählériennes et fibrés holomorphes. *Ann. Sci. École Norm. Sup. (4)*, 12(2):269–294, 1979.

[80] Damien Calaque and Michel Van den Bergh. Hochschild cohomology and Atiyah classes. arXiv:0708.2725, 2007.

[81] Andrei Căldăraru. The Mukai pairing. I. The Hochschild structure. arXiv:math/0308079, 2003.

[82] Andrei Căldăraru. The Mukai pairing. II. The Hochschild-Kostant-Rosenberg isomorphism. *Adv. Math.*, 194(1):34–66, 2005.

[83] P. Candelas, E. Derrick, and L. Parkes. Generalized Calabi-Yau manifolds and the mirror of a rigid manifold. *Nucl. Phys.*, B407:115–154, 1993.

[84] Philip Candelas and Xenia C. de la Ossa. Comments on conifolds. *Nuclear Phys. B*, 342(1):246–268, 1990.

[85] Philip Candelas, Xenia C. de la Ossa, Paul S. Green, and Linda Parkes. A pair of Calabi-Yau manifolds as an exactly soluble superconformal theory. *Nuclear Phys. B*, 359(1):21–74, 1991.

[86] John L. Cardy. Boundary conditions, fusion rules and the Verlinde formula. *Nuclear Phys. B*, 324(3):581–596, 1989.

[87] John L. Cardy. Boundary conformal field theory. In *Encyclopedia of Mathematical Physics*. Academic Press, 2006.

[88] Élie Cartan. *The theory of spinors*. Dover Publications Inc., New York, 1981.

[89] P. Cartier and C. DeWitt-Morette. A rigorous mathematical foundation of functional integration. In *Functional integration (Cargèse, 1996)*, volume 361 of *NATO Adv. Sci. Inst. Ser. B Phys.*, pages 1–50. Plenum, New York, 1997. With an appendix by David Collins.

[90] Ricardo Castaño Bernard and Diego Matessi. Lagrangian 3-torus fibrations. 2006.

[91] Moira Chas and Dennis Sullivan. Closed string operators in topology leading to Lie bialgebras and higher string algebra. In *The legacy of Niels Henrik Abel*, pages 771–784. Springer, Berlin, 2004.

[92] Jeff Cheeger and Detlef Gromoll. On the structure of complete manifolds of nonnegative curvature. *Ann. of Math. (2)*, 96:413–443, 1972.

[93] Jiun-Cheng Chen. Flops and equivalences of derived categories for threefolds with only terminal Gorenstein singularities. *J. Differential Geom.*, 61(2):227–261, 2002.

[94] Shiing Shen Chern. *Complex manifolds without potential theory*. Springer-Verlag, New York, second edition, 1979. With an appendix on the geometry of characteristic classes, Universitext.

[95] Yeuk-Kwan E. Cheung and Zheng Yin. Anomalies, branes, and currents. *Nuclear Phys. B*, 517(1-3):69–91, 1998.

[96] Stefano Chiantese, Florian Gmeiner, and Claus Jeschek. Mirror symmetry for topological sigma models with generalized Kähler geometry. *Internat. J. Modern Phys. A*, 21(11):2377–2389, 2006.

[97] Z. W. Chong, M. Cvetič, G. W. Gibbons, H. Lü, C. N. Pope, and P. Wagner. General metrics of G_2 holonomy and contraction limits. *Nuclear Phys. B*, 638(3):459–482, 2002.

[98] Herb Clemens. Moduli schemes associated to K-trivial threefolds as gradient schemes. *J. Algebraic Geom.*, 14(4):705–739, 2005.

[99] Richard Cleyton and Andrew Swann. Cohomogeneity-one G_2-structures. *J. Geom. Phys.*, 44(2-3):202–220, 2002.

[100] Kevin Costello. Topological conformal field theories and Calabi-Yau categories. *Adv. Math.*, 210(1):165–214, 2007.

[101] David A. Cox and Sheldon Katz. *Mirror symmetry and algebraic geometry*, volume 68 of *Mathematical Surveys and Monographs*. American Mathematical Society, Providence, RI, 1999.

[102] Alastair Craw and Akira Ishii. Flops of G-Hilb and equivalences of derived categories by variation of GIT quotient. *Duke Math. J.*, 124(2):259–307, 2004.

[103] Alastair Craw, Diane Maclagan, and Rekha R. Thomas. Moduli of McKay quiver representations. I. The coherent component. *Proc. Lond. Math. Soc. (3)*, 95(1):179–198, 2007.

[104] Charles W. Curtis and Irving Reiner. *Representation theory of finite groups and associative algebras*. AMS Chelsea Publishing, Providence, RI, 2006.

[105] M. Cvetič, G. W. Gibbons, H. Lü, and C. N. Pope. Supersymmetric M3-branes and G_2 manifolds. *Nuclear Phys. B*, 620(1-2):3–28, 2002.

[106] Jin Dai, R. G. Leigh, and Joseph Polchinski. New connections between string theories. *Mod. Phys. Lett.*, A4:2073–2083, 1989.

[107] Andrew Dancer and Andrew Swann. Hyper-Kähler metrics of cohomogeneity one. *J. Geom. Phys.*, 21(3):218–230, 1997.

[108] B. de Wit, P. G. Lauwers, and Antoine Van Proeyen. Lagrangians of $N = 2$ supergravity-matter systems. *Nucl. Phys.*, B255:569, 1985.

[109] Pierre Deligne, Pavel Etingof, Daniel S. Freed, Lisa C. Jeffrey, David Kazhdan, John W. Morgan, David R. Morrison, and Edward Witten, editors. *Quantum fields and strings: a course for mathematicians. Vol. 1, 2*. American Mathematical Society, Providence, RI, 1999. Material from the Special Year on Quantum Field Theory held at the Institute for Advanced Study, Princeton, NJ, 1996–1997.

[110] Frederik Denef. Quantum quivers and Hall/hole halos. *J. High Energy Phys.*, (10):023, 42, 2002.

[111] Frederik Denef and Gregory W. Moore. Split states, entropy enigmas, holes and halos. 2007. arXiv:hep-th/0702146.

[112] Philippe Di Francesco, Pierre Mathieu, and David Sénéchal. *Conformal field theory*. Graduate Texts in Contemporary Physics. Springer-Verlag, New York, 1997.

[113] Duiliu-Emanuel Diaconescu. D-branes, monopoles and Nahm equations. *Nuclear Phys. B*, 503(1-2):220–238, 1997.

[114] Duiliu-Emanuel Diaconescu, Alberto Garcia-Raboso, Robert L. Karp, and Kuver Sinha. D-brane superpotentials in Calabi-Yau orientifolds. *Adv. Theor. Math. Phys.*, 11(3):471–516, 2007.

[115] Duiliu-Emanuel Diaconescu and Jaume Gomis. Fractional branes and boundary states in orbifold theories. *J. High Energy Phys.*, (10):Paper 1, 44, 2000.

[116] Duiliu-Emanuel Diaconescu, Jaume Gomis, and Michael R. Douglas. Fractional branes and wrapped branes. *J. High Energy Phys.*, (2):Paper 13, 9 pp. (electronic), 1998.

[117] Robbert Dijkgraaf. *A geometrical approach to two-dimensional conformal field theory.* PhD thesis, 1989.

[118] Robbert Dijkgraaf. Intersection theory, integrable hierarchies and topological field theory. In *New symmetry principles in quantum field theory (Cargèse, 1991)*, volume 295 of *NATO Adv. Sci. Inst. Ser. B Phys.*, pages 95–158. Plenum, New York, 1992.

[119] Robbert Dijkgraaf. Fields, strings and duality. In *Symétries quantiques (Les Houches, 1995)*, pages 3–147. North-Holland, Amsterdam, 1998.

[120] Robbert Dijkgraaf, Sergei Gukov, Andrew Neitzke, and Cumrun Vafa. Topological M-theory as unification of form theories of gravity. *Adv. Theor. Math. Phys.*, 9(4):603–665, 2005.

[121] Robbert Dijkgraaf, Cumrun Vafa, Erik Verlinde, and Herman Verlinde. The operator algebra of orbifold models. *Comm. Math. Phys.*, 123(3):485–526, 1989.

[122] Robbert Dijkgraaf, Herman Verlinde, and Erik Verlinde. Notes on topological string theory and 2D quantum gravity. In *String theory and quantum gravity (Trieste, 1990)*, pages 91–156. World Sci. Publ., River Edge, NJ, 1991.

[123] Jacques Distler, Hans Jockers, and Hyuk-Jae Park. D-brane monodromies, derived categories and boundary linear sigma models. 2002. arXiv:hep-th/0206242.

[124] L. J. Dixon. Some world-sheet properties of superstring compactifications,on orbifolds and otherwise. In G. Furlan et al., editors, *Superstrings, Unified Theories, and Cosmology 1987*, pages 67–126. World Scientific, 1988.

[125] Lance J. Dixon, Jeffrey A. Harvey, C. Vafa, and Edward Witten. Strings on orbifolds. *Nucl. Phys.*, B261:678–686, 1985.

[126] I. V. Dolgachev. Mirror symmetry for lattice polarized $K3$ surfaces. *J. Math. Sci.*, 81(3):2599–2630, 1996.

[127] Igor V. Dolgachev and Yi Hu. Variation of geometric invariant theory quotients. *Inst. Hautes Études Sci. Publ. Math.*, (87):5–56, 1998. With an appendix by Nicolas Ressayre.

[128] S. K. Donaldson. Anti self-dual Yang-Mills connections over complex algebraic surfaces and stable vector bundles. *Proc. London Math. Soc. (3)*, 50(1):1–26, 1985.

[129] S. K. Donaldson and P. B. Kronheimer. *The geometry of four-manifolds.* Oxford Mathematical Monographs. The Clarendon Press Oxford University Press, New York, 1990. , Oxford Science Publications.

[130] S.K. Donaldson. Some numerical results in complex differential geometry. arXiv:math/0512625, 2006.

[131] Michael Douglas, Jerome Gauntlett, and Mark Gross, editors. *Strings and geometry*, volume 3 of *Clay Mathematics Proceedings*, Providence, RI, 2004. American Mathematical Society.

[132] Michael R. Douglas. D-branes and discrete torsion. 1998.

[133] Michael R. Douglas. D-branes, categories and $N = 1$ supersymmetry. *J. Math. Phys.*, 42(7):2818–2843, 2001.

[134] Michael R. Douglas. D-branes and $N = 1$ supersymmetry. In *Strings, 2001 (Mumbai)*, volume 1 of *Clay Math. Proc.*, pages 139–152. Amer. Math. Soc., Providence, RI, 2002.

[135] Michael R. Douglas. Dirichlet branes, homological mirror symmetry, and stability. In *Proceedings of the International Congress of Mathematicians, Vol. III (Beijing, 2002)*, pages 395–408, Beijing, 2002. Higher Ed. Press.

[136] Michael R. Douglas and Bartomeu Fiol. D-branes and discrete torsion. II. *J. High Energy Phys.*, (9):053, 22 pp. (electronic), 2005.

[137] Michael R. Douglas, Bartomeu Fiol, and Christian Römelsberger. The spectrum of BPS branes on a noncompact Calabi-Yau. *J. High Energy Phys.*, (9):057, 40 pp. (electronic), 2005.

[138] Michael R. Douglas, Bartomeu Fiol, and Christian Römelsberger. Stability and BPS branes. *J. High Energy Phys.*, (9):006, 15 pp. (electronic), 2005.

[139] Michael R. Douglas, Brian R. Greene, and David R. Morrison. Orbifold resolution by D-branes. *Nuclear Phys. B*, 506(1-2):84–106, 1997.

[140] Michael R. Douglas, Robert L. Karp, Sergio Lukic, and René Reinbacher. Numerical Calabi-Yau metrics. *J. Math. Phys.*, 49(3):032302, 19, 2008.

[141] Michael R. Douglas and Gregory Moore. D-branes, quivers, and ALE instantons. arXiv:hep-th/9603167, 1996.

[142] Yurij A. Drozd and Vladimir V. Kirichenko. *Finite-dimensional algebras.* Springer-Verlag, Berlin, 1994.

[143] Boris Dubrovin. Geometry and analytic theory of Frobenius manifolds. In *Proceedings of the International Congress of Mathematicians, Vol. II (Berlin, 1998)*, number Extra Vol. II, pages 315–326 (electronic), 1998.

[144] E. Elizalde. On the concept of determinant for the differential operators of quantum physics. *J. High Energy Phys.*, (7):Paper 15, 13 pp. (electronic), 1999.

[145] Herbert Federer. *Geometric measure theory.* Die Grundlehren der mathematischen Wissenschaften, Band 153. Springer-Verlag New York Inc., New York, 1969.

[146] Birte Feix. Twistor spaces of hyperkähler manifolds with S^1-actions. *Differential Geom. Appl.*, 19(1):15–28, 2003.

[147] M. Fernández and A. Gray. Riemannian manifolds with structure group G_2. *Ann. Mat. Pura Appl. (4)*, 132:19–45 (1983), 1982.

[148] Gabriele Ferretti, Per Salomonson, and Dimitrios Tsimpis. D-brane probes on G_2 orbifolds. *J. High Energy Phys.*, (3):No. 4, 15, 2002.

[149] Bartomeu Fiol and Marcos Mariño. BPS states and algebras from quivers. *J. High Energy Phys.*, (7):Paper 31, 40, 2000.

[150] Andreas Floer. A relative Morse index for the symplectic action. *Comm. Pure Appl. Math.*, 41(4):393–407, 1988.

[151] Sebastián Franco, Amihay Hanany, Dario Martelli, James Sparks, David Vegh, and Brian Wecht. Gauge theories from toric geometry and brane tilings. *J. High Energy Phys.*, (1):128, 40 pp. (electronic), 2006.

[152] Daniel S. Freed. Higher algebraic structures and quantization. *Comm. Math. Phys.*, 159(2):343–398, 1994.

[153] Daniel S. Freed. *Five lectures on supersymmetry.* American Mathematical Society, Providence, RI, 1999.

[154] Daniel S. Freed. Special Kähler manifolds. *Comm. Math. Phys.*, 203(1):31–52, 1999.

[155] Daniel S. Freed. K-theory in quantum field theory. In *Current developments in mathematics, 2001*, pages 41–87. Int. Press, Somerville, MA, 2002.

[156] Daniel S. Freed and Edward Witten. Anomalies in string theory with D-branes. *Asian J. Math.*, 3(4):819–851, 1999.

[157] Igor Frenkel, James Lepowsky, and Arne Meurman. *Vertex operator algebras and the Monster*, volume 134 of *Pure and Applied Mathematics*. Academic Press Inc., Boston, MA, 1988.

[158] Igor B. Frenkel, Yi-Zhi Huang, and James Lepowsky. On axiomatic approaches to vertex operator algebras and modules. *Mem. Amer. Math. Soc.*, 104(494):viii+64, 1993.

[159] Daniel Friedan. Introduction to Polyakov's string theory. In J. B. Zuber and R. Stora, editors, *Recent advances in field theory and statistical mechanics. Proceedings, 39th Summer school, Les Houches, France, August 2 - September 10, 1982*, Amsterdam, Netherlands, 1984. North-holland (1984).

[160] Daniel Friedan. Nonlinear models in $2+\varepsilon$ dimensions. *Ann. Physics*, 163(2):318–419, 1985.

[161] Jurg Fröhlich, Jurgen Fuchs, Ingo Runkel, and Christoph Schweigert. Kramers-Wannier duality from conformal defects. *Phys. Rev. Lett.*, 93:070601, 2004.

[162] Jurg Fröhlich, Jurgen Fuchs, Ingo Runkel, and Christoph Schweigert. Duality and defects in rational conformal field theory. *Nucl. Phys.*, B763:354–430, 2007.

[163] J. Fuchs, C. Schweigert, and J. Walcher. Projections in string theory and boundary states for Gepner models. *Nuclear Physics B*, 588:110, 2000.

[164] K. Fukaya, Y. Oh, H. Ohta, and K. Ono. Lagrangian intersection Floer theory—anomaly and obstruction.

[165] Kenji Fukaya. Morse homotopy, A^∞-category, and Floer homologies. In *Proceedings of GARC Workshop on Geometry and Topology '93 (Seoul, 1993)*, volume 18 of *Lecture Notes Ser.*, pages 1–102, Seoul, 1993. Seoul Nat. Univ.

[166] Kenji Fukaya. Multivalued Morse theory, asymptotic analysis and mirror symmetry. In *Graphs and patterns in mathematics and theoretical physics*, volume 73 of *Proc. Sympos. Pure Math.*, pages 205–278. Amer. Math. Soc., Providence, RI, 2005.

[167] Kenji Fukaya and Yong-Geun Oh. Zero-loop open strings in the cotangent bundle and Morse homotopy. *Asian J. Math.*, 1(1):96–180, 1997.

[168] Matthias R. Gaberdiel and Barton Zwiebach. Tensor constructions of open string theories. I. Foundations. *Nuclear Phys. B*, 505(3):569–624, 1997.

[169] Matthias R. Gaberdiel and Barton Zwiebach. Tensor constructions of open string theories. II. Vector bundles, D-branes and orientifold groups. *Phys. Lett. B*, 410(2-4):151–159, 1997.

[170] S. J. Gates, Marcus T. Grisaru, M. Rocek, and W. Siegel. Superspace, or one thousand and one lessons in supersymmetry. *Front. Phys.*, 58:1–548, 1983.

[171] J. P. Gauntlett, G. W. Gibbons, G. Papadopoulos, and P. K. Townsend. Hyper-Kähler manifolds and multiply intersecting branes. *Nuclear Phys. B*, 500(1-3):133–162, 1997.

[172] Jerome P. Gauntlett. Branes, calibrations and supergravity. In *Strings and geometry*, volume 3 of *Clay Math. Proc.*, pages 79–126. Amer. Math. Soc., Providence, RI, 2004.

[173] I. M. Gel′fand, M. Kapranov, and A. V. Zelevinsky. *Discriminants, resultants, and multidimensional determinants.* Mathematics: Theory & Applications. Birkhäuser Boston Inc., Boston, MA, 1994.

[174] S. I. Gelfand and Yu. I. Manin. *Homological algebra.* Springer-Verlag, Berlin, 1999. Translated from the 1989 Russian original by the authors, Reprint of the original English edition from the series Encyclopaedia of Mathematical Sciences [*Algebra, V*, Encyclopaedia Math. Sci., 38, Springer, Berlin, 1994;.

[175] Doron Gepner. Space-time supersymmetry in compactified string theory and superconformal models. *Nucl. Phys.*, B296:757, 1988.

[176] Ezra Getzler, John D. S. Jones, and Scott Petrack. Differential forms on loop spaces and the cyclic bar complex. *Topology*, 30(3):339–371, 1991.

[177] G. Gibbons and D. Freedman. Remarks on supersymmetry and Kähler manifolds. In *Superspace and supergravity.* Cambridge University Press, 1981.

[178] G. W. Gibbons, D. N. Page, and C. N. Pope. Einstein metrics on S^3, $\mathbf{R}^3$ and $\mathbf{R}^4$ bundles. *Comm. Math. Phys.*, 127(3):529–553, 1990.

[179] David Gilbarg and Neil S. Trudinger. *Elliptic partial differential equations of second order.* Classics in Mathematics. Springer-Verlag, Berlin, 2001. Reprint of the 1998 edition.

[180] Paul Ginsparg and Gregory Moore. Lectures on 2D gravity and 2D string theory. In *Recent directions in particle theory*. World scientific, 1993.

[181] Victor Ginzburg. Calabi-Yau algebras. 2006. arXiv:math/0612139.

[182] James Glimm and Arthur Jaffe. *Quantum physics. A functional integral point of view*. Springer-Verlag, New York, second edition, 1987.

[183] P. Goddard, A. Kent, and David I. Olive. Virasoro algebras and coset space models. *Phys. Lett.*, B152:88, 1985.

[184] William Goldman and Morris W. Hirsch. The radiance obstruction and parallel forms on affine manifolds. *Trans. Amer. Math. Soc.*, 286(2):629–649, 1984.

[185] Edward Goldstein. Calibrated fibrations on noncompact manifolds via group actions. *Duke Math. J.*, 110(2):309–343, 2001.

[186] G. Gonzalez-Sprinberg and J.-L. Verdier. Construction géométrique de la correspondance de McKay. *Ann. Sci. École Norm. Sup. (4)*, 16(3):409–449 (1984), 1983.

[187] Vassily Gorbounov, Fyodor Malikov, and Vadim Schechtman. Gerbes of chiral differential operators. *Math. Res. Lett.*, 7(1):55–66, 2000.

[188] Vassily Gorbounov, Fyodor Malikov, and Vadim Schechtman. Gerbes of chiral differential operators. III. In *The orbit method in geometry and physics (Marseille, 2000)*, volume 213 of *Progr. Math.*, pages 73–100. Birkhäuser Boston, Boston, MA, 2003.

[189] Vassily Gorbounov, Fyodor Malikov, and Vadim Schechtman. Gerbes of chiral differential operators. II. Vertex algebroids. *Invent. Math.*, 155(3):605–680, 2004.

[190] Michael B. Green, Jeffrey A. Harvey, and Gregory Moore. I-brane inflow and anomalous couplings on D-branes. *Classical Quantum Gravity*, 14(1):47–52, 1997.

[191] Michael B. Green, John H. Schwarz, and Edward Witten. *Superstring theory. Vol. 1*. Cambridge Monographs on Mathematical Physics. Cambridge University Press, Cambridge, 1987. Introduction.

[192] B. R. Greene. String theory on Calabi-Yau manifolds. In *Fields, strings and duality (Boulder, CO, 1996)*, pages 543–726. World Sci. Publishing, River Edge, NJ, 1997.

[193] B. R. Greene and M. R. Plesser. Duality in Calabi-Yau moduli space. *Nuclear Phys. B*, 338(1):15–37, 1990.

[194] Brian R. Greene, Calin-Iuliio Lazaroiu, and Mark Raugas. D-branes on non-abelian threefold quotient singularities. *Nuclear Phys. B*, 553(3):711–749, 1999.

[195] Brian R. Greene, David R. Morrison, and Andrew Strominger. Black hole condensation and the unification of string vacua. *Nuclear Phys. B*, 451(1-2):109–120, 1995.

[196] Brian R. Greene, Alfred Shapere, Cumrun Vafa, and Shing-Tung Yau. Stringy cosmic strings and noncompact Calabi-Yau manifolds. *Nuclear Phys. B*, 337(1):1–36, 1990.

[197] Phillip Griffiths and Joseph Harris. *Principles of algebraic geometry*. Wiley-Interscience [John Wiley & Sons], New York, 1978. Pure and Applied Mathematics.

[198] M. T. Grisaru, A. E. M. van de Ven, and D. Zanon. Four-loop β-function for the $N = 1$ and $N = 2$ supersymmetric nonlinear sigma model in two dimensions. *Phys. Lett. B*, 173(4):423–428, 1986.

[199] Mark Gross. Examples of special Lagrangian fibrations. In *Symplectic geometry and mirror symmetry (Seoul, 2000)*, pages 81–109. World Sci. Publ., River Edge, NJ, 2001.

[200] Mark Gross. Special Lagrangian fibrations. I. Topology. In *Winter School on Mirror Symmetry, Vector Bundles and Lagrangian Submanifolds (Cambridge, MA, 1999)*, volume 23 of *AMS/IP Stud. Adv. Math.*, pages 65–93. Amer. Math. Soc., Providence, RI, 2001.

[201] Mark Gross. Special Lagrangian fibrations. II. Geometry. A survey of techniques in the study of special Lagrangian fibrations. In *Winter School on Mirror Symmetry, Vector Bundles and Lagrangian Submanifolds (Cambridge, MA, 1999)*, volume 23 of *AMS/IP Stud. Adv. Math.*, pages 95–150. Amer. Math. Soc., Providence, RI, 2001.

[202] Mark Gross. Topological mirror symmetry. *Invent. Math.*, 144(1):75–137, 2001.
[203] Mark Gross. Toric degenerations and Batyrev-Borisov duality. *Math. Ann.*, 333(3):645–688, 2005.
[204] Mark Gross and Bernd Siebert. Torus fibrations and toric degenerations. To appear.
[205] Mark Gross and Bernd Siebert. Affine manifolds, log structures, and mirror symmetry. *Turkish J. Math.*, 27(1):33–60, 2003.
[206] Mark Gross and Bernd Siebert. Mirror symmetry via logarithmic degeneration data. I. *J. Differential Geom.*, 72(2):169–338, 2006.
[207] Mark Gross and Bernd Siebert. From real affine geometry to complex geometry. arXiv:math/0703822, 2007.
[208] Mark Gross and Bernd Siebert. Mirror symmetry via logarithmic degeneration data II. 2007. arXiv:0709.2290.
[209] Mark Gross and P. M. H. Wilson. Mirror symmetry via 3-tori for a class of Calabi-Yau threefolds. *Math. Ann.*, 309(3):505–531, 1997.
[210] Mark Gross and P. M. H. Wilson. Large complex structure limits of $K3$ surfaces. *J. Differential Geom.*, 55(3):475–546, 2000.
[211] A. Grothendieck. Éléments de géométrie algébrique. III. Étude cohomologique des faisceaux cohérents. II. *Inst. Hautes Études Sci. Publ. Math.*, (17):91, 1963.
[212] Marco Gualtieri. *Generalized complex geometry*. PhD thesis, University of Oxford, 2003.
[213] Victor Guillemin and Matthew Stenzel. Grauert tubes and the homogeneous Monge-Ampère equation. *J. Differential Geom.*, 34(2):561–570, 1991.
[214] Sergei Gukov and David Tong. D-brane probes of special holonomy manifolds, and dynamics of $\mathcal{N} = 1$ three-dimensional gauge theories. *J. High Energy Phys.*, (4):No. 50, 69, 2002.
[215] Sergei Gukov, Shing-Tung Yau, and Eric Zaslow. Duality and fibrations on G_2 manifolds. *Turkish J. Math.*, 27(1):61–97, 2003.
[216] Christian Haase and Ilia Zharkov. Integral affine structures on spheres and torus fibrations of Calabi-Yau toric hypersurfaces I. 2002.
[217] Amihay Hanany and Johannes Walcher. On duality walls in string theory. *J. High Energy Phys.*, 06:055, 2003.
[218] Amihay Hanany and Edward Witten. Type IIB superstrings, BPS monopoles, and three-dimensional gauge dynamics. *Nuclear Phys. B*, 492(1-2):152–190, 1997.
[219] Dieter Happel, Idun Reiten, and Sverre O. Smalø. Tilting in abelian categories and quasitilted algebras. *Mem. Amer. Math. Soc.*, 120(575):viii+ 88, 1996.
[220] G. Harder and M. S. Narasimhan. On the cohomology groups of moduli spaces of vector bundles on curves. *Math. Ann.*, 212:215–248, 1974/75.
[221] Robin Hartshorne. *Residues and duality*. Lecture notes of a seminar on the work of A. Grothendieck, given at Harvard 1963/64. With an appendix by P. Deligne. Lecture Notes in Mathematics, No. 20. Springer-Verlag, Berlin, 1966.
[222] Robin Hartshorne. *Algebraic geometry*. Springer-Verlag, New York, 1977. Graduate Texts in Mathematics, No. 52.
[223] F. Reese Harvey. *Spinors and calibrations*, volume 9 of *Perspectives in Mathematics*. Academic Press Inc., Boston, MA, 1990.
[224] Jeffrey A. Harvey. TASI 2003 Lectures on anomalies. 2005. arXiv:hep-th/0509097.
[225] Jeffrey A. Harvey and Gregory Moore. On the algebras of BPS states. *Comm. Math. Phys.*, 197(3):489–519, 1998.
[226] Reese Harvey and H. Blaine Lawson, Jr. Calibrated geometries. *Acta Math.*, 148:47–157, 1982.
[227] Mark Haskins. Special Lagrangian cones. *Amer. J. Math.*, 126(4):845–871, 2004.
[228] Yang-Hui He. G_2 quivers. *J. High Energy Phys.*, (2):023, 14, 2003.

[229] Manfred Herbst and Calin-Iuliu Lazaroiu. Localization and traces in open-closed topological Landau-Ginzburg models. *J. High Energy Phys.*, (5):044, 31 pp. (electronic), 2005.

[230] Manfred Herbst, Calin-Iuliu Lazaroiu, and Wolfgang Lerche. Superpotentials, A_∞ relations and WDVV equations for open topological strings. *J. High Energy Phys.*, (2):071, 53 pp. (electronic), 2005.

[231] Nigel J. Hitchin. Polygons and gravitons. *Math. Proc. Cambridge Philos. Soc.*, 85(3):465–476, 1979.

[232] Nigel J. Hitchin. The moduli space of special Lagrangian submanifolds. *Ann. Scuola Norm. Sup. Pisa Cl. Sci. (4)*, 25(3-4):503–515 (1998), 1997.

[233] Nigel J. Hitchin. The geometry of three-forms in six dimensions. *J. Differential Geom.*, 55(3):547–576, 2000.

[234] Nigel J. Hitchin. Lectures on special Lagrangian submanifolds. In *Winter School on Mirror Symmetry, Vector Bundles and Lagrangian Submanifolds (Cambridge, MA, 1999)*, volume 23 of *AMS/IP Stud. Adv. Math.*, pages 151–182. Amer. Math. Soc., Providence, RI, 2001.

[235] Nigel J. Hitchin. Stable forms and special metrics. In *Global differential geometry: the mathematical legacy of Alfred Gray (Bilbao, 2000)*, volume 288 of *Contemp. Math.*, pages 70–89. Amer. Math. Soc., Providence, RI, 2001.

[236] Nigel J. Hitchin. Generalized Calabi-Yau manifolds. *Q. J. Math.*, 54(3):281–308, 2003.

[237] Nigel J. Hitchin. Special holonomy and beyond. In *Strings and geometry*, volume 3 of *Clay Math. Proc.*, pages 159–175. Amer. Math. Soc., Providence, RI, 2004.

[238] Nigel J. Hitchin, A. Karlhede, U. Lindström, and M. Roček. Hyper-Kähler metrics and supergravity. *Comm. Math. Phys.*, 108:535–589, 1987.

[239] Nigel J. Hitchin, A. Karlhede, U. Lindström, and M. Roček. Hyper-Kähler metrics and supersymmetry. *Comm. Math. Phys.*, 108(4):535–589, 1987.

[240] Kentaro Hori. Linear models of supersymmetric D-branes. In *Symplectic geometry and mirror symmetry (Seoul, 2000)*, pages 111–186. World Sci. Publ., River Edge, NJ, 2001.

[241] Kentaro Hori, Amer Iqbal, and Cumrun Vafa. D-branes and mirror symmetry. 2000. arXiv:hep-th/0005247.

[242] Kentaro Hori, Sheldon Katz, Albrecht Klemm, Rahul Pandharipande, Richard Thomas, Cumrun Vafa, Ravi Vakil, and Eric Zaslow. *Mirror symmetry*, volume 1 of *Clay Mathematics Monographs*. American Mathematical Society, Providence, RI, 2003. With a preface by Vafa.

[243] Kentaro Hori and Cumrun Vafa. Mirror symmetry. arXiv:hep-th/0002222, 2000.

[244] Kentaro Hori and Johannes Walcher. D-brane categories for orientifolds—the Landau-Ginzburg case. *J. High Energy Phys.*, (4):030, 36, 2008.

[245] R. Paul Horja. Derived category automorphisms from mirror symmetry. *Duke Math. J.*, 127(1):1–34, 2005.

[246] Yi-Zhi Huang. *Two-dimensional conformal geometry and vertex operator algebras*, volume 148 of *Progress in Mathematics*. Birkhäuser Boston Inc., Boston, MA, 1997.

[247] Yi-Zhi Huang and Antun Milas. Intertwining operator superalgebras and vertex tensor categories for superconformal algebras. I. *Commun. Contemp. Math.*, 4(2):327–355, 2002.

[248] Yi-Zhi Huang and Antun Milas. Intertwining operator superalgebras and vertex tensor categories for superconformal algebras. II. *Trans. Amer. Math. Soc.*, 354(1):363–385 (electronic), 2002.

[249] Daniel Huybrechts. *Complex geometry. An introduction.* Universitext. Springer-Verlag, Berlin, 2005.

[250] Daniel Huybrechts. *Fourier-Mukai transforms in algebraic geometry.* Oxford Mathematical Monographs. Oxford University Press, 2006.

[251] Ken Intriligator and Brian Wecht. The exact superconformal R-symmetry maximizes a. *Nucl. Phys.*, B667:183–200, 2003.

[252] Kenneth A. Intriligator, David R. Morrison, and Nathan Seiberg. Five-dimensional supersymmetric gauge theories and degenerations of Calabi-Yau spaces. *Nucl. Phys.*, B497:56–100, 1997.

[253] Kenneth A. Intriligator and N. Seiberg. Lectures on supersymmetric gauge theories and electric-magnetic duality. *Nucl. Phys. Proc. Suppl.*, 45BC:1–28, 1996.

[254] Kenneth A. Intriligator and N. Seiberg. Mirror symmetry in three-dimensional gauge theories. *Phys. Lett.*, B387:513–519, 1996.

[255] Nobuyuki Ishibashi. The Boundary and Crosscap States in Conformal Field Theories. *Mod. Phys. Lett.*, A4:251, 1989.

[256] Akira Ishii. Representation moduli of the McKay quiver for finite abelian subgroups of $\mathrm{SL}(3,\mathbb{C})$. In *Strings and geometry*, volume 3 of *Clay Math. Proc.*, pages 227–237. Amer. Math. Soc., Providence, RI, 2004.

[257] Y. Ito and I. Nakamura. Hilbert schemes and simple singularities. In *New trends in algebraic geometry (Warwick, 1996)*, volume 264 of *London Math. Soc. Lecture Note Ser.*, pages 151–233. Cambridge Univ. Press, Cambridge, 1999.

[258] Yukari Ito and Hiraku Nakajima. McKay correspondence and Hilbert schemes in dimension three. *Topology*, 39(6):1155–1191, 2000.

[259] Dominic Joyce. Asymptotically locally Euclidean metrics with holonomy $\mathrm{SU}(m)$. *Ann. Global Anal. Geom.*, 19(1):55–73, 2001.

[260] Dominic Joyce. Quasi-ALE metrics with holonomy $\mathrm{SU}(m)$ and $\mathrm{Sp}(m)$. *Ann. Global Anal. Geom.*, 19(2):103–132, 2001.

[261] Dominic Joyce. Singularities of special Lagrangian fibrations and the SYZ conjecture. *Comm. Anal. Geom.*, 11(5):859–907, 2003.

[262] Dominic Joyce. Special Lagrangian submanifolds with isolated conical singularities. V. Survey and applications. *J. Differential Geom.*, 63(2):279–347, 2003.

[263] Dominic Joyce. Special Lagrangian submanifolds with isolated conical singularities. II. Moduli spaces. *Ann. Global Anal. Geom.*, 25(4):301–352, 2004.

[264] Dominic Joyce. Special Lagrangian submanifolds with isolated conical singularities. III. Desingularization, the unobstructed case. *Ann. Global Anal. Geom.*, 26(1):1–58, 2004.

[265] Dominic Joyce. Special Lagrangian submanifolds with isolated conical singularities. IV. Desingularization, obstructions and families. *Ann. Global Anal. Geom.*, 26(2):117–174, 2004.

[266] Dominic Joyce. U(1)-invariant special Lagrangian 3-folds. I. Nonsingular solutions. *Adv. Math.*, 192(1):35–71, 2005.

[267] Dominic Joyce. U(1)-invariant special Lagrangian 3-folds. II. Existence of singular solutions. *Adv. Math.*, 192(1):72–134, 2005.

[268] Dominic Joyce. U(1)-invariant special Lagrangian 3-folds. III. Properties of singular solutions. *Adv. Math.*, 192(1):135–182, 2005.

[269] Dominic Joyce. Configurations in abelian categories. I. Basic properties and moduli stacks. *Adv. Math.*, 203(1):194–255, 2006.

[270] Dominic Joyce. Configurations in abelian categories. II. Ringel-Hall algebras. *Adv. Math.*, 210(2):635–706, 2007.

[271] Dominic Joyce. Configurations in abelian categories. III. Stability conditions and identities. *Adv. Math.*, 215(1):153–219, 2007.

[272] Dominic Joyce. Holomorphic generating functions for invariants counting coherent sheaves on Calabi-Yau 3-folds. *Geom. Topol.*, 11:667–725, 2007.

[273] Dominic Joyce. Configurations in abelian categories. IV. Invariants and changing stability conditions. *Adv. Math.*, 217(1):125–204, 2008.

[274] Dominic D. Joyce. *Compact manifolds with special holonomy*. Oxford Mathematical Monographs. Oxford University Press, Oxford, 2000.

[275] Shamit Kachru, Sheldon Katz, Albion Lawrence, and John McGreevy. Open string instantons and superpotentials. *Phys. Rev. D*, 62(2):026001, Jun 2000.

[276] Shamit Kachru, Sheldon Katz, Albion Lawrence, and John McGreevy. Open string instantons and superpotentials. *Phys. Rev. D (3)*, 62(2):026001, 9, 2000.

[277] Shamit Kachru, Sheldon H. Katz, Albion E. Lawrence, and John McGreevy. Mirror symmetry for open strings. *Phys. Rev.*, D62:126005, 2000.

[278] Shamit Kachru and Eva Silverstein. 4D conformal theories and strings on orbifolds. *Phys. Rev. Lett.*, 80:4855–4858, 1998.

[279] T. V. Kadeishvili. The algebraic structure in the homology of an $A(\infty)$-algebra. *Soobshch. Akad. Nauk Gruzin. SSR*, 108(2):249–252 (1983), 1982.

[280] Hiroshige Kajiura, Kyoji Saito, and Atsushi Takahashi. Matrix factorization and representations of quivers. II. Type ADE case. *Adv. Math.*, 211(1):327–362, 2007.

[281] M. Kapranov. Chow quotients of Grassmannians. I. In *I. M. Gel′fand Seminar*, volume 16 of *Adv. Soviet Math.*, pages 29–110. Amer. Math. Soc., Providence, RI, 1993.

[282] M. Kapranov and E. Vasserot. Kleinian singularities, derived categories and Hall algebras. *Math. Ann.*, 316(3):565–576, 2000.

[283] Anton Kapustin. D-branes in a topologically nontrivial B-field. *Adv. Theor. Math. Phys.*, 4(1):127–154, 2000.

[284] Anton Kapustin. Topological strings on noncommutative manifolds. *Int. J. Geom. Methods Mod. Phys.*, 1(1-2):49–81, 2004.

[285] Anton Kapustin. Chiral de Rham complex and the half-twisted sigma-model. 2005. arXiv:hep-th/0504074.

[286] Anton Kapustin and Yi Li. D-branes in Landau-Ginzburg models and algebraic geometry. *J. High Energy Phys.*, (12):005, 44 pp. (electronic), 2003.

[287] Anton Kapustin and Yi Li. Stability conditions for topological D-branes: A world-sheet approach. *ArXiv High Energy Physics - Theory e-prints*, 2003.

[288] Anton Kapustin and Yi Li. Topological correlators in Landau-Ginzburg models with boundaries. *Adv. Theor. Math. Phys.*, 7(4):727–749, 2003.

[289] Anton Kapustin and Yi Li. Open-string BRST cohomology for generalized complex branes. *Adv. Theor. Math. Phys.*, 9(4):559–574, 2005.

[290] Anton Kapustin and Dmitri Orlov. Remarks on A-branes, mirror symmetry, and the Fukaya category. *J. Geom. Phys.*, 48(1):84–99, 2003.

[291] Anton Kapustin and Dmitri Orlov. Vertex algebras, mirror symmetry, and D-branes: the case of complex tori. *Comm. Math. Phys.*, 233(1):79–136, 2003.

[292] Anton Kapustin and Lev Rozansky. On the relation between open and closed topological strings. *Commun. Math. Phys.*, 252:393–414, 2004.

[293] Anton Kapustin and Alessandro Tomasiello. *Unpublished work.*

[294] Anton Kapustin and Edward Witten. Electric-magnetic duality and the geometric Langlands program. *Commun. Number Theory Phys.*, 1(1):1–236, 2007.

[295] Masaki Kashiwara and Pierre Schapira. *Sheaves on manifolds*, volume 292 of *Grundlehren der Mathematischen Wissenschaften [Fundamental Principles of Mathematical Sciences]*. Springer-Verlag, Berlin, 1994. With a chapter in French by Christian Houzel, Corrected reprint of the 1990 original.

[296] Sheldon Katz. Versal deformations and superpotentials for rational curves in smooth threefolds. In *Symposium in Honor of C. H. Clemens (Salt Lake City, UT, 2000)*, volume 312 of *Contemp. Math.*, pages 129–136. Amer. Math. Soc., Providence, RI, 2002.

[297] Sheldon Katz and Chiu-Chu Melissa Liu. Enumerative geometry of stable maps with Lagrangian boundary conditions and multiple covers of the disc. *Adv. Theor. Math. Phys.*, 5(1):1–49, 2001.

[298] Sheldon Katz and Eric Sharpe. D-branes, open string vertex operators, and Ext groups. *Adv. Theor. Math. Phys.*, 6(6):979–1030 (2003), 2002.

[299] Ralph M. Kaufmann. Orbifolding Frobenius algebras. *Internat. J. Math.*, 14(6):573–617, 2003.

[300] Yujiro Kawamata. D-equivalence and K-equivalence. *J. Differential Geom.*, 61(1):147–171, 2002.

[301] Yujiro Kawamata. Log crepant birational maps and derived categories. *J. Math. Sci. Univ. Tokyo*, 12(2):211–231, 2005.

[302] George R. Kempf. *Complex abelian varieties and theta functions*. Universitext. Springer-Verlag, Berlin, 1991.

[303] Richard P. Kent, IV and David Peifer. A geometric and algebraic description of annular braid groups. *Internat. J. Algebra Comput.*, 12(1-2):85–97, 2002. International Conference on Geometric and Combinatorial Methods in Group Theory and Semigroup Theory (Lincoln, NE, 2000).

[304] A. D. King. Moduli of representations of finite-dimensional algebras. *Quart. J. Math. Oxford Ser. (2)*, 45(180):515–530, 1994.

[305] Igor R. Klebanov and Edward Witten. Superconformal field theory on threebranes at a Calabi-Yau singularity. *Nuclear Phys. B*, 536(1-2):199–218, 1999.

[306] Masanori Kobayashi. A special Lagrangian 3-torus as a real slice. In *Integrable systems and algebraic geometry (Kobe/Kyoto, 1997)*, pages 315–319. World Sci. Publ., River Edge, NJ, 1998.

[307] Ryoichi Kobayashi. Moduli of Einstein metrics on a $K3$ surface and degeneration of type I. In *Kähler metric and moduli spaces*, volume 18 of *Adv. Stud. Pure Math.*, pages 257–311. Academic Press, Boston, MA, 1990.

[308] Anatoly Konechny and Albert Schwarz. Introduction to M(atrix) theory and noncommutative geometry. *Phys. Rep.*, 360(5-6):353–465, 2002.

[309] Maxim Kontsevich. Homological algebra of mirror symmetry. In *Proceedings of the International Congress of Mathematicians, Vol. 1, 2 (Zürich, 1994)*, pages 120–139, Basel, 1995. Birkhäuser.

[310] Maxim Kontsevich. Lecture at Rutgers. 1996.

[311] Maxim Kontsevich and Yan Soibelman. Homological mirror symmetry and torus fibrations. In *Symplectic geometry and mirror symmetry (Seoul, 2000)*, pages 203–263. World Sci. Publishing, River Edge, NJ, 2001.

[312] Maxim Kontsevich and Yan Soibelman. Affine structures and non-Archimedean analytic spaces. In *The unity of mathematics*, volume 244 of *Progr. Math.*, pages 321–385. Birkhäuser Boston, Boston, MA, 2006.

[313] Maxim Kontsevich and Yan Soibelman. Stability structures, motivic Donaldson–Thomas invariants and cluster transformations. 2008. arXiv:0811.2435.

[314] Alexei Kovalev. Twisted connected sums and special Riemannian holonomy. *J. Reine Angew. Math.*, 565:125–160, 2003.

[315] Maximilian Kreuzer and Harald Skarke. Complete classification of reflexive polyhedra in four dimensions. *Adv. Theor. Math. Phys.*, 4(6):1209–1230, 2000.

[316] P. B. Kronheimer. The construction of ALE spaces as hyper-Kähler quotients. *J. Differential Geom.*, 29(3):665–683, 1989.

[317] D. Kutasov, A. Schwimmer, and N. Seiberg. Chiral rings, singularity theory and electric-magnetic duality. *Nuclear Phys. B*, 459(3):455–496, 1996.

[318] J. M. F. Labastida and P. M. Llatas. Topological matter in two dimensions. *Nucl. Phys.*, B379:220–258, 1992.

[319] Serge Lang. *Differential and Riemannian manifolds*, volume 160 of *Graduate Texts in Mathematics*. Springer-Verlag, New York, 1995.

[320] Gerard Laumon. Transformation de Fourier generalisée. alg-geom/9603004, 1996.

[321] Gary Lawlor. The angle criterion. *Invent. Math.*, 95(2):437–446, 1989.

[322] Calin-Iulio Lazaroiu. Instanton amplitudes in open-closed topological string theory. arXiv:hep-th/0011257, 2000.

[323] Calin-Iulio Lazaroiu. Generalized complexes and string field theory. *J. High Energy Phys.*, (6):Paper 52, 32, 2001.

[324] Calin-Iulio Lazaroiu. On the structure of open-closed topological field theory in two dimensions. *Nuclear Phys. B*, 603(3):497–530, 2001.

[325] Calin-Iulio Lazaroiu. Unitarity, D-brane dynamics and D-brane categories. *J. High Energy Phys.*, (12):Paper 31, 14, 2001.

[326] Calin-Iuliu Lazaroiu. On the boundary coupling of topological Landau-Ginzburg models. *J. High Energy Phys.*, (5):037, 19 pp. (electronic), 2005.

[327] J. Le Potier. *Lectures on vector bundles*, volume 54 of *Cambridge Studies in Advanced Mathematics*. Cambridge University Press, Cambridge, 1997. Translated by A. Maciocia.

[328] Tai-Chung Lee. Complete Ricci flat Kähler metric on M_{I}^n, M_{II}^{2n}, M_{III}^{4n}. *Pacific J. Math.*, 185(2):315–326, 1998.

[329] W. Lerche. Introduction to Seiberg-Witten theory and its stringy origin. *Nucl. Phys. Proc. Suppl.*, 55B:83–117, 1997.

[330] Naichung Conan Leung. Einstein type metrics and stability on vector bundles. *J. Differential Geom.*, 45(3):514–546, 1997.

[331] Naichung Conan Leung. Mirror symmetry without corrections. *Comm. Anal. Geom.*, 13(2):287–331, 2005.

[332] David C. Lewellen. Sewing constraints for conformal field theories on surfaces with boundaries. *Nuclear Phys. B*, 372(3):654–682, 1992.

[333] Jun Li and Gang Tian. Virtual moduli cycles and Gromov-Witten invariants of algebraic varieties. *J. Amer. Math. Soc.*, 11(1):119–174, 1998.

[334] Yi Li. *Topological sigma models and generalized geometries*. PhD thesis, California Institute of Technology, Pasadena, California. UMI-31-97353.

[335] Yi Li. Anomalies and graded coisotropic branes. *J. High Energy Phys.*, (3):100, 10 pp. (electronic), 2006.

[336] Ulf Lindström and Maxim Zabzine. $N = 2$ boundary conditions for non-linear sigma models and Landau-Ginzburg models. *J. High Energy Phys.*, (2):006, 42, 2003.

[337] Robert B. Lockhart and Robert C. McOwen. Elliptic differential operators on noncompact manifolds. *Ann. Scuola Norm. Sup. Pisa Cl. Sci. (4)*, 12(3):409–447, 1985.

[338] John Loftin, Shing-Tung Yau, and Eric Zaslow. Affine manifolds, SYZ geometry and the "Y" vertex. *J. Differential Geom.*, 71(1):129–158, 2005.

[339] John C. Loftin. Singular semi-flat Calabi-Yau metrics on S^2. *Comm. Anal. Geom.*, 13(2):333–361, 2005.

[340] Peng Lu. Kähler-Einstein metrics on Kummer threefold and special Lagrangian tori. *Comm. Anal. Geom.*, 7(4):787–806, 1999.

[341] Ernesto Lupercio and Bernardo Uribe. Topological quantum field theories, strings and orbifolds. In *Geometric and topological methods for quantum field theory*, volume 434 of *Contemp. Math.*, pages 73–98. Amer. Math. Soc., Providence, RI, 2007.

[342] E. Macri. Some examples of moduli spaces of stability conditions on derived categories. 2004. arXiv:math.AG/0411613.

[343] Juan Maldacena. The large N limit of superconformal field theories and supergravity. *Adv. Theor. Math. Phys.*, 2(2):231–252, 1998.

[344] Juan Martin Maldacena, Gregory Winthrop Moore, and Nathan Seiberg. D-brane charges in five-brane backgrounds. *J. High Energy Phys.*, 10:005, 2001.

[345] Fyodor Malikov, Vadim Schechtman, and Arkady Vaintrob. Chiral de Rham complex. *Comm. Math. Phys.*, 204(2):439–473, 1999.

[346] Michael A. Mandell. Cochains and homotopy type. *Publ. Math. Inst. Hautes Études Sci.*, (103):213–246, 2006.

[347] Marcos Mariño, Ruben Minasian, Gregory Moore, and Andrew Strominger. Nonlinear instantons from supersymmetric p-branes. *J. High Energy Phys.*, (1):Paper 5, 32, 2000.

[348] Nikita Markarian. The Atiyah class, Hochschild cohomology and the Riemann-Roch theorem. 2006.

[349] S.P. Marshall. *Deformations of special Lagrangian submanifolds*. PhD thesis, University of Oxford, 2002.

[350] Dario Martelli and James Sparks. Toric geometry, Sasaki-Einstein manifolds and a new infinite class of AdS/CFT duals. *Comm. Math. Phys.*, 262(1):51–89, 2006.

[351] Dario Martelli, James Sparks, and Shing-Tung Yau. The geometric dual of a-maximisation for toric Sasaki-Einstein manifolds. *Comm. Math. Phys.*, 268(1):39–65, 2006.

[352] E. Martinec and G. Moore. Noncommutative solitons on orbifolds. arXiv:hep-th/0107199, 2001.

[353] Diego Matessi. Some families of special Lagrangian tori. *Math. Ann.*, 325(2):211–228, 2003.

[354] D. Maulik, N. Nekrasov, A. Okounkov, and R. Pandharipande. Gromov-Witten theory and Donaldson-Thomas theory. I. *Compos. Math.*, 142(5):1263–1285, 2006.

[355] John McKay. Graphs, singularities, and finite groups. In *The Santa Cruz Conference on Finite Groups (Univ. California, Santa Cruz, Calif., 1979)*, volume 37 of *Proc. Sympos. Pure Math.*, pages 183–186. Amer. Math. Soc., Providence, R.I., 1980.

[356] Robert C. McLean. Deformations of calibrated submanifolds. *Comm. Anal. Geom.*, 6(4):705–747, 1998.

[357] Richard B. Melrose. *The Atiyah-Patodi-Singer index theorem*, volume 4 of *Research Notes in Mathematics*. A K Peters Ltd., Wellesley, MA, 1993.

[358] Ruben Minasian and Gregory Moore. K-theory and Ramond-Ramond charge. *J. High Energy Phys.*, (11):Paper 2, 7 pp. (electronic), 1997.

[359] Gregory Moore. http://www.physics.rutgers.edu/~gmoore/clay.html.

[360] Gregory Moore. Lecture at KITP, http://online.itp.ucsb.edu/online/mp01.

[361] Gregory Moore. Lecture at Strings 2000.

[362] Gregory Moore. Some comments on branes, G-flux, and K-theory. *Internat. J. Modern Phys. A*, 16(5):936–944, 2001.

[363] Gregory Moore. K-theory from a physical perspective. In *Topology, geometry and quantum field theory*, volume 308 of *London Math. Soc. Lecture Note Ser.*, pages 194–234. Cambridge Univ. Press, Cambridge, 2004.

[364] Frank Morgan. *Geometric measure theory. A beginner's guide*. Academic Press Inc., San Diego, CA, third edition, 2000.

[365] David R. Morrison. Compactifications of moduli spaces inspired by mirror symmetry. *Astérisque*, (218):243–271, 1993. Journées de Géométrie Algébrique d'Orsay (Orsay, 1992).

[366] David R. Morrison. Mirror symmetry and rational curves on quintic threefolds: a guide for mathematicians. *J. Amer. Math. Soc.*, 6(1):223–247, 1993.

[367] David R. Morrison and M. Ronen Plesser. Summing the instantons: quantum cohomology and mirror symmetry in toric varieties. *Nuclear Phys. B*, 440(1-2):279–354, 1995.

[368] David R. Morrison and M. Ronen Plesser. Non-spherical horizons. I. *Adv. Theor. Math. Phys.*, 3:1–81, 1999.

[369] Paul S. Mostert. On a compact Lie group acting on a manifold. *Ann. of Math. (2)*, 65:447–455, 1957.

[370] S. Mukai. On the moduli space of bundles on $K3$ surfaces. I. In *Vector bundles on algebraic varieties (Bombay, 1984)*, volume 11 of *Tata Inst. Fund. Res. Stud. Math.*, pages 341–413. Tata Inst. Fund. Res., Bombay, 1987.

[371] Shigeru Mukai. Duality between $D(X)$ and $D(\hat{X})$ with its application to Picard sheaves. *Nagoya Math. J.*, 81:153–175, 1981.

[372] David Mumford. *Geometric invariant theory*. Ergebnisse der Mathematik und ihrer Grenzgebiete, Neue Folge, Band 34. Springer-Verlag, Berlin, 1965.

[373] David Mumford. An analytic construction of degenerating abelian varieties over complete rings. *Compositio Math.*, 24:239–272, 1972.

[374] Dennis Nemeschansky and Ashoke Sen. Conformal invariance of supersymmetric sigma models on Calabi-Yau manifolds. *Phys. Lett.*, B178:365, 1986.

[375] Serdar Nergiz and Cihan Saclioğlu. A quasiperiodic Gibbons-Hawking metric and spacetime foam. *Phys. Rev. D (3)*, 53(4):2240–2243, 1996.

[376] Stijn Nevens, Alexander Sevrin, Walter Troost, and Alexander Wijns. Derivative corrections to the Born-Infeld action through beta-function calculations in $N = 2$ "boundary" superspace. *J. High Energy Phys.*, (8):086, 25 pp. (electronic), 2006.

[377] Yong-Geun Oh. Floer cohomology of Lagrangian intersections and pseudo-holomorphic disks. I. *Comm. Pure Appl. Math.*, 46(7):949–993, 1993.

[378] Yong-Geun Oh. Floer cohomology of Lagrangian intersections and pseudo-holomorphic disks. II. $(\mathbf{CP}^n, \mathbf{RP}^n)$. *Comm. Pure Appl. Math.*, 46(7):995–1012, 1993.

[379] So Okada. Stability manifold of $\mathbb{P}^1$. *J. Algebraic Geom.*, 15(3):487–505, 2006.

[380] Christian Okonek, Michael Schneider, and Heinz Spindler. *Vector bundles on complex projective spaces*, volume 3 of *Progress in Mathematics*. Birkhäuser Boston, Mass., 1980.

[381] Andrei Okounkov, Nikolai Reshetikhin, and Cumrun Vafa. Quantum Calabi-Yau and classical crystals. In *The unity of mathematics*, volume 244 of *Progr. Math.*, pages 597–618. Birkhäuser Boston, Boston, MA, 2006.

[382] Hirosi Ooguri, Yaron Oz, and Zheng Yin. D-branes on Calabi-Yau spaces and their mirrors. *Nuclear Phys. B*, 477(2):407–430, 1996.

[383] Hirosi Ooguri and Cumrun Vafa. Summing up Dirichlet instantons. *Phys. Rev. Lett.*, 77(16):3296–3298, 1996.

[384] D. O. Orlov. Equivalences of derived categories and $K3$ surfaces. *J. Math. Sci. (New York)*, 84(5):1361–1381, 1997. Algebraic geometry, 7.

[385] Tommaso Pacini. Deformations of asymptotically conical special Lagrangian submanifolds. *Pacific J. Math.*, 215(1):151–181, 2004.

[386] Rahul Pandharipande and Richard Thomas. Stable pairs and BPS invariants. 2007. arXiv:0711.3899.

[387] Leopoldo A. Pando Zayas and Arkady A. Tseytlin. 3-branes on resolved conifold. *J. High Energy Phys.*, (11), 2000.

[388] Giorgio Patrizio and Pit-Mann Wong. Stein manifolds with compact symmetric center. *Math. Ann.*, 289(3):355–382, 1991.

[389] Henrik Pedersen and Yat Sun Poon. Hyper-Kähler metrics and a generalization of the Bogomolny equations. *Comm. Math. Phys.*, 117(4):569–580, 1988.

[390] Vasily Pestun. Topological strings in generalized complex space. *Adv. Theor. Math. Phys.*, 11:399–450, 2007.

[391] Peter Petersen. *Riemannian geometry*, volume 171 of *Graduate Texts in Mathematics*. Springer-Verlag, New York, 1998.

[392] S. Piunikhin, D. Salamon, and M. Schwarz. Symplectic Floer-Donaldson theory and quantum cohomology. In *Contact and symplectic geometry (Cambridge, 1994)*, volume 8 of *Publ. Newton Inst.*, pages 171–200. Cambridge Univ. Press, Cambridge, 1996.

[393] Joseph Polchinski. Lectures on D-branes. In *Fields, strings and duality (Boulder, CO, 1996)*, pages 293–356. World Sci. Publ., River Edge, NJ, 1997.

[394] Joseph Polchinski. *String theory. Vol. I.* Cambridge Monographs on Mathematical Physics. Cambridge University Press, Cambridge, 2005. An introduction to the bosonic string, Reprint of the 2003 edition.

[395] Joseph Polchinski. *String theory. Vol. II.* Cambridge Monographs on Mathematical Physics. Cambridge University Press, Cambridge, 2005. Superstring theory and beyond, Reprint of 2003 edition.

[396] A. Polishchuk. A_∞-structures on an elliptic curve. *Comm. Math. Phys.*, 247(3):527–551, 2004.

[397] Alexander Polishchuk and Eric Zaslow. Categorical mirror symmetry: the elliptic curve. *Adv. Theor. Math. Phys.*, 2(2):443–470, 1998.

[398] Alexander Polishchuk and Eric Zaslow. Categorical mirror symmetry in the elliptic curve. In *Winter School on Mirror Symmetry, Vector Bundles and Lagrangian Submanifolds (Cambridge, MA, 1999)*, volume 23 of *AMS/IP Stud. Adv. Math.*, pages 275–295. Amer. Math. Soc., Providence, RI, 2001.

[399] Zhenbo Qin. Chamber structures of algebraic surfaces with Kodaira dimension zero and moduli spaces of stable rank two bundles. *Math. Z.*, 207(1):121–136, 1991.

[400] Daniel Quillen. Rational homotopy theory. *Ann. of Math. (2)*, 90:205–295, 1969.

[401] Frank Quinn. Lectures on axiomatic topological quantum field theory, at the graduate summer school on the geometry and topology of manifolds and quantum field theory, Park City, Utah., 1991.

[402] A. Recknagel and V. Schomerus. D-branes in Gepner models. *Nuclear Phys. B*, 531(1-3):185–225, 1998.

[403] Miles Reid. McKay correspondence. alg-geom/9702016, 1996.

[404] Miles Reid. La correspondance de McKay. *Astérisque*, (276):53–72, 2002. Séminaire Bourbaki, Vol. 1999/2000.

[405] Mitchell Rothstein. Sheaves with connection on abelian varieties. *Duke Math. J.*, 84(3):565–598, 1996.

[406] A. N. Rudakov et al., editors. *Helices and Vector Bundles*, volume 148 of *LMS Lecture Notes*. Cambridge, 1990.

[407] Alexei Rudakov. Stability for an abelian category. *J. Algebra*, 197(1):231–245, 1997.

[408] Simon Salamon. *Riemannian geometry and holonomy groups*, volume 201 of *Pitman Research Notes in Mathematics Series*. Longman Scientific & Technical, Harlow, 1989.

[409] Alexander V. Sardo-Infirri. Partial resolutions of orbifold singularities via moduli spaces of HYM-type bundles. 1996.

[410] Stephen Sawin. Direct sum decompositions and indecomposable TQFTs. *J. Math. Phys.*, 36(12):6673–6680, 1995.

[411] Jean-Pierre Schneiders. Quasi-abelian categories and sheaves. *Mém. Soc. Math. Fr. (N.S.)*, (76):vi+134, 1999.

[412] Graeme Segal. Categories and cohomology theories. *Topology*, 13:293–312, 1974.

[413] Graeme Segal. Lectures at Stanford University and KITP Santa Barbara, http://doug-pc.itp.ucsd.edu/online/geom99/, 1999.

[414] N. Seiberg. Electric-magnetic duality in supersymmetric nonabelian gauge theories. *Nucl. Phys.*, B435:129–146, 1995.

[415] N. Seiberg and E. Witten. Electric-magnetic duality, monopole condensation, and confinement in $N = 2$ supersymmetric Yang-Mills theory. *Nuclear Phys. B*, 426(1):19–52, 1994.

[416] Nathan Seiberg and Stephen Shenker. Hypermultiplet moduli space and string compactification to three dimensions. *Phys. Lett. B*, 388(3):521–523, 1996.

[417] Nathan Seiberg and David Shih. Minimal string theory. *C. R. Phys.*, 6(2):165–174, 2005.

[418] Nathan Seiberg and Edward Witten. String theory and noncommutative geometry. *J. High Energy Phys.*, (9):Paper 32, 93 pp. (electronic), 1999.

[419] Paul Seidel. Graded Lagrangian submanifolds. *Bull. Soc. Math. France*, 128(1):103–149, 2000.

[420] Paul Seidel. Homological mirror symmetry for the quartic surface. 2003. arXiv:math/0310414.

[421] Paul Seidel. *Fukaya categories and Picard-Lefschetz theory*. European Mathematical Society, Zürich, 2008.

[422] Paul Seidel and Richard Thomas. Braid group actions on derived categories of coherent sheaves. *Duke Math. J.*, 108(1):37–108, 2001.

[423] Ashoke Sen. Tachyon condensation on the brane antibrane system. *J. High Energy Phys.*, 08:012, 1998.

[424] Eric Sharpe. Kähler cone substructure. *Adv. Theor. Math. Phys.*, 2(6):1441–1462 (1999), 1998.

[425] Eric Sharpe. D-branes, derived categories, and Grothendieck groups. *Nuclear Phys. B*, 561(3):433–450, 1999.

[426] Eric Sharpe. Recent developments in discrete torsion. *Phys. Lett. B*, 498(1-2):104–110, 2001.

[427] Eric Sharpe. Stacks and D-brane bundles. *Nuclear Phys. B*, 610(3):595–613, 2001.

[428] Eric Sharpe. String orbifolds and quotient stacks. *Nuclear Phys. B*, 627(3):445–505, 2002.

[429] Jake Solomon. Intersection theory on the moduli space of holomorphic curves with Lagrangian boundary conditions. 2006.

[430] James Dillon Stasheff. Homotopy associativity of H-spaces. I, II. *Trans. Amer. Math. Soc. 108 (1963)*, pages 275–292; *ibid.* 293–312, 1963.

[431] Andrew Strominger. Special geometry. *Comm. Math. Phys.*, 133(1):163–180, 1990.

[432] Andrew Strominger. Massless black holes and conifolds in string theory. *Nuclear Phys. B*, 451(1-2):96–108, 1995.

[433] Andrew Strominger, Shing-Tung Yau, and Eric Zaslow. Mirror symmetry is T-duality. *Nuclear Phys. B*, 479(1-2):243–259, 1996.

[434] Dennis Sullivan. Infinitesimal computations in topology. *Inst. Hautes Études Sci. Publ. Math.*, (47):269–331 (1978), 1977.

[435] Margaret Symington. Four dimensions from two in symplectic topology. In *Topology and geometry of manifolds (Athens, GA, 2001)*, volume 71 of *Proc. Sympos. Pure Math.*, pages 153–208. Amer. Math. Soc., Providence, RI, 2003.

[436] Balázs Szendrői. Artin group actions on derived categories of threefolds. *J. Reine Angew. Math.*, 572:139–166, 2004.

[437] Balázs Szendrői. Non-commutative Donaldson-Thomas invariants and the conifold. *Geom. Topol.*, 12(2):1171–1202, 2008.

[438] Gerardus 't Hooft, editor. *50 years of Yang-Mills theory.* World Scientific Publishing Co. Pte. Ltd., Hackensack, NJ, 2005.

[439] A. Takahashi. Matrix factorizations and representations of quivers. I.

[440] Joerg Teschner. On boundary perturbations in liouville theory and brane dynamics in noncritical string theories. *J. High Energy Phys.*, 04:023, 2004.

[441] Michael Thaddeus. Toric quotients and flips. In *Topology, geometry and field theory*, pages 193–213. World Sci. Publishing, River Edge, NJ, 1994.

[442] R. P. Thomas and S.-T. Yau. Special Lagrangians, stable bundles and mean curvature flow. *Comm. Anal. Geom.*, 10(5):1075–1113, 2002.

[443] Richard Thomas. *Gauge theory on Calabi-Yau manifolds.* PhD thesis, University of Oxford, 1997.

[444] Richard Thomas. An obstructed bundle on a Calabi-Yau 3-fold. *Adv. Theor. Math. Phys.*, 3(3):567–576, 1999.

[445] Richard Thomas. Derived categories for the working mathematician. In *Winter School on Mirror Symmetry, Vector Bundles and Lagrangian Submanifolds (Cambridge, MA, 1999)*, volume 23 of *AMS/IP Stud. Adv. Math.*, pages 349–361. Amer. Math. Soc., Providence, RI, 2001.

[446] Richard Thomas. Moment maps, monodromy and mirror manifolds. In *Symplectic geometry and mirror symmetry (Seoul, 2000)*, pages 467–498. World Sci. Publ., River Edge, NJ, 2001.

[447] Richard Thomas. Stability conditions and the braid group. *Comm. Anal. Geom.*, 14(1):135–161, 2006.

[448] Gang Tian. Smoothness of the universal deformation space of compact Calabi-Yau manifolds and its Petersson-Weil metric. In *Mathematical aspects of string theory (San Diego, Calif., 1986)*, volume 1 of *Adv. Ser. Math. Phys.*, pages 629–646. World Sci. Publishing, Singapore, 1987.

[449] A. N. Todorov. The Weil-Petersson geometry of the moduli space of $SU(n\geq3)$ (Calabi-Yau) manifolds, I. *Commun. Math. Phys.*, 126:325–246, 1989.

[450] A. A. Tseytlin. On non-abelian generalisation of the Born-Infeld action in string theory. *Nuclear Phys. B*, 501(1):41–52, 1997.

[451] A. A. Tseytlin. Born-Infeld action, supersymmetry and string theory. In *The many faces of the superworld*, pages 417–452. World Sci. Publ., River Edge, NJ, 2000.

[452] Vladimir Turaev. Homotopy field theory in dimension 2 and group-algebras. arXiv.math/9910010, 1999.

[453] K. Uhlenbeck and S.-T. Yau. On the existence of Hermitian-Yang-Mills connections in stable vector bundles. *Comm. Pure Appl. Math.*, 39(S, suppl.):S257–S293, 1986. Frontiers of the mathematical sciences: 1985 (New York, 1985).

[454] Cumrun Vafa. Topological Landau-Ginzburg models. *Modern Phys. Lett. A*, 6(4):337–346, 1991.

[455] Misha Verbitsky and Dmitri Kaledin. *Hyperkahler manifolds*, volume 12 of *Mathematical Physics (Somerville)*. International Press, Somerville, MA, 1999.

[456] Jean-Louis Verdier. Des catégories dérivées des catégories abéliennes. *Astérisque*, (239):xii+253 pp. (1997), 1996. With a preface by Luc Illusie, Edited and with a note by Georges Maltsiniotis.

[457] Jorge Vitória. Mutations vs. Seiberg duality. 2007.

[458] Claire Voisin. *Mirror symmetry*. American Mathematical Society, Providence, RI, 1999.

[459] Johannes Walcher. Stability of Landau-Ginzburg branes. *J. Math. Phys.*, 46(8):082305, 29, 2005.

[460] Julius Wess and Jonathan Bagger. *Supersymmetry and supergravity*. Princeton Series in Physics. Princeton University Press, Princeton, NJ, second edition, 1992.

[461] Wikipedia. D-brane — Wikipedia, the free encyclopedia, 2007. [Online; accessed 20-June-2007].

[462] P. M. H. Wilson. The Kähler cone on Calabi-Yau threefolds. *Invent. Math.*, 107:561–583, 1992. Erratum, *Invent. Math.* 114:231–233, 1993.

[463] E. Witten. Chern-Simons gauge theory as a string theory. In *The Floer memorial volume*, volume 133 of *Progr. Math.*, pages 637–678. Birkhäuser, Basel, 1995.

[464] Edward Witten. Dynamical breaking of supersymmetry. *Nucl. Phys.*, B188:513, 1981.

[465] Edward Witten. Constraints on supersymmetry breaking. *Nuclear Phys. B*, 202(2):253–316, 1982.

[466] Edward Witten. Supersymmetry and Morse theory. *J. Differential Geom.*, 17(4):661–692 (1983), 1982.

[467] Edward Witten. Topological quantum field theory. *Commun. Math. Phys.*, 117:353, 1988.

[468] Edward Witten. Mirror manifolds and topological field theory. In *Essays on mirror manifolds*, pages 120–158. Int. Press, Hong Kong, 1992.

[469] Edward Witten. The N matrix model and gauged WZW models. *Nuclear Phys. B*, 371(1-2):191–245, 1992.

[470] Edward Witten. Solutions of four-dimensional field theories via M-theory. *Nucl. Phys.*, B500:3–42, 1997.

[471] Edward Witten. D-branes and K-theory. *J. High Energy Phys.*, (12):Paper 19, 41 pp. (electronic), 1998.

[472] Edward Witten. Dynamics of quantum field theory. In Pierre Deligne, Pavel Etingof, Daniel S. Freed, Lisa C. Jeffrey, David Kazhdan, John W. Morgan, David R. Morrison, and Edward Witten, editors, *Quantum fields and strings: a course for mathematicians. Vol. 1, 2*, pages Vol 2; pp. 1119–1424, Providence, RI, 1999. American Mathematical Society. Material from the Special Year on Quantum Field Theory held at the Institute for Advanced Study, Princeton, NJ, 1996–1997.

[473] Edward Witten. Index of Dirac operators. In Pierre Deligne, Pavel Etingof, Daniel S. Freed, Lisa C. Jeffrey, David Kazhdan, John W. Morgan, David R. Morrison, and Edward Witten, editors, *Quantum fields and strings: a course for mathematicians. Vol. 1, 2*, page Vol 1, Providence, RI, 1999. American Mathematical Society. Material from the Special Year on Quantum Field Theory held at the Institute for Advanced Study, Princeton, NJ, 1996–1997.

[474] Edward Witten. Two-dimensional models with $(0, 2)$ supersymmetry: perturbative aspects. *Adv. Theor. Math. Phys.*, 11(1):1–63, 2007.

[475] Shing Tung Yau. On the Ricci curvature of a compact Kähler manifold and the complex Monge-Ampère equation. I. *Comm. Pure Appl. Math.*, 31(3):339–411, 1978.

[476] Ilia Zharkov. Limiting behavior of local Calabi-Yau metrics. *Adv. Theor. Math. Phys.*, 8(3):395–420, 2004.

Index